Mechanical Engineering Series

Frederick F. Ling
Series Editor

Springer
*New York
Berlin
Heidelberg
Barcelona
Budapest
Hong Kong
London
Milan
Paris
Santa Clara
Singapore
Tokyo*

Mechanical Engineering Series

Introductory Attitude Dynamics
F.P. Rimrott

Balancing of High-Speed Machinery
M.S. Darlow

Theory of Wire Rope
G.A. Costello

Theory of Vibration
  Vol. I An Introduction
  Vol. II Discrete and Continuous Systems
A.A. Shabana

Laser Machining: Theory and Practice
G. Chryssolouris

Underconstrained Structural Systems
E.N. Kuznetsov

Principles of Heat Transfer in Porous Media, 2nd ed.
M. Kaviany

Mechatronics: Electromechanics and Contromechanics
D.K. Miu

Structural Analysis of Printed Circuit Board Systems
P.A. Engel

**Kinematic and Dynamic Simulation of Multibody Systems:
The Real-Time Challenge**
J. García de Jalón and E. Bayo

**High Sensitivity Moiré:
Experimental Analysis for Mechanics and Materials**
D. Post, B. Han, and P. Ifju

Principles of Convective Heat Transfer
M. Kaviany

Laminar Viscous Flow
V.N. Constantinescu

M. Kaviany

Principles of Heat Transfer in Porous Media

Second Edition

With 202 Illustrations

 Springer

M. Kaviany
Department of Mechanical Engineering
  and Applied Mechanics
The University of Michigan
Ann Arbor, MI 48109-2125 USA

Series Editor
Frederick F. Ling
Ernest F. Gloyna Regents Chair in Engineering
Department of Mechanical Engineering
The University of Texas at Austin
Austin, TX 78712-1063 USA
and
Distinguished William Howard Hart
  Professor Emeritus
Department of Mechanical Engineering,
  Aeronautical Engineering, and Mechanics
Rensselaer Polytechnic Institute
Troy, NY 12180-3590 USA

Library of Congress Cataloging-in-Publication Data
Kaviany, M. (Massoud)
    Principles of heat transfer in porous media / Massoud Kaviany. —
2nd ed.
    p.   cm. — (Mechanical engineering series)
    Includes bibliographical references and index.
    ISBN 0-387-94550-4 (hardcover: alk. paper)
    1. Heat—Transmission—Congresses.   2. Porous materials—Thermal
properties—Congresses.   I. Title.   II. Series: Mechanical
engineering series (Berlin, Germany)
TJ260.K29   1995
621.402′2—dc20   95-4588

Printed on acid-free paper.

© 1995 Springer-Verlag New York, Inc.
All rights reserved. This work may not be translated or copied in whole or in part without the written permission of the publisher (Springer-Verlag New York, Inc., 175 Fifth Avenue, New York, NY 10010, USA), except for brief excerpts in connection with reviews or scholarly analysis. Use in connection with any form of information storage and retrieval, electronic adaptation, computer software, or by similar or dissimilar methodology now known or hereafter developed is forbidden.
The use of general descriptive names, trade names, trademarks, etc., in this publication, even if the former are not especially identified, is not to be taken as a sign that such names, as understood by the Trade Marks and Merchandise Marks Act, may accordingly be used freely by anyone.

Production managed by Natalie Johnson; manufacturing supervised by Jeffrey Taub.
Camera-ready copy prepared from the author's LaTeX files.
Printed and bound by Edwards Brothers, Inc., Ann Arbor, MI.
Printed in the United States of America.

9 8 7 6 5 4 3 2 1

ISBN 0-387-94550-4 Springer-Verlag New York Berlin Heidelberg

To my parents Farideh and Morad

Series Preface

Mechanical engineering, an engineering discipline born of the needs of the industrial revolution, is once again asked to do its substantial share in the call for industrial renewal. The general call is urgent as we face profound issues of productivity and competitiveness that require engineering solutions, among others. The Mechanical Engineering Series is a new series, featuring graduate texts and research monographs, intended to address the need for information in contemporary areas of mechanical engineering.

The series is conceived as a comprehensive one that will cover a broad range of concentrations important to mechanical engineering graduate education and research. We are fortunate to have a distinguished roster of consulting editors, each an expert in one of the areas of concentration. The names of the consulting editors are listed on the first page of the volume. The areas of concentration are applied mechanics, biomechanics, computational mechanics, dynamic systems and control, energetics, mechanics of materials, processing, thermal science, and tribology.

Professor Bergles, the consulting editor for thermal science, and I are pleased to present this volume of the series: *Principles of Heat Transfer in Porous Media, Second Edition*, by Professor Kaviany. The selection of this volume underscores again the interest of the Mechanical Engineering Series to provide our readers with topical monographs as well as graduate texts.

New York, New York   Frederick F. Ling

Mechanical Engineering Series

Frederick F. Ling
Series Editor

Advisory Board

Applied Mechanics	F.A. Leckie University of California, Santa Barbara
Biomechanics	V.C. Mow Columbia University
Computational Mechanics	H.T. Yang University of California, Santa Barbara
Dynamic Systems and Control	K.M. Marshek University of Texas, Austin
Energetics	J. Welty University of Oregon, Eugene
Mechanics of Materials	I. Finnie University of California, Berkeley
Processing	K.K. Wang Cornell University
Thermal Science	A.E. Bergles Rennselaer Polytechnic Institute
Tribology	W.O. Winer Georgia Institute of Technology

Preface

This monograph aims at providing, through integration of available theoretical and empirical treatments, the differential conservation equations and the associated constitutive equations required for the analysis of transport in porous media. Although the empirical treatment of fluid flow and heat transfer in porous media is over a century old, only in the last three decades has the transport in these heterogeneous systems been addressed in sufficient detail. So far, single-phase flow and heat transfer in porous media have been treated or at least formulated satisfactorily. But the subject of two-phase flow and the related heat transfer in porous media is still in its infancy. This monograph identifies the principles of transport in porous media, reviews the available rigorous treatments, and whenever possible compares the available predictions, based on these theoretical treatments of various transport mechanisms, with the existing experimental results. The theoretical treatment is based on the local volume-averaging of the momentum and energy equations with the closure conditions necessary for obtaining solutions. While emphasizing a basic understanding of heat transfer in porous media, the monograph does not ignore the need for the predictive tools. Therefore, whenever a rigorous theoretical treatment of a phenomenon is not available, semiempirical and empirical treatments are given.

The monograph is divided into two parts: Part I deals with single-phase flows and Part II covers two-phase flows. For single-phase flows, all modes of heat transfer are examined first using a single-continuum treatment based on the assumption of a local thermal equilibrium. A two-medium treatment is then made. In Part II, pore-level fluid mechanics and the thermodynamics for the simultaneous presence of both fluid phases in porous media are addressed. Conduction and convection heat transfer are then examined. The heat and mass transfer from surfaces bounding these porous media, which contain both liquid and gaseous phases, is also presented. Since the fluid dynamics of two-phase flow involving phase change is not yet fully understood, specific phase-change processes and their peculiarities are discussed in the last chapter. The contents of each chapter are briefly reviewed here.

The historical and practical aspects of heat transfer in porous media, as well as the length, time, and temperature scales encountered in this field, are reviewed in Chapter 1. The fluid mechanics of the single-phase flow, beginning with the Darcy law and developing along more rigorous treatments based on the local volume averaging, are given in Chapter 2. In this chapter we also examine the porosity variations near the bounding solid surfaces and the hydrodynamic boundary conditions at the interface of the porous plain media. Heat conduction is treated in Chapter 3 and deterministic, stochastic, and semiempirical treatments are examined. Variations of the effective thermal conductivity tensor near the bounding surfaces are also discussed. Hydrodynamic dispersion, which always exists in convective heat transfer in porous media, and the various treatments of it are given in Chapter 4. These include the local volume-averaging treatment for periodic and disordered structures. Again, dispersion near the bounding surfaces is addressed. Radiation heat transfer, which is significant in low-temperature insulation applications as well as in high-temperature combustion applications, is discussed in Chapter 5. The derivation of the radiative properties from the optical properties and the various approximations used in the radiative heat transfer calculations are also considered. The theory of independent scattering is examined for large particles and the inclusion of dependent scattering is formulated and a solution method is presented. Mass transfer in gases, including the low-pressure, small pore size Knudsen regime, the surface diffusion, and chemical reactions are considered in Chapter 6. Part I ends with Chapter 7, where two-medium treatment of transient heat transfer is given. The discussion aims at clarifying the existing misunderstanding about this approach, and several examples are given.

Part II begins with Chapter 8, which discusses two-phase flow and the complexity of the fluid mechanics, including hysteresis. The local volume averaging and semiempirical treatments and the constitutive equations for the capillary pressures and the relative permeability are discussed. In addition, the coefficients for the inertial and surface-tension gradient terms in the momentum equations are presented. In Chapter 9, we address the peculiarities of liquid-vapor coexistence in porous media, including the reduction in vapor pressure and capillary condensation. Conduction and convection heat transfer in two-phase flows are examined in Chapter 10 using both local volume averaging and semiempirical treatments. Heat and mass transfer from partially liquid-saturated permeable surfaces, including the zero Bond number asymptote, are given in Chapter 11. The phase change in the liquid-vapor systems in porous media is discussed in Chapter 12, where specific steady-state and transient processes are considered.

The monograph is written as a detailed description of the fundamentals of heat transfer in porous media. Familiarity with fluid mechanics and heat transfer is assumed. The concepts and physical phenomena are emphasized more than the step-by-step development. When intermediate steps in the derivations are not given, they can either be found in the references or arrived at by the material supplied in the discussion. The symbols used are defined in the nomenclature after Chapter 12. A glossary of the common terms is given following the nomenclature.

Ann Arbor, Michigan   Massoud Kaviany

Preface to Second Edition

In this edition we have updated the references, added new materials to some sections, added some new sections, and corrected typographical errors. Further attempts are made to integrate transport, reaction and phase change in porous media, and we have therefore included discussions of heat transfer-influenced reactions and broader coverage of phase change (including solid–liquid).

Section 4.3.5 includes recent numerical results for the dispersion tensor in periodic structures. Dispersion in oscillating flow is now discussed in Chapter 4. Section 4.10 is updated to include the new numerical results on dispersion near bounding surfaces. New experimental and numerical results on radiative properties—among them the properties in the equation of radiative transfer, in the radiant conductivity model, and in the layered-geometric model—are included. Chapter 6 now provides a more extensive discussion of reaction and local chemical nonequilibrium. An example of local thermal nonequilibrium with exothermic reaction, and comparison between the results of the single- and two-medium treatments and direct simulations, is included in Chapter 7. Section 12.5.3 is on the recent results of vapor flow and condensation front propagation into a dry porous medium. Section 12.6 is on melting/soldification and includes single- and multi-component systems.

We are thankful to those students and colleagues who have made many improvement suggestions. We would also like to thank the Springer-Verlag team, Senior Editor, Dr. Thomas von Foerster, the Thermal Science Editor, Professor Arthur Bergles, and the Senior Production Editor, Ms. Natalie Johnson.

Acknowledgments

I would like to thank my students and collaborators for working in the area of heat transfer in porous media: B. D'Amico, M. Fatehi, K. Hanamura, C.-J. Kim, M. Mittal, A. Oliveira, J. Rogers, M. Sahraoui, B. P. Singh, and Y.-X. Tao. They have been a constant source of ideas and encouragement and have made this monograph possible. I am grateful for their patience and devotion. Special thanks to C.-J. Kim and B. D'Amico for the word processing.

The illustrations for the monograph were drawn by R. Hill, S. Errington, and S. Ackerman of the University of Michigan. I am very thankful to them for their excellent professional services and to other staff members of my department who have provided assistance.

My research in the area of heat transfer in porous media has been sponsored by the Chevron Oil Company, the Department of Energy, the Ford Motor Company, NASA-LeRC, the National Science Foundation, and the Whirlpool Corporation. I am grateful for their financial support and for the encouragement of their project monitors.

Finally, I would like to thank my wife Mitra, who took on many extra responsibilities so that I could attend to this monograph. She remains a constant source of support and strength.

Contents

Series Preface vii

Preface ix

Preface to Second Edition xi

Acknowledgments xiii

1 Introduction 1
  1.1 Historical Background 2
  1.2 Length, Time, and Temperature Scales 7
  1.3 Scope 11
  1.4 References 12

Part I Single-Phase Flow 15

2 Fluid Mechanics 17
  2.1 Stokes Flow and Darcy Equation 17
  2.2 Porosity 20
  2.3 Pore Structure 24
  2.4 Permeability 28
    2.4.1 Capillary Models 29
    2.4.2 Hydraulic Radius Model 32
    2.4.3 Drag Models for Periodic Structures 34
  2.5 High Reynolds Number Flows 45
    2.5.1 Macroscopic Models 45
    2.5.2 Microscopic Fluid Dynamics 48
    2.5.3 Turbulence 51
  2.6 Brinkman Superposition of Bulk and Boundary Effects . . . 52
  2.7 Local Volume-Averaging Method 53
    2.7.1 Local Volume Averages 55
    2.7.2 Theorems 56
    2.7.3 Momentum Equation 58
  2.8 Homogenization Method 61
    2.8.1 Continuity Equation 63
    2.8.2 Momentum Equation 64

	2.9	Semiheuristic Momentum Equations	66
	2.10	Significance of Macroscopic Forces	68
		2.10.1 Macroscopic Hydrodynamic Boundary Layer	69
		2.10.2 Macroscopic Entrance Length	70
	2.11	Porous Plain Media Interfacial Boundary Conditions	71
		2.11.1 Slip Boundary Condition	72
		2.11.2 On Beavers-Joseph Slip Coefficient	75
		2.11.3 Taylor-Richardson Results for Slip Coefficient	78
		2.11.4 Slip Coefficient for a Two-Dimensional Structure	79
		2.11.5 No-Slip Models Using Effective Viscosity	92
		2.11.6 Variable Effective Viscosity for a Two-Dimensional Structure	95
		2.11.7 Variable Permeability for a Two-Dimensional Structure	98
	2.12	Variation of Porosity near Bounding Impermeable Surfaces	101
		2.12.1 Dependence of Average Porosity on Linear Dimensions of System	101
		2.12.2 Local Porosity Variation	102
		2.12.3 Velocity Nonuniformities Due to Porosity Variation	104
		2.12.4 Velocity Nonuniformity for a Two-Dimensional Structure	106
	2.13	Analogy with Magneto-Hydrodynamics	112
	2.14	References	114
3	**Conduction Heat Transfer**		**119**
	3.1	Local Thermal Equilibrium	120
	3.2	Local Volume Averaging for Periodic Structures	121
		3.2.1 Local Volume Averaging	122
		3.2.2 Determination of $\mathbf{b}_f$ and $\mathbf{b}_s$	125
		3.2.3 Numerical Values for $\mathbf{b}_f$ and $\mathbf{b}_s$	126
	3.3	Particle Concentrations from Dilute to Point Contact	127
	3.4	Areal Contact Between Particles Caused by Compressive Force	128
		3.4.1 Effect of Rarefaction	134
		3.4.2 Dependence of Gas Conductivity on Knudsen Number	134
	3.5	Statistical Analyses	135
		3.5.1 A Variational Formulation	136
		3.5.2 A Thermodynamic Analogy	138
	3.6	Summary of Correlations	144
	3.7	Adjacent to Bounding Surfaces	145
		3.7.1 Temperature Slip for a Two-Dimensional Structure	149

		3.7.2	Variable Effective Conductivity for a Two-Dimensional Structure	151
	3.8		On Generalization	152
	3.9		References	153
4	**Convection Heat Transfer**			**157**
	4.1		Dispersion in a Tube—Hydrodynamic Dispersion	157
		4.1.1	No Molecular Diffusion	159
		4.1.2	Molecular Diffusion Included	159
		4.1.3	Asymptotic Behavior for Large Elapsed Times	160
		4.1.4	Turbulent Flow	163
	4.2		Dispersion in Porous Media	164
	4.3		Local Volume Average for Periodic Structures	166
		4.3.1	Local Volume Averaging for $k_s = 0$	167
		4.3.2	Reduction to Taylor-Aris Dispersion	171
		4.3.3	Evaluation of $\mathbf{u}'$ and $\mathbf{b}$	172
		4.3.4	Results for $k_s = 0$ and In-Line Arrangement	174
		4.3.5	Results for $k_s \neq 0$ and General Arrangements	175
	4.4		Three-Dimensional Periodic Structures	183
		4.4.1	Unit-Cell Averaging	184
		4.4.2	Evaluation of $\mathbf{u}'$, $\mathbf{b}$, and $\mathbf{D}$	187
		4.4.3	Comparison with Experimental Results	190
		4.4.4	Effect of Darcean Velocity Direction	191
	4.5		Dispersion in Disordered Structures—Simplified Hydrodynamics	192
		4.5.1	Scheidegger Dynamic and Geometric Models	193
		4.5.2	De Josselin De Jong Purely Geometric Model	196
		4.5.3	Saffman Inclusion of Molecular Diffusion	197
		4.5.4	Horn Method of Moments	200
	4.6		Dispersion in Disordered Structures—Particle Hydrodynamics	205
		4.6.1	Local Volume Averaging	205
		4.6.2	Low Peclet Numbers	207
		4.6.3	High Peclet Numbers	209
		4.6.4	Contribution of Solid Holdup (Mass Transfer)	210
		4.6.5	Contribution Due to Thermal Boundary Layer in Fluid	212
		4.6.6	Combined Effect of All Contributions	213
	4.7		Properties of Dispersion Tensor	215
	4.8		Experimental Determination of $\mathbf{D}$	217
		4.8.1	Experimental Methods	217
		4.8.2	Entrance Effect	225
		4.8.3	Effect of Particle Size Distribution	227
		4.8.4	Some Experimental Results and Correlations	227
	4.9		Dispersion in Oscillating Flow	232

	4.9.1	Formulation and Solution	234
	4.9.2	Longitudinal Dispersion Coefficient	237
4.10	Dispersion Adjacent to Bounding Surfaces	237
	4.10.1	Temperature-Slip Model	240
	4.10.2	No-Slip Treatments	242
	4.10.3	Models Based on Mixing-Length Theory	247
	4.10.4	A Model Using Particle-Based Hydrodynamics	253
	4.10.5	Results of a Two-Dimensional Simulation	254
4.11	References	254

5 Radiation Heat Transfer 259

5.1	Continuum Treatment	260
5.2	Radiation Properties of a Single Particle	264
	5.2.1 Wavelength Dependence of Optical Properties	264
	5.2.2 Solution to Maxwell Equations	268
	5.2.3 Scattering Efficiency and Cross Section	277
	5.2.4 Mie Scattering	278
	5.2.5 Rayleigh Scattering	279
	5.2.6 Geometric- or Ray-Optics Scattering	279
	5.2.7 Comparison of Predictions	283
5.3	Radiative Properties: Dependent and Independent	292
5.4	Volume Averaging for Independent Scattering	302
5.5	Experimental Determination of Radiative Properties	305
	5.5.1 Measurements	305
	5.5.2 Models Used to Interpret Experimental Results	309
5.6	Boundary Conditions	311
	5.6.1 Transparent Boundaries	311
	5.6.2 Opaque Diffuse Emitting/Reflecting Boundaries	311
	5.6.3 Opaque Diffusely Emitting Specularly Reflecting Boundaries	312
	5.6.4 Semitransparent Nonemitting Specularly Reflecting Boundaries	312
5.7	Solution Methods for Equation of Radiative Transfer	313
	5.7.1 Two-Flux Approximations, Quasi-Isotropic Scattering	313
	5.7.2 Diffusion (Differential) Approximation	317
	5.7.3 Spherical Harmonics-Moment (P-N) Approximation	318
	5.7.4 Discrete-Ordinates (S-N) Approximation	322
	5.7.5 Finite-Volume Method	327
5.8	Scaling (Similarity) in Radiative Heat Transfer	327
	5.8.1 Similarity Between Phase Functions	327
	5.8.2 Similarity Between Anisotropic and Isotropic Scattering	331
5.9	Noncontinuum Treatment: Monte Carlo Simulation	333

	5.9.1	Opaque Particles	335
	5.9.2	Semitransparent Particles	337
	5.9.3	Emitting Particles	338
5.10	Geometric, Layered Model	339
5.11	Radiant Conductivity Model	340
	5.11.1	Calculation of F	341
	5.11.2	Effect of Solid Conductivity	343
5.12	Modeling Dependent Scattering	346
	5.12.1	Modeling Dependent Scattering for Large Particles	346
5.13	Summary	359
5.14	References	360

6 Mass Transfer in Gases — 365
6.1	Knudsen Flows	367
6.2	Fick Diffusion	369
6.3	Knudsen Diffusion	370
6.4	Crossed Diffusion	370
6.5	Prediction of Transport Coefficients from Kinetic Theory	371
	6.5.1 Fick Diffusivity in Plain Media	371
	6.5.2 Knudsen Diffusivity for Tube Flows	372
	6.5.3 Slip Self-Diffusivity for Tube Flows	373
	6.5.4 Adsorption and Surface Flux	373
6.6	Dusty-Gas Model for Transition Flows	376
6.7	Local Volume-Averaged Mass Conservation Equation	377
6.8	Chemical Reactions	380
6.9	Evaluation of Total Effective Mass Diffusivity Tensor	383
	6.9.1 Effective Mass Diffusivity	383
	6.9.2 Mass Dispersion Tensor	385
6.10	Evaluation of Local Volume-Averaged Source Terms	385
	6.10.1 Homogeneous Reaction	386
	6.10.2 Heterogeneous Reaction	386
6.11	Local Chemical Nonequilibrium	387
6.12	Modifications to Energy Equation	388
6.13	References	389

7 Two-Medium Treatment — 391
7.1	Local Phase Volume Averaging for Steady Flows	392
	7.1.1 Allowing for Difference in Average Local Temperatures	392
	7.1.2 Evaluation of [b] and [ψ]	394
	7.1.3 Energy Equation for Each Phase	395
	7.1.4 Example: Axial Travel of Thermal Pulses	398
7.2	Interfacial Convective Heat Transfer Coefficient h_{sf}	401
	7.2.1 Models Based on h_{sf}	402
	7.2.2 Experimental Determination of h_{sf}	403

xx Contents

	7.3	Distributed Treatment of Oscillating Flow	404
	7.4	Chemical Reaction	409
		7.4.1 Two-Dimensional Direct Simulation	411
		7.4.2 Volume-Averaged Models	413
		7.4.3 Interfacial Nusselt Number	416
		7.4.4 Comparison of Results of Various Treatments	418
	7.5	References	424

Part II Two-Phase Flow — 425

8 Fluid Mechanics — 427

8.1	Elements of Pore-Level Flow Structure	430
	8.1.1 Surface Tension	432
	8.1.2 Continuous Phase Distribution	442
	8.1.3 Discontinuous Phase Distributions	445
	8.1.4 Contact Line	448
	8.1.5 Thin Extension of Meniscus	455
8.2	Local Volume Averaging	459
	8.2.1 Effect of Surface Tension Gradient	464
8.3	A Semiheuristic Momentum Equation	465
	8.3.1 Inertial Regime	465
	8.3.2 Liquid-Gas Interfacial Drag	466
	8.3.3 Coefficients in Momentum Equations	468
8.4	Capillary Pressure	471
	8.4.1 Hysteresis	471
	8.4.2 Models	474
8.5	Relative Permeability	478
	8.5.1 Constraint on Applicability	479
	8.5.2 Influencing Factors	479
	8.5.3 Models	484
8.6	Microscopic Inertial Coefficient	488
8.7	Liquid-Gas Interfacial Drag	490
8.8	Immiscible Displacement	492
	8.8.1 Interfacial Instabilities	495
	8.8.2 Buckley-Leverett Front	497
	8.8.3 Stability of Buckley-Leverett Front	500
8.9	Fluid-Solid Two-Phase Flow	501
8.10	References	502

9 Thermodynamics — 507

9.1	Thermodynamics of Single-Component Capillary Systems	507
	9.1.1 Work of Surface Formation	507
	9.1.2 First and Second Laws of Thermodynamics	508
	9.1.3 Thickness of Interfacial Layer	510

	9.2	Effect of Curvature in Single-Component Systems	511
		9.2.1 Vapor Pressure Reduction	513
		9.2.2 Reduction of Chemical Potential	513
		9.2.3 Increase in Heat of Evaporation	514
		9.2.4 Liquid Superheat	516
		9.2.5 Change in Freezing Temperature	517
		9.2.6 Change in Triple-Point Temperature	518
	9.3	Multicomponent Systems	519
		9.3.1 Surface Tension of Solution	519
		9.3.2 Vapor Pressure Reduction	520
	9.4	Interfacial Thermodynamics of Meniscus Extension	521
	9.5	Capillary Condensation	523
		9.5.1 Adsorption by Solid Surface	524
		9.5.2 Condensation in a Mesoporous Solid	529
	9.6	Prediction of Fluid Behavior in Small Pores	532
		9.6.1 Phase Transition in Small Pores: Hysteresis	533
		9.6.2 Stability of Liquid Film in Small Pores: Hysteresis	539
	9.7	References	542
10	**Conduction and Convection**	**545**
	10.1	Local Volume Averaging of Energy Equation	545
		10.1.1 Averaging	545
		10.1.2 Effective Thermal Conductivity and Dispersion Tensors	550
	10.2	Effective Thermal Conductivity	551
		10.2.1 Anisotropy	553
		10.2.2 Correlations	554
	10.3	Thermal Dispersion	555
		10.3.1 Anisotropy	557
		10.3.2 Models	558
		10.3.3 Correlations for Lateral Dispersion Coefficient	563
		10.3.4 Dispersion near Bounding Surfaces	569
	10.4	References	569
11	**Transport Through Bounding Surfaces**	**571**
	11.1	Evaporation from Heated Liquid Film	571
		11.1.1 Simple Model for Transition Region	573
		11.1.2 Inclusion of Capillary Meniscus	575
	11.2	Mass Diffusion Adjacent to a Partially Saturated Surface	578
		11.2.1 Large Knudsen Number Model	580
		11.2.2 Small Knudsen Number Model	582
	11.3	Convection from Heterogeneous Planar Surfaces	587
		11.3.1 Mass Transfer from a Single Strip	587
		11.3.2 Simultaneous Heat and Mass Transfer from Multiple Surface Sources	588

11.4 Convection from Heterogeneous Two-Dimensional Surfaces ... 589
    11.4.1 A Simple Surface Model ... 590
    11.4.2 Experimental Observation on Simultaneous Heat and Mass Transfer ... 596
11.5 Simultaneous Heat and Mass Transfer from Packed Beds ... 599
11.6 References ... 601

12 Phase Change ... 603
12.1 Condensation at Vertical Impermeable Bounding Surfaces ... 603
    12.1.1 Thick Liquid-Film Region ($\delta_\ell/d \gg 1$) ... 605
    12.1.2 Thin Liquid-Film Region ($\delta_\ell/d \simeq 1$) ... 613
12.2 Evaporation at Vertical Impermeable Bounding Surfaces ... 614
12.3 Evaporation at Horizontal Impermeable Bounding Surfaces ... 615
    12.3.1 Effect of Bond Number ... 616
    12.3.2 A One-Dimensional Analysis for $Bo \ll 1$ ... 619
12.4 Evaporation at Thin Porous-Layer Coated Surfaces ... 624
12.5 Moving Evaporation or Condensation Front ... 627
    12.5.1 Temperatures Equal to or Larger than Saturation Temperature ... 631
    12.5.2 Temperatures Below Saturation Temperature ... 638
    12.5.3 Condensation Front Moving into Dry Porous Media ... 644
12.6 Melting and Solidification ... 654
    12.6.1 Single-Component Systems ... 656
    12.6.2 Multicomponent Systems ... 658
12.7 References ... 671

Nomenclature ... 677

Glossary ... 683

Citation Index ... 689

Subject Index ... 697

1
Introduction

Examination of *transport*, *reaction*, and *phase change* in natural and engineered porous media relies on the knowledge we have gained in studying these *phenomena* in otherwise plain media. The presence of a *permeable solid* (which we assume to be *rigid* and *stationary*) influences these phenomena significantly. Due to practical limitations, as a general approach we choose to describe these phenomena at a *small* length scale which is yet *larger* than a *fraction* of the linear dimension of the *pore* or the linear dimension of the *solid particle* (for a particle-based porous medium). This requires the use of the local volume averaging theories. Also, depending on the validity, local *mechanical*, *thermal*, and *chemical equilibrium* or *nonequilibrium*, may be imposed between the fluid (liquid and/or gas) and/or solid phases.

Figure 1.1 gives a classification of the transport phenomena in porous media based on the single- or two-phase flow through the pores. Figure 1.2 renders these phenomena at the pore level. Description of transport of species, momentum and energy, *chemical reactions* (endothermic or exothermic) and *phase change* (solid/liquid, solid/gas, and liquid/gas) at the differential, local phase-volume level and the application of the volume averaging theories lead to a relatively accurate and yet solvable local description. In this chapter, a historical background is given first, followed by a review of applications. The length, time, and temperature scales and finally the scope of the monograph are then discussed.

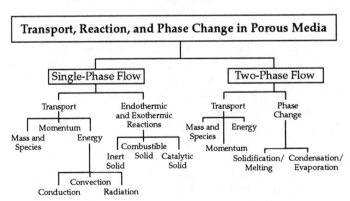

Figure 1.1 Aspects of treatment of transport, reaction, and phase change in porous media.

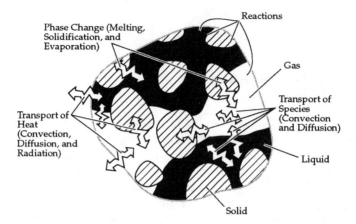

Figure 1.2 A rendering of the pore-level transport, reaction, and phase change in porous media.

1.1 Historical Background

As with any other technological problem, the treatment of fluid flow and heat transfer in porous media has been a combination of direct empirical response on the one hand and a more rigorous first-principles approach on the other hand. While most of the empirical treatments do not use any significant background knowledge base and, accordingly, are rather short-lived, the fundamental approaches build up a knowledge base. A porous medium, being a *heterogeneous* system made of a *solid matrix* with its void filled with fluids, can be treated as a continuum by properly accounting for the role of each phase in transport through this system of phases. In this monograph, we draw upon the existing knowledge base on transport in *homogeneous* systems. This vast knowledge base takes us back to the first observations/reportings of various aspects of fluid flow and heat transfer, a history that is much longer than that documented for the transport in porous media. Darcy's book, which is based on his work on water flow through beds of sands (a water filtration application), was published in 1856. By that time, single-phase flow through pipes was already analyzed by Hagen and Poiseuille, who used the fluid mechanics developed earlier by Navier, Stokes, Poisson, and de Saint Venant. Therefore, the contributions made to our understanding of transport in porous media can be further divided into those that have been *direct* (e.g., Darcy, Carman, Leverett) and those that have been *indirect* (e.g., Hagen, Knudsen, Taylor).

Table 1.1 lists in *chronological* order some of the direct and indirect contributions made to the area of porous media. The list is not intended to be exhaustive. The imposed time limit of two centuries is arbitrary.

1.1 Historical Background   3

TABLE 1.1   CONTRIBUTIONS TO TRANSPORT IN POROUS MEDIA

1800	• Young's characterization of the interface between two fluids in contact and introduction of surface tension (1805)[1].
	• Laplace's condition of static equilibrium for any point on the interface of two contacting fluids (derivation based on the potential theory, 1806)[1].
	• Fresnel's inclusion of phase to the propagation of light which explains diffraction by opaque and semitransparent particles and shows that for ray tracing by geometric optics the particle size should be much larger than the wavelength (1818)[2].
	• Navier (1827)[3]–Stokes' (1845)[3] momentum equation (also obtained by Poisson, 1931; de Saint Venant, 1943)[3].
	• Graham's study of mass diffusion in gases (1829)[4] and measurement of mass diffusivity.
	• Hagen (1839)[3]–Poiseuille (1840)[3] equation for linear flow through pipes.
1850	• Fick's application of Ohm's law for establishing the relationship between the concentration gradient and the diffusion flux (1855)[4].
	• Darcy's introduction of his empirical relation (1856).
	• Maxwell's molecular theory; distribution of velocity in gases (1859)[4], and binary gas mass diffusivity (1860)[4].
	• Marangoni's remark on surface motion of liquids toward regions of higher surface tension (thermocapillary convection or Marangoni effect), when variations of surface tension are present along the liquid surface (1872)[5].
	• Gibbs' thermodynamic treatment of the interface between a liquid and its vapor, the wetting phenomena, and introduction of surface adsorption energy and entropy (1878)[1].
1875	• Rayleigh's investigation of illumination and polarization of sunlit sky (1871)[6].
	• Maxwell's effective conductivity (1873)[7].
	• Reynolds' internal flow transition experiment (1883)[3].
1900	• Prandtl's boundary-layer theory (1904)[3].
	• Schuster (1905)[6] and Schwarzchild's (1906)[6] formulation of division of the radiation field into an outward and an inward stream. *(continued)*

TABLE 1.1 CONTRIBUTIONS TO TRANSPORT IN POROUS MEDIA (CONTINUED)

	• Mie's complete formulation of scattering of light by a homogeneous sphere (1908)[2].
	• Knudsen's experiment with flow of rarefied gases in porous media and observation of the so-called slip flow (1909)[4].
1910	• Observations on simultaneous hydrodynamic dispersion and molecular diffusion (Griffith, 1911)[8].
	• Prandtl's external flow transition experiment (1914)[3].
	• Langmuir's postulate on monolayer adsorption, where in dynamic equilibrium the rate at which gas molecules arrive at a surface and condense on bare sites is equal to the rate at which molecules evaporate from the occupied sites (1916)[9].
	• Solution of the equation of radiative transfer by expansion in spherical harmonics (Jeans, 1917).
1920	• Tollmien's prediction of the critical Reynolds number for flow over a flat plate (1921)[3].
1930	• Carman (1937)-Kozeny's (1927)[10] permeability equation based on specific area.
	• Muskat's description of two-phase (one wetting and one nonwetting) flow in porous media through introduction of an effective permeability for each phase (Muskat and Meres, 1936; Muskat et al., 1937)[11].
1940	• Leverett's introduction of idealized reduced capillary pressure function (Leverett's J function) for correlation of capillary pressure data (Leverett, 1941)[12].
	• Buckley–Leverett's frontal displacement theory for transient two-phase flows with sudden change in inlet saturation (Buckley and Leverett, 1942)[13].
	• Brinkman's modification of the Darcy law and introduction of the effective viscosity (Brinkman, 1947, 1948).
	• Solution of the equation of radiative transfer by the method of discrete ordinate (Wick, 1943; Chandrasekhar, 1950)[14].
1950	• Addition of microscopic inertial effect to the Darcy law (Ergun, 1952).
	• Measurement of the effective thermal conductivity of sintered metals (Grootenhuis et al., 1952).

TABLE 1.1   CONTRIBUTIONS TO TRANSPORT IN POROUS MEDIA (CONTINUED)

	• Introduction of transverse and longitudinal dispersion in porous media (Baron, 1952).
	• Statistical treatment of velocity dispersion (Scheidegger, 1954).
	• Hydrodynamic dispersion in tubes (Taylor, 1953; Aris, 1956).
	• Hydrodynamic and mechanical dispersions in porous media (De Josselin De Jong, 1958; Saffman, 1959).
1960	• Luikov's treatise on two-phase flow and heat transfer in porous media (the original Russian language manuscript, 1961)[15].
	• Measurement of radiative properties of fibrous and foamed insulations and packed beds (Chen and Churchill, 1963).
	• First correlation for the total effective thermal conductivity tensor for two-phase flow in porous media (Weekman and Myers, 1965).
	• A semiempirical interfacial (porous-plain media) boundary condition (Beavers and Joseph, 1967).
	• Development of constitutive equations for diffusion of gases in porous media (Mason et al., 1967; Slattery, 1970)[4].
1970	• Generalization of Aris' moment-analysis technique used for prediction of the dispersion coefficient (Horn, 1971).
	• Introduction of wavelength dependence of transition between independent and dependent radiation scattering from particle suspensions (Hottel et al., 1971).
	• Analysis of two-phase flow in porous media based on volume averaging of the point equations (Slattery, 1970; and Whitaker, 1973).
	• Experimental study of evaporation in a liquid-filled packed bed heated from below (Sondergeld and Turcotte, 1977).
1980	• Application of Horn's general analysis of dispersion to spatially periodic structures (Brenner, 1980).
	• Mapping of independent versus dependent radiation scattering for nonabsorbing dilute suspensions (Brewster and Tien, 1982). (*continued*)

TABLE 1.1   CONTRIBUTIONS TO TRANSPORT IN POROUS MEDIA (CONTINUED)

- Formulation and numerical solution of the dispersion tensor for a two-dimensional medium (Carbonell and Whitaker, 1984).
- Approximate particle-based hydrodynamics used in prediction of dispersion tensor (Koch and Brady, 1985; Koch et al., 1989).
- Dependent scattering in radiative heat transfer in packed beds (Singh and Kaviany, 1991).

---

[1] *Surface Tension and Adsorption*, by Defay and Prigogine, J. Wiley, 1966.
[2] van de Hulst's *Light Scattering by Small Particles*, Dover, 1981.
[3] Schlichting's *Boundary-layer Theory*, McGraw-Hill, 1968.
[4] *Diffusion in Gases and Porous Media*, by Cunningham and Williams, Plenum, 1980.
[5] *Low-Gravity Fluid Mechanics*, by Myshkis et al., Springer-Verlag, 1987.
[6] Chandrasekhar's *Radiative Transfer*, Dover, 1960.
[7] Batchelor–O'Brien 1977 article, Chapter 3.
[8] Taylor 1953 article, Chapter 4.
[9] *Adsorption, Surface Area and Porosity*, by Gregg and Sing, Academic, 1982.
[10] Carman's *Flow of Gases Through Porous Media*, Academic, 1956.
[11] Muskat's *The Flow of Homogeneous Fluids Through Porous Media*, International Human Resources Development, 1937.
[12] Leverett 1941 article, Chapter 9.
[13] Buckley–Leverett 1942 article, Chapter 9.
[14] Davison's *Neutron Transport Theory*, Oxford, 1957.
[15] Luikov's *Heat and Mass Transfer in Capillary-Porous Bodies*, Pergamon, 1966.

As evident, pore-level transport attracts more attention due to the need for a better understanding/prediction of fluid flow and heat and mass transfer. The most rigorous treatments/findings are for simple *unit cells*, which allow detail, pore-level analysis. The pore-level two-phase flow and heat transfer analysis is guided by new findings on *surface forces* and *dynamic wetting/dewetting*. The radiation in porous media, analyzed through dependent scattering, is just taking shape.

As we examine Table 1.1, the evolutionary nature of the progress made in the area of transport phenomena in porous media is clearly evident. The initial macroscopic treatments of 1800s, the early attempts at the microscopic (pore-level) transport in the early 1900s, and the progressively more rigorous approaches (local volume averaging) of the second half of this century are yet to be followed by the treatment of consequences of nonuniformities in the solid phase structure, the pore-level-based analysis of thermal and chemical nonequilibrium between phases, etc.

1.2 Length, Time, and Temperature Scales

The analysis of heat transfer in porous media is required in a large range of applications. The porous media can be *naturally formed* (e.g., rocks, sand beds, sponges, woods) or *fabricated* (e.g., catalytic pellets, wicks, insulations). A review of engineered porous materials is given in Schaefer (1994) and the physics and chemistry of porous media is reviewed by Banavar et al. (1987). The applications are in the areas of chemical, environmental, mechanical, and petroleum engineering and in geology. Table 1.2 gives a list of the applications in these areas. The list is just suggestive and not exhaustive. As expected, the range of *pore sizes* or *particle sizes* (when considering the solid matrix to be made of *consolidated* or *nonconsolidated* particles) is vast and can be of the order of molecular size (ultramicropores with $3 < d < 7$ Å, where d is the average pore size), the order of centimeters (e.g., pebbles, food stuff, debris), or larger. Figure 1.3 gives a classification of the particle size based on *measurement technique, application*, and *statistics*. A review of the particle characteristics for particles with diameters smaller than 1 cm is given by Porter et al. (1984).

Also shown in Figure 1.3 is the *capillary pressure* in a water-air system with the *mean radius of curvature* equal to the particle radius. It is clear that as the particle size spans over many orders of magnitude, the handling of the radiative heat transfer and the significance of forces such as capillarity and gravity also vary greatly.

Other than the *particle dimension* d, the porous medium has a *system dimension* L, which is generally much larger than d. There are cases where L is of the order d such as thin porous layers coated on the heat transfer surfaces. These systems with $L/d \simeq O(1)$ are treated by the examination of the fluid flow and heat transfer through a small number of particles, a treatment we call *direct simulation* of the transport. In these treatments no assumption is made about the existence of the local thermal equilibrium between the finite volumes of the phases. On the other hand, when $L/d \gg 1$ and when the variation of temperature (or concentration) across d is negligible compared to that across L for both the solid and fluid phases, then we can assume that within a distance d both phases are in thermal equilibrium (*local thermal equilibrium*). When the solid matrix structure cannot be *fully* described by the prescription of solid phase distribution over a distance of d, then a *representative elementary volume* with a linear dimension larger than d is needed. We also have to extend the requirement of a negligible temperature (or concentration) variation to that over the *linear dimension of the representative elementary volume* ℓ. For *some* fabricated solid matrices $\ell/d \simeq O(1)$, for natural solid matrices we have $\ell/d \simeq O(10$ or larger$)$. In addition to d, ℓ, and L, a length scale equal to the square root of the permeability is also used. This

TABLE 1.2 EXAMPLES OF VARIOUS INTERESTS IN TRANSPORT THROUGH POROUS MEDIA

CHEMICAL	• Catalytic and inert packed bed reactors (gaseous or aqueous). Filtering. Drying. Trickle bed reactors. Packed bed chromatography. Catalytic converter for air pollution reduction of combustion products. Adsorption/desorption at surfaces. Mass transfer through membranes. Pharmaceutical product formations in bioreactors and in powder and tablet synthesis. Fuel cells.
ENVIRONMENTAL	• Groundwater flow. Contamination migration in groundwater. Injection of grout (plaster) into soil. Air, water-vapor, and water flow through construction materials. Salt water intrusion into coastal aquifers. Waste (radioactive and stable) disposal. Irrigation. Soil cleanup by steam injection. Water percolation in snow. Incineration.
GEOLOGICAL	• Water and mineral migration. Geothermal energy management/harvesting. Thermal cycling of rocks. Glaciological transport.
MECHANICAL	• Single- and two-phase transpiration cooling. Wicked heat pipes. Insulations. Combustion and fire involving pyrolysis of matrices or inert matrices. Drying efficiency. Geothermal energy harvesting. Enhanced heat transfer by surface modification. Tribology and lubrication. Nuclear reactors using gaseous coolants flowing through radioactive pellets. Melting/solidification of binary mixtures. Dehumidifying. Radiant porous burners. Porous preheaters and flame stabilizers. Storage of absorbed solar energy. Catalytic converters and soot traps for automobile emission. Safety as in fires involving porous building materials and in forests. Mold and core-sand formation. Particle sintering by thermal irradiation.
PETROLEUM	• Oil and gas flow in reservoirs. Enhanced oil production. Oil-shale harvesting, including in-situ combustion. Natural gas production.

1.2 Length, Time, and Temperature Scales

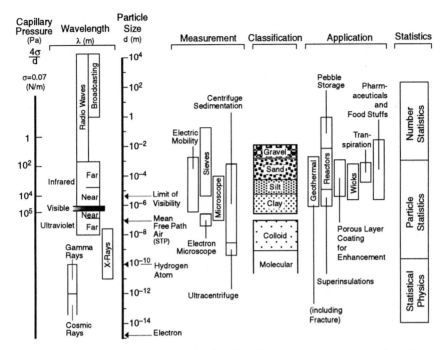

Figure 1.3 Particle sizes and their classifications, measurements, and applications.

length scale (called the *Brinkman screening distance*) $K^{1/2}$ is smaller than d and is generally $O(10^{-2}d)$, where K is the *permeability*. The *four length scales*, with a loose requirement for the presence of local thermal equilibrium (based on the length scale only), are written as

$$K^{1/2} \ll d < \ell \ll L.$$

The ranges for these length scales are given in Table 1.3.

The *time scales* associated with heat *diffusion* across these lengths, i.e., K/α_e, d^2/α_e, ℓ^2/α_e, and L^2/α_e, and those associated with fluid particle *residence* time while passing over these lengths, i.e., $K^{1/2}/u_0$, d/u_0, ℓ/u_0, and L/u_0 are also shown in Table 1.3. The symbols are defined in Nomenclature which follows Chapter 12. A range of typical values for the effective thermal diffusivity α_e and the fluid velocity u_0 are used. As shown, both very *small* diffusion and convection time scales and very *large* time scales are encountered.

The variations of the temperature drop occurring over the four length scales, with a loose requirement for the existence of the local thermal equilibrium imposed, are also shown in Table 1.3. In general, in the presence of heat generation in the solid or fluid phases, the requirements for the local thermal equilibrium assumption are not satisfied. Then a *two-medium*

TABLE 1.3 LENGTH, TIME, AND TEMPERATURE SCALE REQUIREMENT FOR SINGLE CONTINUUM TREATMENT

BRINKMAN SCREENING DISTANCE	PORE SIZE (OR PARTICLE SIZE)	REPRESENTATIVE ELEMENTARY VOLUME	LINEAR DIMENSION OF SYSTEM
LENGTH SCALES			
$K^{1/2}$	d	ℓ	L
$K^{1/2} \ll d$	$d < \ell$	$\ell \ll L$	
$10^{-12} - 10^{-3}$ m	$10^{-10} - 10^{-2}$ m	$10^{-8} - 1$ m	$10^{-6} - 10^{2}$ m
TIME SCALES			
$K^{1/2}$	d	ℓ	L
0	$\dfrac{d^2}{\alpha_e}$, $10^{-15} - 10$ s	$\dfrac{\ell^2}{\alpha_e}$, $10^{-11} - 10^{5}$ s	$\dfrac{L^2}{\alpha_e}$, $10^{-7} - 10^{7}$ s
$\dfrac{K^{1/2}}{u_0}$†, $0 - 10^{3}$ s	$\dfrac{d}{u_0}$, $0 - 10^{4}$ s	$\dfrac{\ell}{u_0}$, $10^{-3} - 10^{6}$ s	$\dfrac{L}{u_0}$, $10^{-1} - 10^{8}$ s
TEMPERATURE SCALES			
$K^{1/2}$	d	ℓ	L
$\Delta T_{K^{1/2}} = 0$	$\Delta T_{K^{1/2}} < \Delta T_d$ 0	$\Delta T_d < \Delta T_\ell \ll \Delta T_L$	ΔT_L‡
0	$0 - 10^{-3}\,°\text{C}$	$0.1 - 10\,°\text{C}$	$1 - 10^{3}\,°\text{C}$

†Velocities as low as 0.1 mm/day and as high as sonic (gases) are encountered.
‡Geothermal gradients are about 0.025°C/m, while in two-phase transpiration cooling and in flames in porous media the gradients can be as high as $10^{4} - 10^{5}\,°\text{C/m}$.

treatment (different *local* solid and fluid temperatures) must be made. Also, when ΔT_d is not much smaller than ΔT_L (such as in flames in porous media), a direct simulation of the transport, which includes the microstructure over a domain of the order of d, must be made.

1.3 Scope

The solutions to the conservation equations, marked by the boundary-layer analyses of Cheng and Minkowycz (1977) and Vafai and Tien (1981) and as reviewed by Nield and Bejan (1991) and as part of the general treatment of convective heat transfer by Kaviany (1994), are not addressed here. Yet, by undertaking the discussion of principles of the fluid flow, conduction, convection, and radiation for both the single- and two-phase flows in porous media, such treatments must lack the depth that these topics deserve. However, as much as the space and our ability permits, we discuss the microscopic (pore-level) and macroscopic (local volume-averaged) analysis of the momentum, heat, and mass transfer in porous media. We show that the Stokes and the laminar-inertial single-phase flow and heat transfer through matrices with simple unit-cell structures are now readily solvable. Then, the *permeability*, the *effective thermal conductivity*, and the *dispersion tensors* can be predicted. The governing conservation equations are reached by the method of local volume averaging and these tensors are evaluated from the pore-level point solution to the momentum and energy equations (or slightly modified versions of them). Turbulent flows and flow and heat transfer through complicated unit cells randomly arranged in the solid matrix are generally approached empirically. Flow nonuniformities and dispersion at and near surfaces bounding the porous media are also discussed in detail.

For radiation heat transfer in porous media, the same unit-cell approach is used and the particles in each cell are treated as scatterers. As with the hydrodynamic interactions among cells, the scattering also becomes dependent when the porosity is not close to unity. The radiation properties are related to the optical properties and the porosity, and both the local volume-averaged and direct simulation treatments are examined.

The two-phase flow in porous media (a three-phase system) is approached from the pore-level fluid mechanics, and the difficulties associated with the analysis of wetting-dewetting and the direct simulation of even the simplest of the flows are discussed. However, the pertinent forces and their expected contributions are examined, and when available, empirical results are used. After arriving at a set of volume-averaged governing conservation and constituitive equations, the boundary conditions at the surfaces bounding the partially saturated porous media, and some liquid-vapor phase change problems, are examined in detail.

Except in Chapter 7, the local thermal equilibrium is assumed to exist between the phases. The two-phase flow treatments are for *nonhygroscopic media*, i.e., the liquid and its vapor coexist and are allowed only in the pore spaces through which they move. A glossary of common words used in transport in porous media, is given in the Glossary.

1.4 References

Aris, R., 1956, "On the Dispersion of a Solute in a Fluid Flowing Through a Tube," *Proc. Roy. Soc. (London)*, A235, 67–77.

Baron, T., 1952, "Generalized Graphic Method for the Design of Fixed Bed Catalytic Reactors," *Chem. Eng. Prog.*, 48, 118–124.

Banavar, J., Koplik, J., and Winkler, K. W., 1987, *Physics and Chemistry of Porous Media*, AIP Conference Proceeding 154, AIP.

Beavers, G. S. and Joseph, D. D., 1967, "Boundary Condition at a Naturally Permeable Wall," *J. Fluid Mech.*, 30, 197–207.

Brenner, H., 1980, "Dispersion Resulting from Flow Through Spatially Periodic Porous Media," *Phil. Trans. Roy. Soc.* (London), 297, 81–133.

Brewster, M. Q. and Tien, C.-L., 1982, "Examination of the Two-Flux Model for Radiative Transfer in Particulate Systems," *Int. J. Heat Mass Transfer*, 12, 1905–1907.

Brinkman, H. C., 1947, "A Calculation of the Viscous Force Exerted by a Flowing Fluid on a Dense Swarm of Particles," *Appl. Sci. Res.*, A1, 27–34.

Brinkman, H. C., 1948, "On the Permeability of Media Consisting of Closely Packed Porous Particles," *Appl. Sci. Res.*, A1, 81–86.

Carbonell, R. G. and Whitaker, S., 1984, "Heat and Mass Transfer in Porous Media," in *Fundamentals of Transport Phenomena in Porous Media*, Bear and Corapcioglu, eds., Martinus Nijhoff Publishers, 121–198.

Carman, P. C., 1937, "The Determination of the Specific Surface Area of Powders. I," *J. Soc. Chem. Ind.*, 57, 225–234.

Chen, J. C. and Churchill, S. W., 1963, "Radiant Heat Transfer in Packed Beds," *AIChE J.*, 9, 35–41.

Cheng, P. and Minkowycz, W. J., 1977, "Free Convection about a Vertical Flat Plate Embedded in a Porous Medium with Application to Heat Transfer from a Dike," *J. Geophys. Res.*, 82, 2040–2044.

Darcy, H., 1856, *Les Fontaines Publiques de la ville de Dijon*, Dalmont, Paris.

De Josselin De Jong, G., 1958, "Longitudinal and Transverse Diffusion in Granular Deposits," *Trans. Amer. Geophys. Union*, 39, 67–74.

Ergun, S., 1952, "Fluid Flow Through Packed Column," *Chem. Eng. Prog.*, 48, 89–94.

Grootenhuis, P., Powell, R. W., and Tye, R. P., 1952, "Thermal and Electrical Conductivity of Porous Metals Made by Powder Metallurgy Method," *Proc. Phys. Soc.*, B65, 502–511.

Horn, F. J. M., 1971, "Calculation of Dispersion Coefficient by Means of Moments," *AIChE J.*, 17, 613–620.

Hottel, H. C., Sarofim, A. F., Dalzell, W. H., and Vasalos, I. A., 1971, "Optical Properties of Coatings, Effect of Pigment Concentration," *AIAA J.*, 9, 1895–1898.

Jeans, J. H., 1917, "The Equation of Radiative Transfer of Energy," *Monthly Notices of Royal Astronomical Society*, 78, 445–461.

Kaviany, M., 1994, *Principles of Convective Heat Transfer*, Springer-Verlag.

Koch, D. L. and Brady, J., 1985, "Dispersion in Fixed Beds," *J. Fluid Mech.*, 154, 399–427.

Koch, D. L., Cox, R. G., Brenner, H., and Brady, J., 1989, "The Effect of Order on Dispersion in Porous Media," *J. Fluid Mech.*, 200, 173–188.

Nield, D. A. and Bejan, A., 1992, *Convection in Porous Media*, Springer–Verlag.

Porter, H. F., Schurr, G. A., Wells, D. F., and Seurau, K. T., 1984, "Solid Drying and Gas–Solid Systems," in *Perry's Chemical Engineer's Handbook*, McGraw-Hill, 20–79.

Saffman, P. G., 1959, "A Theory of Dispersion in a Porous Medium," *J. Fluid Mech.*, 6, 321–349.

Schaefer, D. W., 1994, "Engineered Porous Materials," *MRS Bull.*, XIX, 14–17.

Scheidegger, A. E., 1954, "Statistical Hydrodynamics in Porous Media," *J. Appl. Phys.*, 25, 994–1001.

Singh, B. P. and Kaviany, M., 1991, "Independent Theory versus Direct Simulation of Radiation Heat Transfer in Packed Beds," *Int. J. Heat Mass Transfer*, 34, 2869–2881.

Slattery, J. C., 1970, "Two–Phase Flow Through Porous Media," *AIChE J.*, 16, 345–352.

Sondergeld, C. H. and Turcotte, D. L., 1977, "An Experimental Study of Two–Phase Convection in a Porous Media with Applications to Geological Problems," *J. Geophys. Res.*, 82, 2045–2052.

Taylor, G. I., 1953, "Dispersion of Soluble Matter in Solvent Flowing Slowly Through a Tube," *Proc. Roy. Soc.* (London), A219, 186–203.

Vafai, K. and Tien, C.-L., 1981, "Boundary and Inertia Effects on Flow and Heat Treansfer in Porous Media," *Int. J. Heat Mass Transfer*, 24, 195–243.

Weekman, V. W. and Meyers, J. E., 1965, "Heat Transfer Characteristics of Cocurrent Gas–Liquid Flow in Packed Beds," *AIChE J.*, 11, 13–17.

Whitaker, S., 1973, "The Transport Equations for Multi–Phase Systems," *Chem. Engng. Sci.*, 28, 139–147.

Part I

Single-Phase Flow

2
Fluid Mechanics

In this chapter we examine the *isothermal, single-phase* flow through porous media by starting from the Darcy law. We then examine the permeability tensor and its relation with the matrix structure. The deviation from the Darcy law observed at high velocities and the *pore-level* fluid dynamics are then examined. Attempts at arriving at the Darcy law (the macroscopic momentum equation) from the point description of the flow field (Navier-Stokes equation) by the *local volume-averaging* and *homogenization* techniques are reviewed. Then, a *semiempirical* momentum equation, which includes the *bulk* and *boundary viscous effects*, the *flow development* in porous media, and the *high-velocity effects*, is given. The significance of these terms is assessed by the order-of-magnitude analyses and some estimations. When the porous media are bounded by the fluid occupying them, the *hydrodynamic boundary conditions* on these *interfaces* must be specified. The available *slip velocity model* and the *Brinkman no-slip* (uniform and variable effective viscosity) model for this interface are examined. The dependence of the *slip coefficient* (and the *interfacial effective viscosity*) on the bulk and surface structures of the matrix are studied. The *velocity nonuniformities* observed near the bounding impermeable surfaces and the various theoretical treatments of them are investigated. The chapter ends with an examination of the analogy between porous media- and magnetohydrodynamics.

2.1 Stokes Flow and Darcy Equation

The *bulk*† resistance to flow of an *incompressible* fluid through a solid matrix, as compared to the resistance at and near the surfaces confining this solid matrix, was first measured by Darcy (1856). Since in his experiment the internal surface area (*interstitial area*) was many orders of magnitude larger than the area of the confining surfaces, the *bulk shear stress resistance* was dominant.

His experiment used nearly *uniform-size particles* that were *randomly* and *loosely packed*, i.e., a *nonconsolidated, uniform, rigid,* and *isotropic* solid matrix. The macroscopic flow was steady, one-dimensional, and driven by gravity. A schematic of the flow is given in Figure 2.1. The mass flow

---

† A glossary of common words used in transport in porous media, is given at the end of the monograph.

18 2. Fluid Mechanics

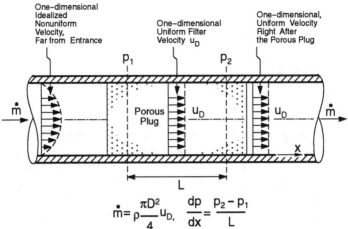

$$\dot{m} = \rho \frac{\pi D^2}{4} u_D, \quad \frac{dp}{dx} = \frac{p_2 - p_1}{L}$$

Figure 2.1 Determination of the filter (or Darcy) velocity.

rate of the liquid $\dot{m}$ was measured and the *filtration* or *filter velocity* u_D was determined by dividing $\dot{m}$ by the product of the fluid density (assumed *incompressible*) and the cross-sectional area A of the channel (which was filled with the particles and then the liquid was flown through it). In applying a volumetric force balance to this flow, he discovered that the bulk resistance can be characterized by the viscosity of the Newtonian fluid (fluid parameter) μ and the *permeability* of the solid matrix (solid matrix parameter) K, such that

$$-\frac{dp}{dx} = \frac{\mu}{K} u_D, \qquad (2.1)$$

where the dimension of K is in square of length, e.g., m². Interpretation of (2.1)† has evolved and one unit still used is *darcy*.‡ One unit of darcy equals 9.87×10^{-13} m². The permeability accounts for the interstitial surface area, the fluid particle path as it flows through the matrix, and other related hydrodynamic characteristics of the matrix. The *Darcy model* has been examined rather extensively and is not closely followed for liquid flows at high velocities and for gas flows at very low and very high velocities.

The fluid flow through the matrix can be viewed as flow *through tortuous conduits* (*capillary* models) or as flow *over objects* (*drag* models). These *phenomenological* models will be reviewed later.

Equation (2.1) states a linear relationship that is characteristic of the *Stokes flow* (viscous force domination over inertial force). This explains the

---

† Numbers in parenthesis refer to equation numbers.
‡ One darcy is the permeability of a matrix when a cubic sample with each side having a width of one cm is used and a fluid with viscosity of one centipoise is flown (one-dimensional) through it, resulting in a pressure drop of one atmosphere (1.013×10^5 Pa).

deviation of the actual flows from (2.1) at high velocities. Noting that in the pores (the void space through which the fluid flows) the velocity is higher than the Darcy velocity by $1/\varepsilon$, where ε is the porosity, then (2.1) can be compared to the Stokes flow counterpart, i.e.,

$$\frac{\mathrm{d}p}{\mathrm{d}x} = \frac{\mu}{\varepsilon}\nabla^2 u_D. \qquad (2.2)$$

At low gas pressures and for small pore size, the mean free path of the gas molecules may be of the order of the pore size and therefore velocity slip occurs (Knudsen effect), resulting in higher permeabilities. This subject will be treated along with mass diffusion in Chapter 6. However, an increase in the permeability due to an increase in gas pressure has been found in some experiments. Scheidegger (1974) discusses the effect of the Knudsen slip, the internal surface roughness, surface absorption, capillary condensation, and molecular diffusion on the measured permeability. By examining these effects at the pore level, it becomes clear that the measured *gas* and *liquid* permeabilities can be noticeably *different*.

For *isotropic media* where the pressure gradient ∇p and the velocity vector $\mathbf{u}_D$ are *parallel*, (2.1) is generalized to

$$-\nabla p = \frac{\mu}{K}\mathbf{u}_D. \qquad (2.3)$$

For *anisotropic media*, in general, these two vectors are *not parallel* and a linear transformation can be made using the *permeability tensor*,† i.e.,

$$-\nabla p = \frac{\mu}{\mathbf{K}}\mathbf{u}_D. \qquad (2.4)$$

The *second-order* tensor $\mathbf{K}$ has nine components, i.e.,

$$\mathbf{K} = \begin{bmatrix} K_{11} & K_{12} & K_{13} \\ K_{21} & K_{22} & K_{23} \\ K_{31} & K_{32} & K_{33} \end{bmatrix}. \qquad (2.5)$$

Since $\mathbf{K}$ is symmetric, then $K_{12} = K_{21}$, etc. Scheidegger has verified this tensor theory of permeability using some experimental results. For *orthotropic* (having three mutually *orthogonal principal axes*) *media*, the tensor is symmetric and a diagonal matrix is formed whenever the coordinate axes are parallel with the *principal axes* along which the components

---

† A review of tensors can be found in *Introduction to Continuum-Mechanics*, 1978, by Lai, Rubin, and Krempl, Pergamon; in *Transport Phenomena*, 1960, by Bird, Stewart, and Lightfoot, Wiley (a modified version is given in *Dynamics of Polymeric Liquids*, Vol. 1, 1987, by Bird, Armstrong, and Hassager, Wiley); in *Momentum, Energy, and Mass Transfer in Continua*, 1981, by Slattery, Krieger; and in *Vectors, Tensors, and the Basic Equations of Fluid Mechanics*, 1989, by Aris, Dover.

of the permeability tensor are $K_{ii} = K_i$ (Dullien, 1979, 215–219).† Then ∇p and $\mathbf{u}_D$ are parallel

$$-\frac{\partial p}{\partial x_i} = \frac{\mu}{K_i} u_{Di}. \tag{2.6}$$

In practice, the permeability is not necessarily measured along the principal axes, but rather in a direction $\mathbf{n}$, which makes angles α, β, and γ with the principal axes. The measured *directional permeability* K_n (Scheidegger)‡ is related to the principal axes values. Defining the directional permeability as

$$-\mathbf{n} \cdot \nabla p = -\frac{\partial p}{\partial x_n} = \mu \left(\mathbf{n} \cdot \mathbf{K}^{-1} \cdot \mathbf{n} \right) u_{Dn} = \frac{\mu}{K_n} u_{Dn}, \tag{2.7}$$

we have

$$\frac{1}{K_n} = \frac{\cos^2 \alpha}{K_1} + \frac{\cos^2 \beta}{K_2} + \frac{\cos^2 \gamma}{K_3}. \tag{2.8}$$

The polar plot of $K_n^{1/2}$ results in an ellipsoid where the axes are the principal axes and their magnitudes are the square root of the principal permeabilities. These treatments are for rigid matrices. In allowing for *compressibility* of the fluid, i.e., $\partial \rho / \partial p \neq 0$, the mass conservation equation allowing for compressibility needs to be used along with the Darcy law. Note that the matrix may also undergo *elastic* or *plastic* deformations. Treatment of the fluid and matrix compression, assuming elastic behaviors, has been reviewed by Scheidegger.

In this chapter it is generally assumed that the fluid can be treated as a *continuum*, i.e., the *Knudsen* number is small and the fluid is *incompressible*. The low pressures or small pore gas flows, where the Knudsen number can be large, is discussed in Chapter 6.

2.2 Porosity

The volume fraction occupied by voids, i.e., the total void volume divided by the total volume occupied by the solid matrix and void volumes, is called the *porosity*. Each void is connected to more than one other pore (*interconnected*), connected only to one other pore (*dead end*), or not connected to any other pore (*isolated*). Fluid flows through the interconnected pores only. The volume fraction of the interconnected pores is called the *effective porosity*. In nonconsolidated media, e.g., particles loosely packed, the effective porosity and porosity are equal. In some consolidated media, the difference between the two can be substantial.

---

† The second numerical entry in the parentheses is the page number.
‡ The year of citation is eliminated after the first reference, unless there are multiple years or when first reference was made in connection with another topic.

For rigid matrices, the porosity ε does not change in the presence of a pressure gradient. In deformable matrices $\partial \varepsilon / \partial p \neq 0$, the extent of change in the porosity depends on the structural properties and elastic versus plastic deformation. Each void is generally represented by one *pore linear length scale d*. This is an *idealization*, because the void shape is not regular. Scheidegger recommends a *distributed linear scale (two-dimensional)* for each void where the linear scale is the diameter of a sphere that as closely as possible occupies the cross section at a given location along the pore. Then a statistical average of this distribution of linear length scales leads to an *average pore diameter* and an *average pore length*. Other definitions of the local pore length scale consider a certain conduit geometry (Dullien, 1992; Adler, 1992).

In general, the voids are nonuniform in their size and in their distribution throughout the matrix. The *nonuniformity* of the void size and the distribution in the bulk of the matrix are usually presented through distribution of statistically averaged local values (along with deviations). These statistical averages are taken over volumes that are larger than the pore volume but much smaller than the volume of the entire solid matrix. This intermediate volume is called the *representative elementary volume*. This is the smallest differential volume that results in statistically meaningful local average properties such as porosity, saturation, and capillary pressure. When the representative elementary volume is appropriately chosen, incremental addition of extra pores (and the solid around them) does not change the magnitude of these local properties. For example, for a matrix made of uniformly sized particles that are regularly arranged, this representative elementary volume includes only a few particles. In a solid matrix with elementary structures that have variations in particle size and are randomly arranged, this may be tens of particles.

The nonuniformities near the boundaries (confining solid or free surfaces) can play a significant role in the transfer rates at the boundaries and should be treated with meaningful local distributions (i.e., different than the bulk of properties). For example, the packing arrangement of spheres near the boundaries is different than the bulk. When the confining boundary is a solid surface, the larger porosities adjacent to this surface result in reduction of the resistance to the flow and, therefore, to an increase in local velocity in this area. This is called *channeling* and occurs over a distance of several particle diameters from the surface. In general, channeling is predicted by using distributions of the average local porosity, which do not totally violate the requirements for the representative elementary volume (i.e., the representative elementary volume is taken such that it gives meaningful values of local porosity). However, when the region over which the channeling occurs is only a few particle diameters in length, then the continuum treatment and application of local porosity fail, and the fluid dynamics (and heat transfer) of the region adjacent to the boundary should

TABLE 2.1   EXAMPLES OF AVERAGE, BULK POROSITIES (ADAPTED FROM SCHEIDEGGER, 1974)

SUBSTANCE	POROSITY
Foam metal (also made of polyurethane and other materials)	0.98
Fiberglas	0.88–0.93
Berl saddles	0.68–0.83
Wire crimps	0.68–0.76
Silica grains	0.65
Black slate powder	0.57–0.66
Raschig rings	0.56–0.65
Leather	0.56–0.59
Catalyst (Fischer-Tropsch, granules only)	0.45
Granular crushed rock	0.44–0.45
Soil	0.43–0.54
Sand	0.37–0.50
Silica powder	0.37–0.49
Spherical packings, well shaken	0.36–0.43
Cigarette filters	0.17–0.49
Brick	0.12–0.34
Hot-compacted copper powder	0.09–0.34
Sandstone (oil sand)	0.08–0.38
Limestone, dolomite	0.04–0.10
Coal	0.02–0.12
Concrete (ordinary mixes)	0.02–0.07

be treated using the point solutions to fields (direct simulations). This will be discussed further in Sections 2.11 and 2.12.

Examples of average bulk porosities are given in Table 2.1 which is an adaptation of the data given by Scheidegger. Measurement of porosity is made using several techniques (Dullien, 1992). Some measure the true porosity and some measure the effective porosity. Table 2.2 names some of the techniques, their principles, the quantity measured, and the porosity obtained (true or effective). The loose packing of spherical particles can be achieved with various porosities for uniform-size spheres. These porosities are independent of the sphere diameter. For uniform-size spheres, the smallest porosity is for the close-pack *rhombohedral* (or *face-centered cubic*) *arrangement*, which gives a porosity of 0.259. The *random* packing of uniformly sized spheres give porosities of 0.38 to 0.41, although 0.37 to 0.43 have also been reported. As will be shown, this variation is in part due to the various ratios of particle diameter to container diameter used in the experiments. When spheres of different diameters are combined, a variety of pore structures and sizes can be constructed.

TABLE 2.2 SOME POROSITY MEASUREMENT TECHNIQUES (ADAPTED FROM DULLIEN, 1979)

PRINCIPLE	QUANTITY MEASURED	POROSITY	POSSIBILITY OF LOCAL POROSITY MEASUREMENT
Direct: Bulk volume compared with crushed volume.	Volume	True	No
Photography (two-dimensional): Sum of areas of solid compared to sum of areas of voids.	Area	True	Yes
Imbibition (wetting liquid): Mass of matrix when fully imbibed with a wetting liquid compared with a mass of dry matrix.	Mass and volume	Effective	No
Mercury injection (nonwetting liquid): Volume of mercury penetrating into the matrix.	Volume	Effective	No
Gas injection: Pressure in contained housing the matrix before and after expansion (via connection to a second container).	Pressure and volume	Effective	No
Attenuation (γ ray): Attenuation in the intensity as the beam travels through the matrix, compared with that through a solid slab of the same thickness.	Intensity	True	Yes

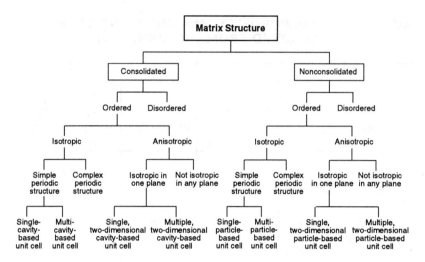

Figure 2.2 A classification of matrix structure based on ease/difficulty of detail hydrodynamic/heat transfer analysis.

Again, it should be noted that the bulk porosity is different from that near the boundaries of the solid matrix. Depending on the preparation of the surface of these boundaries, higher or smaller porosity are found near the surface and through the *penetrative* (or *skin-type*) *depth*.

2.3 Pore Structure

Pores can be very large (often called *caverns*) or very small (of the order of atomic or molecular size, called *micropores* or *ultramicropores*). They are *three-dimensional* and some of them are connected. The scanning electron microscope has been used in the study of the three-dimensional pore structures of micropores. Figure 2.2 gives a classification of the matrix structure. In packed beds of particles, the particles may eventually become *consolidated* due to physical and chemical reactions. The major divisions are based on *ordered* versus *disordered* and *isotropic*. The structures which are most suitable for a *deterministic*, thorough analysis are *simple periodic structures*.

Figure 2.3(a) is an electron micrograph of a tight stainless-steel screen. Flow through screens is encountered in filtering, paper machines, etc. Another application is as a phase separator in cryogenic propellant tanks. The screen is part of the *liquid acquisition device*, which allows liquid withdrawal from the tank under zero gravity. The flow through tight screens follows the Darcy law at low flow rates and the Ergun correlation at high

2.3 Pore Structure 25

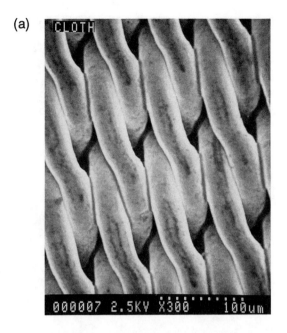

Figure 2.3 (a) Electron micrograph of a stainless screen. The wire counts are 200×1500 (wire counts per inch in each direction). $\varepsilon = 0.358$, $A_o = 6.67 \times 10^4$ m^{-1}, and pore diameter $= 10$ μm. Here the length of the dotted line is 100 μm. (b) A stainless-steel fiber collection (wool) used as a wick.

(a)

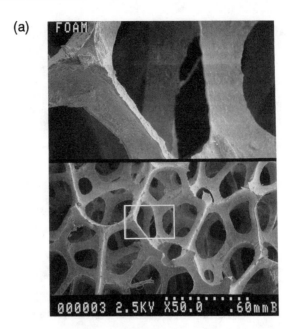

(b)

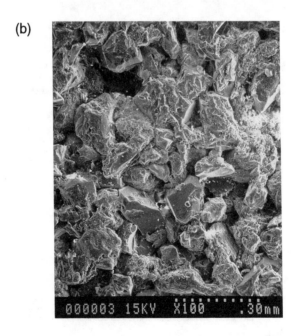

Figure 2.4 (a) Electron micrograph of a polyurethane foam. The number of pores per inch is 100, $\varepsilon = 0.98$, and pore diameter = 20 μm. (b) Electron micrograph of the surface of a cut sandstone piece.

flow rates (Armour and Cannon, 1968). Figure 2.3(b) shows an electron micrograph of a tight stainless-steel fiber collection used as a wick for supply of liquid fuel to a combustion chamber. Figure 2.4(a) is the electron micrograph of a polyurethane foam (used as filter and insulation). Figure 2.4(b) shows the surface structure of a piece of sandstone. The surface particles are different than those inside, because of the surface cuts. Dullien (1979, p. 90) gives three criteria for characterizing the pore structure, namely, *dimensionality* of the network (generally three-dimensional), *pore coordinate number* (the number of connections made to a pore volume), and *microscopic topology* of the network (given by parameters such as *shape, structure*, and *connectivity*).

In the following, we list the matrix structures in order of their simplicity.

- Very long straight cylinders (circular and other cross sections). The simplest land (and *anisotropic*) arrangement is that of two-dimensional structures (long cylinders). The plane can be taken parallel or perpendicular to the axis of the cylinders. The two-dimensional pore structure can be analyzed for any arrangement of cylinders (periodic or nonperiodic arrangement, dense or dilute packing, in-line or staggered, consolidated or nonconsolidated), i.e., the porosity can be very small or very large and a distribution in the particle size can also be incorporated.

- Spheres (and other three-dimensional particles). The isotropic three-dimensional pore structure (pore topology) can be readily analyzed for some cases. The analysis is most difficult for nonperiodic consolidated (dense) packing of polydispersed (variations in particle size) particles.

- Short and long fibers (circular and other cross sections). For certain arrangements, a deterministic analysis of this *anisotropic* structure is possible, but for three-dimensional weaving of fibers including loose-end arrangements, only some stochastic analyses have been performed. The analysis is most difficult for dense packing with natural weaving (anisotropic) with variation in the fiber size (diameter and length).

- Rocks, silicon gels, coal, wood plasters, cements, fabrics, biological tissues, fruits and vegetables, food-stuff, cereals, pharmaceutical products, leathers, papers, thermal insulators, etc., have complex topology and are in general very difficult to analyze.

These complexities have inspired many to study pore structure and to reduce the parameters to a few, such that the problem lends itself to analysis (deterministic or stochastic). Dullien has reviewed this work and has recommended conduit/network models for understanding/interpretation of the

TABLE 2.3 BULK PERMEABILITY OF SOME MATRICES (ADAPTED FROM SCHEIDEGGER, 1974)

MATRIX	PERMEABILITY (m^2)
Sandstone (oil sand)	5.0×10^{-16} to 3.0×10^{-12}
Brick	4.8×10^{-15} to 2.2×10^{-13}
Limestone, dolomite	2.0×10^{-15} to 4.5×10^{-14}
Leather	9.5×10^{-14} to 1.2×10^{-13}
Black slate powder	4.9×10^{-14} to 1.2×10^{-13}
Agar-agar	2.0×10^{-14} to 4.4×10^{-13}
Silica powder	1.3×10^{-14} to 5.1×10^{-14}
Soils	2.9×10^{-13} to 1.4×10^{-11}
Bituminous concrete	1.0×10^{-13} to 2.3×10^{-11}
Fiberglas	2.4×10^{-11} to 5.1×10^{-11}
Sand (loose beds)	2.0×10^{-11} to 1.8×10^{-10}
Hair felt	8.3×10^{-10} to 1.2×10^{-9}
Cork board	3.3×10^{-10} to 1.5×10^{-9}
Wire crimps	3.8×10^{-9} to 1.0×10^{-8}
Cigarette	1.1×10^{-9}
Berl saddles	1.3×10^{-7} to 3.9×10^{-7}

flow characteristics in porous media. He suggests conduits (capillary channels) network models be examined not only in light of the prediction of steady-state single-phase flows, but for two-phase flows in conjunction with the imbibition (invasion) and drainage, including the hysteresis observed in the capillary pressure-saturation curve (these will be discussed in the treatment of two-phase flow in Chapter 8). The capillary-network model requires a pore-size distribution (probability density function giving the distribution of pore volume), which is to be obtained experimentally using the mercury intrusion porosimetry method (or the surface adsorption/desorption measurement), along with microscopy.

More complex geometries are described by fractal geometry (Quiblier, 1984; Adler, 1992).

2.4 Permeability

According to the Darcy law, the permeability K is the measure of the flow conductance of the matrix, i.e., $-\mu u_D/(dp/dx)$. Table 2.3 gives examples of the magnitude of the permeability for some solid matrices (Scheidegger, 1974). The bulk conductance describes the entire bulk hydrodynamic behavior of the flow (when the Darcy law is valid). The bulk hydrodynamic behavior can also be obtained from the application of the first principles (mass and momentum conservation) to flow of viscous fluids at the pore level (including pore surface roughness). However, such an application is a very large undertaking for complex geometries and has only been applied in a few simple structures. This is because the geometry of the fluid conduit is three-dimensional and complex. For complex geometries the paths taken are in application of the Navier-Stokes equation to fluid flow through complex channels (*capillary models*), application of these equations to flow over objects (*drag models*), and a semiheuristic approach of *hydraulic radius*. The first two are being pursued more rigorously as part of modern fluid dynamics.

No general relationship between effective porosity and permeability exists. The few existing empirical, semiempirical, and first-principle-based correlations all have to be used within the restrictions for which they have been developed. Furthermore, as bulk properties, these relationships have not been examined for cases where $\nabla \varepsilon \neq 0$. When used as the *local* permeability, their application to nonuniform matrices must be restricted (if at all applicable) to cases where the gradient of porosity is rather small. In the following we examine the capillary, hydraulic radius, and drag models.

The relationship between the permeability (i.e., momentum transport) and other diffusional-transport effective properties has been discussed by Torquato and Kim (1992).

2.4.1 CAPILLARY MODELS

These models involve application of the Navier-Stokes equation to flow in small-diameter conduits. The flow is assumed to be fully developed (the diameter-based Reynolds number is small enough for the hydrodynamic entrance effects to be negligible) and steady. Ensemble averaging of the pressure drop is used to allow for variation of the conduit diameter.

The average velocity in a conduit is called the *pore velocity*. Each representative elementary volume contains many conduits and, therefore, a statistical average over all local (the pores and the representative elementary volume) velocities must be taken in order for the pore velocity (and average porosity) to be meaningful. Then the filter velocity u_D is related to the pore velocity u_p by $u_p = u_D/\varepsilon$ (this is called *Dupuit-Forchheimer* relation). Scheidegger (1974, 126–135) and Dullien have reviewed the conduit/network models. The simple parallel arrangement of constant and variable cross-section conduits *does not* predict the experimentally observed relationship between the pressure drop and conduit size (and distribution) and porosity. However, the more complex network arrangement (which uses the experimentally obtained pore-entry diameter distribution) has had some success (Dullien, 1979, 181–190).

It should be kept in mind that these models have been proposed for obtaining relationships between permeability (pressure drop) and the matrix property parameters (porosity and other structure variables) for some simple structure such as the matrices made of packed spheres or fibers and, in

(A) PARALLEL/SERIES CONDUIT

For flow along straight tubes of diameter d with the cross-sectional distribution of n tubes per unit area, the porosity ε is $n\pi d^2/4$. Assuming fully developed flow, the pressure drop is found from the integration of the one-dimensional Navier-Stokes equation, which results in the *Hagen-Poiseuille* equation, i.e.,

$$u_p = -\frac{d^2}{32\mu}\frac{dp}{dx}, \tag{2.9}$$

where u_p is the *average pore velocity*. The *Darcean velocity* is

$$u_D = \varepsilon u_p = -\frac{n\pi d^4}{128\mu}\frac{dp}{dx}. \tag{2.10}$$

Note that here the definition of porosity for this arrangement (straight tube) does not create any conflict when either surface flow (flow entering the tubes) or bulk (inside) flow is considered. The volume flow rate into the matrix is $u_D A = \varepsilon u_p A$ at the surface with area A of the matrix and the velocity anywhere inside the matrix is also given by εu_p. When (2.10) is compared with the definition of the permeability (2.1) one obtains

$$K = \frac{n\pi d^4}{128} = \frac{\varepsilon d^2}{32}. \tag{2.11}$$

Since most practical matrices do not comply with the straight-tube model, this relationship between the permeability and porosity does not agree with those found in the experiments. Scheidegger (1974, p. 130) shows that addition of a pore-size distribution $f(d)$, Figure 2.5(a), does not change this relationship between K and ε. Therefore, this parallel-conduits model cannot describe/predict the flow through interstices of packed beds.

Extension to *nonstraight* tubes can be made readily; however, the shortcoming mentioned earlier persists. Other extensions have been made that allow for the variation of the cross sectional area, Figure 2.5(b), and can in principle be applied to conduits that are not straight. An average tube diameter d must be used that is consistent with the definition of the average porosity ε of the matrix and with the treatment of the pressure drop. If an average tube diameter is defined, then the definition of permeability that was given earlier follows for parallel conduits. Scheidegger (1974, 131–133) gives a treatment of the variable-diameter tubes.

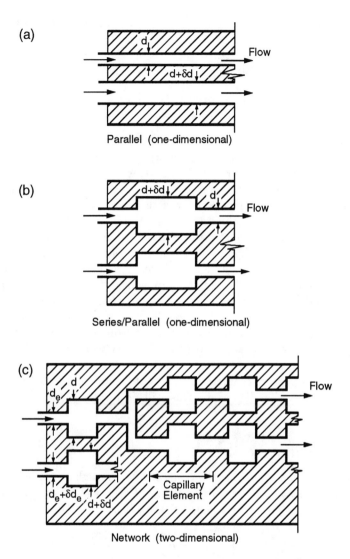

Figure 2.5 Schematic of the (a) parallel, (b) series/parallel, and (c) network arrangement of conduits.

(B) Network of Conduits

Dullien (1979, 181–190) has developed a model for pore structure that consists of capillaries containing segments of variable cross sections, Figure 2.5(c). The experimentally determined variant pore-size distribution (as a function of the pore entrance diameter d_e and the bulk diameter d) is used to assign values for the different rows of capillaries. A cubic network of capillaries is constructed. The different cubic networks (called component networks) are assumed to be independent. His model is a combination of parallel and serial arrangement of capillaries. He obtains the permeability of the three-dimensional network to be (using the cumulative pressure drop, the Hagen-Poiseuille equation, and the Darcy law)

$$K = \frac{\varepsilon}{96} \sum_i \frac{S_i \left(\sum_j V_{ij}/d_j^2\right)^2}{\sum_j V_{ij} \left(\sum_j V_{ij}/d_j^6\right)}, \qquad (2.12)$$

where $i = 1, 2, 3$ stands for the three orthogonal coordinate axes and $j = 1, 2, \cdots, N$ is the index for the capillary segment j; S_i is the fraction of pore volume of the medium represented by the component network i; $V_{ij} = (\pi/4)d_j^2 \ell_{ij}$, where d_j is the diameter of the segment j and ℓ_{ij} is the length of capillary segment j in the i direction.

In these treatments, even for the case of nonuniform cross-sectional area, the one-dimensional (fully developed) flow through circular tubes was used for the determination of the pressure drop. Extensions have been made to include the influence of the radial component of the flow in the expansion and contraction sections, by solving the complete Navier-Stokes equation (Dullien, 1979, 193–194). In light of the many assumptions made in the network theory, the contributions of this two-dimensionality on the permeability (through the pressure drop) is expected to be small. However, for high velocity flows where deviation from the Darcy law occurs, inclusion of the inertial force in the determination of the pressure drop is needed in order to predict this deviation. This will be further discussed in Section 2.4.3 (D).

2.4.2 Hydraulic Radius Model

This is based on a semiheuristic model of flow through solid matrices using the concept of *hydraulic radius* (Carman, 1937), which is often called the *Carman-Kozeny theory* (Dullien, 1979, p.170). The hydraulic diameter is defined as

$$d_h = \frac{4 \times \text{void volume}}{\text{surface area}} = \frac{4\varepsilon}{A_o(1-\varepsilon)}, \qquad (2.13)$$

where A_o is the *volumetric* or *specific surface area based on the solid volume*, i.e., solid surface area divided by the solid volume A_{fs}/V_s where A_{fs} is the interfacial area between the fluid and the solid phases and V_s is the solid volume. The solid volume fraction is $1 - \varepsilon$. The model uses the *tortuosity* (see Glossary) in such a way that it may be interpreted as the correction to the pressure gradient such that a modified pressure is defined as

$$\nabla p_{\text{mod}} = \frac{1}{\tau} \nabla p, \qquad (2.14)$$

where $\tau = 1 + L_t = L_e/L$, L is the length of the straight line between the pressure taps, and L_e is the effective length. Note that L_t, the *excess length* is *also* called the *tortuosity* (Carbonell and Whitaker, 1984). The filter velocity for the Hagen-Poiseuille equation is modified using the tortuosity introduced earlier and adding a shape parameter k_o, which results in

$$\mathbf{u}_p = -\frac{d_h^2}{16k_o \mu \tau} \nabla p, \qquad (2.15)$$

where $k_o = 2$ for a circular capillary and 2.0–2.5 for rectangular, elliptical, and annular shapes (Happel and Brenner, 1986, p. 403). Furthermore, the definition of the pore velocity is modified with

$$\mathbf{u}_D = \mathbf{u}_p \frac{\varepsilon}{\tau} = -\frac{K}{\mu} \nabla p. \qquad (2.16)$$

With these ad hoc modifications, and solving for K, we have

$$K = \frac{\varepsilon d_h^2}{16 k_o \tau^2} = \frac{\varepsilon d_h^2}{16 k_K} = \frac{\varepsilon^3}{k_K (1-\varepsilon)^2 A_o^2}, \qquad (2.17)$$

where $k_K = k_o \tau^2$ is the *Kozeny constant*. Now a *mean particle diameter* d is introduced as the diameter of a hypothetical sphere with the same A_o, i.e.,

$$d = \frac{6}{A_o}, \qquad (2.18)$$

where again $A_o = A_{fs}/V_s$. Then, for a bed made of uniformly sized spheres of diameter d_s, we have $d = d_s$, where subscript s stands for sphere. For cylinders of diameter d_c, we have $d = (3/2)d_c$, where c stands for cylinder. Now we have

$$K = \frac{\varepsilon^3}{36 k_K (1-\varepsilon)^2} d^2. \qquad (2.19)$$

When τ is approximated by $2^{1/2}$ and k_o is taken to be equal to 2.5, then the Kozeny constant k_K is approximated as 5 for packed beds. Happel and Brenner (1986, p. 395) give a discussion of this constant in examining the creeping flow over objects. The semiheuristic Carman-Kozeny model of permeability predicts the permeability of packed beds reasonably well but has limited applicability. This will be discussed in Section 2.4.3. Now (2.19) becomes (the *Carman-Kozeny equation*)

$$K = \frac{\varepsilon^3}{180(1-\varepsilon)^2} d^2. \qquad (2.20)$$

For packed bed of spherical particles with a narrow range of distribution in size, Rumpf and Gupte (1971) show that

$$K = \frac{\varepsilon^{5.5}}{5.6} d^2 \qquad (2.21)$$

gives a better agreement with the experimental results (the experimental range of porosity considered is between 0.35 and 0.67). Dullien (1979, p. 161) reviews other suggested porosity functions. These heuristic and semi-heuristic models have limited range of applicability. As an example, Kyan et al. (1970) show that if (2.19) is applied to fibrous beds, then k_K depends strongly on ε, e.g., for low velocities

$$k_K = \frac{\left[62.3 N_e^2 (1-\varepsilon) + 107.4\right] \varepsilon^3}{16 \varepsilon^6 (1-\varepsilon)^4}, \qquad (2.22)$$

where

$$N_e = \left(\frac{2\pi}{1-\varepsilon}\right)^{1/2} - 2.5 \qquad (2.23)$$

is the effective pore number and is related to the *effective porosity* ε_e by

$$\varepsilon_e = N_e^2 \frac{1-\varepsilon}{2\pi}. \qquad (2.24)$$

For $\varepsilon > 0.95$, which can occur in arrangement of fibers, k_K increases nearly exponentially with ε.

2.4.3 Drag Models for Periodic Structures

In these models, the matrix is envisioned as a collection of objects. The Navier-Stokes equations are solved for flow over these objects. Then the total resistance to the flow is compared with the Darcy resistance and the permeability for these models are found. The *analytical* treatments are limited to *creeping flow* over *periodic arrangement* (*isotropic*) of *dilute concentration* of objects (cylinders and spheres). These treatments are discussed by Clift et al. (1978), Kim and Russell (1985), Happel and Brenner (1986), Tien (1989), Kim and Karrila (1991), and Adler (1992) among others. The numerical treatments extend the results to include the *inertial force* and *asymmetries* such as those near the boundaries of the matrix, bulk anisotropy, and large concentrations including consolidation, e.g., Kim and Karrila.

(A) Approximate Solutions to Creeping Flow over Cylinders

Flow along and perpendicular to assembly of cylinders is generally analyzed by using a unit cell model. Among these are the solutions obtained by Hasimoto (1959) and Kuwabara (1959). Figure 2.6 gives the unit cells used by Sparrow and Loeffler (1959) and by Happel and Brenner (1986, 392–399). In the parallel flow, Figure 2.6(d), analysis of Happel and Brenner (1986, 392–393), closed-form solution is found to the one-dimensional flow

2.4 Permeability

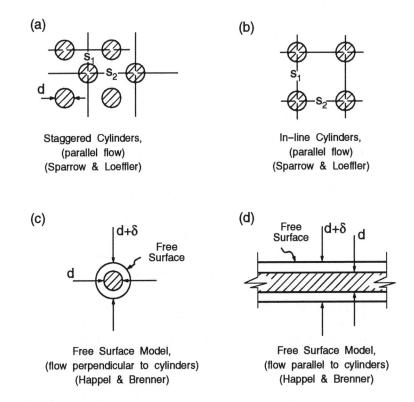

Figure 2.6 Cross-sectional view of the arrangement of cylinders. The unit cells used for the determination of permeability.

in an annulus bounded on the inside by a solid cylinder (diameter d) and on the outside (diameter $d + \delta$) by a shear-free surface (free-surface). This outer cylinder is to represent the extent of the interaction of the solid cylinder with its surrounding. For $\delta \to \infty$, the solution is that of flow along a cylinder. However, because of the one-dimensionality assumption, this limit is not realized. The solution to the one-dimensional Navier-Stokes equation subject to no velocity slip at the solid surface and zero velocity gradient at the free surface is

$$u_p = -\frac{1}{4\mu}\frac{dp}{dx}\left[\frac{d^2}{4} - r^2 + \frac{(d+\delta)^2}{2}\ln\frac{2r}{d}\right]. \tag{2.25}$$

The porosity and the Darcean velocity are

$$\varepsilon = \frac{\delta^2 + 2d\delta}{(d+\delta)^2}, \quad u_D = \overline{u}_p\varepsilon = -\frac{K}{\mu}\frac{dp}{dx}. \tag{2.26}$$

Therefore, we have

$$K = \frac{1}{2(d+\delta)^2} \left[\frac{d^2(d+\delta)^2}{4} - \frac{d^4}{16} - \frac{3(d+\delta)^4}{16} \right.$$
$$\left. + \frac{(d+\delta)^4}{4} \ln \frac{d+\delta}{d} \right]. \quad (2.27)$$

The hydraulic diameter is $d_h = (\delta^2 + 2\delta d)/d = \varepsilon(d+\delta)^2/d$. From the Carman-Kozeny theory, (2.17), for this fluid-solid arrangement, the permeability is given as

$$K = \frac{\varepsilon d_h^2}{16 k_K} = \frac{\varepsilon}{k_K} \frac{(\delta^2 + 2d\delta)^2}{16d^2} = \frac{\varepsilon^3}{k_K} \frac{(d+\delta)^4}{16d^2} = \frac{\varepsilon^3 d^2}{16 k_K (1-\varepsilon)^2}, \quad (2.28)$$

where k_K is again the Kozeny constant. This gives for k_K:

$$k_K = \frac{2\varepsilon^3}{(1-\varepsilon) \left[2\ln \frac{1}{1-\varepsilon} - 3 + 4(1-\varepsilon) - (1-\varepsilon)^2 \right]}. \quad (2.29)$$

For flow perpendicular to the cylinder, it is assumed that far from the cylinder (actually at $d+\delta$) the fluid moves with a velocity of u_∞ with respect to a stationary cylinder. They solve the two-dimensional Navier-Stokes equation, subject to zero shear stress and normal velocity at $(d+\delta)$, and find that *the drag force per unit length f* on the cylinders is

$$f = -\frac{4\pi\mu u_\infty}{\left[\ln \frac{d+\delta}{d} + \frac{d^4}{(d+\delta)^4} - \frac{1}{2} \right]}. \quad (2.30)$$

The pressure gradient balances the volumetric drag force, i.e.,

$$\frac{dp}{dx} = \frac{f}{\pi(d+\delta)^2}. \quad (2.31)$$

The permeability becomes

$$K = \frac{(d+\delta)^2}{16} \left[\ln \frac{d+\delta}{d} - \frac{1}{2} \frac{(d+\delta)^4 - d^4}{(d+\delta)^4 + d^4} \right] \quad (2.32)$$

and the Kozeny constant is

$$k_K = \frac{\frac{2\varepsilon^3}{(1-\varepsilon)}}{\frac{1}{1-\varepsilon} - \frac{1-(1-\varepsilon)^2}{1+(1-\varepsilon)^2}}. \quad (2.33)$$

As evident from (2.29) and (2.33), the Kozeny constant derived from the drag theory is not actually a constant. The analytical-numerical solutions

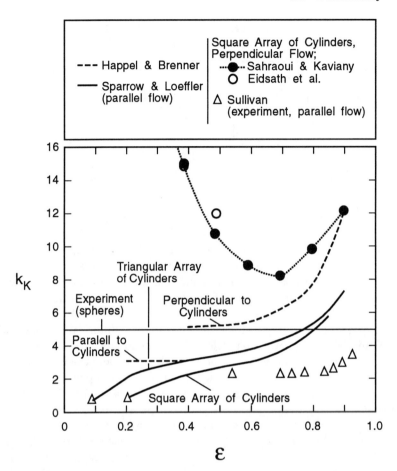

Figure 2.7 Variation of the Kozeny constant with respect to porosity for cylindrical particles and for various flow arrangements.

of Sparrow and Loeffler (1959) for longitudinal fully developed laminar flow between cylinders arranged in triangular or square array are given in Happel and Brenner (1986, pp. 395 and 399). It should be noted that their unit cells are different than those of Happel and Brenner (Figure 2.6). As will be shown, for high porosity these different unit cell models give identical results, i.e., as the interaction between the flows around individual cylinders diminishes (independent flows), flow over a single cylinder is recovered. Figure 2.7 compares the value of the Kozeny constant obtained by the preceding analysis with the experimental results of Sullivan (1942). The agreement between the experimental results and those of Sparrow and Loeffler are very good for low porosities. The smaller experimental values of the Kozeny constant (higher permeability) for high porosity has been

TABLE 2.4   THE KOZENY CONSTANT FOR THE CELL MODELS (HAPPEL AND BRENNER) AND EXPERIMENTS

ε	$\overline{k}_K$ (CYLINDERS)	k_K (SPHERES)	k_K (EXPERIMENT, SPHERE)
0.99	46.25	71.63	
0.90	9.79	11.34	
0.80	6.72	7.22	
0.70	5.60	5.79	5.0
0.60	5.07	5.11	5.0
0.50	4.97	4.74	5.0
0.40	4.66	4.54	5.0

attributed to the higher porosities adjacent to the confining tube in the experiments. This results in the channeling and a lower overall resistance to the flow. Higher porosity adjacent to confining walls is a common problem in packed beds and remedies can be tighter packing or use of fillers in the affected zone.

Since many of the analytical solutions assume a cylindrical or spherical (for cylindrical and spherical particles) shell around the particle, these solutions become less accurate as the porosity decreases and the interparticle interaction increases. Marshall et al. (1994) have developed an improvement over these solutions through a perturbation expansion of the unit-cell geometry.

Numerical (finite element) solution of the Navier-Stokes equation for two-dimensional flow across banks of cylinders has been studied by Eidsath et al. (1983). Their results, in terms of the Kozeny constant, for various arrangements of the cylinders are also shown in Figure 2.7. Additionally shown in Figure 2.7 are the numerical results (finite difference) of a two-dimensional simulation by Sahraoui and Kaviany (1992). The results show that only for $\varepsilon \to 1$ do the results of Happel and Brenner approach the numerical results. The results of this simulation are further discussed in Section (D) later.

The review presented above emphasizes the lack of a universal relationship between permeability and porosity and the limited applicability of the hydraulic radius theory, especially for two-dimensional flows.

(B) Geometric Models for Flow over Spheres

Happel and Brenner (1986, p. 395) report their results for the Kozeny constant for the case of packed spheres. They also develop an average $\overline{k}_K$ for the random orientation of cylinders by defining a $\overline{k}_K$ at an equal void volume and related to the two flow arrangements by

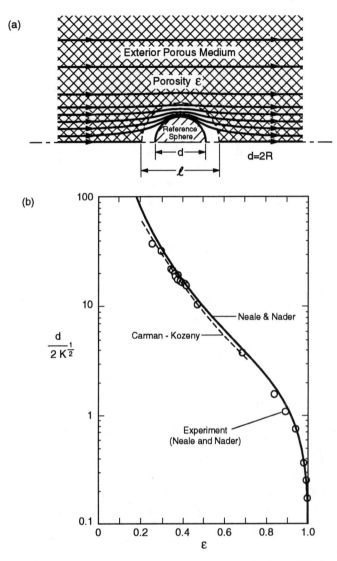

Figure 2.8 (a) A schematic of the two domains used by Neale and Nader for permeability prediction. (b) Variation of inverse of the square root of the normalized permeability with respect to porosity. (From Neale and Nader, reproduced by permission ©1974 AIChE.)

$$\overline{k}_K = \frac{2}{3} k_K (\text{perpendicular flow}) + \frac{1}{3} k_K (\text{parallel flow}). \tag{2.34}$$

Table 2.4 gives their results for k_K for packed spheres and for cylinders. For $\varepsilon > 0.70$, the predicted value of k_K is much higher than those obtained by the extension of the experimental results to $\varepsilon < 0.70$.

Neale and Nader (1974a) extended the formulation of Brinkman (1947, 1948) by modeling the creeping flow through random packing of spheres of diameter d. They take a reference sphere of diameter d as being surrounded by a concentric spherical shell comprising the associated pore space. The sphere and the shell are then embedded in the matrix of porosity ε. Then the creeping flow through the shell and the extended Darcy flow (the Brinkman equation, Section 2.6) through the porous media are solved simultaneously. This model has been successful in predicting the permeability of a packed bed of spheres. Figure 2.8(a) shows the geometry considered and gives typical stream lines. The unit cell is a permeable sphere of diameter ℓ with an interior impermeable sphere of diameter d. Outside of the unit cell is an isotropic homogeneous media of porosity ε. Then the ratio of the diameters is

$$\frac{\ell}{d} = \frac{1}{(1-\varepsilon)^{1/3}}. \tag{2.35}$$

Figure 2.8(b) shows the predicted results along with their and other experimental results and the correlation of Carman-Kozeny, i.e., (2.20). In their analysis, the flow field is represented by

$$\text{plain medium } (R \leq r \leq \frac{\ell}{2}): \quad \mu \nabla^2 \mathbf{u} = \nabla p, \tag{2.36}$$

$$\nabla \cdot \mathbf{u} = 0, \tag{2.37}$$

$$\text{porous medium } (\frac{\ell}{2} \leq r \leq \infty): \quad -\frac{\mu}{\mathbf{K}} \mathbf{u}_D + \mu \nabla^2 \mathbf{u}_D = \nabla p, \tag{2.38}$$

$$\nabla \cdot \mathbf{u} = 0. \tag{2.39}$$

The boundary conditions are the no-slip condition on R and the asymptotic (free stream) velocity for $r \to \infty$. In addition, at the interface between the plain and porous media, they match the *fictitious shear stress* and the pressure. This fictitious shear stress at the plain-porous interface will be further discussed in Section 2.11. The flow field is solved analytically and one of the results is the drag force on the solid sphere, which is given as

$$\mathbf{f} = 6\pi \mu R \mathbf{u}_D \xi \left[\alpha = \frac{R}{K^{1/2}}, \beta = \frac{R}{(1-\varepsilon)^{1/3} K^{1/2}} \right]$$

$$= \mathbf{f}_s \xi \left[\alpha, \beta = \frac{\alpha}{(1-\varepsilon)^{1/3}} \right], \tag{2.40}$$

where the closed-form solution for ξ is

$$\xi(\alpha, \beta) = \frac{4\left(-6\beta^6 - 21\beta^5 - 45\beta^4 - 45\beta^3 + 5\beta^4\alpha^2 + 5\beta^3\alpha^2 + \beta\alpha^5 + \alpha^5\right)}{(-4\beta^6 - 24\beta^5 - 180\beta^4 - 180\beta^3 + 9\beta^5\alpha + 45\beta^4\alpha - 10\beta^3\alpha^3}$$

$$+ 180\beta^3\alpha - 30\beta^2\alpha^3 + 9\beta\alpha^5 - 4\alpha^6 + 9\alpha^5), \tag{2.41}$$

with $\xi(\alpha,\beta) \to 1$ as $\ell/d \to \infty$, and the *Stokes drag force* $\mathbf{f}_s$ is recovered.

A force balance over all the spheres yields

$$2\frac{R^2}{K} - 9(1-\varepsilon)\xi\left[\frac{R}{K^{1/2}}, \frac{R}{K^{1/2}(1-\varepsilon)^{1/3}}\right] = 0, \qquad (2.42)$$

which leads to the $K(\varepsilon, R)$ given in Figure 2.8(b).

One reason for the success of the application of the Brinkman equation in the determination of the permeability is that it predicts a boundary-layer thickness (or the distance over which the velocity disturbance decays), which is of the order of $K^{1/2}$, a distance that is physically reasonable when accounting for the particle-particle hydrodynamic interaction. This distance is called the *Brinkman screening length*.

(C) Creeping Flow over Consolidated and Nonconsolidated Spheres

In the analytical-numerical study of Larson and Higdon (1989), flow through lattices of spheres, including cases where the radius of spheres is allowed to increase past a point of close touching (consolidated media), has been considered. The Stokes flow problem is solved using a collocation method based on the harmonic expansion of the velocity field using the Lamb general solution in spherical coordinate. An asymptotic approximation for low porosities is also developed. Three lattices are considered: *simple cubic* (SC), *body-centered cubic* (BCC), and *face-centered cubic* (FCC).

The solutions to

$$-\nabla p + \mu \nabla^2 \mathbf{u} = 0, \qquad (2.43)$$

$$\nabla \cdot \mathbf{u} = 0, \qquad (2.44)$$

are found as (Happel and Brenner, 1986, 62–71)

$$p = \sum_{i=-\infty}^{\infty} \chi_i, \qquad (2.45)$$

$$\mathbf{u} = \sum_{i=-\infty}^{\infty}\left[\nabla \times \mathbf{r}\,\eta_i + \nabla \xi_i + \frac{(i+3)r^2}{2\mu(i+1)(2i+3)}\nabla \chi_i \right.$$

$$\left. - \frac{i\,\mathbf{r}}{\mu(i+1)(2i+3)}P_i\right], \qquad (2.46)$$

where η_i, ξ_i, and χ_i are the solid spherical harmonics and are written in terms of the associated Legendre function P_i^j. The harmonics are given in the form $r^i P_i^j(\cos\theta)e^{(-1)^{1/2}j\phi}$, where θ is the polar angle and ϕ is the azimuthal angle. The coefficients in the expansion are then determined numerically.

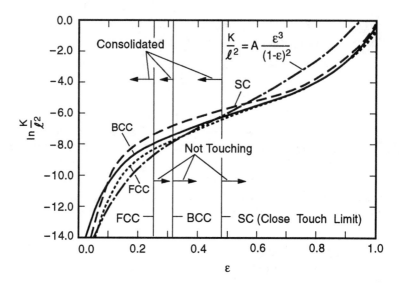

Figure 2.9 The results of Larson and Higdon, and those predicted by the expression given in the text, for variation of the normalized permeability with respect to porosity.

Larson and Higdon have shown that if the interparticle (center to center) distance ℓ in the lattice is used to normalize the permeability, the variation of this normalized permeability with respect to porosity is nearly independent of the lattice structure. Their results are plotted in Figure 2.9. Also plotted is a modified Carman-Kozeny relation based on ℓ, i.e.,

$$\frac{K}{\ell^2} = A\frac{\varepsilon^3}{(1-\varepsilon)^2} \equiv Af(\varepsilon). \quad (2.47)$$

We choose A such that for $\varepsilon = 0.5$, this expression results in a value for K/ℓ^2 close to their predictions. This resulted in $A = 2e^{-6}$. Note that $f(\varepsilon)$ has an inflection point at $\varepsilon = 0.634$.

Instead of the Stokes (and Navier-Stokes for nonzero Reynolds number) equation, the *discretized Boltzmann* equation is used by Ladd (1994) to obtain numerical solutions to flow through periodically arranged particles.

(D) NUMERICAL SOLUTION FOR FLOW OVER CYLINDERS, INCLUDING INERTIAL EFFECT

Sahraoui and Kaviany (1992), by applying the finite difference approximations and an overlaying grid system (cylindrical and Cartesian grids are overlaid), solve the Navier-Stokes equation for flow over various arrangements of cylinders. Their periodic boundary conditions for in-line and staggered arrangements are given in Figure 2.10(a). Using their numerical

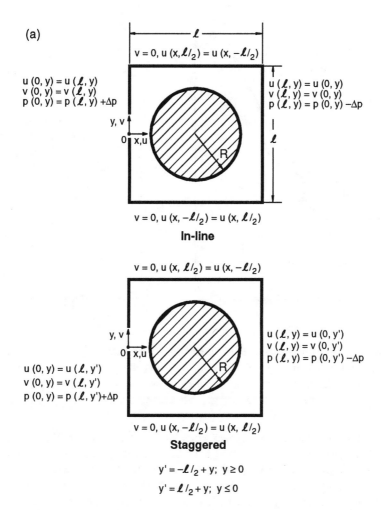

Figure 2.10 (a) Boundary conditions for the unit cell, for in-line and staggered arrangements of cylinders.

scheme the variation of the normalized permeability (using both the particle diameter and the unit cell length) with respect to the porosity, is computed and given in Figure 2.10(b). The results show that, to within a few percent accuracy, a curve fit representation of the form

$$\frac{K}{\ell^2} = 0.0606\,\varepsilon^{5.1} \quad \text{for } 0.4 \leq \varepsilon \leq 0.8 \tag{2.48}$$

or

$$\frac{K}{d^2} = 0.0606\,\frac{\pi}{4}\frac{\varepsilon^{5.1}}{1-\varepsilon} \quad \text{for } 0.4 \leq \varepsilon \leq 0.8 \tag{2.49}$$

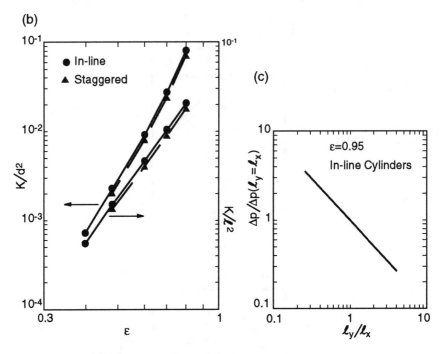

Figure 2.10 (b) Variation of the normalized permeability with respect to the porosity, for in-line and staggered arrangement of cylinders. (c) Effect of anisotropy in structure, for in-line arrangement of cylinders. $\ell_x = \ell_y$ is a square array and flow is along x-axis.

is possible. As evident from these correlations, the Carman-Kozeny representation of the results does not lead to a constant value for the Kozeny constant. This was shown in Figure 2.7, where the Carman-Kozeny relation was used, and as shown, the Kozeny constant varies with porosity. There is an initial decrease in k_K with increase in ε, and then after reaching a minimum, k_K begins to increase with further increase in ε. This failure of the Carman-Kozeny relation to lead to a Kozeny constant, which depends only on the structure and not the porosity, appears to be due to the two-dimensionality of the flow. As was shown by (2.47) and in Figure 2.9, for spherical particles, k_K only depends on the arrangement (and not ε). Figure 2.10(c) shows the effect of anisotropy in the structure on pressure drop. The results are from Marshall et al. (1994). For $\ell_x = \ell_y$, the in-line arrangement is a square array of cylinders. Note the strong dependence on the cell aspect ratio.

The results given in Figures 2.10(b) and (c) are for low Reynolds numbers. As the cell-based Reynolds number increases, using the Darcy law the permeability will become dependent on the flow inertia. Figure 2.10(d) shows how the Kozeny constant for the in-line arrangement with a porosity

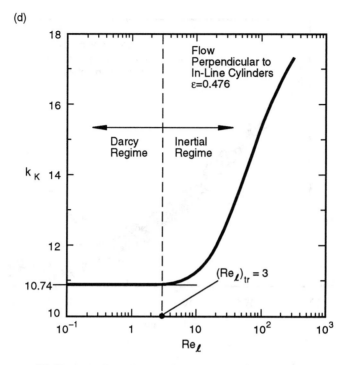

Figure 2.10 (d) Numerical results for the variation of the Kozeny constant as a function of Reynolds number.

of 0.476 changes with $Re_\ell = u_D \ell / \nu$. For Re_ℓ less than the transition $(Re_\ell)_{tr}$ (which is approximately 3), the streamlines around the cylinder are symmetric in form. For $Re_\ell > 3$, the effect of inertia becomes noticeable, and as Re increases, secondary flows are formed. As expected, the apparent permeability and the Kozeny constant both increase as Re_ℓ increases beyond $(Re_\ell)_{tr}$. The effect of weak inertia on flow through a periodically constricted tube is studied by Mei and Auriault (1991), where they find a u_D^3 relation in the inertia regime. The same trend is found by Firdaouss and Guermond (1995).

2.5 High Reynolds Number Flows

2.5.1 MACROSCOPIC MODELS

As the filter velocity is increased, deviations from the Darcy law (2.1) are observed. Based on the capillary (conduit) models, these deviations are due to the inertial (convective) contribution to the momentum balance. At any

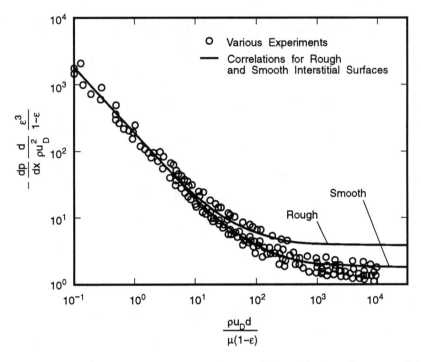

Figure 2.11 Deviations from the Darcy law at high velocities. The two solid lines are correlations given in the text. (From Macdonald et al., reproduced by permission ©1979 Amer. Chem. Soc.)

filter velocity, the sum of the viscous contribution and inertial contribution make up the total resistance to the flow, but the inertial contribution is important only at high velocity and is expected to dominate at very high velocities. The existence of the two asymptotes, namely, viscous (at low velocities) and inertial (at high velocity), suggested by the capillary models has been supported by experiments. Figure 2.11 shows the experimental results of several investigators as carefully analyzed and presented by Macdonald et al. (1979).

In trying to account for the high-velocity regime behavior, Ergun (1952) extends the hydraulic radius theory of Carman-Kozeny. He constructs a correlation, which includes all the fluid parameters (velocity, viscosity, and density) and the matrix parameters (closeness and orientation of packing, size, shape, and surface condition of particles). His correlation is modified by Macdonald et al. as

$$-\frac{dp/dx}{\rho u_D^2} d \frac{\varepsilon^3}{1-\varepsilon} = \frac{180(1-\varepsilon)}{Re_d} + 1.8 \quad \text{(smooth surfaces),} \quad (2.50)$$

$$-\frac{\mathrm{d}p/\mathrm{d}x}{\rho u_D^2} d \frac{\varepsilon^3}{1-\varepsilon} = \frac{180(1-\varepsilon)}{Re_d} + 4 \quad \text{(rough surfaces)}. \tag{2.51}$$

By using the Carman-Kozeny permeability relation, we have for smooth particles

$$-\frac{\mathrm{d}p/\mathrm{d}x}{\rho u_D^2} d = \frac{d^2}{K}\frac{1}{Re_d} + 1.8\frac{1-\varepsilon}{\varepsilon^3}, \tag{2.52}$$

where

$$Re_d = \frac{\rho u_D d}{\mu} \tag{2.53}$$

and d is the appropriate characteristic length for the particle. This is taken as the equivalent mean sphere diameter, which was given before as

$$d = \frac{6}{A_o}. \tag{2.54}$$

Unlike the viscous flow regime, the inertial flow regime is *roughness* dependent. As will be discussed later, this can also be interpreted as the inability of any simple relation to account for the inertial effects. In flow development in the pores, the interpore fluid dynamics depend on several (not only one) length scales. Equation (2.50) can be written as

$$-\frac{\mathrm{d}p}{\mathrm{d}x} = \frac{\mu}{K}u_D + 1.8\frac{1-\varepsilon}{\varepsilon^3}\frac{1}{d}\rho u_D^2 \quad \text{(smooth surfaces)}. \tag{2.55}$$

This equation has been *heuristically* extended to three-dimensional flows (Stanek and Szekely, 1974; Joseph et al., 1984) to become

$$-\nabla p = \frac{\mu}{K}\mathbf{u}_D + 1.8\frac{1-\varepsilon}{\varepsilon^3}\frac{1}{d}\rho |\mathbf{u}_D| \mathbf{u}_D \quad \text{(smooth surfaces)}. \tag{2.56}$$

We note that the above generalization has not been *experimentally* verified. Direct simulation of flow through two-dimensional periodic structures show that for two-dimensional flows, when (2.56) is used, the coefficient of the second term must be *adjusted*. A *large* velocity component *perpendicular* to the direction of interest, would require a *decrease* in this coefficient. However, the advantage of the *vectorial* presentation is that it is *coordinate invariant*.

Extensions to *anisotropic* media can be made, but care must be taken in evaluation of the coefficients. We suggest that for an *isotropic* system with a *Cartesian coordinate*, $|u_{Di}|u_{Di}$ be used instead of $|\mathbf{u}_D|\mathbf{u}_D \cdot \mathbf{s}_i$, where $\mathbf{s}_i$ is the unit vector in the direction i. This is more in accordance with the experimental results. We then have

$$-\frac{\partial p}{\partial x_i} = \frac{\mu}{K}u_{Di} + 1.8\frac{1-\varepsilon}{\varepsilon^3}\frac{1}{d}\rho |u_{Di}|u_{Di} \quad \text{(smooth surfaces)}. \tag{2.57}$$

In order to apply the permeability alone, Ward (1964) recommends

$$-\frac{dp}{dx} = \frac{\mu}{K}u_D + 0.550\frac{1}{K^{1/2}}\rho u_D^2 \qquad (2.58)$$

or

$$-\frac{\partial p}{\partial x_i} = \frac{\mu}{K}u_{Di} + \frac{C_E}{K^{1/2}}\rho |u_{Di}| u_{Di}, \qquad (2.59)$$

where C_E stands for the *Ergun coefficient*. When $K^{1/2}$ is used as the length scale, the experimental results show that deviation from the Darcy law begins at $Re_{K^{1/2}} = u_D K^{1/2}/\nu = 0.2$.

We note that C_E represents *deviation* from the Stokes flow and that in multidimensional flows the extent of the inertial-core formation and flow separation, occuring as the flow passes over the elements of the matrix, would depend on the flow *direction*. Therefore, using the vectorial or the scalar form, C_E would depend on the angle between the direction of the pressure gradient of interest and the flow direction (i.e., C_E becomes a tensor). A discussion of the Ergun equation based on the flow regimes is given by du Plessis (1994).

A more accurate *correlation* for pressure drop in particle beds, up to the Reynolds number of 10^4, is given by Molerus and Schweinzer (1989).

2.5.2 MICROSCOPIC FLUID DYNAMICS

Dybbs and Edwards (1984) have made point measurement of the three-dimensional velocity distribution and have obtained movies of dye streaklines for flow through hexagonal packing of spheres and for complex arrangements of cylinders. They define a Reynolds number that is based on the average pore velocity $\bar{u}_p$ and an average characteristic length scale for the pores $\bar{d}$, i.e.,

$$Re_d = \frac{\rho \bar{u}_p \bar{d}}{\mu}. \qquad (2.60)$$

Based on this Reynolds number, they are able to define *four distinct flow regimes*, discussed here.

- $Re_d < 1$, Darcy or creeping-flow regime:
  The viscous forces dominate over the inertia forces and only the local (pore-level) geometry influences the flow. Adjacent to the walls confining the matrix, the packing is looser and, therefore, channeling occurs with maximum velocities nearly twice that found in the bulk of the matrix. The effects of these walls are confined to only one to two particle diameters. In the region where the flow just enters the matrix, the flow develops in approximately the first three particle diameters (entrance region) and then the axial velocity distribution (pore level) takes on a periodic behavior, i.e., this periodic behavior persists beyond about three particle diameters from the entrance.

- $1\text{–}10 < Re_d < 150$, inertial-flow regime:
  Steady nonlinear (inertial force affected) laminar flow begins between Re of 1 and 10. Boundary layers are more pronounced and core (uniform velocity) regions are present. As Re increases, the cores become larger as the boundary-layer thickness decreases. This is an indication of the dominance of the inertial force over the viscous forces. In each pore, the flow develops (boundary-layer growth), indicating the dependence of the pressure drop on both lateral and longitudinal linear dimensions of the pores. For very narrow and long pores, the effect of the inertial force on the pressure drop is less significant, compared to that for wider pores.

- $150 < Re_d < 300$, unsteady laminar-flow regime:
  Oscillations with frequencies of the order of 1 Hz and amplitudes of the order of one-tenth of the particle diameter are observed. In the dye injection flow visualization, no local (in pore) dispersion of the dye is observed, indicating that the flow remains laminar in this regime. The laminar wake instability may be responsible for the transition from laminar steady flows to unsteady flows.

- $300 < Re_d$, unsteady- and chaotic-flow regime:
  The observed highly unsteady chaotic flow does not appear to be laminar and turbulent type mixing (turbulent dispersion) of dye appears in the pores. The pressure drop measurements (given in connection with the Ergun equation) show that at high Reynolds numbers an asymptotic behavior is found, i.e., the normalized pressure drop does not change with the Reynolds number. From the regime classifications, it appears that this asymptote is in the unsteady- and chaotic-flow regime (which can be the turbulent-flow regime). Examination of the data for various matrices, given in Macdonald et al. (1979), indicates that this *asymptote is reached* for Reynolds numbers between 10^3 and 10^4 (this is also supported by the results of Dybbs and Edwards).

Dybbs and Edwards, also given by Ling (1988), explain the inertial flow regime as a regime in which the flow development through the pores is significant. They analyze the developing flow through tubes of diameter d and determine the pressure drop in the entrance sections. For very long tubes, the average pressure drop is nearly the same as that for fully developed flows. However, as the length decreases and the diameter-to-length ratio becomes of the order of unity, the average pressure drop becomes significantly larger than that for fully developed flows. Since in most porous media the ratio of the three orthogonal linear dimensions are of the order of unity, then, as the Reynolds number (based on tube diameter) increases, this entrance effect is expected to be more and more significant. They find

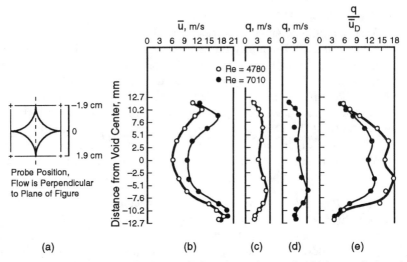

Figure 2.12 The mean velocity, turbulent intensity, and relative turbulent intensity measured by Mickley et al. for two different Reynolds numbers. (From Mickley et al., reproduced by permission ©1965 Pergamon.)

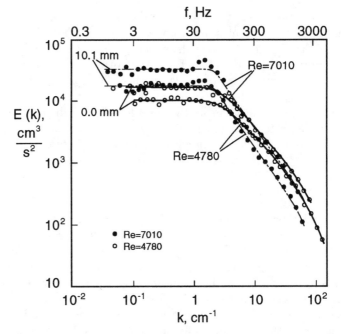

Figure 2.13 The turbulence dissipation spectra for two different Reynolds numbers and at the center of pore, measured by Mickley et al. (From Mickley et al., reproduced by permission ©1965 Pergamon.)

2.5.3 Turbulence

(A) Transition to Turbulence

Early on, the transition to turbulence in flow through matrices was associated with the observed deviation from the Darcy law (e.g., Ward, 1964). This macroscopic deduction was later *disproved* by the pore-level measurement of the velocity fluctuation (e.g., Mickley et al., 1965; Jolls and Hanratty, 1966; van der Merwe and Gauvin, 1971; and Dybbs and Edwards).

In order to have a probe or optical access, the particle size used in the experiments (mostly spherical) is large, and Reynolds numbers

$$Re_d = \frac{\overline{u}_D d}{\nu}, \qquad (2.61)$$

as high as 27,000 (air, $d = 7$ cm, $\overline{u}_D = 20$ m/s, van der Merwe and Gauvin) are achieved.

The transition to turbulence has been found to take place over Reynolds numbers of 110 to 150 (Jolls and Hanratty, deduced from measurements of the local mass transfer rate from an embedded sphere) and by others for Reynolds numbers as large as 300 (laser Doppler anemometry, Dybbs and Edwards). Although there is a variance in the transition Reynolds number, its values are much larger than the transition Reynolds number associated with departure from the Darcy law.

(B) Flow Structure

Mickley et al. use the *rhombohedral* arrangement of 3.8-cm table-tennis balls along with the hot-wire anemometry. Two Reynolds numbers of 4780 and 7010 are achieved using air. Their results for the distributions of the mean longitudinal velocity $\overline{u}$, the turbulent intensity q, and the *relative turbulent intensity* $q/\overline{u}_D$ are given in Figure 2.12. Near symmetry exists between the upper-half and the lower-half of the void. The results show that the mean velocity profile within each half of the pore has two distinct peaks. However, the turbulence intensity is relatively uniformly distributed. The intensity is defined as $q^2 = u'^2 + v'^2 + w'^2$, where u', v', and w' are the fluctuating components of velocity. It was found that at the void center isotropic turbulence occurs, i.e., $u'^2 = v'^2 = w'^2$, but elsewhere the turbulence was anisotropic. The maximum relative turbulence intensity is very high (about 50 percent).

The measured *turbulence spectra* $E(k)$ are given, as a function of the *wave number k*, in Figure 2.13, where

$$k = \frac{2\pi f}{\bar{u}_D}, \qquad \int_0^\infty E(k)\,dk = q^2, \qquad (2.62)$$

and f is the frequency. Since eddies are shed at different frequencies and are associated with a significant kinetic energy, in general the spectra show bumps at shedding frequencies. Figure 2.13 does not show any maxima, indicating no significant eddy shedding. The longitudinal *integral scale*

$$\lambda = \frac{\pi E(0)}{2q^2}, \qquad \text{where } E(0) = \lim_{k \to 0} E(k) \qquad (2.63)$$

is found to be 0.14–0.27 cm ($Re_d = 4780$) and 0.15–0.28 cm ($Re_d = 7010$). This shows that λ/d is 0.037–0.071.

The turbulent spectra measurements of van der Merwe and Gauvin with $d = 7$ cm and for the *simple cubic* arrangement, in contrast to the Mickley et al. data, show that peaks in the spectra exist, indicating eddy shedding. As expected, this shows that pore structure plays a significant role on the structure of turbulence.

2.6 Brinkman Superposition of Bulk and Boundary Effects

In an effort to obtain an expression for the permeability of packed beds made of spheres, Brinkman (1947, 1948) adopted a heuristic momentum equation, which bears his name. His objective was to extend the Stokes drag force on a sphere (a sphere placed in an infinite plain domain) to include the effect of the neighboring spheres. He combines (superimposes) the viscous penetration dominated flow (Stokes flow) with the Darcy flow to obtain

$$\nabla p = -\frac{\mu}{K}\mathbf{u}_D + \mu' \nabla^2 \mathbf{u}_D, \qquad (2.64)$$

where he initially used the Einstein formula for the *effective viscosity* of suspension, given by

$$\mu' = \mu \left[1 + 2.5\left(1 - \varepsilon\right)\right]. \qquad (2.65)$$

The Brinkman approach to flow through solid matrices was that of inclusion of two asymptotes. For $K \to \infty$ (and $\varepsilon \to 1$) the bulk resistance (which he called the *damping force*) diminished, leaving only the force on a single particle due to the external surface shear stress. For $K \to 0$, the damping force (internal surface shear stress) dominates over the external surface shear stress.

An analysis of the Brinkman equation as a model for flow in porous media is given by Durlofsky and Brady (1987). Lundgren (1972) arrives at the Brinkman equation from a statistical formulation and determines a relationship between the effective viscosity μ' and the porosity (for spherical particles). A recent experimental determination of μ' for foam is given by Gilver and Altobelli (1994). We return to this effective viscosity in Section 2.11 in conjunction with the interfacial hydrodynamic condition between a porous and a plain medium. The Brinkman results for the permeability did not agree well with the experimental results. An extension of his work is the unit cell model of Neale and Nader (1974a), which was discussed in Section 2.4.3 (B). The recent analytical treatments of flow through matrices include the interaction between the flow fields around individual objects by directly applying the Navier-Stokes equation to unit cells.

As will be discussed, Brinkman's inclusion of the divergence of the stress tensor (and the resulting boundary value problem) leads to the prediction of the momentum boundary-layer thickness at the walls confining the matrix. This boundary layer is rather thin (experiments and analyses show that this thickness is at most only a few particle diameters). However, the effective viscosity in (2.64) is not equal to the bulk viscosity and varies within this boundary layer. This will be discussed further in Section 2.11.6.

2.7 Local Volume-Averaging Method

The premise of the application of local volume averaging is to arrive at a macroscopic momentum equation by starting from the Navier-Stokes equation, e.g.,

$$\rho_o \left(\frac{\partial \mathbf{u}}{\partial t} + \mathbf{u} \cdot \nabla \mathbf{u} \right) = \nabla \cdot \mathbf{T} + \rho \mathbf{f}, \qquad (2.66)$$

where $\mathbf{T}$ is the second-order stress tensor and $\mathbf{f}$ is the field of the external forces per unit mass. $\mathbf{T} = -p\mathbf{I} + \mathbf{S}$ for an incompressible and Newtonian fluid and $\mathbf{S} = 2\mu\mathbf{D}$, where $\mathbf{D}$ is the deformation rate tensor. Then one arrives at a momentum equation, which has the following features:

- Is based on averages taken over a representative elementary volume, i.e., the smallest volume that represents the local average properties (addition of extra pores and the surrounding solid to this volume does not change this average value).

- Shows the Darcy bulk resistance for viscous regime, $\nabla p = -(\mu/\mathbf{K})\mathbf{u}_D$.

- Shows how $\mathbf{K}$ is related to the more fundamental matrix parameters, i.e., $\mathbf{K} = \mathbf{K}(\text{geometry})$.

- Shows how far the effect of the vorticity generated at walls confining the matrix penetrates into the bulk flow, i.e., the significance of the $\mu'\nabla^2 \mathbf{u}_D$ term.

54   2. Fluid Mechanics

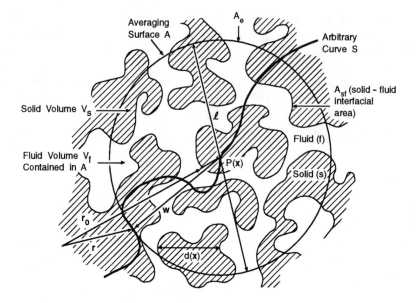

Figure 2.14 A schematic of a representative elementary volume and the position vectors used. The fluid phase is shown as continuous.

- Shows when the inertial term becomes important and what form it should take when it is added to the equation of motion for the bulk flow.
- Can be extended to multiphase flows and can assist in applying the same principles to other conservation equations.
- May be applied at the interface of two homogeneous media (leading to the definition of the interfacial location and conditions).

As will be shown, averaging methods do lead to a more rigorous treatment of transport in porous media. However, a large number of unknowns are introduced, that just as with the semiempirical treatments, require experimental verifications. This is due to the complexity of the flow paths and the interpore and pore-to-pore fluid dynamic interactions. Therefore, although the local volume averaging involves integration of the conservation equations over the representative elementary volume, empiricism to various extents are applied in arriving at the local volume-averaged conservation equations. For example, in arriving at the bulk resistance term, Bear

(1988, p. 104) and Slattery (1981, p. 201) introduce this resistance through a definition, while Whitaker (1969) arrives at it using an arbitrary linear geometrical transformation. A general discussion of volume-averaging is given by Gray et al. (1993). In the following we review the local volume-averaging technique, which in principle gives us the ability to relate the pore-level hydrodynamics to the macroscopic flow behavior.

2.7.1 Local Volume Averages

For single-phase flow through solid matrices around any point $P(\mathbf{x})$ in the matrix, a volume V contained in a surface A can be defined. Figure 2.14 shows this envisioned volume. The volume is occupied by solid (s) and void or fluid (f). A *void distribution function* (or *fluid phase distribution function*) is defined as

$$a(\mathbf{x}) = \begin{cases} 1 & \text{if } \mathbf{x} \text{ is in the void region} \\ 0 & \text{if } \mathbf{x} \text{ is in the solid region.} \end{cases} \quad (2.67)$$

Thus, the *local porosity* is defined as

$$\varepsilon(\mathbf{x}) = \frac{1}{V} \int_V a(\mathbf{x})\, dV = \frac{V_f}{V}, \quad (2.68)$$

where $V = V_f + V_s$.

As was mentioned, the local representative elementary volume V is chosen such that it is the smallest differential volume that results in statistically meaningful local average properties (such as local porosity). When this volume is appropriately chosen, adding extra pores (extra volume) around it will not result in changes to the values of these local properties. This will require that $\ell \gg d$, while at the same time ℓ be small such that $\ell \ll L$. It can be shown (Whitaker, 1969) that using these inequalities for any quantity ψ (tensor of any order), we have

$$\langle\langle\psi\rangle\rangle = \langle\psi\rangle, \quad (2.69)$$

where $\langle\langle\ \rangle\rangle$ is the double averaging, and the *local volume average* is defined as

$$\langle\psi\rangle = \frac{1}{V} \int_V \psi\, dV. \quad (2.70)$$

For any quantity ψ associated with the *fluid*, the *volume averaged* value is

$$\langle\psi\rangle = \frac{1}{V} \int_{V_f} \psi\, dV = \varepsilon \frac{1}{V_f} \int_{V_f} \psi\, dV. \quad (2.71)$$

2.7.2 THEOREMS

In order to locally average the Navier-Stokes equations, the *average of a gradient* (or divergence) must be replaced with the *gradient* (or divergence) *of an average*. Slattery (1969 and 1981, 196–199) develops this transformation, and detailed versions are given by Whitaker (1969) and Gray et al. (1993). We review their steps below. Before embarking on the derivations, two useful relationships that will be needed are given.

The *general transport theorem*, which is the general form of the *Reynolds transport theorem*, states that

$$\frac{d}{dt}\int_{V(t)} \psi \, dV = \int_{V(t)} \frac{\partial \psi}{\partial t} \, dV + \int_{A(t)} \psi \, \mathbf{u}_A \cdot \mathbf{n} \, dA, \qquad (2.72)$$

where $\mathbf{u}_A$ is the velocity of the surface $A(t)$, which encompasses volume $V(t)$ and $\mathbf{n}$ is the unit vector perpendicular to $A(t)$. The velocity $\mathbf{u}_A$ can be different than the fluid velocity on A. When these two velocities are the same, then the general and Reynolds transport theorems are identical.

The second relationship is the *divergence theorem* (the Green transformation), which states

$$\int_{V(t)} \nabla \psi \, dV = \int_{A(t)} \psi \, \mathbf{n} \, dA. \qquad (2.73)$$

Returning to flow through matrices, consider a curved line s passing through the point P in the representative elementary volume shown in Figure 2.14, i.e., P is on s and around P we have local values of $V(s)$, $V_f(s)$, $A(s)$, $A_f(s)$, $\varepsilon(s)$, etc. Then (2.72) can be written for any fluid-related quantity ψ as

$$\frac{d}{ds}\int_{V_f(s)} \psi \, dV = \int_{V_f(s)} \frac{\partial \psi}{\partial s} \, dV + \int_{A_f(s)} \psi \frac{d\mathbf{r}}{ds} \cdot \mathbf{n} \, dA, \qquad (2.74)$$

where $\mathbf{r}$ is the spatial position vector. Since $\psi = \psi[x_i(s), t]$, then

$$\frac{\partial \psi}{\partial s} = 0 \quad \text{and} \quad \frac{d\psi}{ds} = \frac{\partial \psi}{\partial x_i}\frac{dx_i}{ds}. \qquad (2.75)$$

Then (2.74) becomes

$$\frac{d}{ds}\int_{V_f(s)} \psi \, dV = \int_{A_f(s)} \psi \frac{d\mathbf{r}}{ds} \cdot \mathbf{n} \, dA. \qquad (2.76)$$

Since at the solid-fluid interface $d\mathbf{r}/ds$ is perpendicular to $\mathbf{n}$, we can further simplify this equation by dividing the area into $A_f(s) = A_{fs}(s) + A_e(s)$, where A_e represent the entrance and exit surfaces (fluid passing through them) and A_{fs} represent the solid-fluid interfacial area. Now (2.76) becomes

$$\frac{d}{ds}\int_{V_f(s)} \psi \, dV = \int_{A_e(s)} \psi \frac{d\mathbf{r}}{ds} \cdot \mathbf{n} \, dA. \qquad (2.77)$$

2.7 Local Volume-Averaging Method

We now try to remove $d\mathbf{r}/ds$ from the integrand by defining

$$\mathbf{r}(s) = \mathbf{r}_o(s) + \mathbf{w}(s), \tag{2.78}$$

where, as shown in Figure 2.14, $\mathbf{r}_o(s)$ is the position vector for point P and $\mathbf{w}(s)$ is the position vector locating points on $A_f(s)$. We also have the directional derivative given as

$$\frac{d}{ds} = \frac{d\mathbf{r}_o}{ds} \cdot \nabla. \tag{2.79}$$

Noting that $d\mathbf{r}_o/ds$ is not a function of $A_f(s)$, and by using (2.77)–(2.79), we have

$$\frac{d\mathbf{r}_o}{ds} \cdot \left(\nabla \int_{V_f(s)} \psi \, dV - \int_{A_e(s)} \psi \mathbf{n} \, dA \right) = \int_{A_e(s)} \psi \frac{d\mathbf{w}}{ds} \cdot \mathbf{n} \, dA. \tag{2.80}$$

Since the right-hand side of (2.80) is zero, because $d\mathbf{w}/ds$ is tangent to s (in Figure 2.14 we have chosen a portion of s to travel along A_f), we have

$$\nabla \int_{V_f} \psi \, dV = \int_{A_e} \psi \mathbf{n} \, dA, \tag{2.81}$$

where the coefficient $d\mathbf{r}_o/ds$ is dropped because it is an arbitrary vector. Now, using the divergence theorem in the form

$$\int_{V_f} \nabla \psi \, dV = \int_{A_{sf}} \psi \mathbf{n} \, dA + \int_{A_e} \psi \mathbf{n} \, dA \tag{2.82}$$

in (2.81), we have

$$\int_{V_f} \nabla \psi \, dV = \nabla \int_{V_f} \psi \, dV + \int_{A_{sf}} \psi \mathbf{n} \, dA \tag{2.83}$$

or

$$\langle \nabla \psi \rangle = \nabla \langle \psi \rangle + \frac{1}{V} \int_{A_{sf}} \psi \mathbf{n} \, dA. \tag{2.84}$$

In general, ψ can be a *scalar*, a *spatial vector*, or a *second-order tensor* associated with the fluid. When dealing with gradients, we have the *theorem for the volume average of a gradient*.

For a spatial vector or a second-order tensor $\mathbf{b}$, we have

$$\langle \nabla \cdot \mathbf{b} \rangle = \nabla \cdot \langle \mathbf{b} \rangle + \frac{1}{V} \int_{A_{sf}} \mathbf{b} \cdot \mathbf{n} \, dA, \tag{2.85}$$

which is referred to as the *theorem for the volume average of a divergence*.

2.7.3 MOMENTUM EQUATION

The continuity equation is

$$\frac{\partial \rho}{\partial t} + \nabla \cdot \rho \mathbf{u} = 0. \tag{2.86}$$

When volume averaged, it becomes

$$\frac{\partial \langle \rho \rangle}{\partial t} + \frac{1}{V} \int_{V_f} \nabla \cdot \rho \mathbf{u}\, dV = 0. \tag{2.87}$$

Since the velocity on the pore wall is zero and when (2.85) is applied, we have

$$\frac{\partial \langle \rho \rangle}{\partial t} + \nabla \cdot \langle \rho \mathbf{u} \rangle = 0. \tag{2.88}$$

We now examine the derivation of the volume-averaged momentum equation by Slattery (1969) and Whitaker (1969).

(A) DIMENSIONAL ANALYSIS

The momentum equation given for an incompressible and Newtonian fluid, (2.66), can also be averaged and gives (Slattery, 1969)

$$\frac{1}{V} \int_{V_f} \left[\rho_o \left(\frac{\partial \mathbf{u}}{\partial t} + \mathbf{u} \cdot \nabla \mathbf{u} \right) - \nabla \cdot \mathbf{T} - \rho \mathbf{f} \right] dV = 0 \tag{2.89}$$

or

$$\rho_o \left[\frac{\partial \langle \mathbf{u} \rangle}{\partial t} + \langle \mathbf{u} \cdot \nabla \mathbf{u} \rangle \right] = \nabla \cdot \langle \mathbf{T} \rangle + \rho \langle \mathbf{f} \rangle + \frac{1}{V} \int_{A_{sf}} \mathbf{T} \cdot \mathbf{n}\, dA. \tag{2.90}$$

Assuming that $\mathbf{f} = -\nabla \phi$, we have

$$\rho_o \left[\frac{\partial \langle \mathbf{u} \rangle}{\partial t} + \langle \mathbf{u} \cdot \nabla \mathbf{u} \rangle \right] = \nabla \cdot \langle \mathbf{T} \rangle - \rho \nabla \langle \phi \rangle$$

$$+ \frac{1}{V} \int_{A_{sf}} (\mathbf{T} - \rho \phi \mathbf{I}) \cdot \mathbf{n}\, dA \tag{2.91}$$

or

$$\rho_o \left[\frac{\partial \langle \mathbf{u} \rangle}{\partial t} + \langle \mathbf{u} \cdot \nabla \mathbf{u} \rangle \right] = -\nabla \langle \mathcal{P} - p_o \rangle + \nabla \cdot \langle \mathbf{S} \rangle - \langle \mathbf{d} \rangle^{sf}, \tag{2.92}$$

where $\mathcal{P} = p + \rho \phi$ and

$$\langle d \rangle^{sf} = -\frac{1}{V} \int_{A_{sf}} [\mathbf{T} + (p_o - \rho \phi) \mathbf{I}] \cdot \mathbf{n}\, dA \tag{2.93}$$

is the force per unit volume (other than those associated with hydrostatic and ambient pressures) that the fluid exerts on the pore walls. Now, by using

$$\nabla \cdot \langle \mathbf{S} \rangle = 2\mu \nabla \cdot \langle \mathbf{D} \rangle = \mu \nabla \cdot \nabla \langle \mathbf{u} \rangle, \tag{2.94}$$

we have

$$\rho_o \left[\frac{\partial \langle \mathbf{u} \rangle}{\partial t} + \langle \mathbf{u} \cdot \nabla \mathbf{u} \rangle \right] = -\nabla \langle \mathcal{P} - p_o \rangle + \mu \nabla^2 \langle \mathbf{u} \rangle - \langle \mathbf{d} \rangle^{sf}. \tag{2.95}$$

The difficulty is in relating $\langle d \rangle^{sf}$ to the fluid velocity (relative to the solid) and the structural properties of the solid matrix. Slattery uses dimensional analysis, and by noting the difficulties in reducing (2.95) to the Darcy resistance term, he suggests that

$$\langle \mathbf{d} \rangle^{sf} = \frac{\varepsilon \mu}{d^2 k_s} \langle \mathbf{u} \rangle \tag{2.96}$$

for stationary matrices, where d is a characteristic length (pore level) and k_s is dimensionless and is a function of the porosity only. His local volume-averaged equation is

$$\rho_o \left[\frac{\partial \langle \mathbf{u} \rangle}{\partial t} + \langle \mathbf{u} \cdot \nabla \mathbf{u} \rangle \right] = -\nabla \langle \mathcal{P} - p_o \rangle + \mu \nabla^2 \langle \mathbf{u} \rangle - \frac{\varepsilon \mu}{d^2 k_s} \langle \mathbf{u} \rangle. \tag{2.97}$$

The right-hand side of this equation is similar to that recommended by Brinkman.

(B) Transformation Tensor

Subject to negligible inertial effects, Whitaker (1969) uses the substitution $\mathcal{P} = p - p_o + \rho \phi$, and begins with the momentum equation

$$\nabla \mathcal{P} - \mu \nabla^2 \mathbf{u} = 0, \tag{2.98}$$

which leads to $\nabla^2 \mathcal{P} = 0$ and $\nabla^4 \mathbf{u} = 0$. It can be shown that $\langle \mathbf{u} \rangle$ and $\mathbf{u}$ are continuous and that the *transformation* of the average velocity into the point velocity is possible via

$$\mathbf{u} = \mathbf{M}(\mathbf{r}) \cdot \langle \mathbf{u} \rangle. \tag{2.99}$$

It can further be shown that $\nabla^4 \mathbf{M} = 0$ and that the velocity *transformation tensor* $\mathbf{M}$ depends on the pore structure, varies only with the spatial location within the pore, and is independent of $\langle \mathbf{u} \rangle$. Thus the point velocity is a linear vector function of the average velocity. This *closure* is typical of relating the deviations at the microscale to the averaged quantities (as in turbulence closures, Kaviany, 1994, and as in volume averaging of temperature deviation, Section 3.2.1).

Now, referring to an arbitrary curve s (lying entirely in the fluid), which has a tangent τ, we have $\tau \cdot \nabla = d/ds$. The scalar product of τ with the momentum equation gives

$$\tau \cdot \nabla \mathcal{P} = \frac{d\mathcal{P}}{ds} = \mu \tau \cdot \nabla^2 \mathbf{u} = \mu \tau \cdot \nabla^2 \mathbf{M} \cdot \langle \mathbf{u} \rangle, \quad (2.100)$$

where it is assumed that $\nabla^2 \mathbf{M} \cdot \langle \mathbf{u} \rangle$ is much larger than $\nabla^2 \langle \mathbf{u} \rangle \cdot \mathbf{M}$, because $\mathbf{M}$ varies over the small-scale ℓ but $\langle \mathbf{u} \rangle$ varies over L. Integrating from $s = 0$ where $\mathcal{P} = \mathcal{P}_o$, we have

$$\mathcal{P}(s) = \mathcal{P}_o + \mu \left[\int_0^s (\tau \cdot \nabla^2 \mathbf{M}) \, ds \right] \cdot \langle \mathbf{u} \rangle. \quad (2.101)$$

The quantity inside the square bracket is only a function of pore geometry, and the location within the representative elementary volume s and to within a constant can be written as

$$\psi = -\mu \, \mathbf{m} \cdot \langle \mathbf{u} \rangle, \quad (2.102)$$

where $\mathbf{m}$ is the *transformation vector* for $\mathcal{P}$ and depends on the position vector $\mathbf{r}$, in addition to the structure. Then from (2.81) using $\mathcal{P}$ as ψ we have

$$\nabla \int_{V_f} \mathcal{P} \, dV = -\int_{A_e} \mu \mathbf{m} \cdot \langle \mathbf{u} \rangle \mathbf{n} \, dA. \quad (2.103)$$

When both sides are multiplied by $1/V$, we have

$$\nabla \langle \mathcal{P} \rangle = -\mu \mathbf{K}_o \cdot \langle \mathbf{u} \rangle, \quad (2.104)$$

where

$$\mathbf{K}_o = -\frac{1}{V} \int_{A_e} \mathbf{n} \int_0^s (\tau \cdot \nabla^2 \mathbf{M}) \, ds \, dA. \quad (2.105)$$

If $\mathbf{K}_o$ has an inverse $\mathbf{K}_w$ ($\mathbf{K}_w = \mathbf{K}_o^{-1}$), then

$$\langle \mathbf{u} \rangle = -\frac{1}{\mu} \mathbf{K}_w \cdot \nabla \langle \mathcal{P} \rangle. \quad (2.106)$$

Using $\mathcal{P} = p - p_o + \rho \phi$, we have

$$\langle \mathbf{u} \rangle = -\frac{1}{\mu} \mathbf{K}_w \cdot \nabla \left[\langle p - p_o \rangle + \rho \langle \phi \rangle \right]. \quad (2.107)$$

Now using $\phi = -\mathbf{r} \cdot \mathbf{g}$ and $\langle \phi \rangle = -V^{-1} \int_{V_f} (\mathbf{r} \cdot \mathbf{g}) \, dV = -\varepsilon \mathbf{r}_o \cdot \mathbf{g}$, and since the pore pressure $\langle p - p_o \rangle^f = -V^{-1} \int_{V_f} (p - p_o) \, dV = \langle p - p_o \rangle / \varepsilon$, we have

$$\langle \mathbf{u} \rangle = -\frac{1}{\mu} \mathbf{K}_w \cdot \nabla \left\{ \varepsilon \left[\langle p - p_o \rangle^f - \rho \mathbf{r}_o \cdot \mathbf{g} \right] \right\}. \quad (2.108)$$

The gravitational term can be further simplified using

$$\nabla \varepsilon \rho \, \mathbf{r}_o \cdot \mathbf{g} = \varepsilon \rho \mathbf{g} + \rho \mathbf{r}_o \cdot \mathbf{g} \nabla \varepsilon, \qquad (2.109)$$

where $\nabla \mathbf{r}_o$ is the unit tensor. Then (2.108) can be written as

$$\langle \mathbf{u} \rangle = -\frac{1}{\mu} \mathbf{K}_w \cdot \left\{ \varepsilon \left[\nabla \langle p - p_o \rangle^f - \rho \mathbf{g} \right] \right.$$
$$\left. + \left[\langle p - p_o \rangle^f - \rho \mathbf{r}_o \cdot \mathbf{g} \right] \nabla \varepsilon \right\}. \qquad (2.110)$$

When $\nabla \varepsilon \simeq 0$ (nearly uniform matrix), then (2.110) becomes

$$-\nabla \langle p - p_o \rangle^f + \rho \mathbf{g} = \frac{\mu}{\varepsilon \mathbf{K}_w} \langle \mathbf{u} \rangle = \frac{\mu}{\varepsilon \mathbf{K}_w} \mathbf{u}_D. \qquad (2.111)$$

Comparing this with the Darcy law (2.4), when the hydrostatic term is added, we note that this $\mathbf{K}_w$ is different than the Darcy $\mathbf{K}$ by a factor of $1/\varepsilon$. A further refinement of this derivation and a discussion of the closure problem is given in Whitaker (1986).

Bear and Bachmat (Bear, 1988, 92–106) develop an averaging model based on flow through small channels, where the fluid velocity in the void space is in the direction parallel to the channel walls (i.e., one-dimensional flow in a channel, but the channels are not parallel to each other). They also assume that a volumetric resistance force exists and is proportional to the local velocity (i.e., they enter the Darcy resistance into the momentum equation as a replacement for the viscous force term). Later, they develop a more general volume-averaging technique for the multiphase flow in porous media (Bachmat and Bear, 1986).

2.8 Homogenization Method

This is one in the line of methods dealing with the *multiscale systems*. As will be shown, this method is similar to local volume averaging. However, the present treatments are very restrictive and have not been used for other than the hydrodynamics and the effective thermal conductivity of porous media. The method of *homogenization*, as applied to the Stokes flow through periodic structures, was initiated by Sanchez-Palencia (1980, p.129). Later, Ene and Polisevski (1987) expanded and applied the principles to the study of natural convection in porous media. A general discussion of the asymptotic analyses of periodic structures is given by Bensoussan et al. (1978). The method is based on the study of periodic solutions of partial differential equations and the asymptotic behavior of these as the period tends to zero. The hypothesis of periodic structure allows for rigorous treatment of problems such as the fluid flow in porous media and other *microscopically heterogeneous systems*. This is a mathematical method, and the asymptotic solution expansion is made using an *expansion parameter* that is the ratio of the pore (i.e., the representative elementary volume) length scale to the system length scale, i.e.,

$$\delta = \frac{\ell}{L}, \tag{2.112}$$

where ℓ is the *wavelength* of the periodic structure (i.e., the linear dimension of the unit cell). The asymptotic process $\delta \to 0$ represents the *transition* from microscopic to macroscopic phenomena. Homogenization is a simultaneous description (although approximate) of microscopic as well as macroscopic phenomena. In the case of the Darcy law, it will lead to the proof of the *tensorial* character of the permeability and its symmetry. The *convergence* of the homogenization process, for incompressible flows, is discussed in both of the references mentioned earlier.

We start with a Newtonian fluid and a periodically structured matrix bounded in domain V_L, where the structure is three-dimensional and both the fluid and solid phases are *continuous*. In light of the periodicity, the conduit structure has to be *uniform* and *isotropic* in order to deal with only one length scale ratio. For example, organized arrangement of spheres (simple cubic, body-centered cubic, and face-centered cubic) can be used. Three-dimensional interaction of circular tubes, where the fluid is moving on the inside, is another example. It is assumed that the pores have a periodic geometric structure with a period (or wavelength) ℓ, which is much smaller than L (i.e., $\delta \to 0$).

The space variables in V_L are x_i. In V_L, the portion associated with the pore (or fluid) is $V_{L,f}$ and that associated with the solid is $V_{L,s}$. We designate y_i as the stretched coordinate, i.e.,

$$y_i = \frac{x_i}{\delta}. \tag{2.113}$$

Then V_f and V_s are spaces for the variables y_i for fluid and solid phases, respectively. The homogenization proceeds with the following steps.

- Introduction of the two scales, x_i, y_i in the governing equations.

- Integration over the microscale (*mean-value operation*).

- Variational formulation of the expanded (in terms of ℓ) momentum equation, subject to continuity and no-slip constraints.

The mathematical treatment of the variational formulation and the use of the *Hilbert space* are involved and given by Sanchez-Palencia (1980, 130–134). The continuity and momentum equations are

$$\frac{\partial}{\partial x_i}\left(\rho^\delta u_i^\delta\right) = 0, \tag{2.114}$$

$$\rho^\delta u_k^\delta \frac{\partial u_i^\delta}{\partial x_k} = -\frac{\partial p^\delta}{\partial x_i} + \frac{\partial \tau_{ik}^\delta}{\partial x_k} + \rho^\delta f_i, \tag{2.115}$$

in V_f, where

$$\tau_{ik}^\delta = \lambda \delta_{ik} \frac{\partial u_j^\delta}{\partial x_j} + \mu \left(\frac{\partial u_i^\delta}{\partial x_k} + \frac{\partial u_k^\delta}{\partial x_i} \right), \qquad (2.116)$$

where the superscript δ stands for stretched domain, e.g., $\mathbf{u}^\delta(\mathbf{x}) = \mathbf{u}(\mathbf{x}/\delta)$, and λ and μ are the second and first coefficients of the viscosity. The boundary condition on the fluid-solid interfacial area A_{sf} is that of no slip, i.e.,

$$u_i^\delta = 0 \quad \text{on} \quad A_{sf}. \qquad (2.117)$$

Expansion of u_i^δ, p^δ, and ρ^δ in δ gives

$$u_i^\delta(x) = \delta^2 u_i^0(x, y) + \delta^3 u_i^1(x, y) + \ldots, \qquad (2.118)$$

$$p^\delta(x) = p^0(x) + \delta p^1(x, y) + \ldots, \qquad (2.119)$$

$$\rho^\delta(x) = \rho^0(x) + \delta \rho^1(x, y) + \ldots, \qquad (2.120)$$

where $\rho^0 = \rho^0(p^0)$ from the equation of state.

2.8.1 CONTINUITY EQUATION

Now substituting for u_i^δ and ρ^δ in the continuity equation and sorting out terms of order δ^1 and δ^2, we have

$$\frac{\partial}{\partial y_i}(\rho^0 u_i^0) = \rho^0 \frac{\partial u_i^0}{\partial y_i} = \rho^0 \nabla_y \cdot \mathbf{u}^0 = 0, \qquad (2.121)$$

$$\frac{\partial}{\partial x_i}(\rho^0 u_i^0) + \frac{\partial}{\partial y_i}(\rho^0 u_i^1 + \rho^1 u_i^0) = 0, \qquad (2.122)$$

where $\nabla_y \cdot ( )$ is the divergence in the y coordinate. A *mean-value operation*, operating on ψ, is defined as

$$\langle \psi \rangle \equiv \frac{1}{V} \int_V \psi(y) \, dy, \qquad (2.123)$$

which is similar to the local volume average introduced in Section 2.7.1. Then (2.122) becomes

$$\nabla_x \cdot \rho^0 \langle \mathbf{u}^0 \rangle + \frac{1}{V} \int_V \rho^0 \frac{\partial u_i^1}{\partial y_i} \, dy_i + \frac{1}{V} \int_V \frac{\partial \rho^1 u_i^0}{\partial y_i} \, dy_i$$

$$= \nabla_x \cdot \rho^0 \langle \mathbf{u}^0 \rangle = 0, \qquad (2.124)$$

where

$$\int_V \frac{\partial u_i^1}{\partial y_i} \, dy_i = \int_{\partial A} n_i u_i^1 \, ds = \int_{\partial A_{sf}} n_i u_i^1 \, ds + \int_{\partial A_e} n_i u_i^1 \, ds = 0. \qquad (2.125)$$

The last two line integrals (∂A_e is the path on A_e) are zero because $n_i u_i = 0$ on A_{sf} due to no-slip, and on A_e the contributions cancel each other because of the periodicity. The *homogenized or macroscopic continuity equation*, which is (2.124), becomes

$$\nabla_x \cdot \rho^0 \langle \mathbf{u}^0 \rangle = 0. \tag{2.126}$$

Note that because of the uniformity and full-periodicity assumption (uniform, isentropic media) and because no-slip is allowed on A_{sf}, this equation does not include any matrix-related information.

2.8.2 Momentum Equation

Substitution of the expansions for u_i^δ, p^δ, and ρ^δ into (2.115) gives

$$\delta^3 \rho^0 u_k^0 \frac{\partial u_i^0}{\partial y_k} + \ldots = -\frac{1}{\delta} \frac{\partial p^0}{\partial y_i} - \frac{\partial p^0}{\partial x_i} - \frac{\partial p^1}{\partial y_i}$$
$$+ \left[\mu \nabla_y^2 u_i^0 + (\lambda + \mu) \frac{\partial}{\partial y_i} (\nabla_y \cdot \mathbf{u}^0) \right] + \ldots + \rho^0 f_i. \tag{2.127}$$

After sorting and using (2.121), we have

$$\frac{\partial p^0}{\partial y_i} = 0, \quad \text{i.e., } p^0 = p^0(x) \text{ only}, \tag{2.128}$$

$$\frac{\partial p^1}{\partial y_i} = \mu \nabla_y^2 u_i^0 + \rho^0 f_i - \frac{\partial p^0}{\partial x_i}. \tag{2.129}$$

The disappearance of the inertial force was expected from the order of expansion initially selected for the velocity. Next, a *test function* $\mathbf{w}$ is chosen in an appropriate space of y-periodic function $\mathcal{V}_y$. In this space $\mathbf{u}(A_{sf}) = 0$ and $\nabla_y \cdot \mathbf{u} = 0$, and $\mathbf{u}$ is in $\mathbf{h}^1(V_f)$. This space $\mathcal{V}_y$ is the *Hilbert space*† and $\mathbf{h}^1(V_f)$ is its *norm*. The *scalar product* is

$$(\mathbf{u}, \mathbf{w})_{\mathcal{V}_y} = \int_{V_f} \frac{\partial u_i}{\partial y_k} \frac{\partial w_i}{\partial y_k} \, dy. \tag{2.130}$$

By multiplying (2.129) by $\mathbf{w}$ and integrating over V_f, we have

$$\int_{V_f} \frac{\partial p^1}{\partial y_i} w_i \, dy = \int_{\partial A_f} p^1 w_i n_i \, ds = 0, \tag{2.131}$$

$$\mu \int_{V_f} \nabla_y^2 u_i^0 w_i \, dy = \mu \int_{V_f} \frac{\partial}{\partial y_k} \left(\frac{\partial u_i^0}{\partial y_k} w_i \right) - \frac{\partial u_i^0}{\partial y_k} \frac{\partial w_i}{\partial y_k} \, dy$$

---

†Properties of this space are given in *Mathematical Handbook for Scientists and Engineers*, p. 439, edited by Korn and Korn, 2d ed., McGraw-Hill, 1968.

$$= \mu \int_{\partial A_f} \frac{\partial u_i^0}{\partial y_k} w_i n_k \, dy - \mu \int_{V_f} \frac{\partial u_i^0}{\partial y_k} \frac{\partial w_i}{\partial y_k} \, dy$$

$$= -\mu \int_{V_f} \frac{\partial u_i^0}{\partial y_k} \frac{\partial w_i}{\partial y_k} \, dy, \qquad (2.132)$$

where $\mathbf{w}(A_{sf}) = 0$ and the full periodicity was used. After multiplying (2.129) by $\mathbf{w}$, integrating over V_f, and then combining with (2.132), we have

$$(\mathbf{u}^0, \mathbf{w})_{\mathcal{V}_y} = \frac{1}{\mu}\left(\rho^0 f_i - \frac{\partial p^0}{\partial x_i}\right) \int_{V_f} w_i \, dy, \qquad (2.133)$$

where the function in the parenthesis on the right-hand side is not a function of y. This is the extra relation needed in the variational formulation.

The variational problem is to find $\mathbf{u}^0$ in $\mathcal{V}_y$ that satisfies (2.133), or by using the linear property of the function $\mathbf{u}^0$, define $\mathbf{v}^i$ such that

$$\mathbf{u}^0 = \frac{1}{\mu}\left(\rho^0 f_i - \frac{\partial p^0}{\partial x_i}\right)\mathbf{v}^i. \qquad (2.134)$$

Where $\mathbf{v}^i$, $i = 1, 2, 3$, satisfies

$$(\mathbf{v}^i, \mathbf{w})_{\mathcal{V}_y} = \int_{V_f} w_i \, dy. \qquad (2.135)$$

We note that (2.135) is the central relationship needed in transforming the viscous diffusion term into a term proportional to velocity. Now integrating (2.134) over V (volume averaging), we have

$$\frac{1}{V}\int_V \mathbf{u}^0 \, dy = \langle \mathbf{u}^0 \rangle = \frac{1}{\mu}\left(\rho^0 f_i - \frac{\partial p^0}{\partial x_i}\right)\frac{1}{V}\int_V \mathbf{v}^i \, dy \qquad (2.136)$$

or

$$\langle u_j \rangle \simeq \langle u_j^0 \rangle = \frac{K_{ij}}{\mu}\left(\rho^0 f_i - \frac{\partial p^0}{\partial x_i}\right), \qquad (2.137)$$

where

$$K_{ij} = \frac{1}{V}\int v_j^i \, dy. \qquad (2.138)$$

This is the Darcy law, where the gravitational force is included. The permeability tensor K_{ij} is the mean value of component j of v^i. K_{ij} depends only on the *geometry* of the period y. The homogenization method is also used by Rubenstein and Torquato (1989) in order to obtain *rigorous* upper and lower bounds on the permeability for random structures. Each bound is given in terms of various correlation functions that statistically describe the microstructure.

2.9 Semiheuristic Momentum Equations

As much as it is desirable to have one set of governing equations that can describe both the momentum transport through the porous media (K being small) as well as that in the plain media (K being very large), such equations, if they become available, will be too complicated to be of practical use. However, as initiated by Brinkman (1947) with his inclusion of a viscous shear stress term (other than the bulk viscous shear stress), which can take into account the shear stresses initiated at the surfaces bounding the porous media (macroscopic shear), attempts are being made to arrive at an *equivalent* of the Navier-Stokes equation for the description of flow through porous media. Extensions of the Darcy law are being sought to include the following.

- *Flow development in porous media*: As will be shown [Sections 2.10.2 and 2.12.4 (C)], the *macroscopic* entrance length as predicted by the macroscopic equations and by the direct simulations, is rather small. Using the macroscopic equations, the bulk resistance term dampens the velocity in the high flow rate zone and rapidly makes for a uniform velocity field.

- *Boundary generated vorticity (or the boundary effect)*: Similar to the entrance length, the damping of the vorticity (shear stress) generated at the confining walls takes place over a very short distance.

- *High Reynolds number effect*: As was discussed, the microscopic flow regime changes as the Reynolds number (pore level) increases and a pore-level flow development term must be added to the Darcy law to account for the extra pressure drop at high Reynolds numbers.

Among the versions of the extended Darcy law for conservation of momentum (differential) is the one used by Wooding (1957), where he included the inertial force (development term). Later, the boundary term was added and this leads to

$$\frac{\rho_0}{\varepsilon^2}(\mathbf{u}_D \cdot \nabla \mathbf{u}_D) = -\frac{\mu}{K}\mathbf{u}_D - \nabla p + \rho \mathbf{g} + \frac{\mu}{\varepsilon}\nabla^2 \mathbf{u}_D. \tag{2.139}$$

The apparent rationale for the appearance of coefficients $1/\varepsilon^2$ and $1/\varepsilon$ in front of the inertial and boundary viscous terms appears to be that since the pore velocities are larger than the average values (averaging symbol $\langle\ \rangle$ has been dropped), then these forces should be described at conduit scale by multiplying the velocity by $1/\varepsilon$ (and since the inertia term involves product of velocities, $1/\varepsilon^2$ is used). As expected, no rigorous development of (2.139) can be offered.

In an effort to extend the Darcy law (based on the application of local volume averaging), Vafai and Tien (1981) arrive at a semiempirical momentum equation using the Slattery momentum equation (2.95). They argue

2.9 Semiheuristic Momentum Equations

that the term **G** in this equation must include the high Reynolds number effect [the modified Ergun equation (2.57)]. Based on this, they arrive at

$$\frac{\rho_o}{\varepsilon}\left(\frac{\partial u_{Di}}{\partial t} + \mathbf{u}_D \cdot \nabla u_{Di}\right) = -\nabla \langle p \rangle^f + \frac{\mu}{\varepsilon}\nabla^2 u_{Di}$$

$$-\frac{\mu}{K}u_{Di} - \frac{F\varepsilon}{K^{1/2}}\rho|u_{Di}|u_{Di}. \quad (2.140)$$

Note that the viscosity used in both the microscopic and macroscopic viscous terms, is the fluid viscosity. Lundgren (1972), in giving justification to the Brinkman equation (2.64), shows that for spherical particles $\mu' = \mu g(\varepsilon)$, where $0 < g(\varepsilon)$ and $g(1) = 1$. We will discuss μ' further in Section 2.11.6.

For the matrices obeying the *modified Ergun relation* we have

$$\frac{F\varepsilon}{K^{1/2}} = \frac{1.8(1-\varepsilon)}{\varepsilon^3}\frac{1}{d}, \quad \text{where} \quad K = \frac{1}{180}\frac{\varepsilon^3}{(1-\varepsilon)^2}d^2. \quad (2.141)$$

This leads to

$$F = \frac{1.8}{(180\varepsilon^5)^{1/2}} = F(\varepsilon), \quad (2.142)$$

i.e., F is *not* a constant but a function of porosity only. As $\varepsilon \to 1$ (K becomes large), (2.140) reduces to the Navier-Stokes equation.

Based on the Whitaker momentum equation (2.111), when a volumetric body force is added to (2.140), we will have

$$\frac{\rho_o}{\varepsilon}\left(\frac{\partial u_{Di}}{\partial t} + \mathbf{u}_D \cdot \nabla u_{Di}\right) \quad = \quad -\frac{\partial p}{\partial x_i} \quad + \quad \rho f_i$$

(macroscopic inertial force or macroflow-development term)   (pore pressure gradient)   (body force)

$$+ \quad \frac{\mu}{\varepsilon}\nabla^2 u_{Di} \quad - \quad \frac{\mu}{K}u_{Di} \quad - \quad \frac{C_E}{K^{1/2}}\rho|u_{Di}|u_{Di}, \quad (2.143)$$

(macroscopic or bulk viscous shear stress diffusion, also called Brinkman viscous term or bounding surface effect)   (microscopic viscous shear stress, Darcy term)   (microscopic inertial force, also called Ergun inertial term or microflow-development term)

where we have used the Ergun constant C_E as defined in (2.59).

This is a *semiheuristic volume-averaged* treatment of the flow field. As was discussed, the experimental observations of Dybbs and Edwards (1984) show that the macroscopic viscous shear stress diffusion and the flow development (convection) are significant only over a length scale of ℓ from the vorticity generating boundary and the entrance boundary, respectively. However, (2.140) predicts these effects to be confined to distances of the order of $K^{1/2}$ and Ku_D/ν, respectively. We note that $K^{1/2}$ is smaller than d.

Then (2.140) predicts a macroscopic boundary-layer thickness, which is not only smaller than the representative elementary volume ℓ when $\ell \gg d$, but even smaller than the particle size. This will be further discussed in Sections 2.11.3 and 2.11.6. However, (2.140) allows estimation of these macroscopic length scales and shows that for most practical cases, the Darcy law (or the Ergun extension) is sufficient.

2.10 Significance of Macroscopic Forces

Addition of the macroscopic viscous shear stress diffusion and macroscopic inertial force terms to the modified Darcy equation enables us to determine the macroscopic momentum boundary-layer thickness and the flow development length for various flow conditions. The simplest flow that exhibits these features is the flow over a *semi-infinite flat plate* embedded in uniform porous media. Vafai and Tien have treated this problem numerically and Kaviany (1987) has extended their results. His analysis includes the application of the integral method, which will be reviewed later.

Starting from (2.140) and invoking the standard *boundary-layer approximations*, the *two-dimensional*, steady-state, velocity field $[(u,v),(x,y)]$ can be determined from

$$\frac{\partial u_D}{\partial x} + \frac{\partial v_D}{\partial y} = 0, \tag{2.144}$$

$$u_D \frac{\partial u_D}{\partial x} + v_D \frac{\partial u_D}{\partial y} = -\frac{1}{\rho}\frac{dp}{dx} - \frac{\nu \varepsilon}{K} u_D - \frac{C_E \varepsilon}{K^{1/2}} u_D^2 + \nu \frac{\partial^2 u_D}{\partial y^2}, \tag{2.145}$$

where

$$y, \quad x < 0: \quad u_D = u_{D,\infty}, \tag{2.146}$$

$$y = 0, \quad x > 0: \quad u_D = v_D = 0, \tag{2.147}$$

$$y \to \infty, \quad x > 0: \quad u_D = u_{D,\infty}. \tag{2.148}$$

Note that we have used $\langle p \rangle = \varepsilon \langle p \rangle^f$. Since at a sufficiently large distance from the plate the flow field is uniform, the free-stream pressure gradient required for maintaining the velocity $u_{D,\infty}$ must follow

$$0 = -\frac{1}{\rho}\frac{dp}{dx} - \frac{\nu \varepsilon}{K} u_{D,\infty} - \frac{C_E \varepsilon}{K^{1/2}} u_{D,\infty}^2. \tag{2.149}$$

With this, the boundary-layer momentum equation becomes

$$u_D \frac{\partial u_D}{\partial x} + v_D \frac{\partial u_D}{\partial y} = \frac{\nu \varepsilon}{K}(u_{D,\infty} - u_D)$$

$$+ \frac{C_E \varepsilon}{K^{1/2}}(u_{D,\infty}^2 - u_D^2) + \nu \frac{\partial^2 u_D}{\partial y^2}. \tag{2.150}$$

2.10.1 MACROSCOPIC HYDRODYNAMIC BOUNDARY LAYER

Now applying the standard integral method to this equation and constructing a velocity profile based on the finite boundary-layer thickness $\delta(x,y)$, i.e.,

$$\frac{u_D}{u_{D,\infty}} = \frac{3}{2}\left(\frac{y}{\delta}\right) - \frac{1}{2}\left(\frac{y}{\delta}\right)^3, \tag{2.151}$$

and inserting this into the momentum equation, we have

$$\frac{39}{280}\frac{d\delta}{dx} = \frac{3}{2}\frac{x}{Re_x\delta} - \left(\frac{3}{8}\frac{x^2\varepsilon}{KRe_x} + \frac{54}{105}\frac{C_E\varepsilon x}{K^{1/2}}\right)\frac{\delta}{x}$$

$$\equiv \frac{3}{2}\frac{x}{Re_x\delta} - \Gamma_x\frac{\delta}{x}, \tag{2.152}$$

where $Re_x = u_{D,\infty}x/\nu$ and the initial condition for δ is $\delta(0) = 0$. The resistance term Γ_x contains *both of the microscopic resistance terms*. Note that unlike the ever-growing boundary-layer thickness for plain media ($\Gamma_x = 0$), the microscopic resistances for porous media confine the boundary effect to a constant and small distance from the wall. The hydrodynamically fully developed condition is reached when $d\delta/dx \to 0$. This leads to the fully developed momentum boundary-layer thickness given by

$$\frac{\delta}{x} = \left(\frac{3}{2}\frac{1}{Re_x\Gamma_x}\right)^{1/2} = \left(\frac{x^2\varepsilon}{4K} + \frac{36C_E\varepsilon x^2 u_{D,\infty}}{105K^{1/2}\nu}\right)^{-1/2}. \tag{2.153}$$

For the *low Reynolds numbers* (pore level), $C_E = 0$, and we have

$$\delta = 2\left(\frac{K}{\varepsilon}\right)^{1/2}, \tag{2.154}$$

i.e., the boundary-layer thickness is of the order of $K^{1/2}$.

For matrices that obey the Carman-Kozeny relationship we have

$$K = \frac{\varepsilon^3}{180(1-\varepsilon)^2}d^2 \equiv f(\varepsilon)d^2 \tag{2.155}$$

or

$$K^{1/2} = f^{1/2}(\varepsilon)d, \tag{2.156}$$

where $f^{1/2}(\varepsilon)$ is equal to 0.0134 for $\varepsilon = 0.26$, equal to 0.053 for $\varepsilon = 0.5$, equal to 0.145 for $\varepsilon = 0.7$. This shows that $K^{1/2}$ is of the order of $(10^{-2} \sim 10^{-1})d$. Therefore, the *macroscopic boundary-layer thickness* predicted by the Brinkman type extension of the Darcy law is of the order of $(10^{-2} \sim 10^{-1})d$, i.e., *less* than the particle (or pore) size. This is contradictory because the volume-averaging arguments used require that all pore-level variations in the velocity be averaged and represented by u_D.

It should be noted that Lundgren (1972) also points the thinness of the boundary layer by arguing that the thickness of the boundary layer is of the order of $[\mu/(\mu'K)]^{1/2}$. He shows that

$$\frac{\mu}{\mu'} = g(\varepsilon), \quad 0 < g(\varepsilon) < 1.28 \quad \text{and} \quad g(1) = 1. \tag{2.157}$$

From (2.153), when $C_E \neq 0$, we have

$$\delta = \left(\frac{\varepsilon}{4K} + \frac{36 C_E \varepsilon u_{D,\infty}}{105 K^{1/2} \nu}\right)^{-1/2}, \tag{2.158}$$

and δ is further *decreased*.

2.10.2 MACROSCOPIC ENTRANCE LENGTH

The entrance length, i.e., the distance x at which $d\delta/dx \to 0$, is found by numerical integration and for $C_E = 0$, this gives

$$x\left(\frac{d\delta}{dx} \to 0\right) \simeq \frac{K u_{D,\infty}}{\nu} = \frac{K}{d} Re_d \tag{2.159}$$

or

$$\frac{x\left(\frac{d\delta}{dx} \to 0\right)}{d} = \frac{K}{d^2} Re_d = f(\varepsilon) Re_d, \tag{2.160}$$

where $f(\varepsilon)$ is of the order of $10^{-4} \sim 10^{-3}$ for the Carman-Kozeny type media. This results in the entrance lengths of about one particle (or pore) size for $Re_d = 10^3$ to 10^4. Therefore, the entrance length is also *very small*.

This entrance length has been estimated for the Stokes flow by Sangani and Behl (1989), where they found it to be within the first half diameter of the packed bed (for simple, face-centered, and body-centered cubic lattices). The numerical results for two-dimensional media, including the inertial effect, show that the entrance length does depend on Re as given earlier [Section 2.12.2 (C)].

Note that in the computations of the velocity fields, where the linear dimension of the computational domain is L, grid sizes (or if trial functions are used, functions with periods) of the order $0.1K^{1/2}$ are needed for reasonably accurate capturing of the boundary-layer phenomenon. If a uniform mesh is used (or the same trial function is used throughout), then $L/(0.1K^{1/2})$ nodes are needed. Noting that $L \gg \ell \gg d > K^{1/2}$, this requirement becomes rather *impossible* to meet. An alternative will be variable grids along with a two-domain matching similar to matching of inviscid and viscous domains in boundary-layer analysis. Therefore, a straightforward numerical approach to the Brinkman equation fails to capture the boundary layer.

2.11 Porous Plain Media Interfacial Boundary Conditions 71

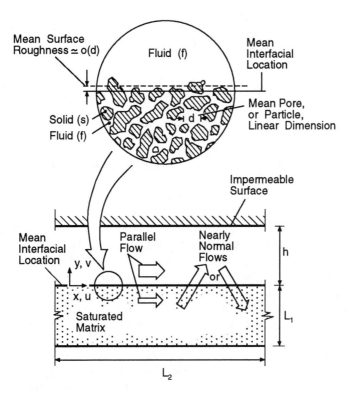

Figure 2.15 A schematic of the interface between a porous layer and its adjacent plain layer.

2.11 Porous Plain Media Interfacial Boundary Conditions

The interface refers to a *layer of thickness* $\delta = O(d)$ where, on the porous media side, the microscopic (bulk) viscous shear stress resistance is very large; and on the plain media side, the macroscopic viscous shear stress is very large. Figure 2.15 gives a schematic of this interface. Through this layer, the pressure and various components of velocity change continuously. When we take $\delta \to 0$, i.e., defining a *massless interface*, then discontinuities take place at this *hypothetical interface*. We can place one boundary layer on each side of this and then in these layers allow for gradual variations such that the asymptotic as well as the interfacial conditions are met.

Ene and Sanchez-Palencia (1975) and Levy and Sanchez-Palencia (1975, 1977) apply expansions around d/L, where L is a large scale characteristic length (L_1 or L_2 in Figure 2.15). Then their application of the Navier-

Stokes equation to the plain media and the Darcy law to the porous media, as $d/L \to 0$, leads to the following.

- Continuity of the pressure.

- Continuity of the tangential velocity (for parallel flows).

- Continuity of the normal velocity and a zero tangential velocity (for nearly normal flows).

The basic difference between *near parallel* and *near normal* flows is that in the former the pressure gradients in both media are of the same order, while in the latter the pressure gradient in the porous media is substantially larger than that in the plain media. Experimental results interpreted macroscopically indicate that for parallel flows, the *tangential interfacial velocity* is not the same as the Darcy velocity, i.e., there is a *discontinuity* in the tangential velocity (Beavers and Joseph, 1967). The first-order approximations mentioned earlier fail to show this phenomena.

Mathematically, the problem in matching the Navier-Stokes equation to the Darcy equation at the interface is that (other than the continuity in pressure) the former requires a velocity boundary condition while the latter does not. In general, this velocity boundary condition can not be taken as the Darcy velocity, and semiempirical replacement for this velocity has been recommended. The semiempirical treatments have been suggested because the very definition of the interface ($\delta \to 0$) is a step away from the rigorous treatment of this region. The rigorous treatment involves application of the Navier-Stokes equation directly to the flow through the interfacial layer, i.e., direct simulation of the interfacial hydrodynamics. This will be discussed in Section 2.11.5.

The following section presents a semiempirical treatment based on the velocity slip and a theoretical treatment for it. Another semiempirical treatment based on an effective viscosity, a direct simulation of parallel flow over a grooved surface, and a general two-dimensional direct simulation, follows.

2.11.1 Slip Boundary Condition

For the special case of steady-state fully developed parallel flow, i.e., $dp/dx = d\langle p \rangle^f/dx$, experiments have been performed by Beavers and Joseph (1967), and Beavers et al. (1970, 1974). Their one-dimensional flow is mathematically given by

(in the fluid layer) $\quad 0 \leq y \leq h: \quad \dfrac{d^2 u}{dy^2} = \dfrac{1}{\mu} \dfrac{dp}{dx},$ (2.161)

(in the porous layer) $\quad -L \leq y \leq 0: \quad \langle u \rangle = u_D = -\dfrac{K}{\mu} \dfrac{d\langle p \rangle^f}{dx},$ (2.162)

2.11 Porous Plain Media Interfacial Boundary Conditions

at $y = h$ and $y = -L$: the surfaces are impermeable. (2.163)

The boundary conditions are

$$u(h) = 0, \qquad (2.164)$$

$$u(0) = u_i, \qquad (2.165)$$

where u_i is the *interfacial velocity* and a boundary condition at $y = -L$ is not needed because the Darcean flow is uniform. Their experimental results show the following features.

- The mass flow rate through the fluid layer is larger when the surface at $y = 0$ is permeable, i.e., the velocity at $y = 0$ is not zero when $K \neq 0$.

- This mass flow rate is larger than the flow rate predicted by the Poiseuille-Couette flow with $u_i = u_D$. This is interpreted as an indication of a *velocity slip* at $y = 0$, i.e., $u_i > u_D$. The velocity distribution in the plain layer is given as

$$u = u_i\left(1 - \frac{y}{h}\right) - \frac{h^2}{2\mu}\frac{\mathrm{d}p}{\mathrm{d}x}\left(1 - \frac{y}{h}\right)\frac{y}{h} \qquad 0 \leq y \leq h. \qquad (2.166)$$

- This slip velocity can be *correlated* with the *square root of the permeability* and with the *velocity gradient* in the fluid layer evaluated at the interface. The idea behind the construction of such a correlation is the notion of continuity in the shear stress across the *interface*. Beavers and Joseph suggest $K^{1/2}$ as being the length scale [as was shown, $K^{1/2}$ is not of the order of d, but of the order $(10^{-2} \sim 10^{-1})d$] over which $u_i - u_D$ occurs. Then we have

$$\left.\frac{\mathrm{d}u}{\mathrm{d}y}\right|_{y=0} = \frac{\alpha}{K^{1/2}}(u_i - u_D), \qquad \text{at } y = 0, \qquad (2.167)$$

where α is the *slip coefficient*. If this boundary condition is used for the fluid layer at $y = 0$, (2.166) becomes

$$u = u_i\left(1 + \frac{\alpha y}{K^{1/2}}\right) + \frac{K}{2\mu}\frac{\mathrm{d}p}{\mathrm{d}x}\left(\frac{2\alpha y}{K^{1/2}} + \frac{y^2}{K}\right) \qquad 0 \leq y \leq h, \qquad (2.168)$$

where the gap size appears in the expression for u_i and this expression is

$$u_i = -\frac{\dfrac{K}{2\mu}\dfrac{\mathrm{d}p}{\mathrm{d}x}\left(\dfrac{h^2}{K} + 2\alpha\dfrac{h}{K^{1/2}}\right)}{\left(1 + \alpha\dfrac{h}{K^{1/2}}\right)} \qquad (2.169)$$

in which $h/K^{1/2}$ is the ratio of *large to small length scales*.

TABLE 2.5   EXPERIMENTAL RESULTS OF BEAVERS AND JOSEPH

MATRIX	BULK POROSITY	BULK PERMEABILITY $m^2 \times 10^9$	$K^{-1/2}$, m	d (MEAN PORE SIZE) mm	α
Foametal A	0.95-0.98	9.68	9.84×10^{-5}	0.407	0.78
Foametal B	0.95-0.98	3.94	1.98×10^{-4}	0.864	1.45
Foametal C	0.95-0.98	81.9	2.86×10^{-4}	1.14	4.0

Since the *interfacial location uncertainty* is of the order of d where d is a mean pore or particle size and since $K^{1/2} < d$ (Table 2.5) it can be expected that $h \gg K^{1/2}$ for any meaningful definition of the length h. Figure 2.15 gives a schematic of these length scales. Note that (2.167) results in $du/dy = 0$ for $h = (2K)^{1/2}$. Therefore, mathematically (2.167) is not valid for $h < (2K)^{1/2}$. However, on the physical grounds discussed earlier, (2.167) is valid only if $h \gg K^{1/2}$ or in terms of d, if $h > d$. This will be further examined in the following sections.

The value of α is determined experimentally from the measured mass flow rate. The fractional increase in the mass flow rate through the channel for $u_i \neq 0$ (permeable surface) over that for $u_i = 0$ is

$$\frac{\dot{m}(K > 0)}{\dot{m}(K = 0)} = \frac{3\left(\dfrac{h}{K^{1/2}} + 2\alpha\right)}{\dfrac{h}{K^{1/2}}\left(1 + \alpha\dfrac{h}{K^{1/2}}\right)}, \quad (2.170)$$

where K is measured separately as a bulk property. Note that the surface conditions (K, ε) are not those of the bulk and significant differences can occur depending on the manufacturing and finishing processes. This will be discussed in Section 2.11.7.

Foametals have a cellular structure consisting of irregularly shaped interconnected pores formed by a lattice construction. Standard selections are 10 to 100 pores per inch. Figure 2.16 shows a photograph of a thin layer of polyurethane foam (the same structure as the foametal). Note that boundaries have roughness of the order of the pore size.

Table 2.5 gives the experimental conditions/results of Beavers and Joseph for three different nickel foametals. For these high-porosity matrices, the value of α changes significantly. Their results show that α *increases* with the *bulk permeability* and the *average pore size*. For another type of matrix with lower bulk permeability than those given earlier, they found a much smaller value for the slip coefficient α.

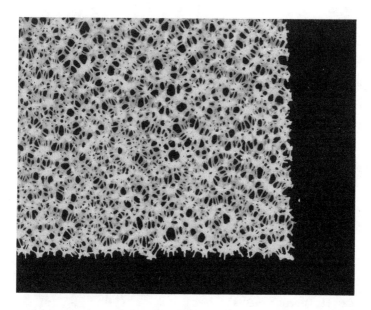

Figure 2.16 Photograph of a thin layer of polyurethane foam, showing the bulk pores and the near surface phase distributions.

2.11.2 ON BEAVERS-JOSEPH SLIP COEFFICIENT

In the classical example of fully developed laminar flow of two superposed immiscible fluids in a channel, the boundary conditions applied at the interface are the continuity of the shear stress (*tangential interfacial stress continuity*) and the tangential velocity (*kinematic condition*). These boundary conditions cannot be readily applied to the interface of porous and plain layers. In addition, with this class of interfaces, the most ambiguous aspect of hydrodynamic interfacial condition is the *prescription* of the *exact location* of the interface.

Saffman (1971) gives a theoretical justification to (2.167). He also shows that α depends on the *interfacial location* and that within an uncertainty of $O(d)$ associated with the specification of this location, α varies *significantly*. He develops an ensemble-averaged, similar to the local volume-average method described before, momentum equation for heterogeneous media. This includes the boundary shear stress term (the Brinkman viscous term). Then he examines the limiting case of a *step function* distribution of the permeability and the porosity (i.e., at the interface of the porous and plain media) using the boundary-layer techniques (inner and outer solutions). His mathematical formulation based on a parallel flow arrangement (similar to the experimental condition of Beavers and Joseph) is as follows (the equations are written for three-dimensional flows however, the analysis is limited to a one-dimensional flow with u along x).

$$\text{Outer I (plain media)}: \quad y > 0 \quad \mu \nabla^2 \mathbf{u}^0 = \nabla p, \qquad (2.171)$$

$$\text{inner (interface)}: \quad \mu \int K_{ij}(\mathbf{x}, \mathbf{r}) u_{D,j}(\mathbf{r}) \, d\mathbf{r}$$
$$= \mu \nabla^2 u_{D,i} - \varepsilon \frac{\partial \langle p \rangle}{\partial x_i}, \qquad (2.172)$$

where $\mathbf{r}$ has three components r_1, r_2, and r_3. The *kernel* has the property

$$K_{ij}(\mathbf{x}, \mathbf{r}) \to 0 \quad y > 0, \qquad (2.173)$$

$$K_{ij}(\mathbf{x}, \mathbf{r}) \to \frac{\varepsilon}{K} \delta_{ij} \delta(\mathbf{x} - \mathbf{r}) \quad y < 0, \qquad (2.174)$$

where δ is the *Dirac delta* and δ_{ij} is the *Kronecker delta*.

$$\text{Outer II (porous media)}: \; y < 0 \; \varepsilon = \varepsilon^0, \; \frac{\mu}{K} \mathbf{u}_D^0 = -\nabla \langle p \rangle. \quad (2.175)$$

As was shown in Section 2.10, the boundary-layer thickness is of the order $K^{1/2}$. Now scaling y with $K^{1/2}$ the boundary conditions are

$$\frac{y}{K^{1/2}} \to \infty: \quad u_D \to u^0, \qquad (2.176)$$

$$y = 0: \quad p = \langle p \rangle, \; \frac{\partial p}{\partial x} = \frac{\partial \langle p \rangle}{\partial x}, \qquad (2.177)$$

$$\frac{y}{K^{1/2}} \to -\infty: \quad u_D = u_D^0, \qquad (2.178)$$

where u^0 and u_D^0 are the asymptotes (outer solutions). For $x_1 = x$, $x_2 = y$, and $u_1 = u$, (2.172) can be written for a constant pressure gradient condition as

$$\mu \int G(y, r_2) \mathbf{u}_D(r_2) \, dr_2 = \frac{\mu}{\varepsilon} \frac{d^2 u_D}{dy^2} - \frac{dp}{dx}, \qquad (2.179)$$

where

$$G(y, r_2) = \frac{1}{\varepsilon} \int K_{11}(\mathbf{x}, \mathbf{r}) \, dr_1 \, dr_3, \qquad (2.180)$$

r_2 is along y, and statistical *homogeneity* parallel to the interface is assumed. The kernel and the porosity satisfy

$$\frac{y}{K^{1/2}} \to \infty: \quad G(y, r_2) \to 0, \; \varepsilon \to 1, \qquad (2.181)$$

$$\frac{y}{K^{1/2}} \to -\infty: \quad G(y, r_2) \to \frac{1}{K} \delta(y, r_2), \; \varepsilon = \varepsilon^0. \qquad (2.182)$$

2.11 Porous Plain Media Interfacial Boundary Conditions

Since the plain medium is assumed to be semi-infinite, the only available length scale is $K^{1/2}$. The velocity can be scaled with $-(K\,dp/dx)/\mu$. Using these, (2.179) becomes

$$\int G^*(y^*, r_2^*) u_D^*(r_2^*)\, dr_2^* - \frac{1}{\varepsilon}\frac{d^2 u_D^*}{dy^{*2}} = 1, \qquad (2.183)$$

where

$$y^* = \frac{y}{K^{1/2}}, \quad u^* = -\frac{\mu u}{K\dfrac{dp}{dx}}, \quad G^* = K^{3/2}G. \qquad (2.184)$$

The boundary conditions are

$$y^* \to \infty: \quad u^* = -\frac{y^{*2}}{2} + Ay^* + B, \quad \varepsilon = 1 \qquad (2.185)$$

$$y^* \to -\infty: \quad u_D^* \to 1, \quad \varepsilon = \varepsilon^0. \qquad (2.186)$$

Saffman *constructs* an asymptotic solution ($y^* \to \infty$) to (2.183) of the form

$$u^* \sim -\frac{y^{*2}}{2} + Ay^* + B + \lambda(y^* + C), \qquad (2.187)$$

where A, B, and C are constants of order of unity determined by the structure G^* and λ is an arbitrary parameter. In order to satisfy (2.183), the class of solutions is a single parameter family. In dimensional form, (2.187) is

$$u \to \frac{1}{\mu}\frac{dp}{dx}\left[\frac{1}{2}y^2 - yK^{1/2}(A+\lambda) - K(B+\lambda C)\right] \quad \text{as } y \to \infty, \qquad (2.188)$$

and when differentiated

$$\frac{du}{dy} \to \frac{1}{\mu}\frac{dp}{dx}\left[y - K^{1/2}(A+\lambda)\right] \quad \text{as } y \to \infty. \qquad (2.189)$$

Now for du/dy to remain finite, λ should be of $O(K^{-1/2})$. Therefore, (2.188) and (2.189) become

$$u \to \frac{1}{\mu}\frac{dp}{dx}\left(\frac{1}{2}y^2 - yK^{1/2}\lambda - K\lambda C\right) = -\frac{1}{\mu}\frac{dp}{dx}K\lambda C \quad \text{on } y=0, \qquad (2.190)$$

$$\frac{du}{dy} \to \frac{1}{\mu}\frac{dp}{dx}\left(y - K^{1/2}\lambda\right) = -\frac{1}{\mu}\frac{dp}{dx}K^{1/2}\lambda \quad \text{on } y=0. \qquad (2.191)$$

Eliminating λ between these two equations, we have

$$u = CK^{1/2}\frac{du}{dy} \quad \text{on } y=0, \qquad (2.192)$$

78   2. Fluid Mechanics

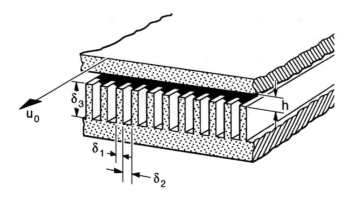

Figure 2.17 The Taylor model for porous-plain interfacial fluid dynamics.

where comparing to the Beavers-Joseph boundary condition (2.167), the slip coefficient becomes $\alpha = 1/C$. As was mentioned, the specification of the location of the interface has an uncertainty of $O(d)$ or at least $O(K^{1/2})$. Then within $O(K^{1/2})$, (2.188) and (2.189) can be written as

$$u = -\frac{1}{\mu}\frac{dp}{dx}\lambda K\left(\frac{y}{K^{1/2}} + C\right), \qquad (2.193)$$

$$\frac{du}{dy} = -\frac{1}{\mu}\frac{dp}{dx}\lambda K^{1/2}, \qquad (2.194)$$

or

$$u = \left(\frac{y}{K^{1/2}} + C\right)K^{1/2}\frac{du}{dy}. \qquad (2.195)$$

Then $\alpha = [(y/K^{1/2}) + C]^{-1}$, i.e., α depends on where the *selection* of the interfacial location [within a distance of $O(K^{1/2})$]. The interfacial slip coefficient has been predicted by Kim and Russell (1985) and by Ochoa-Tapia and Whitaker (1995).

2.11.3 TAYLOR-RICHARDSON RESULTS FOR SLIP COEFFICIENT

In order to study the dependence of α on the matrix structure and in order to examine whether α is independent of the gap size h, Taylor (1971) constructs a model of porous-plain interface. In his model, a stationary disk with a grooved surface resembles a permeable surface, and another flat disk placed a distance h away from this grooved surface is rotated. This results in the *Couette flow* (zero pressure gradient) with a slip velocity u_i at the interface. Richardson (1971) solved the fully developed flow in the gap-groove domain. A schematic of the Taylor model is given in Figure 2.17. For $\delta_3/\delta_2 \to \infty$, the fully developed flow through parallel plates gives the

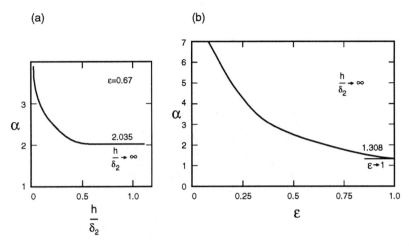

Figure 2.18 Variation of the slip coefficient with respect to (a) gap size and (b) porosity, as computed by Richardson (1971).

permeability as

$$K = \frac{\varepsilon}{12}\delta^2, \qquad (2.196)$$

where $\varepsilon = \delta_2/(\delta_1 + \delta_2)$. Some of the results of their analysis are given here.

- For small gap size (i.e., small h/δ_2), α does depend on the *gap size*. But an asymptotic behavior exists for $h/\delta_2 \to \infty$.

- With an increase in the *porosity*, α decreases and reaches an asymptotic value of 1.308 (i.e, $\partial\alpha/\partial\varepsilon \to 0$ as $\varepsilon \to 1$).

The Taylor-Richardson results for the variation of α with respect to h/δ_2 (for $\varepsilon = 0.67$) and for the variation of α with respect to ε (for $h/\delta_2 \to \infty$) are shown in Figures 2.18 (a) and (b). The asymptotic value of 1.308 instead of the expected 1.0 is due to the *limitation* of their analysis. The dependence of α on the gap size when expressed in terms of $K^{1/2}$ gives the critical value of h beyond which α becomes *independent* of h as

$$h \geq 0.6\,\delta_2 = 2.5K^{1/2}, \quad \varepsilon = 0.67. \qquad (2.197)$$

This is similar to that obtained from the Beavers-Joseph relation which gives $h \geq (2K)^{1/2}$.

2.11.4 SLIP COEFFICIENT FOR A TWO-DIMENSIONAL STRUCTURE

Sahraoui and Kaviany (1992) examine the slip coefficient for a periodic two-dimensional structure made of cylindrical particles. Figures 2.19 (a)

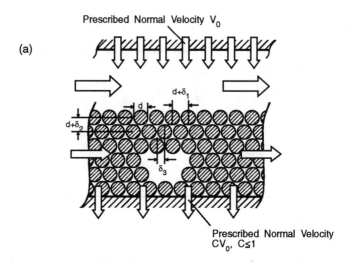

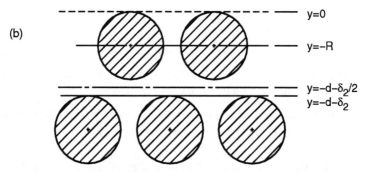

Figure 2.19 The two-dimensional interfacial model used for examination of the interfacial hydrodynamic boundary conditions: (a) arrangement and (b) location of $y = 0$.

and (b) give the parameters specifying the arrangement and define location of the nominal interface ($y = 0$). The point variation of the velocity and pressure are obtained by solving the point conservation equations for mass and momentum, along with the prescribed boundary conditions. In order to relate these point (pore-level or micro) variations to the macro or the Darcean behavior, local area and volume averages are taken. The *local area average* is defined as (for example, for u)

$$\langle u \rangle_A (y) \equiv \frac{1}{\ell} \int_0^\ell u(x, y) \, dx. \qquad (2.198)$$

2.11 Porous Plain Media Interfacial Boundary Conditions

The interfacial, tangential velocity $\langle u \rangle_{A,i}$ is defined as

$$\langle u \rangle_{A,i} = \langle u \rangle_A (y_i), \tag{2.199}$$

where y_i is the selected interfacial location, and is not necessarily the nominal interface. The *local volume average* over a cell [the unit cell is shown in Figure 2.10(a)] is defined as

$$\langle u \rangle_V (y) \equiv \frac{1}{\ell^2} \int_0^\ell \int_{-\ell/2}^{\ell/2} u(x, y + \zeta) \, d\zeta \, dx. \tag{2.200}$$

The pressure is averaged on the fluid phase only, i.e.,

$$\langle p \rangle_V^f = \frac{1}{V_f} \int_{V_f} p \, dV. \tag{2.201}$$

(A) Interface Position

A major difficulty associated with the use of the Beavers and Joseph boundary condition is the choice of the interfacial location where the boundary condition is applied. Beavers and Joseph suggested that the interface be the tangent to the surface of the *outermost* pore. For the arrangement of cylinders considered here, this would be the tangent to the top of the *interfacial layer* of cylinders, which is defined as the *nominal interface* and is assigned as the origin of the y-axis, as shown in Figure 2.19(b).

Because the interfacial effects generally penetrate in the porous medium over distances of the *order* of the Brinkman screening distance $K^{1/2}$ or larger, the magnitude of this distance for a bed of cylinders is considered first. The numerical results of Sahraoui and Kaviany show that the permeability for the in-line arrangement of cylinders is given by (2.48) and (2.49). Based on these, for example, for $\varepsilon = 0.5$ we have $K^{1/2} = 0.0525d$. Note that in practice, for small d, the accurate (to within $K^{1/2}$) determination of the interfacial position is *difficult*. Using the numerical simulation of Sahraoui and Kaviany we can examine the dependence of α on the uncertainty in the assignment of the interfacial location.

Using the numerical results for the area-averaged velocity $\langle u \rangle_A$, the slip coefficient evaluated at an interfacial position y_i (measured from the nominal interface) is determined from (2.167) using

$$\alpha = \frac{d\langle u \rangle_A}{dy} \frac{K^{1/2}}{\langle u \rangle_{A,i} - u_D}. \tag{2.202}$$

Figure 2.20 shows the variation of α with respect to y_i. The results show that for a change in the selected interfacial position from $y_i/\ell = 0$ to 0.04 ($y_i/d = 0.05$), a distance corresponding to $K^{1/2}$, the slip coefficient drops

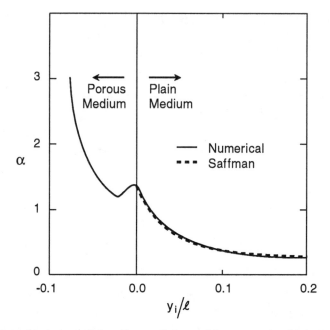

Figure 2.20 Variation of the slip coefficient with respect to the interfacial location. The numerical solution of Sahraoui and Kaviany and the prediction of Saffman are shown.

more than 50 percent. Saffman predicted that the inverse of α is linear with respect to y_i, i.e.,

$$\frac{1}{\alpha} = C + \frac{y_i}{K^{1/2}}. \qquad (2.203)$$

This is also plotted in Figure 2.20, where C is the inverse of the slip coefficient determined from the numerical results at the nominal interface. The prediction of Saffman is in excellent agreement with the numerical results. The variation of the slip coefficient with respect to y_i is not well behaved for $y_i/\ell < 0$. This is because, just below the nominal interface, separation of the flow occurs and wakes are formed. This is shown in Figures 2.21 (a) and (b), where the recirculation region is especially evident for the staggered arrangement.

Larson and Higdon (1986, 1987) consider the Stokes flow through a semi-infinite periodic structure (two-dimensional) made of cylinders with flow either parallel or perpendicular to the cylinder axes. The boundary integral method was used to solve the Stokes equation for flow around different arrangements of cylinders (square and hexagonal arrangements). When the flow is perpendicular to the cylinder axes and parallel to the interfacial plane, they obtained *slip velocities* that are in the direction *opposite* to the velocity in the plain and porous medium for $\varepsilon < 0.9$. This *nonphysical*

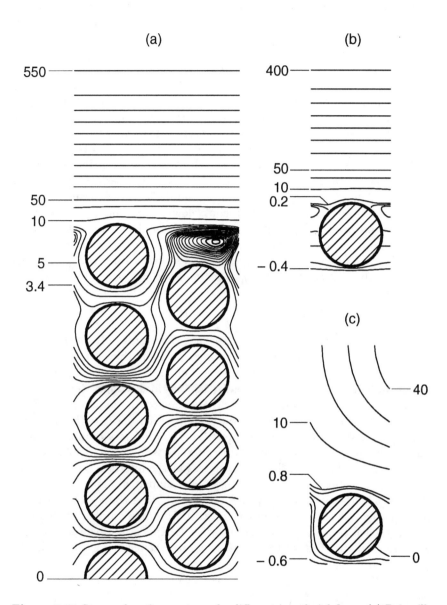

Figure 2.21 Stream function contours for different interfacial flows: (a) Poiseuille flow for staggered arrangement $\psi = 0.2\text{--}3.4$ ($\Delta\psi = 0.2$), 4.25, 10–50 ($\Delta\psi = 20$), 50–550 ($\Delta\psi = 50$); (b) Poiseuille flow for in-line arrangement $\psi = -0.4\text{--}0.2$ ($\Delta\psi = 0.2$), 10–50 ($\Delta\psi = 20$), 50–400 ($\Delta\psi = 50$), and (c) oblique flow for in-line arrangement $\psi = -0.6\text{--}0$ ($\Delta\psi = 0.2$), 0–0.8 ($\Delta\psi = 0.4$), 10–40 ($\Delta\psi = 10$).

84   2. Fluid Mechanics

result is due to their choice of the *interfacial location*. Their slip velocity is calculated using the volumetric flow rate above the interface (for shear flows), i.e.,

$$\frac{\dot{m}_c}{\rho} = \langle u \rangle_{A,i} h + \frac{1}{2}\gamma h^2, \qquad (2.204)$$

where γ is the applied velocity gradient at the impermeable boundary of the channel ($y = h$). In their calculation of the slip velocity, the interfacial location is chosen to be the plane passing through the *axes* of the interfacial row of cylinders. However, the plain medium flow occurs above the surface tangent to the top of the interfacial cylinders. The contribution to $\dot{m}_c$ of the flow between this surface and their defined interface is small for $\varepsilon < 0.9$. This is due to the vortices present between adjacent cylinders. Due to this overestimation of h, a slip velocity opposing the direction of the plain medium flow is obtained using (2.204). They state that due to the ambiguity in the definition of the interfacial location, any reasonable value for α can be correct. However, as was shown α (and $\langle u \rangle_{A,i}$) depends on y_i and the proper choice for y_i is the *nominal interface*.

(B) Inertial Effects

It is generally believed that the slip coefficient depends only on the surface and bulk structural properties of the porous medium. However, we expect α to change as different flow *regimes* are encountered. Here, we only consider *steady-state laminar* flows. The flow at the interface changes due to the *inertial effects* as the Reynolds number (based on flow in the porous medium) increases. Figure 2.22 shows the variation of α with respect to the Reynolds number. For small Reynolds numbers, the viscous forces dominate and α remains constant (the flow field at the interface is invariant). For Reynolds numbers larger than 0.1, the inertial forces at the interface become significant and the slip coefficient increases. The numerical experiments show that for a unit cell located away from the interface (bulk behavior), the inertial effects begin to be important at a Reynolds number of about 3. The slip coefficient begins to decrease for Reynolds numbers larger than 10. Note that the shear at the interface is *extrapolated* using the results for $y > 0$.

(C) Parallel Flows

The interfacial flow also changes for the different flow types (*Poiseuille*, *Couette*, or *oblique*) that can be present in the plain layer. In this section, we study the effect of the two parallel flow types, without introducing the two-dimensional effects (i.e., oblique flows). The Couette flow condition is similar to the experiment and analysis of Taylor and Richardson and to the shear flow over a bed of cylinders studied by Larson and Higdon.

2.11 Porous Plain Media Interfacial Boundary Conditions

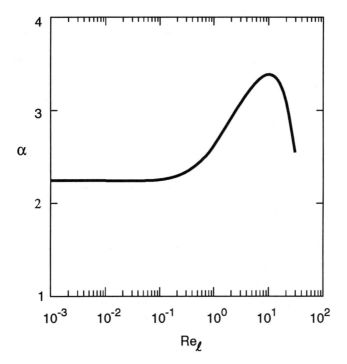

Figure 2.22 Variation of the slip coefficient with respect to the particle Reynolds number.

The numerical results show that α depends on the type of flow considered, as shown in Figure 2.23. This is due to the inertial effects present at the interface for the Poiseuille flow. The effect of increasing the gap size for the Poiseuille flow is very similar to that of increasing the particle Reynolds number. As the gap size is increased, the velocity at the interface increases, and the inertial effects become important. As observed with an increase in the Reynolds number, the separation region behind the cylinder increases in size causing the flow below the interface and the interfacial velocity to decrease. This is presented in Figure 2.23, showing that α increases as the gap size increases. For the Couette flow, the slip coefficient is independent of the gap size, due to the absence of these inertial effects. This agrees with the experimental results of Taylor for flow over a grooved plate. According to Figure 2.23, the assumption that α is the same for the two flow types is correct *when* the channel dimension is between *one* and *two* cell sizes. As the gap size increases, the effect of the inertial forces become more pronounced, and the *difference* in the slip coefficient between the two flows becomes *larger*.

When Beavers et al. performed their experiments, the slip coefficient was computed for different gap sizes. From these different values of the slip coefficient, they compute the *average* value for the specimen considered. In

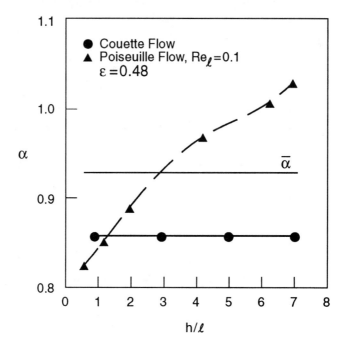

Figure 2.23 Variation of α with respect to the gap size for two parallel flows: (a) a Poiseuille flow, and (b) a Couette flow.

Figure 2.23, the *average slip coefficient* $\bar{\alpha}$ is given for the in-line arrangement of cylinders as it would be found by the procedure used by Beavers et al. For the porous medium made of cylinders with slip coefficient presented in Figure 2.23, the mass flow through the channel was computed for gap size $h/\ell = 0.5$ ($\sigma = 13.5$) using the slip coefficient at $h/\ell = 0.5$. The mass flow was also computed using the average slip coefficient $\bar{\alpha}$. The difference obtained between the two values is about 1 percent. Thus the averaging procedure used for the different gap sizes gives an error well within the experimental uncertainties.

The Couette flow with a *variable porosity* near the interface has been examined by Hsu and Cheng (1991). They find solutions similar to those of Neale and Nader (1974b) for Poiseuille flow and show that variable porosity causes a *decrease* in the slip coefficient.

(D) Oblique Flows

All the available experimental results for the slip coefficient are based on parallel flows. This is because experiments with oblique flows are more difficult to perform and because the relationship between the mass flow rate,

2.11 Porous Plain Media Interfacial Boundary Conditions 87

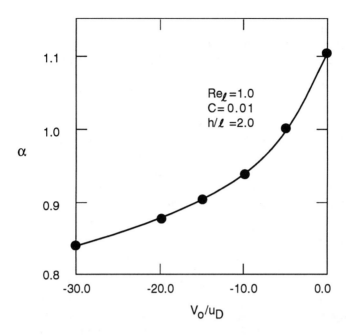

Figure 2.24 Variation of the slip coefficient with respect to the blowing velocity at the impermeable boundary of the channel.

α, and the gap size can not be found in a closed form. The two-dimensional effects are examined here for an oblique flow with an imposed constant pressure gradient along the channel, a blowing velocity V_o at the upper boundary, and a suction velocity CV_o at the lower boundary [Figure 2.19(a)]. Figure 2.24 shows the variation of α with respect to the blowing velocity V_o. The slip coefficient decreases as the flow deviates from the Poiseuille flow ($V_o = 0$) and a 24 percent difference is attained for $V_o/u_D = -30$. This is due to the fact that the flow is two-dimensional, whereas the Beavers and Joseph boundary condition treats the flow as one-dimensional. To account for the two-dimensional effects, Jones (1973) suggested the addition of another term to the slip-boundary condition, giving an expression that is similar to the balance of the shear stress (instead of the velocity gradient across the interface). This expression is

$$\left.\frac{\partial \langle u \rangle_A}{\partial y}\right|_{y_i^+} + \left.\frac{\partial \langle v \rangle_A}{\partial x}\right|_{y_i} = \frac{\alpha}{K^{1/2}}(\langle u \rangle_{A,i} - u_D) \quad \text{at} \quad y_i. \qquad (2.205)$$

When considering a single cell, as shown in Figure 2.21(c), the $\partial \langle v \rangle_A / \partial x$ term in (2.205) is zero due to the periodicity of v. In order to assess the effect of $\partial \langle v \rangle_A / \partial x$, we examined *two* interfacial cells laid side by side. Also, the constant blowing velocity at the upper boundary, giving a periodic flow in each cell, was replaced by a blowing velocity that varies *linearly* with x. These calculations have shown that the change in the calculated α is less than 1 percent, when the second term is included. Thus, we conclude that the second term does not account for the two-dimensional effects, since its contribution is not significant. Generally, the first term is much larger than the second one, since the averaged velocity across the interface varies more significantly than that along the interface. These same observations were made by Fatehi and Kaviany (1990) when they used this two-dimensional boundary condition to analyze the liquid droplet levitation on a heated porous layer. They found that the second term did not noticeably change their prediction of the levitation distance (gap size).

(E) Porosity

In the experiments conducted by Beavers et al. (1967, 1970, 1974), the slip coefficient was evaluated for a wide range of permeabilities spanning over two orders of magnitude. Over this range of permeability, α decreased monotonically as the permeability decreased. This suggests that α has a very strong dependence on the permeability and the porosity. This strong dependence is also observed for the periodic structure considered. The slip coefficient computed for a given Reynolds number and gap size is shown as a function of porosity in Figure 2.25. As expected, when the porosity increases, the slip coefficient increases, showing general agreement with the few experimental data for random (disordered) porous media.

(F) Surface structure

When Beavers and Joseph suggested the slip-boundary condition, they suggested that the slip coefficient would be strongly dependent on the surface structure of the porous medium. Later, Beavers et al. (1974) found that α changed by 85 percent after they machined the surface of the porous slab. In the present model, it is possible to study the surface phase-distribution nonuniformities by using different arrangements of the interfacial row of cylinders. For this purpose, we have taken a surface structure for an in-line arrangement of cylinders, as depicted by Figure 2.26(a), where the variable distance λ represents the surface nonuniformity (e.g., particle dislocation). In random porous media, λ would represent the surface roughness (beyond the nominal surface). The variation of the slip coefficient is shown in Figure 2.26(b) as the distance λ is varied between 0 and $\ell/2$. For a nonzero offset, the fluid flow at the nominal interface increases, giving rise to a higher slip velocity. Therefore, the slip coefficient decreases significantly as the offset increases.

2.11 Porous Plain Media Interfacial Boundary Conditions

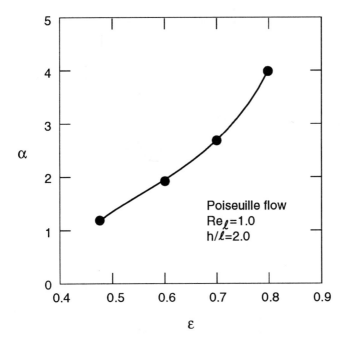

Figure 2.25 Porosity dependence of the slip coefficient for the in-line arrangement of cylinders.

The slip coefficient is also calculated for a staggered arrangement of cylinders. In this configuration, every column of cylinders is offset with respect to the neighboring ones by $\ell/2$, as shown in Figure 2.21(a). When the slip coefficient for the staggered arrangement is compared to the one for the in-line arrangement for an offset distance $\lambda = \ell/2$, a 2 percent difference is obtained. This shows that α is mainly a surface property because the two configurations have the same surface structure but the bulk structures are different. We note that the difference in the permeability for the two cylinder arrangements (for $\varepsilon = 0.48$) is about 10 percent.

(G) Pressure Slip

In the previous sections, we have discussed the velocity slip that occurs at the interface of a porous and a plain medium. The slip is due to the averaging performed in order to obtain the Darcean flow in the porous medium. A pressure slip also occurs when the pressure is averaged in the porous medium. This pressure slip has been investigated by Ene and Sanchez-Palencia (1975) using an order-of-magnitude analysis for low Reynolds number flow in a periodic structure. Through their analysis, they found

90   2. Fluid Mechanics

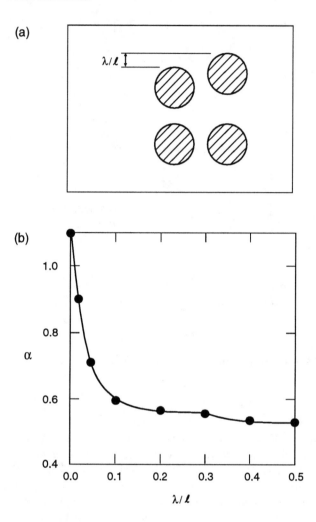

Figure 2.26 Surface nonuniformity effect on the slip coefficient for in-line arrangement: (a) the interfacial arrangement and (b) variation of α with respect to the offset λ.

that the pressure jump across the interface for parallel flows is

$$\frac{\langle p \rangle_v^f - p_p}{\ell} = O\left(\frac{\Delta p}{\ell}\right), \tag{2.206}$$

where $\langle p \rangle_v^f$ is the volume-averaged (over the fluid) pressure in the bulk porous medium, p_p is the average pressure in the plain medium over the cell length, and Δp is the pressure difference applied across the cell to drive

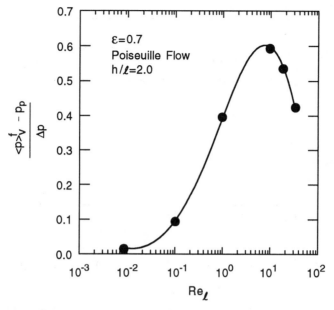

Figure 2.27 Variation of the pressure slip with respect to Reynolds number, for an in-line arrangement.

the flow. Their analysis shows that the *pressure slip* is comparable to the pressure difference across the cell. The interfacial simulation allows for the study of the pressure slip and verification of (2.206). The point solution to the pressure field reveals that this pressure jump is mainly induced by the *inertial effects* present at the interface at high Re_ℓ. This is clearly shown in Figure 2.27 where the pressure slip, normalized using the imposed pressure difference across the unit cell, is given as a function of Re_ℓ. For low Reynolds numbers (less than 0.3), the flow at the interface is Stokean and symmetric with respect to the cylinder axis. Due to this symmetry, the point pressure gradient in the y-direction varies spatially as a nearly *odd function* of y (with respect to the same axis). Thus, when the pressure is averaged in the x-direction (area averaged $\langle p \rangle_A$), the resulting pressure difference across the interface (pressure slip) in the y-direction is negligible. From Figure 2.27, we observe that the pressure slip is very small compared to the pressure difference across the cell and can be neglected. This result is in contradiction with the results of Ene and Sanchez-Palencia due to the approximate nature of their analysis. For higher Reynolds numbers, inertial effects are more important, and as expected, the flow is not symmetric because of the flow separation. Inherent in the nonsymmetrical behavior of the velocity field, the variation of the point pressure gradient in the y-direction does not follow that of an odd function at the interface. For

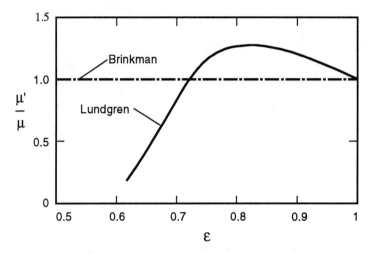

Figure 2.28 Effective viscosity as a function of porosity, computed by Lundgren (1972).

high Reynolds numbers, this pressure difference is of the same order as the pressure difference *across the cell*.

2.11.5 NO-SLIP MODELS USING EFFECTIVE VISCOSITY

Rather than matching the Darcy law with the Navier-Stokes equation via the semiempirical boundary condition of Beavers and Joseph (2.167), the Brinkman extension of the Darcy law can be used to treat the boundary condition at the porous-plain media interface. Since the Brinkman equation is of the second order, the condition of continuity of the tangential shear stress can be invoked. However, if the fluid viscosity is used for the matrix-fluid combination, the results based on the Brinkman extension (2.64) do not agree with the existing experimental results. This has been remedied by the *renormalization* of the viscosity for the *matrix-fluid* combination, i.e., defining an *effective viscosity* μ', which depends on the *fluid viscosity* as well as the *structural properties of the matrix*. However, this effective viscosity is defined as a bulk property (i.e., assumed to have the same value adjacent to the surface as it has in the bulk). Lundgren (1972) gives justifications for the Brinkman equation and finds that the ratio of the effective viscosity μ' in (2.64) to the fluid viscosity is not always greater than unity [unlike the Einstein relation for suspensions, (2.65) and the result of Sahraoui and Kaviany (1992) which will be presented in Section 2.11.6]. Figure 2.28 gives the effective viscosity calculated by Lundgren for a random bed of spheres. His analysis is for dilute concentration of spheres, and therefore, the decrease in the effective viscosity for larger concentrations may be the result of the violation of this restriction (*diluteness*). His determination of

μ' involves a nonuniform flow passing through a dilute bed of spheres. Then the *drag* or the mean traction at a point is calculated (by approximations given in Lundgren) and ensemble (volume) averages of $\langle \mathbf{u} \rangle$ and $\nabla^2 \langle \mathbf{u} \rangle$ are evaluated. Then, through the Brinkman equation the coefficient μ' is determined. Experimental determination of μ' has been made by Gilver and Altobelli (1994) for a foam inserted in a tube. They find that μ' is about 10 μ. This experiment is in the Ergun regime (i.e., large Re_d). Further analysis of the viscosity renormalization is given by Koplik et al. (1983).

Assuming a constant μ'/μ, similar to that given earlier, Neale and Nader (1974b) show that the Beavers-Joseph constant α is related to μ'/μ. Before reviewing their work, note that μ'/μ is a bulk property and, therefore, from their analysis, α should be a *bulk property*. However, one expects that depending on the surface manufacturing and finish, a large variation in α can occur. Therefore, the Neale and Nader analysis is restricted to uniform matrices (including the region adjacent to surfaces). They consider the parallel flow of Beavers and Joseph and solve the one-dimensional two-domain problem given here.

$$\text{Plain media}: \quad 0 < y < h \quad \frac{dp}{dx} = \mu \frac{d^2 u}{dy^2}, \ u(h) = 0, \quad (2.207)$$

$$\text{interface}: \quad u = u_i, \ \mu \frac{du}{dy} = \mu' \frac{du_D}{dy}, \quad (2.208)$$

$$\text{porous media}: \quad -\infty < y \leq 0 \quad \frac{dp}{dx} = \mu' \frac{d^2 u_D}{dy^2} - \frac{\mu}{K} u_D, \quad (2.209)$$

$$u_D(y \to -\infty) = -\frac{K}{\mu}\frac{dp}{dx}$$

$$\simeq u_D(-\infty). \quad (2.210)$$

The boundary conditions for $y \to -\infty$ are based on the presumption that a boundary-layer regime exists inside the porous media and immediately below the interface. This formulation is different than that of Beavers-Joseph in that it allows for matching of the shear stress at the interface, and it allows for the presence of a boundary layer in the porous media. The solution to the velocity distribution in the porous media is

$$u_D = u_i + [u_D(-\infty) - u_i]\left\{1 - \exp\left[\left(\frac{\mu}{\mu' K}\right)^{1/2} y\right]\right\}, \quad (2.211)$$

which results in

$$\frac{du_D}{dy}(0) = [u_i - u_D(-\infty)]\left(\frac{\mu}{\mu' K}\right)^{1/2}. \quad (2.212)$$

When matching shear stresses, we have

$$\frac{du}{dy}(0) = \frac{(\mu'/\mu)^{1/2}}{K^{1/2}}[u_i - u_D(-\infty)]. \quad (2.213)$$

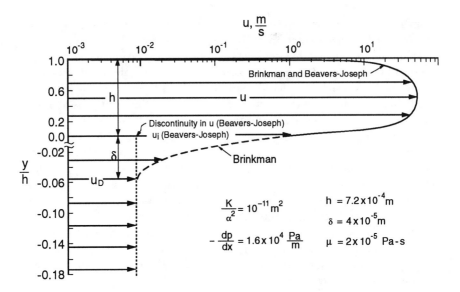

Figure 2.29 Comparison between the velocity distribution found using the Beavers and Joseph semiempirical boundary condition and that based on the Brinkman extension of the Darcy law.

This is identical to the Beavers-Joseph boundary condition, if we take $\alpha = (\mu'/\mu)^{1/2}$. The flow in the plain media is exactly as before, i.e., (2.168).

A boundary-layer thickness δ can be defined such that $u_D(-\delta) = 1.01\, u_D(-\infty)$. Then we have

$$\delta = K^{1/2}\left(\frac{\mu'}{\mu}\right)^{1/2} \ln \frac{50\left(\frac{h^2}{K} - 2\right)}{1 + \frac{h}{K^{1/2}}\left(\frac{\mu'}{\mu}\right)^{1/2}}, \qquad (2.214)$$

where, as was found before for impermeable bounding surfaces, the boundary-layer thickness is of the order of $K^{1/2}$. Figure 2.29 shows the velocity profiles obtained using the Beavers-Joseph or Neale-Nader formulation. Neale and Nader recommend that the ratio μ'/μ be taken as unity. This simplification is not justified considering the large variations in α found in the experiments. Vafai and Thiyagaraja (1987) also apply the Brinkman equation with $\mu = \mu'$ to a general class of parallel interfaces, including the case of two adjacent porous media (jump in the permeability). The variation of α with respect to ε (and K) were examined in Section 2.11.3.

2.11.6 VARIABLE EFFECTIVE VISCOSITY FOR A TWO-DIMENSIONAL STRUCTURE

When the no-slip boundary condition is used, special attention should be given to the choice of the averaging volume. The averaging volume must be small enough so that the velocity variations near the interface are *not masked*. Moreover, the averaging volume should be taken in such a way that the *interfacial velocity* (as obtained from the local simulation) and the *Darcean velocity* (in the bulk of the porous medium) are recovered. If a unit cell is taken as the averaging volume, it will *mask* the velocity variation over the Brinkman screening distance. The numerical results show that for the range of porosities considered, the bulk flow is recovered at the *lower half* of the interfacial row of cylinders. Thus the volume-averaged velocity should not make this boundary-layer effect penetrate beyond one cell into the porous medium.

Based on these, we choose an averaging volume that has an infinitesimal thickness at the interface (guaranteeing the continuity of velocity). For points away from the interface, the averaging volume increases with the distance from the interface (to a cell size) so that the Darcean velocity is recovered. In the numerical computations, the averaging volume is the *grid size* at the nominal interface and the *cell size* at $y \leq -\ell/2$. For any point y existing between 0 and $-\ell/2$ the averaging volume is taken as $-2y\ell$ and the volume-averaged velocity is defined as

$$\langle u \rangle_V (y) \equiv \frac{-1}{2y\ell} \int_{-2y}^{0} \int_{0}^{\ell} u(x, y') \, \mathrm{d}x \, \mathrm{d}y'. \qquad (2.215)$$

(A) BRINKMAN MODEL

This Brinkman model is commonly used in the analysis of flow and heat transfer in composite (porous-plain) media. As was discussed in Section 2.9, some investigators have added, in an ad hoc manner, a macroscopic shear term to the Darcy law (with $\mu' = \mu$) to allow for the variation of the velocity near the boundary. Others [e.g., Lundgren (1972)] have formally proved the validity of the Brinkman equation for *dilute* concentration of particles. In applying (2.64), the viscosity μ' needs to be prescribed. Brinkman suggested $\mu' = \mu$, while later Lundgren showed that $\mu' = \mu'(\mu, \varepsilon)$.

Here, the results of the one-dimensional Brinkman model are compared to the volume-averaged point solution (direct simulation) in an attempt to examine the validity of this model. We begin by using the results of Neale and Nader for a parallel flow, which gives $\mu'/\mu = \alpha^2$. Therefore, we guarantee that the interface velocity obtained from the Brinkman equation is the same as that obtained from the local simulation (within a small error). The boundary conditions used here are similar to the Beavers and Joseph model discussed earlier, except at the interface where the continuity

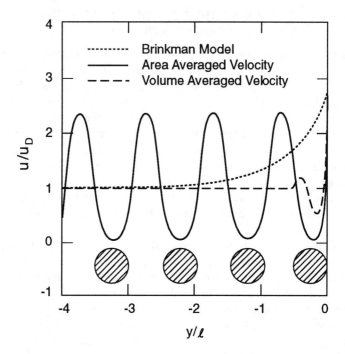

Figure 2.30 Comparison of the velocity distributions obtained from the Brinkman solution, the area-averaged local solution, and the volume-averaged (variable volume) local solution, using $\langle u \rangle_V (y) = -1/(2y\ell) \times \int_{-2y}^{0} \int_{0}^{\ell} u(x, y') \, dx \, dy'$.

conditions are used, i.e.,
$$u(0) = \langle u \rangle_V (0), \tag{2.216}$$
and
$$\mu \frac{du}{dy}\bigg|_{0+} = \mu' \frac{d\langle u \rangle_V}{dy}\bigg|_{0-}. \tag{2.217}$$

In the porous medium, the velocity distribution is (Neale and Nader)
$$\langle u \rangle_V (y) = -\frac{K}{\mu} \frac{dp}{dx} \left[1 + \frac{\sigma^2 - 2}{2(1 + \sigma\alpha)} \exp\left(\frac{1}{\alpha} \frac{y}{K^{1/2}}\right) \right]. \tag{2.218}$$

This velocity profile is presented in Figure 2.30 and compared with the volume-averaged point solution. These results show that the volume- [given by (2.215)] and area-averaged solutions reach the bulk behavior within the

2.11 Porous Plain Media Interfacial Boundary Conditions

interfacial cell, whereas the *Brinkman* model reaches the bulk behavior beyond the second cell. The Brinkman model, using a small decay factor in (2.218), *underestimates* the resistance to the flow at the interface and makes the boundary effect penetrate further into the porous medium. The resistance at the interface is also *underestimated* using the boundary conditions (2.217), which establishes continuity of the shear stress at the interface. For $\varepsilon = 0.8$, we have $\mu'/\mu \simeq 16$, which gives a *small velocity gradient* at the interface (on the porous medium side). This smaller velocity gradient makes the boundary effect penetrate *more* into the porous medium (compared to the averaged solution). Note that the screening length is $K^{1/2} = 0.267d$ (for $\varepsilon = 0.8$). The Brinkman model results in a velocity profile that is much closer to the local simulation for $\varepsilon = 0.48$, where the ratio of the viscosities is very close to unity. However, for a staggered cylinder arrangement, as shown in Figure 2.21(a), the Brinkman model gives a *faster* decay than the point solution. Thus, we conclude that the Brinkman model with a constant effective viscosity predicts the correct slip velocity but generally does *not* result in the correct velocity profile in the porous medium near the interface. This shortcoming might be overcome by using a variable effective viscosity model, as discussed later.

(B) VARIABLE EFFECTIVE VISCOSITY MODEL

In the variable effective viscosity model, $\mu'(y)$ is readily obtained by using the volume-averaged point distribution by integration of

$$-\frac{\mu}{K}\langle u\rangle_v + \frac{d}{dy}\left[\mu'(y)\frac{d\langle u\rangle_v}{dy}\right] = \frac{d\langle p\rangle_v^f}{dx}, \qquad (2.219)$$

which gives

$$\mu'(y) = \frac{\int_0^y \left[\frac{d\langle p\rangle_v^f}{dx} + \frac{\mu}{K}\langle u\rangle_v(y')\right]dy' + \mu'(0)\left.\frac{d\langle u\rangle_v}{dy}\right|_0}{\left.\frac{d\langle u\rangle_v}{dy}\right|_y}, \qquad (2.220)$$

where $\mu'(0)$ is taken as the fluid viscosity μ, which will give a continuous velocity gradient on both sides of the interface as the boundary condition. In Figure 2.31, the ratio of the local effective to the fluid viscosity is given as a function of y/ℓ. The results are only for $y/\ell > -0.2$, because further into the porous medium $d\langle u\rangle_v/dy \to 0$ and $\mu'(y)$ is undetermined (large oscillations of μ' occur). As expected and shown in Figure 2.31, the local effective viscosity decreases with an increase in porosity (because the pore size becomes larger and the fluid encounters less resistance).

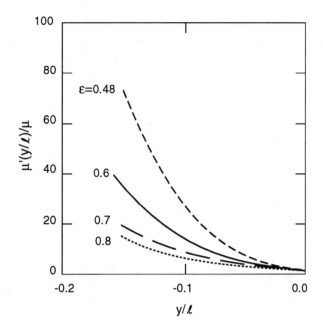

Figure 2.31 Local effective viscosity distribution for in-line arrangement of cylinders with different porosities.

2.11.7 VARIABLE PERMEABILITY FOR A TWO-DIMENSIONAL STRUCTURE

The presence of the interface could also be modeled through the permeability variation, i.e.,

$$-\frac{\mu}{K(y)}\langle u\rangle_v + \mu\frac{d^2\langle u\rangle_v}{dy^2} = \frac{d\langle p\rangle_v^f}{dx}. \qquad (2.221)$$

This model has been used in predicting the shear stress and the heat transfer rate at the impermeable surfaces bounding packed beds (Vafai, 1984; Cheng and Vortmeyer, 1988). Sangani and Behl (1989) used a combination of the variable permeability and viscosity models [i.e., a combination of (2.219) and (2.221)] to solve for a shear flow over a porous medium made of spheres. They used the areal void distribution of their bed of spheres to model the variations in the local permeability and viscosity. They assume inverse and direct proportionality between the local K and μ' and the local planar void fraction, respectively. Their predicted interfacial velocity is in good agreement with the results from their local simulation. It should be mentioned that if such arbitrary relationships for the local K and μ' are

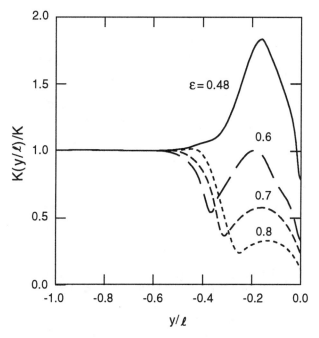

Figure 2.32 Local permeability distribution for in-line arrangement of cylinders with different porosities.

not assumed, then the simultaneous determination of these relations, using the results from local simulation, is not possible.

Commonly, an area-averaged void distribution $\varepsilon(y)$ is prescribed based on the experimental results for random packing of spheres. Then, the permeability is calculated using the Carman-Kozeny relation and this $\varepsilon(y)$.

Here, as was done earlier for the variable effective viscosity model, we find the variation in the permeability near the interface from the computed velocity field and by using (2.221), i.e.,

$$K(y) = \frac{\mu \langle u \rangle_v}{\mu \dfrac{d^2 \langle u \rangle_v}{dy^2} - \dfrac{d\langle p \rangle_v^f}{dx}}. \qquad (2.222)$$

The computed variation of the permeability near the interface is shown in Figure 2.32, for the in-line arrangement for various porosities. Intuitively, we expect the permeability to be higher at the interface because of the lower resistance to the flow (compared to the bulk permeability). However, the results show that the surface permeability is lower than that in bulk except for $\varepsilon = 0.48$. This decrease in the permeability is due to the dominating effect of the shear stress term in (2.221). From these results, we infer

100   2. Fluid Mechanics

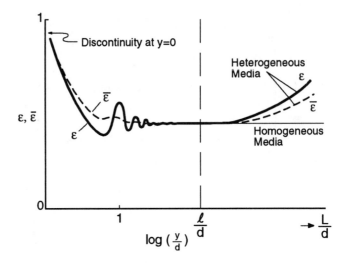

Figure 2.33 Variation in the local and average porosity with respect to the averaging volume (volume represented by linear dimension y).

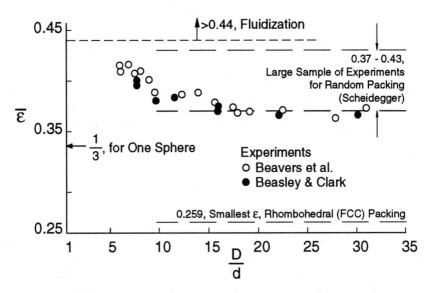

Figure 2.34 Variation of porosity with respect to the ratio of the diameter of the cylindrical container to the particle diameter. The results are for uniform-size spheres randomly packed (nonconsolidated).

that the permeability at the interface *cannot* be modeled using $\varepsilon(y)$ and the Carman-Kozeny relation. The variable permeability model has been suggested by Saleh et al. (1993), based on their experimental observation of the flow near the interface.

2.12 Variation of Porosity near Bounding Impermeable Surfaces

The significant flow field *nonuniformities* near the bounding solid surfaces are mostly associated with matrices that are made of *nonconsolidated, monosized* particles, although it can also occur in manufactured consolidated matrices or be caused in the presence of critical stresses (compacting, fracture, etc.). These are generally confined to distances of order of d, and as expected, when d/L is very small, they are not significant, where L is the smallest linear dimension of the system. When d/L is not very small, these nonuniformities must be included in the analysis. In general, the packing at and near the solid surfaces is *looser* than the bulk and the increase in local porosity results in higher velocities in this region—a phenomenon called *channeling*. Note that defining porosity for systems where d/L is not very small will not be in accord with the definitions of the representative elementary volume. Therefore, the inclusion of local variation of ε into the extended forms of the Darcy law may not be very meaningful for systems in which d/L is not very small. In the following, this point is elaborated.

2.12.1 DEPENDENCE OF AVERAGE POROSITY ON LINEAR DIMENSIONS OF SYSTEM

Designating the system volume as V_L, where L is a characteristic linear dimension of the system, we define an *average porosity* as

$$\overline{\varepsilon} = \frac{1}{V_L} \int_{V_L} a(\mathbf{x}) \, dV, \qquad (2.223)$$

where a is defined in (2.67) and we only required that $L \geq d$. This is different than the *local porosity* is defined in (2.68) as

$$\varepsilon(\mathbf{x}) = \frac{1}{V_\ell} \int_{V_\ell} a(\mathbf{x}) \, dV, \qquad (2.224)$$

where the representative elementary volume V_ℓ has a linear dimension ℓ and we required that $\ell \ll L$ and $\ell \gg d$. Figure 2.33 shows the expected variation of the local porosity with respect to $\log(y/d)$ where y is a coordinate axis [similar presentation is made in Bear (1988, p.17)]. For uniform porosity, $\overline{\varepsilon} \to \varepsilon$ as the averaging volume increases and reaches that of the representative elementary volume.

For monosized spheres, randomly packed, Beavers et al. (1973) and Beasley and Clark (1984) have measured the average porosity $\bar{\varepsilon}$. Their experimental results for packing in cylindrical containers of *diameter D* are presented, along with that accumulated by Scheidegger, in Figure 2.34. Note that D is *equivalent* to y used in Figure 2.33.

The point to be made is that only nonuniformities in ε that persist beyond length scale ℓ can be meaningfully represented as $\varepsilon(\mathbf{x})$ in a momentum equation that is based on scaling arguments used in the volume averaging. For systems with $y/d < 1$, the least rigorous treatment would be that of using $\bar{\varepsilon}$ (a constant).

2.12.2 LOCAL POROSITY VARIATION

Although not experimentally tested valid, we consider (2.108), which is a volume-averaged momentum equation, and assume negligible body force and isotropic permeability. When ε changes over ℓ, we have

$$\langle \mathbf{u} \rangle = -\frac{\varepsilon K_w}{\mu} \nabla \langle p - p_o \rangle^f - \frac{\varepsilon K_w}{\mu} \langle p - p_o \rangle^f \nabla \varepsilon, \qquad (2.225)$$

$$\text{order}: \quad O\left(\frac{d^2}{\mu} \frac{\Delta p}{\ell}\right) \qquad O\left(\frac{d^2}{\mu} \Delta p \frac{1}{\ell}\right).$$

To show that changes of ε over d are not consistent with the volume-average formulation, introduce $(\Delta \varepsilon)/(\Delta x_i) \simeq 1/d$, which leads to

$$\frac{\varepsilon K_w}{\mu} \nabla \langle p \rangle^f \quad \text{is} \quad O\left(\frac{d^2}{\mu} \frac{\Delta p}{\ell}\right) \qquad (2.226)$$

$$\frac{K_w}{\mu} \langle p \rangle^f \nabla \varepsilon \quad \text{is} \quad O\left(\frac{d^2}{\mu} \frac{\Delta p}{d}\right) \gg O\left(\frac{d^2}{\mu} \frac{\Delta p}{\ell}\right). \qquad (2.227)$$

This is, as expected, a *departure* from the Darcy flows, and the concept of bulk or microscopic viscous resistance no longer holds. This example and the arguments given in the last section show that only porosity changes over distances larger than ℓ can be meaningfully included in the extensions of the Darcy law. As will be discussed later the challenges in simulation of the channeling are in the reduction of the observed variation in void fraction distributions to variations in *local porosity* and in predicting reasonably accurate flow fields. Random placement of monosized particles at and above planar solid surfaces (and for curved surfaces with radius of curvature much larger than that of the particles) results in *nonuniformity* of packing adjacent to these surfaces. If a planar (or areal) void fraction, i.e., A_f/A_t, was used with A in the x-z plane and with y being the coordinate axis perpendicular to the bounding surface and originated at the surface, then $(A_f/A_t)(y)$ has a damping, sinusoidal variation with y. However, with a three-dimensional void fraction (porosity), the sampling size in the third

2.12 Variation of Porosity near Bounding Impermeable Surfaces

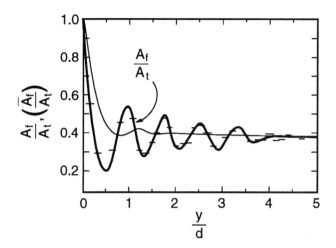

Figure 2.35 The local and average area void fraction. (From Benenati and Brosilow, reproduced by permission ©1962 AIChE.)

dimension must be prescribed. Depending on what is chosen for this size (e.g., a fraction of the particle diameter, one or several particle diameters), the porosity distribution obtained is substantially different. We review the experimental results of Benenati and Brosilow (1962). They pour monosized spherical lead particles into a cylindrical container and then fill all the interstices with a liquid epoxy resin. Upon curing of the resin, the solid cylinder is *machined*, in stages, to successively smaller diameters, each time reducing the diameter by δ, where δ is about $d/6$. The weight and diameter of the cylinder are measured after each machine cut, and knowing the densities, are used to determine *local void fraction*. In order to avoid confusion with the porosity, we present their experimental results as *areal void fraction*. This implies that we assume that A_f/A_t does not change over δ, which is only true when $\delta/d \ll 1$. However, it is important not to adapt ε as the quantity they measured because ε has meaning only if $\delta \gg d$. Figure 2.35 gives their experimental results for A_f/A_t.

Martin (1978) *approximated* this *area fraction variation* as

$$\frac{A_f}{A_t} = \left.\frac{A_f}{A_t}\right|_{\min} + \left(1 - \frac{A_f}{A_t}\right)\left(\frac{y}{R} - 1\right)^2 \quad \text{for } y \leq R, \quad (2.228)$$

$$\frac{A_f}{A_t} = \left.\frac{A_f}{A_t}\right|_{\infty} + \left(\left.\frac{A_f}{A_t}\right|_{\min} - \left.\frac{A_f}{A_t}\right|_{\infty}\right)$$
$$\times \exp\left(-\frac{y}{4R} + \frac{1}{4}\right) \cos\left[\frac{\pi}{0.810}\left(\frac{y}{R} - 1\right)\right] \quad \text{for } y > R, \quad (2.229)$$

where he uses $(A_f/A_t)_\infty = 0.39$ and $(A_f/A_t)_{\min} = 0.23$.

Also shown in Figure 2.35 is an *average* given by

$$\overline{\left(\frac{A_f}{A_t}\right)} = \frac{1}{y}\int_0^y \frac{A_f}{A_t}\,dy. \tag{2.230}$$

The horizontal bars in the figure are the values for δ. As expected, the experimental results show nonuniformity in A_f/A_t. However, for purposes of flow analysis, areal void fraction, or selection of a third dimension which is too small compared to the linear dimension of the representative elementary volume, is *not* acceptable. The alternatives are the direct simulation of the nonuniform region (microtreatment) or adaptation of a local porosity variation to be introduced into the volume-averaged momentum equation (a combined macro- and microtreatment).

2.12.3 Velocity Nonuniformities Due to Porosity Variation

Using the local measurement techniques, it is desirable to make high resolution point measurements and to average the data over the pore (or unit cell) in which the measurements are made. This is because in our macroscopic treatments a pore is the smallest unit we can use in conjunction with the modified Darcy law. However, the planar averaged velocity $\overline{u}(y)$, where y is as given in the last section, illustrates the channeling *most* clearly. This is because the channeling is generally confined to a few diameter lengths; and therefore, the volume-averaged velocities (over one or more pores) *masks* the extent of the velocity variation. Schwartz and Smith (1953), Schertz and Bischoff (1969), Newel and Standish (1973), and Dybbs and Edwards (1984), among others, have made local velocity measurements and have shown that due to the larger void fraction, the velocity peaks near the solid surface (as much as 30 to 100 percent *larger* than the average velocity in the matrix). The reported location of the velocity maximum is within *one* to *three* particle diameters from the solid surface.

The dependence of the average porosity on the ratio of D/d (Figure 2.33) indicates that the local permeability also changes with D/d. Figure 2.36 shows the experimental results of Chu and Ng (1989) for permeability for various bed diameters and particle sizes. The results for $K\,(D/d \to \infty)$ are from the Carman-Kozeny relation with the constant 180 *replaced* by 150. As expected, in general the larger porosity associated with small D/d results in larger permeabilities. However, Chu and Ng argue that the presence of the tube wall has *two counteracting* effects on the fluid flow. A *higher porosity promotes* the flow along the walls, but *higher surface area* per unit volume (a result of the presence of the wall) *hinders* it. The packing determines whether K is less than or greater than the bulk $K\,(D/d \to \infty)$. The Chu and Ng model for the fluid flow confirms those tendencies. More discussion on the effect of the bed size, including the high Reynolds number flows, is given by Beavers et al. (1973).

2.12 Variation of Porosity near Bounding Impermeable Surfaces

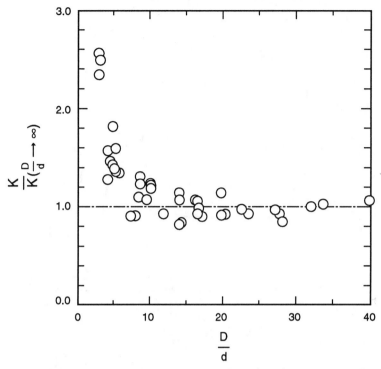

Figure 2.36 Experimental results of Chu and Ng (1989) for permeability (normalized) as a function of the relative tube size. (From Chu and Ng, reproduced by permission ©1989 AIChE.)

Consider the variation of velocity near a bounding solid surface, a problem also studied by Vafai (1984). We take $C_E = 0$ in (2.143), and by assuming a one-dimensional flow with a prescribed variation of the porosity perpendicular to the flow and by including the macroscopic viscous term, we will have

$$\mu \frac{d^2 u_D}{dy^2} - \frac{\mu \, \varepsilon(y) u_D}{K(y)} - \varepsilon(y) \frac{d\langle p \rangle^f}{dx} = 0, \qquad (2.231)$$

where $d\langle p \rangle^f/dx$ is a constant. The boundary conditions are $u_D(0) = 0$ and $u_D(\infty) = u_{D,\infty} = -[K(\infty)/\mu \, \varepsilon(\infty)] \, d\langle p \rangle^f/dx$.

An *estimate* of the boundary-layer thickness can be made assuming an *exponential* decay and this leads to

$$\int_0^\delta \left[\frac{\varepsilon(y)}{K(y)}\right]^{1/2} dy \simeq 4. \qquad (2.232)$$

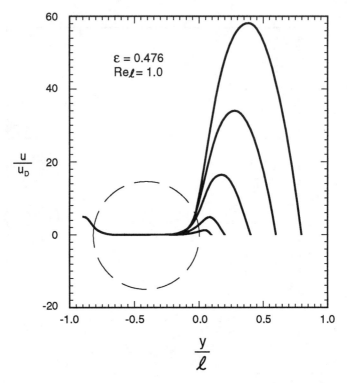

Figure 2.37 Channeling of flow adjacent to an impermeable surface. The distance between the nominal interface and the impermeable surface is varied (as a fraction of lattice spacing).

From this we note that in order to account for variation of ε in the boundary layer, ε must *change* over $K^{1/2}$, i.e., $\nabla \varepsilon = o\left(K^{-1/2}\right)$. Chandrasekhar and Vortmeyer (1979) use $\varepsilon(y)$ and $K(y)$ variation over a distance of d, using (2.228) and (2.229). In the following section, the point solutions to the velocity field are compared with those predicted based on these models.

2.12.4 Velocity Nonuniformity for a Two-Dimensional Structure

The numerical solution for the two-dimensional periodic arrangement of cylinders using the numerical method described by Sahraoui and Kaviany (1992) are used to examine the validity of one-dimensional modeling of channeling, boundary, and entrance effects. The results are given here.

2.12 Variation of Porosity near Bounding Impermeable Surfaces

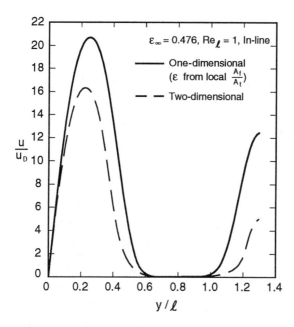

Figure 2.38 One-dimensional approximation made to flow in a periodic structure bounded by an impermeable surface. The two-dimensional results are also shown.

(A) Channeling

In this two-dimensional simulation, we refer to channeling as the high-velocity region that exists adjacent to the impermeable bounding surface (in flows that are parallel to this surface). Figure 2.37 shows how the distance between the nominal interface ($y = 0$) and the location of the impermeable wall ($y = h$) influences the normalized (using the Darcean velocity) velocity in this region. The results are for $0.1 \leq h/\ell \leq 0.8$. For $h/\ell \geq 0.2$, the *maximum* velocity in the *channel* becomes greater than the *maximum* velocity in the *pore*. The maximum velocity occurs near *middistance* between the nominal interface and the wall. As h/ℓ increases, u/u_D becomes very large and the flow can become *unsteady* and *turbulent* in the channel while remaining *laminar* in the bed far from the surface.

(B) Boundary Effect

We now examine the one-dimensional approximations made to the velocity variation near an impermeable bounding surface. We examine two approximations; the *first* uses an *effective permeability* that varies with the local *planar void fraction*. We use the bulk permeability $K = K(\varepsilon)$ for the in-line arrangement, found numerically and given by (2.48). Then we use $\varepsilon = A_f/A_t$ for an in-line arrangement with $\varepsilon_\infty = 0.476$ far from the nominal interface. Then we use this $K = K(A_f/A_t)$ in the one-dimensional Darcy equation. The results are given in Figure 2.38.

The velocity distribution obtained from the two-dimensional simulation and averaged along the x-direction is also given in Figure 2.38. Note that using the local areal void fraction A_f/A_t leads to a *higher* maximum velocity and a *larger* penetration distance. The periodic flow in the bed is also over-predicted using this local variation of porosity. We conclude that using the local areal void fraction is *not* suitable for the one-dimensional modeling of the *velocity variation* near the impermeable bounding surface. This conclusion is also expected to be valid for the *spherical* particles, i.e., (2.229) should *not* be used for determination of the local permeability.

The *second* approximation uses the *averaged* local areal void fraction, i.e., (2.230). Figure 2.39(a) shows the variation of the averaged local areal void fraction with respect to the distance from the wall. Since the magnitude of the periodic oscillation of $\overline{(A_f/A_t)}$ decays *exponentially* and *many* periods have to be included, an approximate *exponential* curve fit is generally used. Noting that as $y \to 0$, we have $\varepsilon \to 1$ and as $y \to \infty$, we have $\varepsilon \to \infty$, and using only one constant a_1, this approximation is also shown in Figure 2.39(a). Here a_1 is chosen such that the variation of $\overline{(A_f/A_t)}$ for $0 \leq y/\ell \leq 2$ is *best* presented. When this exponential variation in the local area fraction is used in $K = K(\varepsilon)$, the velocity distribution shown in Figure 2.39(b) is obtained. Also shown in this figure are the numerical results for the two-dimensional flow (averaged along the x-axis). Note that this one-dimensional approximation *underpredicts* the *maximum velocity* and *overpredicts* the *penetration* distance.

By choosing a value of a_1 that best matches the variation of $\overline{(A_f/A_t)}$ over larger y/ℓ, such as the choice of a_1 made in Figure 2.40(a), the predicted maximum velocity will become larger [as shown in Figure 2.40(b)]. However, the penetration distance also increases substantially. We note that the penetration distance δ found from the constant (bulk) permeability and the inclusion of the Brinkman term, i.e., (2.232) results in a δ of the order of $K^{1/2}$ that is much *less than* the particle diameter. We conclude that using the bulk permeability and the Brinkman extension *does not* lead to the accurate prediction of the penetration depth (screening distance). Also, when variable permeability with simple exponential-type distribution of the local porosity is used, accurate velocity distributions and penetration depths are not found. Sangani and Behl (1989) use simultaneous and rather arbitrarily prescribed variations in K and μ' and find a good agreement between their one-dimensional approximation and their numerical results for packed beds of spheres.

2.12 Variation of Porosity near Bounding Impermeable Surfaces

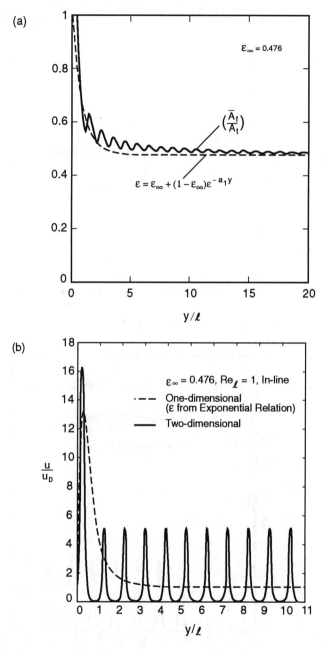

Figure 2.39 (a) Variation of the averaged local areal void fraction and an approximate exponential curve fit. (b) Variation in the local velocity found using the averaged local areal void fraction. Also shown are the two-dimensional results.

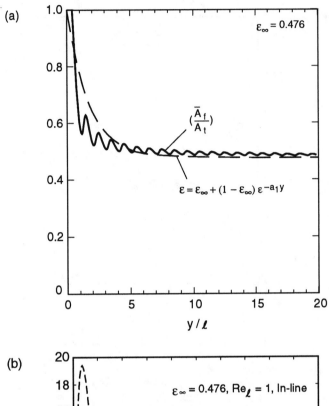
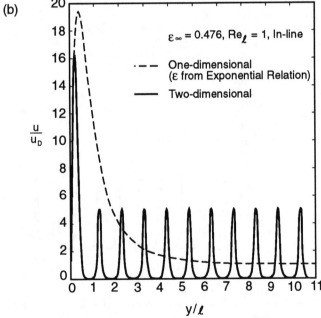

Figure 2.40 Same as Figures 2.39 (a) and (b), except the value of a_1 is altered.

2.12 Variation of Porosity near Bounding Impermeable Surfaces

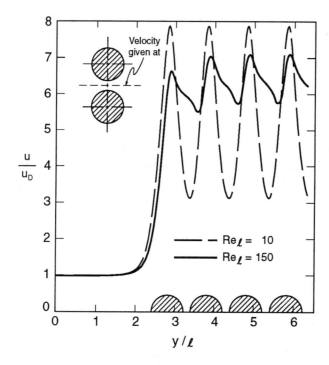

Figure 2.41 Velocity variation along the centerline between two adjacent cylinders. The results are for $Re_\ell = 10$ and 150 and show that flow becomes periodic after one particle length for $Re_\ell = 10$ and after two particle lengths for $Re_\ell = 150$.

(C) Entrance Effect

In the preceding discussion we considered flow *parallel* to the interface of a two-dimensional periodic structure. Here we consider flow *perpendicular* to such interfaces. This is the problem of *entrance* into a bed and was discussed in Section 2.10.2. Figure 2.41 gives the velocity along the *centerline* between two adjacent cylinders in an in-line arrangement. The results show that the *entrance length* is about one particle long for low Reynolds numbers and becomes larger as Re_ℓ increase. The dependence on Re_ℓ was predicted by the inclusion of the macroscopic inertial effect. However, the predicted entrance length, i.e., (2.159) is significantly *less*. This difference between the prediction by the macroscopic model and that by the point solution is similar to that found for the boundary effect. An improvement can be made to the macroscopic model by allowing for the variation of the permeability. This is not attempted. We conclude that the entrance length is of the order of the particle size and, as expected, *depends* on the Reynolds number.

The *exit* flow from a high porosity slab (ceramic foam) has been measured by Hall and Hiatt (1994). The results show *recirculation* and large turbulence intensity in the region *immediately* downstream of the porous slab.

2.13 Analogy with Magneto-Hydrodynamics

Flow through matrices experiences a volumetric resistance that is proportional to the first power of velocity. A similar volumetric resistance is experienced in the flow of charged fluid particles in the presence of a magnetic field. The volumetric *electromagnetic force* $\mathbf{f}_e$ acting on the field is

$$\mathbf{f}_e = \mathbf{j}_e \times \mathbf{b} = \sigma_e (\mathbf{e} + \mathbf{u} \times \mathbf{b}) \times \mathbf{b}$$
$$= \sigma_e \mathbf{e} \times \mathbf{b} + \sigma_e (\mathbf{u} \times \mathbf{b}) \times \mathbf{b}, \quad (2.233)$$

where

$$\mathbf{j}_e = \text{current density (A/m}^2\text{)}, \quad (2.234)$$
$$\mathbf{e} = \text{electric field intensity (V/m)}, \quad (2.235)$$
$$\mathbf{b} = \text{magnetic induction (Tesla, or V-s/m}^2\text{)}, \quad (2.236)$$
$$\sigma_e = \text{electric conductivity (1/ohm-m)}, \quad (2.237)$$

and N/m^3 is equal to J/m^4 = W-s/m^4 = V^2-s/ohm-m^4. Part of this force is present even if the field is at rest, while the other is due to the current induced by the motion through the magnetic field $\mathbf{b}$. The force $\sigma_e \mathbf{e} \times \mathbf{b}$ *accelerates* or *decelerates* (assisting or resisting) the flow, depending on the directions of $\mathbf{e}$, $\mathbf{b}$, and $\mathbf{u}$. However, $\sigma_e \mathbf{e} \times \mathbf{b}$ provides a current whose interaction with $\mathbf{b}$ always *decelerates* the flow (resistive).

Based on the continuum treatment of the charged fields, the conservation of linear momentum when scaled using u_o, L, ρ_o, σ_{e_o}, b_o, and e_o for velocity, length, density, electrical conductivity, magnetic induction, and electric field intensity, respectively, will become (Romig, 1964)

$$\rho_o^* \left(\frac{\partial \mathbf{u}^*}{\partial t} + \mathbf{u}^* \cdot \nabla \mathbf{u}^* \right) = -\nabla p^* + \frac{1}{Re} \nabla \cdot \mathbf{S}^* + \Omega \kappa (\sigma_e^* \mathbf{e}^* \times \mathbf{b}^*)$$
$$+ \Omega \sigma_e^* (\mathbf{u}^* \times \mathbf{b}^*) \times \mathbf{b}^* + \frac{Gr}{Re^2} T^*, \quad (2.238)$$

where an asterisk (∗) indicates that the variable is *dimensionless* and the thermal buoyancy term is included for the completeness. The dimensionless parameters are

$$\Omega = \text{magnetic interaction parameter}$$

$$= \frac{\text{ponderomotive force}}{\text{inertia force}} = \frac{\sigma_{e_o} b_o^2 L}{\rho_o u_o} \geq 0, \qquad (2.239)$$

$$\kappa = \text{generator coefficient}$$
$$= \frac{\text{applied electric field}}{\text{induced electric field}} = \frac{e_o}{u_o b_o} \geq \text{or} < 0, \qquad (2.240)$$

$$Re = \frac{\rho_o u_o L}{\mu_o}, \qquad (2.241)$$

$$Gr = \frac{\rho_o^2 g \beta_o (T_h - T_c) L^3}{\mu_o^2}. \qquad (2.242)$$

In the *absence* of the gravitational and imposed electric fields, for a uniform magnetic field $\mathbf{b}_o$ perpendicular to the free stream, and under the boundary-layer flow assumptions, (2.238) becomes

$$u^* \frac{\partial u^*}{\partial x} + v^* \frac{\partial u^*}{\partial y} = -\frac{\partial p^*}{\partial x^*} + \frac{1}{Re} \frac{\partial^2 u^*}{\partial y^{*2}} - \Omega u^*. \qquad (2.243)$$

The x-component of the momentum equation is found by the dot product with the unit vector $\mathbf{s}_x$ in that direction, i.e., $\mathbf{s}_x \cdot [(\mathbf{u} \times \mathbf{b}) \times \mathbf{b}] = -ub_o^2$, and we have taken $\sigma = \sigma_o$. For *low* Reynolds number flows the inertial force is negligible and when $\Omega Re > 1$, i.e., $\Omega > 1$, then (2.243) reduces to the Darcy law, i.e., $-\partial p^*/\partial x^* = \Omega$. Rossow (1957) has solved the boundary-layer problem (2.243) using a first-order regular expansion around Ω. As expected, the boundary-layer thickness and other features of this problem are exactly as those given in Section 2.6 for the Brinkman equation.

Features such as the boundary layers (associated with the Brinkman flow or small and moderate Ω), velocity slip (associated with Darcy flow or large Ω), and channeling (associated with variations in Ω) are common in both porous-media hydrodynamics and magneto-hydrodynamics. The magneto-hydrodynamics is not limited to creeping flows, because both liquids and gases are used as fluids and, for example, associated with reentry of space vehicles very high-speed flows are considered. Inclusion of the body force (e.g., natural convection) has been extensively studied in magnetic-hydrodynamics (Romig gives a review). Examples are Lykoudis (1962), Cramer (1963), Poots (1961), and Sparrow and Cess (1961). A review of electro- and magnetohydrodynamics and plasmas is given by Kaviany (1994). Their results can be readily related to flow in a porous media.

The review by Romig includes the effect of the Joule heating which results in the local variation in the properties, which in turn result in interesting local velocity variations near the bounding surfaces. Reviews are also given in Cramer and Pai (1973). In the hydrodynamic analogy between those two flows, the dimensionless parameters that play an analogous role are

$$\frac{\mu L}{\rho_o K u_o} \quad : \quad \frac{\sigma_{e_o} b_o^2 L}{\rho u_o}, \qquad (2.244)$$

or dimensionally

$$\frac{\mu_o}{K}\left(\frac{\text{N-s}}{\text{m}^4}\right) \quad : \quad \sigma_{e_o} b_o^2 \left(\frac{\text{N-s}}{\text{m}^4}\right). \qquad (2.245)$$

2.14 References

Adler, P. M., 1992, *Porous Media: Geometry and Transport*, Butterworth-Heinemann, Stoneham, MA.

Armour, J. C. and Cannon, J. N., 1968, "Fluid Flow Through Woven Screens," *AIChE J.*, 14, 415–420.

Bachmat, Y. and Bear, J., 1986, "Macroscopic Modeling of Transport Phenomena in Porous Media," *Transport in Porous Media*, 1, 241–269.

Bear, J., 1988, *Dynamics of Fluids in Porous Media*, Dover.

Beasley, D. E. and Clark, J. A., 1984, "Transient Response of Packed Beds for Thermal Energy Storage," *Int. J. Heat Mass Transfer*, 27, 1659–1669.

Beavers, G. S. and Joseph, D. D., 1967, "Boundary Condition at a Naturally Permeable Wall," *J. Fluid Mech.*, 30, 197–207.

Beavers, G. S., Sparrow, E. M., and Magnuson, R. A., 1970, "Experiments on Coupled Parallel Flow in a Channel and a Bounding Porous Medium," *J. Basic Eng.*, 92 D, 843–848.

Beavers, G. S., Sparrow, E. M., and Masha, B. A., 1974, "Boundary Condition at a Porous Surface Which Bounds a Fluid Flow," *AIChE J.*, 20, 596–597.

Beavers, G. S., Sparrow, E. M., and Rodenz, D. E., 1973, "Influence of Bed Size on the Flow Characteristics and Porosity of Randomly Packed Beds of Spheres," *J. Appl. Mech.*, 40, 655–660.

Benenati, R. F. and Brosilow, C. B., 1962, "Void Fraction Distribution in Packed Beds," *AIChE J.*, 8, 359–361.

Bensoussan, A., Lions, J. L., and Papanicolaou, G., 1978, *Asymptotic Analysis for Periodic Structures*, North-Holland.

Brinkman, H. C., 1947, "A Calculation of the Viscous Force Exerted by a Flowing Fluid on a Dense Swarm of Particles," *Appl. Sci. Res.*, A1, 27–34.

Brinkman, H. C., 1948, "On the Permeability of Media Consisting of Closely Packed Porous Particles," *Appl. Sci. Res.*, A1, 81–86.

Carbonell, R. G. and Whitaker, S., 1984, "Heat and Mass Transfer in Porous Media," in *Fundamentals of Transport Phenomena in Porous Media*, Bear and Corapcioglu, eds., Martinus Nijhoff Publishers, 121–198.

Carman, P. C., 1937, "The Determination of the Specific Surface Area of Powder I," *J. Soc. Chem. Ind.*, 57, 225–234.

Chandrasekhar, B. C. and Vortmeyer, D., 1979, "Flow Model for Velocity Distribution in Fixed Porous Beds under Isothermal Conditions," *Warme-und Stoffubertrigung*, 12, 105–111.

Cheng, P. and Vortmeyer, D., 1988, "Transverse Thermal Dispersion and Wall Channelling in a Packed Bed with Forced Convective Flow," *Chem. Eng. Sci.*, 43, 2523–2532.

Chu, C. F. and Ng, K. M., 1989, "Flow in Packed Tubes with a Small Tube to Particle Diameter Ratio," *AIChE J.*, 35, 148–158.

Clift, R., Grace, J. R., and Weber, M. E., 1978, *Bubbles, Drops, and Particles*, Academic.

Cramer, K. R., 1963, "Several Magneto–Hydrodynamic Free–Convection Solutions," *ASME J. Heat Transfer*, 85, 35–39.

Cramer, K. R. and Pai, S.-I., 1973, *Magneto Fluid Dynamics for Engineers and Applied Physicists*, McGraw-Hill.

Darcy, H., 1856, *Les Fontaines Publiques de la ville de Dijon*, Dalmont, Paris.

du Plessis, J. P., 1994, "Analytical Quantification of Coefficients in the Ergun Equation for Fluid Friction in a Packed Bed," *Transp. Porous Media*, 16, 189–207.

Dullien, F. A. L., 1979, *Porous Media: Fluid Transport and Pore Structure*, Academic.

Dullien, F. A. L., 1992, *Porous Media: Fluid Transport and Pore Structure*, Second Edition, Academic.

Durlofsky, L. and Brady, J. F., 1987, "Analysis of the Brinkman Equation as a Model for Flow in Porous Media," *Phys. Fluids*, 11, 3329–3341.

Dybbs, A. and Edwards, R. V., 1984, "A New Look at Porous Media Fluid Mechanics—Darcy to Turbulent," in *Fundamentals of Transport Phenomena in Porous Media*, Bear and Corapcioglu, eds., Martinus Nijhoff Publishers, 199–254.

Eidsath, A., Carbonell, R. G., Whitaker, S., and Herrmann, L. R., 1983, "Dispersion in Pulsed Systems—III, Comparison between Theory and Experiments in Packed Beds," *Chem. Engng. Sci.*, 38, 1803–1816.

Ene, H. I. and Polisevski, D., 1987, *Thermal Flow in Porous Media*, D. Reidel.

Ene, H. I. and Sanchez–Palencia, E., 1975, "Equations et Phenomenes de Surface Pour L'ecoulement dans un Modelle de Milieu Poreux," *J. de Meca.*, 14, 73–107.

Ergun, S., 1952, "Fluid Flow Through Packed Column," *Chem. Eng. Prog.*, 48, 89–94.

Fatehi, M. and Kaviany, M., 1990, "Analysis of Levitation of Saturated Liquid Droplets on Permeable Surfaces," *Int. J. Heat Mass Transfer*, 33, 983–994.

Firdaouss, M. and Guermond, J.-L., 1995, "Sur L'Homogénéisation des Équations de Navier-Stokes à Faible Nombre de Reynolds," *C. R. Acad. Sci. Paris*, 320, 245–251.

Gray, W. A., Leijnse, A., Kolar, R. L., and Blain, C. A., 1993, *Mathematical Tools for Changing Spatial Scales in the Analysis of Physical Systems*, CRC.

Gilver, R. C. and Altobelli, S. A., 1994, "A Determination of the Effective Viscosity for the Brinkman–Forchhiemer Flow Model," *J. Fluid Mech.*, 258, 355–370.

Hall, M. J. and Hiatt, J. P., 1994, "Exit Flows from Highly Porous Media," *Phys. Fluids*, 6, 469–479.

Happel, J. and Brenner, H., 1986, *Low Reynolds Number Hydrodynamics*, Martinus Nijhoff Publishers.

Hsu, C. T. and Cheng, P., 1991, "A Singular Perturbation Solution for Couette Flow over a Semi–Infinite Porous Bed," *ASME J. Fluid Engineering*, 113, 137–142.

Jolls, K. R. and Hanratty, T. J., 1966, "Transition to Turbulence for Flow Through a Dumped Bed of Spheres," *Chem. Engng. Sci.*, 21, 1185–1190.

Jones, I. P., 1973, "Low Reynolds Number Flow Past a Porous Spherical Shell," *Proc. Camb. Phil. Soc.*, 73, 231–238.

Joseph, D. D., Nield, D. A., and Papanicolaou, G., 1984, "Nonlinear Equation Governing Flow in a Saturated Porous Medium," *Wat. Resour. Res.*, 18, 1049–1052.

Kaviany, M., 1987, "Boundary–Layer Treatment of Forced Convection Heat Transfer from a Semi–Infinite Flat Plate Embedded in Porous Media," *ASME J. Heat Transfer*, 109, 345–349.

Kaviany, M., 1994, *Principles of Convective Heat Transfer*, Springer–Verlag.

Kim, S. and Russell, W. B., 1985, "Modelling of Porous Media by Renormalization of the Stokes Equation," *J. Fluid Mech.*, 154, 269–286.

Kim, S. and Karria, S. J., 1991, *Microhydrodynamics*, Butterworth–Heinemann, Stoneham, MA.

Koplik, J., Levine, H., and Zee, A., 1983, "Viscosity Renormalization in the Brinkman Equation," *Phys. Fluids*, 26, 2864–2870.

Kyan, C. P., Wasan, D. T., and Kintner, R. C., 1970, "Flow of Single–Phase Fluids Through Fibrous Beds," *Ind. Eng. Chem. Fund.*, 9, 596–603.

Ladd, A. J., 1994, "Numerical Simulation of Particulate Suspensions via a Discretized Boltzmann Equation, Part 1 and 2," *J. Fluid Mech.*, 271, 285–339.

Larson, R. E. and Higdon, J. J. L., 1986, "Microscopic Flow Near the Surface of Two–Dimensional Porous Media, Part 1: Axial Flow," *J. Fluid. Mech.*, 166, 449–472.

Larson, R. E. and Higdon, J. J. L., 1987, "Microscopic Flow Near the Surface of Two–Dimensional Porous Media, Part 2: Transverse Flow," *J. Fluid Mech.*, 178, 119–136.

Larson, R. E. and Higdon, J. J. L., 1989, "A Periodic Grain Consolidation Model of Porous Media," *Phys. Fluids*, A1, 38–46.

Levy, Th. and Sanchez–Palencia, E., 1975, "On Boundary Conditions for Fluid Flow in Porous Media," *Int. J. Eng. Sci.*, 13, 923–940.

Levy, Th. and Sanchez–Palencia, E., 1977, "Equation and Interface Conditions for Acoustic Phenomena in Porous Media," *J. Math. Anal. Appl.*, 61, 813–834.

Ling, J.-X., 1988, *Physical Model for Heat Transfer in Porous Media*, Ph.D. thesis, Case Western Reserve University, 74–78.

Lundgren, T. S., 1972, "Slow Flow Through Stationary Random Beds and Suspensions of Spheres," *J. Fluid Mech.*, 51, 273–299.

Lykoudis, P. S., 1962, "Natural Convection of an Electrically Conducting Fluid in the Presence of a Magnetic Field," *Int. J. Heat Mass Transfer*, 4, 23–34.

Macdonald, I. F., El–Sayed, M. S., Mow, K., and Dullien, F. A. L., 1979, "Flow Through Porous Media—Ergun Equation Revisited," *Ind. Eng. Chem. Fund.*, 18, 199–208.

Marshall, H., Sahraoui, M., and Kaviany, M., 1994, "An Improved Analytic Solution for Analysis of Particle Trajectories in Fibrous, Two–Dimensional Filters," *Phys. Fluids*, 6, 507–520.

Martin, H., 1978, "Low Peclet Number Particle–To–Fluid Heat and Mass Transfer in Packed Beds," *Chem. Engng. Sci.*, 33, 913–919.

Mei, C. C. and Auriault, J.-L., 1991, "The Effect of Weak Inertia on Flow Through a Porous Medium," *J. Fluid Mech.*, 222, 647–663.

Mickley, H. S., Smith, K. A., and Korchak, E. I., 1965, "Fluid Flow in Packed Beds," *Chem. Engng. Sci.*, 20, 237–246.

Molerus, O. and Schweinzer, J., 1989, "Resistance of Particle Beds at Reynolds Numbers up to $Re \sim 10^4$," *Chem. Engng. Sci.*, 44, 1071–1079.

Neale, G. H. and Nader, W. K., 1974a, "Prediction of Transport Process within Porous Media: Creeping Flow Relative to a Fixed Swarm of Spherical Particles," *AIChE J.*, 20, 530–538.

Neale, G. H. and Nader, W. K., 1974b, "Practical Significance of Brinkman Extension of Darcy's Law: Coupled Parallel Flows within a Channel and a Bounding Porous Medium," *Can. J. Chem. Eng.*, 52, 475–478.

Newell, R. and Standish, N., 1973, "Velocity Distribution in Rectangular Packed Beds and Non–Ferrous Blast Furnaces," *Metall. Trans.*, 4, 1851–1857.

Ochoa–Tapia, J. A. and Whitaker, S., 1995, "Momentum at the Boundary between a Porous Medium and a Homogeneous Fluid, I: Theoretical Development, II: Comparison with Experiment," *Int. J. Heat Mass Transfer*, to appear.

Poots, G., 1961, "Laminar Natural Convection Flow in Magneto–Hydrodynamics," *Int. J. Heat Mass Transfer*, 3, 1–25.

Quiblier, J. A., 1984, "A New Three–Dimensional Modeling Technique for Studying Porous Media," *J. Colloid Interface Sci.*, 98, 84–102.

Richardson, S., 1971, "A Model for the Boundary Condition of a Porous Material: Part 2," *J. Fluid Mech.*, 49, 327–336.

Romig, M. F., 1964, "The Influence of Electric and Magnetic Fields on Heat Transfer to Electrically Conducting Fluids," *Adv. Heat Transfer*, 1, 267–354, Academic.

Rossow, V. J., 1957, "On Flow of Electrically Conducting Fluids of a Transverse Magnetic Field," *NASA Technical Note*, 3971.

Rubenstein, J. and Torquato, S., 1989, "Flow in Random Porous Media: Mathematical Formulation, Variational Principles, and Rigorous Bounds," *J. Fluid Mech.*, 206, 25–46.

Rumpf, H. and Gupte, A. R., 1971, "Einfüsse und Korngrössenverteilung in Widerstandsdesetz der Porenströmung," *Chemie–Ing. Techn.*, 43, 367–375.

Saffman, P. G., 1971, "On the Boundary Condition at the Surface of a Porous Medium," *Stud. Appl. Math.*, 50, 93–101.

Sahraoui, M. and Kaviany, M., 1992, "Slip and No–Slip Boundary Condition at Interface of Porous, Plain Media," *Int. J. Heat Mass Transfer*, 35, 927–943.

Sanchez–Palencia, E., 1980, *Non–Homogeneous Media and Vibration Theory*, Lecture Notes in Physics, 127, Springer–Verlag.

Sangani, A. S. and Behl, S., 1989, "The Planar Singular Solutions of Stokes and Laplace Equations and Their Application to Transport Processes near Porous Surfaces," *Phys. Fluids*, A1, 21–37.

Saleh, S., Thovert, J. F., and Adler, P. M., 1993, "Flow Along Porous Media by Particle Image Velocimetry," *AIChE J.*, 39, 1765–1776.

Scheidegger, A. E., 1974, *The Physics of Flow Through Porous Media*, Third Edition, University of Toronto Press.

Schertz, W. W. and Bischoff, K. B., 1969, "Thermal and Material Transport in Nonisothermal Packed Beds," *AIChE J.*, 15, 597–604.

Schwartz, C. E. and Smith, J. M., 1953, "Flow Distribution in Packed Beds," *Ind. Engng. Chem.*, 45, 1209–1218.

Slattery, J. C., 1969, "Single–phase Flow Through Porous Media," *AIChE J.*, 15, 866–872.

Slattery, J. C., 1981, *Momentum, Energy, and Mass Transfer in Continua*, Second Edition, R. F. Krieger.

Sparrow, E. M. and Cess, R. D., 1961, "The Effect of a Magnetic Field on Free Convection Heat Transfer," *Int. J. Heat Mass Transfer*, 3, 267–274.

Sparrow, E. M. and Loeffler, A. L. Jr., 1959, "Longitudinal Laminar Flow Between Cylinders Arranged in Regular Array," *AIChE J.*, 5, 325–330.

Stanek, V. and Szekely, J., 1974, "Three–Dimensional Flow of Fluids Through Nonuniform Packed Beds," *AIChE J.*, 20, 974–980.

Sullivan, R. R., 1942, "Specific Surface Measurement on Compact Bundles of Parallel Fibers," *J. Appl. Phys.*, 13, 725–730.

Taylor, G. I., 1971, "A Model for the Boundary Condition of a Porous Material: Part 1," *J. Fluid Mech.*, 49, 319–326.

Tien, C., 1989, *Granular Filteration of Aerosols and Hydrosols*, Butterworth–Heinemann, Stoneham, MA.

Torquato, S. and Kim, I. C., 1992, "Cross–Property Relations for Momentum and Diffusional Transport in Porous Media," *J. Appl. Phys.*, 72, 2612–2619.

Vafai, K., 1984, "Convective Flow and Heat Transfer in Variable–Porosity Media," *J. Fluid Mech.*, 147, 233–259.

Vafai, K. and Thiyagaraja, R., 1987, "Analysis of Flow and Heat Transfer at the Interface Region of a Porous Medium," *Int. J. Heat Mass Transfer*, 30, 1391–1405.

Vafai, K. and Tien, C.-L., 1981, "Boundary and Inertia Effects on Flow and Heat Transfer in Porous Media," *Int. J. Heat Mass Transfer*, 24, 195–203.

van der Merwe, D. F. and Gauvin, W. H., 1971, "Velocity and Turbulence Measurement of Air Flow Through a Packed Bed," *AIChE J.*, 17, 519–528.

Ward, J. C., 1964, "Turbulent Flow in Porous Media," *J. Hyd. Div. ASCE*, 90, HY5, 1–12.

Whitaker, S., 1969, "Advances in Theory of Fluid Motion in Porous Media," *Ind. Engng. Chem.*, 61, 14–28.

Whitaker, S., 1986, "Flow in Porous Media I: A Theoretical Derivation of Darcy's Law," *Transp. Porous Media*, 1, 3–25.

Wooding, R. A., 1957, "Steady State Free Thermal Convection of Liquid in a Saturated Permeable Medium," *Phys. Fluids*, 12, 273–285.

3
Conduction Heat Transfer

Heat conduction through *fully saturated* matrices (i.e., a *single-phase fluid* occupying the pores), as with heat conduction through any *heterogeneous* media, depends on the structure of the *matrix* and the *thermal conductivity* of *each phase*. One of the most difficult aspects of the analysis of heat conduction through a porous medium is the *structural modeling*. This is because the representative elementary volumes are *three-dimensional* and have complicated structures that vary greatly among different porous media. Since the thermal conductivity of the solid phase is generally *larger* than that of the fluid, the manner in which the solid is *interconnected* influences the conduction significantly. Even when dealing with the *nonconsolidated* particles, the *contact* between the particles plays a significant role.

For the analysis of the macroscopic heat flow through heterogeneous media, the local volume-averaged (or *effective*) properties such as the *effective thermal conductivity* $\langle k \rangle = k_e$ are used. These local effective properties such as the *heat capacity* $\langle \rho c_p \rangle$, thermal conductivity $\langle k \rangle$, and radiation absorption and scattering coefficients $\langle \sigma_a \rangle$, $\langle \sigma_s \rangle$ need to be arrived at from the application of the *first principles* to the volume over which these local properties are averaged, i.e., the representative elementary volume.

As will be shown, the local average $\langle \rho c_p \rangle$ is obtained by *simple* volume averaging. However, we reiterate that the effective thermal conductivity is expected to depend on the following.

- The *thermal conductivity* of *each phase*, i.e., the relative magnitude of k_s/k_f is important.

- The *structure* of the solid matrix, i.e., the extent of the *continuity* of the solid phase is very important.

- The *contact resistance* between the nonconsolidated particles, i.e., the solid surface *oxidation* and other *coatings* are all important.

- For gases, when the ratio of the *mean free path* and the *average linear pore dimension* (i.e., the *Knudsen number*) becomes large, the *bulk* gas conductivity *cannot* be used for the fluid phase.

In the following, after reviewing the requirements for the validity of the assumption of the *local thermal equilibrium*, the various attempts at *prediction* of the effective thermal conductivity are reviewed along with some comparison with *experimental* results. The treatment based on the volume averaging of the point conduction equation for solid matrices with *periodic* structures is given along with the numerical results for a two-dimensional *isotropic periodic* structure. In this treatment the *intercell contact* is obtained empirically. The interparticle contact area for beds of spheres subject to *compressive* force is then studied for *elastic* contact deformations. The statistical treatments are then reviewed. The existing closed-form analytical treatments and the empirical correlations are also given and comparisons are made with the experimental results. Finally, the *nonuniformity* of the local effective thermal conductivity near *bounding* fluid and *solid surfaces* is examined.

3.1 Local Thermal Equilibrium

In principle, determination of the thermal conductivity of saturated porous media involves application of the point conduction (energy) equation to a point in the representative elementary volume of the matrix and the integration over this volume. In doing so, we realize that at the pore level there will be a *difference* ΔT_d between the temperature at a point in the solid and in the fluid. Similarly, across the representative elementary volume, we have a maximum temperature difference ΔT_ℓ. However, we assume that these temperature differences are much smaller than those occurring over the system dimension, ΔT_L. Thus, we impose the assumption of *local thermal equilibrium* by requiring that

$$\Delta T_{K^{1/2}} < \Delta T_d < \Delta T_\ell \ll \Delta T_L. \tag{3.1}$$

With this assumed negligible local temperature difference between the phases, we assume that within the representative elementary volume $V = V_f + V_s$ the solid and fluid phases are in *local thermal equilibrium*, i.e.,

$$\frac{1}{V_f} \int_{V_f} T_f \, dV = \frac{1}{V_s} \int_{V_s} T_s \, dV = \frac{1}{V} \int_V T \, dV. \tag{3.2}$$

Note that, although $\Delta T_{K^{1/2}}$, ΔT_d, and ΔT_ℓ are small, their gradients in their respective length scales are not small. Based on an order-of-magnitude analysis, the *criteria* for the *validity* of local thermal equilibrium approximation is given here (Carbonell and Whitaker, 1984).

- The *time scale t* must *satisfy*

$$\frac{\varepsilon(\rho c)_f \ell^2}{t} \left(\frac{1}{k_f} + \frac{1}{k_s} \right) \ll 1 \tag{3.3}$$

and

$$\frac{(1-\varepsilon)(\rho c)_s \ell^2}{t} \left(\frac{1}{k_f} + \frac{1}{k_s} \right) \ll 1. \tag{3.4}$$

- The *length scales* must *satisfy*

$$\frac{\varepsilon k_f \ell}{A_o L^2}\left(\frac{1}{k_f}+\frac{1}{k_s}\right) \ll 1 \qquad (3.5)$$

and

$$\frac{(1-\varepsilon)k_s \ell}{A_o L^2}\left(\frac{1}{k_f}+\frac{1}{k_s}\right) \ll 1, \qquad (3.6)$$

where A_o is the specific surface area.

For very fast transients and when heat generation exists in the solid or fluid phase, inequality (3.1) may not be satisfied and a *two-temperature treatment* should be made. Glatzmaier and Ramirez (1988) have discussed this in relation to the application of the transient hot-wire method for the measurement of the effective thermal conductivity. This will be further discussed in Chapter 7 in connection with the two-medium treatment.

Note that in the analysis of the unit cells, we use ΔT_d and ΔT_ℓ, and in dealing with the macroscopic heat transfer of the saturated matrix, we deal with ΔT_L.

3.2 Local Volume Averaging for Periodic Structures

Taking the local volume average of the point energy equation has been discussed by Slattery (1981, 406–413). He applies his averaging theorems, and by making no assumptions about the structure of the matrix, he arrives at a differential local volume-averaged energy equation, which includes the effective thermal conductivity $\langle k \rangle$, where $\langle k \rangle$ depends on the phase conductivities and the structure of the solid matrix through a function that has to be determined empirically. Carbonell and Whitaker (1984) (by limiting their scope to *periodic structures*) have related this function to the geometry of the unit cell. The parameters of the unit cells are k_s/k_f, ε, and the solid-phase distribution within a cell, including interparticle contact. Therefore, for periodic structures, Carbonell and Whitaker have *reduced* the *empiricism* to the level of specification of this *contact*. In the following, their treatments are reviewed, including comparison with the experimental results.

The point energy equation for transient conduction, written for each phase, is

$$(\rho c_p)_s \frac{\partial T_s}{\partial t} = \nabla \cdot k_s \nabla T_s, \qquad (3.7)$$

$$(\rho c_p)_f \frac{\partial T_f}{\partial t} = \nabla \cdot k_f \nabla T_f. \qquad (3.8)$$

The boundary conditions on the interfacial area A_{fs} are the continuity of temperature and heat flux, i.e.,

$$T_f = T_s \quad \text{on} \quad A_{fs}, \tag{3.9}$$

$$\mathbf{n}_{fs} \cdot k_f \nabla T_f = \mathbf{n}_{fs} \cdot k_s \nabla T_s \quad \text{on} \quad A_{fs}. \tag{3.10}$$

3.2.1 LOCAL VOLUME AVERAGING

In dealing with any quantity ϕ that has nonzero values in *both phases*, the *intrinsic volume-averaged* value, defined by

$$\langle \phi_f \rangle^f = \frac{1}{V_f} \int_{V_f} \phi_f \, dV = \langle \phi \rangle^f, \tag{3.11}$$

is more meaningful than the average value defined by (2.71). For example, by using (3.11) under the local thermal equilibrium condition, we have $\langle T \rangle^s = \langle T \rangle^f = \langle T \rangle$. As before, the representative elementary volume is $V = V_f + V_s$. Then the fluid (or solid) temperature within V is decomposed using

$$T_f = \langle T \rangle^f + T'_f, \tag{3.12}$$

where T'_f is the *spatial deviation* component. Assuming that over V the thermal conductivities and heat capacities remain constant and by taking the intrinsic volume averages, we have

$$\frac{1}{V}\int_{V_s} (\rho c_p)_s \frac{\partial T_s}{\partial t} \, dV = \frac{V_s}{V}(\rho c_p)_s \frac{\partial}{\partial t}\left(\frac{1}{V_s}\int_{V_s} T_s \, dV\right)$$

$$= (1-\varepsilon)(\rho c_p)_s \frac{\partial \langle T \rangle^s}{\partial t}. \tag{3.13}$$

Now using the Slattery theorem (2.85) we have

$$\langle \nabla \cdot k_s \nabla T_s \rangle = \nabla \cdot \langle k_s \nabla T_s \rangle + \frac{1}{V}\int_{A_{fs}} \mathbf{n}_{sf} \cdot k_s \nabla T_s \, dA$$

$$= \nabla \cdot [(1-\varepsilon)k_s \nabla \langle T \rangle^s + k_s \langle T \rangle^s \nabla(1-\varepsilon)]$$

$$+ \nabla \cdot k_s \frac{1}{V}\int_{A_{fs}} \mathbf{n}_{sf} T_s \, dA$$

$$+ \frac{1}{V}\int_{A_{fs}} \mathbf{n}_{sf} \cdot k_s \nabla T_s \, dA. \tag{3.14}$$

Next they show (Carbonell and Whitaker, 1984, 135–139), through order-of-magnitude arguments and further assumptions, that

$$\frac{1}{V}\int_{A_{fs}} \mathbf{n}_{sf} T_s \, dA \simeq \frac{1}{V}\int_{A_{fs}} \mathbf{n}_{sf} T'_s \, dA - [\nabla(1-\varepsilon)]\langle T \rangle^s. \tag{3.15}$$

3.2 Local Volume Averaging for Periodic Structures

This is rather central in the developments that follow because under the assumption of local thermal equilibrium, the structure specific features can be included only through T'. The temperature deviation T'_s can also be introduced by direct substitution of $T_s = \langle T \rangle^s + T'_s$ in (3.14).

By using (3.15) in (3.14) we have

$$\langle \nabla \cdot k_s \nabla T_s \rangle = \nabla \cdot \left[(1-\varepsilon) k_s \nabla \langle T \rangle^s + k_s \frac{1}{V} \int_{A_{fs}} \mathbf{n}_{sf} T'_s \, dA \right]$$
$$+ \frac{1}{V} \int_{A_{fs}} \mathbf{n}_{sf} \cdot k_s \nabla T_s \, dA. \tag{3.16}$$

Then the volume-averaged energy equation for the solid phase becomes

$$(1-\varepsilon)(\rho c_p)_s \frac{\partial \langle T \rangle^s}{\partial t} = \nabla \cdot \left[(1-\varepsilon) k_s \nabla \langle T \rangle^s + k_s \frac{1}{V} \int_{A_{fs}} \mathbf{n}_{sf} T'_s \, dA \right]$$
$$+ \frac{1}{V} \int_{A_{fs}} \mathbf{n}_{sf} \cdot k_s \nabla T_s \, dA. \tag{3.17}$$

Similarly,

$$\varepsilon (\rho c_p)_f \frac{\partial \langle T \rangle^f}{\partial t} = \nabla \cdot \left(\varepsilon k_f \nabla \langle T \rangle^f + k_f \frac{1}{V} \int_{A_{fs}} \mathbf{n}_{fs} T'_f \, dA \right)$$
$$+ \frac{1}{V} \int_{A_{fs}} \mathbf{n}_{fs} \cdot k_f \nabla T_f \, dA. \tag{3.18}$$

Now noting that $\mathbf{n}_{fs} = -\mathbf{n}_{sf}$ and by adding (3.17) and (3.18), we have

$$\varepsilon(\rho c_p)_f \frac{\partial \langle T \rangle^f}{\partial t} + (1-\varepsilon)(\rho c_p)_s \frac{\partial \langle T \rangle^s}{\partial t}$$
$$= \nabla \cdot \left[\varepsilon k_f \nabla \langle T \rangle^f + (1-\varepsilon) k_s \nabla \langle T \rangle^s \right]$$
$$+ \nabla \cdot \left[\frac{1}{V} \int_{A_{fs}} \mathbf{n}_{fs} \left(k_f T'_f - k_s T'_s \right) dA \right]$$
$$+ \frac{1}{V} \int_{A_{fs}} \mathbf{n}_{fs} \cdot (k_f \nabla T_f - k_s \nabla T_s) \, dA. \tag{3.19}$$

From the assumption of local thermal equilibrium, we have

$$\langle T \rangle^f = \langle T \rangle^s = \langle T \rangle, \tag{3.20}$$

and on A_{fs}, we have

$$T'_f = T'_s. \tag{3.21}$$

However, inside V_f and V_s, the distributions of deviations T'_f and T'_s differ, and these distributions depend on the matrix structure. Then, by applying the boundary conditions (3.9)–(3.10), the local volume-averaged energy equation becomes

$$\left[\varepsilon\left(\rho c_p\right)_f+(1-\varepsilon)\left(\rho c_p\right)_s\right] \frac{\partial \langle T \rangle}{\partial t}$$
$$= \nabla \cdot \left\{ [\varepsilon k_f + (1-\varepsilon)k_s] \nabla \langle T \rangle + \frac{k_f - k_s}{V} \int_{A_{fs}} \mathbf{n}_{fs} T'_f \, dA \right\}. \quad (3.22)$$

Slattery (1981, pp. 408 and 413) and Chang and Slattery (1988) discuss the modeling of the last term using some symmetry arguments and dimensional analyses. Nozad et al. (1985) proceed with a set of *closure constitutive equations* (or *tranformations*) given by

$$T'_f = \mathbf{b}_f \cdot \nabla \langle T \rangle, \quad (3.23)$$

$$T'_s = \mathbf{b}_s \cdot \nabla \langle T \rangle. \quad (3.24)$$

The required choice of a *transformation vector* **b** *instead of a scalar*, also satisfies the *tensorial* character of the effective thermal conductivity (similar to permeability). Note that $\mathbf{b} = \mathbf{b}(\mathbf{x})$, where $\mathbf{x}$ is within the representative elementary volume, transforms the gradient of local volume-averaged temperature (changing over the length scale L) into deviations changing over length scale ℓ. The *effective thermal conductivity tensor* $\mathbf{K}_e$ is defined, and by using (3.23) in (3.22), we have

$$\left[\varepsilon\left(\rho c_p\right)_f+(1-\varepsilon)\left(\rho c_p\right)_s\right] \frac{\partial \langle T \rangle}{\partial t} = \nabla \cdot (\mathbf{K}_e \cdot \nabla \langle T \rangle), \quad (3.25)$$

where $\mathbf{K}_e$ is given by

$$\mathbf{K}_e = [\varepsilon k_f + (1-\varepsilon)k_s]\mathbf{I} + \frac{k_f - k_s}{V} \int_{A_{fs}} \mathbf{n}_{fs} \mathbf{b}_f \, dA$$

$$= [\varepsilon k_f + (1-\varepsilon)k_s]\mathbf{I} + (k_f - k_s)\varepsilon \frac{1}{V_f} \int_{A_{fs}} \mathbf{n}_{fs} \mathbf{b}_f \, dA. \quad (3.26)$$

A product (of two vectors) such as $\mathbf{n}_{fs}\mathbf{b}_f$ is called a *dyad product* and is a special form of the second-order tensors. Each vector is decomposed as

$$\mathbf{n}_{fs} = n_{fs1}\,\mathbf{s}_1 + n_{fs2}\,\mathbf{s}_2 + n_{fs3}\,\mathbf{s}_3, \quad (3.27)$$

where $\mathbf{s}_i$ is the unit vector in the i-direction. Then the dyad product is given by

$$\mathbf{n}_{fs}\mathbf{b} = \begin{bmatrix} n_{fs1}\,b_1 & n_{fs1}\,b_2 & n_{fs1}\,b_3 \\ n_{fs2}\,b_1 & n_{fs2}\,b_2 & n_{fs2}\,b_3 \\ n_{fs3}\,b_1 & n_{fs3}\,b_2 & n_{fs3}\,b_3 \end{bmatrix}. \quad (3.28)$$

Also, the unit tensor used in (3.26) is

$$\mathbf{I} = \begin{bmatrix} 1 & 0 & 0 \\ 0 & 1 & 0 \\ 0 & 0 & 1 \end{bmatrix}. \quad (3.29)$$

3.2.2 DETERMINATION OF $\mathbf{b}_f$ AND $\mathbf{b}_s$

This is done for uniform matrices (where ε is constant). The treatment is *not* restricted to *isotropic* structures. The general behavior of the effective thermal conductivity tensor for a relatively large class of periodic structures can be predicted with reasonable accuracies.

As discussed by Sahraoui and Kaviany (1993), $\mathbf{K}_e$ can be determined from $[\langle T \rangle]$ and $[T']$ equations. Here we follow the formulation of Carbonell and Whitaker in arriving at expressions to be used for the evaluation of [**b**]. This is done by developing the equations for T'_s and T'_f and then using (3.23) and (3.24).

With the assumption of uniform ε, (3.7) can be written as

$$(1-\varepsilon)(\rho c_p)_s \frac{\partial T_s}{\partial t} = \nabla \cdot [(1-\varepsilon)k_s \nabla T_s]. \tag{3.30}$$

Noting that $T'_s = T_s - \langle T_s \rangle^s$, (3.30), when subtracted from (3.17), gives

$$(1-\varepsilon)(\rho c_p)_s \frac{\partial T'_s}{\partial t} = (1-\varepsilon)k_s \nabla^2 T'_s, \tag{3.31}$$

where the surface integrals have been dropped through an order-of-magnitude analysis.

One further assumption is that T'_f and T'_s have *quasi-steady* fields. This requires that for a *time scale* of t

$$\frac{k_s t}{(\rho c_p)_s \ell^2} \gg 1 \quad \text{and} \quad \frac{k_f t}{(\rho c_p)_f \ell^2} \gg 1. \tag{3.32}$$

These state that *no* heat *diffusion* boundary layer (or penetration) exists within the representative elementary volume. These restrictions are of the same order as those given for the requirement for the local thermal equilibrium. Based on these quasi-steady states, we have

$$(1-\varepsilon)k_s \nabla^2 T'_s = 0 = \nabla^2 T'_s \tag{3.33}$$

and

$$\varepsilon k_f \nabla^2 T'_f = 0 = \nabla^2 T'_f. \tag{3.34}$$

Using $T = \langle T \rangle + T'$ in the boundary condition (3.10), we have

$$\mathbf{n}_{fs} \cdot k_f \nabla \left(\langle T \rangle + T'_f \right) = \mathbf{n}_{fs} \cdot k_s \nabla \left(\langle T \rangle + T'_s \right) \tag{3.35}$$

or

$$\mathbf{n}_{fs} \cdot k_f \nabla T'_f = \mathbf{n}_{fs} \cdot k_s \nabla T'_s + \mathbf{n}_{fs} \cdot (k_s - k_f) \nabla \langle T \rangle. \tag{3.36}$$

Now substituting in (3.33)–(3.36) and (3.21) for T' using (3.23) and (3.24), we have the following equations for $\mathbf{b}_f$ and $\mathbf{b}_s$

$$\nabla^2 \mathbf{b}_f = 0, \quad \nabla^2 \mathbf{b}_s = 0, \tag{3.37}$$

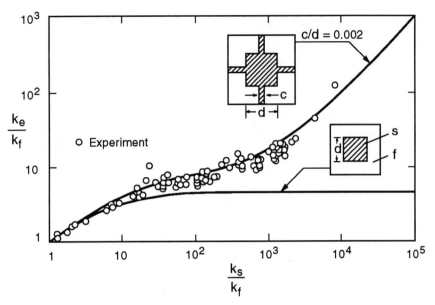

Figure 3.1 Comparison of the numerical predictions of Sahraoui and Kaviany, with the available experimental results. (Adapted from Nozad et al., reproduced by permission ©1985 Pergamon.)

$$k_f \mathbf{n}_{fs} \cdot \nabla \mathbf{b}_f = k_s \mathbf{n}_{fs} \cdot \nabla \mathbf{b}_s + \mathbf{n}_{fs}(k_s - k_f) \quad \text{on} \quad A_{fs}, \quad (3.38)$$

$$\mathbf{b}_f = \mathbf{b}_s \quad \text{on} \quad A_{fs}, \quad (3.39)$$

$$\mathbf{b}_f(\mathbf{x} + \boldsymbol{\ell}_i) = \mathbf{b}_f(\mathbf{x}), \quad \mathbf{b}_s(\mathbf{x} + \boldsymbol{\ell}_i) = \mathbf{b}_s(\mathbf{x}), \quad (3.40)$$

where $\nabla \mathbf{b}_f$ and $\nabla \mathbf{b}_s$ are scaled with $\nabla \langle T \rangle$ and $\boldsymbol{\ell}_i$ is the *period* (cell length) in direction i. The properties of these equations are discussed by Ryan et al. (1980), including the uniqueness of the solution.

3.2.3 NUMERICAL VALUES FOR $\mathbf{b}_f$ AND $\mathbf{b}_s$

A two-dimensional periodic structure with an equilateral unit cell and the solid phase distribution as shown in Figure 3.1 has been considered by Nozad et al. (1985). The structure considered is isotropic. Numerical solutions to the system (3.37)–(3.40) have been obtained. Sahraoui and Kaviany (1993) repeat the computation of Nozad et al. and find an error which has also been noted by Shonnard and Whitaker (1989). The results of Sahraoui and Kaviany will be further discussed in Section 3.7. Two predictions are shown: one with the fluid phase continuous and $\varepsilon = 0.36$, the other with the fluid phase discontinuous with $\varepsilon = 0.36$. The symmetric contacts are used on all four sides with the contact arm width taken as $c/d = 0.002$. This selection of c/d gives the closest agreement with the experiments. The results of Sahraoui and Kaviany for the effective thermal conductivity are given in Figure 3.1, along with some experimental results reported by Nozad et al. The agreement with the experiment is fairly good. Futher discussions of the effective conductivity of periodic structures are given by Quintard and Whitaker (1993) and Ochoa-Tapia et al. (1994).

3.3 Particle Concentrations from Dilute to Point Contact

A review of prediction of $\mathbf{K}_e$ using three-dimensional treatments (along with the existing symmetries) reveals the evolutionary path taken in this field. The historical progression of the analyses has been along the following steps.

- Analytical solutions based on very *dilute concentrations* of particles—*no* particle/particle *interactions*. The field in and around a particle is analyzed and an effective conductivity is defined. Mathematically stated,

    given: k_s, k_f, and particle geometry;

    then $\mathbf{q} = -k\nabla T$, $\nabla \cdot \mathbf{q} = 0 = \nabla^2 T$ in each domain,

    $\mathbf{n} \cdot k_s \nabla T = \mathbf{n} \cdot k_f \nabla T$ on interfacial area A_{fs},

    $\langle \mathbf{q} \rangle = -k_e \langle \nabla T \rangle$ for isotropic media.

- Analytical solutions for *larger concentrations*, while allowing for *no particle contacts*.

- Analytical-numerical solutions for *touching particles*. Batchelor and O'Brien (1977) apply ensemble averages over a large volume (similar to a representative elementary volume) and express the heat flux as

$$\langle \mathbf{q} \rangle = \frac{1}{V} \int_{V_f} \mathbf{q} \, dV + \frac{1}{V} \sum_i \int_{V_{si}} \mathbf{q} \, dV$$

$$= -\frac{k_f}{V} \int_{V_f} \nabla T \, dV - \frac{k_s}{V} \sum_i \int_{V_{si}} \nabla T \, dV$$

$$= -k_f \langle \nabla T \rangle - \frac{(k_s/k_f - 1)}{V} k_f \sum_{i=1}^{n} \int_{V_{si}} \nabla T \, dV$$

$$= -k_f \langle \nabla T \rangle + n \left(1 - \frac{k_f}{k_s}\right) \left\langle \int_{A_{fs}} \mathbf{x} (\mathbf{q} \cdot \mathbf{n}) \, dA \right\rangle, \quad (3.41)$$

where V_f is the volume of the fluid, V_{si} is the volume of *each* solid particle, $V = V_f + \sum_i V_{si}$, n is the *number density*, and $\langle \ \rangle$ denotes an average over many particles in one realization. Note that $\langle \nabla T \rangle$ is equal to $(1/V) \int_V \nabla T \, dV$, $\mathbf{x}$ is the position vector of the *interface*, and A_{fs} is the interfacial area. The dependence of $\langle \mathbf{q} \rangle$ on k_s/k_f for various arrangements is examined. Special attention is given to the volume near the contact points. Asymptotes for $k_s/k_f \to \infty$, $\mathbf{x} \to$ constant (nearly planar interface) near the contact point and for small contact area are also found. Various analytical-numerical methods are used [e.g., Sangani and Acrivos (1983) use the method of generalized functions].

- Analytical solution for periodic structures using the *homogenization technique* (multiple-scale expansion) for simple geometries is used by Chang (1982).

- A *semiempirical* approach is taken by Hadley (1986), where the existing bounds and the experimental studies are used to arrive at a generalized correlation.

Table 3.1 gives some of the available closed-form solutions for the isotropic effective thermal conductivity for beds of spherical particles. The results are plotted for $\varepsilon = 0.38$ (nearly that of random packing of spheres) in Figure 3.2. Figure 3.3 shows the range of the effective thermal conductivity encountered when various solid matrix materials (polymers with k_s of about 0.3 W/m-K, ceramics 2 to 50 W/m-K and metals 10 to 500 W/m-K) and saturating fluids (including vacuum) are used. In general, the solid conductivity is larger than that of the fluid. However, in liquid applications, the conductivity of the solid can be smaller. The insulation examples are from Tien and Cunningham (1976), Stark and Frickle (1993), and Hrubesh and Pekala (1994).

3.4 Areal Contact Between Particles Caused by Compressive Force

In order to predict the effective thermal conductivity for beds made of spherical particles where the particles are subject to a *compressive* load (this could be in part due to the *weight* of particles), the areal contact must be included. Except for the effect of the surface *roughness* (additional contact resistance) and the presence of *coatings* on the surfaces, the problem is that of dealing with the *consolidated* particles. Here, the contact area is predictable for the *elastic* deformation of the particles. Chen and Tien (1973), Ogniewicz and Yovanovich (1978), and Batchelor and O'Brien (1977) have analyzed the conductive heat flow through regular

3.4 Areal Contact Between Particles Caused by Compressive Force

TABLE 3.1   SOME PREDICTIONS FOR EFFECTIVE THERMAL CONDUCTIVITY FOR PACKED BEDS

APPROACH	EXPRESSION
Temperature field within a particle is unaffected by the presence of other particles—called *Maxwell lower bound* (Woodside and Messer, 1961)	Dilute particle concentration ($\varepsilon \simeq 1$) $$\frac{k_e}{k_f} = \frac{2\varepsilon + (3 - 2\varepsilon)k_s/k_f}{3 - \varepsilon + \varepsilon k_s/k_f}$$
Variational formulation *Hashin and Shtrikman* (1962) *upper bound*	$k_s/k_f \geq 1$, geometry represented by ε only $$\frac{k_e}{k_f} = \frac{k_s}{k_f}\left[1 + \frac{3\varepsilon(1 - k_s/k_f)}{(1-\varepsilon) + k_s/k_f(2+\varepsilon)}\right]$$ (same as Maxwell upper bound)
Variational formulation *Hashin and Shtrikman* (1962) *lower bound*	$k_s/k_f \geq 1$, geometry represented by ε only $$\frac{k_e}{k_f} = \left[1 + \frac{3(1-\varepsilon)(k_s/k_f - 1)}{3 + \varepsilon(k_s/k_f - 1)}\right]$$ (same as Maxwell lower bound)
Variational formulation plus structural statistics *Miller* (1969) *upper bound*	$k_s/k_f \geq 1$, symmetry in unit cell, $1/9 \leq G \leq 1/3$, $G = 1/9$ for spherical cell shape, $G = 1/3$ for plate cell shape $$\frac{k_e}{k_f} = \left[1 + (1-\varepsilon)\left(\frac{k_s}{k_f} - 1\right)\right] \cdot F_u\left(\varepsilon, \frac{k_s}{k_f}\right) \dagger$$
Variational formulation plus structural statistics *Miller* (1969) *lower bound*	Same as above $$\frac{k_e}{k_f} = \frac{k_s}{k_f}\left[\frac{k_s}{k_f} - (1-\varepsilon)\left(\frac{k_s}{k_f} - 1\right)\right.$$ $$\left. - \frac{\frac{4}{3}(k_s/k_f - 1)^2(1-\varepsilon)\varepsilon}{1 + k_s/k_f + 3(1-2\varepsilon)(k_s/k_f - 1)G}\right]^{-1}$$

$$\dagger F_u\left(\varepsilon, \frac{k_s}{k_f}\right) = 1 - \frac{(1-\varepsilon)\left(\frac{k_s}{k_f} - 1\right)^2 \varepsilon}{3\left[1 + (1-\varepsilon)\left(\frac{k_s}{k_f} - 1\right)\right]\left[1 + \{(1-\varepsilon) + 3(2\varepsilon - 1)G\}\left(\frac{k_s}{k_f} - 1\right)\right]}$$

TABLE 3.1 SOME PREDICTIONS FOR EFFECTIVE THERMAL CONDUCTIVITY FOR PACKED BEDS (CONTINUED)

Maxwell (Batchelor and O'Brien, 1977)	Dilute particle concentration $$\frac{k_e}{k_f} = 1 + \frac{3(k_s/k_f - 1)}{k_s/k_f + 2}(1 - \varepsilon)$$
Solution of Laplace equation in solid and fluid domains, including *point contact* (Batchelor and O'Brien, 1977)	For special case of point contact and high k_s/k_f, adjustment of a constant using the experimental data, random arrangement, $k_s/k_f > 100$ $$\frac{k_e}{k_f} = 4\ln\frac{k_s}{k_f} - 11$$
Parallel arrangement	$$\frac{k_e}{k_f} = \varepsilon + (1 - \varepsilon)\frac{k_s}{k_f}$$
Series arrangement	$$\frac{k_e}{k_f} = \frac{k_s/k_f}{\varepsilon k_s/k_f + 1 - \varepsilon}$$
Geometric mean	$k_e = k_f^\varepsilon k_s^{1-\varepsilon}$ i.e., $k_e/k_f = (k_s/k_f)^{1-\varepsilon}$
Homogenization of diffusion equation (Chang, 1982)	Periodic structure, two-dimensional unit cell $$\frac{k_e}{k_f} = \frac{(2-\varepsilon)k_s/k_f + 1}{2 - \varepsilon + k_s/k_f}$$
Solid body containing dilute inclusion of fluid—called *Maxwell upper bound* (Hadley, 1986)	Dilute fluid inclusion $$\frac{k_e}{k_f} = \frac{2(k_s/k_f)^2(1-\varepsilon) + (1+2\varepsilon)k_s/k_f}{(2+\varepsilon)k_s/k_f + 1 - \varepsilon}$$
Weighted average (using α_0) of the Maxwell upper bound with an expression obtained by introduction of an adjustable function f_0 into a weighted averaged expression (Hadley, 1986)	Periodic structure, $f_0 = 0.8 + 0.1\varepsilon$, $\alpha_0 = \alpha_0(\varepsilon)$ $$\frac{k_e}{k_f} = (1 - \alpha_0)\frac{\varepsilon f_0 + k_s/k_f(1 - \varepsilon f_0)}{1 - \varepsilon(1 - f_0) + k_s/k_f \varepsilon(1 - f_0)}$$ $$+ \alpha_0 \frac{2(k_s/k_f)^2(1-\varepsilon) + (1+2\varepsilon)k_s/k_f}{(2+\varepsilon)k_s/k_f + 1 - \varepsilon}$$ $\log\alpha_0 = -4.898\varepsilon$, $\quad 0 \leq \varepsilon \leq 0.0827$ $\log\alpha_0 = -0.405$ $\quad -3.154(\varepsilon - 0.0827)$, $\quad 0.0827 \leq \varepsilon \leq 0.298$ $\log\alpha_0 = -1.084$ $\quad -6.778(\varepsilon - 0.298)$, $\quad 0.298 \leq \varepsilon \leq 0.580$

3.4 Areal Contact Between Particles Caused by Compressive Force 131

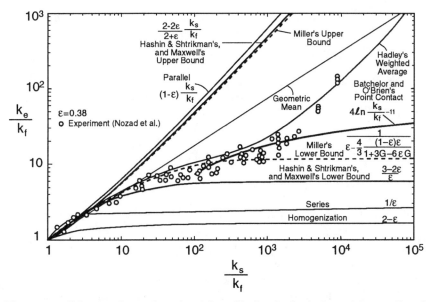

Figure 3.2 Effective thermal conductivity of beds of spherical particles predicted by various analyses. Also shown are the experimental results given by Nozad et al.

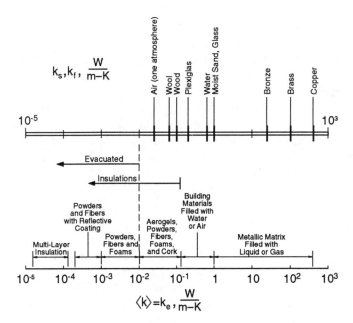

Figure 3.3 Thermal conductivity of plain and porous media (around room temperature).

arrangement of spherical particles. When the fluid is a gas, its contribution near the contact zone is also included, except in Chen and Tien where microspheres (solid, hollow, and coated) in vacuum, as in *super-insulations*, are analyzed.

The *Hertzian* elastic deformation gives the radius of the contact area between two spheres of radius R with the *Poisson* ratio μ_p and the *Young* modules E_s and subjected to a collinear force f as (Timoshenko and Goodier, 1970, p. 412)

$$R_c = \left[\frac{3(1-\mu_p^2)}{4E_s}fR\right]^{1/3}, \quad R_c \ll R. \tag{3.42}$$

For $k_f = 0$ (vacuum), Chen and Tien analyzed the conductive heat flow (assuming *constant heat flow* across the contact area, which gives results very close to the *isothermal* contact region boundary condition) for *simple* cubic, *face-centered* cubic, and *body-centered* cubic close-pack arrangements. Each layer is assumed to be isothermal normal to the direction of the heat flow, therefore, across each particle the thermal resistance is a group of *parallel* resistances. The contact resistances are much *larger* than that experienced through the bulk of the particles. The analytical results of Chen and Tien give the contact resistance R_1 (*constriction resistance*) for solid spheres as

$$R_1 = \frac{0.64}{k_s R_c}. \tag{3.43}$$

Assuming that the sphere-sphere resistance is *twice* the constriction resistance of a sphere in contact (through an areal contact of radius R_c) with a *semi-infinite slab*, the sphere-sphere constriction resistance is given as $\sim 0.5/(k_s R_c)$ for the *isothermal* contact area and $\sim 0.54/(k_s R_c)$ for the *constant heat flux* boundary condition. Therefore, the constant heat flux approximation, which is easier to apply, is justifiable because the isothermal boundary condition is expected to give similar results (the actual boundary condition is *neither* of these two).

The *overall* conductance of the matrix is given as

$$\frac{k_e}{k_s} = \left[\frac{3(1-\mu_p^2)}{4E_s}fR\right]^{1/3}\frac{1}{0.531S}\left(\frac{N_A}{N_L}\right), \tag{3.44}$$

where N_A and N_L are the number of particles per unit area and length, respectively. The constants S, N_A, and N_L are given in Table 3.2 for the three close-pack arrangements. The arrangements are shown in Figure 3.4. The force f is related to the external pressure p through $f = pS_F/N_A$, where S_F is also given in Table 3.2.

3.4 Areal Contact Between Particles Caused by Compressive Force

TABLE 3.2 MAGNITUDE OF STRUCTURAL PARAMETERS FOR DIFFERENT CLOSE PACKINGS OF SPHERICAL PARTICLES

PARAMETER	SIMPLE CUBIC	BODY-CENTERED CUBIC	FACE-CENTERED CUBIC
ε	0.476	0.32	0.26
N_L	$\dfrac{1}{2R}$	$\dfrac{3^{1/2}}{2R}$	$\dfrac{(3/8)^{1/2}}{R}$
N_A	$\dfrac{1}{4R^2}$	$\dfrac{3}{16R^2}$	$\dfrac{1}{2(3)^{1/2}R^2}$
S	1	0.25	$\dfrac{1}{3}$
S_F	1	$\dfrac{3^{1/2}}{4}$	$\dfrac{1}{6^{1/2}}$
Total number of contact points	6	8	12
Points used in the heat flow analysis	2	8	6

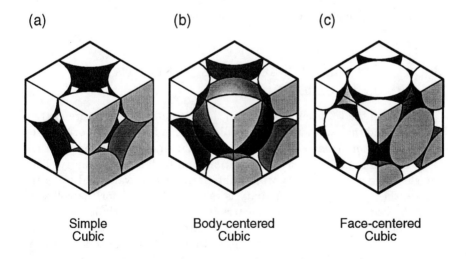

Figure 3.4 The three close-pack cubic arrangements.

3.4.1 EFFECT OF RAREFACTION

When the *gas pressure* is high enough and the pore linear dimension is not very small, so that the heat conduction in the gas phase becomes important or when a *liquid* is present, their contributions should be included. For em gases, the variation of the thermal conductivity with the gap size (in the contact region) has been accounted for in arriving at the contribution of the gas conduction (Ogniewicz and Yovanovich, 1978). They assume *no* thermal *interactions* between the gas and the solid (i.e., one-dimensional heat flow), and therefore, place the gas conductivity in *parallel* with that of the solid; the solid contribution is given by (3.44). As expected, since the solid-solid contact area is rather small, the fluid can conduct a significant amount of heat if k_f/k_s is *not* very small. Their effective conductivity, when used along with the analysis of Chen and Tien, leads to

$$k_e = \left[\frac{3(1-\mu_p^2)}{4E_s}fR\right]^{1/3} \frac{1}{0.531S}\left(\frac{N_A}{N_L}\right)(k_s + k_f\, In), \qquad (3.45)$$

where k_f is the gas thermal conductivity under standard temperature-pressure condition. The term In is an *integral function* that includes the following parameters: *Knudsen number* ($Kn = \lambda/C$ where λ is the mean distance between two molecular collisions, the *mean free path*, and C is the average linear pore size or interparticle distance; the effects of rarefaction on transport will be further discussed in Chapter 6), R_c/R, k_s/k_f, *packing geometry*, and the *gas pressure* and *temperature*. The gas conduction contribution is also addressed by Cunningham and Tien (1978), where the model suggested by Luikov et al. (1968) for the effective thermal conductivity is used.

3.4.2 DEPENDENDENCE OF GAS CONDUCTIVITY ON KNUDSEN NUMBER

Because of the traditional interest in the *evacuated* (low pressure gas) *insulations* and the recent interest in *aerogels*, i.e., a network of short-bonded chains of 2–5 nm (made of organic or inorganic materials) fused together, the Knudsen number in these applications can be larger than unity. Then the gas thermal conductivity will deviate from its bulk value and using the kinetic theory, the gas conductivity is modified to allow for the collisions with the pore surface. The parameters in the analysis include the surface accommodation factor as well as the geometric (pore) and transport (i.e., Prandtl number) parameters. The rarefaction heat transfer is discussed by Springer (1971) and by Giedt and Willis (1985). Applications to aerosols are discussed by Hrubesh and Pekala (1994) and Zeng et al. (1994). A general discussion is given by Wolf and Strieder (1994) and the application to fibrous insulations is discussed by Stark and Frickle (1993) and Zheng and Strieder (1994).

3.5 Statistical Analyses

Due to the statistical nature of most solid matrix microstructures, the regular (periodic) microstructures only correspond to *discrete* values of porosity in an otherwise continuous variation in porosity. Therefore, the statistical treatments that are based on the microstructure of the matrix should be able to predict the periodic structures while also being able to accommodate microstructural *defects* and other statistics. In the *existing* treatments, the medium treated is macroscopically

- uniform, and

- isotropic (spheres), or

- anisotropic (fibers)—the effective thermal conductivity perpendicular to the fibers (transverse conductivity) and parallel to the fibers (axial conductivity) are found. The axial conductivity is simply found by the *parallel arrangement* of resistances (Table 3.1).

The existing statistical treatments of effective thermal conductivity include various extents of information about the macro- and microstructure. Some of the treatments are given here.

- *Simplest statistical* information, i.e., the porosity and thermal conductivity of each phase (one-point correlation function) are used. These analyses establish the *most restrictive* upper and lower bounds.

- The preceding plus a *symmetrical unit-cell* model of the microstructure (a three-point correlation function). The length scale for each phase within this unit cell is given by a probability distribution.

- The preceding extended to *unit-cell mixtures* (various packings) and *asymmetric unit cells*.

In the following, we review a statistical treatment and its extension, which are based on the application of the *variational principles*. Then a treatment similar to that applied in the *statistical thermodynamics* is reviewed. The variational formulation, along with the microstructure statistics, leads to the lower and upper *bounds* for the effective thermal conductivity, while the thermodynamic analogy, which does not directly include any extremum, predicts the *absolute values*.

3.5.1 A VARIATIONAL FORMULATION

(A) SIMPLEST STATISTICS

Prediction of the effective *magnetic* permeability of the macroscopically uniform and isotropic multiphase materials through the application of the variational theorems is done by Hashin and Shtrikman (1962). Because of the mathematical analogy, the results can be applied to the *heat* and *electrical* conductivity as well. The formulation leads to the conditions for the *absolute maximum* and the *absolute minimum* for the magnetostatic energy stored in a multiphase medium. These establish bounds for the effective magnetic permeability of the medium. Their analysis uses the parameters ε, k_s, and k_f. Their prediction, as rearranged by Miller (1969), is given in Table 3.1 and plotted along with the other predictions in Figure 3.2.

As an illustration of the *inadequacy* of the simple statistics method in given realistic (usable) predictions of the effective thermal conductivity, the *measured* k_e for an air-saturated *polyurethane foam* is shown in Figure 3.5. These foams are manufactured with 10 to 100 pores per inch, but the porosity for all of them is about 0.98. Based on the simple statistics model, k_e should be independent of the structure as long as ε remains constant. This is not observed in the measurement. Therefore, extra care must be taken in using any prediction that does not account for the *microstructure*.

(B) THREE-POINT CORRELATION FUNCTION

As an application of use of additional statistical information, the *three-point correlation function* (Miller, 1969) leads to the development of improved bounds on the effective properties. Beran (1965) derives *bounds* for the effective conductivity (electrical permeability), using two variational principles. As *trial* functions, he uses perturbation expansion of the fields. The first-order perturbation effect is found in terms of the expressions involving *third-order* correlation functions of the fluctuating part of the conductivity. The three-point correlation function is a difficult statistic to obtain. However, for the broad class of *periodic structures*, Miller (1969) shows that these three-point correlation functions, which appear in the effective thermal conductivity bounds, are *simply constants* for each phase.

The n-point correlation function $\gamma_n(\mathbf{x}_1, \mathbf{x}_2, \ldots \mathbf{x}_n)$ is defined as

$$\gamma_n = \int k(\mathbf{x}_1) \ldots k(\mathbf{x}_n) \, dF_{1\ldots n}(\mathbf{x}_1, \mathbf{x}_2, \ldots, \mathbf{x}_n), \qquad (3.46)$$

where $F_{1\ldots n}$ is the n-point distribution function. For a two-phase medium with the thermal conductivity k_s in the volume V_s and k_f in V_f [i.e., $k(\mathbf{x}_s) = k_s$, $k(\mathbf{x}_f) = k_f$], the n-point correlation function γ_n is related to the *probability* that n points $\mathbf{x}_1, \mathbf{x}_2, \ldots, \mathbf{x}_n$, thrown at random into the

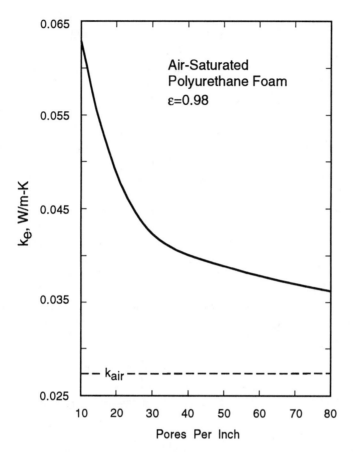

Figure 3.5 Variation of the measured effective thermal conductivity with the cell size. The results are for a constant porosity of 0.98 (from Scott Paper Co., Chester, PA).

medium, all lie in the same phase (s or f). This can be extended to cases where n points fall in all possible combinations of V_s and V_f. If $P_{1,\ldots 1,1}$ is the probability of n points being in phase s, and $P_{1,\ldots,1,2}$ is the probability of $n-1$ points being in phase s and point $\mathbf{x}_n$ in phase f, etc., we have

$$\gamma_n = \sum_{\alpha_n=1}^{2} \cdots \sum_{\alpha_1=1}^{2} P_{\alpha_1,\ldots,\alpha_n}(\mathbf{x}_1, \mathbf{x}_2, \ldots, \mathbf{x}_n) k(\mathbf{x}_1) \ldots k(\mathbf{x}_n). \quad (3.47)$$

Miller shows that for $n = 1$ (*one-point correlation*)

$$\gamma_1(\mathbf{x}) = \varepsilon k_f + (1-\varepsilon)k_s = \langle k \rangle, \quad (3.48)$$

i.e., *no* statistical information about the distribution of phases (structure of matrix) is used.

Miller applies those summations to the three-point correlation formulas for the upper and lower bounds of the effective thermal conductivity, and his final results are given in Table 3.1. These results are also plotted in Figure 3.2, along with other bounds. Torquato (1991) gives a summary of the recent work and improved bounds.

Vafai (1980) incorporates the probability of the occurrence of different packing of spheres found from experiments into the analysis and finds the proper bounds for the effective thermal conductivity of this mixture of packings. He also extends the results of Schulgasser (1976) for needle-shaped fibers placed in parallel (two-dimensional) and random (three-dimensional) arrangements (fibrous porous media) and finds the bounds for the *transverse* effective thermal conductivity.

Hadley (1986) combines the *Maxwell* upper bound with an expression obtained by the introduction of an *adjustable* function $f_o(\varepsilon)$. In combining these, he uses a *weighting function* $\alpha_o(\varepsilon)$ which, along with $f_o(\varepsilon)$, is found from the *experimental* results. While $f_o(\varepsilon)$ changes *slightly* with ε, α_o is *very sensitive* to changes in ε. Figure 3.6 shows the results for $\alpha_o(\varepsilon)$ obtained from Hadley's experiments with powders and disks. The range of ε in his experiments are also shown. His complete expression for the effective thermal conductivity is given in Table 3.1. The plotted results for $\alpha_o(\varepsilon)$ are also given in algebraic forms. These equations are repeated here.

$$\frac{k_e}{k_f} = (1-\alpha_o)\frac{\varepsilon f_o + \frac{k_s}{k_f}(1-\varepsilon f_o)}{1-\varepsilon(1-f_o) + \frac{k_s}{k_f}\varepsilon(1-f_o)}$$

$$+\alpha_o o\frac{2\left(\frac{k_s}{k_f}\right)^2(1-\varepsilon) + (1+2\varepsilon)\frac{k_s}{k_f}}{(2+\varepsilon)\frac{k_s}{k_f}+1-\varepsilon} \qquad (3.49)$$

$$f_o = 0.8, \qquad (3.50)$$

where

$\log \alpha_o = -4.898\varepsilon$ $\qquad 0 \leq \varepsilon \leq 0.0827,$

$\log \alpha_o = -0.405 - 3.154(\varepsilon - 0.0827)$ $\qquad 0.0827 \leq \varepsilon \leq 0.298,$ $\quad (3.51)$

$\log \alpha_o = -1.084 - 6.778(\varepsilon - 0.298)$ $\qquad 0.298 \leq \varepsilon \leq 0.580.$

3.5.2 A Thermodynamic Analogy

While the analysis for regular arrangements of spherical particles leads to the determination of the constriction resistance and the effective thermal conductivity, the results are for discrete values of porosity. This limitation can be slightly lifted by assuming *lattice defect* and applying the same

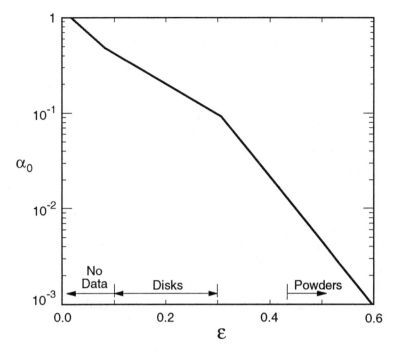

Figure 3.6 Hadley's empirical results for $\alpha_0(\varepsilon)$.

deterministic techniques. However, it is desirable to obtain results over a relatively *large* and *continuous* range of porosities. Nayak and Tien (1978) applied the principles of statistical thermodynamics to the problem of determining the effective thermal conductivity of *randomly* arranged (and as a subset, the regularly arranged) spherical particles in *vacuum*. The idea of extending the statistical mechanics to the description of such a macroscopic system has been suggested by Lienhard and Davis (1971).

This involves *specification* of the variation of the *average number* of *contact points* with respect to the *porosity*, as well as the frequency distribution of those contact points for a given packing (or porosity). We will next review the steps that lead to the determination of the effective thermal conductivity.

(A) POINTS OF CONTACT

The *points* of *contact* between a given sphere and adjacent spheres (and in some cases the angular distribution of these points) must be determined. For regular packings, the number of such points, called the *coordination number*, is indicative of the packing employed; simple-cubic, ortho-rhombic (body-centered-cubic), *tetragonal-spherominal*, and rhombohedral (face-centered-cubic) packings have 6, 8, 10, and 12 contact points, respectively.

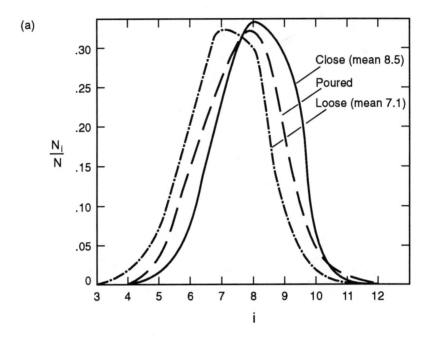

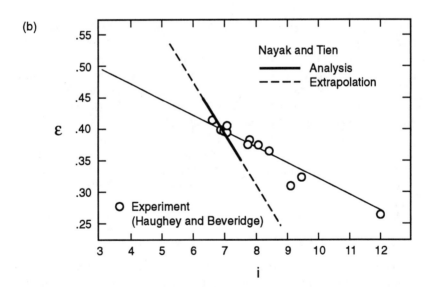

Figure 3.7 (a) The frequency distribution of the number of contact points as a function of the number of contact points for various packings and (b) distribution of the number of the contact points as a function of porosity. (From Nayak and Tien, reproduced by permission ©1978 Pergamon.)

3.5 Statistical Analyses

The frequency distribution of the number of contact points N_i/N, where N_i is the number of samples that have the number of contact points i and N is the total number of samples, is measured and given by Haughey and Beveridge (1969). The results are plotted in Figure 3.7(a). Also given in the figure are the mean values of the contact points for loose and close random packing. The probability distribution is nearly symmetric (normal about the mean value).

The number of contact points can be related to the bulk porosity. This is done by Haughey and Beveridge and they recommend

$$i = 22.47 - 39.39\varepsilon, \quad 0.259 \leq \varepsilon \leq 0.5. \tag{3.52}$$

This is a curve fit to the experimental data of several investigators. The experimental results along with their correlations, are given in Figure 3.7(b). A monotonic increase in i with decrease in ε is expected, as more contact (without deformation) is required for denser (small ε) packing.

(B) Equilibrium Distribution

This is obtained (Lienhard and Davis, 1971) by furnishing the following.

- *Constraints*—in this case these are the *number constraint*, stating that

$$\sum_i N_i = N, \tag{3.53}$$

  and the *volume constraint*, which states that

$$\sum_i \frac{N_i}{1 - \varepsilon_i} = \frac{N}{1 - \varepsilon}. \tag{3.54}$$

- A *probability distribution*—which is the generalized *Boltzmann* statistic (Boltzons); Nayak and Tien give

$$P = \prod_i \frac{g_i^{N_i}}{N_i!}, \tag{3.55}$$

  where g_i is the number of different ways the i contact points can be made, and as before, N_i is the number of particles that have i contact points.

Then the distribution N_i, which maximizes P and hence $\ln P$, is found as the *equilibrium distribution*. By setting $\mathrm{d}(\ln P) = 0$ and solving (3.55) along with constraint (3.53) and (3.54) and using the method of *Lagrangian multiplier*, the equilibrium distribution is found. Nayak and Tien assume a relationship between i and ε based on their analysis of regular packing of spheres, and this is given by

$$i = 11.6(1 - \varepsilon_i), \quad 0.35 < \varepsilon_i < 0.44. \tag{3.56}$$

This relationship is also given in Figure 3.7(b), and since they apply this to $0.26 < \varepsilon < 0.52$, the extrapolation of this correlation is also shown in that figure. Note that the *extrapolations* are *not* in agreement with the experimental results. Based on (3.56), the constraint in (3.54) becomes

$$\sum_i \frac{N_i}{i} = \frac{N}{11.6(1-\varepsilon)}. \tag{3.57}$$

Now the equilibrium distribution, subject to (3.54) and (3.57), is

$$\frac{N_i}{N} = \frac{g_i e^{-\beta/i}}{\sum_i g_i e^{-\beta/i}}, \tag{3.58}$$

where β is the Lagrangian multiplier. As was mentioned, the *degeneracy* g_i associated with i contacts on a particle is the number of different ways i contact points can be made, and Nayak and Tien suggest

$$g_i = \frac{12!}{i!(12-i)!}; \tag{3.59}$$

when (3.59) is used in (3.58), the expression for the mean porosity is

$$\frac{1}{1-\varepsilon} = 11.6 \sum_i \frac{\frac{g_i}{i} e^{-\beta/i}}{\sum_i g_i e^{-\beta/i}}. \tag{3.60}$$

Equation (3.60) is used to obtain $\beta(\varepsilon)$ numerically. As expected, the equilibrium distribution is close to those given in Figure 3.7(a).

(C) Effective Thermal Conductivity

As was given in connection with the constriction resistance per contact point, the contact resistance is nearly $R_i = 0.5(k_s R_c)^{-1}$, where R_c (the radius of contact area) depends on the applied force. The solid area of a random cross-sectional slice through a particle of diameter d is shown to be equal to $\pi d^2/6$. The force at any point is related to the pressure through

$$f = \frac{\pi}{3} \frac{p\, d^2}{1-\varepsilon} \frac{1}{i \cos\theta}, \tag{3.61}$$

where θ is the angular location of contact referenced to the loading direction, $0 \leq \theta \leq \pi/2$ (for $\theta = \pi/2$ no load on the contact point). Then, using (3.42), we have

$$R_i = \frac{1}{k_s}\left[\frac{(1-\mu_p^2)}{E_s}\frac{2\pi p\, d^3}{(1-\varepsilon)i\cos\theta}\right]^{-1/3} \quad (3.62)$$

Each resistance R_i is associated with an average area of $\pi d^2/[3i(1-\varepsilon)]$ and an average layer height of $(d/2)\cos\theta$. The equivalent thermal resistance over the bulk area and per unit length is

$$R^*_{e,i} = \frac{2\pi d R_i}{3(1-\varepsilon_i)i\cos\theta}. \quad (3.63)$$

The effective thermal conductivity per contact is

$$k_{e,i} = \frac{1}{R^*_{e,i}} = \frac{3(1-\varepsilon_i)ik_s\cos\theta}{2\pi}\left[\frac{(1-\mu_p^2)p}{E_s(1-\varepsilon)}\frac{2\pi}{i\cos\theta}\right]^{1/3} \quad (3.64)$$

The effective lattice conductivity is found by averaging over the distribution of contact point numbers and their angular distribution, i.e.,

$$k_e = \frac{2}{\pi}\sum_i \frac{N_i}{N}\int_0^{\pi/2} k_{e,i}\, d\theta, \quad (3.65)$$

$$\frac{k_e}{k_s}\left[\frac{(1-\mu_p^2)p}{E_s}\right]^{-1/3} = \frac{\frac{3}{\pi^2}(2\pi)^{1/3}\int_0^{\pi/2}(\cos\theta)^{2/3}\, d\theta}{(1-\varepsilon)^{1/3}}$$

$$\times \sum_i \frac{N_i}{N}i^{2/3}(1-\varepsilon), \quad (3.66)$$

where $\int_0^{\pi/2}(\cos\theta)^{2/3}\, d\theta = 1.1186$. Then

$$\frac{k_e}{k_s}\left[\frac{(1-\mu_p^2)p}{E_s}\right]^{-1/3} = \frac{C}{(1-\varepsilon)^{1/3}}\frac{\sum_i g_i(1-\varepsilon_i)_i^{2/3}e^{-\beta/i}}{\sum_i g_i e^{-\beta/i}}, \quad (3.67)$$

where $C = 0.65$. Their numerical results for (3.67), along with the results based on the analysis of the lattice defect, which gives porosities around those found by regular arrangements, are given in Figure 3.8. Good agreement is found between their statistical model and their deterministic analysis.

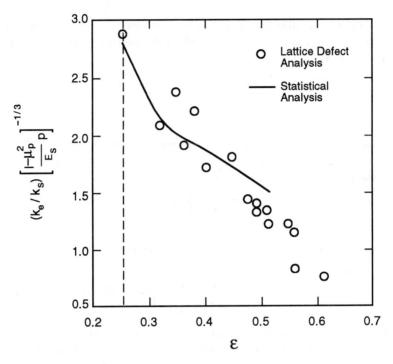

Figure 3.8 Comparison of the Nayak and Tien statistical and lattice defect analyses. (From Nayak and Tien, reproduced by permission ©1978 Pergamon.)

3.6 Summary of Correlations

For packed beds of spherical particles and for the entire range of values of k_s/k_f (larger and smaller than unity) some *empirical* correlations for the effective thermal conductivity, in addition to those given in the previous sections, are available. Three of these are constructed by Krupiczka (1967), Kunii and Smith (1960), and Zehnder and Schlünder (1970). An extensive review of the literature on the effective thermal conductivity prior to 1960 is given by Krupiczka. The prediction of Krupiczka gives

$$\frac{k_e}{k_f} = \left(\frac{k_s}{k_f}\right)^{+0.280 - 0.757 \log \varepsilon - 0.057 \log(k_s/k_f)}. \tag{3.68}$$

The prediction of Kunii and Smith (for $0.260 \leq \varepsilon \leq 0.476$) gives

$$\frac{k_e}{k_f} = \varepsilon + \frac{(1-\varepsilon)}{\phi_2 + 4.63(\varepsilon - 0.26)(\phi_1 - \phi_2) + \frac{2}{3}\frac{k_f}{k_s}}, \tag{3.69}$$

where $\phi_1 = \phi_1(k_f/k_s)$ and $\phi_2 = \phi_2(k_f/k_s)$ and they are monotonically decreasing functions of k_s/k_f. The prediction of Zehnder and Schlünder is

$$\frac{k_e}{k_f} = 1 - (1-\varepsilon)^{1/2} + \frac{2(1-\varepsilon)^{1/2}}{1 - \frac{k_f}{k_s}B}$$

$$\times \left[\frac{\left(1 - \frac{k_f}{k_s}\right)B}{\left(1 + \frac{k_f}{k_s}B\right)^2} \ln \frac{1}{\frac{k_f}{k_s}B} - \frac{B+1}{2} - \frac{B-1}{1 - \frac{k_f}{k_s}B} \right], \quad (3.70)$$

where

$$B = 1.25 \left(\frac{1-\varepsilon}{\varepsilon}\right)^{10/9}. \quad (3.71)$$

A modification to the Zehnder-Schlünder correlation is given by Hsu et al. (1994).

Prasad et al. (1989) performed experiments for k_s/k_f larger and smaller than unity (but for only one order magnitude to either side of unity) and examined the accuracy of the preceding three predictions. In Figure 3.9, the experimental results of Prasad et al., Nozad (1983), and other experimental results reported by her are plotted against these predictions and that of Hadley. While for moderate values of k_s/k_f the four predictions are in agreement, for high values *only* the Hadley correlation can predict the effective thermal conductivity correctly. Another correlation that uses k_s/k_f and ε as parameters is that of Nimick and Leith (1992) with accuracies similar to that of Hadley. A correlation for the effective conductivity of rocks based on a geometric model of nonintersecting, oblate spherical pores, has been developed by Zimmerman (1989). Correlations for fibrous insulations are given by Stark and Frickle (1993). For two-dimensional structures, some correlations for the bulk effective thermal conductivity are given by Sahraoui and Kaviany (1993) and Ochoa-Tapia et al. (1994) and by Hsu et al. (1995) for two- and three dimensional periodic structures.

3.7 Adjacent to Bounding Surfaces

The phase distributions adjacent to the surfaces that bound porous media is generally not uniform. This porosity nonuniformity is expected to influence the local effective thermal conductivity. Therefore the *bulk* effective thermal conductivity, which was reviewed in the preceding sections, is not expected to be applicable *near the bounding surfaces*. The factors contributing to the difference between the bulk and the near boundary k_e are given here.

- When the bounding medium is a *solid*, *point* or *areal* contacts occur between the elements of matrix (e.g., particles) and the bounding surface. Then the thermal conductivity of the bounding surface, in addition to the thermal conductivity of the solid matrix and the fluid, should be *included* in the analysis (Figure 3.10). We can define a *local area-averaged* (averaged parallel to the bounding surface) effective thermal conductivity which, within a distance of $O(d)$, changes continuously as the bounding surface is approached. The effect of the conductivity of the bounding surface may extend to a short distance from the bounding surface. However, when *local volume-averaged* k_e is used, this continuous variation is replaced with a *discontinuity*, where k_e is averaged over the representative elementary volume. The discontinuity in k_e does not pose any mathematical problem, but a *significant error* in temperature field near the surface and, therefore, in the heat flow can result if k_e is averaged over ℓ.

- When the bounding medium is a fluid, the *ambiguity* is in the location of the interface. This ambiguity is confined to $O(d)$. Over a distance of order $O(d)$ from the interface, the *area-averaged* k_e changes continuously. Again, volume-averaged k_e masks these variations. When significant heat flows perpendicular to the interface, the ambiguity in the interfacial location and the use of the volume-averaged k_e can result in significant errors.

The variation of the local effective thermal conductivity adjacent to bounding rigid surfaces is addressed by Ofuchi and Kunii (1965) for packed beds of spherical particles. Their semiempirical relation for the *local* effective thermal conductivity, which is taken as *uniform* over a distance $d/2$ from the surface, is

$$\frac{k_e}{k_f} = \varepsilon_w + \frac{1-\varepsilon_w}{2\phi_w + \frac{2}{3}\frac{k_f}{k_s}} \quad \text{for a distance } \frac{d}{2} \text{ from the wall,} \quad (3.72)$$

where $\varepsilon_w = 0.4$, which is the closest packing of spherical particles on a *planar surface*. Their *bulk* porosity varies between 0.338 and 0.603. They give $0.026 < \phi_w\,(k_s/k_f) < 0.16$ and the function ϕ_w is given by them in *graphical* form for $0.3 < k_s/k_f < 10^4$. When $k_f/k_s > 1$, the wall region effective thermal conductivity is *larger* than the bulk, and vice versa. This expression is valid only for a *loose packing* (nonconsolidated) of monosized spherical particles.

Also hidden in ϕ_w is the effect of thermal conductivity of the *bounding solid* surface (in their experiment a *marble* plate, $k = 2.08$ W/m-K, was used). Therefore, general expressions for the local effective thermal conductivity adjacent to bounding surfaces would naturally be very complex, while application of the bulk values can lead to significant errors in the predicted heat flow rate. This point will also be elaborated where dealing with

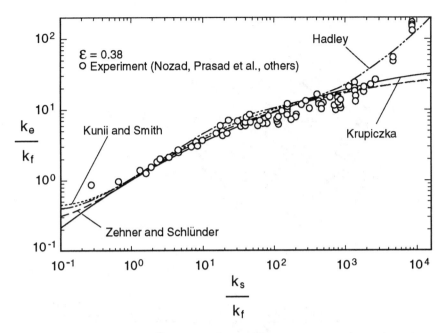

Figure 3.9 Comparison of several correlations for the effective thermal conductivity with experimental results from several sources.

convective heat flow near the bounding surfaces (Section 4.10). Effective thermal conductivity for packed beds of spheres, pellets and rings in beds with the ratio of radial size to particle radius of the order of unity, is studied by Melanson and Dixon (1985). As a part of their results, correlations based on the experimental results are given.

Figure 3.10 gives a schematic of the phase distributions adjacent to a bounding planar surface. We consider two limiting cases of very large thermal conductivity for the bounding surface $k_{s_b} \to \infty$ and very low thermal conductivity $k_{s_b} \to 0$. When $k_{s_b} \to \infty$, i.e, isothermal bounding surface, the heat transfer through the bounding solid surface (made of $A_{s_b f}$ and $A_{s_b s}$) is

$$Q_b = -k_f A_{s_b f} \frac{\partial T}{\partial n} - k_s A_{s_b s} \frac{\partial T}{\partial n} \quad \text{on } A_b \quad \text{for} \quad k_{s_b} \to \infty, \quad (3.73)$$

$$T_b \text{ is constant} \quad \text{on } A_b \quad \text{for} \quad k_{s_b} \to \infty. \quad (3.74)$$

When in addition $k_s \gg k_f$, we have

$$Q_b = -k_s A_{s_b s} \frac{\partial T}{\partial n} \quad \text{for} \quad k_{s_b} \to \infty. \quad (3.75)$$

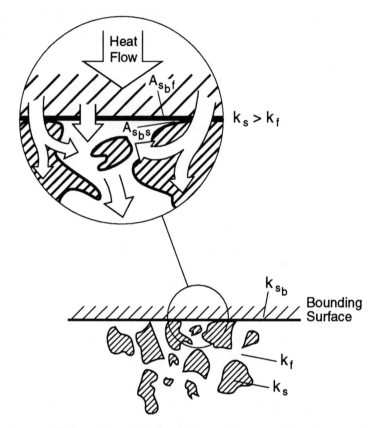

Figure 3.10 The coupling of the bounding surface thermal conductivity and the effective thermal conductivity of the porous media at and near the bounding solid surface.

However, when $k_{s_b} \to 0$, even for $k_s \gg k_f$, the heat flow through the bounding surface-fluid interface $A_{s_b f}$ is not negligible, because T_b (the bounding surface temperature) will be much higher over $A_{s_b f}$ than it is over $A_{s_b s}$.

A local volume-averaged treatment of the effective thermal conductivity at the bounding surface is given by Prat (1989, 1990). He also examines the numerical results for a *two-dimensional* structure made of *square cylinders*. In the following we examine the results of a numerical simulation for variation of $\mathbf{K}_e$ near the bounding surfaces and examine the resulting slip in the interfacial temperature when a uniform $\mathbf{K}_e$ is used.

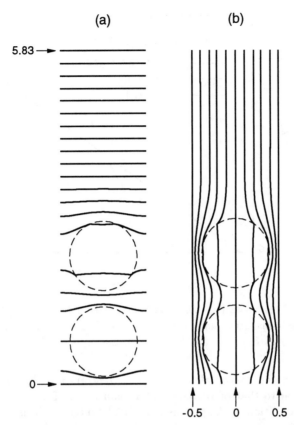

Figure 3.11 Constant temperature contours for (a) transverse (y-direction) heat flow and (b) longitudinal (x-direction) heat flow. The results are for $k_s/k_f=5$, and $k_{s_b}/k_f = 1$.

3.7.1 Temperature Slip for a Two-Dimensional Structure

Variation of $\mathbf{K}_e$ in the interfacial region between a two-dimensional periodic structure made of circular cylinders and a plain medium (solid or fluid phase) has been examined by Sahraoui and Kaviany (1993). Figure 3.11 shows their computed temperature field for heat flow (a) perpendicular and (b) parallel to this interface. The results are for $k_s/k_f = 5$, $\varepsilon = 0.5$, and for the plain medium being an *extension* of the fluid saturating the porous medium ($k_{s_b}/k_f = 1$). As expected the temperature field is different near the interface (compared to that in the plain medium and the bulk of the porous medium).

150   3. Conduction Heat Transfer

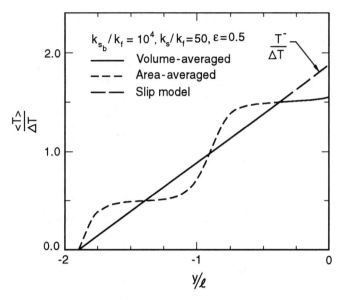

Figure 3.12 Volume- and area-averaged temperature distributions for heat conduction in an in-line arrangement of circular cylinders with a perfectly conducting solid bounding surface ($\varepsilon = 0.5$, $k_s/k_f = 50$, and $k_{s_b}/k_f = 10^4$).

Sahraoui and Kaviany also construct a *temperature slip boundary* condition similar to that of Beavers and Joseph (1967) for the *hydrodynamic* boundary condition at the porous-plain media interface (this was discussed in Section 2.11.1). *The temperature slip* boundary condition is defined as

$$\left.\frac{d\langle T\rangle_v}{dy}\right|_{y=0^-} = \frac{\alpha_T}{\lambda}(T^- - T^+), \qquad (3.76)$$

where α_T is the dimensionless *slip coefficient*, T^- is the *porous medium interfacial temperature*, T^+ is the *plain medium interfacial temperature*, and λ is a pore-level length scale. The temperature gradient is evaluated on the porous medium side. However, it can be written in terms of the temperature gradient on the plain medium side, but α_T will be different. Using the bulk k_e, and when the plain medium is the fluid, the continuity of the heat flux gives

$$k_e \left.\frac{d\langle T\rangle_v}{dy}\right|_{y=0^-} = k_f \left.\frac{dT}{dy}\right|_{y=0^+}. \qquad (3.77)$$

The temperature slip coefficient α_T calculated using (3.76) for $\lambda = \ell$ is

$$\alpha_T = \frac{\ell \left.\dfrac{d\langle T\rangle_v}{dy}\right|_{y=0^-}}{(T^- - T_{s_b})}. \qquad (3.78)$$

Figure 3.12 shows the distribution of the volume- and area-averaged temperatures for $k_s/k_f = 50$ and $\varepsilon = 0.5$. The temperature $T^-/\Delta T$, shown in Figure 3.12, is found by the extrapolation of the temperature distribution using the results away from the boundary. A temperature gradient $\Delta T/\ell$ is imposed in the bed. The extrapolated temperature at the interface is higher than the actual temperature of the interface. Note that in the available *experimental* results for packed beds of spheres, the extrapolated $T^-/\Delta T$ is *lower* than the actual temperature. This is due to the different behaviors in $\langle \varepsilon \rangle_A (y)$ for *circular cylinders* and for *spheres*.

For the packed beds of spheres, the porosity near the wall is *higher* than in the bulk and the local effective conductivity is *lower*. For this reason a *higher* temperature gradient exists near the wall. For an arrangement of circular cylinders, the porosity near the wall is *lower* than in the bulk and this results in a *higher* local effective conductivity and a *smaller* temperature gradient near the interface. The packed beds of spheres can be *simulated* by moving the first layer of cylindrical particles *away* from the solid surface. When the interface is placed at a distance of $\ell/2$ away from the center of the particle, the slip in temperature is *zero*. When the interface is placed at a distance further than $\ell/2$, a temperature slip similar to the one in the packed beds of spheres is obtained.

When the particle is in *direct contact* (i.e., point contact) with a bounding solid surface, the particle conductivity k_s/k_f significantly *influences* the heat flux at the wall. This is shown in Table 3.3, where the results show that the *slip* in the temperature increases as k_s/k_f increases.

3.7.2 Variable Effective Conductivity for a Two-Dimensional Structure

The calculated local effective conductivity, by Sahraoui and Kaviany (1993), is given in Figures 3.13 and 3.14 (for $k_{e\perp}$ and $k_{e\|}$), for $\varepsilon = 0.5$ and several k_s/k_f. The volume averages are made using progressively larger volumes as the distance from the interface increases (or y decreases). This is similar to that used in Section 2.11.6. Figure 3.13 shows that the *transverse* effective conductivity first *increases* and then begins to *decrease* over a distance of $\ell/2$, where ℓ is the unit-cell linear dimension. This is a direct result of the solid phase distribution. Figure 3.14 shows the results for the *longitudinal* component and since the *nonuniformity* of the phase distributions are *less severe* in this direction, the local longitudinal effective conductivity does not reach as large of a maximum as does $k_{e\perp}/k_f$. In the bulk (far away from the interface) an *isotropic* behavior is expected and this is found in Figures 3.13 and 3.14 for $y < 0.5\ell$.

TABLE 3.3   EFFECT OF THE PARTICLE CONDUCTIVITY ON THE SLIP COEFFICIENT FOR CIRCULAR CYLINDERS IN CONTACT WITH A SOLID BOUNDING MEDIUM OF HIGH CONDUCTIVITY ($k_{s_b}/k_f = 10^4$, $\varepsilon = 0.5$)

$\dfrac{k_s}{k_f}$	$\dfrac{k_e}{k_f}$	α_T	$\dfrac{T^- - T_{s_b}}{\Delta T}$
5	2.01	7.11	0.12
10	2.42	4.58	0.20
50	2.92	2.90	0.34

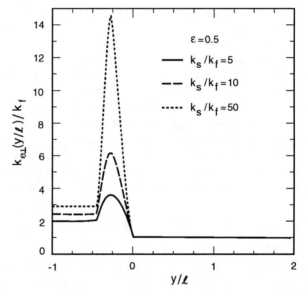

Figure 3.13 Distribution of the local transverse effective conductivity for an in-line arrangement of circular cylinders for different solid conductivities. The results are for $\varepsilon = 0.5$, $k_{s_b}/k_f = 1$.

3.8   On Generalization

Heat conduction through multiphase media is handled through imposition of the first law of thermodynamics along with the conduction flux law. When gases are involved, rarefication (for small pore size and near constrictions) should be attended. The thermal conduction of each phase is temperature-dependent (and for gases, also pressure-dependent), and that should also be considered.

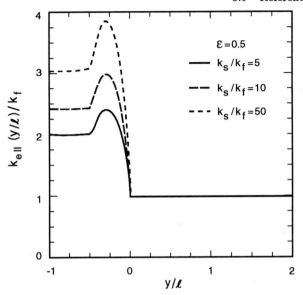

Figure 3.14 Distribution of the longitudinal local effective conductivity for an in-line arrangement of circular cylinders, for different solid conductivities ($\varepsilon = 0.5$, $k_{s_b}/k_f = 1$).

A generalization based on the porosity alone is not possible. Solid matrices with the same porosity but with various cellular structures have different effective thermal conductivities (Figure 3.5 gives an example for polyurethane foam). Also some high conductivity consolidated solid matrices have conductivities that are nearly *independent* of porosity (at low porosities). Examples are given by Koh and Fortinn (1973), where for oxygen-free, high-conductivity copper sintered powders with porosity between 0 and 0.30, they find nearly no change in k_e with respect to ε for a given average temperature.

In using the available results for k_e, special care must be taken to assure that the results are applicable to the porous medium considered. Moreover, the effective thermal conductivity near the bounding surfaces can be significantly different than the bulk values.

3.9 References

Batchelor, G. K. and O'Brien, R. W., 1977, "Thermal or Electrical Conduction Through a Granular Material," *Proc. Roy. Soc.* (London), A355, 313–333.

Beran, M., 1965, "Use of Variational Approach to Determine Bounds for Effective Permittivity in Random Media," *Il Nuovo Cimento*, 38, 771–787.

Beavers, G. S. and Joseph, D. D., 1967, "Boundary Conditions at a Naturally Permeable Wall," *J. Fluid Mech.*, 30, 197–207.

Carbonell, R. G. and Whitaker, S., 1984, "Heat and Mass Transfer in Porous Media," in *Fundamentals of Transport Phenomena in Porous Media*, Bear and Corapcioglu, eds., Martinus Nijhoff, 121–198.

Chang, H.-C., 1982, "Multi-Scale Analysis of Effective Transport in Periodic Heterogeneous Media," *Chem. Eng. Comm.*, 15, 83–91.

Chang, S.-H. and Slattery, J., 1988, "A New Description for Dispersion," *Transport in Porous Media*, 3, 515–527.

Chen, C. K. and Tien, C.-L., 1973, "Conductance of Packed Spheres in Vacuum," *ASME J. Heat Transfer*, 95, 302–308.

Cunningham, G. R. and Tien, C.-L., 1978, "Heat Transfer in Microsphere Insulation in the Presence of a Gas," in *Proceedings of 15th International Thermal Conductivity Conference*, Plenum, 325–337.

Giedt, W. H. and Willis, D. R., 1985, Rarefied Gases, *Handbook of Heat Transfer, Fundamentals*, Rohsenow, W. M. et al., Editors, Second Edition, McGraw-Hill.

Glatzmaier, G. C. and Ramirez, W. F., 1988, "Use of Volume Averaging for the Modeling of Thermal Properties of Porous Materials," *Chem. Engng. Sci.*, 43, 3157–3169.

Hadley, G. R., 1986, "Thermal Conductivity of Packed Metal Powders," *Int. J. Heat Mass Transfer*, 29, 909–920.

Hashin, Z. and Shtrikman, S., 1962, "A Variational Approach to the Theory of the Effective Magnetic Permeability of Multiphase Materials," *J. Appl. Phys.*, 10, 3125–3131.

Haughey, D. P. and Beveridge, G. S. G., 1969, "Structural Properties of Packed Beds—A Review," *Can. J. Chem. Eng.*, 47, 130–140.

Hrubesh, L. W. and Pekala, R. W., 1994, "Thermal Properties of Organic and Inorganic Aerosols," *J. Mat. Sci.*, 9, 731–738.

Hsu, C. T., Cheng, P. and Wong, K. W., 1994, "Modified Zehnder–Schlünder Models for Stagnant Thermal Conductivity of Porous Media," *Int. J. Heat Mass Transfer*, 37, 2751–2759.

Hsu, C. T., Cheng, P. and Wong, K. W., 1995, "A Lumped Parameter Model for Stagnant Thermal Conductivity of Spatially Periodic Porous Media," *ASME J. Heat Transfer*, to appear.

Koh, J. C. Y. and Fortinn, A., 1973, "Prediction of Thermal Conductivity and Electrical Resistivity of Porous Metallic Materials," *Int. J. Heat Mass Transfer*, 16, 2013–2022.

Krupiczka, R., 1967, "Analysis of Thermal Conductivity in Granular Materials," *Int. Chem. Engng.*, 7, 122–144.

Kunii, D. and Smith, J. M., 1960, "Heat Transfer Characteristics of Porous Rocks," *AIChE J.*, 6, 71–78.

Lienhard, J. H. and Davis, L. B., 1971, "An Extension of Statistical Mechanics to the Description of a Broad Class of Macroscopic Systems," *J. Appl. Math. Phys.*, 22, 85–96.

Luikov, A. V., Shashkov, A. G., Vasiliev, L. L., and Fraiman, Yu. E., 1968, "Thermal Conductivity of Porous Systems," *Int. J. Heat Mass Transfer*, 11, 117–140.

Melanson, M. M. and Dixon, A. G., 1985, "Solid Conduction in Low d_t/d_p Beds of Spheres, Pellets and Rings," *Int. J. Heat Mass Transfer*, 28, 383–394.

Miller, M., 1969, "Bounds for Effective Electrical, Thermal and Magnetic Properties of Heterogeneous Materials," *J. Math. Phys.*, 10, 1988–2004.

Nayak, L. and Tien, C.-L., 1978, "A Statistical Thermodynamic Theory for Coordination–Number Distribution and Effective Thermal Conductivity of Random Packed Beds," *Int. J. Heat Mass Transfer*, 21, 669–676.

Nimick, F. B. and Leith, J. R., 1992, "A Model for Thermal Conductivity of Granular Porous Media," *ASME J. Heat Transfer*, 114, 505–508.

Nozad, I., 1983, *An Experimental and Theoretical Study of Heat Conduction in Two- and Three-Phase Systems*, Ph.D. thesis, University of California, Davis.

Nozad, I., Carbonell, R. G., and Whitaker, S., 1985, "Heat Conduction in Multi-Phase Systems I: Theory and Experiments for Two-Phase Systems," *Chem. Engng. Sci.*, 40, 843–855.

Ochoa-Tapia, J. A., Stroeve, P., and Whitaker, S., 1994, "Diffusion Transport in Two-Phase Media: Spatially Periodic Models and Maxwell's Theory for Isotropic and Anisotropic Systems," *Chem. Engng. Sci.*, 49, 709–726.

Ofuchi, K. and Kunii, D., 1965, "Heat Transfer Characteristics of Packed Beds with Stagnant Fluids," *Int. J. Heat Mass Transfer*, 8, 749–757.

Ogniewicz, Y. and Yovanovich, M. M., 1978, "Effective Conductivity of Regularly Packed Spheres: Basic Cell Model with Constriction," *Prog. Astronaut. Aeronaut.*, 60, 209–228.

Prasad, V., Kladas, N., Bandyopadhaya, A., and Tian, Q., 1989, "Evaluation of Correlations for Stagnant Thermal Conductivity of Liquid-Saturated Porous Beds of Spheres," *Int. J. Heat Mass Transfer*, 32, 1793–1796.

Prat, M., 1989, "On the Boundary Conditions at the Macroscopic Level," *Transp. Porous Media*, 4, 259–280.

Prat, M., 1990, "Modelling of Heat Transfer by Conduction in a Transition Region Between a Porous Medium and an External Fluid," *Transp. Porous Media*, 5, 71–95.

Quintard, M. and Whitaker, S., 1993, "Transport in Ordered and Disordered Porous Media: Volume–Averaged Equations, Closure Problems, and Comparison with Experiment," *Chem. Engng. Sci.*, 48, 2537–2564.

Ryan, D., Carbonell, R. G., and Whitaker, S., 1980, "Effective Diffusivities for Catalyst Pellets under Reactive Conditions," *Chem. Engng. Sci.*, 35, 10–16.

Sahraoui, M. and Kaviany, M., 1993, "Slip and No-Slip Temperature Boundary Conditions at Interface of Porous, Plain Media: Conduction," *Int. J. Heat Mass Transfer*, 36, 1019–1033.

Sangani, A. S. and Acrivos, A., 1983, "The Effective Conductivity of a Periodic Array of Spheres," *Proc. Roy. Soc.* (London), A386, 263–275.

Schulgasser, K., 1976, "On the Conductivity of Fiber-Reinforced Materials," *J. Math. Phys.*, 17, 382–387.

Shonnard, D. R. and Whitaker, S., 1989, "The Effective Thermal Conductivity for a Point Contact Porous Medium: An Experimental Study," *Int. J. Heat Mass Transfer*, 32, 503–512.

Slattery, J. C., 1981, *Momentum, Energy, and Mass Transfer in Continua*, Second Edition, R. F. Krieger.

Springer, G. S., 1971, "Heat Transfer in Rarefied Gases," *Advan. Heat Transfer*, 7, 163–218.

Stark, C. and Frickle, J., 1993, "Improved Heat Transfer Models to Fibrous Insulations," *Int. J. Heat Mass Transfer*, 36, 617–625.

Tien, C.-L. and Cunningham, G. R., 1976, "Glass Microsphere Cryogenic Insulation," *Cryogenics*, 16, 583–586.

Timoshenko, S. P. and Goodier, J. N., 1970, *Theory of Elasticity*, article 140, McGraw-Hill.

Torquato, S., 1991, "Random Heterogeneous Media: Microstructure and Improved Bounds on Effective Properties," *Appl. Mech. Rev.*, 44, 37–76.

Vafai, K., 1980, "Some Fundamental Problems in Heat and Mass Transfer Through Porous Media," Ph.D. thesis, University of California–Berkeley.

Wolf, J. R. and Strieder, W. C., 1994, "Pressure–Dependent Gas Heat Transport in a Spherical Pore," *AIChE J.*, 40, 1287–1296.

Woodside, W. and Messer, J. H., 1961, "Thermal Conductivity of Porous Media: I. Unconsolidated Sands" and "II. Consolidated Rocks," *J. Appl. Phys.*, 32, 1688–1706.

Zehnder, P. and Schlünder, E. U., 1970, "Thermal Conductivity of Granular Materials at Moderate Temperatures (in German), *Chemie. Ingr.-Tech.*, 42, 933–941.

Zeng, S. Q., Hurt, A. J., Cao, W., and Grief, R., "Pore Size Distribution and Apparent Gas Conductivity of Silica Aerogel," *ASME J. Heat Transfer*, 116, 756–759.

Zheng, L. and Strieder, W., 1994, "Knudsen Void Gas Heat Transport in Fibrous Media," *Int. J. Heat Mass Transfer*, 37, 1433–1440.

Zimmerman, R. W., 1987, "Thermal Conductivity of Fluid-Saturated Rocks," *J. Pet. Sci. Eng.*, 3, 219–227.

4

Convection Heat Transfer

As we consider *simultaneous* fluid flow and heat transfer in porous media, the role of the *macroscopic* (Darcean) and *microscopic* (pore-level) velocity fields on the temperature field needs to be examined. Experiments have shown that the *mere* inclusion of $\mathbf{u}_D \cdot \nabla \langle T \rangle$ in the energy equation does not accurately account for *all* the *hydrodynamic effects*. The *pore-level* hydrodynamics also influence the temperature field. Inclusion of the effect of the pore-level velocity nonuniformity on the temperature distribution (called the *dispersion effect* and generally included in the diffusion transport) is the *main* concern in this chapter.

We begin with Taylor's insightful look at the hydrodynamic dispersion in *tubes* and then examine this phenomenon in *porous media*. Over the last three decades continuous contributions have been made to this area: the pore-level fluid mechanics has been treated with and without the inertial effects; the treatment for the ordered and disordered arrangement of spheres in the bulk (i.e., away from the bounding surfaces of the medium) has advanced significantly, and the treatment of dispersion near the bounding surfaces has begun.

After reviewing Taylor's treatment of dispersion in tubes, we examine the local volume averaging of the energy equation. The available treatments and closure conditions for periodic structures and the predicted dispersion tensor are examined and comparisons with the experimental results are made. The treatment for random structures including the earlier and less-rigorous treatments are then considered and again the closed-form results for the dispersion tensor are compared with the experimental results. We then look at the experimental methods used for the determination of the dispersion coefficients. We end with examination of the anisotropy and nonuniformity of the dispersion tensor near the bounding surfaces of porous media.

4.1 Dispersion in a Tube—Hydrodynamic Dispersion

In order to *illustrate* the role of the velocity *nonuniformities* in the distribution of the heat content of fluids, consider the introduction of a temperature *nonuniformity* to a fluid flowing inside a circular tube. The tube wall thermal conductivity is assumed to be zero (i.e., no radial and axial conduction in the tube wall). Figure 4.1(a) shows a schematic of the problem. The

(a) Initial Disturbance (b) Purely Hydrodynamic Dispersion

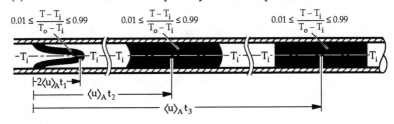

(c) Molecular Diffusion - Hydrodynamic Dispersion

(d) Distribution of Asymptotic Temperature and Velocity Deviations

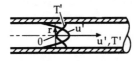

Figure 4.1 Spreading of the heat content in a tube caused by a net fluid flow, the presence of a radial velocity nonuniformity, and by molecular diffusion.

laminar velocity field is assumed to be *fully developed* and given by the Poiseuille-Hagen distribution, i.e.,

$$u = 2\langle u \rangle_A \left(1 - \frac{r^2}{R^2}\right), \tag{4.1}$$

where $\langle u \rangle_A$ is the *average axial velocity* in the x-direction and R is the radius of the tube. For $t < 0$, the temperature throughout is T_i. At $t = 0$, the temperature of the region $0 \leq r \leq R$, $x_o \leq x \leq x_o + \delta$ is suddenly raised to $T_o > T_i$. The temperature distribution within the fluid for $t > 0$, $T(r, x, t)$, is of interest. This problem was considered by Taylor (1953, 1954a) and later by Aris (1956) and reviews are given by Nunge and Gill (1969), Fried and Combarnous (1971), and Quintard and Whitaker (1993). Recent extensions of the Taylor dispersion in tubes are discussed by Yuan et al. (1991) and Batycky et al. (1993). We consider first the case of $k_f = 0$ and then examine the role of $k_f \neq 0$.

4.1.1 NO MOLECULAR DIFFUSION

When no molecular diffusion is present, the region of $T = T_o$, which was initially a disk of thickness δ, will be distorted into a *paraboloid* located downstream and between $x = 2\langle u \rangle_A t \left(1 - r^2/R^2\right)$ and $x = 2\langle u \rangle_A t \left(1 - r^2/R^2\right) + \delta$. Taylor (1953) examines this case for $\delta/R \ll 1$. Figure 4.1(b) shows this paraboloid, which results from a *purely hydrodynamic dispersion*.

The relative heat content of the cylindrical shell (integrated axially) between r and $r + \Delta r$ *remains* constant and is given by $2\pi r (\rho c_p)_f (T_o - T_i) \delta \Delta r$. This is due to lack of any radial velocity or molecular diffusion. However, when not integrated in the axial direction, a radial nonuniformity in temperature develops that was *not* initially present. Note that this nonuniformity does *not* develop for *plug flow*.

The axial distribution of the heat content can be examined through the mean relative temperature. On the paraboloid, a small displacement Δr around r results in an axial displacement of $\Delta x = (-2\langle u \rangle_A t/R^2) 2r \Delta r$. The heat content between $x + \Delta x$, for $\Delta x \gg \delta$ is $2\pi r (\rho c_p)_f (T_o - T_i) \delta (dr/dx) \Delta x$. Next, the *area-* and *segment-averaged relative* temperature is found for the volume $\pi R^2 \Delta x$. This is

$$\langle\langle T \rangle_A - T_i \rangle_{\Delta x} = \overline{T} - T_i = \frac{1}{-\pi R^2 \Delta x} \int_0^R 2\pi r (T_o - T_i) \delta \left(\frac{-R^2 \Delta x}{4\langle u \rangle_A t\, r}\right)$$

$$= \frac{(T_o - T_i)\delta}{2\langle u \rangle_A t}. \quad (4.2)$$

This shows that the heat content is *uniformly spread* over $0 \le x \le 2\langle u \rangle_A t$.

This example demonstrates how an upstream introduced, radially uniform disturbance can be *distorted (dispersed)* by the presence of a velocity nonuniformity. The *discrete* initial temperature distribution chosen is *not* required for the occurrence of dispersion, and other initial temperature distributions can also be used. When dealing with *both* molecular *diffusion* and *hydrodynamic dispersion*, the initial discontinuities are used to illustrate the role of the molecular diffusion in distorting these discontinuities. The role of molecular diffusion is considered next.

4.1.2 MOLECULAR DIFFUSION INCLUDED

For $k_f \ne 0$, the energy equation becomes

$$(\rho c_p)_f \frac{\partial T}{\partial t} + 2(\rho c_p)_f \langle u \rangle_A \left(1 - \frac{r^2}{R^2}\right) \frac{\partial T}{\partial x}$$

$$= k_f \left(\frac{\partial^2 T}{\partial r^2} + \frac{1}{r}\frac{\partial T}{\partial r} + \frac{\partial^2 T}{\partial x^2}\right). \quad (4.3)$$

This describes the behavior of $T(r,x,t)$. We note that the *Peclet number* $Pe = (\rho c_p)_f \langle u \rangle_A / (k_f / R)$ is the important dimensionless parameter signifying the relative strength of convective and molecular diffusion transports. Figure 4.1(c) gives a schematic of the evolution of the temperature field obtained from (4.3) for a given Pe for the given initial temperature disturbance. Note that since $k_f \neq 0$ the condition of $\partial T/\partial r(r = R) = 0$ due to $k_s = 0$ causes *radial* changes in T near $r = R$. The paraboloid just discussed for $k_f = 0$ now *spreads* both *axially* and *radially* because of the molecular diffusion ($t = t_1$). Later, these lateral and axial diffusions change the paraboloid into a *diluted plug*. At $t = t_2$, asymmetry around $x = \langle u \rangle_A t_2$ is still present, while for larger elapsed time t_3 this asymmetry *disappears*. For $t \geq t_3$, the initial disk-like disturbance is returned except that the initial thickness δ is turned into a much larger *ever-increasing* thickness (the initial discontinuity in T disappeared due to diffusion). This is the *asymptotic* behavior for large elapsed times. We now examine this asymptote.

4.1.3 Asymptotic Behavior for Large Elapsed Times

Taylor and Aris have found expressions for the rate of spreading of the thick disk (or plug) observed for $t \geq t_3$. Nunge and Gill solve (4.3) numerically and show that the required elapsed time for the asymptotic behavior is nearly the diffusion time for the distance R, i.e.,

$$t_3 \simeq \frac{0.8 R^2}{\left(\dfrac{k}{\rho c_p}\right)_f}. \tag{4.4}$$

However, the experimental results of Han et al. (1985) suggest a value much larger than unity (instead of 0.8). The asymptotic behavior has some very interesting features, a few of which are discussed here.

- The *center* of the temperature disturbance (this is the region confined between $\beta \leq (T - T_i)/(T_o - T_i) \leq 1 - \beta$ say $\beta = 0.01$) moves at the *mean velocity* of the *fluid* $\langle u \rangle_A$.

- Since the *centerline* velocity is twice that of the mean, the fluid particles moving at the center *arrive* in the disturbed region with temperature T_i, initially *undergo* an increase in T, begin to lose (for $T_i < T_o$) their internal energy, and finally *leave* again at temperature T_i.

- The temperature distribution in the disturbed region is *symmetrical* around the center of this region. This is interesting, considering that the flow is *unidirectional*.

Under the assumptions of negligible axial conduction ($\partial^2/\partial x^2 = 0$) and the existence of an asymptotic behavior ($\partial/\partial t \to 0$), Taylor (1953) solves (4.3).

4.1 Dispersion in a Tube—Hydrodynamic Dispersion   161

The axial coordinate is *transformed* according to $x_1 = x - \langle u \rangle_A t$, and then (4.3), subject to the assumptions mentioned earlier, becomes

$$2\left(\frac{\rho c_p}{k}\right)_f \langle u \rangle_A \left(\frac{1}{2} - \frac{r^2}{R^2}\right) \frac{\partial T}{\partial x_1} = \frac{1}{r}\frac{\partial}{\partial r}\left(r \frac{\partial T}{\partial r}\right). \tag{4.5}$$

He further assumes that $\partial T/\partial x_1 = constant$ and finds the solution to the differential equation subject to

$$T(r = 0, x_1) = T(x_1) \tag{4.6}$$

and

$$\frac{\partial T}{\partial r}(r = R, x_1) = 0. \tag{4.7}$$

The solution is

$$T = T(x_1) + T'(r) = T(x_1) + \frac{R^2 \langle u \rangle_A}{4\left(\frac{k}{\rho c_p}\right)_f} \frac{\partial T}{\partial x_1}\left(\frac{r^2}{R^2} - \frac{1}{2}\frac{r^4}{R^4}\right)$$

$$= T(x_1) + f(r)\frac{\partial T}{\partial x_1}. \tag{4.8}$$

This radial *deviation* of the temperature $T' = T'(r)$ which, although small, is required for the realization of the features of the disturbed region. Note that as with the *transformation* (i.e., *closure*) used for volume averaging discussed in Section 3.2.1, here the gradient of the *area-averaged* temperature is used along with a transformation function f to describe the deviation in temperature. The radial distributions of u' and T' are also shown in Figure 4.1.

The heat flow across x_1 is

$$\frac{Q}{(\rho c_p)_f} = 2\pi \int_0^R u' T' r\, dr 4\pi \langle u \rangle_A \int_0^R [T - T(x_1)]\left(\frac{1}{2} - \frac{r^2}{R^2}\right) r\, dr$$

$$= -\frac{\pi R^4 \langle u \rangle_A^2}{48\left(\frac{k}{\rho c_p}\right)_f} \frac{\partial T(x_1)}{\partial x_1} \simeq -\frac{\pi}{48} \frac{R^4 \langle u \rangle_A^2}{\left(\frac{k}{\rho c_p}\right)_f} \frac{\partial \langle T \rangle_A}{\partial x_1}, \tag{4.9}$$

where the *area-averaged* temperature was defined as

$$\langle T \rangle_A = \frac{2}{R^2} \int_0^R T r\, dr. \tag{4.10}$$

We have also assumed that the radial variation of T is small.

The heat flux is given as

$$\frac{Q}{(\rho c_p)_f \pi R^2} = -\frac{R^2 \langle u \rangle_A^2}{48\left(\frac{k}{\rho c_p}\right)_f} \frac{\partial \langle T \rangle_A}{\partial x_1}. \tag{4.11}$$

This corresponds to an axial diffusive transport with the *axial dispersion coefficient* (or *Taylor* dispersion coefficient) D_{xx}^d given by

$$D_{xx}^d = \frac{R^2 \langle u \rangle_A^2}{48 \left(\frac{k}{\rho c_p}\right)_f}. \tag{4.12}$$

In dimensionless form using the fluid molecular diffusivity $\alpha_f = (k/\rho c_p)_f$, we have

$$\frac{D_{xx}^d}{\alpha_f} = \frac{1}{48} Pe^2. \tag{4.13}$$

For $Pe > 48^{1/2}$, the dispersion contribution is *larger* than the molecular diffusion, where $Pe = Re \, Pr = \langle u \rangle_A R/\alpha_f$ and $Re \leq 1150$ for laminar flow. The penetration of the disturbed region on both sides of the x_1 plane can be *remarkably* fast. Consider molecular diffusion of heat in a semi-infinite medium subject to a sudden change in the temperature at location $x_1 = 0$. Then the penetration depth based on $(T - T_i)/(T_o - T_i) = 0.99$ is $3.64(\alpha_f t)^{1/2}$. The corresponding molecular diffusion-hydrodynamic dispersion results in the penetration depth of $3.64(Pe/48^{1/2})(\alpha_f t)^{1/2}$. The *attraction* of representing dispersion as an enhancement in diffusion is apparent. Aris shows that the *total axial diffusion coefficient* for the tube is

$$\frac{D_{xx}}{\alpha_f} = \underbrace{1}_{\text{molecular diffusion}} + \underbrace{\frac{1}{48} Pe^2}_{\substack{\text{molecular diffusion-}\\ \text{hydrodynamic}\\ \text{dispersion}}} = 1 + \frac{D_{xx}^d}{\alpha_f}. \tag{4.14}$$

This Pe^2 relationship does not depend on a specific velocity nonuniformity. However, the proportionality constant decreases as the velocity distribution becomes more uniform, and, as expected, this constant is zero for the plug flow.

Other solutions to (4.3) reported by Nunge and Gill show that allowing for the development of the velocity field, superimposing small amplitude pulsations, and using channels (flow through parallel plates) or concentric/eccentric annuli does not change the Pe^2 dependency.

The effect of the velocity nonuniformities in the enhancement of diffusion (or spreading) of heat content has been demonstrated through the preceding example, where analytical treatments are possible. Dispersion occurring in flow through disordered or periodic structures is based on the same interactions between the velocity and temperature nonuniformities. However, due to the *multidimensionality* of the temperature and velocity fields, the thermal dispersion coefficient must be presented as a *tensor*, and numerical treatments (at least partly) are required.

4.1.4 TURBULENT FLOW

Since dispersion in a tube for laminar flows is partially controlled by the lateral molecular diffusion, Taylor (1954b) considered the effect of an *additional lateral* mixing present in the turbulent flows. He uses the same formulation as that used for laminar flow along with the experimentally determined *mean velocity distribution*. A brief review of his analysis is given later.

Using the *turbulence closure* and the use of the *thermal eddy diffusivity* α_t, the energy equation for internal fully developed turbulent flows with *negligible* axial conduction (he shows that this makes only a small contribution) is

$$\frac{\partial \overline{T}}{\partial t} + \overline{u}\frac{\partial \overline{T}}{\partial x} = \frac{1}{r}\frac{\partial}{\partial r}\left[(\alpha_f + \alpha_t)r\frac{\partial \overline{T}}{\partial r}\right], \tag{4.15}$$

where $r = 0$ is at the center of the tube and $\overline{u}$ *is the mean velocity* (as compared to the *fluctuating* component of the velocity) and $\overline{T}$ is the *mean temperature*. He assumes *full* analogy between turbulent *heat* and *momentum* transfer and uses the momentum turbulence closure to arrive at

$$\alpha_f + \alpha_t = \frac{\overline{\tau}}{\rho\left|\frac{\partial \overline{u}}{\partial r}\right|} = \frac{\overline{q}}{(\rho c_p)_f \left|\frac{\partial \overline{T}}{\partial r}\right|}, \tag{4.16}$$

where $\overline{\tau}$ is the mean shear stress at radius r. Then he uses the correlation available from the *measured mean velocity distribution*, i.e.,

$$\frac{\overline{u}(0) - \overline{u}}{\overline{u}_\tau} = f\left(\frac{r}{R}\right), \tag{4.17}$$

where $\overline{u}(0)$ is the *mean velocity at the centerline*, and $\overline{u}_\tau$ is the *friction velocity* $(\overline{\tau}_R/\rho)^{1/2}$, and $\overline{\tau}_R$ is the *wall shear stress*. In the distribution of mean velocity near the wall, $f(r/R)$ becomes the *von Karman logarithmic law*. The *area-average mean velocity* is given by $\langle \overline{u}\rangle_A = \overline{u}(0) - 4.25\overline{u}_\tau$. Also, the mean shear stress is distributed according to

$$\overline{\tau} = \overline{\tau}_R \frac{r}{R}. \tag{4.18}$$

Then, from (4.16) and (4.17), we have

$$\alpha_f + \alpha_t = \frac{r\overline{u}_\tau}{R\left|\frac{df}{dr}\right|}. \tag{4.19}$$

Using (4.17) and (4.19), the energy equation (4.15) becomes

$$\frac{R}{\overline{u}_\tau}\frac{\partial \overline{T}}{\partial t} - R\left[f\left(\frac{r}{R}\right) - \frac{\overline{u}(0)}{\overline{u}_\tau}\right]\frac{\partial \overline{T}}{\partial x} = \frac{1}{r}\frac{\partial}{\partial r}\left(\frac{r^2}{\left|\frac{df}{dr}\right|}\frac{\partial \overline{T}}{\partial r}\right). \tag{4.20}$$

Now, as before, we define $x = x_1 + \langle \bar{u} \rangle_A t$ and take $\partial \overline{T}/\partial t \to 0$ for the long-time asymptote. We also assume that $\partial T/\partial x_1$ is independent of x and r, i.e., is a constant. Then, we have

$$\frac{\partial \overline{T}}{\partial x_1} = \text{constant} = \frac{1}{R\left[f\left(\frac{r}{R}\right) - 4.25\right]} \frac{1}{r} \frac{\partial}{\partial r}\left(\frac{r^2}{\left|\frac{df}{dr}\right|} \frac{\partial \overline{T}}{\partial r}\right). \quad (4.21)$$

Taylor numerically integrates this using the correlation for $f(r/R)$ and the boundary conditions given by (4.6) and (4.7). Following the same approach as in the previous section, he defines the dispersion coefficient as in (4.14) and arrives at

$$\frac{D_{xx}^d}{\alpha_f} = 10.06 \frac{R \bar{u}_\tau}{\alpha_f}, \quad (4.22)$$

where $u_\tau = u_\tau (Re = \langle \bar{u} \rangle_A R/\nu) = \langle \bar{u} \rangle_A [\gamma(Re)/2]^{1/2}$ and $\gamma(Re)$ is the *resistance coefficient*. He shows that this coefficient takes on values in the range $0.047 \leq \gamma^{1/2} \leq 0.118$ for $1.4 \times 10^3 \leq 2Re \leq 5.4 \times 10^6$. Then the dispersion coefficient can be written in terms of the average mean velocity as

$$\frac{D_{xx}^d}{\alpha_f} = 7.14 \frac{R \langle \bar{u} \rangle_A}{\alpha_f} \gamma^{1/2} = 7.14 Pe \gamma^{1/2}. \quad (4.23)$$

His experiments verify this relation as well as the predicted spread rate (not discussed here, but found in Taylor's paper).

The difference between dispersion in laminar and turbulent flows (leading to Pe^2 and Pe dependence, respectively) is associated with the additional lateral mixing due to the turbulent eddy diffusion. Note that the results of Aris show that the velocity profile does not change the power of Pe for laminar flow. As will be shown (observed first experimentally and then developed theoretically) for *ordered, in-line arranged* matrices nearly simulating a tube flow, the dispersion coefficient is *proportional* to Pe^2, while for *ordered, staggered arrangements* and *disordered* matrices this becomes nearly Pe. A phenomenon *similar* to that for laminar and turbulent flows. The enhanced lateral transport in the disordered media (as in turbulent flows) is responsible for this difference.

4.2 Dispersion in Porous Media

Dispersion in porous media is different than that just reviewed for the hydrodynamically fully-developed flows in regularly shaped constant cross-section-area straight channels. A schematic of an ideal unit cell, representing an element of a solid matrix through which fluid flows with an axial temperature gradient present across it, is given in Figure 4.2. We expect the dispersion in porous media to have the following features.

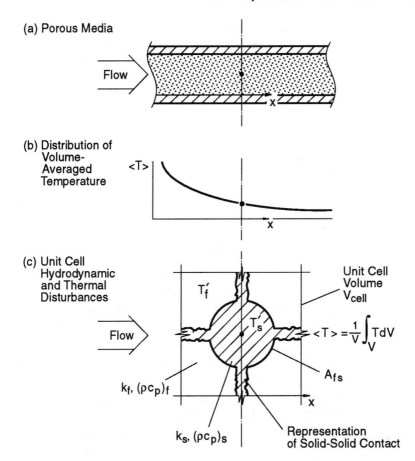

Figure 4.2 Macroscopic and microscopic renderation of flow and heat transfer in an ordered porous medium. A schematic of a unit cell in a periodic structure is shown along with the local volume-averaged and the local deviation of temperature.

- **Pr_f**: At first, we may expect that dispersion would depend on the Prandtl number ν/α_f, which signifies the ratio of the momentum to thermal boundary-layer thickness. However, in contrast to heat transfer in external flows, no simultaneous boundary-layer growth occurs around the particles. Therefore, the Prandtl number dependence is expected to be *weak*.

- **Pe, Re, ϵ, structure**: Dispersion should depend on the pore-level hydrodynamics. The pore structure, pore velocity, and upstream conditions determine whether they are recirculation zones (closed streamlines), dead ends, flow reversals, etc. The classification of the structure to *ordered* and *disordered* media allows for further specification of the pore-level hydrodynamics. In ordered and isotropic media and when simple unit-cell structures with convenient symmetries are present, the flow field can be analyzed for various pore Reynolds numbers. The bulk of the available experimental results show that the Re dependency is rather weak and $Pe = Re\, Pr$ can *approximately* express the pore-level hydrodynamics and heat transfer. This is at least surprising because *several flow regimes* exist in the pore depending on Re. This Pe dependence is characteristic of *fully-developed* velocity and temperature fields.

- k_s/k_f: Dispersion should depend on the ratio of the molecular thermal resistances because the temperature field in the pore is influenced by the solid conductivity. If the ratio of the thermal penetration time into the solid phase to the residence time in the pore, $(d^2/\alpha_s)/(d/\varepsilon u_D)$ $= \varepsilon u_D d/\alpha_s = \varepsilon Pe\, \alpha_f/\alpha_s$ is large, the solid contribution is expected to be small and vice versa. Similarly, the conduction-to-convection time-scale ratio in the fluid leads to $(d^2/\varepsilon^2 \alpha_f)/(d/\varepsilon u_D) = u_D d/(\varepsilon \alpha_f)$ $= (1/\varepsilon)Pe$. Therefore, α_f/α_s (or k_s/k_f for steady-state) is expected to influence the dispersion tensor.

- $(\rho c_p)_s/(\rho c_p)_f$: Dispersion should depend on the ratio of the volumetric heat capacities. The transient temperature distribution in the unit cell (pore plus solid) depends on the extent of the ability of the solid to store/release heat. Large values of $(\rho c_p)_s$ dampen the temporal temperature disturbances.

Based on this, we expect the dispersion tensor to have a functional form

$$\frac{\mathbf{D}^d}{\alpha_f} = \frac{\mathbf{D}^d}{\alpha_f}\left(Re,\ Pr,\ \varepsilon,\ \text{structure},\ \frac{k_s}{k_f},\ \text{and}\ \frac{(\rho c_p)_s}{(\rho c_p)_f}\right). \tag{4.24}$$

In the following, various analyses of the dispersion phenomenon in porous media are discussed. The local volume averaging applied to fluid flow and conduction in earlier chapters, will be applied here to convection. Alternatively the *homogenization theory* discussed in Chapter 2 can be applied, as done by Mei (1992). The numerical results by Lee et al. (1995), using this method, are in good agreement with the experimental results as are those obtained from the application of the local volume averaging. Next the experimental methods used for the determination of the dispersion tensor are reviewed and some of these experimental results are presented.

4.3 Local Volume Average for Periodic Structures

The local volume-averaging technique can be applied to *ordered* and *disordered* porous media. As with conduction, the ratio k_s/k_f can also be any arbitrary number. Here we begin by reviewing an *existing* formulation and numerical results for the case of *periodic structures* with $k_s = 0$. Since dispersion occurring in heat transfer is *analogous* to that in *mass transfer*, the case of $k_s = 0$ corresponds to an impermeable (to species) solid phase. Since mass transfer is also of interest and experiments can be readily performed, this case ($k_s = 0$) has been studied *extensively*.

4.3.1 LOCAL VOLUME AVERAGING FOR $k_s = 0$

The principle of volume averaging and the requirement of the existence of the local thermal equilibrium between the fluid and solid phases were discussed in Section 3.1. In addition to the diffusion time and length scale requirements for the existence of the local thermal equilibrium, the residence time scales (the time it takes for a fluid particle to cover the length scales, $K^{1/2}$, d, ℓ, and L) must be included in the length and time scale requirements given in Section 3.1.

The following are based on the development given by Carbonell and Whitaker (1983) and that given by Slattery (1981, p. 406). The objective is to average the energy equation over the fluid phase (here $k_s = 0$, then the solid phase energy equation is not needed) so that the various contributions of the fluid flow including dispersion can be expressed as integrals over the representative elementary volume.

We begin by writing (4.3) in the *vectorial* form, i.e.,

$$\frac{\partial T}{\partial t} + \nabla \cdot \mathbf{u} T = \nabla \cdot \alpha_f \nabla T \quad \text{in} \quad V_f, \tag{4.25}$$

where $(\rho c_p)_f$ is assumed to be constant. For $k_s = 0$, we have

$$\mathbf{u} = 0 \quad \text{and} \quad \mathbf{n} \cdot \nabla T = 0 \quad \text{on} \quad A_{fs}, \tag{4.26}$$

where, as defined before, V_f is the volume occupied by the fluid and A_{fs} is the interfacial area between the fluid and solid phases.

In accordance with the definition of the representative elementary volume, by using (2.71) the phase average of (4.25) becomes

$$\frac{\partial \langle T \rangle}{\partial t} + \langle \nabla \cdot \mathbf{u} T \rangle = \langle \nabla \cdot \alpha_f \nabla T \rangle. \tag{4.27}$$

Using the theorem (2.85), and by applying the boundary condition (4.26), we have

$$\frac{\partial \langle T \rangle}{\partial t} + \nabla \cdot \langle \mathbf{u} T \rangle = \nabla \cdot \langle \alpha_f \nabla T \rangle. \tag{4.28}$$

Now, using (2.84) we obtain

$$\frac{\partial \langle T \rangle}{\partial t} + \nabla \cdot \langle \mathbf{u} T \rangle = \nabla \cdot \left(\alpha_f \nabla \langle T \rangle + \frac{\alpha_f}{V} \int_{A_{fs}} \mathbf{n} T \, dA \right). \tag{4.29}$$

As in (3.12), using the *intrinsic phase average* defined by (3.11), we introduce

$$T = \langle T \rangle^f + T', \quad (4.30)$$

$$\mathbf{u} = \langle \mathbf{u} \rangle^f + \mathbf{u}', \quad (4.31)$$

where we have

$$\frac{1}{V_f} \int_{V_f} T' \, dV = \langle T' \rangle^f = 0. \quad (4.32)$$

The operant in the convective term can be written as

$$\frac{1}{V} \int_V \mathbf{u} T \, dV = \varepsilon \frac{1}{V_f} \int_{V_f} \mathbf{u} T \, dV = \varepsilon \langle \mathbf{u} \rangle^f \langle T \rangle^f + \varepsilon \langle \mathbf{u}' T' \rangle^f. \quad (4.33)$$

Then we have

$$\frac{\partial \varepsilon \langle T \rangle^f}{\partial t} + \nabla \cdot [\varepsilon \langle \mathbf{u} \rangle^f \langle T \rangle^f] = \nabla \cdot \left[\varepsilon \alpha_f \nabla \langle T \rangle^f + \frac{\alpha_f}{V} \int_{A_{fs}} \mathbf{n} T' \, dA \right]$$
$$- \nabla \cdot \left[\varepsilon \langle \mathbf{u}' T' \rangle^f \right]. \quad (4.34)$$

As was done in the treatment of conduction (Section 3.2) we now introduce a *closure constitutive equation* (based on an analogy with the treatment of turbulence) as

$$T' = \mathbf{b}(\mathbf{x}) \cdot \nabla \langle T \rangle^f, \quad (4.35)$$

where $\mathbf{b}(\mathbf{x})$ is a *vector function* that *transforms* the gradient of the intrinsic phase-averaged temperature into the local variation of the deviation from the averaged temperature.

When $\nabla \varepsilon$ is small such that locally ε can be taken as constant and by using $\nabla \cdot \langle \mathbf{u} \rangle^f = 0$ [from (2.88)], we have

$$\frac{\partial \langle T \rangle^f}{\partial t} + \langle \mathbf{u} \rangle^f \cdot \nabla \langle T \rangle^f = \nabla \cdot \left[\alpha_f \left(\mathbf{I} + \frac{1}{V_f} \int_{A_{fs}} \mathbf{n} \mathbf{b} \, dA \right) \cdot \nabla \langle T \rangle^f \right]$$
$$- \nabla \cdot \left[\langle \mathbf{u}' \mathbf{b} \rangle^f \cdot \nabla \langle T \rangle^f \right], \quad (4.36)$$

where we have used

$$\langle \mathbf{u}' T' \rangle^f = \left\langle \mathbf{u}' \mathbf{b} \cdot \nabla \langle T \rangle^f \right\rangle^f = \langle \mathbf{u}' \mathbf{b} \rangle^f \cdot \nabla \langle T \rangle^f, \quad (4.37)$$

which uses the fact that $\nabla \langle T \rangle^f$ is constant in the representative elementary volume.

Next, Carbonell and Whitaker define the *tortuosity tensor* as

$$\mathbf{L}_t^* = \frac{1}{V_f} \int_{A_{fs}} \mathbf{n} \mathbf{b} \, dA. \quad (4.38)$$

Note that the concept of *tortuosity* has meaning only when $k_s = 0$, i.e., the tracer can only travel through the fluid phase. Also, since $\mathbf{b} = \mathbf{b}(\mathbf{u})$, the tortuosity is a *function* of the velocity field. In order to be consistent with the definition given in the treatment of conduction, we assume that $\mathbf{L}_t^* = \mathbf{L}_t^*(\mathbf{u} = 0)$ so that *no hydrodynamic effects are included in* $\mathbf{L}_t^*$.

Next, the *dispersion tensor* is defined as

$$\mathbf{D}^d = -\frac{1}{V_f} \int_{V_f} \mathbf{u}' \mathbf{b} \, dV. \tag{4.39}$$

Then (4.36) becomes

$$\frac{\partial \langle T \rangle^f}{\partial t} + \langle \mathbf{u} \rangle^f \cdot \nabla \langle T \rangle^f = \nabla \cdot \left\{ \underbrace{\left[\alpha_f (\mathbf{I} + \mathbf{L}_t^*) + \mathbf{D}^d \right]}_{} \cdot \nabla \langle T \rangle^f \right\}, \tag{4.40}$$

where the underlined quantity is designated as $\mathbf{D}$ and is the *total effective thermal diffusivity tensor*, for $k_s = 0$.

A generalization of (4.40) can be made for $k_s \neq 0$ and under the assumption of the local thermal equilibrium and by adding the solid- and fluid-phase energy equations. Then, we have

$$\left[\varepsilon (\rho c_p)_f + (1 - \varepsilon)(\rho c_p)_s \right] \frac{\partial \langle T \rangle}{\partial t} + (\rho c_p)_f \mathbf{u}_D \cdot \nabla \langle T \rangle$$

$$= (\rho c_p)_f \nabla \cdot (\mathbf{D} \cdot \nabla \langle T \rangle), \tag{4.41}$$

where $\langle T \rangle^s = \langle T \rangle^f = \langle T \rangle$, and the *total diffusivity tensor* is

$$\mathbf{D} = \frac{\mathbf{K}_e}{(\rho c_p)_f} + \varepsilon \mathbf{D}^d. \tag{4.42}$$

Note that for the case of $k_s = 0$ under study, we have

$$\mathbf{K}_e = \varepsilon k_f (\mathbf{I} + \mathbf{L}_t^*) \quad \text{for} \quad k_s = 0, \ (\rho c_p)_s = 0 \tag{4.43}$$

and (4.40) is recovered.

The knowledge of $\mathbf{b}$ leads to the determination of $\mathbf{L}_t^*$ and $\mathbf{D}^d$, which then can be used in the energy equation (4.40). The vector $\mathbf{b}$, which is a function of position only, has a magnitude of the order of the representative elementary volume ℓ and is determined from the differential equation and boundary conditions for T'.

The equation for T' is found by using (4.34), (4.25), (4.30), and (4.31). The result is

$$\frac{\partial T'}{\partial t} + \mathbf{u}\cdot\nabla T' + \mathbf{u}'\cdot\nabla\langle T\rangle^f - \alpha_f\nabla^2 T'$$
$$= -\nabla\cdot\left(\alpha_f\frac{A_{fs}}{V_f}\langle \mathbf{n}\,T'\rangle^{fs}\right) + \nabla\cdot\langle \mathbf{u}'T'\rangle^f, \quad (4.44)$$

where
$$\langle \mathbf{n}\,T'\rangle^{fs} = \frac{1}{A_{fs}}\int_{A_{fs}}\mathbf{n}\,T'\,\mathrm{d}A. \quad (4.45)$$

The boundary condition (4.26) becomes
$$-\mathbf{n}\cdot\nabla T' = \mathbf{n}\cdot\nabla\langle T\rangle^f \quad \text{on } A_{fs}. \quad (4.46)$$

Through the following order-of-magnitude arguments, it can be shown that the right-hand side of (4.44) makes a negligible contribution. From (4.46)
$$\frac{\Delta T'}{\ell} \simeq \frac{\langle\Delta T\rangle^f}{L} \quad \text{or} \quad \Delta T' \simeq \frac{\ell}{L}\langle\Delta T\rangle^f, \quad (4.47)$$

which is consistent with the requirement for the local thermal equilibrium. We also have $\nabla\cdot\langle \mathbf{u}'T'\rangle^f \simeq u_p\Delta T'/L$ and $\mathbf{u}'\cdot\nabla\langle T\rangle^f \simeq u_p\Delta T_L/L$, which show that $\nabla\cdot\langle \mathbf{u}'T'\rangle^f$ is much smaller than $\mathbf{u}'\cdot\nabla\langle T\rangle^f$. Also

$$\nabla^2 T' \simeq \frac{\Delta T'}{\ell^2} \quad \text{while} \quad \nabla\cdot\frac{A_{fs}}{V_f}\langle \mathbf{n}\,T'\rangle_{fs} \simeq \frac{\Delta T'}{\ell L} \ll \frac{\Delta T'}{\ell^2}. \quad (4.48)$$

Furthermore, for the asymptotic behavior (large elapsed time), the time derivative is also small. This simplification reduces (4.44) to
$$\mathbf{u}\cdot\nabla T' + \mathbf{u}'\cdot\nabla\langle T\rangle^f = \alpha_f\nabla^2 T'. \quad (4.49)$$

Note that the same arguments used earlier leading to (4.49) are used by Taylor (1953) in his treatment of the dispersion in a tube.

Now, by substituting (4.35) for **b** in (4.49) and (4.46), we have
$$\mathbf{u}\cdot\nabla\left(\mathbf{b}\cdot\nabla\langle T\rangle^f\right) + \mathbf{u}'\cdot\nabla\langle T\rangle^f = \alpha_f\nabla^2\left(\mathbf{b}\cdot\nabla\langle T\rangle^f\right) \quad (4.50)$$

and
$$-\mathbf{n}\cdot\nabla\left(\mathbf{b}\cdot\nabla\langle T\rangle^f\right) = \mathbf{n}\cdot\nabla\langle T\rangle^f \quad \text{on } A_{fs}. \quad (4.51)$$

Next, the quantities inside the parentheses are expanded and approximated using an order-of-magnitude argument, i.e., that the derivatives of the averaged quantities are smaller than the derivative of the disturbed quantities (e.g., **b**). The results are

$$\mathbf{u}' + \mathbf{u}\cdot\nabla\mathbf{b} = \alpha_f\nabla^2\mathbf{b} \quad \text{in } V_f \quad (4.52)$$

and
$$-\mathbf{n} \cdot \nabla \mathbf{b} = \mathbf{n} \quad \text{on } A_{fs}. \tag{4.53}$$

Also, for periodic structures, we have
$$\mathbf{b}(\mathbf{x} + \boldsymbol{\ell}_i) = \mathbf{b}(\mathbf{x}), \quad i = 1, 2, 3, \tag{4.54}$$

where $\mathbf{x}$ is located on the surface of the unit cell and $\boldsymbol{\ell}_i$ is the *period* in the i direction.

Determination of $\mathbf{L}_t^*$ and $\mathbf{D}^d$ requires knowledge of $\mathbf{u}$ (and then $\mathbf{u}' = \mathbf{u} - \langle \mathbf{u} \rangle^f$) and $\mathbf{b}$ on the *unit-cell* (or *pore*) *level*. Solution of the Navier-Stokes equation over the cell leads to $\mathbf{u}$ and $\mathbf{u}'$. Then (4.52)–(4.54) are solved for $\mathbf{b}$.

4.3.2 REDUCTION TO TAYLOR-ARIS DISPERSION

For fully-developed laminar flow through a tube, (4.40) for a one-dimensional heat flow becomes

$$\frac{\partial \langle T \rangle^f}{\partial t} + \langle u \rangle^f \frac{\partial \langle T \rangle^f}{\partial x} = [\alpha_f (1 + L_{t,xx}^*) + D_{xx}^d] \frac{\partial^2 \langle T \rangle^f}{\partial x^2}, \tag{4.55}$$

where
$$L_{t,xx}^* = \frac{1}{V_f} \int_{A_{fs}} n_x b_x \, dA, \tag{4.56}$$

$$D_{xx}^d = -\frac{1}{V_f} \int_{V_f} u' b_x \, dV. \tag{4.57}$$

Since $n_x = 0$ ($\mathbf{n} = \mathbf{s}_r$ in a tube with $r = 0$ on the tube surface), $L_{t,xx}^* = 0$. From (4.52), we have

$$u' + u \frac{\partial b_x}{\partial x} = \alpha_f \left[\frac{\partial^2 b_x}{\partial x^2} + \frac{1}{r} \frac{\partial}{\partial r} \left(r \frac{\partial b_x}{\partial r} \right) \right]. \tag{4.58}$$

Also from (4.53), we have

$$-[(\mathbf{n} \cdot \nabla) \mathbf{b}]_x = -\frac{\partial b_x}{\partial r} = n_x = 0 \quad \text{on } A_{fs} \tag{4.59}$$

or
$$\frac{\partial b_x}{\partial r} = 0 \quad \text{at } r = R. \tag{4.60}$$

Since no axial periodicity exists, condition (4.54) can be interpreted as

$$\frac{\partial b_x}{\partial x} = 0 = \frac{\partial^2 b_x}{\partial x^2}, \tag{4.61}$$

implying an infinite period.

172   4. Convection Heat Transfer

An extra constraint is found from (4.35), i.e., since no axial periodicity exists, the volume integral becomes an area integral. This leads to

$$\langle b_x \rangle^f = 0 = \frac{2}{R^2} \int_0^R b_x r \, dr. \qquad (4.62)$$

By using (4.61), (4.58) becomes

$$u' = \alpha_f \frac{1}{r} \frac{\partial}{\partial r} \left(r \frac{\partial b_x}{\partial r} \right). \qquad (4.63)$$

From the Poiseuille-Hagen distribution, we have for velocity derivation

$$u' = u - \langle u \rangle^f = \langle u \rangle^f \left(1 - 2 \frac{r^2}{R^2} \right). \qquad (4.64)$$

Note that in comparing (4.64) to (2.99), here the transformation tensor is only function of r. Now by solving for b_x in (4.63) using (4.64) and by determining the constants of the integration using (4.60) and (4.62), we have

$$b_x(r) = \frac{\langle u \rangle^f R^2}{4\alpha_f} \left(\frac{r^2}{R^2} - \frac{1}{2} \frac{r^4}{R^4} - \frac{1}{3} \right). \qquad (4.65)$$

When this is used in (4.57), the result is

$$\frac{\partial \langle T \rangle^f}{\partial t} + \langle u \rangle^f \frac{\partial \langle T \rangle^f}{\partial x} = \left(\alpha_f + \frac{\langle u \rangle^f \langle u \rangle^f R^2}{48 \alpha_f} \right) \frac{\partial^2 \langle T \rangle^f}{\partial x^2}. \qquad (4.66)$$

This is the Taylor-Aris result, with $D_\parallel^d = \langle u \rangle^f \langle u \rangle^f R^2/(48\alpha_f)$. Therefore, the Carbonell-Whitaker formulation for porous media reduces to the Taylor-Aris formulation for tubes.

4.3.3 EVALUATION OF u' AND b

Eidsath et al. (1983) consider a periodic structure made of cylindrical particles with in-line and staggered arrangements. These two-dimensional structures simulate porous media made of fibers. Regardless of the applicability of the models, we *expect* the numerical solutions for these *periodic* models to show the Pe^2 dependence of the dispersion coefficient.

(A) HYDRODYNAMICS

Determination of *velocity deviation* u' involves the solution of the Navier-Stokes equation for steady laminar flow through a two-dimensional periodic arrangement made of cylindrical particles (Figure 4.3). The boundary conditions are no-slip on A_{fs} and periodicity of the velocity field on the boundaries of the unit cell (this was discussed in connection with the

4.3 Local Volume Average for Periodic Structures   173

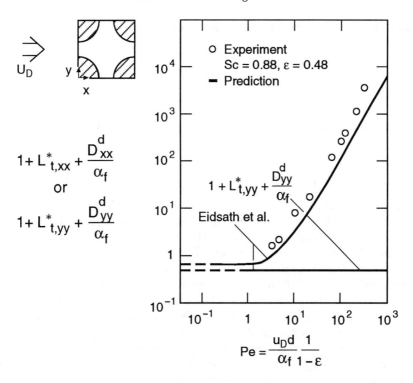

Figure 4.3 The predicted results of Eidsath et al. compared with the experimental results of Gunn and Pryce. (From Eidsath et al., reproduced by permission ©1983 Pergamon.)

determination of the Kozeny constant in Section 2.4.3). For flow over cylinders, there is a *strong dependence* on the Reynolds number (based on the cylinder diameter), i.e., the flow departs significantly from that of the creeping (Stokes) flow when Re_d is large.

(B) DETERMINATION OF b

This requires solution of (4.52)–(4.54) which in turn requires the solution for $\mathbf{u}'$. The Prandtl number $Pr = \nu/\alpha_f$ is introduced after the Reynolds number (the only dimensionless number in the momentum equations) is introduced. Then Pr appears along with $Re_d = \langle u \rangle^f d/\nu$ and as product $Pr\, Re_d = Pe = \langle u \rangle^f d/\alpha_f$. Therefore, *specification* of both Re_d and Pr is *required*.

Eidsath et al. solve the **b**-equation numerically (finite element approximations). In their two-dimensional domain (x, y) on the symmetry line (with the main flow in the x-direction), they use

$$\mathbf{n} \cdot \nabla b_x = 0 \tag{4.67}$$

and

$$b_y = 0. \tag{4.68}$$

The arrangements are that of flow around in-line and staggered cylinders (same as that for $\mathbf{u}'$).

4.3.4 RESULTS FOR $k_s = 0$ AND IN-LINE ARRANGEMENT

The numerical results of Eidsath et al. for the *in-line* arrangement of *cylinders* with $\varepsilon = 0.37$ are given in Figure 4.3, along with the experimental results of Gunn and Pryce (1969). The experiments are for in-line (i.e., simple cubic) arrangement of *spherical* particles of 0.37 to 6.0 mm in diameter and $Re_d = u_D d/\nu$ between 2 and 250. In their experiments both simple-cubic ($\varepsilon = 0.476$) and rhombohedral ($\varepsilon = 0.26$) arrangements were used. However, the results for the rhombohedral arrangement are slightly more scattered, and the trends are different than those for the simple cubic arrangement (this is discussed in Sections 4.3.6 and 4.8.4). In their analogous mass transfer experiments, argon gas was injected into air (Schmidt number near unity). Further discussion of the experimental and predicted results are given by Quintard and Whitaker (1993). Note that $\langle u \rangle^f$ is equal to u_D/ε. The results for both dimensionless *longitudinal total thermal diffusivity*

$$\frac{\langle \alpha_{e\,xx} \rangle^f}{\alpha_f} + \frac{D^d_{xx}}{\alpha_f} = \frac{\langle k_{e\,xx} \rangle^f}{k_f} + \frac{k^d_{xx}}{k_f} = 1 + L^*_{t,xx} + \frac{D^d_{xx}}{\alpha_f} \tag{4.69}$$

and *transverse total thermal diffusivity*

$$\frac{\langle \alpha_{e\,yy} \rangle^f}{\alpha_f} + \frac{D^d_{yy}}{\alpha_f} = \frac{\langle k_{e\,yy} \rangle^f}{k_f} + \frac{k^d_{yy}}{k_f} = 1 + L^*_{t,yy} + \frac{D^d_{yy}}{\alpha_f} \tag{4.70}$$

are given. Note that the factor ε is needed when the description over both s and f phases are used. For $\langle u \rangle^f = 0$, the terms $1 + L^*_{t,xx}$ and $1 + L^*_{t,yy}$ are the dimensionless stagnant fluid phase *effective diffusivity* components (for $k_s = 0$), and, as expected, $\langle \alpha_{e\,xx} \rangle^f / \alpha_f = 1 + L^*_{t,xx} \leq 1$ and $\langle \alpha_{e\,yy} \rangle^f / \alpha_f = 1 + L^*_{t,yy} \leq 1$. No data are available for the transverse total diffusivity. Although not stated, apparently Prandtl (Schmidt) numbers of near unity have been used in the numerical simulations. In this case, for $Re_d = 250$, the flow will be *unsteady*. The agreement between the two-dimensional prediction and the experimental results are good in light of the many simplifications assumed in the simulations. The Peclet number dependency at high Pe is $Pe^{1.7}$, a power *lower* than 2 predicted by Taylor and Aris. The results of Taylor-Aris with $d = 6/A_o = 1.5 d_{\text{tube}}$, can also be shown using

$$\frac{\langle u \rangle^f d}{\alpha_f} \frac{\varepsilon}{1-\varepsilon} \equiv \frac{\langle u \rangle^f d}{1.5 \alpha_f} = Pe \qquad (4.71)$$

which leads to

$$1 + \frac{D_{xx}^d}{\alpha_f} = 1 + \frac{Pe^2}{432}. \qquad (4.72)$$

For $\mathbf{u} = 0$ and using the same method, Ryan et al. (1980) have computed $L_{t,xx}^*$ for square cells with square particles placed inside in a symmetrical arrangement. As expected, they also found that the isotropic effective thermal diffusivity ($\mathbf{u} = 0$) is

$$\frac{\langle \alpha_{e\,xx} \rangle^f}{\alpha_f} = 1 + L_{t,xx}^* \leq 1, \quad k_s = 0, \qquad (4.73)$$

with the equality reached only as $\varepsilon \to 1$.

4.3.5 Results for $k_s \neq 0$ and General Arrangements

In the two-dimensional numerical simulations of Sahraoui and Kaviany (1994), heat diffusion through the solid and periodic arrangements other than the in-line are considered and their results are reviewed below. Figure 4.4 shows typical constant, dimensionless b_x and b_y contours.

The structure parameters used are k_s/k_f, ε and the particle arrangement, and the flow parameters are $Pe_\ell = u_D \ell / \alpha_f$, $Re_\ell = u_D \ell / \nu_f$, and the flow direction with respect to the principal axes of the structure. First the results for the longitudinal total diffusivity D_{xx} are represented, followed by the transverse total diffusivity D_{yy}.

(A) Longitudinal Component

The total dispersion tensor is examined for a bed of circular and square cylinders. There are two mechanisms that contribute to the hydrodynamic dispersion. The first mechanism is due to the velocity gradient in the pore caused by the no velocity slip occurring at the particle surface and also due to the tortuous fluid particle path caused by the solid particle arrangement. The second mechanism is due to the presence of a flow recirculation region, i.e., presence of closed streamlines. The heat transfer out of this region occurs only by molecular diffusion. This dispersion mechanism is especially important for some periodic structures, where vortices can exist between adjacent cylinders.

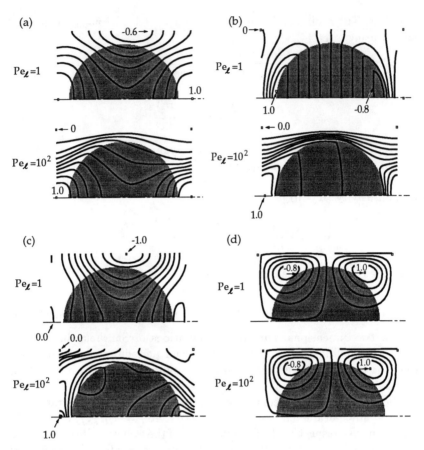

Figure 4.4 (a) Contours of constant $b_x/|b_{x_{max}}|$ for the in-line arrangement of cylinders and for $Pe_\ell = 1$ and 10^2 ($\varepsilon = 0.5$, $k_s/k_f = 1$, and $Re\ell = 0.01$. (b) Effect of k_s/k_f on $b_x/|b_{x_{max}}|$ for the in-line arrangement of cylinders, with $k_s/k_f = 100$ and for $Pe_\ell = 1$ and 10^2 ($\varepsilon = 0.5$ and $Re_\ell = 0.01$). (c) Effect of the Peclet number on $b_x/|b_{x_{max}}|$ for the staggered arrangement of cylinders and for $Pe_\ell = 1$ and 10^2 ($\varepsilon = 0.5, k_s/k_f = 1$, and $Re_\ell = 0.01$). (d) Contours of constant $b_y/|b_{y_{max}}|$ for the in-line arrangement of cylinders and for $Pe_\ell = 1$ and 10^2 ($\varepsilon = 0.5$, $k_s/k_f = 1$, and $Re_\ell = 0.01$).

(i) Porosity

Since the recirculation region covers a larger portion of the pore volume and the velocity gradients become more pronounced as the spacing between the particle decreases, D^d is expected to increase with a decrease in porosity. The increase in D^d_{xx}/α_f with decrease in porosity, for in-line arrangement of cylinders, is shown in Figure 4.5(a). Note that the variation

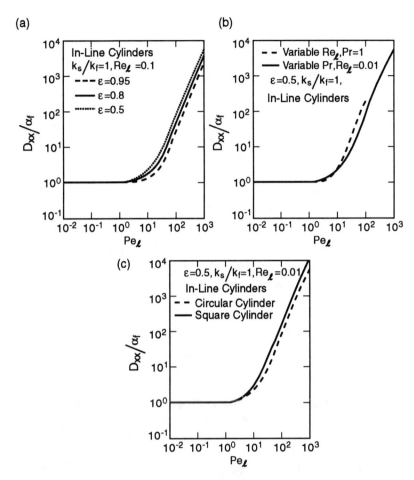

Figure 4.5 (a) Effect of the porosity variation on D_{xx}/α_f for in-line arrangement of circular cylinders. (b) Variation of D_{xx}/α_f as a function of the Peclet number for constant Pr and constant Re_ℓ for in-line arrangement of circular cylinders. (c) Comparison of the variation of D_{xx}/α_f with respect to Pe_ℓ for in-line arrangement of square and circular cylinders.

in $D_{xx}/\alpha_f = 1 + \varepsilon D_{xx}^d/\alpha_f$ with respect to Pe_ℓ is shown in Figure 4.5(a). These results show that the high Pe_ℓ asymptotic behavior is independent of the porosity, i.e., the results for all porosities show that the change in D_{xx}/α_f, caused by the porosity variation, is not very significant compared to the variation with respect to Pe_ℓ. For example, the difference in D_{xx}^d/α_f between $\varepsilon = 0.5$ and 0.95 is only about 60%. As it will be discussed below, the particle arrangement or the flow direction can change D_{xx}^d/α_f by orders of magnitude.

(ii) Reynolds and Prandtl Numbers

So far, the Peclet number $Pe_\ell = Re_\ell Pr$ is varied by by changing the Prandtl number while keeping the Reynolds number the same. In order to evaluate the effect of the flow inertia on D_{xx}/α_f, we now vary the Reynolds number while keeping the Prandtl number the same. The results are shown in Figure 4.5(b) for $Pe_\ell \leq 10^2$. Note that the flow becomes unsteady for $Re_\ell \geq 150$. Also shown in Figure 4.5(b) are the results of varying Pr while keeping Re_ℓ the same. The results show that at high Peclet numbers the power a in the Pe_ℓ^a relation is the same for both variable Pr and Re_ℓ. However, due to the inertial effects, varying the Reynolds number results in a higher D_{xx}/α_f. The inertial effects on the flow field for in-line arrangement of circular cylinders were discussed in Section 2.5.1. For lower Reynolds numbers, separation occurs behind the tip of the cylinder and the streamlines curve around the cylinder. For higher Reynolds numbers, the flow separates before the tip of the cylinder making the flow field nearly rectilinear. This earlier flow separation causes an increase in the extent of the recirculation region and as the Reynolds number increases, D_{xx}^d/α_f increases.

(iii) Particle Shape

The effect of the particle shape on D_{xx}^d/α_f is examined using circular and square cylinders. The results for square and circular cylinders are compared in Figure 4.5(c). The results show that due to a larger recirculation region between the cylinders, D_{xx}^d/α_f for square cylinders and at high Pe_ℓ, is larger. For the square cylinders, the flow is partly rectilinear and can be compared to the flow between two parallel plates. However, for the square cylinders, D_{xx}^d/α_f is greatly affected by recirculation, while for a straight channel the dispersion is only caused by the velocity gradient in the channel. A channel having the same size as the gap between the square cylinders will have a D_{xx}^d/α_f which is lower by about 80 percent.

(iv) Particle Conductivity

The effect of solid to fluid conductivity ratio k_s/k_f on D_{xx}/α_f has been recently studied by Yuan et al. (1991). In their model, a thick-wall capillary tube is used in order to evaluate this effect. They obtain the same Pe_ℓ dependency as that of Taylor. Moreover, they find that at high Peclet numbers, D_{xx}/α_f decreases with increasing k_s/k_f while at low Peclet numbers it increases. The variation of D_{xx}/α_f with respect to k_s/k_f for $Pe_\ell = 1$ and 10^3 are shown in Figure 4.6(a) and (b). For low Peclet number flows, i.e., $Pe_\ell < 10$, D_{xx}/α_f increases as shown in Figure 4.6(a). This is expected because for low Peclet numbers, the hydrodynamic effects are not very significant and the transport is diffusion controlled. In this regime

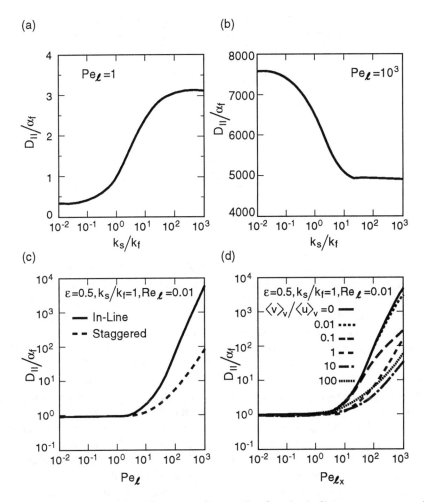

Figure 4.6 (a) and (b): Effect of k_s/k_f on D_{xx}/α_f for in-line arrangement of circular cylinders ($\varepsilon = 0.5$ and $Re_\ell = 0.01$) for $Pe_\ell = 10^3$, respectively. (c) Effect of the particle arrangement on D_{xx}/α_f for variable Pe_ℓ using in-line and staggered arrangements of circular cylinders. (d) Effect of the flow tilt on D_{xx}/α_f for in-line arrangement of circular cylinders.

the effective conductivity initially increases with an increase in k_s/k_f and then reaches an asymptote. As the Peclet number increases, convection dominates and the effect of k_s/k_f on $D_{\parallel}/\alpha_f$ is noticeably different. The transition between the high and low Peclet number regimes occurs around $Pe_\ell = 10$. For higher Peclet numbers ($Pe_\ell > 10$), D_{xx}^d/α_f is enhanced by lowering k_s/k_f, as shown in Figure 4.6(b) for $Pe_\ell = 10^3$. This is consistent with the results of Yuan et al.

TABLE 4.1   EFFECT OF PARTICLE ARRANGEMENT ON THE COEFFICIENTS IN $D_{xx}/\alpha_f = a_1 Pe_\ell^{a_2}$, FOR $k_s/k_f = 1$, AND $Re_\ell = 0.01$. THE RESULTS ARE FOR $10^2 \leq Pe_\ell \leq 10^3$

ε	IN-LINE		STAGGERED	
	a_2	a_1	a_2	a_1
0.50	1.71	0.048	1.26	0.018
0.60	1.68	0.049	1.37	0.013
0.70	1.68	0.044	1.43	0.011
0.80	1.67	0.039	1.49	0.0080
0.95	1.86	0.0076	1.54	0.0042

(v) Particle Arrangement

As the pore geometry changes, so does the flow field, and therefore, the dispersion tensor changes. Here, we study the effect of the particle arrangement on D_{xx}/α_f by examining the in-line and the staggered arrangements of particles. In the staggered arrangement, two adjacent columns of in-line cylinders are shifted with respect to each other by a distance of half a cell size. The results for D_{xx}^d/α_f are shown in Figure 4.6(c). At $Pe_\ell = 10^3$ and for the staggered arrangement, D_{xx}^d/α_f is lower by about two orders of magnitude for the same values of ε, k_s/k_f, and Re_ℓ. This difference is attributed to two effects. First, for the staggered arrangement the recirculation region, that is present between the two adjacent particles for the in-line arrangement, is not present resulting in a lower D_{xx}/α_f. The second effect, which is more significant, is due to the interruptions made to the motion of fluid particles by the staggered solid particles. For the in-line arrangement, the velocity distribution does not change significantly along the flow direction. For the staggered arrangement, the fluid particles follow a tortuous path and undergo periodic and substantial change in direction and magnitude of their velocity. This results in a lower value for D_{xx}/α_f compared with the in-line arrangement. The substantial change in the velocity of the fluid particles also occurs in the disordered porous media. From the results shown in Figure 4.6(c) for high Pe_ℓ, the exponent a_2 in $D_{xx}/\alpha_f = a_1 Pr_\ell^{a_2}$ is 1.26 for $\varepsilon = 0.5$. This trend is also found in the experimental results of Gunn and Pryce (1969) for the rhombohedral arrangement of spherical particles. In the existing literature, these experimental results have *not* been compared to the predictions. This is because a Pe_ℓ^2 relation had been expected for *all* periodic arrangements of particles. The results show that a Pe_ℓ^2 relation is not found for any periodic structure. Table 4.1 gives the coefficients a_1 and a_2 in $D_{xx}/\alpha_f = a_1 Pe_\ell^{a_2}$, for different ε and for the staggered and the in-line arrangements of cylinders. The results show that as ε increases a_2 increases and approaches a value of 2, as expected for a periodic structure. Note that D_{xx}/α_f is significantly lower for the staggered compared to the in-line arrangement.

TABLE 4.2 EFFECT OF FLOW DIRECTION ON THE COEFFICIENTS IN $D_{xx}/\alpha_f = a_1 Pe_\ell^{a_2}$, **FOR** $k_s/k_f = 1$, $Re_\ell = 0.01$, **and** $\epsilon = 0.5$. **THE RESULTS ARE FOR** $10^2 \le Pe_\ell \le 5 \times 10^3$

$\langle v \rangle_V / \langle u \rangle_V$	a_2	a_1	$\langle v \rangle_V / \langle u \rangle_V$	a_2	a_1
0	1.71	0.048	0.9	1.17	0.045
0.01	1.25	0.440	1	1.17	0.049
0.05	0.97	0.807	2	1.15	0.031
0.10	0.97	0.403	4	1.13	0.024
0.30	1.15	0.050	6	1.12	0.020
0.50	1.30	0.010	10	1.10	0.018
0.70	1.25	0.020	100	0.98	0.074

(vi) Flow Direction

So far, D_{xx}/α_f has only been examined for Darcean flows along the principal axes of the solid matrix (or bed). The off-principal axes flows have been examined by Koch et al., for the in-line arrangement of circular cylinders (discussed in Section 4.4.4). Their results show that with a slight deviation from the principal axes, instead of a monotomic increase in D_{xx}/α_f with respect to Pe_ℓ, an asymptote is reached for high Peclet numbers (e.g., for $Pe_\ell \ge 10^3$ and a tilt angle of $0.2°$). This trend is not found here, as shown in Figure 4.6(d), is the variation of D_{xx}^d/α_f with respect to Pe_{ℓ_x} (i.e., the Peclet number based on the flow along x-principal axis) for the in-line arrangement of cylinders. The results are for several tilt angles from $0°$ up to $89.4°$. At high values of Pe_{ℓ_x}, as the tilt angle increases, D_{xx}/α_f decreases. This is because the y-direction flow eliminates the recirculation region between the cylinders and creates a more tortuous fluid particle path. Note that a tilt in Darcean flow corresponds to the staggering of the particles.

Figure 4.6(d) also shows that at high Peclet numbers, D_{xx}/α_f approaches an asymptotic behavior of the form $D_{xx}/\alpha_f = a_1 Pe_\ell^{a_2}$. The coefficients a_1 and a_2 are computed for different tilt angles and are shown in Table 4.2. These results show that the exponent a_2 decreases drastically for small tilt angles and becomes nearly unity for $\langle v \rangle_V / \langle u \rangle_V = 0.05$ to 0.1. As the tile angle further increases, the exponent first increases, also seen in Figure 4.6(d), and then decreases again. From Table 4.2, we find that for $\langle v \rangle_V / \langle u \rangle_V = 0.9$ and 1, D_{xx}/α_f is larger compared to the results for 0.7. This is because as the tilt angle changes, the tortuosity and the magnitude of the velocity gradient change to enhance D_{xx}^d/α_f. For $\langle v \rangle_V / \langle u \rangle_V = 10^2$, D_{xx}/α_f is larger compared to $\langle v \rangle_V / \langle u \rangle_V = 10$. This is because the overall Peclet number, based on the total velocity, is about 10 times larger compared to that for $\langle v \rangle_V / \langle u \rangle_V = 10$, while the tilt angles are not very different. The experimental results of Gunn and Pryce (1969) for the rhombohedral arrangement of spheres shows a behavior similar to that for the staggered arrangement (i.e., $\langle v \rangle_V / \langle u \rangle_V = 1$).

In order to simulate random porous media we use the ensemble averaging over the flow direction (i.e., tilt angle) for the in-line arrangement of cylinders. We use a uniform probability distribution function for the flow direction distribution and take the ensemble average of D_{xx}/α_f, for the same Peclet number, using the results for the different tilt angles shown in Table 4.2. Using these averaged values, the computed coefficients a_1 and a_2 are 0.062 and 1.17, respectively. This a_2 is close to that found for $\langle v \rangle_V / \langle u \rangle_V = 1$, i.e., the average tilt angle and it is also close to the experimental results for random arrangement of spheres (i.e., $1 \leq a_2 \leq 1.2$).

(B) Transverse Component

The variables that most noticeably affect D_{yy}^d/α_f are the Peclet number, the particle arrangement, and the flow direction. The numerical result shows that for a given Pe_ℓ, D_{yy}^d/α_f is independent of whether Pr or Re_ℓ is varied. Also, the porosity and the particle shape (i.e., sqaure versus circular cylinders) do not affect D_{yy}^d/α_f (but the effective conductivity does depend on these parameters for $k_s/k_f \neq 1$).

(i) Particle Arrangement and Peclet Number

In the staggered arrangement, the fluid particles follow a more tortuous path and this enhances D_{yy}^d/α_f. This is evident in Figure 4.7(a), where for the staggered arrangement, D_{yy}/α_f increases more rapidly with the Peclet number (as compared to in-line). For the nearly rectilinear flow fields of the in-line arrangement, the heat transfer in the transverse direction occurs only by diffusion. The only dispersion mechanism of heat transfer in the transverse direction is the recirculation which provides some mixing of the flow. For the in-line arrangement and at high Pe_ℓ ($Pe_\ell = 10^3$), the increase in D_{yy}/α_f due to this mixing is about 35 percent. Note that D_{yy}^d/α_f for the staggered arrangement is yet small compared to D_{xx}^d/α_f, because there is no net flow in the y direction.

(ii) Flow Direction

The effect of flow direction (with respect to the principal axes) on D_{yy}^d/α_f, for the in-line arrangement of circular cylinders is shown in Figure 4.7(b). For a given Pe_{ℓ_x}, and when the velocity in the transverse direction is small compared to that in the longitudinal direction (e.g., $\langle v \rangle_V / \langle u \rangle_V = 0.1$), D_{yy}^d/α_f is slightly larger than that for the zero tilt angle. As the velocity in the transverse direction increases, D_{yy}^d/α_f is increased further. For $\langle v \rangle_V / \langle u \rangle_V = 1$ a substantial increase in the transverse hydrodynamic dispersion is found, as expected, since the in-line arrangement is a staggered arrangement for the oblique direction of the flow.

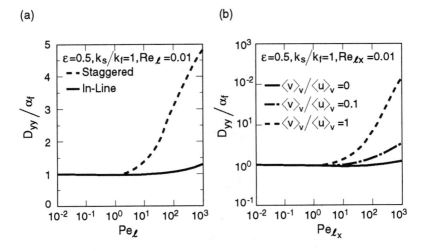

Figure 4.7 (a) Effect of the particle arrangement on D_{yy}/α_f for in-line and staggered arrangements of cylinders. (b) Effect of the flow tilt angle on D_{yy}/α_f for the in-line arrangement of circular cylinders.

4.4 Three-Dimensional Periodic Structures

Closed-form expressions for the dispersion tensor have been obtained by Koch et al. (1989) for packed beds of spherical particles periodically arranged. Their development is similar to that of Brenner (1980). Some of the general features are also given by Koch and Brady (1987). The restrictions applied in the analysis are given here.

- The *Stokes flow* is assumed (and the velocity disturbance caused by the particles is approximated to that due to *point sources*).

- The *point force* used is that of the viscous drag on a *single particle*. [The force is actually *larger* when the interactions with the other particles are included (Hasimoto, 1959).]

- The velocity field does *not* satisfy the boundary conditions on the particle surface.

- *Dilute* suspension of particles is assumed.

- *Equal* thermal conductivities and heat capacities are assumed for the fluid and solid phases.

- For $\mathbf{u} = 0$, the results of the analysis do *not* predict the effective thermal conductivity.

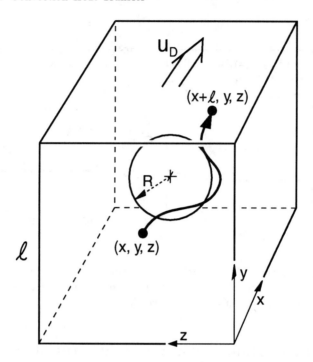

Figure 4.8 A cubic unit cell with a spherical particle located at its center. The flow is perpendicular to the front face and is periodic.

Despite these restrictions, the advantages are that closed-form solutions, which also show the expected asymptotic behavior at high Pe, are obtained. Also, the three-dimensionality of the flow and other interesting features of the dispersion *tensor* are maintained. A schematic of the unit cell is given in Figure 4.8.

The development is based on the *large elapsed-time* asymptotic behavior (along with the development of Taylor-Aris) of disturbances introduced into the fluid flowing through periodic structures. The local volume averaging and the closure condition used for the determination of the disturbance fields are the same as those used by Carbonell and Whitaker.

4.4.1 UNIT-CELL AVERAGING

The energy equation for any point in the solid ($\mathbf{u} = 0$) or fluid phase can be written in terms of the *heat flux* (molecular plus convection) $\mathbf{q} = (\rho c_p)_f \mathbf{u} T - k_f \nabla T$, where it is assumed that $(\rho c_p)_f = (\rho c_p)_s$ and $k_f = k_s$. The energy equation becomes

$$\rho c_p \frac{\partial T}{\partial t} + \nabla \cdot \mathbf{q} = 0. \tag{4.74}$$

The average of (4.74) over a unit cell (a schematic of the unit-cell volume V is given in Figure 4.8) is taken as

$$\rho c_p \frac{\partial \langle T \rangle}{\partial t} + \nabla \cdot \langle \mathbf{q} \rangle = 0, \qquad (4.75)$$

where

$$\langle \mathbf{q} \rangle = \rho c_p \langle \mathbf{u} T \rangle - k \langle \nabla T \rangle = \frac{1}{V} \int_V \mathbf{q} \, dV. \qquad (4.76)$$

Introducing $T' = T - \langle T \rangle$ and $\mathbf{u}' = \mathbf{u} - \langle \mathbf{u} \rangle$, they arrive at

$$\langle \mathbf{q} \rangle = \rho c_p \langle \mathbf{u} T \rangle - k \langle \nabla T \rangle = \rho c_p [\langle \mathbf{u} \rangle \langle T \rangle + \langle \mathbf{u}'T' \rangle] - k \nabla \langle T \rangle. \qquad (4.77)$$

Note that when the theorem for the volume average of a gradient (or a divergence) is used [i.e., (2.84)–(2.85)], the integral term for the gradient is not zero. However, in the derivation of Koch et al., unlike that of Carbonell and Whitaker, (4.29), the integral term is dropped. This integral term leads to the tortuosity tensor. However, as the results of Section 3.2 show, the integral term *vanishes* under the condition of *equal properties* for the solid and fluid phases (as assumed here).

The equation for the temperature disturbance is obtained by subtracting the decomposed energy equation from (4.74), and this gives

$$\frac{\partial T'}{\partial t} + \nabla \cdot \mathbf{u} T' - \alpha \nabla^2 T' = -\mathbf{u}' \cdot \nabla \langle T \rangle + \nabla \cdot \langle \mathbf{u}'T' \rangle. \qquad (4.78)$$

Under the assumption of nearly constant average temperature gradient, which makes $\mathbf{u}'$ and T' stationary and $\langle \mathbf{u}'T' \rangle$ independent of the position, the last term in (4.78) becomes zero. Note that the same argument led to (4.49).

We now introduce a *transformation* similar to that used by Carbonell and Whitaker and by Brenner, i.e.,

$$T' = \mathbf{b}(\mathbf{x}) \cdot \nabla \langle T \rangle, \qquad (4.79)$$

which is the same as the $\mathbf{b}$ introduced in (4.35). When this is used in (4.77) we have

$$\frac{1}{\rho c_p} \langle \mathbf{q} \rangle = \langle \mathbf{u} \rangle \langle T \rangle - \mathbf{D} \cdot \nabla \langle T \rangle, \qquad (4.80)$$

where

$$\mathbf{D} = \alpha \mathbf{I} - \langle \mathbf{u}'\mathbf{b} \rangle \qquad (4.81)$$

is the *sum* of the *molecular diffusivity* and the *dispersion* tensors. Noting that $\mathbf{b} = \mathbf{b}(\mathbf{x})$ only, and inserting (4.79) into (4.78) without the first and last terms, which can be shown to be negligibly small, we have the **b** equation as

$$\nabla \cdot (\mathbf{u} \mathbf{b} - \alpha_f \nabla \mathbf{b}) = -\mathbf{u}'. \qquad (4.82)$$

Next, we place the centers of the particles at

$$\mathbf{x}_n = n_1 \boldsymbol{\ell}_1 + n_2 \boldsymbol{\ell}_2 + n_3 \boldsymbol{\ell}_3 \quad (n_1, n_2, n_3 = 0, \pm 1, \pm 2, \ldots), \tag{4.83}$$

where $\mathbf{x}_n$ is the position vector of the particles, and $\boldsymbol{\ell}_1$, $\boldsymbol{\ell}_2$, and $\boldsymbol{\ell}_3$ are the basic vectors (in the physical plane) determining the unit cell of the array.

Now, *scaling* the length with the lattice spacing $\ell = V^{1/3}$ (i.e., assuming a *simple cubic lattice*), the velocity with the filter velocity u_D, and the diffusivity with α, the symmetric part of $\mathbf{D}^*$ in (4.81) is

$$\mathbf{D}_s^* = \mathbf{I} - \frac{Pe}{2} \int_V \left(\mathbf{u}^{*'} \mathbf{b} + \mathbf{b} \mathbf{u}^{*'} \right) d\mathbf{x}^*, \tag{4.84}$$

where $Pe = u_D \ell / \alpha$, an *asterisk* indicates that the variable is dimensionless and $\mathbf{b}$ is scaled with ℓ.

Substituting for $\mathbf{u}'$ from (4.82), we have

$$\mathbf{D}_s^* = \mathbf{I} + \frac{Pe}{2} \int_V \left[\nabla \cdot \left(\mathbf{u}^* \mathbf{b} - \frac{1}{Pe} \nabla \mathbf{b} \right) \mathbf{b} \right.$$

$$\left. + \mathbf{b} \nabla \cdot \left(\mathbf{u}^* \mathbf{b} - \frac{1}{Pe} \nabla \mathbf{b} \right) \right] d\mathbf{x}^*. \tag{4.85}$$

Integration by parts following application of the divergence theorem (2.73) gives

$$\mathbf{D}_s^* = \mathbf{I} + \int_V (\nabla \mathbf{b} \cdot \nabla \mathbf{b}^\dagger) \, d\mathbf{x}^* + \frac{Pe}{2} \int_A \mathbf{n} \cdot \left[\mathbf{u}^* \mathbf{b} \mathbf{b} - \frac{1}{Pe} \nabla (\mathbf{b} \mathbf{b}) \right] dA^*, \tag{4.86}$$

where $\dagger$ stands for the *transpose* and $\mathbf{n}$ is the normal unit vector pointing outward from the cell boundaries. The area integral is zero because of the symmetry at the opposite faces of the cell. This leads to

$$\mathbf{D}_s^* = \mathbf{I} + \int_V (\nabla \mathbf{b} \cdot \nabla \mathbf{b}^\dagger) \, d\mathbf{x}^*. \tag{4.87}$$

The velocity field $\mathbf{u}^{*'}$ is available in the literature (Hasimoto) for flow through dilute suspension of particles. This field is available in the transformed form. Therefore, the Fourier transform of the $\mathbf{D}^*$ and $\mathbf{b}$ are used.

Now we expand $\mathbf{u}^*(\mathbf{x}^*)$ in terms of the Fourier series as

$$\mathbf{u}^*(\mathbf{x}^*) = \sum_{\mathbf{k}^*} e^{-2\pi i \mathbf{k}^* \cdot \mathbf{x}^*} \widehat{\mathbf{u}}^*(\mathbf{k}^*), \tag{4.88}$$

where the summation is over the integral values of n_1, n_2, and n_3. Also

$$\mathbf{k} = n_1 \mathbf{c}_1 + n_2 \mathbf{c}_2 + n_3 \mathbf{c}_3 \tag{4.89}$$

are the vectors in reciprocal lattice which satisfy

$$\mathbf{k} \cdot \boldsymbol{\ell}_i = n_i \quad i = 1, 2, 3. \tag{4.90}$$

Based on these two relations, the basic vectors $\mathbf{c}_1$, $\mathbf{c}_2$, and $\mathbf{c}_3$ in the reciprocal lattice are

$$\mathbf{c}_1 = \frac{\boldsymbol{\ell}_2 \times \boldsymbol{\ell}_3}{V}, \quad \mathbf{c}_2 = \frac{\boldsymbol{\ell}_3 \times \boldsymbol{\ell}_1}{V}, \quad \mathbf{c}_3 = \frac{\boldsymbol{\ell}_1 \times \boldsymbol{\ell}_2}{V}, \tag{4.91}$$

and

$$V = \boldsymbol{\ell}_1 \cdot (\boldsymbol{\ell}_2 \times \boldsymbol{\ell}_3). \tag{4.92}$$

For the *simple cubic lattice*, we have

$$\begin{bmatrix} \boldsymbol{\ell}_1 \\ \boldsymbol{\ell}_2 \\ \boldsymbol{\ell}_3 \end{bmatrix} = \ell \begin{bmatrix} (1,0,0) \\ (0,1,0) \\ (0,0,1) \end{bmatrix}, \tag{4.93}$$

$$V = \ell^3, \tag{4.94}$$

$$\begin{bmatrix} \mathbf{c}_1 \\ \mathbf{c}_2 \\ \mathbf{c}_3 \end{bmatrix} = \frac{1}{\ell} \begin{bmatrix} (1,0,0) \\ (0,1,0) \\ (0,0,1) \end{bmatrix}. \tag{4.95}$$

Note that for *body-centered cubic* lattice $V = \ell^3/2$, and for *face-centered cubic* lattice $V = \ell^3/4$, where ℓ is the *lattice spacing*.

Then the $\mathbf{D}^*$ and $\mathbf{b}$ equations, (4.87) and (4.82), become

$$\mathbf{D}_s^* = \mathbf{I} + \sum_{\mathbf{k}^*} 4\pi^2 k^{*2} \widehat{\mathbf{b}}(\mathbf{k}^*) \widehat{\mathbf{b}}(-\mathbf{k}^*), \tag{4.96}$$

$$k^{*2} = \mathbf{k}^* \cdot \mathbf{k}^*, \tag{4.97}$$

$$4\pi^2 \frac{k^{*2}}{Pe} \widehat{\mathbf{b}}(\mathbf{k}^*) + 2\pi \sum_{\mathbf{k}^{*\prime}} \left[\widehat{\mathbf{u}}^* \left(\mathbf{k}^* - \mathbf{k}^{*\prime} \right) \cdot i\mathbf{k}^{*\prime} \right] \widehat{\mathbf{b}} \left(\mathbf{k}^{*\prime} \right) = -\widehat{\mathbf{u}}^{*\prime}(\mathbf{k}^*). \tag{4.98}$$

4.4.2 EVALUATION OF $\mathbf{u}'$, $\mathbf{b}$, AND $\mathbf{D}$

For *dilute* concentration of particles of radius R and lattice spacing ℓ, the velocity disturbance caused by the particle is approximated as that due to a point source. This velocity field is determined by Hasimoto, and his formulation is briefly given later. This formulation is based on the Stokes flow (with a point force at the center of the particle). The equations of motion and continuity are

$$\frac{1}{V} \mathbf{f} \sum_n \delta(\mathbf{x} - \mathbf{x}_n) = -\nabla p + \mu \nabla^2 \mathbf{u}, \tag{4.99}$$

$$\nabla \cdot \mathbf{u} = 0, \tag{4.100}$$

where $\mathbf{f}$ is the force acting on one of the particles and $\delta(\mathbf{x} - \mathbf{x}_n)$ is the Dirac delta function. Note that the force $\mathbf{f}$ acting on one of the particles in the bed is larger than the drag force on an isolated sphere. Because of the periodicity of the fields, $\mathbf{u}$ and $-\nabla p$ are expanded in the Fourier series. The $\mathbf{u}$ expansion is already given in (4.88), and for ∇p we have

$$-\nabla p(\mathbf{x}) = -\sum_{\mathbf{k}} e^{-2\pi i(\mathbf{k} \cdot \mathbf{x})} \widehat{\nabla p}(\mathbf{k}). \tag{4.101}$$

By multiplying (4.99) and (4.100) by $e^{2\pi i(\mathbf{k} \cdot \mathbf{x})}$, using (4.99) and (4.101), and integrating over the cell, we have

$$\frac{\mathbf{f}}{V} = -\widehat{\nabla p}(\mathbf{k}) - 4\pi^2 \mu k^2 \widehat{\mathbf{u}}(\mathbf{k}), \tag{4.102}$$

$$\mathbf{k} \cdot \widehat{\mathbf{u}}(\mathbf{k}) = 0, \tag{4.103}$$

where $\widehat{\nabla p}(\mathbf{k})$ satisfies $\widehat{\nabla p}(\mathbf{k}) \times \mathbf{k} = 0$. Also

$$\int_V \delta(\mathbf{x} - \mathbf{x}_n) \, d\mathbf{x} = \begin{cases} 1 & \text{for } \mathbf{x}_n \text{ in } V \\ 0 & \text{for } \mathbf{x}_n \text{ not in } V \end{cases} \tag{4.104}$$

when $\mathbf{k} = 0$, i.e., considering the particle only, we have

$$-\widehat{\nabla p}(\mathbf{k} = 0) = \frac{\mathbf{f}}{V}. \tag{4.105}$$

This states that the force acting on the particles is balanced by the mean pressure gradient of the fluid. Taking the scalar product of (4.102) with $\mathbf{k}$, and using (4.103) and (4.105), we have for $\mathbf{k} \neq 0$

$$-\mathbf{k} \cdot \widehat{\nabla p}(\mathbf{k}) = \frac{1}{V}(\mathbf{k} \cdot \mathbf{f}) = -\mathbf{k} \cdot \widehat{\nabla p}(\mathbf{k} = 0). \tag{4.106}$$

This can be written as

$$-\widehat{\nabla p}(\mathbf{k}) = \frac{1}{V} \frac{(\mathbf{k} \cdot \mathbf{f})\mathbf{k}}{k^2}, \quad \mathbf{k} \neq 0. \tag{4.107}$$

Using (4.107) in (4.102), we have

$$\widehat{\mathbf{u}}(\mathbf{k}) = \frac{1}{4\pi^2 \mu V} \left[\frac{(\mathbf{k} \cdot \mathbf{f})\mathbf{k}}{k^4} - \frac{\mathbf{f}}{k^2} \right]. \tag{4.108}$$

Hasimoto shows that the ratio of $\mathbf{f}$ found for cubic lattices (simple, face-centered, or body-centered cubic arrangement) to that for the Stokes flow over a single sphere is *greater* than unity and does *not* differ much between these three arrangements. He finds

$$\mathbf{f} = 11\mathbf{f}_s, \quad \text{where } \mathbf{f}_s = 6\pi \mu R u_D \tag{4.109}$$

4.4 Three-Dimensional Periodic Structures

for the *simple cubic lattice and* $\varepsilon = 0.784$ where $\mathbf{f}_s$ is the Stokes drag force. Note that although different approaches are used, the geometric model of Neale and Nader (1974a, see Chapter 2 for reference) discussed in Section 2.4.3 (B), leads to similar results for $\mathbf{f}$.

When in (4.108) the contents of the square bracket are written in terms of *tensors* and then quantities are made dimensionless, we have

$$\widehat{\mathbf{u}}^{*'}(\mathbf{k}^*) = -\frac{1}{4\pi^2}\frac{R}{\ell}\frac{\mathbf{f}^* \cdot \left(\mathbf{I} - \frac{\mathbf{k}^*\mathbf{k}^*}{k^{*2}}\right)}{k^{*2}}, \qquad \mathbf{k}^* \neq 0 \qquad (4.110)$$

$$\widehat{\mathbf{u}}^{*'}(\mathbf{k}^*) = 0, \qquad \mathbf{k}^* = 0. \qquad (4.111)$$

Note that the force $\mathbf{f}$ exerted by the sphere on the fluid is made dimensionless using $\mu\ell u_D$. When this drag force is taken to be the *Stokes* drag, i.e., $6\pi R \mathbf{u}_D^*/\ell$, then (4.110) becomes

$$\widehat{\mathbf{u}}^{*'} = -\frac{3}{2\pi}\frac{R}{\ell}\frac{\mathbf{u}_D^* \cdot \left(\mathbf{I} - \frac{\mathbf{k}^*\mathbf{k}^*}{k^{*2}}\right)}{k^{*2}}, \qquad \mathbf{k}^* \neq 0, \qquad (4.112)$$

except near the particle (another *restriction* is $R/\ell \ll 1$). Also in the $\mathbf{b}$ equation (4.98), it is assumed that $\mathbf{u}^* = \mathbf{u}_D^*$, such that from (4.98) we have

$$\widehat{\mathbf{b}}(\mathbf{k}^*) = -\frac{\mathbf{u}^{*'}(\mathbf{k}^*)}{4\pi^2\dfrac{k^{*2}}{Pe} + 2\pi i \mathbf{u}_D^* \cdot \mathbf{k}^*}. \qquad (4.113)$$

Using this in (4.96), we have

$$\mathbf{D}_s^* = \mathbf{I} + \sum_{\mathbf{k}^*}\frac{\widehat{\mathbf{u}}^{*'}(\mathbf{k}^*)\widehat{\mathbf{u}}^{*'}(-\mathbf{k}^*)k^{*2}}{4\pi^2\dfrac{k^{*4}}{Pe^2} + (\mathbf{u}_D^* \cdot \mathbf{k}^*)^2}. \qquad (4.114)$$

Using (4.112) for the velocity disturbance in (4.114), we have for the *hydrodynamic dispersion* only, i.e., by removing $\mathbf{I}$ in (4.114), the following

$$\mathbf{D}_s^{*\,d} = \frac{9}{4\pi^2}\left(\frac{R}{\ell}\right)^2\sum_{\mathbf{k}^*\neq 0}\frac{\left[\mathbf{u}_D^* \cdot \left(\mathbf{I} - \dfrac{\mathbf{k}^*\mathbf{k}^*}{k^{*2}}\right)\right]^2}{k^{*2}\left[4\pi^2\dfrac{k^{*4}}{Pe^2} + (\mathbf{u}_D^* \cdot \mathbf{k}^*)^2\right]}. \qquad (4.115)$$

The two-dimensional analog of the simple cubic array of spheres is a *square array (doubly periodic)* of cylinders of *infinite* length. Koch et al. also obtained a solution for this medium by replacing the drag force ($6\pi\ell\mathbf{u}_D^*/R$ for a sphere) by the force on a cylinder per unit length, i.e., $4\pi\mathbf{u}_D^*/[\ln(\ell/R)-$

1.3015] (a leading-order solution found by Hasimoto) and by using **k** as the two-dimensional reciprocal lattice vector. Their result is

$$\mathbf{D}_s^{*\,d} = \frac{1}{16\pi^4}\left(\frac{4\pi}{\ln\frac{\ell}{R} - 1.3015}\right)^2 \sum_{\mathbf{k}^*\neq 0} \frac{\left[\mathbf{u}_D^* \cdot \left(\mathbf{I} - \frac{\mathbf{k}^*\mathbf{k}^*}{k^{*2}}\right)\right]^2}{k^{*2}\left[4\pi^2\frac{k^{*4}}{Pe^2} + (\mathbf{u}_D^* \cdot \mathbf{k}^*)^2\right]}. \tag{4.116}$$

At low values of $Pe = u_D\ell/\alpha_f$, the term $(\mathbf{u}_D^* \cdot \mathbf{k}^*)^2$ in the denominator makes a negligible contribution, and, therefore, $\mathbf{D}_s^*$ is proportional to Pe^2 (similar to the Taylor-Aris dispersion).

4.4.3 Comparison with Experimental Results

Koch et al. have added to (4.115) the *purely* conductive contribution (same as $\mathbf{K}_e$ in the Carbonell and Whitaker treatment) determined previously by Sangani and Acrivos (1983). The combined results are given in Figure 4.9 along with the experimental results of Gunn and Pryce. The total dimensional effective diffusivity is

$$\mathbf{D} = \frac{\mathbf{K}_e}{(\rho c_p)_f} + \varepsilon\, \mathbf{D}_s^d. \tag{4.117}$$

Note that the experiment of Gunn and Pryce is for $k_s = 0$, while $k_s = k_f$ is assumed in the analysis. $D_\parallel^d$ is the component of the dispersion tensor, (4.115), in the direction of flow $\mathbf{u}_D$, and $D_\perp^d$ is that in the direction perpendicular to flow (transverse dispersion).

As is expected, complete agreement between the experimental and predicted results does not exist, but the trend predicted in (4.115) is similar to that observed in the experiment. This is interesting in light of the major differences between the model and the experiments.

- The experiments are for $\varepsilon = 0.48$, while the analysis is, in principle, for $\varepsilon \to 1$ but evaluated at $\varepsilon = 0.48$.

- The experiments are for $0.02 < Re_d < 250$, while the solution of Hasimoto for the velocity is for Stokes flow. For $Re_d > 4$, flow *separation* begins, and as Re increases, the flow will deviate substantially from his solution. This difference in the flow field may be masked because of the *counteracting* effects of the *thinning* of the boundary layer and the tail *separation* (and shedding that occurs at high Re). Also, when there exists an inertial core, the assumptions such as $\mathbf{u} \simeq \mathbf{u}_D$ made in the **b** equation may be more justifiable.

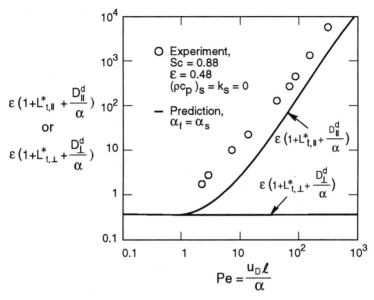

Figure 4.9 The prediction of Koch et al. (for simple cubic arrangement of spheres) for longitudinal and transverse total effective thermal diffusivity, compared with the experimental results of Gunn and Pryce ($Sc = 0.88$). (From Koch et al., reproduced by permission ©1989 Cambridge University.)

4.4.4 Effect of Darcean Velocity Direction

For off-principal axes flows, i.e., when the Darcean velocity is *not* perpendicular to any vector in the reciprocal lattice $\mathbf{k}$, i.e., $\mathbf{u}_D \cdot \mathbf{k} \neq 0$, then Koch et al. show that axial and transverse diffusions are *independent* of Pe, for $Pe \gg 1$. In a *square array* (two-dimensional, x, y), this implies that $\mathbf{u}_D \cdot \mathbf{s}_x \neq 0$ and $\mathbf{u}_D \cdot \mathbf{s}_y \neq 0$, where $\mathbf{s}$ is the unit vector. They argue that (4.116) gives the contribution to the total effective diffusivity arising from the bulk fluid away from the boundary layers. Thus the structure dependence weakens and the results will be similar to that for the *disordered* media. They suggest that the boundary layer and $(\rho c_p)_s$ contributions be *added* to (4.116) in order to *completely* account for all mechanisms. These two contributions are discussed in their treatment of the disordered media (Koch and Brady, 1985), which will be reviewed shortly. For the bulk fluid contribution, given by (4.116), their results for the specific *flow direction* given by

$$\mathbf{u}_D^* = \frac{1}{\pi}\mathbf{s}_x + \left(1 - \frac{1}{\pi^2}\right)^{1/2}\mathbf{s}_y, \qquad (4.118)$$

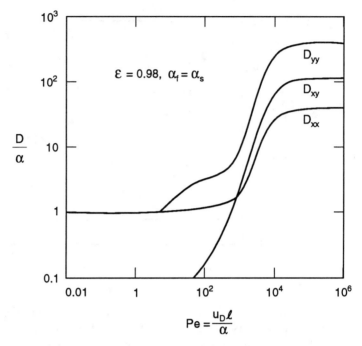

Figure 4.10 The effect of Darcean velocity direction on the dispersion tensor, as predicted by Koch et al., for the specific flow direction given in the text (two-dimensional flow). (From Koch et al., reproduced by permission ©1989 Cambridge University.)

and for $\varepsilon = 0.98$ (very dilute particle concentration) are given in Figure 4.10. The results show that the total effective thermal diffusivity is *non-diagonal* relative to the x, y axes and to $\mathbf{u}_D$. Therefore, $D_{xy} = D_{yx} \neq 0$. Also, the high Pe asymptote is observed. No experimental verification is available. We note that unlike those discussed in Section 4.3.5, their results also show that for a very small deviation from $\mathbf{s}_x$ or $\mathbf{s}_y$, i.e., small tilt angle, the axial dispersion coefficient becomes *very* large. Therefore, although nonzero values for nondiagonal elements are expected for the *off-axis* flows, the magnitude of $D_\parallel$ is not expected to change by orders of magnitude for small tilt angles. Note that this in part may be due to the assumptions of a very dilute particle concentration.

4.5 Dispersion in Disordered Structures—Simplified Hydrodynamics

The present analytical-numerical capabilities have enabled more *exact* analyses of the dispersion phenomena. However, review of the earlier more *ap-*

4.5 Dispersion in Disordered Structures—Simplified Hydrodynamics

proximate analyses helps in the construction of a perspective of the field of convective heat and mass transfer in porous media and acts as an instructional tool in the *evolutionary* modeling. In this light, some of the earlier attempts in modeling the dispersion in porous media are reviewed later. While some of the techniques have limited expandability, i.e., cannot be made to include more realistic features of the transport, others like the Horn method of moments have recently been applied to periodic structure, along with inclusion of more realistic cell-based hydrodynamics (Brenner). The first three methods reviewed later are *stochastic*. The fourth is *deterministic* with an extremum hypothesis (the maximum of a functional is sought). Bear (1988, 587–616) gives a thorough review of the statistical treatments, the properties of the dispersion coefficient, and the related *medium dispersivity tensor* (relates the mean tracer displacement to the dispersion coefficient).

4.5.1 SCHEIDEGGER DYNAMIC AND GEOMETRIC MODELS

Scheidegger (1954) obtained relationships between the velocity dispersion coefficient and the Darcean velocity by using a stochastic model (*the random walk model*). Since *no* allowance was made for the molecular diffusion, his results have been interpreted as that for *purely mechanical*† dispersion of heat in porous media. He uses *ensemble averages* over observations (states) where each observation is based on a microdynamic equation (such as the Darcy law) completely describing the observation. The ensemble-averaged values of the observations are called the *expectation value*. For steady-state processes, it is assumed that the ensemble averages and the time averages are interchangeable (the *ergodic hypothesis*). His approach, which is similar to that applied to the *Brownian motion*, is based on the following construction.

- The *path-line* of fluid particles flowing through channels and undergoing random knocks from the channel walls is followed. The ensemble is that of all possible random walks in time interval $(0, t)$ from a given point.

- It is assumed that during the observation time interval $t = N\tau$, where N = number of steps, what happens in each time step τ is independent of the prior experience (*Markov process*).

- The particle will have various velocities along its path and the mean velocity will be the Darcy velocity (or during microdynamic transition

---

† As indicated in Brenner (1980), any model that does not allow for the molecular participation in the transport [such as the purely random walk treatment of Scheidegger (1954) and De Josselin De Jong (1958)] does not realistically represent the pore-level conduction and convection, and, therefore, is not expected to be valid.

between each time step, the mean velocity at any time is the Darcy velocity).

Again, although many advances have been made since the introduction of this analysis [see, e.g., Sahimi et al. (1983)], it is reviewed here to offer a perspective on the treatment of disordered systems.

(A) PROBABILITY DENSITY FUNCTION

In every time step, the particle is displaced to position $\mathbf{x}$ (initially at $\mathbf{x} = 0$). The probability of displacement $\mathbf{x}$ taking place in τ is $\psi(\mathbf{x})$, such that

$$\int \psi(\mathbf{x}) \, d\mathbf{x} = 1. \tag{4.119}$$

The *ensemble-averaged displacement* over τ is

$$\langle \mathbf{x} \rangle = \int \mathbf{x}\,\psi(\mathbf{x}) \, d\mathbf{x}. \tag{4.120}$$

Now we define $\mathbf{x}' = \mathbf{x} - \langle \mathbf{x} \rangle$ such that $\langle \mathbf{x}' \rangle = 0$, and then use $P(\mathbf{x}')$ as the *probability of displacement* $\mathbf{x}'$. Because of the assumed uniformity and isotropy, $P(\mathbf{x}') = P(x'_1)P(x'_2)P(x'_3)$, and the probabilities are the same for any τ. After a large number of trials N, the *central limit theorem* states that P is *Gaussian*. For σ being the *standard deviation* of P, we have

$$P(\mathbf{x}') = \frac{1}{(2\pi N \sigma^2)^{3/2}} \exp\left(-\frac{x'^2_1 + x'^2_2 + x'^2_3}{2N\sigma^2}\right). \tag{4.121}$$

Now define the *dispersion coefficient* as the Einstein equation used in analysis of the Brownian motion

$$D^d = \frac{\sigma^2}{2\tau} = \frac{N\sigma^2}{2t}, \tag{4.122}$$

such that (4.121) by using (4.122) and $\mathbf{x}' = \mathbf{x} - \langle \mathbf{x} \rangle$ becomes

$$P(\mathbf{x}, t) = \frac{1}{(4\pi D^d t)^{3/2}}$$

$$\times \exp\left[-\frac{(x_1 - \langle x_1 \rangle)^2 + (x_2 - \langle x_2 \rangle)^2 + (x_3 - \langle x_3 \rangle)^2}{4D^d t}\right]. \tag{4.123}$$

This describes the motion of a fluid particle through the porous medium.

(B) MICRODYNAMIC EQUATION

Over τ, the external force $-\nabla p$ drives the flow along the displacement $\Delta \mathbf{x}$ making an angle θ with $-\nabla p$. Then using K'/ε' as the local conductance, we have

$$\frac{\Delta \mathbf{x}}{\tau} = -\frac{K'}{\varepsilon'}\frac{1}{\mu}\nabla p \cos\theta = \frac{\mathbf{x}}{t}. \tag{4.124}$$

By using (4.120), the average of $\Delta \mathbf{x}$ over the ensemble is found. The average of $\Delta \mathbf{x}$ will be along $-\nabla p$ (uniform and isotropic media), and its magnitude is equal to the average component of $\Delta \mathbf{x}$ in the direction of $-\nabla p$, i.e., $|\Delta \mathbf{x}|\cos\theta$. With this, we have

$$\frac{\langle \Delta \mathbf{x}\rangle}{\tau} = -\left\langle\frac{K'}{\varepsilon'}\right\rangle\frac{1}{\mu}\nabla p \langle\cos^2\theta\rangle = \mathbf{u}_p = \frac{\langle\mathbf{x}\rangle}{t}. \tag{4.125}$$

Substituting this in (4.123), we have

$$P(\mathbf{x},t) = \frac{1}{(4\pi D^d t)^{3/2}}\exp\left[-\frac{\left(\mathbf{x}+\left\langle\frac{K'}{\varepsilon'}\right\rangle\frac{t}{\mu}\nabla p\langle\cos^2\theta\rangle\right)^2}{4D^d t}\right]. \tag{4.126}$$

The *square-length* $\langle \mathbf{x}^2\rangle$ is (from the general *theory of Gaussian distribution*)

$$\langle \mathbf{x}^2\rangle = 6D^d t + \left(\left\langle\frac{K'}{\varepsilon'}\right\rangle\frac{t}{\mu}\nabla p\langle\cos^2\theta\rangle\right)^2. \tag{4.127}$$

Evaluation of D^d requires further specification of $\langle \mathbf{x}^2\rangle$. This is done by a *dynamic* and a *geometric* treatment.

(C) DYNAMIC AND GEOMETRIC TREATMENTS

The *dynamic model* is constructed by assuming a uniform flow across each channel, an *acceleration* term $\rho(d^2\mathbf{x}/dt^2)$ can be added to the Darcy law such that (4.124) can be written as

$$\frac{K'}{\varepsilon'}\frac{1}{\mu}\left(-\nabla p\cos\theta - \rho\frac{d^2\mathbf{x}}{dt^2}\right) = \frac{d\mathbf{x}}{dt}. \tag{4.128}$$

When (4.128) is multiplied by $\mathbf{x}$ and $\mathbf{x}(d^2\mathbf{x}/dt^2) = (d^2\mathbf{x}^2/dt^2)/2 - (d\mathbf{x}/dt)^2$ is used and then the average over the ensemble averaging is taken, a diffusional equation for $\langle \mathbf{x}^2\rangle$ emerges giving

$$\langle \mathbf{x}^2\rangle = 2\left\langle\frac{K'}{\varepsilon'}\right\rangle\frac{\rho}{\mu}\left[\left\langle\frac{K'}{\varepsilon'}\frac{1}{\mu}\nabla p\cos\theta\right\rangle^2 - \left(\left\langle\frac{K'}{\varepsilon'}\right\rangle\frac{1}{\mu}\nabla p\langle\cos^2\theta\rangle\right)^2\right]t$$
$$+ \left(\left\langle\frac{K'}{\varepsilon'}\right\rangle\frac{1}{\mu}\nabla p\langle\cos^2\theta\rangle\right)^2 t^2, \tag{4.129}$$

which, when compared to (4.127), gives

$$D^d = \left\langle \frac{K'}{\varepsilon'} \right\rangle \frac{\rho}{3\mu} \left[\left\langle \frac{K'}{\varepsilon'} \frac{1}{\mu} \nabla p \cos^2 \theta \right\rangle^2 \right.$$

$$\left. - \left(\left\langle \frac{K'}{\varepsilon'} \right\rangle \frac{1}{\mu} \nabla p \left\langle \cos^2 \theta \right\rangle \right)^2 \right]. \quad (4.130)$$

For a given K'/ε' and θ, we have $\langle K'/\varepsilon' \rangle \langle \cos^2 \theta \rangle \equiv K/\varepsilon$, such that as expected from (4.125) we have $-(K/\varepsilon)(1/\mu)\nabla p = \mathbf{u}_p$. Then

$$D^d = \frac{\rho a_1}{\mu^3} (\nabla p)^2 = a_2 u_p^2 \quad \text{or} \quad \frac{D^d}{\alpha_f} = a_3 Pe^2, \quad (4.131)$$

where a_1 represents the porous medium, a_2 and a_3 represent both the porous medium and the fluid, and $Pe = u_p K^{1/2}/\alpha_f = u_D K^{1/2}/(\varepsilon \alpha_f)$. Note that it is only *accidental* that the Pe^2 relationship found is similar to that of the Taylor-Aris treatment, because no direct allowance for the molecular diffusion was made in the development leading to (4.131). However, it has been assumed that in each channel the velocity is uniform (indicative of *infinite* lateral diffusion) when (4.128) was developed.

In his *geometric treatment*, Scheidegger assumes that $P \neq P(\mathbf{u}_p)$ allowing for *no streamline* crossing of the fluid particles, and finds

$$D^d = a_4 u_p \quad \text{or} \quad \frac{D^d}{\alpha_f} = a_5 Pe, \quad (4.132)$$

which has been found to be the high Pe asymptote for the disordered media (the mechanical dispersion, independent of molecular diffusion). In a further examination of this method, Todorovic (1970) finds that D^d is *not* Gaussian and that it depends on *two parameters*, which in turn depend on the hydrodynamics of the channel flow. An expression similar to above has been obtained by Georgiadis and Catton (1988).

4.5.2 DE JOSSELIN DE JONG PURELY GEOMETRIC MODEL

This analysis (De Josselin De Jong, 1958), which is similar to that of Scheidegger, is concerned with the velocity dispersion in disordered media. The basic assumptions are given here.

- The medium is made of spheres with *elementary channels* of tetrahedral cross sections of uniform length ℓ (but the specific geometry is *never* used in the analysis).

- The velocity in the channels is uniform (based on a very small channel length).

4.5 Dispersion in Disordered Structures—Simplified Hydrodynamics

- This uniform velocity is proportional to $\cos\theta$, where θ is the angle between the channel axis and $\mathbf{u}_D/u_D$. (This is different than that considered in the Brownian motion.)

- At any location (r,θ,ϕ), the choice of the direction between θ, ϕ and $\theta+d\theta$, $\phi+d\phi$ is distributed proportional to the *fraction* (of total flow) of fluid flowing in each of these directions.

The results of the analysis for $Pe = u_D\ell/(\varepsilon\alpha_f)$ are

$$\frac{D_\perp^d}{\alpha_f} = \frac{3}{16} Pe, \tag{4.133}$$

$$\frac{D_\parallel^d}{\alpha_f} = \frac{1}{6}\left(\lambda + \frac{3}{4} - 0.577\right) Pe, \tag{4.134}$$

where λ is found from the *transcendental* equation $(\lambda - 3/2 - 0.577)N_o = 3L_o/(2\ell) = e^{2\lambda}/2$. For spherical particles, $\ell = R/6$, where R is the radius of the sphere. L_o is the length along $\mathbf{u}_D$ associated with the *maximum* number of fluid particles traveling this distance at a given time t_o. This makes L_o the length of the medium along $\mathbf{u}_D$. Therefore, (4.133) and (4.134) relate D^d to $\ln L_o$. This relation shows a length *dependence* without any asymptotic (large length) behavior. A similar treatment is given by Haring and Greenkorn (1970), except distributions in the pore radius and the pore length are allowed using a *two-parameter* probability distribution function.

4.5.3 SAFFMAN INCLUSION OF MOLECULAR DIFFUSION

Saffman (1959a,b, 1960) analyzed the dispersion in porous media as being analogous to the Brownian motion. He also used the ensemble-averaged random displacement (or motion) of the fluid particles. This motion is the sum of the Brownian motion and a net motion due to the presence of a pressure gradient. The underlying assumptions are given here.

- The Stokes flow is assumed in the channels.

- Circular channel of uniform radius R and uniform length ℓ ($R/\ell \simeq 1/10$ to $1/5$) is assumed.

- Permeability is assumed to be given as $K = \varepsilon R^2/24$.

- The results are for $Pe = u_D\ell/(\varepsilon\alpha_f) < 8\ell^2/R^2 = 200$ to 800.

- The axial (along the channels) total diffusivity is that of Taylor-Aris, i.e., $\alpha_f + R^2 u_c^2/(48\alpha_f)$, where u_c is given later.

The Darcean flow is given by

$$u_D = -\frac{K}{\mu}\frac{dp}{dx}. \tag{4.135}$$

He begins by taking the x-axis to be the direction of the Darcean flow $\mathbf{u}_D$, θ as the angle that $\boldsymbol{\ell}$ makes with x, and ϕ as the azimuthal angle. The average velocity u_c in any circular channel is proportional to the pressure drop between the ends of the channel divided by the channel length, i.e., the pressure gradient along the pore is $\nabla p \cos\theta$. Then u_c is found from the Hagen-Poiseuille equation, (2.9), as

$$u_c = -\frac{R^2}{8\mu}\frac{dp}{dx}\cos\theta. \qquad (4.136)$$

Also, since K is taken to be $\varepsilon R^2/24$ (note that this is only accurate for $\varepsilon \to 1$), we have from (4.135)–(4.136)

$$u_c = \frac{3}{\varepsilon}\cos\theta\, u_D. \qquad (4.137)$$

Note that this channel velocity is not the same as the pore velocity. The factor 3 is just the result of the relationship assumed for the permeability. The *ensemble-averaged Lagrangian correlation function* (or covariance) of the velocity u at time t and a later time t' is $\langle u(t)u(t')\rangle$. The displacement is $x(t) = \int_0^t u(t')\,dt'$, and we have

$$\frac{d}{dt}\langle x^2(t)\rangle = 2\langle x(t)u(t)\rangle = 2\int_0^t \langle u(t)u(t')\rangle\,dt'$$
$$= 2\int_0^t R(t-t')\,dt' = 2\int_0^t R(\tau)\,d\tau, \qquad (4.138)$$

where $R(\tau)$ is the *correlation function of the velocity u* for a time interval τ, and the coordinate axes are chosen such that $\langle u\rangle = \langle x\rangle = 0$. Then (4.138) can be written as

$$\langle x^2\rangle = 2\int_0^t (t-\tau)R(\tau)\,d\tau. \qquad (4.139)$$

When $t \gg \tau$, the dispersion is given as

$$\frac{\langle x^2\rangle}{2t} = \int_0^t R(\tau)\,d\tau = D \qquad (4.140)$$

with $t \to \infty$.

Since u and x are taken to move with the mean velocity, we have

$$u = (u_c - u_B)\cos\theta - \frac{u_D}{\varepsilon} = \frac{u_D}{\varepsilon}(3\cos^2\theta - 1) - u_B\cos\theta, \qquad (4.141)$$

where $\theta = \theta(t)$ and u_B is the *random* part of the velocity along the capillary due to a diffusion process with a *total* diffusivity equal to $\alpha_f + R^2 u_c^2/(48\alpha_f)$,

4.5 Dispersion in Disordered Structures—Simplified Hydrodynamics

i.e., the Taylor-Aris diffusion (or diffusion plus axial dispersion). The motion u_B is similar to the Brownian motion in the molecular diffusion with the properties

$$\langle u_B \rangle = 0, \qquad \int_0^\infty \langle u_B(0)u_B(\tau) \rangle \, d\tau = \alpha_f + \frac{R^2 u_c^2}{48\alpha_f}. \qquad (4.142)$$

The ensemble-averaged covariance, using (4.141) and noting that $\langle u_B \rangle = 0$, is

$$\langle u(\theta)u(\theta') \rangle = \frac{u_D^2}{\varepsilon^2} \langle (3\cos^2\theta - 1)(3\cos^2\theta' - 1) \rangle$$
$$+ \langle \cos\theta \cos\theta' u_B(\theta) u_B(\theta') \rangle. \qquad (4.143)$$

The ensemble average in the last term on the right-hand side is taken over θ only. This is because only during the period where the fluid particle is in the channel, is the Lagrangian correlation meaningful. Then

$$\langle \cos\theta \cos\theta' u_B(\theta) u_B(\theta') \rangle = \int_0^1 \cos^2\theta \langle u_B(\theta) u_B(\theta') \rangle \, d\cos\theta, \qquad (4.144)$$

where $\langle u_B(\theta) u_B(\theta') \rangle$ depends on τ only. Now using (4.142) and (4.137), (4.144) must be integrated as shown in (4.140) to give the contribution of the last term in (4.143) to the diffusivity. This leads to

$$\int_0^t \langle \cos\theta \cos\theta' u_B(\theta) u_B(\theta') \rangle \, d\tau = \int_0^1 \left(\alpha_f + \frac{R^2 u_c^2}{48\alpha_f} \right) \cos^2\theta \, d\cos\theta$$
$$= \frac{\alpha_f}{3} + \frac{3}{80} \frac{R^2 u_D^2}{\varepsilon^2 \alpha_f}. \qquad (4.145)$$

The first term is equivalent to the *effective thermal conductivity* (for $k_s = 0$), and this topic was discussed in Section 3.1. This predicted effective thermal conductivity *does not* agree with the experimental results *or* with the correlations and predictions of others. This is due to the many simplifications made in the spatial and temporal integrations, where *no structural* information was included in the analysis. The second term makes a Pe^2 contribution. The only contribution to the first term on the right-hand side of (4.143) is from those fluid particles that begin their travel in the channel and during τ remain in the channel. The probability for their occurrence is constructed by allowing for the transient transport of the tracers by diffusion (Taylor-Aris) and by convection (u_c) in the channel. When this probability is determined from the transport differential equation (*Kolmogorov equation*) and used in the ensemble, (4.143), and then temporal, (4.140), averages are performed and the contributions added to (4.145), the result for the total dispersion (*not* including the effective thermal conductivity) is

$$\frac{D_\parallel^d}{\alpha_f} = \frac{Pe}{6} \left(\ln 1.5 Pe - \frac{17}{12} - \frac{1}{200} Pe^2 \right), \qquad (4.146)$$

where $Pe = u_D \ell/(\varepsilon \alpha_f)$ and $\ell/R = 5$ is used. This $Pe \ln Pe$ relation is also found in a more exact analysis of Koch and Brady (1985), which will be discussed in Section 4.6.

The lateral dispersion is found along the same line (for y- and z-axis being perpendicular to $\mathbf{u}_D$, and for ϕ being the angle between the y-axis and the projection of the channel on the y-z plane). The velocity component parallel to the y-axis is $u_c \sin\theta \cos\phi$. Then the counterpart of (4.140) for v (y-component) is

$$v = \frac{3u_D}{\varepsilon} \cos\theta \sin\theta \cos\phi + u_B \sin\theta \cos\phi \qquad (4.147)$$

and the lateral dispersion is $\int_0^\infty \langle v(0)v(\tau)\rangle \, d\tau$. Substitution of (4.147), expansion of the terms, and integration, using the same approach as that used for the axial dispersion, lead to

$$\frac{D_\parallel^d}{\alpha_f} = \frac{3}{16} Pe + \frac{1}{1000} Pe^2, \qquad (4.148)$$

for $\ell/R = 5$. The predicted results for $D_\parallel^d$ are *larger* than the experimental results. This could be due to lack of any constraints on the lateral displacements.

4.5.4 HORN METHOD OF MOMENTS

Horn (1971) expands upon the formulations of Aris (1956) and develops a general method of moments, which is briefly reviewed here. The developments are along the following lines.

- Dispersion is related to the width of the spread in the temperature disturbance—much like the displacements discussed in the previous sections. The width of the spread is expressed in terms of the *moments* of the *temperature* field taken along the direction of the flow.

- The analysis leads to the determination of the contribution to the dispersion tensor due to the Taylor-type interaction between *velocity* and *temperature disturbances*. It *allows* for anisotropic molecular diffusivity, however, it is taken to be *isotropic*.

- In order to allow for the cases where the conductivity of one of the phases is *infinite*, the interfacial heat flux is expressed in terms of a *prescribed* heat transfer coefficient.

- A solution for the case of $k_f \to \infty$ and a packing of spherical particles is found, although the method is more general.

The porous medium is generated by translating the region X in the y-z plane (in the case of a two-dimensional unit cell, X is three-dimensional)

4.5 Dispersion in Disordered Structures—Simplified Hydrodynamics

along the x-axis where the velocity component is u_p (the pore velocity). The cross section X is occupied by the different phases (an example would be a unit cell with the particle located in its center). The boundaries of X are *impermeable* and *adiabatic*. The energy equation written in terms of the *volumetric enthalpy* is

$$\frac{\partial i_f}{\partial t} + \mathbf{u}_f \cdot \nabla i_f = \alpha_f \nabla^2 i_f \quad \text{in } V_f, \tag{4.149}$$

$$\frac{\partial i_s}{\partial t} = \alpha_s \nabla^2 i_s \quad \text{in } V_s. \tag{4.150}$$

At the solid-fluid interface we have

$$-\mathbf{n} \cdot \mathbf{q} = \frac{h_{sf}}{(\rho c_p)_f}\left(i_f - \frac{i_s}{C}\right) \quad \text{on } A_{fs}, \tag{4.151}$$

where $C = (\rho c_p)_s/(\rho c_p)_f$, h_{sf} is the *average* (around the particle) *interstitial heat transfer coefficient*, and $\mathbf{n}$ points from the fluid phase into the solid phase. The volume-average symbol $\langle\ \rangle$ has been *dropped*. A single equation can be written representing both (4.149) and (4.150) by using $i, \mathbf{u}$, and α such that they take on the solid- or fluid-phase values depending on whether the *phase distribution function* $a(\mathbf{x})$ lies in V_s or V_f. This leads to

$$\frac{\partial i}{\partial t} + u\frac{\partial i}{\partial x} + \mathbf{u} \cdot \nabla_1 i = \alpha\frac{\partial^2 i}{\partial x^2} + \alpha\nabla_1^2 i, \tag{4.152}$$

where $\mathbf{u} \cdot \nabla_1$, ∇_1^2, and $\mathbf{x}_1$ denote the *components* in the X space. In the pores we have $\mathbf{u} = \mathbf{u}_p$, and in the solid phase we have $\mathbf{u} = 0$, where $\mathbf{u}_p$ in general varies across the pore but in the following is treated as a constant.

(A) MOMENT TRANSFORMATION OF VOLUMETRIC ENTHALPY

The volumetric enthalpy *disturbance* is introduced at $t = 0$ over a *finite* region. The time-dependent distribution of this disturbance is called the *peak*. The peak is characterized by

$$\mu^n(\mathbf{x}_1, t) = \int_{-\infty}^{\infty} x^n i(\mathbf{x}_1, x, t)\, dx, \tag{4.153}$$

which is an integration along the flow direction x with a *weighting* function x^n. This is called the *moment transformation*, and the power n determines the *order*. Only the first *three* ($n = 0, 1, 2$) moments will be needed. When combined with the integration over the X space, we have

$$M^n(t) = \int_X \int_{-\infty}^{\infty} x^n i(\mathbf{x}_1, x, t)\, dV, \tag{4.154}$$

where dV denotes a volume element of X. The *center* X^* of the peak is defined as

$$X^* = \frac{M^1(t)}{M^0(t)} = u^*t, \qquad (4.155)$$

where u^* is the *asymptotic center velocity*. The *width* s of the peak is defined by

$$s^2(t) = \frac{M^1(t)}{M^0(t)} - X^{*2}(t) = \frac{M^{*2}(t)}{M^0(t)} = 2D^*t, \qquad (4.156)$$

where M^{*2} is the second moment about the center of the peak and D^* is the *asymptotic dispersion coefficient*. If u, ρc_p, and α are position-independent in X, then the peak becomes a *Gaussian* distribution.

The *moment* equations are obtained from (4.152) and (4.153) and are

$$\frac{\partial \mu^0}{\partial t} = -\mathbf{u} \cdot \nabla_1 \mu^0 + \alpha \nabla_1^2 \mu^0, \qquad (4.157)$$

$$\frac{\partial \mu^1}{\partial t} = u\mu^0 - \mathbf{u} \cdot \nabla_1 \mu^1 + \alpha \nabla_1^2 \mu^1, \qquad (4.158)$$

$$\frac{\partial \mu^2}{\partial t} = 2u\mu^1 + 2\alpha \mu^0 - \mathbf{u} \cdot \nabla_1 \mu^2 + \alpha \nabla_1^2 \mu^2, \qquad (4.159)$$

with the *boundary conditions*

$$-\alpha \mathbf{n} \cdot \nabla \mu^n = \frac{h}{(\rho c_p)_f} \left(\mu_f^n - \frac{\mu_s^n}{C} \right) \quad \text{on } A_{fs}, \qquad (4.160)$$

where we have used $\mathbf{u} = 0$ on A_{fs}.

(B) Solution for Longitudinal Dispersion Coefficient

Since the boundaries of X are impermeable and adiabatic, we have $\mathbf{u} \cdot \mathbf{n} = 0$, and $\partial i/\partial n = 0$. Then

$$\int_X \mathbf{u} \cdot \nabla \mu^n \, dV = \int_{A_X} \mathbf{n} \cdot \mathbf{u} \mu^n \, dA = 0 \qquad (4.161)$$

and

$$\int_X \alpha \nabla^2 \mu^n \, dV = \int_{A_X} \alpha \mathbf{n} \cdot \nabla \mu^n \, dA = 0. \qquad (4.162)$$

Because (4.157) has no source term, the asymptotic equilibrium solution is

4.5 Dispersion in Disordered Structures—Simplified Hydrodynamics

$$\mu^0 = a_1 \frac{\rho c_p}{(\rho c_p)_f}, \qquad (4.163)$$

where a_1 is a constant that is taken as unity, and we have

$$\mu^0 = \frac{\rho c_p}{(\rho c_p)_f}. \qquad (4.164)$$

In (4.158) $u\mu^0 = u\rho c_p/(\rho c_p)_f$ is the *sink* term for transport in the X space. Since the boundaries of X are impermeable and adiabatic, i.e., (4.161)–(4.162) hold, then the volumetric enthalpy must be *time-dependent*. The solution is [details are given in Horn (1971)]

$$\mu^1 = \frac{\rho c_p}{(\rho c_p)_f} \overline{u} t + \psi, \qquad (4.165)$$

where ψ is a *function* of position (X space) *only* and $\overline{u}$ is the average of u over X with $\rho c_p/(\rho c_p)_f$ as the weight, i.e.,

$$\overline{u} = \frac{\displaystyle\int_X \frac{\rho c_p}{(\rho c_p)_f} u \, dV}{\displaystyle\int_X \frac{\rho c_p}{(\rho c_p)_f} \, dV} = \frac{\varepsilon}{\varepsilon + \frac{(\rho c_p)_s}{(\rho c_p)_f}(1-\varepsilon)} u_p, \qquad (4.166)$$

since $u = 0$ in V_s. Using (4.161)–(4.162), (4.165), and (4.155), we have

$$\frac{M^1}{M^0} = \overline{u} t + \frac{\overline{\psi}}{k}. \qquad (4.167)$$

Using (4.161)–(4.162), (4.159) becomes

$$\frac{dM^2}{dt} = 2\int_X u\mu^1 \, dV + 2\int_X \alpha \mu^0 \, dV. \qquad (4.168)$$

Now inserting (4.164)–(4.165) in the preceding and dividing by M^0, we have

$$\frac{1}{M^0} \frac{dM^2}{dt} = 2\overline{u}^2 t + 2\left(\overline{\frac{u\psi}{k}}\right) + \overline{\alpha}. \qquad (4.169)$$

Integration leads to

$$\frac{M^2}{M^0} = \overline{u}^2 t^2 + 2\left[\left(\overline{\frac{u\psi}{k}}\right) + \overline{\alpha}\right] t + a_2, \qquad (4.170)$$

where a_2 is a constant which can be taken to be zero. Using this, we have

$$\frac{M^2}{M^0} = 2\left[\left(\overline{\frac{u\psi}{k}}\right) - \overline{u}\left(\overline{\frac{\psi}{k}}\right) + \overline{\alpha}\right] t = 2D^* t, \qquad (4.171)$$

where (4.155) and (4.156) have been used. Then, solving for D^*, we have

$$D^* = \overline{[(u-\overline{u})\psi/k]} + \overline{\alpha} = D_\parallel^d + \overline{\alpha}, \qquad (4.172)$$

where $D_\parallel^d + \overline{\alpha}$ is equivalent to the Taylor-Aris contribution to the total effective thermal diffusivity. The equation for ψ is obtained by expressing the *longitudinal total effective thermal diffusivity* $D_\parallel^d + \overline{\alpha}$ as the maximum of a functional [details are given in Horn (1971)]. This involves selection of *trial* functions and maximization with respect to parameters of these functions.

(C) SOLUTION FOR $k_s \to \infty$

The $\rho c_p/(\rho c_p)_f$ *weighted* average axial velocity is given by (4.166). From this

$$\frac{\rho c_p}{(\rho c_p)_f}(u-\overline{u}) = \begin{cases} \dfrac{C(1-\varepsilon)}{\varepsilon + C(1-\varepsilon)} u_p & \text{in } V_f \\[2mm] \dfrac{\varepsilon}{\varepsilon + C(1-\varepsilon)} u_p & \text{in } V_s. \end{cases} \qquad (4.173)$$

The solution to the ψ equation is

$$\psi = \begin{cases} \dfrac{\varepsilon u_p}{\varepsilon + C(1-\varepsilon)}\left[\dfrac{R^2}{6\alpha_s} + \dfrac{CR}{3h_{sf}/(\rho c_p)_f}\right] & \text{in } V_f \\[3mm] \dfrac{C\varepsilon u_p}{\varepsilon + C(1-\varepsilon)}\dfrac{r^2}{6D} & \text{in } V_s, \end{cases} \qquad (4.174)$$

where $C = (\rho c_p)_s/(\rho c_p)_f$. Now using (4.172) with (4.173) and (4.174) and completing the integration (averaging), we have

$$D_\parallel^d = \frac{C^2\varepsilon^2(1-\varepsilon)}{[\varepsilon + C(1-\varepsilon)]^3}\left[\frac{Pe^2}{15C}\left(\frac{\alpha_f}{\alpha_s}\right) + \frac{Pe}{3}\frac{u_D(\rho c_p)_f}{\varepsilon h_{sf}}\right], \quad k_s \to \infty, \qquad (4.175)$$

where $Pe = u_D R/(\varepsilon \alpha_f)$ and $u_D = \varepsilon u_p$.

For *finite* values of k_s, the ψ equation has to be solved for the fluid domain and construction of the trial functions can be fairly involved and are not attempted by Horn. In the preceding, u_p was taken to be uniform across the pore; therefore, the sink term in the ψ equation was uniform in the fluid domain. If this restriction was not made, determination of ψ would have been *even* more complicated, especially if the full Navier-Stokes equations had to be solved. This would have involved numerical solutions such as that of Eidsath et al. or the analytical treatment of Koch and Brady and Koch et al.

The generalization of the Horn method of moments has been developed by Brenner where a general class of periodic structures is reviewed.

4.6 Dispersion in Disordered Structures—Particle Hydrodynamics

Comparing to the earlier treatments of disordered structures, a rigorous description of the flow in the pores is included in the analysis of Koch and Brady (1985). They obtain an ensemble-averaged energy equation by using the concept of the representative elementary volume. The averaging can be viewed as averaging over an ensemble of *realizations* of the packed bed, each realization having a different *arrangement* of particles. The requirements for the representative elementary volume was given earlier. As was done in the investigation of the periodic structures, they are able to obtain closed-form solutions for the dispersion tensor. This is achieved by imposing the following simplifications/approximations.

- Long-time asymptotic behavior (Taylor-Aris dispersion).
- Dilute concentration of particles (only fluid phase is *continuous*).
- Single-particle-based velocity disturbance found as the solution to the *Brinkman* momentum equation. This solution is used along with the point-force treatment discussed in Section 4.5.
- Slowly varying (spatial) average temperature fields.

Their development, which is subject to these conditions, is reviewed, and then their predictions are compared with the available experimental results.

4.6.1 LOCAL VOLUME AVERAGING

Allowing for differences in k and in ρc_p between the fluid and the solid phases and by using the volumetric enthalpy $i = \rho c_p T$, the heat flux equation becomes

$$\mathbf{q} = \begin{cases} \mathbf{u} i - \alpha_f \nabla i & \text{fluid,} \\ -\alpha_s \nabla i & \text{solid,} \end{cases} \quad (4.176)$$

or

$$\mathbf{q} = \mathbf{u} i - \alpha \nabla i, \quad (4.177)$$

where $\mathbf{u} = 0$ in the solid phase and α will take the fluid- or solid-phase values depending on where $\mathbf{x}$ lies. The boundary condition at the solid-fluid interface is

$$\alpha_s \mathbf{n} \cdot \nabla i = \alpha_f \mathbf{n} \cdot \nabla i. \quad (4.178)$$

Note that for *mass transfer* C is the *ratio of the solubilities* of the solute in the solid and fluid such that there is a *jump* in the concentration across the interface. Since we will use the results of Koch and Brady for both heat and mass transfer, we will include C in the analysis. Assuming that the local thermal equilibrium exists (although, as before, at the pore-level, a small difference between the disturbance temperatures of the phases is allowed in arriving at the expressions for the effective thermal conductivity and the dispersion tensor) and averaging over the representative elementary volume $V = V_f + V_s$, the energy equation becomes

$$\frac{\partial}{\partial t}\langle i \rangle + \nabla \cdot \langle \mathbf{q} \rangle = 0, \tag{4.179}$$

where again the *integral* (which accounts for conduction across the interface) emerging from the application of the theorem on the average of a divergence, (2.85), is not included in (4.179). However, *later* they include this conduction contribution in an ad hoc manner by making use of the effective thermal conductivity relation developed by Jeffrey (1973).

Next, the disturbance in $\mathbf{u}, i$, and α is introduced as

$$\mathbf{u} = \langle \mathbf{u} \rangle + \mathbf{u}', \tag{4.180}$$

$$i = \langle i \rangle + i', \tag{4.181}$$

$$\alpha = \langle \alpha \rangle + \alpha'. \tag{4.182}$$

Then (4.177) becomes

$$\langle \mathbf{q} \rangle = \langle \mathbf{u} \rangle \langle i \rangle + \langle \mathbf{u}' i' \rangle - \langle \alpha \rangle \nabla \langle i \rangle - \langle \alpha' \nabla i' \rangle, \tag{4.183}$$

where $\langle \mathbf{u} \rangle = \mathbf{u}_D$.

From the results of Jeffrey, Koch and Brady make the following substitution

$$-\langle \alpha \rangle \nabla \langle i \rangle - \langle \alpha' \nabla i' \rangle = -\alpha_f \nabla \langle i \rangle - \alpha_f \left(\frac{k_s}{k_f} - 1 \right)$$

$$\times \int_{|\mathbf{x} - \mathbf{r}_1| \leq R} P(\mathbf{r}_1) \nabla \langle i(\mathbf{x}|\mathbf{r}_1) \rangle_1 \, d\mathbf{r}_1, \tag{4.184}$$

where $P(\mathbf{r}_1)$ is the *probability density function* for finding a particle at $\mathbf{r}_1$, which is taken to be *uniform* and equal to $3(1 - \varepsilon)/(4\pi R^3)$, and where R is the radius of the particle. $\langle i(\mathbf{x}|\mathbf{r}_1) \rangle_1$ is the *conditional ensemble average* at point $\mathbf{x}$ *conditional* on the presence of a particle centered at $\mathbf{r}_1$. The integral is over the volume of the particle centered at $\mathbf{r}_1$. Note that in dealing with the effective thermal conductivity through volume averaging, an expression similar to (4.184) was arrived at in Section 3.2. The average enthalpies in the solid and fluid phases are

$$\langle i \rangle_f = \frac{\langle i \rangle}{1 - (1 - \varepsilon)(1 - C)}, \quad \langle i \rangle_s = C \langle i \rangle_f. \tag{4.185}$$

4.6 Dispersion in Disordered Structures—Particle Hydrodynamics

The heat flux vector (4.183) is written as

$$\langle \mathbf{q} \rangle = (1+\gamma) \langle \mathbf{u} \rangle \langle i \rangle - \mathbf{D} \cdot \nabla \langle i \rangle, \qquad (4.186)$$

where

$$\gamma = \frac{(1-\varepsilon)(1-C)}{1-(1-\varepsilon)(1-C)}. \qquad (4.187)$$

The *correction* γ made to the convective heat flux is equivalent to replacing the enthalpy flow averaged over the bed with the average enthalpy over the fluid and using the *intrinsic* fluid-phase average velocity. The diffusion-dispersion tensor, i.e., the total thermal diffusivity, can be written as

$$\mathbf{D} = \underbrace{\alpha_f \mathbf{I}}_{\substack{\text{fluid} \\ \text{molecular} \\ \text{diffusivity } \mathbf{D}^f,}} + \underbrace{\frac{\alpha_f \left(\frac{k_s}{k_f}-1\right)}{\nabla \langle i \rangle} \int_{|\mathbf{x}-\mathbf{r}_1|\leq R} P(\mathbf{r}_1) \nabla \langle i(\mathbf{x}|\mathbf{r}_1)\rangle_1 \, d\mathbf{r}_1}_{\substack{\text{solid contribution to} \\ \text{molecular diffusivity } \mathbf{D}^s,}}$$

$$-\underbrace{\frac{\langle \mathbf{u}' i' \rangle - \gamma \langle \mathbf{u} \rangle \langle i \rangle}{\nabla \langle i \rangle}}_{\substack{\text{contribution} \\ \text{from} \\ \text{dispersion } \varepsilon \mathbf{D}^d}} \qquad (4.188)$$

$$\mathbf{D} = \alpha_f \mathbf{I} + \mathbf{D}^s + \varepsilon \mathbf{D}^d = \frac{\mathbf{K}_e}{(\rho c_p)_f} + \varepsilon \mathbf{D}^d. \qquad (4.189)$$

The expression for $\mathbf{K}_e$ is taken from Jeffrey and that for $(\rho c_p)_f / (\rho c_p)_s \to \infty$ and is

$$\mathbf{K}_e = \left[-1.5(1-\varepsilon) + 0.588(1-\varepsilon)^2\right] \mathbf{I}. \qquad (4.190)$$

In the following, $\mathbf{D}^d$ will be evaluated for various asymptotic conditions.

4.6.2 Low Peclet Numbers

For $(1-\varepsilon) \to 0$ and $Pe = u_D R/\alpha_f \ll 1$, approximate expressions for $\mathbf{u}'$ and i' are developed later. The dispersion contribution in (4.189) can be written as

$$\mathbf{D}^d \cdot \nabla \langle i \rangle = -\langle \mathbf{u}' i' \rangle + \gamma \langle \mathbf{u} \rangle \langle i \rangle. \qquad (4.191)$$

Using the definition of the conditional average and considering low Peclet number flows where the second term makes a negligible contribution, (4.191) becomes

$$\mathbf{D}^d \cdot \nabla \langle i \rangle = -\int_{|\mathbf{x}-\mathbf{r}_1|\leq R} P(\mathbf{r}_1) \langle \mathbf{u}' \rangle_1 \langle i' \rangle_1 \, d\mathbf{r}_1, \qquad (4.192)$$

208   4. Convection Heat Transfer

where $\langle \mathbf{u}' \rangle_1 = \langle \mathbf{u} \rangle_1 - \langle \mathbf{u} \rangle = \langle \mathbf{u} \rangle_1 - \mathbf{u}_D$ and $\langle i' \rangle_1 = \langle i \rangle_1 - \langle i \rangle$. The solution for $\langle \mathbf{u} \rangle_1$ is found by solving the Brinkman equation

$$-\nabla \langle p \rangle_1 + \mu \nabla^2 \langle \mathbf{u} \rangle_1 - \frac{\mu}{K} \langle \mathbf{u} \rangle_1 = 0. \tag{4.193}$$

The solution is given by Acrivos et al. (1980), which satisfies the no-slip condition on the particle and tends to $\mathbf{u}_D$ for $r > o(K^{1/2})$. This solution is

$$\langle \mathbf{u} \rangle_1 = \mathbf{u}_D \left[1 + o(K^{1/2}) \right]. \tag{4.194}$$

The permeability is *approximated* as $K = (2/9)[R^2/(1-\varepsilon)]$ and is expected to be valid only when $\varepsilon \to 1$. Note that this is significantly different than the Carman-Kozeny equation. Next, an equation for $\langle i' \rangle_1$ is found as

$$\langle \mathbf{u} \rangle \cdot \nabla \langle i' \rangle_1 - \alpha_f \nabla^2 \langle i' \rangle_1 + \langle \mathbf{u}' \rangle_1 \cdot \nabla \langle i \rangle = 0. \tag{4.195}$$

Transformation $\mathbf{R} = K^{-1/2}(\mathbf{x} - \mathbf{r}_1)$ *places* the dominant convection contribution at $\mathbf{R} = o(1)$. Then (4.195) becomes (in *dimensionless* form)

$$Pe K^{1/2} \langle \mathbf{u} \rangle_1 \cdot \nabla_R \langle i' \rangle_1 - \nabla_R^2 \langle i' \rangle_1 = -Pe K \langle \mathbf{u}' \rangle_1 \cdot \nabla \langle i \rangle. \tag{4.196}$$

Using (4.194) for $\langle \mathbf{u} \rangle_1$ and solving (4.196) by Fourier transform gives

$$\langle \widehat{i'}(\mathbf{s}) \rangle_1 = -Pe K \frac{\langle \widehat{\mathbf{u}'}(\mathbf{s}) \rangle_1 \cdot \nabla \langle i \rangle}{s^2 - i Pe K^{1/2} \langle \mathbf{u} \rangle \cdot \mathbf{s}}, \tag{4.197}$$

where $\mathbf{s}$ is the *transform variable* of $\mathbf{R} \langle \widehat{\mathbf{u}}'(\mathbf{s}) \rangle_1$. The Fourier transform of the point-force velocity disturbance is given by Saffman (1973) as

$$\langle \widehat{\mathbf{u}'}(\mathbf{s}) \rangle_1 = \frac{6\pi}{K^{1/2}} \frac{\langle \mathbf{u} \rangle \cdot \left(\mathbf{I} - \frac{\mathbf{ss}}{s^2} \right)}{s^2 + 1}. \tag{4.198}$$

Returning to (4.192) and transforming this equation to $\mathbf{R}$-variable, we have

$$\mathbf{D}^d \cdot \nabla \langle i \rangle = -\frac{3Pe(1-\varepsilon)K^{3/2}}{4\pi} \int \langle \mathbf{u}'(\mathbf{R}) \rangle_1 \langle i'(\mathbf{R}) \rangle_1 \, d\mathbf{R}. \tag{4.199}$$

Using the convolution theorem in $\mathbf{s}$, we have

$$\mathbf{D}^d \cdot \nabla \langle i \rangle = -\frac{3Pe(1-\varepsilon)K^{3/2}}{4\pi(2\pi)^3} \int \langle \mathbf{u}'(-\mathbf{s}) \rangle \langle i'(\mathbf{s}) \rangle_1 \, d\mathbf{s}. \tag{4.200}$$

Since $\nabla \langle i \rangle$ is a constant, $\mathbf{D}^d$ is also a constant and is given by

$$\mathbf{D}^d = \frac{3}{4\pi^2} Pe^2 K^{1/2} \int \frac{\langle \mathbf{u} \rangle \cdot \left(\mathbf{I} - \frac{\mathbf{ss}}{s^2} \right) \langle \mathbf{u} \rangle \cdot \left(\mathbf{I} - \frac{\mathbf{ss}}{s^2} \right)}{(s^2+1)^2 (s^2 - i Pe K^{1/2} \langle \mathbf{u} \rangle \cdot \mathbf{s})} \, d\mathbf{s}. \tag{4.201}$$

4.6 Dispersion in Disordered Structures—Particle Hydrodynamics

Integration of (4.201) shows that the *off-diagonal* elements of the tensor are zero. The *transverse* component of $\mathbf{D}^d$ is

$$D_\perp^d = K^{-1/2}\left[\frac{1}{4} + \frac{3}{4PeK^{1/2}} - \frac{3}{4Pe^2K}\right.$$

$$\left. + \left(\frac{3}{2Pe^3K^{3/2}} - \frac{3}{4PeK^{1/2}}\right)\ln\left(PeK^{1/2}+1\right)\right], \quad (4.202)$$

when Pe is taken as a *definite* positive. The *longitudinal* component is

$$D_\parallel^d = K^{-1/2}\left[\frac{3}{4}PeK^{1/2} - 2 - \frac{3}{2PeK^{1/2}} + \frac{3}{Pe^2K}\right.$$

$$\left. + \left(\frac{3}{PeK^{1/2}} - \frac{3}{Pe^2K}\right)\ln\left(PeK^{1/2}+1\right)\right], \quad (4.203)$$

where the *dimensionless* permeability is $2/[9(1-\varepsilon)]$. Note that $D_\parallel^d$ and $D_\perp^d$ are made *dimensionless* using α_f.

For low Pe, (4.202) and (4.203) are given as

$$\left.\begin{array}{l} D_\perp^d = \dfrac{\sqrt{2}}{60}\dfrac{Pe^2}{(1-\varepsilon)^{1/2}} \\[2ex] D_\parallel^d = \dfrac{\sqrt{2}}{15}\dfrac{Pe^2}{(1-\varepsilon)^{1/2}} \end{array}\right\} \quad Pe^2 \ll 1 - \varepsilon \ll 1. \quad (4.204)$$

It should be noted that since the analysis is for $(1-\varepsilon) \to 0$, the dispersion contribution can be significant even at low Pe. This is inherent in the assumption that convection extends a distance of $2^{1/2}R/[3(1-\varepsilon)]$, resulting in the spread of the heat much *beyond* a distance of a particle diameter associated with the *pure conduction* contribution. For dense packings, one does not expect the dispersion contribution to be as large (for low Pe), and the permeability-porosity relation will be *substantially* different than that assumed earlier.

4.6.3 HIGH PECLET NUMBERS

By neglecting the molecular conduction, (4.195) becomes

$$\langle\mathbf{u}\rangle_1 \cdot \nabla_R \langle i'\rangle_1 = -K^{1/2}\langle\mathbf{u}'\rangle_1 \cdot \nabla\langle i\rangle, \quad (4.205)$$

with the solution given as

$$\langle i'\rangle_1 = -K^{1/2}\int_{-\infty}^{\eta}\frac{\langle\mathbf{u}'(\psi,\eta,\theta)\rangle_1 \cdot \nabla\langle i\rangle}{h_\eta|\langle\mathbf{u}(\psi,\eta,\theta)\rangle_1|}\,d\eta, \quad (4.206)$$

where η is the coordinate along the *streamlines* of the conditionally averaged velocity field $\langle \mathbf{u} \rangle_1$, which is taken to be along the bulk flow with the *metric coefficient* h_η. Also θ is the coordinate in the direction of rotational *invariance* and ψ is the third orthogonal coordinate. Also implied is that $\langle i' \rangle_1 \to 0$ as $\eta \to -\infty$. $\langle \mathbf{u}' \rangle_1$ is obtained from the integration of the Brinkman equation (4.193). Using (4.206) and the preceding in (4.200), with $\langle \mathbf{u}' \rangle_1$ transformed to the Fourier space, and using the convolution theorem leads to

$$\left. \begin{array}{l} D_\parallel^d = \dfrac{3}{4} Pe \\[6pt] D_\perp^d = 0 \end{array} \right\} \qquad Pe^{1/2} \gg 1,\ \varepsilon \to 1, \qquad (4.207)$$

where the result for $D_\parallel^d$ is in accord with the asymptote of (4.203), and supports that *molecular diffusion* plays no role at large Pe (pure mechanical dispersion caused by *stochastic* velocity field in random fixed beds). $D_\perp^d$ is not in accord with (4.202) because (4.207) is integrated along a streamline, and the transverse component of the velocity disturbance along any streamline is *zero*, as it must be for any streamline that possesses symmetry along the flow. By making a diffusion correction, the asymptote of (4.202) for *large* Pe is found and is

$$D_\perp^d = \frac{3\sqrt{2}}{8}(1-\varepsilon)^{1/2}, \qquad Pe \gg 1,\quad (1-\varepsilon) \ll 1, \qquad (4.208)$$

which is *independent* of Pe because transverse transport is only *by* molecular diffusion. Figure 4.11 shows the combined results of (4.202)–(4.203), the low Pe asymptote (4.204), and (4.207) and (4.208), which are for high Pe. Plotted are the *normalized* parameters

$$D^d K^{1/2} = D^d \frac{\sqrt{2}}{3(1-\varepsilon)^{1/2}}, \qquad Pe K^{1/2} = \frac{u_D R}{\alpha_f} \frac{\sqrt{2}}{3(1-\varepsilon)^{1/2}}. \qquad (4.209)$$

The high Pe asymptotes, i.e., $D_\parallel^d$ proportional to Pe and $D_\perp^d = $ constant, are those discussed earlier.

4.6.4 Contribution of Solid Holdup (Mass Transfer)

Since the developments of Koch and Brady are applicable to both heat and mass transfer except for the condition of jump across the solid-fluid interface (due to the difference in the solubility), in the following, we add their analysis for the holdup of solute in the particle. At high Pe the particle contributes to dispersion by trapping portions of diffusing species. We will not change the variables to those associated with mass transfer and,

4.6 Dispersion in Disordered Structures—Particle Hydrodynamics

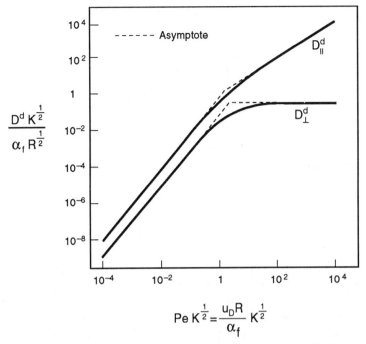

Figure 4.11 Low and high Peclet number behavior of the longitudinal and transverse thermal dispersion coefficients for disordered media. (From Koch and Brady, reproduced by permission ©1985 Cambridge University.)

therefore, we will have a jump in i across A_{sf}, but this jump is in the *concentration*.

The leading order effect of the solid thermal diffusivity is given by (4.192), and by allowing for a difference between the conditionally averaged fluid and solid enthalpies, we have

$$\mathbf{D}^d \cdot \nabla \langle i \rangle = -\frac{3Pe(1-\varepsilon)}{4\pi} \int_{|\mathbf{x}-\mathbf{r}_1|\leq 1} \langle \mathbf{u}' \rangle_1 (\langle i \rangle_1 - \langle i \rangle_s) \, d\mathbf{r}_1, \qquad (4.210)$$

where $\langle i \rangle_s$ is the enthalpy at equilibrium.

Using $\langle \mathbf{u}' \rangle = -\langle \mathbf{u} \rangle = -\langle \mathbf{u}_D \rangle$ inside the particle and the expression for $\langle i \rangle_s$ given right before (4.186), we have

$$\mathbf{D}^d \cdot \nabla \langle i \rangle = \frac{3Pe(1-\varepsilon)\langle \mathbf{u}_D \rangle}{4\pi} \int (\langle i \rangle_1 - C(1+\gamma)\langle i \rangle) \, d\mathbf{r}_1. \qquad (4.211)$$

The solid conditionally averaged enthalpy on the surface is

$$\langle i \rangle_1 = C \langle i \rangle_f = C(1+\gamma)\langle i \rangle \quad \text{at } |\mathbf{x} - \mathbf{r}_1| = 1, \qquad (4.212)$$

where γ is given by (4.187). The particle disturbance enthalpy is

$$\langle i' \rangle_1 = \langle i \rangle_1 - C(1+\gamma)\langle i \rangle. \qquad (4.213)$$

Then the solid energy equation becomes

$$\nabla^2 \langle i' \rangle_1 = -CPe(1-\gamma)^2 \langle \mathbf{u}_D \rangle \cdot \nabla \langle i \rangle \qquad (4.214)$$

$$\text{with } \langle i' \rangle = 0 \quad \text{at } |\mathbf{x} - \mathbf{r}_1| = 1. \qquad (4.215)$$

When the solution of (4.214) and (4.215), with the right-hand side being a constant, is found and substituted in (4.211), the result is

$$\left.\begin{array}{l} D^d_\parallel = \dfrac{1}{15} C(1+\gamma)^2 \dfrac{\alpha_f}{\alpha_s} Pe^2 (1-\varepsilon), \\[2mm] D^d_\perp = 0, \end{array}\right\} \quad Pe \gg 1. \qquad (4.216)$$

The effect of a solid heat capacitance is also referred to as the *holdup effect*. Reis et al. (1979) have also solved the combined fluid-solid problem by assuming a uniform velocity and the *lack* of the local thermal equilibrium. Their analysis includes a constant *interfacial heat transfer coefficient*, which appears as an extra term in (4.211). Their result, which is based on assumptions similar to those made by Taylor and by Koch et al., is the same as (4.216) except for the additional term due to that interstitial heat transfer coefficient at A_{sf}.

4.6.5 Contribution Due to Thermal Boundary Layer in Fluid

The fluid enthalpy distribution determined using the point force solution is *singular* at the particle surface. This is because the thermal diffusion in the *boundary layer* (which is important even though this boundary layer becomes thinner as Pe increases) has *not* been included. Solution of the boundary-layer convection-diffusion equation obtained by Koch and Brady leads to the extra *contribution* of the boundary layer as

$$D^d_\parallel = \frac{\pi^2}{6}(1-\varepsilon) Pe \ln Pe,$$

$$D^d_\perp = 0. \qquad (4.217)$$

This contribution is due to the no-slip condition at the solid surface and is present in any porous media. The $Pe \ln Pe$ behavior is *independent* of the details of the *microstructure*, but the coefficient *varies*. The $Pe \ln Pe$ contribution was also found by Saffman (Section 4.5.3).

4.6.6 COMBINED EFFECT OF ALL CONTRIBUTIONS

The various contributions made to the dispersion coefficient are summarized in Table 4.3 for various Peclet number regimes. Figure 4.12 shows these predictions of Koch and Brady (for $\varepsilon = 0.5$) along with the experimental results reported in Fried and Combarnous (1971), which were obtained from mass transfer experiments using impermeable particles. All the experimental data are for years prior to 1963 and the experimental results of Gunn and Pryce (1969) are *not* included. The experimental results and the *data reduction* method presented by Fried and Combarnous are for $(\rho c_p)_s = k_s = 0$.

The experiments involve introduction of a step change in the concentration of a *dilute solute* and observation of the resultant concentration distribution *downstream*. Then the data are fitted to the exponential distribution which is the solution to (4.179) and (4.186). This solution is (Fried and Combarnous)

$$\frac{\langle i \rangle}{\langle i(t=0) \rangle} = \frac{1}{2} \left\{ \operatorname{erfc} \left[\frac{x - (1+\gamma)u_D t}{(2D_\| t)^{1/2}} \right] \right.$$
$$\left. + \exp\left[\frac{(1+\gamma)u_D x}{D_\|}\right] \operatorname{erfc}\left[\frac{x + (1+\gamma)u_D t}{2(D_\| t)^{1/2}}\right] \right\}. \quad (4.218)$$

For the *transverse* dispersion, the solution is found to be

$$(1+\gamma)u_D \frac{\partial \langle i \rangle}{\partial x} = D_\perp \frac{\partial^2 \langle i \rangle}{\partial y^2}, \quad (4.219)$$

which leads to

$$\frac{\langle i \rangle}{\langle i(t=0) \rangle} = \frac{1}{2}\left[1 + \operatorname{erf}\left\{\frac{y}{2}\left[\frac{D_\perp x}{(1+\gamma)u_D}\right]^{-1/2}\right\}\right]. \quad (4.220)$$

Note that in Fried and Combarnous, the Peclet number is based on $(1+\gamma)u_D$ and $2R$, while Koch and Brady use u_D and R. A more complete review of the experimental methods is given later (Section 4.8).

The high Pe prediction of Koch and Brady is chosen to apply at $Pe \geq 1$. The values for $\mathbf{D}^s$ (which is *isotropic*) are those given by (4.190). The hydrodynamic treatment of Koch and Brady involves the leading order analysis of the Stokes (creeping) flow. Therefore, no *Reynolds number* effects are included. With this in mind, we should interpret their large Pe predictions as being for large Pr (or Sc), not for large Re. This *low Re* and *high Pe restriction* is not satisfied in *all* the experimental data given by Fried and Combarnous.

The agreement between the predictions and the experiments are good except for $Pe \leq 20$. The prediction of Jeffrey (1973) for $\mathbf{D}^s$ does not agree with the experimental results. Note that the improvement over the Maxwell

TABLE 4.3   THE LEADING BEHAVIOR OF THE DISPERSION COEFFICIENT FOR VARIOUS PECLET NUMBER REGIMES (KOCH AND BRADY, 1985)

$D_\parallel^d/\alpha_f$	$D_\perp^d/\alpha_f$
(a) $\dfrac{3\,(k_s/k_f - 1)}{k_s/k_f + 2}(1-\varepsilon)$ $+\dfrac{2^{1/2}}{15}\dfrac{Pe^2}{(1-\varepsilon)^{1/2}}$	(a) $\dfrac{3\,(k_s/k_f - 1)}{k_s/k_f}(1-\varepsilon)+$ $\dfrac{2^{1/2}}{60}\dfrac{Pe^2}{(1-\varepsilon)^{1/2}}$
(b) $\left[\dfrac{3}{4}\dfrac{PeK^{1/2}}{R^{1/2}} - 2 - \dfrac{3}{2}\dfrac{R^{1/2}}{PeK^{1/2}}\right.$ $+\dfrac{3R}{Pe^2 K} + \left(\dfrac{3R^{1/2}}{PeK^{1/2}} - \dfrac{3R^{3/2}}{Pe^3 K^{3/2}}\right)$ $\left. \times \ln\left(\dfrac{PeK^{1/2}}{R^{1/2}}+1\right)\right]\left(\dfrac{R}{K}\right)^{1/2}$	(b) $\left[\dfrac{1}{4} + \dfrac{3}{4}\dfrac{R^{1/2}}{PeK^{1/2}} - \dfrac{3}{4}\dfrac{R}{Pe^2 K}\right.$ $+\left(\dfrac{3}{2}\dfrac{R^{3/2}}{Pe^3 K^{3/2}} - \dfrac{3}{4}\dfrac{R^{1/2}}{PeK^{1/2}}\right)$ $\left. \times \ln\left(\dfrac{PeK^{1/2}}{R^{1/2}}+1\right)\right]\left(\dfrac{R}{K}\right)^{1/2}$
(c) $\dfrac{3}{4}Pe$	(c) $\dfrac{3(2)^{1/2}}{8}(1-\varepsilon)^{1/2}$
(d) $\dfrac{3}{4}Pe + \dfrac{1}{6}\pi^2(1-\varepsilon)Pe\ln Pe$ $+\dfrac{C}{15}(1+\gamma)^2\dfrac{\alpha_f(1-\varepsilon)Pe^2}{\alpha_s}$	(d) $\dfrac{63(2)^{1/2}}{320}(1-\varepsilon)^{1/2}Pe$

$Pe = u_D R/\alpha_f$, C and γ defined in text.
(a) For $Pe \ll (1-\varepsilon)^{1/2} \ll 1$, $PeK^{1/2}/R^{1/2} \ll 1$; (b) For $(1-\varepsilon)^{3/4} \ll Pe \ll 1$; (c) For $(1-\varepsilon)^{1/2} \ll Pe \ll 1$, $PeK^{1/2}/R^{1/2} \gg 1$; (d) For $Pe \gg 1$, the last term in $D_\parallel^d/\alpha_f$ is for mass transfer only.

prediction, i.e., the Jeffrey results for $\mathbf{K}_e/(\rho c_p)_f = \varepsilon \alpha_f \mathbf{I} + \mathbf{D}^s$, is for *dilute* concentration, i.e., (4.190) is valid only for $\varepsilon \to 1$.

Also note that up to $Pe = 1$, the effective molecular diffusion dominates. Therefore, the very low Pe results given in Table 4.3 do not make any significant contribution to the total diffusivity. With this in mind, and for Pr characteristics of gases (< 1), the significant dispersion contribution will be associated with $Re > 1$. This, in principle, requires inclusion of the inertial term in the momentum equation. However, as will be shown, the experimental results indicate that even for the inertial regime, Pe remains the only parameter on which $\mathbf{D}^d$ depends (Section 4.8.4).

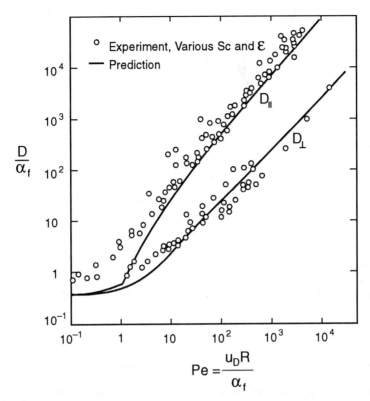

Figure 4.12 The prediction of Koch and Brady (1985) (for random arrangement of spheres) for longitudinal and transverse total effective thermal diffusivity, compared with the experimental results of many investigators as presented in Fried and Combarnous. (From Koch and Brady, reproduced by permission ©1985 Cambridge University.)

4.7 Properties of Dispersion Tensor

Given here are some of the properties of the dispersion tensor.

- The conduction-convection energy equation is given as

$$\left[\varepsilon\,(\rho c_p)_f + (1-\varepsilon)\,(\rho c_p)_s\right]\frac{\partial T}{\partial t} + (\rho c_p)_f\,\mathbf{u}_D\cdot\nabla T$$
$$= (\rho c_p)_f\,\nabla\cdot\mathbf{D}\cdot\nabla T, \tag{4.221}$$

where

$$\mathbf{D} = \frac{\mathbf{K}_e}{(\rho c_p)_f} + \varepsilon\,\mathbf{D}^d, \tag{4.222}$$

$\mathbf{K}_e$ is the *effective thermal conductivity tensor*, $\mathbf{D}^d$ is the *thermal dispersion tensor*, and $\mathbf{D}$ is called the *total thermal diffusivity* tensor.

- For $k_s = 0$, the conduction-convection energy equation can be written as

$$\frac{\partial \langle T \rangle^f}{\partial t} + \langle \mathbf{u} \rangle^f \cdot \nabla \langle T \rangle^f = \nabla \cdot \mathbf{D} \cdot \nabla \langle T \rangle^f, \quad k_s = 0 \quad (4.223)$$

where

$$\mathbf{D} = \frac{\mathbf{K}_e}{(\rho c_p)_f} + \mathbf{D}^d, \quad k_s = 0, \quad (4.224)$$

and $\mathbf{K}_e$ is the *effective thermal conductivity of the fluid phase*, which is different than k_f, because of the presence of the solid phase.

- $\mathbf{D}$ is a *second-rank* tensor.

- $\mathbf{D}$ is *positive-definite*.

- $\mathbf{D}$ is *symmetrical* for random structures, i.e., $D_{ij}^d = D_{ji}^d$.

- The *off-diagonal* elements of $\mathbf{D}$ are zero for *isotropic* media, i.e., $D_{ij}^d = 0$ for $i \neq j$.

- $\mathbf{D}$ is *invariant* to the origin of the coordinate system.

- For *isotropic* media, the *longitudinal* dispersion is given as $D_\parallel^d = (\mathbf{u}_D/u_D) \cdot \mathbf{D}^d \cdot (\mathbf{u}_D/u_D)$.

- For *isotropic* media, there is a *transverse* isotropy in $\mathbf{D}^d$, such that the transverse dispersion for $\mathbf{n}$ perpendicular to $\mathbf{u}_D$ is $D_\parallel = \mathbf{n} \cdot \mathbf{D}^d \cdot \mathbf{n}$.

- The last two can be *combined* to give $\mathbf{D}^d = \mathbf{nn} D_\parallel^d + (\mathbf{I} - \mathbf{nn}) D_\perp^d$.

- For *periodic* structures, $D_\perp^d$ is small (but its magnitude depends on the direction of flow with *respect* to the symmetry axis of the unit cell).

- It is *expected* that

$$\mathbf{D}^d = \mathbf{D}^d \left(\text{structure}, \varepsilon, \frac{(\rho c_p)_s}{(\rho c_p)_f}, \frac{k_s}{k_f}, Pr, Re_d \right). \quad (4.225)$$

We have discussed some predictions for $\mathbf{D}$ and we now consider the experimental results in order to examine the effect of the *variables* given in (4.225) on $\mathbf{D}$.

4.8 Experimental Determination of **D**

4.8.1 EXPERIMENTAL METHODS

Extraction of $D_\parallel$ and $D_\perp$ from the heat (or mass) transfer data requires having a relationship between these unknowns and the quantities measured. The measured quantity is in general the temperature (or concentration) measured at various *locations*. In order to plan the experiment and to provide this relationship, the temperature field and its dependence on $D_\parallel$ and $D_\perp$ must be *predicted*. This requires the solution to the *macroscopic* energy equation containing the dispersion coefficients. The heat (or solute) input required for causing measurable temperature (or concentration) disturbances can be applied either at the *confining boundaries* of the porous medium or *inside* the medium.

The following is a *classification* of the heat input *strategies*. Also given are the *solutions* to the resulting disturbed temperature fields. The unknowns of the experiment $D_\parallel$ and $D_\perp$ appear in these solutions and can be determined using the measured temperature field. Table 4.4 gives some of the strategies used.

(A) STEADY-STATE HEAT ADDITION/REMOVAL AT BOUNDING SURFACES

Experimental evaluation of the axial and transverse dispersion coefficients as well as the surface heat transfer coefficient are made from the measured temperature in the porous medium at various radial locations by Gunn and Khalid (1975). The solution is sought to the steady-state two-dimensional *constant-property* energy equation written as

$$u_D \frac{\partial T}{\partial x} = D_\perp \frac{1}{r} \frac{\partial}{\partial r}\left(r \frac{\partial T}{\partial x}\right) + D_\parallel \frac{\partial^2 T}{\partial x^2}. \tag{4.226}$$

This is subject to the confining boundaries being maintained at two different temperatures T_{s_1} and T_{s_2} with $T_{s_2} > T_{s_1}$. In the porous medium at the bounding surface, allowance is made for a *jump in temperature*. This is done by introducing a *surface convective heat transfer coefficient* h_s, a treatment similar to that made for the hydrodynamic interfacial boundary condition (Section 2.11) and the interfacial conduction boundary condition (Section 3.7).

This slip condition is made to mask the variation of $D_\perp$ near the bounding surfaces. We will examine this variation further in Section 4.9. With the introduction of h_s, the boundary conditions have become

$$-(\rho c_p)_f D_\perp \frac{\partial T}{\partial r} = h_s (T - T_{s_1}) \quad \text{at} \quad r = R \text{ for } x < 0, \tag{4.227}$$

$$-(\rho c_p)_f D_\perp \frac{\partial T}{\partial r} = h_s (T - T_{s_2}) \quad \text{at} \quad r = R \text{ for } x \geq 0, \tag{4.228}$$

TABLE 4.4　VARIOUS EXPERIMENTAL-ANALYTICAL TECHNIQUES FOR DETERMINATION OF $D_\parallel$ and $D_\perp$

METHOD	MEASUREMENT	QUANTITY DETERMINED
(a) Steady state, upstream heat addition through finite width source	Temperature at various locations, downstream (on and off the centerline)	$D_\perp$
(b) Point source suddenly activated and then ceased immediately	Temperature at various locations	$D_\parallel$, $D_\perp$
(c) Frequency response to upstream, sinusoidal heat addition (within the bed)	Time variation of the inlet and exit temperature	$D_\parallel$, or h_{fs} (interstitial heat transfer coefficient)
(d) Steady state, heat addition at bounding surfaces	Temperature at various radial locations	$D_\parallel$, $D_\perp$, and h_s (heat transfer coefficient at the bounding surface)
(e) Upstream pulse and step rise in temperature	Time variation of temperature measured at various downstream locations	$D_\parallel$
(f) Steady state, upstream heat addition over half of the bed cross section	Temperature at various radial locations	$D_\perp$

(a) Gunn and Pryce (1969); f(b) Lawson and Elrick (1972); (c) Gunn and De Souza (1974); (d) Gunn and Khalid (1975); (e) Han et al. (1985); (f) Han et al. (1985).

$$\frac{\partial T}{\partial r} = 0 \quad \text{at} \quad r = 0, \qquad (4.229)$$

$$T \to T_{s_2} \quad \text{as} \quad x \to \infty, \qquad (4.230)$$

$$T \to T_{s_1} \quad \text{as} \quad x \to -\infty, \qquad (4.231)$$

T and $\dfrac{\partial T}{\partial x}$ are continuous at $x = 0$. (4.232)

Now by defining a dimensionless temperature $T^* = (T - T_{s_1})/(T_{s_2} - T_{s_1})$ and applying the separation-of-variables techniques with the eigenvalues designated by λ_i, we have the solution in the forms of the Bessel function of the zeroth order and the first kind (r-direction) J_0 and the exponential function (x-direction). Since the measurements are made with *finite* size probes and the radial temperature gradient is the *largest*, the temperature is *averaged* over an annulus of inside radius R_i and outside radius R_o. The solution is given by Gunn and Khalid as

$$\overline{T}^* = 1 + \sum_{j=1}^{\infty} \frac{2h_s R^2}{(\rho c_p)_f D_\perp} \left(\frac{D_\parallel}{D_\perp}\right)^{1/2} \left[1 + \frac{\dfrac{Ru_d}{2D_\parallel}}{\left(\dfrac{R^2 u_D^2}{4D_\parallel} + \lambda_j^2 \dfrac{D_\perp}{D_\parallel}\right)^{1/2}}\right]$$

$$\times \left[\frac{R_o J_1\left(\lambda_j \dfrac{R_o}{R}\right) - R_i J_1\left(\lambda_j \dfrac{R_i}{R}\right)}{(R_o^2 - R_i^2) J_0(\lambda_j)\left(\dfrac{R^2 h_s^2 D_\parallel}{(\rho c_p)_f^2 D_\perp^3} + \lambda_j^2\right)}\right]$$

$$\times \exp\left\{-\frac{2x D_\parallel}{u_D}\left[(1+A)^{1/2} - 1\right]\right\} \quad \text{for } x > 0, \quad (4.233)$$

where $A = 4\lambda_j D_\perp D_\parallel / (R^2 u_D^2) \geq 0$, and the eigenvalues are found from

$$\lambda J_1(\lambda) - \frac{h_s R}{(\rho c_p)_f (D_\perp D_\parallel)^{1/2}} J_0(\lambda) = 0. \quad (4.234)$$

The measured dimensionless temperature $\overline{T}^*_{\text{exp}}$ is first used in (4.233) along with the available data on $D_\parallel$ (Gunn and De Souza, 1974), and then the values of $D_\perp$ and h_s are determined simultaneously, while minimizing the difference between the measured and predicted $\overline{T}^*$. Next, the minimization is repeated, except this time $D_\parallel$ is *also* evaluated using the previously determined $D_\perp$ and h_s as the initial guess. Since the number of radially averaged (over $R_o - R_i$) temperature measurements is larger than three, the least-square approximation method can be applied.

(B) Heat Addition Inside Porous Media

These are *dominantly* transient experiments and vary in the radial extent of the heat source used, as well as in the imposed temporal variation of the heat input.

(i) Radially Uniform Step Rise in Temperature (Transient)

A common method used for determination of $D_\parallel$ is that of the sudden introduction of a *jump* in the temperature (concentration) and then *maintenance* of the amplitude of this disturbance (Fried and Combarnous, 1971; Han et al. 1985). The solution is sought to the following energy equation ($k_s = 0$, $D_\parallel = \alpha_e^f + D_\parallel^d$, where α_e^f is the effective thermal diffusivity of the fluid phase, which is different than α_f)

$$\frac{\partial \langle i \rangle^f}{\partial t} + \langle u \rangle^f \frac{\partial \langle i \rangle^f}{\partial x} = D_\parallel \frac{\partial^2 \langle i \rangle^f}{\partial x^2}, \qquad (4.235)$$

which is for the case of $(\rho c_p)_s = 0$ and is subject to the conditions

$$t = 0: \quad \langle i \rangle^f = 0, \quad \text{for } 0 \leq x < \infty, \qquad (4.236)$$

$$t > 0: \quad \langle i \rangle^f = i_o, \quad \text{at } x = 0, \qquad (4.237)$$

$$t > 0: \quad \langle i \rangle^f \to 0, \quad \text{as } x \to \infty. \qquad (4.238)$$

Taking the Laplace transform of (4.235)–(4.238), we have

$$p \langle \hat{i} \rangle + u_p \frac{d \langle \hat{i} \rangle}{dx} = D_\parallel \frac{d^2 \langle \hat{i} \rangle}{dx^2} \qquad (4.239)$$

with

$$\langle \hat{i} \rangle (x = 0) = \frac{i_o}{p}, \qquad (4.240)$$

$$\langle \hat{i} \rangle (x \to \infty) = 0, \qquad (4.241)$$

where $\langle \hat{i} \rangle = \int_0^\infty e^{-pt} \langle i \rangle^f \, dt$ is the Laplace transform of $\langle i \rangle^f$ and $u_p = \langle u \rangle^f$. The solution to (4.239) subject to the preceding conditions is

$$\langle \hat{i} \rangle = i_o \exp\left(\frac{u_p x}{2 D_\parallel}\right) \frac{1}{p} \exp\left[-\frac{x}{D_\parallel^{1/2}} \left(\frac{u_p^2}{4 D_\parallel} + p\right)^{1/2}\right], \qquad (4.242)$$

the inverse of which is

$$\frac{\langle i \rangle^f}{i_o} = \frac{T}{T_o} = \frac{1}{2} \operatorname{erfc}\left[\frac{x - u_p t}{2 (D_\parallel t)^{1/2}}\right] + \frac{1}{2} \operatorname{erf}\left[\frac{x + u_p t}{2 (D_\parallel t)^{1/2}}\right] \exp\left(\frac{u_p x}{D_\parallel}\right). \qquad (4.243)$$

The second term is generally negligible. This equation is used for the evaluation of $D_\parallel$.

(ii) Radially Uniform Pulsed Change in Temperature (Transient)

The pulse response has been used in the measurements of Han et al. for the determination of $D_\|$. The condition at $x = 0$ is

$$\langle i \rangle^f = F(t) \quad \text{at } x = 0, \tag{4.244}$$

where $F(t)$ is the *temporal* variation of the *amplitude* of the volumetric enthalpy disturbance. For the determination of $D_\|$, (4.235) is solved subject to (4.236)–(4.238) and (4.244) by using the Laplace transform. The transformed solution is similar to (4.242), i.e.,

$$\langle \hat{i} \rangle = \langle F \rangle (p) \exp\left(\frac{u_p x}{2D_\|}\right) \exp\left[-\frac{x}{D_\|^{1/2}} \left(\frac{u_p^2}{4D_\|} + p\right)^{1/2}\right], \tag{4.245}$$

where $u_p = \langle u \rangle^f$. The *moment* of the volumetric enthalpy is taken as its time integral, i.e.,

$$\mu^n = \int_0^\infty t^n \langle i \rangle^f \, dt \quad n = 0, 1, 2. \tag{4.246}$$

One of the properties of the Laplace transform is that

$$\mu^n = (-1)^n \frac{d^n}{dp^n} \langle \hat{i} \rangle. \tag{4.247}$$

The first absolute moment is μ^1/μ^0, which is the *mean residence time* for the pulse at a given axial position. This is found by using (4.245) in (4.247), which gives

$$\frac{\mu^1(x)}{\mu^0(x)} = \frac{\mu^1(0)}{\mu^0(0)} + \frac{x}{u_p}, \tag{4.248}$$

where $\mu^1(0)/\mu^0(0)$ is the mean residence time for pulses introduced at $x = 0$ (undispersed). The second moment can be modified to be a measure of the *average pulse spread* relative to the mean residence time. This requires that we use $t - \mu^1/\mu^0$. Then

$$\frac{\mu^2}{\mu^0} = \frac{1}{\mu^0} \int_0^\infty \left(t - \frac{\mu^1}{\mu^0}\right)^2 \langle i \rangle^f \, dt. \tag{4.249}$$

Upon evaluation, we have

$$\frac{\mu^2(x)}{\mu^0(x)} = \frac{\mu^2(0)}{\mu^0(0)} + \frac{2D_\|}{x^2} \left[\frac{\mu^1(x)}{\mu^0(x)} - \frac{\mu^1(0)}{\mu^0(0)}\right]^3, \tag{4.250}$$

which, when (4.248) is used, gives

$$\frac{\mu^2(x)}{\mu^0(x)} = \frac{\mu^2(0)}{\mu^0(0)} + \frac{2D_\| x}{u_p}, \qquad (4.251)$$

where $\mu^2(0)/\mu^0(0)$ is the *spread* of the pulse introduced at $x = 0$. From the measurement of $\mu^2(x)$, $D_\|$ can be calculated from (4.251).

(iii) Radially Uniform Sinusoidal Change in Temperature (Transient)

One method for the experimental determination of $D_\|$ is the introduction of *sinusoidal (radially uniform)* temperature disturbances and then the downstream measurement of its amplitude and phase. Gunn and Pryce (1969) and Gunn and De Souza have used this method, and a general treatment is given in Gunn (1970).

The solution to (4.235) is sought subject to

$$\langle i \rangle = i_o e^{i\omega t}, \quad x = 0, \quad \langle i \rangle \to 0 \ \text{ as } \ x \to \infty, \qquad (4.252)$$

where ω is the angular frequency. Then (4.234) becomes

$$i\omega + u_p \frac{d\langle i \rangle^f}{dx} = D_\| \frac{d^2 \langle i \rangle^f}{dx^2}. \qquad (4.253)$$

Now, for any location x, the solution of (4.253) is

$$\frac{\langle i \rangle^f}{i_o} = \frac{T}{T_o}$$

$$= \exp\left\{\left[1 - \left(\frac{1}{2}\left[1 + (4\omega D_\|/u_p^2)^2\right]^{1/2} + \frac{1}{2}\right)^{1/2}\right]\frac{u_p x}{2D_\|}\right\}. \qquad (4.254)$$

The evaluation of $D_\|$, is done by comparing the measured *spatial* (or more conveniently the *temporal*) variation of the amplitude of the disturbance to that for the solution of (4.253). Then a least-square approximation is made to the differences between the two. The frequency modulation is used by Gunn and Pryce for the evaluation of $D_\|$.

(iv) Radially Nonuniform Temperature Distribution (Steady State)

For determination of $D_\perp$, Han et al. perform experiments in channels with *rectangular* cross sections with a *large* aspect ratio, such that two-dimensional descriptions of the energy (or species) equation is permissible. For this case we have

$$\langle u \rangle^f \frac{\partial \langle i \rangle^f}{\partial x} = D_\| \frac{\partial^2 \langle i \rangle^f}{\partial x^2} + D_\perp \frac{\partial^2 \langle i \rangle^f}{\partial y^2}. \qquad (4.255)$$

For $\partial^2/\partial y^2 \gg \partial^2/\partial x^2$, which is a reasonable approximation where the upstream ($x = 0$) y-direction distribution is a step-change, the initial condition for (4.255) can be written as

$$\text{at } x = 0 \quad \langle i \rangle^f = \begin{cases} i_o & 0 \leq y \leq \dfrac{w}{2} \\ 0 & -\dfrac{w}{2} \leq y < 0. \end{cases} \quad (4.256)$$

Note that $y = 0$ is at the *center* of a channel of $2w$ width. Then (4.255) reduces to

$$u_p \frac{\partial \langle i \rangle^f}{\partial x} = D_\perp \frac{\partial^2 \langle i \rangle^f}{\partial y^2}, \quad (4.257)$$

where $\langle u \rangle^f = u_p$.

Now the discontinuity given by (4.256) can be treated as that of a disturbance introduced at $y = 0$, which then *penetrates* into a semi-infinite medium. A similarity solution exists and is given by

$$\frac{\langle i \rangle^f}{i_o} = \frac{T}{T_o} = \frac{1}{2}\text{erfc}\left[\frac{u}{2\left(xD_\parallel/u_p\right)^{1/2}}\right]. \quad (4.258)$$

Then measurement of $\langle T \rangle^f(y) = T(y)$ leads to determination of $D_\perp$.

(v) Line or Point Heat Source

The line source disturbance has been used by Lawson and Elrick (1972) for the determination of $D_\parallel$ and $D_\perp$. The energy equation is

$$\left[\varepsilon\left(\rho c_p\right)_f + (1-\varepsilon)\left(\rho c_p\right)_s\right] \frac{\partial T}{\partial t} + \left(\rho c_p\right)_f u_D \frac{\partial T}{\partial x}$$

$$= \left(\rho c_p\right)_f D_\parallel \frac{\partial^2 T}{\partial x^2} + \left(\rho c_p\right)_f D_\perp \frac{\partial^2 T}{\partial y^2} + \left(\rho c_p\right)_f D_\perp \frac{\partial^2 T}{\partial z^2}, \quad (4.259)$$

or

$$C_1 \frac{\partial T}{\partial t} + u_D \frac{\partial T}{\partial x} = D_\parallel \frac{\partial^2 T}{\partial x^2} + D_\perp \frac{\partial^2 T}{\partial y^2} + D_\perp \frac{\partial^2 T}{\partial z^2}, \quad (4.260)$$

where

$$C_1 = \frac{\varepsilon\left(\rho c_p\right)_f + (1-\varepsilon)\left(\rho c_p\right)_s}{\left(\rho c_p\right)_f}, \quad (4.261)$$

which is equal to $\varepsilon + (1-\varepsilon)C$. Subject to a line (or point) source *introduced* at $x = y = z = 0$ with the medium initially at $T = 0$.

(a) Instantaneous Line or Point Source

The solution is found from the formulation of Carslaw and Jaeger (1986, 256–268), which uses the *coordinate transformation* $x - u_p t/C_1$. The solution for a *line source* given by Lawson and Elrick is

$$T = \frac{C_1 A_2}{4\pi t \left(D_\| D_\perp\right)^{1/2}} \exp\left\{-\frac{1}{4t}\left[\frac{C_1 (x - u_D t/C_1)^2}{D_\|} + \frac{C_1 y^2}{D_\perp}\right]\right\}, \quad (4.262)$$

where

$$A_2 = \frac{1}{\varepsilon (\rho c_p)_f + (1-\varepsilon)(\rho c_p)_s} \int_{-\infty}^{\infty}\int_{-\infty}^{\infty} (\rho c_p) T \, dx \, dy. \quad (4.263)$$

Note that ρc_p becomes $\varepsilon (\rho c_p)_f + (1-\varepsilon)(\rho c_p)_s$ when integrated over *both* phases and A_2 is a normalized heat content. For a *point source*, the same formulation is used and the solution is

$$T = \frac{A_3}{8 (\pi t)^{3/2} \left(D_\| D_\perp^2\right)^{1/2}}$$

$$\times \exp\left[-\frac{1}{4t}\left(\frac{C_1 (x - u_D t/C_1)^2}{D_\|} + \frac{C_1 y^2}{D_\perp} + \frac{C_1 z^2}{D_\perp}\right)\right], \quad (4.264)$$

where

$$A_3 = \frac{1}{\varepsilon (\rho c_p)_f + (1-\varepsilon)(\rho c_p)_s} \int_{-\infty}^{\infty}\int_{-\infty}^{\infty}\int_{-\infty}^{\infty} (\rho c_p) T \, dx \, dy \, dz. \quad (4.265)$$

For a *line source* at any location (x, y) except for $(0,0)$, the temperature will undergo a *maximum* with respect to time. This time is designated as t_m, and from (4.262) we have

$$x^2 + \frac{D_\|}{D_\perp} y^2 - \frac{u_D^2 t_m^2}{C_1^2} - \frac{4 t_m D_\|}{C_1} = 0, \quad (4.266)$$

or at $y = 0$

$$D_\| = \frac{x^2 - u_p^2 t_m^2 / C_1^2}{4 t_m / C_1}. \quad (4.267)$$

(b) Continuous Line Source

This can be treated by replacing t in (4.262) or (4.264) with $t - t'$ and integrating these expressions with respect to t. For example, (4.262) becomes

$$T = \frac{C_1 B_2}{4\pi \left(D_\| D_\perp\right)^{1/2}} \int_0^t \exp\left[-\frac{1}{4(t-t')}\right.$$

$$\times \left(\frac{C_1}{D_\parallel} C_1 \left[x - \frac{u_D(t-t')}{C_1}\right]^2 + \frac{C_1 y^2}{D_\perp}\right)\right] \frac{dt'}{t-t'}. \quad (4.268)$$

There exists a *steady-state asymptote*. At this asymptote and for $y = 0$, we have

$$D_\perp = \frac{C_1^4 B_2^2}{4\pi x u_D T^2}, \quad (4.269)$$

where B_2 is related to the *heat source strength* and T is the *mean* temperature at $(x, 0)$.

4.8.2 ENTRANCE EFFECT

A local disturbance introduced at $x = 0$ upstream from the probe location x_p undergoes *lateral* and *axial* diffusion-convection with the net convective flux expressed in terms of $\mathbf{u}_D \cdot \nabla T$ and the molecular plus the dispersive flux given by $\nabla \cdot \mathbf{D} \cdot \nabla T$. In the analysis/prediction of $\mathbf{D}$ (for both capillary tubes and porous media), it was assumed that lateral spreading of the heat content has reached its *asymptotic* limit (for fully developed flows, lateral implies the direction in which *no* net convective transport occurs).

This *evolution* toward the asymptote requires a sufficiently *large* elapsed time (or *length*, downstream of $x = 0$). This length can be determined for capillary tubes (through numerical solutions), but for porous media, where even for the case of regular arrangement of particles the direct simulation would require inclusion of many particles in the model, the *prediction* of this length is almost computationally unrealistic.

Han et al. performed experiments using randomly packed spherical particles, evaluated $D_\parallel$ and $D_\perp$ at several downstream locations, and observed *asymptotic* values for these quantities as the probing location moved *further* and *further* downstream. Figure 4.13 shows the experimental (step change in concentration) results for $D_\parallel$. Their results show that as $Pe = u_D d/\alpha_f$ *increases*, the probe location for the observation of the asymptotic behavior must move *further* downstream. They also present data for probe locations in between the two presented in this figure, supporting this asymptotic behavior (to within the experimental accuracy). The predictions of Koch and Brady for $D_\parallel$ are also shown in Figure 4.13 and good agreement is found with the experimental results.

Han et al. correlated their experimental results and found the required *entrance length* L_e to be

$$\frac{L_e}{d^2}\frac{\alpha_f}{u_p} \geq 0.3 \quad \text{or} \quad \frac{L_e}{d} \geq 0.3\frac{u_D d}{\alpha_f} = 0.3\frac{Pe}{\varepsilon}. \quad (4.270)$$

Han et al. review the available experimental results and conclude that the violation of the above inequality has led to the lower values of $D_\parallel$ *reported* by others for *large* Pe. This entrance effect was previously mistakenly

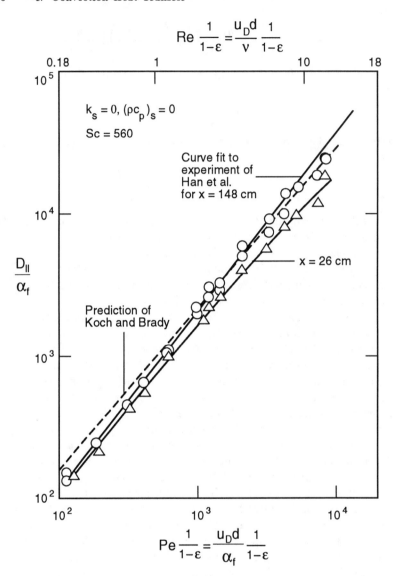

Figure 4.13 Entrance effect on normalized $D_\parallel$, experimental results of Han et al. for two different axial locations, and the prediction of Koch and Brady for the asymptotic behavior.

associated with the *transition to turbulence* even though for the *high* $Pr(Sc)$ data the Reynolds numbers are much *smaller* than the *transitional* value of about 150–300.

4.8.3 EFFECT OF PARTICLE SIZE DISTRIBUTION

In general, the *spread* of the *particle size* for a given average porosity and particle size distribution causes an increase in $D_\|$ and $D_\perp$. Han et al. performed experiments in packed beds of spheres using various particle size spreads. They found that for a given $\bar{d}$ and $\bar{\varepsilon}$, when the *spread* was due to a *small* number of *much* larger particles (*non-Gaussian distribution*), the values of $D_\|$ and $D_\perp$ were much *larger* than when the spread was due to a *large* number of particles having *slightly* different diameters than $\bar{d}$ (*Gaussian distribution*). When the particle size varies substantially, significant variation in the pore velocity is found, and this random redistribution of the flow between the pores enhances dispersion.

Carbonell (1979) studies the effect of the pore size distribution (i.e., tube diameter distribution in the parallel arrangement of tubes) on $D_\|$. He uses the Taylor-Aris dispersion coefficient and finds the effect of the *standard deviation* of the pore size distribution (Gaussian distribution) on the *mean dispersion coefficient*. He shows that the mean dispersion coefficient *increases* as the dimensionless standard deviation σ^2/d^2 *increases*. Note that since the Taylor dispersion coefficient is proportional to Pe^2, the Carbonell model does *not* agree with the experimental results (and predictions) for randomly packed spheres, which show a Pe proportionality.

The nonuniformity of the average pore velocity in the bed associated with the porosity nonuniformity has been identified as the *cause* of the decrease in the average (over the bed) fluid-particle heat (mass) transfer rate. Martin (1978) and Schlünder (1978) examine this flow *maldistribution* in the bulk as well as near the wall (as discussed in Chapter 2) and its effects on the local and average *particle-bed transfer coefficients*. (This subject, lack of local thermal equilibrium, will be discussed in Chapter 7.) As was mentioned, dispersion, which for disordered media can in part be viewed as the fluid particle displacement, is enhanced due to the *maldistribution*.

4.8.4 SOME EXPERIMENTAL RESULTS AND CORRELATIONS

In the treatment of dispersion in *ordered* and *disordered* media, we reviewed some of the closed-form solutions for the longitudinal and transverse dispersion coefficients. These are summarized in Table 4.5. We also examined the experimental techniques used for the evaluation of the total effective thermal conductivity tensor. These experimental results have been correlated, and the available correlations for the longitudinal and transverse dispersion coefficients are given in Table 4.6.

For the *random* arrangement of spherical particles, the closed-form solution of Koch and Brady accurately *predicts* the experimental results for $D_\perp^d$ and $D_\|^d$. For K_e, we recommend the correlations given in Chapter 3 and then the use of (4.221) and (4.222) or (4.223) and (4.224), whichever is *applicable*.

TABLE 4.5   SOME CLOSED-FORM SOLUTIONS FOR TOTAL DIFFUSIVITY D AND DISPERSION COEFFICIENT D^d

CONSTRAINTS	$D_\parallel/\alpha_f$	$D_\perp/\alpha_f$
(a) $Re \gg 1$, disordered media	$\dfrac{D_\parallel}{\alpha_f} = \dfrac{Pe}{2}$ $(Pe = 2u_D R/\alpha_f)$	$\dfrac{D_\perp}{\alpha_f} = \dfrac{Pe}{11}$
(b) Disordered media	$\dfrac{D_\parallel^d}{\alpha_f} = \left(\lambda + \dfrac{3}{4} - 0.577\right)\dfrac{Pe}{6}$ $(Pe = u_D \ell/\varepsilon\alpha_f;\ \lambda$ depends on the bed length)	$\dfrac{D_\perp^d}{\alpha_f} = \dfrac{3Pe}{16}$
(c) Stokes flow, $\ell/R = 5$, disordered media	$\dfrac{D_\parallel^d}{\alpha_f} = \dfrac{Pe}{6}\left(\ln 1.22 Pe - \dfrac{17}{12}\right.$ $\left. - \dfrac{1}{200}Pe\right)$ $\dfrac{D_\parallel^d}{\alpha_f} = \dfrac{Pe^2}{15},\ Pe \ll 1$ $(Pe = u_D\ell/\varepsilon\alpha_f;\ \ell$ is the average channel length)	$\dfrac{D_\perp^d}{\alpha_f} = \dfrac{3}{16}Pe + \dfrac{1}{1000}Pe^2$ $\dfrac{D_\perp^d}{\alpha_f} = \dfrac{1}{40}Pe^2,\ Pe \ll 1$
(d) $k_f \to \infty$ (uniform T in fluid, but not in solid), disordered media	$\dfrac{D_\parallel^d}{\alpha_f} = \dfrac{C^2\varepsilon^2(1-\varepsilon)}{[\varepsilon + C(1-\varepsilon)]^3}$ $\times\left[\dfrac{Pe^2}{15C}\dfrac{\alpha_f}{\alpha_s} + \dfrac{Pe}{3}\dfrac{u_D(\rho c_p)_f}{\varepsilon h_{sf}}\right]$ $(Pe = u_D R/(\varepsilon \alpha_f),$ $C = (\rho c_p)_s/(\rho c_p)_f,\ h_{sf}$ is the interstitial heat transfer coefficient)	
(e) Table 4.3, Stokes flow, disordered media	Table 4.3	Table 4.3
(f) $\alpha_f = \alpha_s$, Stokes flow, disordered media	Section 4.4.2 $(Pe = u_D\ell/\alpha_f)$	Section 4.4.2

(a) Aris and Amundson (1957); (b) De Josselin De Jong (1958); (c) Saffman (1960); (d) Horn (1971); (e) Koch and Brady (1985); (f) Koch et al. (1989).

TABLE 4.6 VARIOUS CORRELATIONS FOR TOTAL DIFFUSIVITY COEFFICIENT D OR DISPERSION COEFFICIENT D^d

CONSTRAINTS	$D_\parallel/\alpha_f$	$D_\perp/\alpha_f$
(a) Disordered media		$D_\perp/\alpha_f = Pe/(5\sim 15)$ $(Pe = u_D d/\alpha_f)$
(b) Disordered media	$D_\parallel/\alpha_f = k_s/k_f + (0.7\sim 0.8)Pe$ $(Pe = u_D d/\alpha_f)$	$D_\perp/\alpha_f = k_e/k_f + (0.1\sim 0.3)Pe$ $(Pe = u_D d/\alpha_f)$
(c) $Re < 50$ Disordered media	$D_\parallel/\alpha_f = 0.73 + \dfrac{0.5Pe}{1+(0.97/Pe)}$ $[Pe = u_D d/(\varepsilon \alpha_f)]$	
(d) Disordered media	$D_\parallel/\alpha_f = k_e/k_f + 0.8Pe$ $(Pe = u_D d/\alpha_f)$	
(e) Disordered media	$D_\parallel/\alpha_f = k_e/k_f + 0.5Pe^{(1\sim 1.2)}$ for $Pe \leq 10^4$; $D_\parallel/\alpha_f = 1.8Pe$ for $Pe \geq 10^4$ $(Pe = u_D d/\alpha_f)$	
(f) Two-dimensional ordered media	$D_\parallel/\alpha_f = 1 + 0.128(Pe_\ell - 1)$ for $1 < Pe_\ell < 10$, $\varepsilon = 0.8$; $D_\parallel/\alpha_f = 1 + 0.071(Pe_\ell - 1)$ for $1 < Pe_\ell < 10$, $\varepsilon = 0.9$; $D_\parallel/\alpha_f = 0.019 Pe_\ell^{1.82}$ for $10 \leq Pe_\ell \leq 10^3$, $\varepsilon = 0.8$; $D_\parallel/\alpha_f = 0.009 Pe_\ell^{1.86}$ for $10 \leq Pe_\ell \leq 10^3$, $\varepsilon = 0.9$	

(a) Baron (1952); (b) Yagi et al. (1960), also Schertz and Bischoff (1969);
(c) Edwards and Richardson (1968); (d) Vortmeyer (1975); (e) Bear (1988);
(f) Sahraoui and Kaviany (1994).

We now proceed to examine the existing experimental data and point out the *dependence* of $\mathbf{D}^d$ on the various *dimensionless* parameters.

(A) A Close Examination of Experimental Results

Examination of the experimental results of Gunn and Pryce for the regular arrangement of particles shows the following features.

230   4. Convection Heat Transfer

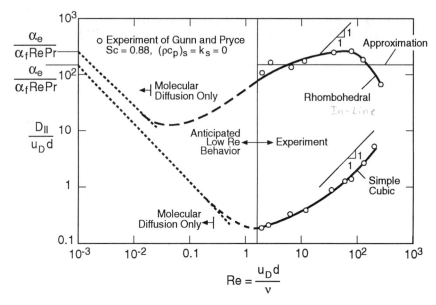

Figure 4.14 Experimental results of Gunn and Pryce for normalized $D_\parallel$ along with the extrapolation for low Reynolds numbers.

- There is a *large difference* between the *predicted* and *measured* relative amplitude of the disturbance (for different frequencies) in the experimental results for the *rhombohedral arrangement*. But for the *simple cubic arrangement*, this difference is *negligibly* small.

- Because air was used as the fluid and, therefore, the resistance to the flow is rather small, the resulting velocities (and flow rates) are *large*. Therefore, the *low Pe* asymptote (molecular diffusion only) was *never* achieved.

Figure 4.14 shows their experimental results for both of the arrangements. The pure diffusion ($Re \to 0$) asymptote α_e/α_f is chosen to include the porosity effect but is otherwise arbitrary. The results show that for the *rhombohedral* arrangement, the high Pe asymptotic behavior, i.e., $D_\parallel$ being proportional to Pe^2, is not observed. Instead a Pe proportionality appears more representative. As was discussed in Section 4.3.5, the staggered regular arrangements result in a Pe dependency similar to that for random arrangements.

(B) Trends

In general, $D_\parallel$ and $D_\perp$ are plotted versus Pe, i.e., Re and Pr are combined together (e.g., Figure 4.12). Based on the experience with internal and external flows and heat transfer, we expect *both* Re and Pr to influence

4.8 Experimental Determination of D

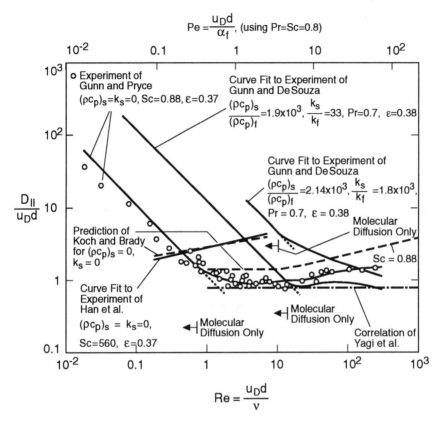

Figure 4.15 Presentation of normalized $D_\parallel$ versus Re for different Pr, k_s/k_f, and $(\rho c_p)_s/(\rho c_p)_f$. Both the predictions and the experimental results are shown.

the dispersion coefficient. In Figure 4.15, the results for $D_\parallel$ for different Pr (or Sc) are plotted versus Re. The following trends are observed.

- **Pr**: For a given Re, the effect of Pr is *best* represented through Pe, i.e., as Pr *increases*, the *molecular regime* occurs at a *smaller* Re.

- **Re**: The pore hydrodynamics represented by Re does *not* influence $D_\parallel$ as much as $Re\, Pr = Pe$.

- **k_s/k_f**: For $Pr \simeq 0.8$ (experiments of Gunn and Pryce and of Gunn and De Souza), the effect of k_s/k_f on the value of $D_\parallel$ is rather *significant*. The *molecular regime* persists to *higher* Re as k_s/k_f increases.

- **$(\rho c_p)_s/(\rho c_p)_f$**: For steady state (large elapsed time or behavior at large distances from the entrance) this ratio is *not* expected to be significant. No isolation of the effect of this ratio in Figure 4.10 is possible.

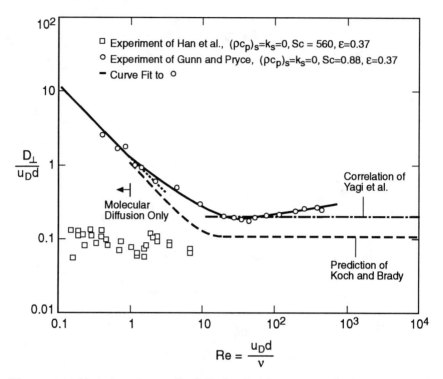

Figure 4.16 Variation of normalized $D_\perp/u_D d$ with respect to Re for two different $Pr(Sc)$.

The predictions of Koch and Brady, Table 4.3, are in good agreement with the experiments for low and high Pr. The predictions of Yagi et al. (1960) are not accurate at high Pr.

We now examine the *transverse* dispersion. Figure 4.16, which is a plot of normalized $D_\perp$ versus Re, shows a trend similar to that found for $D_\parallel$. It shows that the high Re hydrodynamics does *not* significantly influence the dispersion. The same results are plotted as normalized $D_\perp$ versus Pe in Figure 4.17. Again we observe that Pe is the parameter of choice for the dispersion.

4.9 Dispersion in Oscillating Flow

So far we have addressed steady flow through porous media. Periodic flows are also of practical interest and the hydrodynamic dispersion can also be examined for this class of flows. We begin by examining the periodic flow through capillary tubes. At *high* frequencies the velocity distribution in the tube differs greatly from that of the Hagen-Poiseuille profile. By

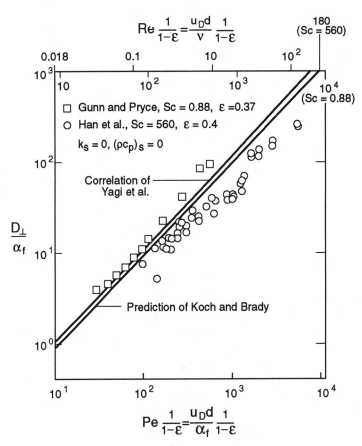

Figure 4.17 Variation of normalized $D_\perp$ with respect to Pe for two different $Pr(Sc)$. The experimental results are the same as those used in Figure 4.16.

neglecting the hydrodynamic and thermal entrance effects, *closed-form* expressions can be found for the temperature and velocity fields. An interesting feature of the high-frequency flow and inlet temperature oscillation is that the *fluid phase* is also used for the storage/release of the heat. This requires that the fluid- and solid-phase thermal diffusivity be of the *same* order of magnitude. This transient problem has been considered by Chatwin (1975), Watson (1983), Kurzweg (1985), and Kaviany (1986, 1990). Since the fluid volume is also used for the storage/release of heat, the inside radius of the tube should be chosen in accord with the viscosity and frequency so that the viscous boundary-layer thickness is nearly equal to radius. When the fluid occupying the capillary tubes *connecting* two reservoirs of different temperatures oscillates, a *large* amount of heat is transferred between the two reservoirs by the oscillating fluid. Since the heat *continuously* flows between the reservoirs, a *mean* temperature gradient is established along the tubes. At any *axial* location the temperature in each phase oscillates around the local mean value. This mean temperature gradient is *uniform* along the tube, when the entrance effects are negligible.

4.9.1 Formulation and Solution

The *pressure gradient* across the tube (or collection of tubes) oscillates with an amplitude $|dp^*/dx^*|$. The dimensionless momentum and energy equations are

$$\frac{\partial u^*}{\partial t^*} = \left|\frac{dp^*}{dx^*}\right|_0 \cos t^* + \frac{1}{\gamma^2}\frac{1}{r^*}\frac{\partial}{\partial r^*}\left(r^*\frac{\partial u^*}{\partial r^*}\right), \qquad (4.271)$$

$$\frac{\partial T_f^*}{\partial t^*} + u^*\frac{d\langle T^*\rangle_A}{dx^*} = \frac{1}{\gamma^2 Pr}\frac{1}{r^*}\frac{\partial}{\partial r^*}\left(r^*\frac{\partial T_f^*}{\partial r^*}\right) \qquad 0 \leq r^* \leq 1, \qquad (4.272)$$

$$\frac{\partial T_s^*}{\partial t^*} = \frac{\sigma_s}{\sigma_f}\frac{1}{\gamma^2 Pr}\frac{1}{r^*}\frac{\partial}{\partial r^*}\left(r^*\frac{\partial T_s^*}{\partial r^*}\right) \qquad 1 \leq r^* \leq \frac{R_o}{R_i}, \qquad (4.273)$$

where it is assumed that there exists a uniform and constant gradient of the time- and *area-averaged temperature*, i.e., $d\langle T^*\rangle_A/dx^*$ is uniform and constant. This is similar to the Taylor-Aris dispersion formulation (Section 4.1) except for the *sinusoidal* variations and the *direct participation* of the solid phase. The *dimensionless* variables and parameters are

$$x^* = \frac{x}{R_i}, \qquad r^* = \frac{r}{R_i}, \qquad u^* = \frac{u}{R\omega}, \qquad t^* = t\omega, \qquad (4.274)$$

$$p^* = \frac{p}{\rho R^2 \omega^2}, \qquad T^* = \frac{T - T_{f\,\min}}{T_{f\,\max} - T_{f\,\min}}, \qquad (4.275)$$

$$\gamma = \frac{R\omega^{1/2}}{\nu^{1/2}} \text{ (Stokes or Womersley number)}, \qquad Pr = \frac{\nu}{\alpha_f}, \qquad (4.276)$$

$$u_c^* = \left|\frac{dp}{dx}\right|_0 \frac{1}{\rho_f R_i \omega^2} \text{ (characteristic velocity)}, \qquad (4.277)$$

where, again, ω is the angular frequency and is equal to $2\pi f$.

The boundary conditions are

$$u^*(1, t^*) = \frac{\partial u^*}{\partial r^*}(0, t^*) = \left.\frac{\partial T_f^*}{\partial r^*}\right|_0 = \left.\frac{\partial T_s^*}{\partial r^*}\right|_{R_i/R_o} = 0, \qquad (4.278)$$

$$T_f^*(1, t^*) = T_s^*(1, t^*), \qquad (4.279)$$

$$\left.\frac{k_f}{k_s}\frac{\partial T_f^*}{\partial r^*}\right|_1 = \left.\frac{\partial T_s^*}{\partial r^*}\right|_1. \qquad (4.280)$$

The dimensionless average displacement of the fluid along the tube axis over half the period is called the *tidal displacement* and is given by

$$\Delta x^* = 2 \left| \int_{-\pi/2}^{\pi/2} \int_0^1 u^* r^* \, dr^* \, dt^* \right|. \tag{4.281}$$

Since no axial diffusion is allowed in (4.271) and (4.273) and because there is no net fluid motion, the axial heat transfer can be expressed in terms of an *axial hydrodynamic dispersion* $D_\parallel^d$. As will be shown, $D_\parallel^d/\alpha_f$ is very large when the frequency is *relatively large*. This dispersion coefficient is given by

$$\frac{D_\parallel^d}{R^2 \omega} = -\frac{1}{\pi} \int_0^{2\pi} \int_0^1 \text{Real}(u^*) \text{Real}(T^*) r^* \, dr^* \, dt^*. \tag{4.282}$$

The average dimensionless *work* done per unit time is

$$W = \frac{L}{2\pi R_i} \int_0^{2\pi} \int_0^1 u^* \left| \frac{dp^*}{dx^*} \right|_0 \cos t^* r^* \, dr^* \, dt^*. \tag{4.283}$$

The solution for the velocity field is available in Chatwin (1975), and Schlichting (1979), and is

$$u^* = iu_c^* \left[\frac{I_0\left(i^{1/2}\gamma r^*\right)}{I_0\left(i^{1/2}\gamma\right)} - 1 \right] e^{it^*} \equiv iu_c^* F e^{it^*}, \tag{4.284}$$

where $I_0(i^{1/2}z) = \text{ber}(z) + i\,\text{bei}(z)$, ber and bei are the Kelvin functions (Oliver, 1972), and $i = (-1)^{1/2}$. The solutions to the temperature fields are based on the form

$$T^* = \frac{d\overline{\langle T^*\rangle}_A}{dx^*} x^* + \frac{d\overline{\langle T^*\rangle}_A}{dx^*} \theta^{*\prime}(r^*) e^{it^*}, \tag{4.285}$$

i.e., there is a time- and area-averaged component $\overline{\langle T^*\rangle}_A$ and a spatially fluctuating component $(d\overline{\langle T^*\rangle}_A/dx^*)\theta^{*\prime}$, where only the latter is a function of the radial position (with periodic time variation given by the exponential product).

The solutions to (4.271) and (4.273), which are of the form (4.285) and subject to (4.280), are (for $Pr \neq 1$) found as (Kaviany, 1990)

$$T_f^* = \frac{d\overline{T}^*}{dx^*} x^* + \frac{d\overline{T}^*}{dx^*} \left\{ \frac{u_c^*}{1-Pr} + \frac{Pr\, u_c^*}{1-Pr} \left[\frac{I_0\left(i^{1/2}\gamma r^*\right)}{I_0\left(i^{1/2}\gamma\right)} - 1 \right] \right.$$

$$\left. + a_1 I_0 \left[i^{1/2} \left(\frac{\alpha_f}{\alpha_s} Pr \right)^{1/2} r^* \right] \right\} e^{it^*}, \tag{4.286}$$

$$T_s^* = \frac{d\overline{T}^*}{dx^*}x^* + \frac{d\overline{T}^*}{dx^*}\left\{a_2 I_0\left[i^{1/2}\gamma\left(\frac{\alpha_f}{\alpha_s}Pr\right)^{1/2}r^*\right]\right.$$
$$\left.+a_3 K_0\left[i^{1/2}\gamma\left(\frac{\alpha_f}{\alpha_s}Pr\right)^{1/2}r^*\right]\right\}e^{it^*}, \qquad (4.287)$$

where

$$a_1 = \frac{a_2 I_0(i^{1/2}z_2) + a_3 K_0(i^{1/2}z_2) - u_c^*(1-Pr)^{-1}}{I_0(i^{1/2}z_1)}, \qquad (4.288)$$

$$a_2 = \frac{\dfrac{Pr}{(1-Pr)}u_c^*\gamma\dfrac{I_0'(i^{1/2}\gamma)}{I_0(i^{1/2}\gamma)} - \dfrac{\gamma Pr^{1/2}I_0'(i^{1/2}z_1)u_c^*}{I_0(i^{1/2}z_1)(1-Pr)}}{\left[-z_1\dfrac{I_0'(i^{1/2}z_1)}{I_0(i^{1/2}z_1)}\left\{I_0(i^{1/2}z_2) - K_0(i^{1/2}z_2)\dfrac{I_0'(i^{1/2}z_3)}{K_0'(i^{1/2}z_3)}\right\}\right.}$$
$$\left.+z_2\left(\dfrac{k_s}{k_f}\right)\left\{I_0'(i^{1/2}z_2) - K_0'(i^{1/2}z_2)\dfrac{I_0'(i^{1/2}z_3)}{K_0'(i^{1/2}z_3)}\right\}\right], \qquad (4.289)$$

$$a_3 = a_2 \frac{I_0'(i^{1/2}z_3)}{K_0'(i^{1/2}z_3)}, \qquad (4.290)$$

with

$$z_1 = \gamma Pr^{1/2}, \qquad z_2 = z_1(\alpha_f/\alpha_s)^{1/2}, \qquad z_3 = z_2(R_o/R_i). \qquad (4.291)$$

The following functions are given by Oliver (1972) and are available as computer library subroutines (e.g., IMSL).

$$I_0'\left(i^{1/2}z\right) = \text{ber}'(z) + i\,\text{bei}'(z), \qquad (4.292)$$

$$K_0\left(i^{1/2}z\right) = \text{ker}(z) + i\,\text{kei}(z), \qquad (4.293)$$

$$K_0'\left(i^{1/2}z\right) = \text{ker}'(z) + i\,\text{kei}'(z). \qquad (4.294)$$

For a porosity ε, the heat flux is

$$q = \varepsilon(\rho c_p)_f D_\parallel^d \frac{dT}{dx} = \varepsilon(\rho c_p)_f D_\parallel^d \frac{T_{f\,\max} - T_{f\,\min}}{L}, \qquad (4.295)$$

where

$$\varepsilon = \frac{\pi}{2(3)^{1/2}\left[1 + 4(R_o - R_i)^2 + 4(R_o - R_i)\right]}. \qquad (4.296)$$

Note that when considering many parallel tubes packed closely, the adiabatic boundary condition is applied at the *middistance* between the inside and outside radius (Kaviany, 1990). Here R_o refers to the *position* of the adiabatic boundary. The *transition to turbulence* occurs for $\Delta x(\omega/\nu)^{1/2} > 400$

to 800 where these critical values are determined experimentally by Grassmann and Tuma (1979). The measurements of Kurzweg et al. (1989) show that the onset of turbulence occurs at large γ ($\gamma > 5$) when $\Delta x \, (\omega/\nu)^{1/2}$ is less than 700, but this onset condition is delayed to ever-increasing values when $\gamma \to 0$. In applications such as those discussed earlier, the Womersley number γ is generally less than 5, and, therefore, the flow is laminar.

4.9.2 LONGITUDINAL DISPERSION COEFFICIENT

Results for the *glass-water* system with $R_i = 0.4$ mm, $L = 10^3 R_i$, $R_o - R_i = 0.125 R_i$, and the thermophysical properties evaluated at 60°C ($Pr = 3.0$) are given in Figures 4.18–4.20. The details can be found in Kaviany (1990). Figure 4.18 shows the distribution of the velocity and temperature fluctuations for every 1/12 increment of the period. The results are for $\Delta x = 20$ cm and $f = 1$ Hz. Note that the velocity distribution has its *maximum* at the *centerline*, i.e., the viscous boundary layer fills the entire fluid cross section. The Womersley number γ is 1.48. The velocity and temperature fields are *not* in phase.

Figure 4.19 shows the velocity field for $\Delta x = 20$ cm and for various frequencies. Note that when $\gamma > 5$, the viscous boundary-layer thickness becomes *smaller* than the tube radius and an *inertial core* is present. The core becomes *larger* as the frequency *increases*. The velocity is scaled with respect to $f \Delta x$. Since the presence of the viscous and thermal boundary layers are responsible for the heat storage/release in the fluid, as these boundary layers become thinner the heat transfer efficiency decreases (i.e., the required pressure gradient *increases* while the axial heat flux does not increase in the same proportion).

Figure 4.20 shows how the longitudinal dispersion coefficient increases in proportion to f^2 at *low frequencies*, where the boundary layers (the thermal boundary layer is also important) *fill* the entire fluid flow cross section. At higher frequencies $f > f_c$, the dispersion coefficient becomes proportional to $f^{1/2}$. The results are for several tidal displacements. It also shows that the required pressure gradient also increases as f increases, but unlike $D_\parallel^d$ the required pressure gradient increases substantially for $f > f_c$.

4.10 Dispersion Adjacent to Bounding Surfaces

The nonuniformities in the phase distributions at and near the bounding surface and its effects on the fluid flow and heat transfer are most significant if the primary heat transfer is through these surfaces. In analyzing the *variation* of the dispersion coefficient *at* and *near* these surfaces, the following should be considered.

238   4. Convection Heat Transfer

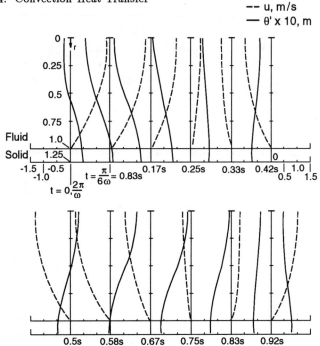

Figure 4.18 Distribution of velocity and temperature fluctuations for various elapsed times. The results are for $\Delta x = 20$ cm and $f = 1$ Hz and for a glass-water system.

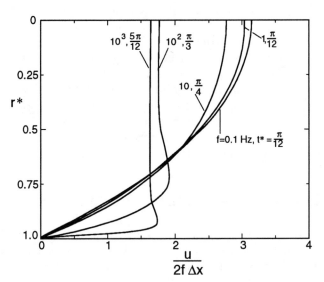

Figure 4.19 Distribution of velocity for $\Delta x = 20$ cm and several oscillating frequencies.

4.10 Dispersion Adjacent to Bounding Surfaces 239

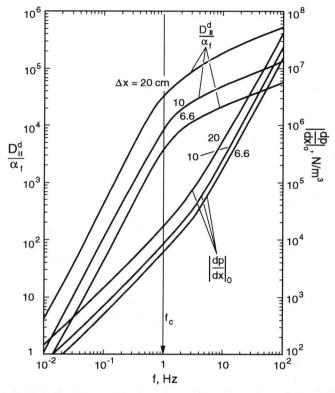

Figure 4.20 Variation of the dispersion coefficient with respect to frequency. The critical frequency f_c is also shown.

- Both $D_\perp^d$ and $D_\parallel^d$ are expected to *vanish* at the bounding solid surface.

- Near the bounding surface (fluid or solid), the porosity distribution is nonuniform.

- For packed beds of spheres the porosity is generally *larger* near the bounding solid surface, resulting in an *increase* in the local velocity (component parallel to the surface) and the local Reynolds number.

- Since the *Darcean* velocity is taken to be uniform, it does *not* allow for the *local* variations in $D_\parallel(Pe)$ and therefore does *not* lead to vanishing $D_\parallel$ and $D_\perp$ on the solid surface. Inclusion of the *porosity variation* near the surface results in an *increase* in $D_\perp(Pe)$ and $D_\parallel(Pe)$. Inclusion of the *macroscopic* shear stress (Brinkman) term $\mu'\nabla^2\mathbf{u}$ *insures* that $D_\parallel = D_\perp = 0$ at the surface. But since the Brinkman screening length is much smaller than the pore size, this inclusion does *not* account for the *actual* effect of the pore-level hydrodynamics on the local variation of $D_\parallel$ and $D_\perp$.

- Inclusion of the *pore-level* (or particle-based) hydrodynamics along with the appropriate volume averaging allows for the inclusion of the local variation of $D_\perp$ and $D_\parallel$ into the energy equations. In principle, these variations can only be included if the change from the bulk value to zero at the surface takes place over *several* representative elementary volumes. Otherwise it will not be in accord with the volume averaging.

- This leaves the rigorous continuum treatment of the near-surface hydrodynamics heat transfer to be rather impossible. The alternative has been the use of various *area* averages introduced into the volume-averaged equations. These area-averaging-continuum descriptions do *not* describe the flow and heat transfer *accurately* (Section 2.12.4), but contain a few adjustable constants, which enable them to *match specific experimental* data.

- An alternative is the *direct simulation* of the flow and heat transfer at and near the bounding surfaces. Because of the computational limitations, only *simple periodic structures* can be analyzed.

In the following we consider solid (impermeable) and fluid bounding surfaces and review the available treatments of the variation (anisotropy and nonuniformity) of the dispersion tensor near the solid and fluid bounding surfaces. Both the temperature slip and no-slip treatments are reviewed.

4.10.1 TEMPERATURE-SLIP MODEL

Since generally the variation od $D_\perp$ near the interface is not known, the *empirical slip* boundary condition is used (as discussed in Sections 2.11.1 and 3.7.1). This boundary condition uses the extrapolation of the temperature fields away from the interface and an *empirical slip coefficient* α_T. The slip boundary condition, based on the temperature gradient in the porous medium, is

$$\left.\frac{d\langle T^*\rangle_V}{dy^*}\right|_{y^*=0^-} = \alpha_T(T^{*-} - T^{*+}), \qquad (4.297)$$

where the temperatures T^{*+} and T^{*-} at the interface are found by the extrapolation of the temperature fields away from the interface (i.e., where the boundary-layer effects are not present). The slip coefficient is calculated using

4.10 Dispersion Adjacent to Bounding Surfaces

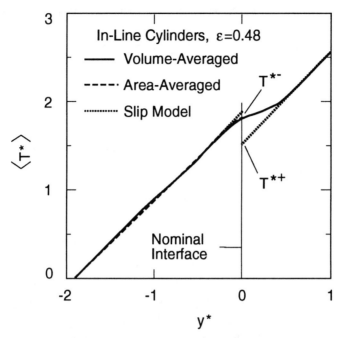

Figure 4.21 Distribution of the volume- and area-averaged temperature and the distribution of the extrapolated temperatures for the slip model ($\varepsilon = 0.48$, $Re_\ell = 0.1$, $Pe_\ell = 10^2$, and $k_s/k_f = 1$).

$$\alpha_T = \frac{\left.\dfrac{\mathrm{d}\langle T^*\rangle_V}{\mathrm{d}y^*}\right|_{y^*=0^-}}{(T^{*-} - T^{*+})}. \tag{4.298}$$

For a two-dimensional porous medium, made of periodic arrangement of circular cylinders, Sahraoui and Kaviany (1994) have computed the variation of the dispersion tensor near the bounding surfaces. The computed area- and volume-averaged local temperatures are shown in Figure 4.21 for $Pe_\ell = 10^2$, $k_s/k_f = 1$, and $\varepsilon = 0.48$. The thermal boundary layer in the fluid bounding medium contributes to the temperature jump more significantly, compared to the boundary layer in the porous medium. The thermal boundary layer in the plain medium is due to the local two dimensionality of the flow. As Pe_ℓ increases, this boundary layer effect becomes more significant. Due to the mixing in the recirculation region between the cylinders, the flow in the bed side also contributes, but slightly, to the slip in temperature. The results for $T^{*-} - T^{*+}$ and α_T are given in Table 4.7, for different Peclet numbers and for a given gap size h^*. The results show

TABLE 4.7 EFFECT OF THE PECLET NUMBER ON SLIP COEFFICIENT ($k_s/k_f = 1.0$, $Re_\ell = 0.1$, $\epsilon = 0.48$, AND $h^* = 2$)

Pe_ℓ	$T^{*-} - T^{*+}$	α_T
10	0.10	10.62
10^2	0.72	1.55
10^3	1.55	0.89

TABLE 4.8 EFFECT OF THE GAP SIZE ON SLIP COEFFICIENT ($k_s/k_f = 1.0$, $Re_\ell = 0.1$, $\epsilon = 0.48$, AND $Pe_\ell = 10^2$)

h^*	$T^{*-} - T^{*+}$	α_T
1	0.35	3.09
2	0.72	1.55
3	0.77	1.45
4	0.82	1.37

that as the Peclet number increases, the slip in the temperature becomes noticeable and can be larger than the temperature difference across one cell (in the bulk of the porous medium). Thus, if a uniform D_{yy}/α_f is used along with the no-slip condition, a significant error results in the computed interfacial heat flux. From the results on conduction heat transfer, we expect this error to become larger as k_s/k_f becomes significantly different than unity. Note that for small Peclet numbers, i.e., $Pe_\ell < 10$, conduction dominates the lateral heat transfer and the results for the slip conditions discussed in Section 3.7 apply.

The effect of h^* on α_T is similar to that of Pe_ℓ. This is because as h^* increases, the velocity near the interface also increases. A more detailed discussion of the effect of the gap size on the local flow near the interface is given in Section 2.11.4, where the effect of h^* on the hydrodynamic slip boundary condition was examined. The effect of h^* on the total temperature slip is demonstrated in Table 4.8 for $\epsilon = 0.48$ and $Pe_\ell = 10^2$. For $h^* > 3$, $T^{*-} - T^{*+}$ reaches an asymptotic value near unity. This is because the penetration of the boundary effect in the plain medium reaches an asymptote for $h > 3$. The slip coefficient decreases with increasing h^*, because the temperature slip increases while the gradient of the volume-averaged temperature remains the same.

4.10.2 NO-SLIP TREATMENTS

An early attempt at the inclusion of the *wall effect* allowed for the *variation* of the local effective thermal conductivity resulting from the local porosity variation near the bounding surface. Yagi and Kunii (1960) and Ofuchi and Kunii (1965) have found a *semiempirical* relation for the *local* effective thermal *conductivity* near the walls. As was discussed in Section 3.7, it is expected that the wall properties also influence the local effective thermal conductivity. Since the variation in the local *lateral* dispersion was not addressed by this group, their results are given in terms of a *wall* (or *surface*) *heat transfer coefficient* h_s and a constant value for the *near* wall *lateral dispersion coefficient*.

The use of the wall heat transfer coefficient allows for a *temperature slip* at the wall. This slip condition, as with the velocity slip on permeable surfaces (Section 2.11), is introduced to mask the variation in $D_\perp$ near the surface. The near wall lateral dispersion coefficient is given as $D_\perp/\alpha_f = Pe/50$, $Pe = u_D d/\alpha_f$, which gives a $D_\perp$ that is *smaller* than the bulk value.

These *empirical* correlations do *not* provide the needed $D_\perp(\mathbf{x})$ and $D_\parallel(\mathbf{x})$ for the accurate prediction of the temperature distribution within porous media. When the primary heat transfer is through the walls, the temperature distribution in the transverse direction is generally the *most* significant, and the transverse temperature gradients are *much* larger than the axial ones. *Smaller* effective thermal conductivities and *larger* temperature gradients near the walls have been reported by all the investigators. This larger-than-bulk total effective thermal *resistance* is related in part to the effective thermal conductivity variation near the wall. It is also related to the near-the-wall hydrodynamics, which is different than that of the bulk, because of the porosity variation and the no-slip condition at the wall (this was discussed in Section 2.12).

(A) LAYERED MODELS

Assuming that $D_\parallel$ and $D_\perp$ only depend on the local velocity and the velocity in turn depends on the local porosity (and permeability) through the Darcy law, Carbonell (1980) examines the effect of this velocity nonuniformity on the *axial* temperature (concentration) distribution. He considers radially uniform internal heat (solute) addition in a bed with a *radial variation* of porosity $\varepsilon(r)$.

The problem becomes that of *laterally averaging* (*area* averaging) the velocity, temperature, and $D_\parallel$. He considers a tube of radius R and a porosity distribution (and the associated velocity and dispersion coefficient distributions) given by a *jump* at radius r_o, i.e.,

$$\varepsilon = \varepsilon_0, \ u = u_0, \ D_\parallel = D_{\parallel,0} \qquad 0 \leq \frac{r}{R} \leq \frac{r_o}{R}, \qquad (4.299)$$

$$\varepsilon = \varepsilon_1, \ u = u_1, \ D_\parallel = D_{\parallel,1} \qquad \frac{r_o}{R} \leq \frac{r}{R} \leq 1, \qquad (4.300)$$

where the distribution can be envisioned as that of an outer high porosity cylindrical shell *surrounding* a low porosity core (the distribution that leads to the *channeling* phenomena). The deviations from the laterally averaged

$\langle i \rangle$, $\langle u \rangle$, and $\langle D_\| \rangle$ are given as

$$i = \langle i \rangle + i', \tag{4.301}$$

$$u = \langle u \rangle^f + u', \tag{4.302}$$

$$D_\| = \langle D_\| \rangle + D_\|', \tag{4.303}$$

where $\langle i \rangle = 2R^{-2}\int_0^R ir\,dr$. It is also assumed that $D_\perp = \gamma D_\|$. Then the steady state one-dimensional energy equation (for $k_s = 0$) becomes

$$\langle u \rangle^f \frac{d\langle i \rangle}{dx} = \langle D_\| \rangle \frac{d^2\langle i \rangle}{dx^2} - \frac{d}{dx}\langle u'i' \rangle + \frac{d^2}{dx^2}\langle D_\|' i' \rangle, \tag{4.304}$$

where

$$\langle u \rangle^f = u_0 \frac{r_o^2}{R} + u_1\left(1 - \frac{r_o^2}{R^2}\right), \tag{4.305}$$

$$\langle D \rangle = D_0 \frac{r_o^2}{R} + D_1\left(1 - \frac{r_o^2}{R^2}\right), \tag{4.306}$$

and

$$u' = -(u_1 - u_0)\left(1 - \frac{r_o^2}{R^2}\right) \qquad 0 \leq \frac{r}{R} \leq \frac{r_o}{R}, \tag{4.307}$$

$$D_\|' = -(D_{\|,1} - D_{\|,0})\left(1 - \frac{r_o^2}{R^2}\right), \tag{4.308}$$

$$u' = (u_1 - u_0)\frac{r_o^2}{R^2} \qquad \frac{r_o}{R} \leq \frac{r}{R} \leq 1, \tag{4.309}$$

$$D_\|' = (D_{\|,1} - D_{\|,0})\frac{r_o^2}{R^2}. \tag{4.310}$$

Evaluation of i' is made along the treatment made by Taylor, i.e., by using the transformation $x_u = x - \langle u \rangle^f t$ and *neglecting* the temporal and axial diffusion terms. This results in

$$u' \frac{\partial \langle i \rangle}{\partial x_u} = \frac{\gamma}{r}\frac{d}{dr}\left(rD_\| \frac{\partial i'}{\partial r}\right). \tag{4.311}$$

This is subject to the *adiabatic* and *symmetry* boundary conditions and the averaging condition. These are

$$\frac{\partial i'}{\partial r} = 0 \quad \text{at} \quad r = 0, R \tag{4.312}$$

$$\langle i' \rangle = 0. \tag{4.313}$$

4.10 Dispersion Adjacent to Bounding Surfaces

The solution is

$$i' = \frac{1}{\gamma}\left[\phi(r) - \frac{2}{R^2}\int_0^R \phi(r)r\,dr\right]\frac{\partial\langle i\rangle}{\partial x_u} \equiv f(r)\frac{\partial\langle i\rangle}{\partial x_u} = f(r)\frac{\partial\langle i\rangle}{\partial x}, \quad (4.314)$$

where

$$\phi(r) \equiv \int_0^r \frac{dr}{D_\|(r)r}\int_0^r u'(r)r\,dr. \quad (4.315)$$

Using (4.314) and (4.310) in (4.306), we have

$$\langle u\rangle^f \frac{d\langle i\rangle}{dx} = [(\rho c_p)\langle D_\|\rangle - \langle u'f(r)\rangle]\frac{d^2\langle i\rangle}{dx^2} + \langle D'_\| f(r)\rangle\frac{d^3\langle i\rangle}{dx^3}. \quad (4.316)$$

For $f(r) = 0$, the original, one-dimensional energy equation is recovered. For $f(r) < 0$, the *axial dispersion* is *enhanced* by the presence of the velocity (porosity) nonuniformity. The last term is generally negligible.

If radial instead of axial transport is of interest, then (4.313), along with $-D_\perp \partial\langle i\rangle/\partial t = -k_e \partial T/\partial r = q$ at $r = R$, must be used. Note that $D_\perp^d$ for this case, using the same one-dimensional treatment given earlier, does *vanish* at the wall. Then its behavior near the wall has to be prescribed. The dependence of $D_\perp^d$ on the radial location is addressed later.

(B) Results for a Two-Dimensional Structure

In the no-slip temperature boundary condition, a variable D_{yy} is used in order to model the nonconformity near the interface. Through a local simulation, $D_{yy}(y)$ has been computed by Sahraoui and Kaviany (1994) using the volume-averaged transverse heat flow and the gradient of the volume-averaged temperature, i.e.,

$$-\frac{D_{yy}(y^*)}{\alpha_f}\frac{d\langle T^*\rangle_V}{dy^*} = \left\langle Pe_\ell v^* T^* - \frac{k}{k_f}\frac{\partial T^*}{\partial y^*}\right\rangle_V. \quad (4.317)$$

The transverse heat flow is the same as the volume-averaged heat flow, i.e.,

$$\left\langle Pe_\ell v^* T^* - \frac{k}{k_f}\frac{\partial T^*}{\partial y^*}\right\rangle_V = \left\langle -\frac{\partial T^*}{\partial y^*}\right\rangle_{A_x} (y^* = h^*). \quad (4.318)$$

Note that the upper boundary is $v(x,h) = 0$. Then,

$$\frac{D_{yy}(y^*)}{\alpha_f} = \frac{\left\langle \dfrac{\partial T^*}{\partial y^*}\right\rangle_{A_x}(y^* = h^*)}{\dfrac{d\langle T^*\rangle_V}{dy^*}(y^*)}. \quad (4.319)$$

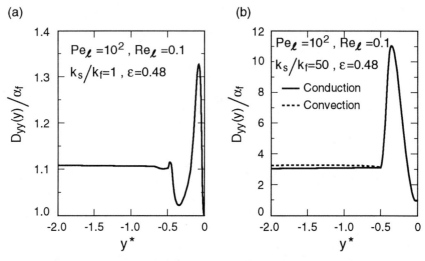

Figure 4.22 (a) Distribution of $D_{yy}(y^*)/\alpha_f$ near the solid bounding surface. (b) Same as (a), but showing the effect of k_s/k_f on $D_{yy}(y^*)/\alpha_f$ near the solid bounding surface.

(i) Dispersion Near a Solid Bounding Surface

We first examine $D_{yy}(y)^*/\alpha_f$ for a solid bounding surface and for $k_s/k_f = 1$. Since the gap size is part of the porous medium, h^* is chosen such that the last cylinder row also has a porosity of 0.48 (i.e., corresponding to the simple cubic arrangement for spheres). Here the interface is taken as the bounding surface (instead of the surface tangent to the tip of the cylinders). The result for $D_{yy}(y^*)/\alpha_f$ presented in Figure 4.22(a), for $Pe_\ell = 10^2$ and $k_s/k_f = 1$, show that the boundary effect only penetrates half of a cell size into the porous medium. Near the bounding surface, $D_{yy}(y^*)/\alpha_f$ undergoes a peak and further away from the bounding surface, and before reaching the bulk value, $D_{yy}(y^*)/\alpha_f$ decreases to values lower than the bulk value. The decrease in $D_{yy}(y^*)/\alpha_f$ is due to the recirculation region between the cylinders.

The effect of k_s/k_f on $D_{yy}(y^*)/\alpha_f$ is shown in Figure 4.22(b) for $k_s/k_f = 50$ and $Pe_\ell = 10^2$. For small $|y^*|$ where the averaging volume does not enclose any solid, $D_{yy}(y^*)/\alpha_f$ is unity. As $|y^*|$ increases, the contribution of the transverse effective conductivity $k_{e_{yy}}(y^*)/k_f$ becomes more significant compared to the hydrodynamic effect. The results presented in Figure 4.22(b) for convection and conduction are decomposed and the only significant difference between the two exists for $y^* < -0.4$. This significant *dominance of conduction* is also evident in the experiments of Yagi and Kunii. Their results show that the temperature slip, for their random packed bed of spheres, is nearly the same for the stagnant and flowing of air. Therefore, accurate modeling of $k_{e_{yy}}(y^*)/k_f$ is more important in predicting the heat transfer across the bounding surface. In the previous studies mentioned above, $k_{e_{yy}}(y^*)/k_f$ is found using the local porosity and an effective conductivity-porosity correlation (for packed beds of spherical particles). Using the model for $k_{e_{yy}}(y^*)/k_f$ used by Cheng and Hsu, in the one-dimensional energy equation, and comparing the results with the experimental results of Yagi and Kunii, we find that this local effective conductivity predicts a lower heat flux at the interface. Therefore, in the previous studies such as Cheng and Hsu, the empirical constant introduced in modeling $D_{yy}(y^*)/\alpha_f$ is used mostly to correct the deficiency in predicting $k_{e_{yy}}(y^*)/k_f$.

(ii) Dispersion Near a Bounding Channel Flow

For a porous medium bounded by a fluid (plain medium), the nonuniformity in $D_{yy}(y^*)/\alpha_f$ is influenced by the flow in the fluid channel. This is shown in Figure 4.23(a), where for $k_s/k_f = 1$ the effect of the thermal boundary layer in the plain medium (which depends on Pe_ℓ) is rather dominant. Away from the interface, in the porous medium, this Peclet number dependence is very weak.

As the Peclet number vanishes, $D_{yy}(y^*)/\alpha_f$ becomes uniform for $k_s/k_f = 1$. For $k_s/k_f \neq 1$, the nonuniformity in $D_{yy}(y^*)/\alpha_f$ is present on both sides of the interface, as shown in Figure 4.23(b). The nonuniformity in $D_{yy}(y^*)/\alpha_f$ in the porous medium depends mostly on the magnitude of k_s/k_f and again modeling of $k_{e_{yy}}/k_f$ becomes more important than D_{yy}^d/α_f.

The effect of h^* on the variation of D_{yy}/α_f is shown in Figure 4.23(c), where an increase in h^* increases the local Peclet number near the interface and then gives an increase in D_{yy}/α_f. For $h^* > 3$, as shown in Figure 4.23(c), the nonuniformity in D_{yy}/α_f extends to about one cell size in the plain medium. The penetration depth in the plain medium is independent of h^*, however, the magnitude of $D_{yy}(y^*)/\alpha_f$ increases with h^*.

4.10.3 Models Based on Mixing-Length Theory

In noting that the value of $D_\perp$ and $D_\parallel$ must vanish at the bounding surface, several investigators have proposed inclusion of a *wall effect* similar to that used for the *turbulent flows*. The observed variation in porosity discussed in Section 2.12, along with a prescription of $D_\perp(\mathbf{x})$, has been used by Cheng and Hsu (1986a,b), Cheng and Zhu (1987), and Cheng and Vortmeyer (1988). A similar approach is used by Tobis and Ziolkowski (1988). A *geometric-based* prediction of the variation of $D_\perp$ near the surface is done by Kuo and Tien (1989). These are reviewed here.

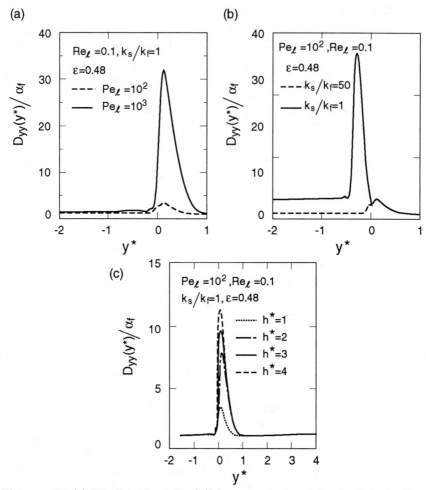

Figure 4.23 (a) Distribution of $D_{yy}(y^*)/\alpha_f$ near the interface of a bed of cylinders with the fluid bounding medium for $Pe_\ell = 10^2$ and $Pe_\ell = 10^3 (h^* = 1)$. (b) Effect of k_s/k_f on the distribution of $D_{yy}(y^*)/\alpha_f$, for the solid bounding medium with $k_s/k_f = 1$ and $k_s/k_f = 50 (h^* = 2)$. (c) Effect of the gap size on the distribution of $D_{yy}(y^*)/\alpha_f$, for $h^* = 1, 2, 3, 4$.

(A) APPLICATION OF TURBULENCE WALL FUNCTION

Cheng and co-workers have analyzed the fully developed (hydrodynamic and thermal) heat transfer through the bounding surfaces (parallel plate channels, circular tubes, and annuli) of packed beds and have introduced the following features in their analysis.

(i) Hydrodynamics

- They consider the *random* arrangement of spherical particles with the momentum balance given by the Darcy-Ergun equation (2.57). In some cases the Brinkman extension is also used. The Carman-Kozeny permeability equation is used for $K = K(\varepsilon, d)$.

- The *areal* void fraction distribution of Section 2.12.1 with an *exponential* function curve fit was used as the *local* porosity distribution. This local porosity distribution is taken as $\varepsilon = \varepsilon_\infty + (1 - \varepsilon_\infty)e^{-a_1 y/d}$ with y measured from the wall and a_1 as a constant. Then the velocity distribution is determined from the Darcy-Ergun-Brinkman equation.

- The preceding hydrodynamic formulation is an attempt to model the pore-level flow using a unidirectional flow. When the Brinkman extension is used, it results in a two-dimensional flow over a distance of $o\left(K^{1/2}\right)$, which is generally much *smaller* than the particle size. This *masks* the *lateral* flow and, therefore, *underpredicts* the momentum and heat flow in that direction. The result of this formulation is the simulation of the familiar channeling phenomenon adjacent to the bounding surface.

(ii) Heat Transfer

The *lateral* dispersion coefficient is taken as

$$\frac{D_\perp^d}{\alpha_f} = a_1 Pe_{u_D} \frac{u_D(y)}{u_D} \frac{\ell(y)}{d}, \qquad (4.320)$$

where $Pe_{u_D} = u_D d/\alpha_f$, u_D is the local velocity obtained from the formulation, and ℓ is taken as

$$\frac{\ell}{d} = 1 - e^{-a_2 y/d}, \qquad (4.321)$$

which is the *van Driest wall function* for turbulent flows. Note that since both u_D and ℓ change with y, then $D_\perp$ does *not* decrease *monotonically* with y.

The *local* effective thermal conductivity is taken to vary with the porosity according to the available correlations for the *bulk* effective property of packed beds of *spheres* given in the general form

$$\frac{k_e}{k_f} = \frac{k_e}{k_f}\left(\frac{k_s}{k_f}, \varepsilon\right). \qquad (4.322)$$

The *larger* experimentally observed temperature gradients near the bounding surfaces can then be predicted by using specific values for a_2 which result in good agreement with the experimentally determined temperature distribution. Then if $k_e(y=0)$ is *also* correctly prescribed, the heat flux at the wall can also be correctly calculated.

In general, a_2 depends on the *local Re* and Pr. For large values of k_s/k_f, the value of $k_e(y=0)$ *cannot* be accurately extrapolated from the bulk k_e correlations (Section 3.7). Even the *near*-wall value of Yagi and Kunii is *averaged* over a distance of $d/2$, and over this distance, k_e changes *significantly*. As was mentioned in Section 3.7, the thermal conductivity of the bounding surface also influences $k_e(y=0)$.

Tobis and Ziolkowski use the concept of the *eddy viscosity* of turbulence and allow for a *plain* boundary layer (i.e., *without any matrix*) *forming* next to the surface. They assume the same uniform pressure gradient along this plain boundary layer and the remaining porous medium. The form of the momentum equation used is

$$\frac{d}{dy}\left[\mu(1+a_3 y)\right]\frac{du}{dy} = \frac{dp}{dx} = -\frac{\mu}{K}u_D - \rho\frac{C_E \varepsilon}{K^{1/2}}u_D^2, \qquad (4.323)$$

where $\mu_t = \mu a_3 y$ is the *eddy viscosity*. They also introduce a *thermal eddy diffusivity* as

$$\frac{\alpha_t}{\alpha_f} = \frac{\mu_t}{\mu} Pr = a_3 Pr\, y, \qquad (4.324)$$

such that the heat flux becomes

$$q = -k_f(1+a_3 Pr\, y)\frac{dT}{dy}. \qquad (4.325)$$

They envision that within the momentum boundary layer the velocity *increases* from the value of zero at the surface to its maximum (at $du/dy=0$), which takes place *nearly* at a distance of $d/2$ from the surface. The local porosity ε_m at $y=d/2$ is assumed to be known ($\varepsilon_m=0.26$, i.e., closest regular packing). Then the *maximum* velocity is taken as $u_{\max} = u_D \varepsilon/\varepsilon_m$, where ε is the bulk porosity. These *three* boundary conditions allow for solution of (4.323) and the *exact position* of the velocity maximum. As expected, this boundary-layer thickness is the same as the Brinkman screening distance and a result similar to (2.153) emerges.

Their result for $C_E=0$ is

$$\delta = \left(\frac{K_s}{\varepsilon_m}\right)^{1/2}, \qquad (4.326)$$

where $K_s = K_s(\varepsilon_s)$ and ε_s is the *average porosity* near the surface such that $\varepsilon > \varepsilon_s > \varepsilon_m$.

By comparing the results of their experiments and predictions for the temperature distribution and the Nusselt number, they find that a_1 *depends* on the Reynolds number. This is expected because the hydrodynamics of the region extending from the surface to several particle diameters (a few in

4.10 Dispersion Adjacent to Bounding Surfaces

the case of regular arrangements) into the bed is *strongly* Reynolds-number-dependent. This is because boundary layers grow on the particle and on the bounding surface, and their thickness depends on the local Reynolds number. These have been discussed in Section 2.12.4 for a two-dimensional periodic structure.

The treatment of Tobis and Ziolkowski is based on *two-* and *three-layer* models which require the introduction of empiricism (based on the comparison between the results of the model and the experiments) and generally lacks universality. However, these models give predictions/correlations for the solid bounding-surface Nusselt number.

(B) A GEOMETRICAL MODEL FOR MIXING-LENGTH

In order to provide a *closure* for $\langle \mathbf{u}'T' \rangle$, rather than directly solving for this as was done for the simple periodic structures and for the disordered media, e.g., (4.44), (4.78), and (4.183), Kuo and Tien apply a *mixing-length* method to find expressions for the lateral dispersion coefficient in the bulk as well as near the bounding surface.

(i) Bulk Transverse Dispersion Coefficient

The volume-averaged lateral dispersion is related to the lateral temperature gradient through

$$D_\perp = -\varepsilon \langle v'T' \rangle = \varepsilon \langle |v'|\ell \rangle \frac{d\langle T \rangle}{dy}, \qquad (4.327)$$

where v' is the *disturbance* component of the *lateral* velocity and $\langle v' \rangle = 0$. The *mixing-length* ℓ depends on the local structure.

The porous medium is assumed to be made of *randomly arranged spheres* of diameters d. In this model, each sphere is *replaced* with a cylinder with its axis going through the center of the sphere and *along* the flow direction. The *average* distance between the centers of the two adjacent spheres (uniform distribution) is

$$\ell_c = \frac{d}{3(1-\varepsilon)}. \qquad (4.328)$$

The *local mixing length*, the distance the fluid particle deflects from a sphere located at (r, θ), is given as

$$\ell = \left[w + \left(\frac{d}{2} - r \right) \right] \sin \theta, \qquad (4.329)$$

where w is the *average* annulus gap between any two spheres replaced by two cylinders and is given as

$$w = \frac{d}{2} \left[\frac{1}{(1-\varepsilon)^{1/3}} - 1 \right]. \qquad (4.330)$$

The lateral velocity *deviation* is taken as

$$|v'| = \frac{\ell}{\ell_c/2} \frac{u_D}{\varepsilon}. \tag{4.331}$$

When (4.328)–(4.331) are used in (4.327) and integrated over $0 < r < d/2$ and $0 < \theta < \pi/2$, the result is given as

$$D_\perp = f_\infty(\varepsilon) \langle u_D \rangle d, \tag{4.332}$$

where $f_\infty(\varepsilon)$ is nearly equal to 0.075, for $0.35 \leq \varepsilon \leq 0.5$.

(ii) Adjacent to Bounding Surfaces

It should be mentioned that the *volume* averaging of any mixing length *nullifies* its introduction. This is because the mixing length is assumed to be a local (*point*) quantity, and near the wall its significant variation over only a few particle lengths is sought. Near the bounding surface, it is *assumed* that

$$|v'| = a_1 \frac{y}{d} u_D, \tag{4.333}$$

with y measuring the distance from the surface. The temperature deviation is taken as

$$T' = -a_2 d \frac{\mathrm{d}\langle T \rangle}{\mathrm{d}y}. \tag{4.334}$$

Then from (4.332) and (4.327) with y treated as a constant (i.e., *volume* integral is *replaced* with an *areal* integral in order to maintain the *locality*), we have

$$f(\varepsilon) = \varepsilon \frac{y}{d} \langle a_1 a_2 \rangle = a_3 \left(\frac{y}{d}\right)^2, \tag{4.335}$$

where geometrical arrangements and expansion of ε in terms of y/d is used in arriving at (4.335). In order to *match* with the bulk value, they use (in line with the *van Driest wall function*)

$$f(\varepsilon) = f_\infty(\varepsilon) \left[1 - e^{-a_4 (y/d)^2} \right]. \tag{4.336}$$

The constant a_4 is determined using the available *experimental* results for the temperature distribution and by using (4.336) in the energy equation. However, their predicted Nusselt number does *not* agree with the experimental results. This could be due to the value of $k_e(y = 0)$ used and, as expected, a_4 *depends* on the Reynolds number.

The preceding attempt at the introduction of some local variations (*geometrical*) is ad hoc and heuristic, and in general, is along the lines of the other empirical mixing-length prescriptions.

4.10.4 A MODEL USING PARTICLE-BASED HYDRODYNAMICS

In an effort to obtain a $D_\perp(y)$ that contains the local hydrodynamic and thermal features but is still consistent with the averaging requirement, Hsu and Cheng (1988 and 1990) considered evaluation of $\langle u'T' \rangle$. However, their pore hydrodynamics is based on *dilute* particle concentrations (similar to that of Koch and Brady and Koch et al.) and does *not* include any influence of the bounding surface. Instead, the variation of the *local area-averaged* void fraction is introduced in order to account for the *lower* values of $D_\perp$ near the surface. This is the same approach they used above in the application of the mixing-length concept.

In principle, the T' equation (4.44) can be solved if $\mathbf{u}'$ is known, similar to the methods discussed in Sections 4.3.3 and 4.6.1. Hsu and Cheng consider *two flow regimes*, namely, a *boundary-layer* and a *creeping* flow regime and then recommend

$$T' = d\,\mathbf{b}_{T1}(\mathbf{x}) \cdot \nabla \langle T \rangle^f, \qquad (4.337)$$

$$\mathbf{u}' = \langle u \rangle^f \left[\mathbf{b}_{u1}(\mathbf{x}) + \mathbf{b}_{u2}(\mathbf{x}) Re^{-1/2} \right] \qquad \text{for } Re \gg 10, \qquad (4.338)$$

and

$$T' = \frac{d^2 \langle u \rangle^f}{\alpha_f} \mathbf{b}_{T2}(\mathbf{x}) \cdot \nabla \langle T \rangle^f, \qquad (4.339)$$

$$\mathbf{u}' = \langle u \rangle^f \left[\mathbf{b}_{u3}(\mathbf{x}) + \mathbf{b}_{u4}(\mathbf{x}) Re \right] \qquad \text{for } Re \ll 10. \qquad (4.340)$$

Insertion of (4.338) and (4.340) into (4.39) and the subsequent integration gives a Pe^2 *proportionality* for the creeping flow regime and a Pe *proportionality* for the boundary-layer flow regime for $D_\perp$. The constants are to be determined from the *experimental* results. Note that the more exact analysis of Koch and Brady and the numerical solution of Carbonell and Whitaker do not contain any constants. Although not an easy task, one approach that is still semiempirical is the following. The vector functions $\mathbf{b}_{ui}$ can be constructed to include the effect of the bounding solid surface. Then, by determining $\langle v'T' \rangle$ either by area averaging or volume averaging over a small lateral distance Δy, a local variation can be introduced. This, in principle, eliminates the need for the introduction of any constant. However, this approach has not yet been developed rigorously.

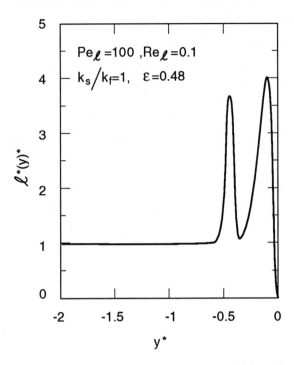

Figure 4.24 Distribution of the mixing length near a solid bounding surface.

4.10.5 RESULTS OF A TWO-DIMENSIONAL SIMULATION

The numerical results of Sahraoui and Kaviany (1994) for the distribution of the mixing length $\ell^* = \ell^*(y^*)$ is given in Figure 4.24. The application of (4.320) gives

$$\ell^*(y^*) = \frac{D_{yy}(y^*)}{D_{yy}(y^* \to \infty)} \frac{\langle \varepsilon \rangle_A (y^*)}{\langle u^* \rangle_A (y^*)}. \tag{4.341}$$

Using the computed area-averaged quantities, the distribution of ℓ^* is found. Their results show that $\ell^* = \ell^*(y^*)$ has *two* peaks, one at $\ell^* = 0.1$ and one at $y^* = -0.4$. This shows that the simple, *integral concept* (i.e., a monotonic decay near the bounding solid surface), for the mixing length does *not* directly apply to the variations near the bounding surface.

4.11 References

Acrivos, A., Hinch, E. J., and Jeffrey, D. J., 1980, "Heat Transfer to a Slowly Moving Fluid from a Dilute Bed of Heated Spheres," *J. Fluid Mech.*, 101, 403–421.

4.11 References

Aris, R., 1956, "On the Dispersion of a Solute in a Fluid Flowing Through a Tube," *Proc. Roy. Soc.* (London), A235, 67–77.

Aris, R. and Amundson, N. R., 1957, "Some Remarks on Longitudinal Mixing or Diffusion in Fixed Beds," *AIChE J.*, 3, 280–282.

Baron, T., 1952, "Generalized Graphic Method for the Design of Fixed and Catalytic Reactor," *Chem. Eng. Prog.*, 48, 118–124.

Batycky, R. P., Edwards, D. A., and Brenner, H., 1993, "Thermal Taylor Dispersion in an Insulated Circular Cylinder – I. Theory, II. Applications," *Int. J. Heat Mass Transfer*, 36, 4317–4333.

Bear, J., 1988, *Dynamics of Fluids in Porous Media*, Dover.

Brenner, H., 1980, "Dispersion Resulting from Flow Through Spatially Periodic Porous Media," *Phil. Trans. Roy. Soc.* (London), 297, 81–133.

Carbonell, R. G., 1979, "Effect of Pore Distribution and Flow Segregation on Dispersion in Porous Media," *Chem. Engng. Sci.*, 34, 1031–1039.

Carbonell, R. G., 1980, "Flow Nonuniformities in Packed Beds, Effect on Dispersion," *Chem. Engng. Sci.*, 35, 1347–1356.

Carbonell, R. G. and Whitaker, S., 1983, "Dispersion in Pulsed Systems—II. Theoretical Developments for Passive Dispersion in Porous Media," *Chem. Engng. Sci.*, 38, 1795–1802.

Carslaw, H. S. and Jaeger, J. C., 1986, *Conduction of Heat in Solids*, Oxford University.

Chatwin, P. C., 1975, "On the Longitudinal Dispersion of Passive Containment in Oscillatory Flows in Tubes," *J. Fluid Mech.*, 71, 513–527.

Cheng, P. and Hsu, C. T., 1986a, "Fully-Developed, Forced Convective Flow Through an Annular Packed Bed–Sphere Bed with Wall Effects," *Int. J. Heat Mass Transfer*, 29, 1843–1853.

Cheng, P. and Hsu, C. T., 1986b, "Application of van Driest's Mixing Length Theory to Transverse Thermal Dispersion in Forced Convection Flow Through a Packed Bed," *Int. Comm. Heat Mass Transfer*, 13, 613–625.

Cheng, P. and Vortmeyer, D., 1988, "Transverse Thermal Dispersion and Wall Channeling in a Packed Bed with Forced Convective Flow," *Chem. Engng. Sci.*, 43, 2523–2532.

Cheng, P. and Zhu, H., 1987, "Effect of Radial Thermal Dispersion on Fully-Developed Forced Convective in Cylindrical Packed Tubes," *Int. J. Heat Mass Transfer*, 30, 2373–2383.

De Josselin De Jong, G., 1958, "Longitudinal and Transverse Diffusion in Granular Deposits," *Trans. Amer. Geophys. Union*, 39, 67–74.

Eidsath, A., Carbonell, R. G., Whitaker, S., and Herman, L. R., 1983, "Dispersion in Pulsed Systems—III. Comparison between Theory and Experiment for Packed Beds," *Chem. Engng. Sci.*, 38, 1803–1816.

Edwards, M. F. and Richardson, J. E., 1968, "Gas Dispersion in Packed Beds," *Chem. Engng. Sci.*, 23, 109–123.

Fried, J. J. and Combarnous, M. A., 1971, "Dispersion in Porous Media," *Advances in Hydro. Science*, 7, 169–282.

Georgiadis, J. G. and Catton, I., 1988, "An Effective Equation Governing Transport in Porous Media," *ASME J. Heat Transfer*, 110, 635–641.

Grassmann, P. and Tuma, M., 1979 "Critical Reynolds Number for Oscillating and Pulsating Tube Flow," (in German), *Thermo-Fluid Dyn.*, 12, 203–209.

Gunn, D. J., 1970, "The Transient and Frequency Response of Particles and Beds of Particles," *Chem. Engng. Sci.*, 25, 53–66.
Gunn, D. J. and De Souza, J. F. C., 1974, "Heat Transfer and Axial Dispersion in Packed Beds," *Chem. Engng. Sci.*, 29, 1363–1371.
Gunn, D. J. and Khalid, M., 1975, "Thermal Dispersion and Wall Transfer in Packed Bed," *Chem. Engng. Sci.*, 30, 261–267.
Gunn, D. J. and Pryce, C., 1969, "Dispersion in Packed Beds," *Trans. Instn. Chem. Engrs.*, 47, T341–T350.
Han, N.-W., Bhakta, J., and Carbonell, R. G., 1985, "Longitudinal and Lateral Dispersion in Packed Beds: Effect of Column Length and Particle Size Distribution," *AIChE J.*, 31, 277–288.
Haring, R. E. and Greenkorn, R. A., 1970, "A Statistical Model of a Porous Media with Nonuniform Pores," *AIChE J.*, 16, 477–483.
Hasimoto, H., 1959, "On the Periodic Fundamental Solutions of the Stokes Equation and Their Application to Viscous Flow Past a Cubic Array of Spheres," *J. Fluid Mech.*, 5, 317–328.
Horn, F. J. M., 1971, "Calculation of Dispersion Coefficient by Means of Moments," *AIChE J.*, 17, 613–620.
Hsu, C. T. and Cheng, P., 1988, "Closure Schemes of the Macroscopic Energy Equation for Convective Heat Transfer in Porous Media," *Int. Comm. Heat Mass Transfer*, 15, 689–703.
Hsu, C. T. and Cheng, P., 1990, "Thermal Dispersion in a Porous Medium," *Int. J. Heat Mass Transfer*, 33, 1587–1597.
Jeffrey, D. J., 1973, "Conduction Through a Random Suspension of Spheres," *Proc. Roy. Soc.* (London), A335, 355–367
Kaviany, M., 1986, "Some Aspects of Heat Diffusion in Fluids by Oscillation," *Int. J. Heat Mass Transfer*, 29, 2002–2006.
Kaviany, M., 1990, "Performance of a Heat Exchanger Based on Enhanced Heat Diffusion in Fluids by Oscillation: Analysis," *ASME J. Heat Transfer*, 112, 110–116.
Koch, D. L. and Brady, J. F., 1985, "Dispersion in Fixed Beds," *J. Fluid Mech.*, 154, 399–427.
Koch, D. L. and Brady, J. F., 1987, "The Symmetry Properties of the Effective Diffusivity Tensor in Anisotropic Porous Media," *Phys. Fluids*, 30, 642–650.
Koch, D. L., Cox, R. G., Brenner, H., and Brady, J. F., 1989, "The Effect of Order on Dispersion in Porous Media," *J. Fluid Mech.*, 200, 173–188.
Kuo, S. M. and Tien, C.-L., 1989, "Transverse Dispersion in Packed-Sphere Beds," *Proceedings of National Heat Transfer Conference*, 1, 629–634.
Kurzweg, U. H., 1985, "Enhanced Heat Conduction in Oscillatory Viscous Flows within Parallel-Plate Channels," *J. Fluid Mech.*, 156, 291–300.
Lawson, D. W. and Elrick, D. E., 1972, "A New Method for Determining and Interpreting Dispersion Coefficient in Porous Media," in *Proceedings of the Second Symposium on Fundamentals of Transport Phenomena in Porous Media*, Ontario, Canada, 753–779.
Lee, C. K., Sun, C. C., and Mei, C. C., 1995, "Micromechanical Theory of Periodic Porous Media by Homogenization Method, II. Computation of Dispersivities of Heat and Solute," *Wat. Resour. Res.*, to appear.

Martin, H., 1978, "Low Peclet Number Particle–to–Fluid Heat and Mass Transfer in Packed Beds," *Chem. Engng. Sci.*, 33, 913–919.

Mei, C. C., 1992, "Method of Homogenization Applied to Dispersion in Porous Media," *Transp. Porous Media*, 9, 262–274.

Nunge, R. J. and Gill, W. N., 1969, "Mechanisms Affecting Dispersion and Miscible Displacement," in *Flow Through Porous Media*, American Chemical Society Publication, 180–196.

Ofuchi, K. and Kunii, D., 1965, "Heat Transfer Characteristics of Packed Beds with Stagnant Fluids," *Int. J. Heat Mass Transfer*, 8, 749–757.

Oliver, F. W. J., 1972, "Bessel Functions of Integer Order," in *Handbook of Mathematical Functions*, Abramowitz and Stegun, eds., NSRDS–NBS, p. 379.

Quintard, M. and Whitaker, S., 1993, "Transport in Ordered and Disordered Porous Media: Volume–Averaged Equations, Closure Problems, and Comparison with Experiments," *Chem. Engng. Sci.*, 48, 2537–2564.

Reis, J. F. G., Lightfoot, E. N., Noble, P. T., and Chiang, A. S., 1979, "Chromatography in Bed of Spheres," *Sep. Sci. Tech.*, 14, 867–894.

Ryan, D., Carbonell, R. G., and Whitaker, S., 1980, "Effective Diffusivities for Catalyst Pellets under Reactive Conditions," *Chem. Engng. Sci.*, 35, 10–16.

Saffman, P. G., 1959a, "Dispersion in Flow Through a Network of Capillaries," *Chem. Engng. Sci.*, 11, 125–129.

Saffman, P. G., 1959b, "A Theory of Dispersion in a Porous Medium," *J. Fluid Mech.*, 6, 321–349.

Saffman, P. G., 1960, "Dispersion Due to Molecular Diffusion and Macroscopic Mixing in Flow Through a Network of Capillaries," *J. Fluid Mech.*, 7, 194–208.

Saffman, P. G., 1973, "On the Settling of Free and Fixed Suspensions," *Stud. Appl. Math.*, 52, 115–127.

Sahimi, M., Hughes, B. D., Scriven, L. E., and Davis, H. T., 1983, "Stochastic Transport in Disordered Systems," *J. Chem. Phys.*, 78, 6849–6864.

Sahraoui, M. and Kaviany, M., 1994, "Slip and No–Slip Temperature Boundary Conditions at Interface of Porous, Plain Media: Convection," *Int. J. Heat Mass Transfer*, 37, 1029–1044.

Sangani, A. S. and Acrivos, A., 1983, "The Effective Conductivity of a Periodic Array of Spheres," *Proc. Roy. Soc.* (London), A386, 263–275.

Scheidegger, A. E., 1954, "Statistical Hydrodynamics in Porous Media," *J. Appl. Phys.*, 25, 994–1001.

Schertz, W. W. and Bishoff, K. G., 1969, "Thermal and Material Transfer in Nonisothermal Packed Beds," *AIChE J.*, 4, 597–604.

Schlichting, H., 1979, *Boundary–Layer Theory*, Seventh Edition, McGraw–Hill.

Schlünder, E. U., 1978, "Transport Phenomena in Packed Bed Reactors," in *Chemical Reaction Engineering Reviews—-Houston*, ACS Symposium Series, No. 72, American Chemical Society, 110–161.

Slattery, J. C., 1981, *Momentum, Energy, and Mass Transfer in Continua*, 2nd ed., R. F. Krieger.

Taylor, G. I., 1953, "Dispersion of Soluble Matter in Solvent Flowing Slowly Through a Tube," *Proc. Roy. Soc.* (London), A219, 186–203.

Taylor, G. I., 1954a, "Condition under which Dispersion of a Solute in a Stream of Solvent Can be Used to Measure Molecular Diffusion," *Proc. Roy. Soc.* (London), A225, 473–477.

Taylor, G. I., 1954b, "The Dispersion of Matter in Turbulent Flow in a Pipe," *Proc. Roy. Soc.* (London), A223, 446–468.

Tobis, J. and Ziolkowski, D., 1988, "Modelling of Heat Transfer at the Wall of a Packed–Bed Apparatus," *Chem. Engng. Sci.*, 43, 3031–3036.

Todorovic, P., 1970, "A Stochastic Model of Longitudinal Diffusion in Porous Media," *Wat. Resources Res.*, 6, 211–222.

Vortmeyer, D., 1975, "Axial Heat Dispersion in Packed Beds," *Chem. Engng. Sci.*, 30, 999–1001.

Watson, E. J., 1983, "Diffusion in Oscillatory Pipe Flow," *J. Fluid Mech.*, 133, 233–244.

Yagi, S. and Kunii, D., 1960, "Studies of Heat Transfer Near Wall Surface in Packed Beds," *AIChE J.*, 6, 97–104.

Yagi, S., Kunii, D., and Wakao, N., 1960, "Studies on Axial Effective Thermal Conductivities in Packed Beds," *AIChE J.*, 6, 543–546.

Yuan, Z., Somerton, W., and Udell, K. S., 1991, "Thermal Dispersion in Thick Wall Tubes as a Model of Porous Media," *Int. J. Heat Mass Transfer*, 34, 2715–2726.

5
Radiation Heat Transfer

In this chapter, heat transfer *by* radiation in porous media is examined. The medium may be treated either as a single *continuum* or as a *collection of particles* (i.e., *scatterers*). In the *particle-based* analysis, the interaction of radiation with a collection of elements of the solid matrix (e.g., particles in a packed bed) is considered. On the other hand, the continuum treatment attempts to obtain the *effective/radiative properties* of the *medium* by using the element-based interaction along with a local volume-averaging procedure. This volume averaging is greatly simplified if it is assumed that the *interaction* of a particle with radiation is *not* affected by the presence of neighboring particles [i.e., the scattering (or absorption) is *independent*]. In case the assumption of independent scattering fails, the volume averaging must include *dependent* effects.

In this chapter, we first discuss the fundamentals of the continuum treatment leading up to the equation of radiative transfer. Then we examine the *spectral* radiative behavior of a single *isolated* particle. *Asymptotic* approaches based on *large* and *small* particle size are discussed and their predictions are compared. Following this, we establish the range of validity of the *theory of independent scattering* and give the volume-averaging procedure to obtain the *independent properties* of the system from the radiative properties of an *individual particle*. Then some experiments on *packed beds* and *suspensions* are examined. We discuss some *methods* to solve the equation of radiative transfer. A *noncontinuum* (e.g., *Monte Carlo*) method of solution for *large particles* is also discussed. Since this method does not use properties obtained from volume averaging, it models *dependent* scattering in the large particle regime. *Continuum methods* of modeling *dependent* scattering for *large* and *small* particle size limits are also discussed. Finally, we offer some *guidelines* on how to approach a radiation problem in porous media.

The inclusion of the contribution of radiative heat transfer in the energy equation is guided by the following.

- In the presence of a matrix, the *absorption/emission/scattering* from *gases* is generally *masked* by the absorption/emission/scattering from the matrix (Mengüc and Viskanta, 1986). If the gas contribution is significant, it will be included as if *independent* of the matrix contributions, i.e., the spectral or band contributions will be *superimposed*, e.g.,

$$\langle \sigma_{\lambda a} \rangle = \langle (\sigma_{\lambda a})_s \rangle + \langle (\sigma_{\lambda a})_g \rangle. \qquad (5.1)$$

This relation is strictly true only if the assumption of independent scattering holds.

- For *liquids*, the temperature difference existing in the system is generally *small* and the liquids are highly absorbing in the *infrared* wavelength range. Therefore, in general, *no* radiation heat transfer is considered in dealing with *fully* or *partially liquid saturated* matrices. However, when it is necessary to include radiation heat transfer with liquids, the *relative index of refraction n* used for the properties is n_s/n_f (i.e., for gases $n_f \simeq 1$, but *not* for liquids).

- Because the source of thermal radiation energy is generally *nonpolarized* (gas or solid surface) and because the waves undergo substantial *reflections* in the interstices of the matrix, the thermal radiation is conceived as being *nonpolarized*.

5.1 Continuum Treatment

The fundamentals of radiation heat transfer in absorbing/emitting/scattering media have been given by Chandrasekhar (1960), Ozisik (1985), Siegel Howell (1981, 1992), Brewster (1992), and Modest (1993). Some of the principles are briefly given herein. Their approach treats the solid-fluid phases as a single continuum. Therefore, the following applies to *heterogeneous* (solid and fluid phases are present simultaneously) differential elements. A schematic showing the coordinate system for a *plane-parallel* geometry (which is the geometry used through most of this chapter) is given in Figure 5.1. The *unit vector* in the beam direction is given by s and the *length* of the position vector is given by S. The *incident* beam is shown with subscript i and the incident *solid angle* is shown by $d\Omega_i$.

- It is assumed that the *particle size* is much *smaller* than the linear size of the system. Then the radiative properties are averaged over a representative elementary volume with a linear dimension ℓ, such that $d \ll \ell \ll L$.

- The matrix-fluid system is treated as a continuum by assuming that the *local thermal equilibrium* (as discussed in Section 3.1) exists in accord with the treatment of conduction and convection.

- *Azimuthal symmetry* is assumed so that $I_\lambda(\theta, \phi) = I_\lambda(\theta)$.

- The *spectral* (indicated by subscript λ) *radiation intensity* I_λ is the radiation energy in the direction θ *per unit time, per unit projected area, per unit solid angle*, and *per interval* $d\lambda$ around λ. Then I_λ is given in W/(m^2-sr-μm).

5.1 Continuum Treatment

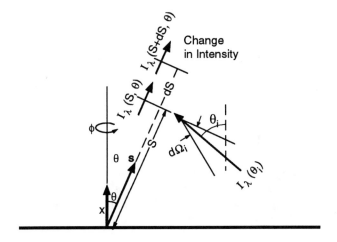

Figure 5.1 A schematic of the coordinate system.

- The *solid angle differential* is $d\Omega = \sin\theta\, d\phi\, d\theta$, when θ is the *polar angle* and ϕ is the *azimuthal angle*. Since an azimuthal symmetry is assumed in the following, the differential solid angle is taken as $d\Omega = 2\pi \sin\theta\, d\theta = -2\pi\, d\cos\theta$.

- The *absorbed* energy is $dI_\lambda = -\sigma_{\lambda a} I_\lambda\, dS$, where again S is the path length and $\sigma_{\lambda a}$ is the *absorption coefficient* (1/m). Figure 5.1 shows dS and $dI_\lambda = I_\lambda(S + dS, \theta) - I_\lambda(S, \theta)$.

- The *scattered energy* is $dI_\lambda = -\sigma_{\lambda s} I_\lambda\, dS$, where $\sigma_{\lambda s}$ is the *scattering coefficient* (1/m).

- Based on the assumption of local thermal equilibrium, the *spontaneously emitted energy* is $dI_\lambda = \sigma_{\lambda a} I_{\lambda b}\, dS$. The *spectral blackbody intensity* $I_{\lambda b}$ is related to the *spectral blackbody emissive power* by $I_{\lambda b} = E_{\lambda b}/\pi$. The spectral blackbody emissive power is given by *Planck's law* for emission into *vacuum*

$$E_{\lambda b} = \frac{2\pi C_1}{\lambda^5 \left(e^{C_2/\lambda T} - 1\right)}, \tag{5.2}$$

where $C_1 = 2\pi h_p c_o^2 = 0.59544 \times 10^8$ W-μm^4/m^2, $C_2 = h_p c_o/k_B = 1.4388 \times 10^5$ μm-K, h_p (Planck constant) $= 6.626 \times 10^{-34}$ J-s, k_B (Boltzmann constant) $= 1.381 \times 10^{-23}$ J/K, and $c_o = 2.9979 \times 10^8$ m/s. The *total blackbody emissive power* is equal to σT^4 and σ (Stephan-Boltzmann constant) $= 2\pi^5 C_1/(15 C_2^4) = 5.6696 \times 10^{-8}$ W/m^2-K^4.

- The *increase* in intensity in the direction (θ), because of *in-scattering* from a direction (θ_i) over a length dS, is

$$\mathrm{d}I_\lambda(S,\theta) = \langle\sigma_{\lambda s}\rangle\, I_\lambda(S,\theta_i)\,\mathrm{d}S\frac{\langle\Phi_\lambda\rangle\,(\theta_o)}{2}, \qquad (5.3)$$

where θ_o is the angle between the incident and scattered beam and the *phase function* $\langle\Phi_\lambda\rangle$ has the property of

$$\int_{-1}^{1}\langle\Phi_\lambda\rangle\,(\theta_o)\,\mathrm{d}\cos\theta_o = 2 \quad\text{and}\quad \int_{-1}^{1}\mathrm{d}\cos\theta_o = 2. \qquad (5.4)$$

For a plane-parallel slab geometry, the scattering angle θ_o is related to the polar and azimuthal angles of the *incident* beam (θ_i,ϕ_i) and the polar and azimuthal angles of the *scattered* beam (θ,ϕ) by

$$\cos\theta_o = \cos\theta_i\cos\theta + \sin\theta_i\sin\theta\cos(\phi-\phi_i). \qquad (5.5)$$

The phase function is generally expressed as a *finite* series of Legendre polynomials

$$\langle\Phi\rangle\,(\cos\theta_o) = 1 + \sum_{i=1}^{N} a_i P_i(\cos\theta_o). \qquad (5.6)$$

When there is azimuthal symmetry, using (5.5), the *summation theorem* of Legendre polynomials and integration (Ozisik, 1985, p.260) leads to

$$\langle\Phi\rangle\,(\cos\theta_i \to \cos\theta) = 1 + \sum_{i=1}^{N} a_i P_i(\cos\theta) P_i(\cos\theta_i). \qquad (5.7)$$

- To account for the in-scattering into the direction (θ,ϕ) from all directions, we integrate

$$\begin{aligned}
\mathrm{d}I_\lambda &= \langle\sigma_{\lambda s}\rangle\frac{\mathrm{d}S}{4\pi}\int_{-1}^{1}\int_{0}^{2\pi} I_\lambda(\theta_i)\,\langle\Phi_\lambda\rangle\,[\theta_o(\theta_i,\phi_i \to \theta,\phi)]\,\mathrm{d}(\phi-\phi_i)\,\mathrm{d}\cos\theta_i \\
&= \langle\sigma_{\lambda s}\rangle\frac{\mathrm{d}S}{2}\int_{-1}^{1} I_\lambda(\theta_i)\,\langle\Phi_\lambda\rangle\,(\theta_i \to \theta)\,\mathrm{d}\cos\theta_i,
\end{aligned} \qquad (5.8)$$

where

$$\langle\Phi_\lambda\rangle\,(\theta_i \to \theta) = \frac{1}{2\pi}\int_{0}^{2\pi}\langle\Phi_\lambda\rangle\,[\theta_o(\theta_i,\phi_i \to \theta,\phi)]\,\mathrm{d}(\phi-\phi_i). \qquad (5.9)$$

- Based on this, the *equation of radiative transfer* for radiation in a direction θ becomes

$$\begin{aligned}
\frac{\partial I_\lambda(S)}{\partial S} = &-\langle\sigma_{\lambda a}\rangle\, I_\lambda(S) + \langle\sigma_{\lambda a}\rangle\, I_{\lambda b}[T(S)] - \langle\sigma_{\lambda s}\rangle\, I_\lambda(S) \\
&+ \frac{\langle\sigma_{\lambda s}\rangle}{2}\int_{-1}^{1} I_\lambda(S,\theta_i)\,\langle\Phi_\lambda\rangle\,(\theta_i \to \theta)\,\mathrm{d}\cos\theta_i.
\end{aligned} \qquad (5.10)$$

The *arrow* indicates from the *incident* to the *scattered* direction and θ_o is the angle between the incident (θ_i) and scattered (θ) beam. When rewritten, we have

$$\frac{\partial I_\lambda}{\partial S} = \langle \sigma_{\lambda a} \rangle I_{\lambda b} - (\langle \sigma_{\lambda a} \rangle + \langle \sigma_{\lambda s} \rangle) I_\lambda$$

$$+ \frac{\langle \sigma_{\lambda s} \rangle}{2} \int_{-1}^{1} I_\lambda \langle \Phi_\lambda \rangle (\mu_i \to \mu) \, d\mu_i, \tag{5.11}$$

where we have used $\mu = \cos\theta$.

- The spectral *radiative heat flux* in the direction normal to the parallel slab faces is found from the directional spectral intensity $I_\lambda(\theta)$ by noting that I_λ is per unit projected area ($dA\cos\theta$) and is in the θ-direction. Then the contribution from all directions to the normal heat flux is

$$q_{\lambda r} = 2\pi \int_{-1}^{1} I_\lambda \cos\theta \, d\cos\theta, \tag{5.12}$$

where $I_\lambda \cos\theta$ is the *spectral directional emissive power*. In vectorial form, we have

$$\mathbf{q}_{\lambda r} = 2\pi \int_{-1}^{1} \mathbf{s} I_\lambda(S, \mathbf{s}) \, d\mu. \tag{5.13}$$

- The *divergence of the total radiative heat flux*, which is used in the energy equation, is found from the radiative transfer equation by its integration over $\int_{4\pi} d\Omega$ and $\int_0^\infty d\lambda$. This gives

$$\int_0^\infty \int_{4\pi} \nabla \cdot \mathbf{s} I_\lambda \, d\Omega \, d\lambda$$

$$= \nabla \cdot \mathbf{q}_r = 4\pi \int_0^\infty \langle \sigma_{\lambda a} \rangle I_{\lambda b}[T(S)] \, d\lambda$$

$$- 2\pi \int_0^\infty \langle \sigma_{\lambda a} \rangle \int_{-1}^{1} I_\lambda(S, \theta) \, d\cos\theta \, d\lambda \tag{5.14}$$

or

$$\nabla \cdot \mathbf{q}_r = 4\pi \int_0^\infty \langle \sigma_{\lambda a} \rangle I_{\lambda b}(S) \, d\lambda - 2\pi \int_0^\infty \langle \sigma_{\lambda a} \rangle \int_{-1}^{1} I_\lambda(S, \theta) \, d\mu \, d\lambda. \tag{5.15}$$

When *no* other *mode of heat transfer* is present and the emitted and absorbed energy are equal, then $\nabla \cdot \mathbf{q}_r = 0$, and the *state of radiative equilibrium* exists.

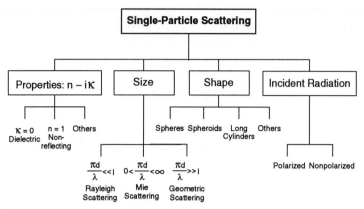

Figure 5.2 Parameters influencing scattering from a single particle.

5.2 Radiation Properties of a Single Particle

In this section, we treat scattering from a *single* (i.e., single sctatterers) particle. In Section 5.3, we will relate the radiation properties obtained for the individual particles to that for the *collection* of such elements in the representative elementary volume. Figure 5.2 gives a *classification* for single-particle scattering, where variation in *optical properties, size, shape*, and the *incident radiation* are considered. In terms of the theoretical treatments, the *most* important distinction, which also leads to significant variation in the rigor of treatment, is that based on *size of the spherical particles*. For large *size parameter*, $\alpha_R = \pi d/\lambda$, geometric optics can be used. The applicability of Rayleigh, Mie, and geometric optic treatments will be discussed in Section 5.2.7. In approaching the theoretical treatment of scattering from particles, we consider cases where the following simplifications can apply (a) constant optical properties, n_s and κ_s, *within* the scatterers, where n_s and κ_s are the solid *index of refraction* and *extinction*, respectively, and, (b) smooth scattering surface. The *optical properties* for solids are strongly *wavelength*-dependent. We will examine this spectral behavior before considering the incidence of electromagnetic waves upon an isolated solid particle.

5.2.1 WAVELENGTH DEPENDENCE OF OPTICAL PROPERTIES

The *relationship* between the optical properties n and κ, and the other *molecular-crystalline* properties of the solid are discussed by Siegel and

5.2 Radiation Properties of a Single Particle

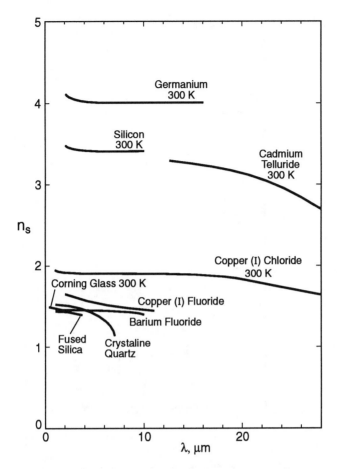

Figure 5.3 Variation of index of refraction of some solids used in optics (visible and infrared), with respect to wavelength. This variation is generally negligible for these materials. (Adapted from Driscoll and Vaughan, 1978.)

Howell. The theoretical treatments of the prediction of these properties are also discussed by them. Here we examine some of the limited experimental data on $n(\lambda)$ and $\kappa(\lambda)$ for *solids*.

Figure 5.3 shows the measured wavelength-dependence of n_s for materials used in *visible* and *infrared optics*. We have included these as examples (and because their spectral behavior is studied most extensively), and *not* because of their common use in heat transfer in porous media. For these optical materials, the wavelength-dependence is *not* very strong (in as far as the radiative heat transfer is concerned), except for *cadmium, telluride*, and *crystalline quartz*. Because of their *near*-room-temperature applications, the measured values are for 300 K.

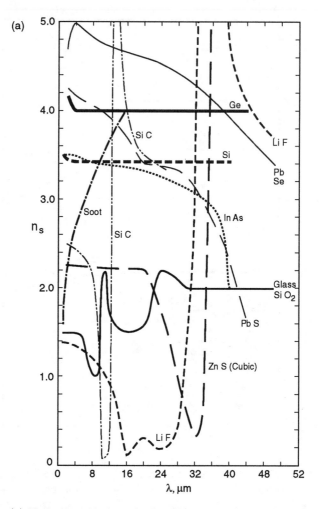

Figure 5.4 (a) Variation of index of refraction for some nonmetallic solids, with respect to the wavelength.

Figures 5.4 (a) and (b) give $n_s(\lambda)$ and $\kappa_s(\lambda)$ for *nonmetallic solids*. Again most of the data is for near room temperature. For $n_s(\lambda)$, other than Ge and Si, the other materials shown exhibit *strong* wavelength-dependence [Figure 5.4(a)]. The variations in the region $\lambda > 10$ μm, corresponding to *below-room-temperature* applications, are in general as significant as they are for $\lambda < 10$ μm. The high-temperature applications require data for $\lambda \simeq 1$ μm. Note the variations in $\kappa_s(\lambda)$ shown in Figure 5.4(b). For *large* particles, the values of κ_s as small as 10^{-5} can result in *significant* absorption because the attenuation is a function of the *product* $\kappa_s \alpha_R$. Therefore, *reliable* data

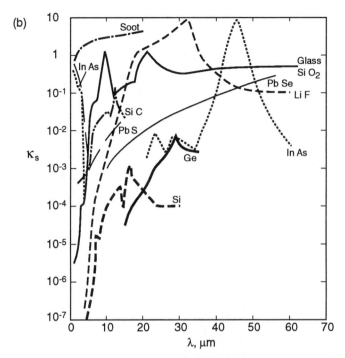

Figure 5.4 (b) Variation of index of extinction for some nonmetallic solids, with respect to the wavelength. (Adapted from Palik, 1985.)

for $\kappa_s(\lambda)$ are *very* important in determining the transmission through *beds*. The wavelength-dependence of κ_s is *very* strong and should be *included* in the radiative heat transfer analysis.

Figures 5.5 (a) and (b) give $n_s(\lambda)$ and $\kappa_s(\lambda)$ for *metallic solids*. Note that n_s does *not* increase *monotonically* with λ (through the spectrum shown) for *all* the materials in Figure 5.5(a), even though Al, Cu, Ag, and Au do show *monotonic* increase of κ_s with λ. Therefore, for *high*-temperature applications where $\lambda = O(1 \ \mu m)$, *extreme* care should be used in order to properly account for variation of n_s with λ. Figure 5.5(b) shows that *except* for *titanium* below $\lambda = 5 \ \mu m$, κ_s *increases* monotonically with λ for metals. *Extensive* documentation of the optical properties of metals can be found in Weaver (1981).

As was mentioned, when the fluid phase is a liquid, we do not expect the radiation heat transfer to be significant, although there are exceptions for some high-temperature applications. However, in Table 5.1, the *index of refraction* of some *liquids* are given for the sake of completeness. In treatment of scattering from particles, the *relative index of refraction*, i.e.,

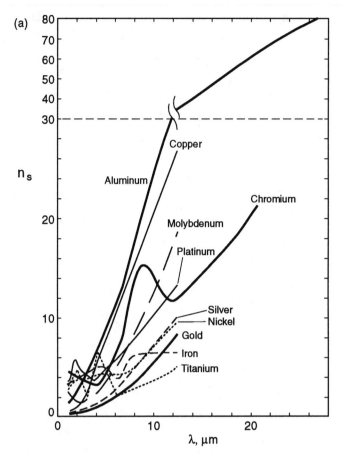

Figure 5.5 (a) Variation of index of refraction for some metallic solids, with respect to the wavelength.

n_s/n_f, is the significant parameter, and as expected when dealing with liquids, this ratio can be substantially *different* from n_s.

5.2.2 SOLUTION TO MAXWELL EQUATIONS

The review given later is based on the treatment of propagation of electromagnetic waves in *heterogeneous* media given by Lorrain and Corson (1970, interfacial aspects), van de Hulst (1981, comprehensive treatment of scattering)†, Siegel and Howell (1981, thermal radiation aspects and interfacial conditions), Kerker (1969, including stratified spheres and different

---

† Note that the units in van de Hulst are not the rational MKS units adapted by Lorrain and Corson.

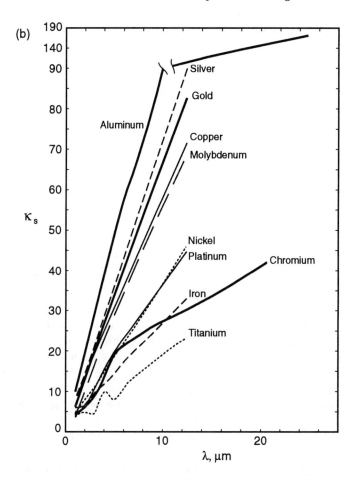

Figure 5.5 (b) Variation of index of extinction for some metallic solids, with respect to the wavelength. (Adapted from Weaver, 1981.)

anisotropies), and Bohren and Huffman (1983, electromagnetic and scattering theories).

In Figures 5.6 through 5.9, we show the *nomenclature* used in the following analysis, as well as the *geometrical* parameters and *definitions*, and aspects of *reflection, refraction,* and *transmission* of electromagnetic waves incident on a particle. Figure 5.6 shows the direction of the *incident unit vector* $\mathbf{s}_i$, which is along the cross product of the *electric field intensity* $\mathbf{e}$ and *magnetic field intensity* $\mathbf{h}$ vectors. The *magnetic induction vector* $\mathbf{b}$ and *electric displacement vector* $\mathbf{d}$ are also shown. The fluid has indices of refraction and extinction n_f and κ_f, as well as *electric conductivity* σ_{ef} and a dielectric constant ε_{df}. The particle surface A_{fs} has a *unit normal*

TABLE 5.1   INDEX OF REFRACTION OF LIQUIDS (FROM WEAST, 1987)

SUBSTANCE	n_ℓ
Pure elements	
H_2	1.10974
Br_2	1.061
Cl_2	1.385
N_2	1.2053
O_2	1.221
Most inorganic compounds	1.4–2.8
Most organic compounds	1.3–1.6
Water	1.317–1.333
Most aqueous solutions	1.34–1.44

vector **n** and *tangent vector* **τ** at a location on the surface the *position* $\mathbf{r}_{fs}$ of which is given with respect to a fixed-coordinate system. The *particle radius* is R and its *optical and electrical properties* are n_s, κ_s, σ_{es}, and ε_{ds}. The angle that the incident beam makes with the normal to the surface A_{fs} in the plane of incidence is θ_i, i.e., *the incidence angle*.

It is more *convenient* to decompose the field vectors into *parallel* (in the plane of incidence) and *perpendicular* (normal to the plane of incidence). For example, the incidence electric field intensity vector $\mathbf{e}_i$ can be decomposed into $\mathbf{e}_{\|i}$ and $\mathbf{e}_{\perp i}$. Figure 5.7 shows this *decomposition* and the azimuthal angle ϕ between **e** and $\mathbf{s}_i$. The *reflection angle* θ_o is the angle between $\mathbf{s}_i$ and $\mathbf{s}_r$. The incident electric field changes direction upon reflection from the solid surface and the components of the reflected field are $\mathbf{e}_{\|r}$ and $\mathbf{e}_{\perp r}$.

The incident radiation can *penetrate* through the interface. In Figure 5.8 the *reflected* and *transmitted* radiation are shown. In this particular example, $\mathbf{h}_i$ is *normal* to the plane of incidence and upon reflection *undergoes a change* in direction. The transmitted **h** has the *same* direction as $\mathbf{h}_i$. The unit vector in the direction of transmitted beam is $\mathbf{s}_t$. In Figure 5.8, the incident radiation is in the x-z plane.

Figure 5.9(a) shows an *incident* radiation with $\mathbf{e}_i$ being *perpendicular* to the plane of incidence. Figure 5.9(b) shows the *direction* of reflected and transmitted radiation for $n_f < n_s$. Note that the *reflected* waves *interfere* with the *incident* waves. The *transmitted* waves are also shown. The *dark lines* with gray contrasts *indicate* the *crests* of the waves. In Figure 5.9(c), normal incidence on a surface is shown. Note that the incidence energy *divides* between the portion reflected and that transmitted.

5.2 Radiation Properties of a Single Particle

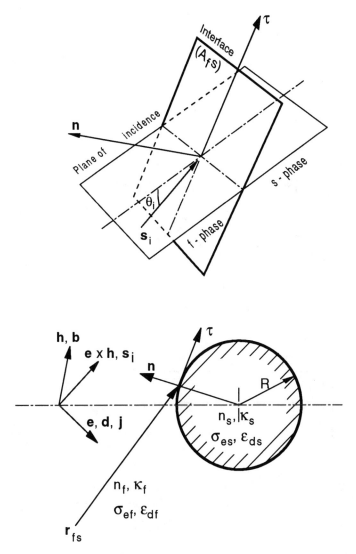

Figure 5.6 A schematic of the incident electromagnetic wave and the fluid-solid interface.

In the following, we briefly review some of the fundamental results of the theoretical treatment of electromagnetic wave/solid surface *interactions*. The details are available in the preceding references.

- The *complex amplitude scattering matrix* $[F(\theta_o)]$ describes the amplitude and the phase of the scattered waves (as was mentioned

272   5. Radiation Heat Transfer

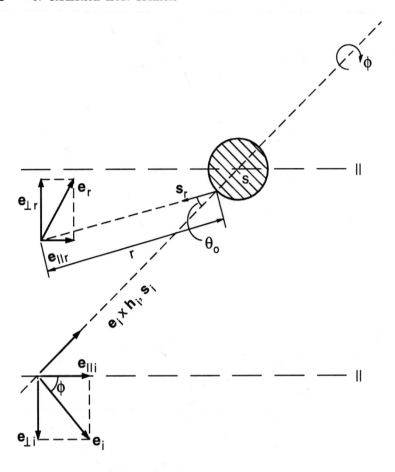

Figure 5.7 A schematic of the incident wave upon the particle and the reflected wave. The parallel (∥) plane contains both the incident beam and the particle.

θ_o is the angle between the incident and the scattered beam), i.e.,

$$\mathbf{e}_s = \frac{e^{-i2\pi r/\lambda}}{i2\pi r/\lambda}[F]\mathbf{e}_i \quad \text{for } r \gg R, \tag{5.16}$$

where again $\mathbf{e}_i$ is the incident electric field intensity.

- For a sphere of uniform properties, the scattered electric field intensity is given by

5.2 Radiation Properties of a Single Particle

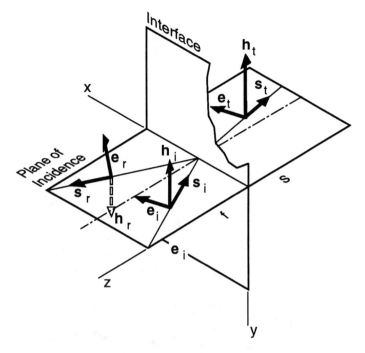

Figure 5.8 Incident, reflected, and transmitted electromagnetic waves. (From Lorrain and Corson, reproduced by permission ©1970 Freeman and Co.)

$$\begin{bmatrix} e_\| \\ e_\perp \end{bmatrix} = \frac{e^{-i2\pi r/\lambda}}{i2\pi r/\lambda} \begin{bmatrix} F_\| & 0 \\ 0 & F_\perp \end{bmatrix} \begin{bmatrix} e_{\|i} \\ e_{\perp i} \end{bmatrix}$$

$$= e_i \frac{e^{-i2\pi r/\lambda}}{i2\pi r/\lambda} \begin{bmatrix} F_\|(\theta_o) & 0 \\ 0 & F_\perp(\theta_o) \end{bmatrix} \begin{bmatrix} \cos\phi \\ \sin\phi \end{bmatrix}, \quad (5.17)$$

where again $\|$ stands for parallel and $\perp$ for perpendicular and ϕ is the azimuthal angle.

- The *intensity* of the scattered radiation is given as

$$\begin{bmatrix} I_{\lambda\|} \\ I_{\lambda\perp} \end{bmatrix} = \frac{I'_{\lambda i}}{4\pi^2 r^2/\lambda^2} \begin{bmatrix} F_\|^2(\theta_o) & 0 \\ 0 & F_\perp^2(\theta_o) \end{bmatrix} \begin{bmatrix} \cos^2\phi \\ \sin^2\phi \end{bmatrix}. \quad (5.18)$$

The *intensity parameter* is defined as $i_{\lambda\|} = |F_\||^2$ and $i_{\lambda\perp} = |F_\perp|^2$ and for nonpolarized radiation $i_\lambda = (i_{\lambda\|} + i_{\lambda\perp})/2$.

274   5. Radiation Heat Transfer

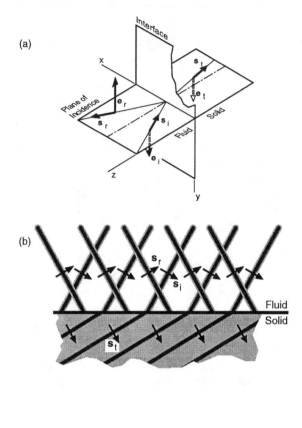

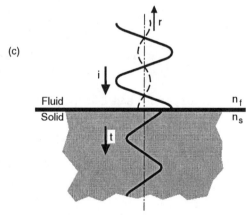

Figure 5.9 (a) Incidence of an electromagnetic wave upon a surface and the consequence change in direction of the reflected **e**. Also shown are the crests of **e** (b), and the reflection and transmission of a normal incidence (c). (From Lorrain and Corson, reproduced by permission ©1970 Freeman and Co.)

5.2 Radiation Properties of a Single Particle

- *Determination* of $[F]$ is through the solution to the Maxwell equations subject to the interfacial boundary conditions at the interface of the surrounding fluid (f-phase) and the scatterer (s-phase) (Lorrain and Corson).

- The magnetic inductance **b** is related to the magnetic field intensity **h** through a scalar μ_m (*magnetic permeability*), the electric displacement **d** is related to electric field intensity **e** through ε_d (dielectric constant), and the *current density* **j** is related to the electric field intensity **e** through σ_e (electric conductivity). The units are given in the nomenclature. These three parameters μ_m, ε_d, and σ_e define the *electric-magnetic properties* of each phase. Under the preceding *proportionality* assumptions, the *complex index of refraction* in each phase is related to these properties as

$$m = \left[\mu_m c_o^2 \left(\varepsilon_d - \frac{i\sigma_e}{\omega}\right)\right]^{1/2}. \tag{5.19}$$

However, in practice $m = n - i\kappa$ is *measured* independently. In the preceding, c_o is the speed of light in a vacuum and ω is the angular frequency (rad/s).

- *Time periodic* behavior is assumed such that $\mathbf{e} = \mathbf{e}(\mathbf{x})e^{i\omega t}$ and $\mathbf{h} = \mathbf{h}(\mathbf{x})e^{i\omega t}$ with $\omega = 2\pi f$ and f the frequency (Hz).

- In passing through the solid, or in general, when encountering a change in the complex index of refraction $m = n - i\kappa$, the frequency of the incident beam *does not* change. However, the wavelength *does* change according to $\lambda_s = \lambda_f/n$, where $n = n_s/n_f$.

- Under these simplifications, and assuming that there are *no* free charges in the solid (this results in $\nabla \cdot \mathbf{d} = 0$), the *Maxwell equations* reduce to

$$\nabla \times \mathbf{e} = -i\mu_m \omega \mathbf{h}, \tag{5.20}$$

$$\nabla \times \mathbf{h} = (\sigma + i\omega\varepsilon_d)\mathbf{h}. \tag{5.21}$$

- The boundary conditions at the interface between the fluid and solid phases are continuity of the *tangential components* of **e** and **h** and continuity of *normal components* of **d** and **b**. The interfacial conditions are

$$\boldsymbol{\tau} \cdot \mathbf{h}_f = \boldsymbol{\tau} \cdot \mathbf{h}_s \quad \text{on} \quad A_{fs}, \tag{5.22}$$

$$\boldsymbol{\tau} \cdot \mathbf{e}_f = \boldsymbol{\tau} \cdot \mathbf{e}_s \quad \text{on} \quad A_{fs}, \tag{5.23}$$

$$(\mathbf{s}_i - \mathbf{s}_r) \cdot \mathbf{r}_{fs} = 0 \quad \text{or} \quad \theta_i = \theta_r, \tag{5.24}$$

$$\frac{\sin \theta_t}{\sin \theta_i} = \frac{n_f}{n_s}, \tag{5.25}$$

where $\mathbf{r}_{fs}$ is the interfacial location vector; $\mathbf{s}_i$, and $\mathbf{s}_r$ are the unit vectors of the incident and reflected beams; and θ_i, θ_r, and θ_t are the *incident*, *reflection*, and *transmission* angles.

- For *perfect conductors*, the wave penetrates essentially a short distance along the normal to the surface (the *concept of skin depth*), no matter what the incident angle. Also for perfect conductors,

$$\frac{\mathbf{e}_r}{\mathbf{e}_i} \simeq -1 \quad \text{and} \quad \frac{\mathbf{e}_t}{\mathbf{e}_i} \simeq 0. \tag{5.26}$$

- Equations (5.20) and (5.21) for **e** and **h** can be further simplified to yield

$$\nabla^2 \begin{bmatrix} \mathbf{e} \\ \mathbf{h} \end{bmatrix} = m^2 k^2 \begin{bmatrix} \mathbf{e} \\ \mathbf{h} \end{bmatrix}, \tag{5.27}$$

where $k = 2\pi/\lambda$ is the *wave number*.

- The above field equations are solved along with the boundary conditions mostly for spheres of uniform properties and the solutions are in the series form. The results for the *two elements* of the amplitude scattering matrix for *uniform property spheres* of radius R are

$$F_\parallel(\theta_o) = \sum_{i=1}^{\infty} \frac{2i+1}{i(i+1)} \left[a_i \pi_i(\cos\theta_o) + b_i \tau_i(\cos\theta_o) \right], \tag{5.28}$$

$$F_\perp(\theta_o) = \sum_{i=1}^{\infty} \frac{2i+1}{i(i+1)} \left[a_i \tau_i(\cos\theta_o) + b_i \pi_i(\cos\theta_o) \right], \tag{5.29}$$

where in terms of the *size parameter* $\alpha_R = 2\pi R/\lambda$, we have

$$a_i(\alpha_R, m) = \frac{\psi_i(\alpha_R) A_i - m\psi_i'(\alpha_R)}{\zeta_i(\alpha_R) A_i - m\zeta_i'(\alpha_R)}, \tag{5.30}$$

$$b_i(\alpha_R, m) = \frac{m\psi_i(\alpha_R) A_i - \psi_i'(\alpha_R)}{m\zeta_i(\alpha_R) A_i - \zeta_i'(\alpha_R)}, \tag{5.31}$$

where $A_i = \psi_i'(\alpha_R m)/\psi_i(\alpha_R m)$, ψ_i and ζ_i are the *Ricatti-Bessel functions* (and are related to the Bessel functions J), and prime indicates differentiation. For $z = \alpha_R$ or $\alpha_R m$, we have the relationship between ψ_i, ζ_i, and J_i as

$$\psi_i(z) = \left(\frac{\pi z}{2}\right)^{1/2} J_{i+\frac{1}{2}}(z), \tag{5.32}$$

$$\zeta_i(z) = \left(\frac{\pi z}{2}\right)^{1/2} J_{i+\frac{1}{2}}(z) + (-1)^n i J_{-i-\frac{1}{2}}(z), \tag{5.33}$$

where $J_{i+\frac{1}{2}}(z)$ is the *Bessel function of the first kind of half-integral* (of complex argument for $J_{i+\frac{1}{2}}(\alpha_R m)$, where $\kappa \neq 0$). In general, high-order accurate *approximations* are used for evaluation of ψ_i and ζ_i.

We will discuss the computation of these functions in Section 5.2.7. The *angular distribution functions* are given as

$$\pi_i = \frac{dP_i(\cos\theta_o)}{d\cos\theta_o}, \qquad (5.34)$$

$$\tau_i = \cos\theta_o \pi_i(\cos\theta_o) - \sin^2\theta_o \frac{d^2 P_i(\cos\theta_o)}{d(\cos\theta_o)^2}, \qquad (5.35)$$

where $P_i(\cos\theta_o)$ is the Legendre polynomial. For $\cos\theta_o = 1$, $\tau_i = \pi_i = i(i+1)/2$, then

$$F_\|(0) = F_\perp(0) = \frac{1}{2}\sum_{i=1}^{\infty}(2i+1)(a_i + b_i). \qquad (5.36)$$

The representation of a_i, b_i, ψ_i, ζ_i, τ_i, and π_i in terms of recurrence relations is discussed by Crosbie and Davidson (1985).

- The *asymmetry factor* g or $\overline{\cos\theta}$ is a measure of scattering asymmetry ($g = -1$ for *completely backward* and $g = 1$ for *completely forward* scattering; $g = 0$ for *isotropic* scattering) and is given by

$$\overline{\cos\theta} = \overline{\mu} = g = \frac{\int_{-1}^{1} I_\lambda(\mu)\mu\, d\mu}{\int_{-1}^{1} I_\lambda(\mu)\, d\mu}. \qquad (5.37)$$

5.2.3 Scattering Efficiency and Cross Section

For incidence of a planar radiation upon a spherical particle, the *spectral power* (W/μm) arriving at the sphere is $\pi R^2 I_{\lambda i}$. The fraction that is scattered can be found by integrating the *local scattered spectral intensity* $I_{\lambda s}$ over a sphere with a radius larger than the particle radius (since the intensity decays as r^{-2}, the location r is irrelevant), i.e., $\int_{4\pi} I_{\lambda s} r^2\, d\Omega$. Then a *spectral scattering efficiency* $\eta_{\lambda s}$ is defined as

$$\eta_{\lambda s} = \frac{\int_{4\pi} I_{\lambda s} r^2\, d\Omega}{\pi R^2 I_{\lambda i}}, \qquad (5.38)$$

where, from (5.18), we have

$$I_{\lambda s} = \frac{I_{\lambda i}\left[F_\|^2(\theta_o) + F_\perp^2(\theta_o)\right]}{8\pi^2 r^2/\lambda^2}. \qquad (5.39)$$

The *degree of polarization* is

$$\frac{F_\|^2 - F_\perp^2}{F_\|^2 + F_\perp^2}. \qquad (5.40)$$

The *spectral scattering cross section* is defined as

$$A_{\lambda s} = \eta_{\lambda s} \pi R^2. \tag{5.41}$$

Similarly the *spectral absorption efficiency* and cross section are defined as

$$\eta_{\lambda a} = \frac{\int_{4\pi} I_{\lambda a} r^2 \, d\Omega}{\pi R^2 I_{\lambda i}}, \tag{5.42}$$

$$A_{\lambda a} = \eta_{\lambda a} \pi R^2. \tag{5.43}$$

The measurement of $I_{\lambda a}$ is *difficult*, but it can be predicted through analysis such as the *Mie theory*. Finally, the *spectral extinction efficiency* and *spectral extinction cross section* are defined as

$$\eta_{\lambda \, \text{ex}} = \eta_{\lambda s} + \eta_{\lambda a}, \tag{5.44}$$

$$A_{\lambda \, \text{ex}} = A_{\lambda s} + A_{\lambda a}. \tag{5.45}$$

5.2.4 Mie Scattering

The spectral extinction cross section is evaluated based on the energy transmitted at $\theta_o = 0$. This relationship is discussed in van de Hulst (1981, 30–31) and is given in terms of the *Mie solution* for scattering from a sphere. At $\theta_o = 0$, both $F_\parallel(0)$ and $F_\perp(0)$ have the *same* magnitude $F(0)$, as given by (5.36). Then we have

$$A_{\lambda \, \text{ex}} = \frac{\lambda^2}{\pi} \text{Real}\,[F(0)] \tag{5.46}$$

or

$$A_{\lambda \, \text{ex}} = \frac{\lambda^2}{2\pi} \sum_{i=1}^{\infty} (2i+1) \text{Real}\,[a_i(\alpha_R, \alpha_R m) + b_i(\alpha_R, \alpha_R m)]. \tag{5.47}$$

The spectral scattering cross section as defined by (5.41) is given by

$$\begin{aligned} A_{\lambda s} &= \frac{\lambda^2}{4\pi} \int_0^\pi \left[F_\parallel^2(\theta_o) + F_\perp^2(\theta_o) \right] \sin\theta_o \, d\theta \\ &= \frac{\lambda^2}{2\pi} \sum_{i=1}^{\infty} (2i+1)(|a_i|^2 + |b_i|^2) \end{aligned} \tag{5.48}$$

or

$$A_{\lambda s} = \frac{\lambda^2}{2\pi} \sum_{i=1}^{\infty} (2i+1) \left[|\,a_i(\alpha_R, \alpha_R m)|^2 + |b_i(\alpha_R, \alpha_R m)|^2 \right]. \tag{5.49}$$

Note that $m = n_s/n_f - i\kappa_s/n_f = n - i\kappa$. Approximate solutions are discussed extensively by van de Hulst, where some *tabulations* of the coefficients are also given. The phase function is in general *fairly* complicated and must be determined for a given $(\alpha_R, \alpha_R m)$.

5.2.5 RAYLEIGH SCATTERING

For particles of arbitrary shapes with *linear* dimensions *small* compared to λ (i.e., λ_f and λ_s), the scattering of the waves is done by the oscillating *induced dipole moment*. The problem was formulated by Rayleigh and is reviewed in Chandrasekhar (1960) and van de Hulst (1981).

For spherical particles, the solution has been found by Lorentz. The *induced dipole moment* **p** is related to the imposed electric field intensity $\mathbf{e}_i$ through the polarizability parameter α_p as

$$\mathbf{p} = \alpha_p \mathbf{e}_i, \tag{5.50}$$

where α_p is a *scalar* for *isotropic* distribution of dipoles. The scattering matrix is then given as

$$[F] = i\left(\frac{2\pi R}{\lambda}\right)^3 \left(\frac{m^2-1}{m^2+2}\right)\begin{bmatrix} \cos\theta_o & 0 \\ 0 & 1 \end{bmatrix}. \tag{5.51}$$

Then $F(0) = i(2\pi R/\lambda)^3 (m^2-1)/(m^2+2)$. The scattering cross section for the *electric dipole scattering* is developed by van de Hulst (1981, 64–70) and is

$$A_{\lambda s} = \frac{128\pi^5}{3}\frac{R^6}{\lambda^4}\left|\frac{m^2-1}{m^2+2}\right|^2. \tag{5.52}$$

Thus, Rayleigh scattering shows a λ^{-4} *dependence* (assuming independent scattering), i.e., *larger* wavelengths are scattered *less* significantly. The absorption cross section is

$$A_{\lambda a} = -\frac{8\pi^2}{\lambda}\mathrm{Real}\left(i\frac{m^2-1}{m^2+2}R^3\right) = -\frac{8\pi^2 R^3}{\lambda}\mathrm{Im}\left(\frac{m^2-1}{m^2+2}\right). \tag{5.53}$$

The Phase function is

$$\Phi(\theta_o) = \frac{3}{4}\left(1+\cos^2\theta_o\right), \tag{5.54}$$

which is *symmetric* around the $\theta_o = \pi/2$ plane. The range of validity of the Rayleigh scattering has been investigated by Kerker et al. (1978), Ku and Felske (1984), Selamet (1989), and Selamet and Arpaci (1989).

5.2.6 GEOMETRIC- OR RAY-OPTICS SCATTERING

The method involves ray tracing and reviews are given in van de Hulst (1981, 200–226), and Born and Wolf (1988, also on diffraction and other optical phenomena). The restrictions in applying geometric-optics are the following.

- The *size parameter* must be *large* $\alpha_R \gg 1$.

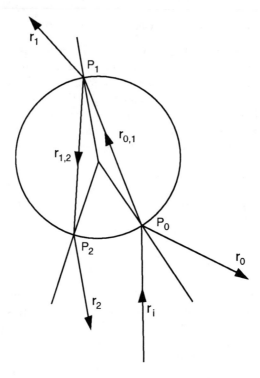

Figure 5.10 A schematic of ray tracing for multiple internal reflection in a sphere.

- The *phase shift* given as $(2\pi/\lambda)d(n-1) = 2\alpha_R(n-1)$, i.e., the change of the phase of a light ray passing through the sphere along the diameter, must be *large*.

- Geometric-optics *fails* at *extreme incident* angles.

- Only half of the total scattering, i.e., that due to *reflection* and *refraction*, is considered.

- The other half arising from *diffraction* around the object must be *included separately* leading to the *Fraunhofer* diffraction pattern. Note that the Mie theory includes reflection, refraction, and diffraction and that the Rayleigh theory does *not* distinguish between these three.

Generally, only up to three internal reflections are included (Liou and Hansen, 1971) in order to account for 99 percent of the scattered energy. Figure 5.10 gives a schematic of the ray tracing in a sphere where *P stands* for the *number of internal reflections* before the ray *leaves* the sphere.

5.2 Radiation Properties of a Single Particle

The *angle of refraction* θ_r is related to the angle of incidence and the index of refraction through the *Snell law*. This states that

$$n_f \cos\theta_i = n_s \cos\theta_r \quad \text{for} \quad \kappa_s \to 0 \tag{5.55}$$

or

$$\cos\theta_i = n\cos\theta_r, \quad n = \frac{n_s}{n_f}. \tag{5.56}$$

In general, the polarization is decomposed into components parallel and perpendicular to the plane of incidence. For each direction the *directional, spectral specular reflectivity* is found, i.e., $\rho'_{\parallel\lambda}$ and $\rho'_{\perp\lambda}$. For *nonpolarized irradiation*, the *directional spectral specular reflectivity* is

$$\rho'_\lambda = \frac{1}{2}\left(\rho'_{\perp\lambda} + \rho'_{\parallel\lambda}\right), \tag{5.57}$$

where in terms of the angles θ_i and θ_r we have (Siegel and Howell, 1981, pp. 95 and 101)

$$\rho'_{\parallel\lambda} = \left[\frac{\tan(\theta_i - \theta_r)}{\tan(\theta_i + \theta_r)}\right]^2 \quad \text{and} \quad \rho'_{\perp\lambda} = \left[\frac{\sin(\theta_i - \theta_r)}{\sin(\theta_i + \theta_r)}\right]^2 \quad \text{for} \quad \kappa \to 0 \tag{5.58}$$

or in terms of θ_i and $n_s/n_f = n$ as

$$\rho'_\lambda = \frac{1}{2}\left[\frac{(n^2 - \cos^2\theta_i)^{1/2} - \sin\theta_i}{(n^2 - \cos^2\theta_i)^{1/2} + \sin\theta_i}\right]^2$$

$$+ \frac{1}{2}\left[\frac{n^2\sin\theta_i - (n^2 - \cos^2\theta_i)^{1/2}}{n^2\sin\theta_i + (n^2 - \cos^2\theta_i)^{1/2}}\right]^2 \quad \text{for} \quad \kappa \to 0. \tag{5.59}$$

Thus the reflected parts of energy are $\rho'_{\parallel\lambda}$ and $\rho'_{\perp\lambda}$. The refracted parts are $1 - \rho'_{\parallel\lambda}$ and $1 - \rho'_{\perp\lambda}$. Then the *energy carried by various rays* is (van de Hulst, 1981, p.204)

$$\beta_\parallel = \rho'_{\parallel\lambda} \qquad \text{for } P=0, \tag{5.60}$$

$$\beta_\parallel = (1 - \rho'_{\parallel\lambda})^2 (\rho'_{\parallel\lambda})^{P-1} \quad \text{for } P = 1, 2, 3, \ldots \quad \text{if } \kappa = 0, \tag{5.61}$$

and

$$\beta_\parallel = (1 - \rho'_{\parallel\lambda})^2 (\rho'_{\parallel\lambda})^{P-1} \exp(-4\kappa P \alpha_R \sin\theta_r)$$

$$\text{for } P = 1, 2, 3, \ldots \quad \text{if } \kappa \neq 0. \tag{5.62}$$

For the other polarization replace $\parallel$ with $\perp$. However when κ_s is *not* small, then (5.58) should *not* be used to calculate the reflectivity. Instead, an exact analysis should be followed (Siegel and Howell, 1981, p.100).

The total *deviation* from the original direction is (Figure 5.10)

$$\theta' = 2\theta_i - 2P\theta_r. \tag{5.63}$$

The *scattering angle* in the interval $(0, \pi)$ is given by

$$\theta_o = k2\pi + q\theta', \tag{5.64}$$

where k is an *integer* and $q = +1$ or -1. Differentiation and use of the Snell law leads to

$$\frac{d\theta'}{d\theta_i} = 2 - 2p\frac{\tan\theta_i}{\tan\theta_r}, \tag{5.65}$$

$$d\theta_o = \left|\frac{d\theta'}{d\theta_i}\right| d\theta_i. \tag{5.66}$$

The emergent pencil *spreads* into an area $r^2 \sin\theta_o \, d\theta_o d\phi$, where r is a large distance from the sphere. Dividing the *emergent flux* by this area, we obtain the *intensity*

$$I_\parallel(p,\theta_i) = \frac{R^2 \beta_\parallel I_i \cos\theta_i \sin\theta_i \, d\theta_i \, d\phi}{r^2 \sin\theta_o \, d\theta_o \, d\phi} = \frac{R^2}{r^2} I_i \beta_\parallel D, \tag{5.67}$$

where

$$D = \frac{\sin\theta_i \cos\theta_i}{\sin\theta_o \left|\dfrac{d\theta'}{d\theta_i}\right|}, \tag{5.68}$$

and similarly for $I_\perp(p,\theta_i)$.

Defining the *gain* G relative to the isotropic scattering as the ratio of *scattered* intensity to the intensity that would be found *in any direction* if the sphere scattered the entire incident energy *isotropically*, we have

$$G_\parallel(\theta_o) = \frac{I_\parallel(4\pi r^2)}{I_i \pi R^2} = 4\beta_\parallel D. \tag{5.69}$$

The gain for *nonpolarized* incident radiation is

$$G = \frac{1}{2}(G_\parallel + G_\perp). \tag{5.70}$$

The sum of the gains in a particular direction resulting from various values of P gives the phase function Φ for a *nonabsorbing* sphere. For an absorbing sphere, the resulting values of the phase function must be *divided* by the *scattering efficiency*.

The fraction of energy scattered, or the scattering efficiency, can be calculated as

$$\eta_{\lambda s} = -\frac{1}{4\pi}\int_0^{2\pi}\int_{-1}^{1}\sum_{p=0}^{n} G(P,\theta_o)\, d\cos\theta_o \, d\phi$$

$$= -\frac{1}{2}\int_{-1}^{1}\sum_{p=0}^{n} G(P,\theta_o)\, d\cos\theta_o. \tag{5.71}$$

As was mentioned earlier, a value of $n = 3$ is generally *sufficient*. The integral over θ_o is replaced by a summation for carrying out the calculation. The absorption efficiency is given by

$$\eta_{\lambda a} = 1 - \eta_{\lambda s}. \tag{5.72}$$

The phase function for *large opaque specularly* reflecting spheres is obtained by Siegel and Howell (1981, 578–581) and is

$$\Phi_{\lambda r}(\theta_o) = \frac{\rho'_\lambda}{\rho_\lambda}, \tag{5.73}$$

where $\rho'_\lambda[(\pi - \theta_o)/2]$ is the *directional* and ρ_λ is the *hemispherical* specular reflectivity.

For *diffuse* reflection, they show

$$\Phi_{\lambda r}(\theta_o) = \frac{8}{3\pi}(\sin\theta_o - \theta_o \cos\theta_o). \tag{5.74}$$

The *diffraction* component of the phase function given by van de Hulst (1981, p. 108) is $F(\theta) = \alpha_R J_1(\alpha_R \sin\theta_o)/\sin\theta_o$, which leads to

$$\Phi_{\lambda d}(\theta_o) = \frac{F^2(\theta_o)}{\alpha_R^2} = \frac{4J_1^2(\alpha_R \sin\theta_o)}{\sin^2\theta_o}, \tag{5.75}$$

where J_1 is the Bessel function of the first order and the first kind. This has a *very* strong forward component (lobe around $\theta_o = 0$) with *lobes* for $\theta_o > 0$ *decreasing* exponentially in *strength*.

The scattering and absorption efficiencies for specularly or diffusely scattering, large spheres are also given in Siegel and Howell (1981, pp. 575 and 582) and along with that of the diffraction, are given here:

specular or diffuse reflection: $\quad \eta_{\lambda s} = \rho_\lambda, \quad \eta_{\lambda a} = 1 - \rho_\lambda, \quad$ (5.76)

diffraction : $\quad \eta_{\lambda s d} = 1,$ (5.77)

where ρ_λ is the *hemispherical spectral reflectivity*.

Since the *diffraction* scattering is dominantly forward for *large* particles, it is *customary* to *exclude* the diffraction contribution from the *phase function* and the *scattering coefficient simultaneously*.

5.2.7 COMPARISON OF PREDICTIONS

We expect the Mie theory to be applicable for all values of n, κ, and size parameter α_R. The Rayleigh theory is applicable for *small* α_R and *small* values of $|m\alpha_R|$ and the geometric treatment is expected to be valid for $\alpha_R \gg 1$. Here, we consider *spherical* particles only. The *optical properties* are expected to be wavelength-dependent. Van de Hulst (1981, 131–134) gives the *classifications* for the case of $\kappa_s = 0$. His results are plotted in a diagram

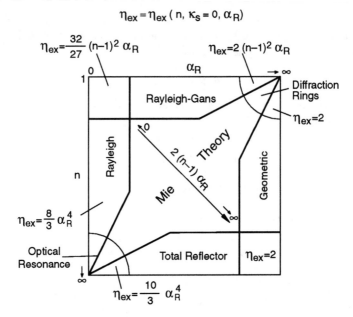

Figure 5.11 The α_R-n plane (van de Hulst diagram) showing the various asymptotes for prediction of the extinction efficiency based on the Mie theory. The results are for $\kappa_s = 0$. (From van de Hulst, reproduced by permission ©1981 Dover.)

named after him and this diagram is shown in Figure 5.11. He gives the *asymptotic* relations for η_{ex} so that the necessity of carrying out the full Mie solution can be *avoided*. Some of these asymptotes are shown in the van de Hulst diagram. However, because of faster computers and improved subroutines, carrying out a *full Mie solution* is *no* longer as prohibitive a task as it once was. The problem lies more in making practical use of it, because no method of solution can handle the *sharp* forward peak produced for *large* particles. Thus this peak has to be *truncated* for geometric size particles and the phase function *renormalized* to ensure energy conservation. The computation involved *increases* with *increasing* α_R. However, for very large values of α_R, the theory of *geometric scattering* provides a convenient alternative.

For small particles, the Rayleigh theory can be used. Although this does *not* result in a substantial savings in computation over the Mie theory, it provides a *closed-form solution*. Here, we compute $\eta_{\lambda s}(\lambda)$, $\eta_{\lambda a}(\lambda)$, $g_\lambda(\lambda)$, and $\Phi_\lambda(\lambda)$ for a 0.2-mm sphere using the available *experimental results* for $n_s(\lambda)$ and $\kappa_s(\lambda)$ for *glass* and *iron* (carbon steel). The computations are based on the Rayleigh, Mie, and geometric treatments. Then comparisons are made among the results of these three theories, and the limit of applicability of the Rayleigh and geometric treatments for these examples are discussed.

5.2 Radiation Properties of a Single Particle

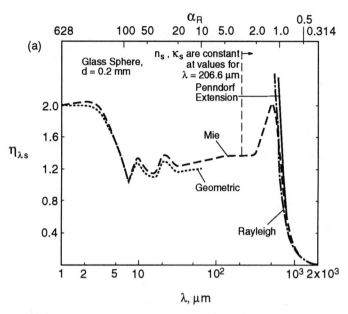

Figure 5.12 (a) Variation of the spectral scattering efficiency with respect to wavelength for a glass spherical particle of diameter 0.2 mm. When appropriate, the Rayleigh, Mie, and geometrical treatments are shown. Also shown is the Penndorf extension.

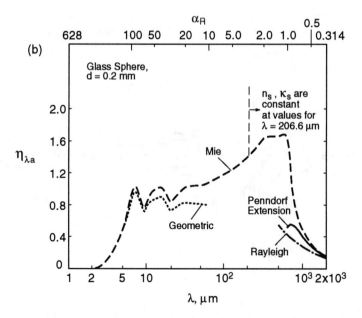

Figure 5.12 (b) Same as Figure 5.12(a), except for the variation of the spectral absorption efficiency.

286   5. Radiation Heat Transfer

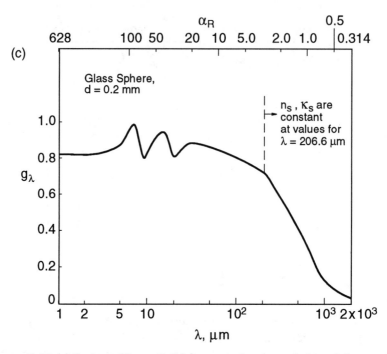

Figure 5.12 (c) Same as Figure 5.12(a), except for the variation of the spectral asymmetry factor.

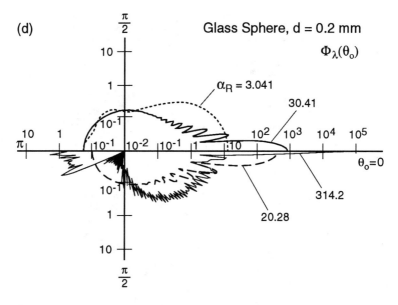

Figure 5.12 (d) Same as Figure 5.12(a), except for the distribution of the phase function.

5.2 Radiation Properties of a Single Particle

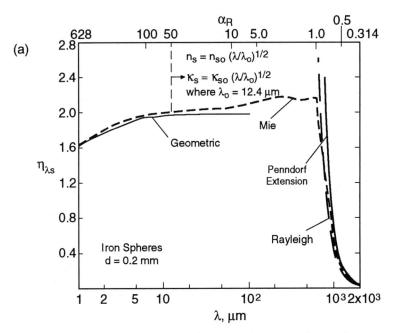

Figure 5.13 (a) Same as Figure 5.12(a), except for iron spheres of 0.2 mm in diameter.

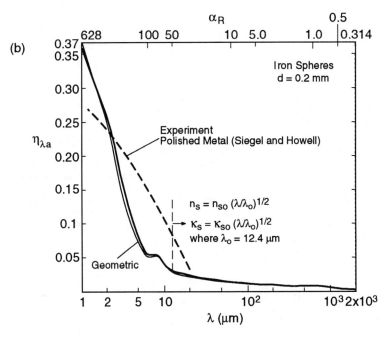

Figure 5.13 (b) Same as Figure 5.12(b), except for iron.

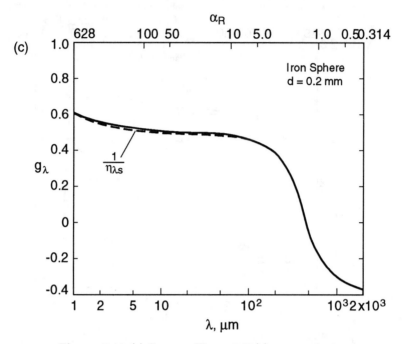

Figure 5.13 (c) Same as Figure 5.12(c), except for iron.

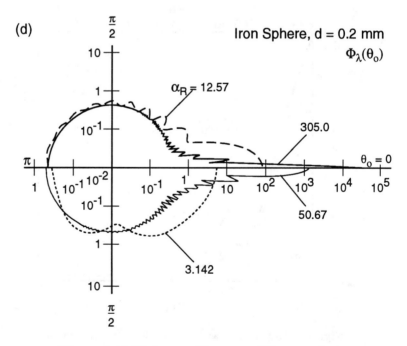

Figure 5.13 (d) Same as Figure 5.12(d), except for iron.

5.2 Radiation Properties of a Single Particle

The results of *single* particle scattering for 0.2-mm glass and steel spheres are shown in Figures 5.12 (a)–(d) and Figures 5.13 (a)–(d), respectively. The optical properties of glass and iron are taken from Hsieh and Su (1979) and Weast (1987). The data for iron are *not* available for wavelengths greater than 12.4 μm. Therefore, beyond this wavelength, the values at 12.4 μm are used along with the *Hagen-Rubens law* to *extrapolate* to higher wavelengths.

$$n_s = n_{s_{s0}}\sqrt{\lambda/\lambda_0}, \tag{5.78}$$

$$\kappa_s = \kappa_{s0}\sqrt{\lambda/\lambda_0}, \tag{5.79}$$

where $\lambda_0 = 12.4$ μm.

For glass, the optical properties are available for wavelengths *less* than 206.6 μm. Since the optical properties do *not* show much change near this limit, the values at $\lambda = 206.6$ μm are used for *higher* wavelengths.

The experimentally obtained optical constants for metals may be *greatly in error* (Siegel and Howell, 1981, 107–108), e.g., Weast gives for iron and for $\lambda = 0.587$ μm the indices as $n_s = 1.51$ and $\kappa_s = 1.63$. The corresponding values obtained from the 68th edition are $n_s = 2.80$ and $\kappa_s = 3.34$.

For glass, the values of n_s are fairly *well* documented (Palik, 1985). However, the value of κ for the wavelength range where glass is almost *transparent* ($\lambda < 2.5$ μm) is difficult to measure and may contain *large* experimental errors (Palik, 1985). *Small* glass spheres in this range may be treated as transparent. However, as the sphere size increases, *absorption* becomes significant. Also, because of differences in composition, the properties vary with the *type* of glass used.

For Figures 5.12 and 5.13, the Mie scattering calculations are done as explained in Sections 5.2.2–5.2.4 using the Mie theory subroutine of Bohren and Huffman with minor modifications. In particular, if this subroutine is to be used for *very large* values of the size parameters, double precision must be used.

For *computation*, (5.30) and (5.31) may be rewritten as

$$a_n = \frac{\left[D_i(m\alpha_R)/m + \dfrac{i}{\alpha_R}\right]\psi_i(\alpha_R) - \psi_{i-1}(\alpha_R)}{\left[mD_i(m\alpha_R) + \dfrac{i}{\alpha_R}\right]\zeta_i(\alpha_R) - \zeta_{i-1}(\alpha_R)}, \tag{5.80}$$

$$b_n = \frac{\left[mD_i(m\alpha_R) + \dfrac{i}{\alpha_R}\right]\psi_i(\alpha_R) - \psi_{i-1}(\alpha_R)}{\left[mD_i(m\alpha_R) + \dfrac{i}{\alpha_R}\right]\zeta_i(\alpha_R) - \zeta_{i-1}(\alpha_R)}, \tag{5.81}$$

where $D_i(z)$ is the *logarithmic derivative*

$$D_i(z) = \frac{d}{dz}\ln\psi_i(z) \tag{5.82}$$

and

$$\psi'_i(\alpha_R) = \psi_{i-1}(\alpha_R) - \frac{n\psi_i(\alpha_R)}{\alpha_R}, \tag{5.83}$$

$$\zeta'_i(\alpha_R) = \zeta_{i-1}(\alpha_R) - \frac{i\zeta_i(\alpha_R)}{\alpha_R}, \tag{5.84}$$

are *recurrence* relations used to eliminate ψ'_i and ζ'_i.

The logarithmic derivative *satisfies* the *recurrence* relation

$$D_{n-1} = \frac{n}{z} - \frac{1}{D_n - n/z}, \tag{5.85}$$

which is used to calculate $D_n(m\alpha_R)$ by downward recurrence using the preceding recurrence relation and starting with a value of n *sufficiently* higher than the number of terms needed for convergence. *Convergence* is assumed to occur after $\alpha_R + 4\alpha_R^{1/3} + 2$ terms.

Both ψ_i and ζ_i ($= \psi_i + i\chi_i$) satisfy

$$\psi_{i+1}(\alpha_R) = \frac{2i+1}{\alpha_R}\psi_i(\alpha_R) - \psi_{i-1}(\alpha_R) \tag{5.86}$$

and are computed by this *upward* recurrence relation. The initial terms are

$$\psi_{-1}(\alpha_R) = \cos\alpha_R, \tag{5.87}$$

$$\psi_0(\alpha_R) = \sin\alpha_R, \tag{5.88}$$

$$\chi_{-1}(\alpha_R) = -\sin\alpha_R, \tag{5.89}$$

$$\chi_0(\alpha_R) = \cos\alpha_R. \tag{5.90}$$

The Rayleigh and Rayleigh-Penndorf scattering and absorption efficiencies are computed as follows (Selamet and Arpaci, 1989). In the Rayleigh limit,

$$(\eta_{\lambda s})_R = \frac{8}{3}\alpha_R^4 \left|\frac{m^2-1}{m^2+2}\right|^2 \tag{5.91}$$

and

$$(\eta_{\lambda a})_R = -4\alpha_R \operatorname{Im}\left(\frac{m^2-1}{m^2+2}\right). \tag{5.92}$$

Inserting $m = n - i\kappa = n_s/n_f - i\kappa_s/n_f$ into these equations

$$(\eta_{\lambda s})_R = \frac{8}{3}\left(1 - 3\frac{M_2}{M_1}\right)\alpha_R^4, \tag{5.93}$$

$$(\eta_{\lambda a})_R = 12\left(\frac{N_1}{M_1}\right)\alpha_R, \tag{5.94}$$

5.2 Radiation Properties of a Single Particle

where

$$M_1 = N_1^2 + (2 + N_2)^2, \tag{5.95}$$

$$M_2 = 1 + 2N_2, \tag{5.96}$$

$$N_1 = 2n\kappa, \tag{5.97}$$

$$N_2 = n^2 - \kappa^2. \tag{5.98}$$

The Rayleigh limit can be extended to *higher* particle sizes by using the *Penndorf extension* (Penndorf, 1962). The Penndorf extension can be expressed as

$$(\eta_{\lambda s})_P = (\eta_{\lambda s})_R \left[1 + 2\frac{\alpha_R^2}{M_1}\left(\frac{3}{5}M_3 - 2N_1\alpha_R\right)\right], \tag{5.99}$$

$$(\eta_{\lambda e})_P = (\eta_{\lambda a})_R + 2\alpha_R^3\left[N_1\left(\frac{1}{15} + \frac{5}{3}\frac{1}{M_4} + \frac{6}{5}\frac{M_5}{M_1^2}\right) + \frac{4}{3}\frac{M_6}{M_1^2}\alpha_R\right], \tag{5.100}$$

and

$$(\eta_{\lambda a})_P = (\eta_{\lambda e})_P - (\eta_{\lambda s})_P, \tag{5.101}$$

where

$$M_3 = N_3 - 4, \tag{5.102}$$

$$M_4 = N_1^2 + (3 + 2N_2)^2, \tag{5.103}$$

$$M_5 = 4(N_2 - 5) + 7N_3, \tag{5.104}$$

$$M_6 = (N_2 + N_3 - 2)^2 - 9N_1^2, \tag{5.105}$$

and

$$N_3 = (n^2 + \kappa^2)^2 = N_1^2 + N_2^2. \tag{5.106}$$

For *glass*, Figures 5.12 (a) and (b) show the results computed from these equations as well as the *exact* Mie calculation. For *iron*, Figure 5.13, *even at small size parameters*, the particle does not lie in the Rayleigh limit because of the very *high* refraction index. The condition for Rayleigh scattering is not only that $\alpha_R \ll 1$ but also $|m\alpha_R| \ll 1$ (van de Hulst, 1981, p. 75). Even though this is *clearly* violated for iron, *some agreement* with Mie calculation is seen for the scattering efficiency. The absorption efficiencies for iron given by Rayleigh scattering and the Penndorf extension are *highly inaccurate*.

The geometric scattering calculations are done as explained in Section 5.2.6. The value of θ_i is varied in *discrete* steps, and the *ray* is traced through the sphere. The values of θ_o and $G(P, \theta_o)$ are calculated for $P = 0, 1, \ldots$.

The tracing is *stopped* at some value of P depending upon the accuracy required. Although $P = 2$ or $P = 3$ is accurate enough for most purposes, higher accuracies can be obtained by continuing to trace to about $P = 6$.

Two *different* types of points of *singularity* are encountered in these calculations. *Glory* occurs when $\sin\theta_o = 0$ but $\sin\theta_i \cos\theta_i \neq 0$. *Rainbow* occurs when $|d\theta'/d\theta_i| = 0$. Both of these make the denominator on the right-hand side of (5.68) zero. However, the solid angle affected is extremely small, and by making the step size for θ_i small enough, a *fairly* accurate computation can be carried out.

Figures 5.12 (a) and (b), and Figures 5.13 (a) and (b) show the scattering and absorption efficiencies for a 0.2-mm glass and iron sphere, respectively. Also plotted are the Rayleigh, Rayleigh-Penndorf, and geometric approximations. In general, changes in efficiencies at *smaller* size parameters ($\alpha_R < 10$) are due to changes in *size* parameter while changes in efficiencies at *larger* size parameters ($\alpha_R > 10$) are mainly due to variation in *optical properties* (n_s and κ_s) with the wavelength.

Figures 5.12(c) and 5.13(c) show the asymmetry parameter g_λ for glass and iron. Also plotted on Figure 5.13(c) is $1/\eta_{\lambda s}$, which shows good agreement with g_λ for large values of α_R. This follows from the assumption of isotropic reflection and assuming diffraction to be totally forward. Then

$$g_\lambda = \frac{\int_{-1}^{1} I_\lambda(\mu)\mu\, d\mu}{\int_{-1}^{1} I_\lambda(\mu)\, d\mu} = \frac{1}{\eta_{\lambda s}}. \tag{5.107}$$

Figures 5.12(d) and 5.13(d) show some phase functions for glass and iron spheres at different size parameters. The extremely forward character of diffraction at large size parameters *justifies* the *neglect* of diffraction for larger size particles.

Most *porous media applications* involve particles even *larger* than the 0.2-mm diameter considered here. Also, in some applications the wavelengths are generally in the combustion (1–6 μm) range. This results in very large size parameters. The *computation* required for Mie calculations *increases* with *increasing size parameters* and is generally *substantial*. The diffraction peak that is included in the Mie phase function has to be removed because the methods for solution of the *equation of radiative transfer cannot* handle the *extremely* sharp peaks produced by large particles. *Geometric optics* provides a *convenient* alternative where the computation required does *not* increase with size.

5.3 Radiative Properties: Dependent and Independent

The properties of an isolated *single* particle were discussed in the previous section. However the equation of radiative transfer requires knowledge of

5.3 Radiative Properties: Dependent and Independent

the radiative properties of the medium, i.e., $\langle\sigma_a\rangle$, $\langle\sigma_s\rangle$, and $\langle\Phi\rangle$. The scattering and absorption are called *dependent* if the scattering and absorbing characteristics of a particle in a medium are *influenced by neighboring particles* and are called *independent* if the presence of neighboring particles has *no* effect on absorption and scattering by a single particle. The assumption of independent scattering greatly simplifies the task of obtaining the radiative properties of the medium. Also, many important applications lie in the independent regime; therefore, the independent theory and its limits will be examined in detail in this section.

In obtaining the properties of a packed bed, the independent theory assumes the following.

- *No interference* occurs between the scattered waves (*far-field effects*). This leads to a limit on the *minimum* value of C/λ, where C is the average interparticle clearance. However, most packed beds are made up of large particles and can therefore be assumed to be *above* any such limit.

- *Point scattering* occurs, i.e., the distance between the particles is large compared to their size. Thus a representative elementary volume containing many particles can be found in which there is *no multiple scattering* and each particle scatters as if it were *alone*. Then this small volume can be treated as a *single scattering volume*. This leads to a *limit* on the *porosity*.

- The *variation of intensity* across this elemental volume is *not* large.

Then the radiative properties of the particles can be averaged across this small volume by *adding* their scattering (absorbing) *cross sections*. The total scattering (absorbing) cross sections divided by this volume gives the scattering (absorbing) *coefficient*. The phase function of the single scattering volume is the *same* as that for a single particle.

Using the *number of the scatterers* per unit volume N_s (particles/m³) and assuming independent scattering from each scatterer, the *spectral scattering coefficient* for uniformly distributed *monosize* scatterers is defined as

$$\langle\sigma_{\lambda s}\rangle = N_s A_{\lambda s}. \tag{5.108}$$

Similarly $\langle\sigma_{\lambda a}\rangle = N_s A_{\lambda a}$ and $\langle\sigma_{\lambda\mathrm{ex}}\rangle = \langle\sigma_{\lambda s}\rangle + \langle\sigma_{\lambda a}\rangle$. For spherical particles the volume of each particle is $4\pi R^3/3$ and in terms of porosity ε, we have

$$\frac{4}{3}\pi N_s R^3 = 1-\varepsilon \quad\text{or}\quad N_s = \frac{3}{4\pi}\frac{1-\varepsilon}{R^3}. \tag{5.109}$$

Then we have

$$\langle\sigma_{\lambda s}\rangle = \frac{3}{4\pi}\frac{(1-\varepsilon)}{R^3}A_{\lambda s} \tag{5.110}$$

or
$$\langle \sigma_{\lambda s} \rangle = \frac{3}{4} \frac{(1-\varepsilon)}{R} \eta_{\lambda s}. \tag{5.111}$$

When the particle diameter is not uniform, we can describe the distribution $N_s(R)\,\mathrm{d}R$, i.e., the *number of particles with a radius between $R + \mathrm{d}R$ per unit volume (number density)*. Note that $N_s(R)\,\mathrm{d}R$ has a dimension of particles/m^3. Then assuming independent scattering, we can define the *average spectral scattering coefficient* as

$$\langle \sigma_{\lambda s} \rangle = \int_0^\infty \eta_{\lambda s}(R) \pi R^2 N_s(R)\,\mathrm{d}R. \tag{5.112}$$

A similar treatment is given to the absorption and scattering coefficients. The *volumetric size distribution function* satisfies

$$N_s = \int_0^\infty N_s(R)\,\mathrm{d}R, \tag{5.113}$$

where N_s is the *average number of scatterers per unit volume*.

Whenever the particles are placed close to each other, it is expected that they *interact*. One of these interactions is the *radiation* interaction. In particular the extent to which the scattering and absorption of radiation by a particle is *influenced* by the presence of the neighboring particles. This influence is classified by *two mechanisms*: the *coherent addition*, which accounts for the *phase difference* of the superimposed far field scattered radiations and the *disturbance of the internal field* of the individual particle due to the presence of other particles (Kumar and Tien, 1990). These interactions among particles can in principle be determined from the *Maxwell equations* along with the particle arrangement and interfacial conditions. However, the complete solution is very difficult, and, therefore, *approximate* treatments, i.e., *modeling* of the interactions, have been performed. This analysis leads to the prediction of the *extent* of interactions, i.e., *dependency* of the scattering and absorption of individual particles on the presence of the other particles. One possible approach is to solve the problem of scattering by a collection of particles and attempt to obtain the radiative properties of the medium from it. However the collection *cannot* in general be assumed to be a single scattering volume. For closely packed particles, even a small collection of particles is *not* a single scattering volume. Thus, some sort of a *regression* method might be required to obtain the dependent properties of the medium. For Rayleigh scattering-absorption of dense concentration of small particles, the interaction has been analyzed by Ishimaru and Kuga (1982), Cartigny et al. (1986), and Drolen and Tien (1987).

Hottel et al. (1971) were among the first to examine the interparticle radiation *interaction* by measuring the *bidirectional* reflectance and transmittance of *suspensions* and comparing them with the predictions based on

Mie theory, i.e., by examining $(\eta_{\lambda\,\text{ex}})_{\exp}/(\eta_{\lambda\,\text{ex}})_{\text{Mie}}$. They used visible radiation and a small concentration of small particles. An *arbitrary criterion* of 0.95 has been assigned. Therefore, if this ratio is *less* than 0.95 the scattering is considered *dependent* (because the interference of the surrounding particles is expected to *redirect* the scattered energy back to the *forward* direction).

Hottel et al. (1971) identified the *limits of independent scattering* as $C/\lambda > 0.4$ and $C/d > 0.4$ (i.e., $\varepsilon > 0.73$). Brewster and Tien (1982a) and Brewster (1983) also considered *larger* particles (maximum value of $\alpha_R = 74$). Their results indicated that no dependent effects occur as long as $C/\lambda > 0.3$, *even* for a close pack arrangement ($\varepsilon = 0.3$). It was suggested by Brewster that the *point* scattering assumption is only an *artifice* necessary in the derivation of the theory and is *not* crucial to its application or validity. Thereafter, the C/λ criteria for the applicability of the theory of independent scattering was verified by Yamada et al. (1986, $C/\lambda > 0.5$), and Drolen and Tien. However, Ishimaru and Kuga note *dependent* effects at *much* higher values of C/λ. In sum, these experiments seem to have developed confidence in application of the theory of independent scattering in packed beds consisting of large particles, where C/λ almost always has a value *much larger* than the mentioned limit of the theory of independent scattering. Thus, the approach of obtaining the radiative properties of the packed beds from the independent properties of an individual particle has been applied to packed beds *without* any regard to their *porosity* (Brewster; Drolen and Tien). However, as will be shown later, all these experiments were similar in design and most of these experiments used *suspensions* of *small transparent latex* particles. Only in the Brewster experiment was a close packing of large semitransparent spheres considered.

Figure 5.14 shows a map of independent/dependent scattering for packed beds and *suspensions* of spherical particles (Tien and Drolen, 1987). The map is developed based on available experimental results. The experiments are from *several* investigators, and *some* of the experiments will be reviewed later. The results show that for relatively *high temperatures* in most packed beds the scattering of thermal radiation can be considered *independent*.

The *rhombohedral lattice arrangement* gives the *maximum* concentration for a given interparticle spacing. This is *assumed* in arriving at the relation between the *average interparticle clearance* C and the *porosity*. This relation is

$$\frac{C}{d} = \frac{0.905}{(1-\varepsilon)^{1/3}} - 1 \quad \text{or} \quad \frac{C}{\lambda} = \frac{\alpha_R}{\pi}\left[\frac{0.905}{(1-\varepsilon)^{1/3}} - 1\right], \qquad (5.114)$$

where $C/\lambda > 0.5$ (some suggest 0.3) has been *recommended* for *independent scattering* (based on the experimental results). The total interparticle clearance should *include* the average distance from a point on the surface of one particle to the nearest point on the surface of the *adjacent* particle in a close pack. This *average close-pack separation* should be *added* to the

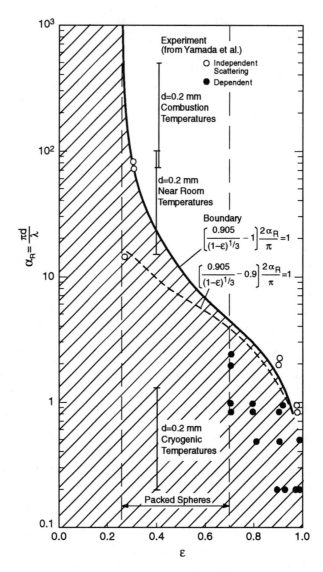

Figure 5.14 Experimental results for dependent versus independent scattering shown in the α_R-ε plane. Also shown are two empirical boundaries separating the two regimes.

interparticle clearance C obtained when the *actual* packing is referred to a rhombohedral packing ($\varepsilon = 0.26$). This separation can be *represented* by $a_1 d$ where a_1 is a constant ($a_1 \simeq 0.1$). Therefore we suggest that the *condition* for independent scattering be modified to

5.3 Radiative Properties: Dependent and Independent

TABLE 5.2 SIZE PARAMETER $\alpha_R = \pi d/\lambda$, FOR A 0.2-mm-DIAMETER PARTICLE

λT (μm-K) $(F_{0-\lambda T})^\dagger$	1888 (0.05)	2898 (0.25)	12555 (0.95)
T (K) = 4	$\lambda = 472(\mu m)/\frac{\pi d}{\lambda} = 1.33$	724/0.867	3139/0.200
300	$6.29/10^2$	9.66/65	41.9/15
1500	$1.26/2.00 \times 10^2$	$1.93/3.25 \times 10^2$	8.37/75

$^\dagger F_{0-\lambda T} = \dfrac{15}{\pi^4}\sum_{i=1}^{4}\dfrac{e^{-ix}}{i}\left(x^3 + \dfrac{3x^2}{i} + \dfrac{6x}{i^2} + \dfrac{6}{i^3}\right),$

for $x = \dfrac{14,388(\mu\text{m-K})}{\lambda T}$, Chang and Rhee (1984).

$$C + 0.1d > 0.5\lambda, \qquad (5.115)$$

where C is given earlier. This is *also* plotted in Figure 5.14. As expected for $\varepsilon \to 1$ this *correction* is small, while for $\varepsilon \to 0.26$, it becomes *significant*.

In Figure 5.14 the size parameters associated with randomly packed bed of 0.2-mm-diameter spheres at *very high* (*combustion*), intermediate (*room temperature*), and *very low* (*cryogenic*) temperatures are also given. Note that based on (5.115) only the first temperature range falls into the *dependent* scattering regime (for $d = 0.2$ mm and $\varepsilon = 0.4$). Table 5.2 gives the range of *temperatures, wavelengths*, and *size parameters* for the 0.2-mm sphere considered.

Singh and Kaviany (1991) examine *dependent* scattering in beds consisting of *large* particles (*geometric range*) by carrying out Monte Carlo simulations. They argue that the C/λ criterion *only accounts* for the *far-field effects* and that the porosity of the system is of *critical importance* if *near-field effects* are to be considered. According to the regime map shown in Figure 5.14, a packed bed of *large particles* should lie in the independent range. This is because a very large diameter *ensures* a large value of C/λ *even* for small porosities. However, Singh and Kaviany show *dependent* scattering for very large particles in systems with low porosity. Figure 5.15(a) shows the *transmittance* through a medium consisting of large (geometric range) *totally reflecting* spheres. The scattering is assumed to be *specular*. The transmittance through packed beds of different porosities and at different values of τ_{ind} was calculated by the method of discrete ordinates using a 24-point Gaussian quadrature. It is clear from Figure 5.15(a)

298 5. Radiation Heat Transfer

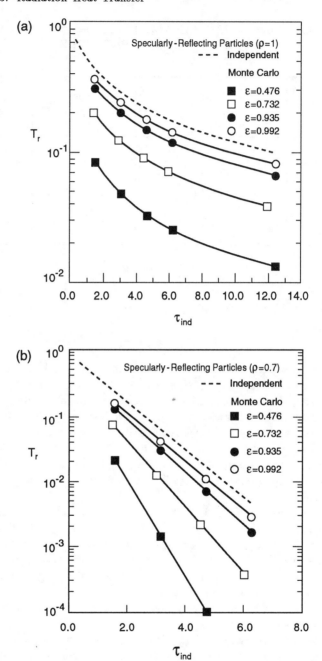

Figure 5.15 (a) Variation of transmittance with respect to the bed optical thickness for totally reflecting spheres ($\rho = 1$) and several porosities. (b) Same as (a), except for $\rho = 0.7$.

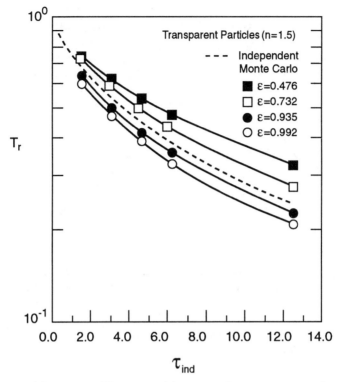

Figure 5.16 (a) Same as Figure 5.15(a), except for transparent spheres ($n = 1.5, \eta_{\lambda a} = 0$).

that the independent theory *fails* for low porosities. As the porosity is *increased*, the Monte Carlo solution begins to approach the independent theory solution. For $\varepsilon = 0.992$, the agreement obtained is good. The *bulk* behavior (away from the bounding surface) predicted by the Monte Carlo simulations for $\varepsilon = 0.992$ and the results of the independent theory are in very close agreement. A small difference occurs at the *boundaries*, where the bulk properties are no longer valid. However, although this difference occurs at the boundary, the commonly made assumption that the prediction by the continuum treatment will *improve* with increase in the optical thickness is *not* justifiable because this offset is carried over to larger optical thicknesses.

Figure 5.15(b) shows the effect of the porosity on the bed transmittance for absorbing particles ($\rho = 0.7$). Again, the independent theory fails for low porosities although the agreement for dilute systems is good. Thus, the transmittance for a packed bed of *opaque* particles can be *significantly less* than that predicted by the independent theory. This is due to *multiple*

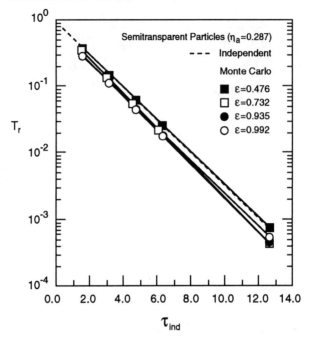

Figure 5.16 (b) Same as Figure 5.15(a), except for semitransparent spheres ($n = 1.5, \eta_{\lambda a} = 0.287$).

scattering in a representative elementary volume, so that the effective cross section presented by a particle is *more* than its independent cross section.

Figures 5.16(a)–(c) show the effect of change in the porosity on the transmittance through a medium of *semitransparent* particles. The particles considered are large spheres with $n = 1.5$. For these particles, the only parameter that determines the radiative properties of a particle is the product $\kappa \alpha_R$ (as long as κ is *not* too large). Figure 5.16(a) is plotted for the case of $\kappa = 0$ (*transparent* spheres). Differences from opaque particles (Figure 5.15) are obvious. The violation of the independent theory results in a decrease in the transmittance for opaque spheres, but for transparent spheres, it results in an *increase* in the transmittance. This results because the change in the *optical thickness* across one particle in a packed bed is large. Therefore, a transparent particle while transmitting the ray through it, also *transports* it across a substantial optical thickness. In a *dilute* suspension, a particle, while allowing for transmission through it, does not result in a substantial transport.

Figure 5.16(b) is plotted for semitransparent particles with $\kappa \alpha_R = 0.1$, which gives $\eta_{\lambda a} = 0.287$. The absorption decreases this effect (transportation across a layer of substantial optical thickness) to the extent that it is exactly *balanced* by the decrease due to *multiple* scattering in the

5.3 Radiative Properties: Dependent and Independent

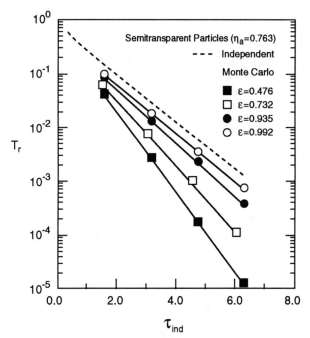

Figure 5.16 (c) Same as Figure 5.15(a), except for semitransparent spheres ($n = 1.5, \eta_{\lambda a} = 0.763$).

elementary volume for $\varepsilon = 0.476$. As a result, the Monte Carlo prediction for $\varepsilon = 0.476$ shows very good agreement with the prediction from the independent theory. The results for *dilute* systems are exactly as expected: giving slightly less transmittance than the independent theory solution but showing the same *bulk* behavior. Therefore, due to these two *opposing* effects, the magnitude of deviation from independent theory for packed beds of transparent and semitransparent particles is smaller than that for opaque spheres.

Figure 5.16(c) shows the effect of variation in porosity on transmittance through a medium of highly absorbing semitransparent particles ($\kappa \alpha_R = 0.5, \eta_{\lambda a} = 0.763$). Here, the *multiple scattering* effect clearly *dominates over* the *transportation* effect. The predicted transmittance for low porosities by the Monte Carlo method is far less than that predicted by the independent theory, while the most dilute system ($\varepsilon = 0.992$) again shows good agreement with the independent theory. It is encouraging to note that the $\varepsilon = 0.992$ system matched the independent theory results for all cases considered. However, the effect of the porosity on transmittance is *noticeable even* for relatively high porosities ($\varepsilon = 0.935$).

As seen earlier, the failure is *more* drastic for transmission through a bed of *opaque* spheres than for *transparent* and semitransparent spheres with low absorption. Also, the deviation from the independent theory is shown to increase with a decrease in the porosity. This deviation can be significant for porosities as high as 0.935. The independent theory gives good predictions for the bulk behavior of highly porous systems ($\varepsilon \geq 0.992$) for all cases considered. Two distinct dependent scattering effects were identified. The *multiple scattering* of the reflected rays *increases* the effective scattering and absorption cross sections of the particles. This results in a *decrease* in transmission through the bed. The transmission through a particle in a packed bed results in a *decrease* in the effective cross sections, resulting in an *increase* in the transmission through a bed. For opaque particles, *only* the multiple scattering effect is found while for transparent and semitransparent particles, *both* of these effects are found and tend to oppose each other.

In conclusion, we note that *both* the C/λ *criterion* and the *porosity* criterion must be satisfied before the independent theory can be used with confidence.

5.4 Volume Averaging for Independent Scattering

We have already referred to the problem of scattering by a collection of objects in Section 5.3. If the scatterers behave *independently*, a *simple* volume integration over the particle concentration distribution results. We now look back at the scattering property of spherical particles as predicted by *Rayleigh*, *Mie*, and *geometric* analyses. We assume independent scattering. Assuming that a continuum treatment of radiation in solid-fluid systems using independent scattering is possible, we use volume averaging over a representative elementary volume to average over the scatterers as in (5.112).

Table 5.3 gives some *approximations* for $\langle \sigma_{\lambda s} \rangle$ $\langle \sigma_{\lambda a} \rangle$ and $\langle \Phi_\lambda \rangle (\theta_o)$ along with the applicable *constraints*. The wavelength λ is that for the wave traveling in the fluid, and if $n_f \neq 1$, then $\lambda = \lambda_o / n_f$, where λ_o is for travel in vacuum. Table 5.4 shows the various *approximations* used to represent the phase function $\langle \Phi_\lambda \rangle (\theta_o)$ in terms of *Legendre polynomials*.

The scattering-absorption of incident beams by a long circular *cylinder* has also been studied by van de Hulst (1981, 297–326). He considers other particle shapes (329–339). Wang and Tien (1983), Tong and Tien (1983), and Tong et al. (1983) consider *fibers* used in insulations. They use the efficiencies derived by van de Hulst and examine the effects of κ_s and d on the overall performance of the insulations. The effect of fiber *orientation* on the scattering phase function of the medium is discussed by Lee (1990). The *effective* radiative properties of a fiber-sphere composite is predicted by Lee et al. (1994).

TABLE 5.3 VOLUME AVERAGING OF RADIATIVE PROPERTIES: INDEPENDENT SCATTERING

CONSTRAINTS	$\langle \sigma_{\lambda s} \rangle$ (1/m), $\langle \sigma_{\lambda a} \rangle$ (1/m), $\langle \Phi_\lambda \rangle (\theta_o)$ • λ IS FOR WAVES TRAVELING IN THE FLUID • $m = n - i\kappa = n_s/n_f - i\kappa_s/n_f$ • $\theta_o = 0$ FOR FORWARD SCATTERED BEAM AND π FOR BACKWARD
(a) Large opaque specularly reflecting spherical particles $2\pi R/\lambda > 5$	$\langle \sigma_{\lambda s r} \rangle = \pi \rho_\lambda \int_0^\infty R^2 N_s(R)\,dR$ $N_s(R)\,dR$ is the number density of particles having radius between R and $R + dR$, ρ_λ is the hemispherical reflectivity $\langle \sigma_{\lambda a r} \rangle = \pi(1 - \rho_\lambda) \int_0^\infty R^2 N_s(R)\,dR$ $\langle \Phi_{\lambda r} \rangle (\theta_o) = \dfrac{\rho'_\lambda}{\rho_\lambda}$ $\rho'_\lambda [(\pi - \theta_i)/2]$ is the directional specular reflectivity for incident angle θ_i
(b) Large opaque diffusely reflecting spherical particles $2\pi R/\lambda > 5$	$\langle \sigma_{\lambda s r} \rangle = \pi \rho_\lambda \int_0^\infty R^2 N_s(R)\,dR$ $\langle \sigma_{\lambda a r} \rangle = \pi(1 - \rho_\lambda) \int_0^\infty R^2 N_s(R)\,dR$ $\langle \Phi_{\lambda r} \rangle (\theta_o) = \dfrac{8}{3\pi}(\sin \theta_o - \theta_o \cos \theta_o)$
(c) Large spherical particles, diffraction contribution, $2\pi R/\lambda > 20$	$\langle \sigma_{\lambda s d} \rangle = \pi \int_0^\infty R^2 N_s(R)\,dR$ $\langle \Phi_{\lambda d} \rangle (\theta_o) = \dfrac{4 J_1^2 [(2\pi R/\lambda) \sin \theta_o]}{\sin^2 \theta_o}$ J_1 is the Bessel function of first order and first kind, diffraction contribution. (*continued*)

(a) Siegel and Howell (1981, p. 575); (b) Siegel and Howell (1981, p. 582); (c) van de Hulst (1981, p. 108).

TABLE 5.3 VOLUME AVERAGING OF RADIATIVE PROPERTIES: INDEPENDENT SCATTERING (CONTINUED)

(d) Small spherical particles (extension of Rayleigh's scattering) limits are given in Ku and Felske (1984) and Selamet and Arpaci (1989)	$\langle \sigma_{\lambda s} \rangle = \dfrac{128\pi^5}{3z_1^2 \lambda^4} \left\{ \left[(n^2+\kappa^2)^2 + n^2 - \kappa^2 - 2\right]^2 + 36n^2\kappa^2 \right\}$ $\times \left\{ \displaystyle\int_0^\infty R^6 N_s(R)\,dR + \dfrac{24\pi^2}{5z_1\lambda^2}\left[(n^2+\kappa^2)^2 - 9\right] \right.$ $\left. \times \displaystyle\int_0^R R^8 N_s(R)\,dR - \dfrac{64n\kappa\pi^3}{z_1\lambda^3}\displaystyle\int_0^R R^9 N_s(R)\,dR \right.$ $\langle \sigma_{\lambda e} \rangle = \dfrac{48n\kappa\pi^2}{z_1^2 \lambda}\displaystyle\int_0^\infty R^3 N_s(R)\,dR + \left\{\dfrac{4}{15} + \dfrac{20}{3z_2}\right.$ $\left. + \dfrac{4.8}{z_1^2}\left[7(n^2+\kappa^2) + 4(n^2-\kappa^2-5)\right]\right\}$ $\times \dfrac{8nk\pi^4}{\lambda^3}\displaystyle\int_0^\infty R^5 N_s(R)\,dR + \dfrac{128\pi^5}{3z_1^2\lambda^4}$ $\times \left\{\left[(n^2+\kappa^2)^2 + n^2 - \kappa^2 - 2\right]^2 - 36n^2\kappa^2\right\}$ $\times \displaystyle\int_0^R R^6 N_s(R)\,dR$ where $z_1 = (n^2+\kappa^2)^2 + 4(n^2-\kappa^2) + 4$ $z_2 = 4(n^2+\kappa^2)^2 + 12(n^2-\kappa^2) + 9$
(e) Small spherical particles, $2\pi R/\lambda < 0.6/n$ (*Rayleigh scattering*)	$\langle \sigma_{\lambda s} \rangle = \dfrac{128\pi^5}{3\lambda^4}\left|\dfrac{m^2-1}{m^2+2}\right|^2 \displaystyle\int_0^\infty R^6 N_s(R)\,dR$ $\langle \Phi_\lambda \rangle(\theta_o) = \dfrac{3}{4}(1+\cos^2\theta_o)$
(f) d < interparticle spacing $< \lambda$, $2\pi R/\lambda \ll 1$ (*Lorentz-Lorenz scattering*)	$\langle \sigma_{\lambda s} \rangle = \dfrac{24\pi^3}{\lambda^4 N_s}\left|\dfrac{m^2-1}{m^2+2}\right|^2$ Independent of R, N_s is the number density of scatterers

(d) Penndorf (1962); (e) van de Hulst (1981, p. 70); (f) Siegel and Howell (1981, p. 587).

5.4 Volume Averaging for Independent Scattering

TABLE 5.3 VOLUME AVERAGING OF RADIATIVE PROPERTIES: INDEPENDENT SCATTERING (CONTINUED)

(g) Spherical particles, interparticle spacing $\gg \lambda$, $2\pi R/\lambda \ll 1$, random arrangement	$\langle \sigma_{\lambda s} \rangle = \dfrac{8}{3} \dfrac{\pi^3}{\lambda^4 N_s} \lvert m^2 - 1 \rvert^2$ Independent of R
Same with $n \to \infty$	$\langle \sigma_{\lambda s} \rangle = \dfrac{160 \pi^5}{3 \lambda^4} \int_0^\infty R^6 N_s(R)\, dR$ $\quad + \dfrac{256 \pi^7}{5 \lambda^6} \int_0^\infty R^8 N_s(R)\, dR$ $\langle \Phi \rangle (\theta_o) = \dfrac{3}{5}\left[\left(1 - \dfrac{1}{2}\cos\theta_o\right)^2 + \left(\cos\theta_o - \dfrac{1}{2}\right)^2\right]$ Small spherical particles such that R^8 term is negligible
(h) Nonspherical (Rayleigh-ellipsoid approximation)	$\langle \sigma_{\lambda a} \rangle = \dfrac{\pi}{\lambda}\left(\dfrac{\overline{V}}{\overline{A}}\right) \operatorname{Im}\left[\dfrac{2n}{n-1}(\log n - i\kappa)\right]$ $\overline{V}/\overline{A}$ is the average diameter

(g) van de Hulst (1981, p. 83); (h) Bohren and Huffman (1983).

For *small* particles, a simplified approach to modeling the spectral scattering and absorption coefficient is given by Mengüc and Viskanta (1985a).

5.5 Experimental Determination of Radiative Properties

5.5.1 MEASUREMENTS

We consider only those experiments that use beds and suspensions, rather than individual particles. The experimental results are generally compared with the prediction based on the Mie theory of scattering and volume averaging under the assumption of *independent* scattering. The source is either a *laser* with very *limited* and *discrete* wavelengths or a *blackbody* cavity that is used in conjunction with a *mechanical chopper*.

One of the first experiments is that of Chen and Churchill (1963) in which *isothermal* packed beds (3 mm $\leq d \leq$ 5 mm, $0.35 < \varepsilon < 0.49$) of *glass*,

TABLE 5.4 PHASE FUNCTION IN TERMS OF LEGENDRE POLYNOMIALS

APPROXIMATIONS FOR $\langle \Phi_\lambda \rangle (\theta_o)$

(a) Spherical particles, independent scattering

$$\langle \Phi_\lambda \rangle (\theta_o) = \sum_{i=0}^{N} (2i+1) A_i P_i(\cos\theta_o)$$

where $\cos\theta_o = \cos\theta_i \cos\theta + \sin\theta_i \sin\theta \cos(\phi - \phi_i)$,
$A_i = (1/2) \int_0^\pi \Phi_\lambda(\theta_o) P_i(\cos\theta_o) \, d\cos\theta_o$, P_i is the Legendre polynomial of degree i, and $\Phi_\lambda(\theta_o)$ is the exact phase function obtained through the Mie scattering analysis

- For isotropic scattering $A_0 = 1$, $A_i = 0$
- For Rayleigh scattering $A_0 = 1$, $A_2 = 1/10$, $A_i = 0$
- For linear-isotropic scattering $A_0 = 1$, $-1/3 \leq A_1 \leq 1/3$, $A_i = 0$

Strong forward scattering

$$\langle \Phi_\lambda \rangle (\theta_o) = 2 f_\lambda \delta (1 - \cos\theta_o) + (1 - f_\lambda) \sum_{i=0}^{M} (2i+1) \widehat{A}_i P_i(\cos\theta_o)$$

where $f_\lambda = A_{M+1}$, $\widehat{A}_i = (A_i - f_\lambda)/(1 - f_\lambda)$, $i = 0, 1, \ldots, M$ and δ is the Dirac delta, $(M+1)/2$ is the order of approximation for a spike in the forward direction $(\delta - M)$ approximation

(b) Strongly forward scattering

$$\langle \Phi_\lambda \rangle (\theta_o) = 2 f_\lambda \delta (1 - \cos\theta_o) + (1 - f_\lambda) \left(1 + 3\widehat{A}_1 \cos\theta_o \right)$$

is the Delta-Eddington approximation

$$f_\lambda = \begin{cases} A_2, & A_2 \geq (3A_1 - 1)/2 \\ (A_3 - 1)/2, & \text{else} \end{cases} \quad \text{with } \widehat{A}_1 = \frac{A_1 - A_2}{1 - A_2}$$

(c) Not very accurate (Lee and Buckius, 1982). For the two-flux model an approximation (linear isotropic) is

$$\text{forward scattered, } f_\lambda = \frac{1}{2} \int_0^{\pi/2} \langle \Phi_\lambda \rangle (\theta_o) \, d\cos\theta_o \simeq \frac{1}{2} + \frac{1}{2} \sum_{i=0}^{\infty} \frac{(-1)^i A_{2i+1}(2i)!}{2^{2i+1} i!(i+1)!}$$

$$\text{backward scattered, } b_\lambda = \frac{1}{2} \int_{-\pi/2}^{0} \langle \Phi_\lambda \rangle (\theta_o) \, d\cos\theta_o = 1 - f_\lambda$$

TABLE 5.4 PHASE FUNCTION IN TERMS OF LEGENDRE POLYNOMIALS (CONTINUED)

(d) Strongly forward scattering

$$\langle \Phi_\lambda \rangle (\cos\theta_o) = 2 f_\lambda \delta (1 - \cos\theta_o) + (1 - f_\lambda) \frac{1 - g_\lambda^2}{(1 + g_\lambda^2 - 2\cos\theta_o)^{3/2}}$$

is the δ-Henyey-Greenstein approximation,

$$g_\lambda = \frac{A_1 - A_2}{1 - A_1}, \text{ and } f_\lambda = \frac{A_2 - A_1^2}{1 - 2A_1 + A_2}$$

(a) Chu and Churchill (1955); (b) Wiscombe (1977), McKellar and Box (1981); (c) Lee and Buckius (1982); (d) McKellar and Box (1981).

aluminum oxide, carbon steel, and *silicon carbide* particles are used with air as the fluid. A variable-temperature blackbody source is used (685 K $\leq T_b \leq$ 1352 K) and modulated to a 10-Hz square wave in order to *eliminate* the steady background contributions.

The measured transmission was used along with the solution to the one-dimensional radiative transfer equation (using the two-flux approximation and *isotropic* scattering). Then the *absorption* and *back scattering* coefficients σ_a and $B\sigma_s$ (where B is the *fraction of energy back scattered* in a plane parallel slab geometry) were determined by the least-square approximation method. No comparison was made with the predicted values of σ_a and σ_s (from Mie or other theories).

Hottel et al. (1970, 1971) use *nonabsorbing* spheres (*latex* particles in water with $0.705 \leq \varepsilon \leq 1$) and two monochromatic sources of 0.426 and 0.546 μm and measure the bidirectional reflectance and *transmittance* from single particles as well as packed beds. The results of the Mie theory for $\sigma_{\lambda s}$ and $\Phi_\lambda(\theta_o)$ used in the equation of radiative transfer are compared to the measurement. They found that if the average *clearance* between the particles C is larger than $0.3\lambda_o(0.4\lambda)$, then the Mie theory predicts these properties correctly, i.e., the *scattering is independent*.

Extensions of the Hottel et al. study are made by Brewster and Tien (1982a) and Brewster (1983) for nonabsorbing as well as absorbing (dyed) particles, and by Yamada et al. (1986) and Drolen and Tien (1987). They all confirm that for $C/\lambda < 0.5$ the measured values of the transmittance are different (by 5 percent or more for $\langle \eta_{\lambda s} \rangle$) than the predicted values based on the independent Mie theory equation of radiative transfer. Ishimaru and Kuga (1982), however, report dependent effects for much higher values of C/λ. The latex particles used had n_s ranging from 1.58 to 1.61 (depending on the wavelength). For the absorbing case considered by Brewster, κ_s was 0.0013, which for a size parameter (α_R) of 73.7 gives $\omega_a = 0.76$.

Nelson et al. (1986) used the same particles and fluid to measure the *back-scattered* energy. They noted that in the Rayleigh scattering, allowance must be made for the *index of refraction* of the fluid (as stated in Table 5.3), and in the equation of radiative transfer, allowance must be made for the *liquid absorption*. For small porosities (large *optical thicknesses*), they also note a deviation from the independent scattering.

Experimental results of Papini (1989) show a strong wavelength-dependence of *the directional-hemispherical spectral reflectance* for natural fibers (cotton and wool). For $\lambda \geq 0.6$ μm, this reflectance increases substantially over that for $\lambda \leq 0.6$ μm.

When the interparticle interactions in radiative transfer are significant, then modification to the efficiencies can be made. For spherical particles, correlations such as those which will be discussed in Section 5.12 have been found to be satisfactory for the experimental results of Chu et al. (1988). They use silicon oxide particles with average diameters of 70 and 500 Å. A collection of these particles is called *aerosol* and is used as insulation. At STP, the mean free path of air is about 200 Å, resulting in a *smaller* conductivity for the gas phase in the silicon oxide-air system. Source wavelengths of 0.532, 0.633, and 1.053 μm (monochromatic sources), 0.8–3.2 μm (a spectrophotometer), and 1.6–1.7 μm (silicon carbide rod heated to 1100 K) are used. The *minimum* porosity used is 0.9.

The spectral variation of the extinction coefficient, a single wavelength scattering coefficient, and the phase function for *polyurethane foam* (cellular structure) and for *fiberglass insulation* have been studied by Glicksman et al. (1987). A wavelength range of 2.5–40 μm available in a spectrophotometer (a critical examination of the use of this instrument is given in their paper), is used for measurement of $\langle \sigma_{\lambda\,ex} \rangle$. A single wavelength of 9.64 μm (CO_2 laser) is used for the measurement of $\langle \sigma_{\lambda\,s} \rangle$ and $\langle \Phi_\lambda \rangle$. Due to the finite divergence of the laser beam, it is *impossible* to distinguish unscattered radiation from that scattered with $\theta_o < 10°$. Therefore, radiation scattered with $\theta_o < 10°$ was assumed to be a part of the *transmitted* radiation and measurement of the phase function was only made for $\theta_o > 10°$. In order to obtain the phase function for $\theta_o > 10°$, it is assumed that the source of the scattered energy is the *diverging* laser beam (i.e., no emission or in-scattering from the rest of the volume). Their results show the following:

- The extinction coefficient for a single cell wall (polyurethane) $\sigma_{\lambda\,ex}$ is *strongly* spectral, while the measured transmission through several cells $\langle \sigma_{\lambda\,ex} \rangle$ behaves as if it belonged to a *gray* body. This has been associated with the different *chemical* composition of the cell *wall junctions*.

- The total extinction coefficient is found to be 15–25 cm^{-1} (density of the polyurethane foams used is 29–55 kg/m^3). This shows that for the sample thickness *larger* than 7 mm, the optical thickness is *greater* than 10.

- For the foam, the absorption coefficient is about 2.5 times the scattering coefficient ($\omega_a = 0.29$).

- For both materials, the scattering is strongly *forward* ($\langle \Phi_\lambda \rangle = 0$ for $\theta_o > 40°$).

- For the fiberglass (density of 10 kg/m^3), the absorption and scattering coefficients are *nearly* the same ($\omega_a = 0.5$), and the extinction coefficient is about 7 cm^{-1}.

- Polyurethane insulations are not opaque to infrared radiation, and radiation can be a significant *mode* of heat transfer in these insulations.

Effective spectral absorption and scattering coefficients and spectral scattering phase functions have been indirectly measured for some *ceramic foams* by Hendricks and Howell (1994). The measured quantities, the effective spectral hemispherical reflectance and transmittance, are used in a discrete-ordinates approximated radiation model to calculate the above radiative properties. The phase functions are given in terms of the Henyey-Greenstein parameters. Their experimental results for the effective *total* extinction coefficient is given in Figure 5.17, as a function of the pore diameter (a linear pore dimension). A correlation showing an *inverse* proportionality with the pore diameter, is also shown.

5.5.2 MODELS USED TO INTERPRET EXPERIMENTAL RESULTS

(A) HIGHLY FORWARD SCATTERING MATRICES

When the in-scattering term is almost the entire scattered energy (large specularly reflecting particles), then the normal (along x) intensity is approximated by

$$\frac{dI_{\lambda n}}{dx} = -\langle \sigma_{\lambda a} \rangle I_{\lambda n} + \langle \sigma_{\lambda a} \rangle I_{\lambda b}. \tag{5.116}$$

This is the approximation used by Grosshandler and Monterio (1982). With the normal incident intensity $I_{\lambda i}$ as the boundary condition for $x = 0$, we have

$$\langle \sigma_{\lambda a} \rangle = -\frac{1}{L} \ln \frac{I_{b\lambda} - I_{\lambda n L}}{I_{b\lambda} - I_{\lambda i}}, \tag{5.117}$$

where $I_{\lambda n L}$ is the measured normal intensity at $x = L$. It should be noted that the fraction of energy scattered in the included angle of the collector should be estimated to verify the applicability of (5.117). Otherwise a proper correction must be applied.

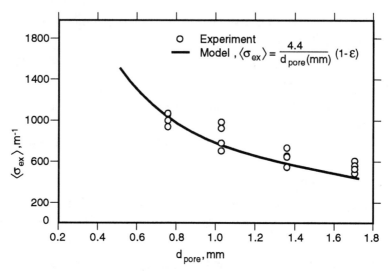

Figure 5.17 Experimental results for the effective, total extinction coefficient for some ceramic foams, as a function of the pore diameter. (From Hendricks and Howell, reproduced by permission ©1994 American Society of Mechanical Engineers.)

(B) No Emission with Scattering from Limited Incident Angles

For a highly localized incident radiation (intensity I_0) entering a porous media (an unexpanded laser beam impinging upon a porous layer) at $x = 0$, Glicksman et al. (1987) divided the polar angle distribution of the phase function into two regions. In $0 \leq \theta_o \leq \theta_{o\ell}$, where $\theta_{o\ell}$ is the divergence angle of the laser beam, no in-scattering was considered, i.e.,

$$\frac{I_\lambda(x, \theta_o = 0)}{I_0} = e^{-\langle\sigma_\lambda \text{ex}\rangle x} \quad \text{for } 0 \leq \theta_o \leq \theta_{o\ell}, \quad (5.118)$$

$$\cos\theta \frac{\partial I_\lambda(x, \theta_o)}{\partial x} = -\langle\sigma_{\lambda\text{ex}}\rangle I_\lambda(x, \theta_o) + \frac{\langle\sigma_{\lambda s}\rangle}{2} I_\lambda(x, \theta_o = 0)$$
$$\times \langle\Phi_\lambda(\theta_o)\rangle \sin\theta_{o\ell}\Delta\theta_{o\ell} \quad \text{for } \theta_{o\ell} \leq \theta_o \leq \pi, \quad (5.119)$$

where $\sin\theta_{o\ell}\Delta\theta_{o\ell}$ is the *solid angle* of the laser source. In their measurement, (5.118) is used for determination of $\langle\sigma_{\lambda a}\rangle + \langle\sigma_{\lambda s}\rangle = \langle\sigma_{\lambda\text{ex}}\rangle$. The solution to (5.119) using (5.118) and $I_{\lambda L}(\theta_o)$ (the measured boundary condition at $x = L$) is

$$\frac{I_{\lambda L}(\theta_o)}{I_0 \sin\theta_{o\ell}\Delta\theta_{o\ell}} = \frac{\langle\sigma_{\lambda s}\rangle \langle\Phi_\lambda\rangle (\theta_o)}{2 \langle\sigma_{\lambda\text{ex}}\rangle (1 - \cos\theta_o)} \left(e^{-\sigma_\lambda \text{ex} L} - e^{-\langle\sigma_\lambda \text{ex}\rangle L / \cos\theta_o} \right)$$
$$\text{for } \theta_{o\ell} \leq \theta_o \leq \frac{\pi}{2}, \quad (5.120)$$

$$\frac{I_{\lambda L}(\theta_o)}{I_0 \sin\theta_o \ell \Delta\theta_o \ell} = \frac{\langle\sigma_{\lambda s}\rangle \langle\Phi_{\lambda s}\rangle(\theta_o)}{2\langle\sigma_{\lambda \mathrm{ex}}\rangle(1-\cos\theta_o)}\left[1 - e^{-\langle\sigma_\lambda \mathrm{ex}\rangle L\left(1-\frac{1}{\cos\theta_o}\right)}\right]$$

$$\text{for } \frac{\pi}{2} \leq \theta_o \leq \pi. \quad (5.121)$$

Then (5.120) and (5.121) are used for determination of the product $\langle\sigma_{\lambda s}\rangle\langle\Phi_\lambda\rangle(\theta_o)$.

Since

$$\frac{1}{2}\int_\pi^{\Delta\theta_o \ell} \langle\sigma_{\lambda s}\rangle\langle\Phi_\lambda\rangle(\theta_o)\sin\theta_o\,d\theta_o = \langle\sigma_{\lambda s}\rangle, \quad (5.122)$$

then $\langle\sigma_{\lambda s}\rangle$ and $\langle\Phi_\lambda\rangle(\theta_o)$ are found from the product after normalization of $\langle\Phi_\lambda\rangle(\theta_o)$ over all angles.

5.6 Boundary Conditions

For plane-parallel geometry (azimuthal symmetry) and for a porous medium matrix confined between a surface 1 ($x = 0$, $\tau = 0$, $\varepsilon_{\lambda 1}$) and a surface 2 ($x = L$, $\tau = \tau_L$, $\varepsilon_{\lambda 2}$) with the directional specular intensities at the boundaries given by $I(0,\mu)$, $\mu > 0$ and $I(\tau_L, \mu)$, $\mu < 0$, the following equations (Ozisik, 1985, 274–275) describe the boundary conditions.

5.6.1 Transparent Boundaries

External sources are $I^+_{\lambda i}(\mu, x = 0)$, $I^-_{\lambda i}(\mu, x = L)$, and we have

$$I_\lambda(0, \mu) = I^+_{\lambda i}(\mu) \quad \text{for } \mu > 0, \quad (5.123)$$

$$I_\lambda(\tau_L, \mu) = I^-_{\lambda i}(\mu) \quad \text{for } \mu < 0. \quad (5.124)$$

5.6.2 Opaque Diffuse Emitting/Reflecting Boundaries

Then we have

$$I_\lambda(0,\mu) = \varepsilon_{\lambda 1} I_{\lambda b}(T_1) + \rho_{\lambda 1}\frac{\int_{-1}^0 I_\lambda(0,\mu')\mu'\,d\mu'}{\int_{-1}^0 \mu'\,d\mu'}$$

$$= \varepsilon_{\lambda 1} I_{\lambda b}(T_1) + 2\rho_{\lambda 1}\int_0^1 I_\lambda(0,-\mu')\mu'\,d\mu'$$

$$= \varepsilon_{\lambda 1} I_{\lambda b}(T_1) + 2(1-\varepsilon_{\lambda 1})\int_0^1 I_\lambda(0,-\mu')\mu'\,d\mu' \quad \text{for } \mu > 0 \,(5.125)$$

and

$$I_\lambda(\tau_L, \mu) = \varepsilon_{\lambda 2} I_{\lambda b}(T_2) + 2(1 - \varepsilon_{\lambda 2}) \int_0^1 I_\lambda(\tau_L, \mu') \mu' \, d\mu' \quad \text{for } \mu < 0. \tag{5.126}$$

5.6.3 Opaque Diffusely Emitting Specularly Reflecting Boundaries

We have

$$I_\lambda(0, \mu) = \varepsilon_{\lambda 1} I_{\lambda b}(T_1) + \rho^s_{\lambda 1} I_\lambda(0, -\mu) \quad \text{for } \mu > 0, \tag{5.127}$$

$$I_\lambda(0, \mu) = \varepsilon_{\lambda 2} I_{\lambda b}(T_2) + \rho^s_{\lambda 2} I_\lambda(\tau_L, -\mu) \quad \text{for } \mu < 0. \tag{5.128}$$

5.6.4 Semitransparent Nonemitting Specularly Reflecting Boundaries

For planar boundaries with the spectral specular reflectivity $\rho^s_{\lambda B}$ with a *significant* difference between the index of refraction of the bounding medium and the porous medium, Hottel et al. (1968) point out the significant *internal reflections*. The boundary conditions are

$$I_\lambda(0, \mu) = \rho_{\lambda B} I_\lambda(0, -\mu) + (1 - \rho_{\lambda B}) I^+_{\lambda i}(\mu) \quad \text{for } \mu > 0, \tag{5.129}$$

$$I_\lambda(\tau_L, \mu) = \rho_{\lambda B} I_\lambda(\tau_L, -\mu) + (1 - \rho_{\lambda B}) I^-_{\lambda i}(\mu) \quad \text{for } \mu < 0. \tag{5.130}$$

Furthermore, if the intensity of the radiation emerging (leaving) from the porous medium is desired, this is related to the intensities $I_\lambda(0, \mu)$ and $I_\lambda(\tau_L, \mu)$ by (Hottel et al., 1968)

$$\frac{I_\lambda(\tau_L, \mu_e)}{I_\lambda(\tau_L, \mu)} = \frac{I_\lambda(0, \mu_e)}{I_\lambda(0, \mu)} = \frac{1 - \rho^s_{\lambda B}}{(n_f/n_B)^2}, \tag{5.131}$$

where the Snell law gives the direction of refraction as

$$n_B \sin \theta_e = n_f \sin \theta, \tag{5.132}$$

where n_f is the index of refraction of the fluid in the porous medium and n_B is that of the bounding medium. If $n_f > n_B$, then the refracted part of the energy may *not* leave the porous media and may undergo *total* internal refraction. In that case, the equations in this section should be modified to include that contribution.

5.7 Solution Methods for Equation of Radiative Transfer

The approximate solution methods are reviewed by Davison (1957, *spherical harmonics* and *discrete ordinate methods*), Sparrow and Cess (1978, plane-parallel, exact and kernel approximation), Siegel and Howell (1981), Ozisik (1985, Chapter 9), and Raithby and Chui (1990, finite volume).

The integration of the equation of radiative transfer is made *difficult* when this integro-differential equation includes the following effects.

- *Emission* is significant and *coupling* of this equation with the energy equation is required.

- Both *absorption* and *scattering* are significant.

- *Scattering* is highly *anisotropic*.

- The coefficients are *highly wavelength-dependent*.

- Radiation in *more* than one dimension must be considered.

- Boundary conditions include *emission, reflection* (diffuse and specular), and *transmission*.

The approximate solution to the equation of radiative transfer has been (and continues to be) attempted using various mathematical techniques. Following are a *few* examples of the solution methods.

5.7.1 Two-Flux Approximations, Quasi-Isotropic Scattering

The *two-flux approximation* (or *Schuster-Schwarzchild approximation*) for plane-parallel geometry has been discussed by Chandrasekhar (1960, p. 55), Ozisik (1985, 330–332), Vortmeyer (1978), and Brewster and Tien (1982b), among others. The principle is the *division* of the radiation field into the *forward* I_λ^+ and *backward* I_λ^- components. The two-flux method is based on the assumption of *hemispherical isotropy* and fails to give good results whenever this assumption is violated. This results in the case of a highly *anisotropic* phase function as noted by Brewster and Tien (1982b) and by Mengüc and Viskanta (1982). Singh and Kaviany (1991) note that hemispherical anisotropy is destroyed in a nonemitting but absorbing bed and show that under these conditions, the two-flux model *fails* even for an isotropic phase function. Under the two-flux approximation the radiative transfer equation (5.11) is written as

$$\cos\theta \frac{\partial I_\lambda^+(x,\theta)}{\partial x} = -(\langle\sigma_{\lambda s}\rangle + \langle\sigma_{\lambda a}\rangle) I_\lambda^+(x,0) + \langle\sigma_{\lambda a}\rangle I_{\lambda b}(x)$$

$$+ \frac{\langle\sigma_{\lambda s}\rangle}{4\pi} \int_0^{2\pi} \left[\int_0^1 I_\lambda^+(x,\theta_i) \langle\Phi_\lambda\rangle(\theta_o) \,\mathrm{d}\cos\theta_i \right.$$

$$\left. + \int_{-1}^0 I_\lambda^-(x,\theta_i) \langle\Phi_\lambda\rangle(\theta_o) \,\mathrm{d}\cos\theta_i \right] \mathrm{d}\phi. \qquad (5.133)$$

Note that, because of the assumed *azimuthal symmetry*, integration over 2π will not be performed from this point on. A similar equation can be written for I_λ^-.

Integration of this over the forward direction gives

$$\frac{\mathrm{d}}{\mathrm{d}x} \int_0^1 \cos\theta I_\lambda^+(x,\theta) \,\mathrm{d}\cos\theta$$

$$= -(\langle\sigma_{\lambda s}\rangle + \langle\sigma_{\lambda a}\rangle) \int_0^1 I_\lambda^+(x,\theta) \,\mathrm{d}\cos\theta + \langle\sigma_{\lambda a}\rangle I_{\lambda b}(x)$$

$$+ \frac{\langle\sigma_{\lambda s}\rangle}{2} \left[\int_0^1 I_\lambda^+(x,\theta) \langle\Phi_\lambda\rangle(\theta_i \to \theta) \,\mathrm{d}\cos\theta_i \right.$$

$$\left. + \int_{-1}^0 I_\lambda^-(x,\theta) \langle\Phi_\lambda\rangle(\theta_i \to \theta) \,\mathrm{d}\cos\theta \right]. \qquad (5.134)$$

When $\int_0^1 I_\lambda^+ \,\mathrm{d}\cos\theta = I_\lambda^+$ is used and the scattering is assumed *quasi-isotropic*, we have

$$\frac{1}{2} \frac{\mathrm{d} I_\lambda^+}{\mathrm{d}x} = -(\langle\sigma_{\lambda s}\rangle + \langle\sigma_{\lambda a}\rangle) I_\lambda^+ + \langle\sigma_{\lambda a}\rangle I_{\lambda b}$$

$$+ \langle\sigma_{\lambda s}\rangle \left[(1-B) I_\lambda^+ + B I_\lambda^- \right] \qquad (5.135)$$

or

$$\frac{\mathrm{d} I_\lambda^+}{\mathrm{d}x} = -(2\langle\sigma_{\lambda s}\rangle B + 2\langle\sigma_{\lambda\mathrm{ex}}\rangle) I_\lambda^+ + 2\langle\sigma_{\lambda a}\rangle I_{\lambda b} + 2\langle\sigma_{\lambda s}\rangle B I_\lambda^-, \qquad (5.136)$$

where

$$B = \frac{1}{2} \int_0^1 \int_{-1}^0 \langle\Phi_\lambda\rangle(\theta_i \to \theta) \,\mathrm{d}\cos\theta_i \,\mathrm{d}\cos\theta. \qquad (5.137)$$

By defining $\overline{\sigma}_{\lambda s} = 2\langle\sigma_{\lambda s}\rangle B$ and $\overline{\sigma}_{\lambda a} = 2\langle\sigma_{\lambda a}\rangle$, we have

$$\frac{\mathrm{d} I_\lambda^+}{\mathrm{d}x} = -(\overline{\sigma}_{\lambda s} + \overline{\sigma}_{\lambda a}) I_\lambda^+ + \sigma_{\lambda a} I_{\lambda b} + \overline{\sigma}_{\lambda s} I_\lambda^-, \qquad (5.138)$$

$$-\frac{\mathrm{d} I_\lambda^-}{\mathrm{d}x} = -(\overline{\sigma}_{\lambda s} + \overline{\sigma}_{\lambda a}) I_\lambda^- + \sigma_{\lambda a} I_{\lambda b} + \overline{\sigma}_{\lambda s} I_\lambda^+. \qquad (5.139)$$

(A) NONEMITTING MEDIUM

The *transmittance* for a *nonemitting* bed is found by applying the two-flux approximation to the equation of radiative transfer. Chen and Churchill (1963) use

$$\frac{dI_\lambda^+}{dx} = -(\overline{\sigma}_{\lambda s} + \overline{\sigma}_{\lambda a}) I_\lambda^+ + \overline{\sigma}_s I_\lambda^-, \qquad (5.140)$$

$$-\frac{dI_\lambda^-}{dx} = -(\overline{\sigma}_{\lambda s} + \overline{\sigma}_{\lambda a}) I_\lambda^- + \overline{\sigma}_s I_\lambda^+, \qquad (5.141)$$

with the boundary conditions

$$I_\lambda^+ = I_{\lambda i} \quad \text{at } x = 0, \qquad (5.142)$$

$$I_\lambda^- = 0 \quad (\textit{infinite radiation absorption}) \text{ at } x = L. \qquad (5.143)$$

The solution for the *transmittance* T_r is

$$T_r^{-1} = \frac{I_{\lambda i}}{I_\lambda} = \cosh\left(\overline{\sigma}_{\lambda a}^2 + 2\overline{\sigma}_{\lambda a}\overline{\sigma}_{\lambda s}\right)^{1/2} L + \frac{\overline{\sigma}_{\lambda s} + \overline{\sigma}_{\lambda a}}{\left(\overline{\sigma}_{\lambda a}^2 + 2\overline{\sigma}_{\lambda a}\overline{\sigma}_{\lambda s}\right)^{1/2}}$$

$$\times \sinh\left(\overline{\sigma}_{\lambda a}^2 + 2\overline{\sigma}_{\lambda a}\overline{\sigma}_{\lambda s}\right)^{1/2} L \qquad (5.144)$$

and the *reflectance* R_r is given by

$$R_r = T_r \frac{\overline{\sigma}_{\lambda s}}{\left(\overline{\sigma}_{\lambda a}^2 + 2\overline{\sigma}_{\lambda s}\overline{\sigma}_{\lambda a}\right)^{1/2}} \sinh\left(\overline{\sigma}_{\lambda a}^2 + 2\overline{\sigma}_{\lambda a}\overline{\sigma}_{\lambda s}\right)^{1/2} L. \qquad (5.145)$$

In order to obtain $\langle \Phi_\lambda \rangle (\theta_i \to \theta)$ from the solutions for the phase function available from the Mie theory, namely, $\langle \Phi_\lambda \rangle (\theta_i, \phi_i \to \theta, \phi)$, the integration over ϕ must be taken. Brewster and Tien (1982b) suggest that instead of using a large of number of terms in the expansion in terms of the Legendre polynomial, for large particles where strong peaks are present, the integration should be carried through semianalytically as

$$\langle \Phi_\lambda \rangle (\theta_i \to \theta) = \frac{1}{\pi} \int_0^\pi \langle \Phi_\lambda \rangle \left[\theta_o(\theta_i, \phi_i \to \theta, \phi)\right] d(\phi - \phi_i)$$

$$= \frac{1}{\pi} \int_{\theta_1}^{\theta_2} \frac{\langle \Phi_\lambda \rangle \left[\theta_o(\theta_i, \phi_i \to \theta, \phi)\right] \sin \theta_o \, d\theta_o}{\left[\sin^2 \theta \sin^2 \theta_i - (\cos \theta_o - \cos \theta \cos \theta_i)^2\right]^{1/2}}, \qquad (5.146)$$

where $\cos \theta_o = \cos \theta \cos \theta_i + \sin \theta \sin \theta_i \cos(\phi - \phi_i)$ is used with $\phi_i = 0$, because of the azimuthal symmetry. The bounds are found by setting $\phi - \phi_i$ equal to 0 and π, i.e.,

$$\cos \theta_1 = \cos \theta \cos \theta_i + \sin \theta \sin \theta_i, \qquad (5.147)$$

$$\cos \theta_2 = \cos \theta \cos \theta_i - \sin \theta \sin \theta_i. \qquad (5.148)$$

Then over $\Delta\theta$, $\langle \Phi_\lambda \rangle$ is taken as constant, i.e., over $\Delta\theta$, the integration is done analytically.

(B) EMITTING MEDIUM

When the medium is *emitting*, Tong and Tien (1983) formulate the problem as given here. Under *radiation and local thermal equilibrium* (no other mode of heat transfer, i.e., evacuated with $k_e = 0$) assumptions, we have $\nabla q_{\lambda r} = 0$ or from (5.14)

$$4\pi \langle \sigma_{\lambda a} \rangle I_{\lambda b} = 2\pi \int_{-1}^{1} \langle \sigma_{\lambda a} \rangle I_\lambda \, d\mu = 2\pi \langle \sigma_{\lambda a} \rangle (I_\lambda^+ + I_\lambda^-) \qquad (5.149)$$

or

$$I_{\lambda b} = \frac{1}{2}(I_\lambda^+ + I_\lambda^-). \qquad (5.150)$$

The two-flux approximation with this replacement leads to

$$\frac{dI_\lambda^+}{dx} = -[2\langle \sigma_{\lambda s} \rangle B + \langle \sigma_{\lambda a} \rangle]I_\lambda^+ + [2\langle \sigma_{\lambda s} \rangle B + \langle \sigma_{\lambda a} \rangle]I_\lambda^-, \qquad (5.151)$$

$$\frac{dI_\lambda^-}{dx} = -[2\langle \sigma_{\lambda s} \rangle B + \langle \sigma_{\lambda a} \rangle]I_\lambda^- + [2\langle \sigma_{\lambda s} \rangle B + \langle \sigma_{\lambda a} \rangle]I_\lambda^+. \qquad (5.152)$$

Defining

$$\beta = \frac{2\langle \sigma_{\lambda s} \rangle B + \langle \sigma_{\lambda a} \rangle}{\langle \sigma_{\lambda a} \rangle + \langle \sigma_{\lambda s} \rangle} \qquad (5.153)$$

and

$$\tau = (\langle \sigma_{\lambda s} \rangle + \langle \sigma_{\lambda a} \rangle)x, \qquad (5.154)$$

we have

$$\frac{dI_\lambda^+}{d\tau} = -\beta(I_\lambda^+ - I_\lambda^-), \qquad (5.155)$$

$$\frac{dI_\lambda^-}{d\tau} = -\beta(I_\lambda^- - I_\lambda^+) = -\frac{dI_\lambda^+}{d\tau}. \qquad (5.156)$$

The boundary conditions are (1 is for the *lower* surface and 2 is for the *upper* surface)

$$I_\lambda^+(0) = \varepsilon_{\lambda 1} I_{\lambda b1} + (1 - \varepsilon_{\lambda 1}) I_\lambda^-(0), \qquad (5.157)$$

$$I_\lambda^-(L) = \varepsilon_{\lambda 2} I_{\lambda b2} + (1 - \varepsilon_{\lambda 2}) I_\lambda^+(L). \qquad (5.158)$$

The solution to these is given as

$$I_\lambda^+ - I_\lambda^- = \frac{(I_{\lambda b1} - I_{\lambda b2})}{\dfrac{1}{\varepsilon_{\lambda 1}} + \dfrac{1}{\varepsilon_{\lambda 2}} - 1 + \beta \tau_L}. \qquad (5.159)$$

The net *radiative spectral heat flux* is given by (5.12), i.e.,

$$q_{\lambda r} = 2\pi \int_{-1}^{1} I_\lambda \cos\theta \, d\cos\theta$$

$$= 2\pi \int_{0}^{1} I_\lambda^+ \mu \, d\mu + 2\pi \int_{-1}^{0} I_\lambda^- \mu \, d\mu, \qquad (5.160)$$

and for a quasi-isotropic scattering, using the definition of I_λ^+ and I_λ^-, we have

$$q_{\lambda r} = 2\pi \left(\frac{1}{2}I_\lambda^+ - \frac{1}{2}I_\lambda^-\right) = \pi\left(I_\lambda^+ - I_\lambda^-\right), \quad (5.161)$$

which gives

$$q_{\lambda r} = \frac{(E_{\lambda b1} - E_{\lambda b2})}{\dfrac{1}{\epsilon_{\lambda 1}} + \dfrac{1}{\epsilon_{\lambda 2}} - 1 + \beta\tau_L}. \quad (5.162)$$

5.7.2 Diffusion (Differential) Approximation

When $\tau_L \gg 1$, the medium is called *optically thick* and the approximate form of the radiative heat flux q_r, given below, is called the *Rosseland or diffusion (local gradient dependence only) approximation*. The derivation is based on expansion of I_λ with respect to τ_L^{-1} as the expansion parameter and is given in Deissler (1964), Ozisik (1985, 316–319), and Siegel and Howell (1981, 498–501). The results for the spectral local radiative heat flux in the plane-parallel geometry is

$$q_{\lambda r} = -\frac{4\pi}{3(\langle\sigma_{\lambda s}\rangle + \langle\sigma_{\lambda a}\rangle)}\frac{dI_{\lambda b}}{dx} = -\frac{4}{3(\langle\sigma_{\lambda s}\rangle + \langle\sigma_{\lambda a}\rangle)}\frac{dE_{\lambda b}}{dx}$$

for $\tau_L \gg 1$ and x away from the boundaries. (5.163)

Because of the diffusion (local gradient) approximation, the surface radiosity/irradiation must be related to the gradient of the medium emissive power. This is done for surface 1 (at $T_{s\,1}$) at $x = 0$ (and surface 2 at $x = L$) as

$$(E_{\lambda b1} - E_{\lambda bs1}) = \left(\frac{1}{\epsilon_{\lambda s 1}} - \frac{1}{2}\right)q_{\lambda r 1}$$

$$+ \frac{1}{2(\langle\sigma_{\lambda s}\rangle + \langle\sigma_{\lambda a}\rangle)^2}\left.\frac{d^2 E_{\lambda b}}{dx^2}\right|_1. \quad (5.164)$$

This is the so-called *jump boundary condition*. When radiative equilibrium does not exist, the $dq_{\lambda r}/dx \neq 0$ and $d^2 E_{\lambda b}/dx^2 \neq 0$. A similar equation can be written for the boundary condition at surface 2. The second order term on the right-hand side is of the order of $1/\tau_L^2$ and is generally negligible (Sparrow and Cess, 1978, p. 224).

Anisotropic scattering has been included in the optically thick approximation by Modest and Azad (1980). Since for $\tau_L \gg 1$, the intensity deviates slightly from an isotropic distribution, they arrive at

$$q_{\lambda r} = -\frac{4}{3\left[\langle\sigma_{\lambda a}\rangle + \left(1 - \frac{A_1}{3}\right)\langle\sigma_{\lambda s}\rangle\right]} \frac{dE_{\lambda b}}{dx}. \quad (5.165)$$

For *semitransparent* boundaries, Azad (1985) gives modified boundary conditions that improve the accuracy of the differential approximation for cases where τ_L is not large.

When (5.163) is integrated over all wavelengths, we have

$$q_r = -\frac{4}{3} \frac{\sigma}{\overline{\sigma}_{ex\,R}} \frac{dT^4}{dx}, \quad (5.166)$$

where the *Rosseland mean extinction coefficient* $\overline{\sigma}_{ex\,R}$ is

$$\frac{1}{\overline{\sigma}_{ex\,R}} = \frac{\displaystyle\int_0^\infty \frac{1}{(\langle\sigma_{\lambda s}\rangle + \langle\sigma_{\lambda a}\rangle)} \frac{dE_{\lambda b}}{dE_b} d\lambda}{\displaystyle\int_0^\infty \frac{dE_{\lambda b}}{dE_b} d\lambda}$$

$$= \int_0^\infty \frac{1}{(\langle\sigma_{\lambda s}\rangle + \langle\sigma_{\lambda a}\rangle)} \frac{dE_{\lambda b}}{dE_b} d\lambda, \quad (5.167)$$

where $E_b = n_f^2 \sigma T^4$ and

$$E_{\lambda b}(E_b) = \frac{2\pi C_1 n_f^2}{\lambda^5 \exp\left[C_2/\lambda (\sigma/E_b)^{1/4}\right] - 1}. \quad (5.168)$$

5.7.3 SPHERICAL HARMONICS-MOMENT (P-N) APPROXIMATION

This method has been applied and described by Jeans (1917), Cheng (1969), Bayazitoğlu and Higenyi (1979), Ratzel and Howell (1983), Mengüc and Viskanta (1985b), and Tong and Swathi (1987). The steps are as follows.

- Directional intensity $I(\tau, \mu)$ is expressed in terms of the spherical harmonics, which are related to the associated Legendre functions P_i^j for angular variations, and the spatial dependence is separated and expressed in terms of position-dependent coefficients η_i^j.

- The equation of radiative transfer is multiplied by *powers of the directional cosines* and integrated over the 4π solid angle. The *integrated weighted directional intensity* (or *moments*) of the directional intensity has some physical significance, i.e.,

$$\text{zeroth moment: } I_0(\mathbf{x}) = \int_{4\pi} I(\mathbf{x}, \Omega) \, d\Omega$$

is the average radiation energy intensity,

5.7 Solution Methods for Equation of Radiative Transfer

first moment: $I_k(\mathbf{x}) = \int_{4\pi} (\mathbf{s}_k \cdot \mathbf{s}_x) I(\mathbf{x}, \Omega) \, d\Omega$

is the radiative heat flux along the k-axis,

where $\mathbf{s}_k$ is the *unit vector* in the k-direction and $\mathbf{s}_x$ is the position unit vector. *Higher*-order moments use products of the direction cosines as the weight function.

- Instead of solving for the unknown position coefficients η_i^j, the moments (which are related to η_i^j) are treated as unknowns. The outcome is a set of coupled differential equation for the moments.

- Since the series representation of the local directional intensity is truncated and only the leading terms are used, a *closure condition* must be imposed. This requires approximation of the moments that are one order higher than those maintained. These higher moments are represented by the lower-order moments.

- Generally *constant properties, gray media*, and *isotropic scattering* is assumed. However, these are *not* the restrictions of the method.

- The treatment of two-dimensional enclosure is available in Ratzel and Howell, and the three-dimensional formulation is given by Mengüc and Viskanta. In the following, only the one-dimensional results are given for the sake of clarity.

(A) Expansion

The directional intensity is expanded in terms of the spherical harmonics as

$$I(\mathbf{x}, \mu) = \sum_{i=1}^{\infty} \sum_{j=-i}^{i} \left[\frac{2i+1}{4\pi} \frac{(i-j)!}{(i+j)!} \right]^{1/2} e^{(-1)^{1/2} j \phi} \eta_i^j(\mathbf{x}) P_i^j(\mu), \quad (5.169)$$

where $\eta_i^j(\mathbf{x})$ is the position dependent coefficient and P_i^j is the associated Legendre polynomial of the first kind. Expansion in terms of the ordinary Legendre polynomials is discussed in Ozisik (1985, p. 334) and applied by Tong and Swathi. The properties of P_i^j are

$$P_i^j(\mu) = \frac{(1-\mu^2)^{j/2}}{2^i i!} \frac{d^{i+j}}{d\mu^{i+j}} (\mu^2 - 1)^i, \quad (5.170)$$

$$P_i^{-j} = P_i^j (-1)^j \frac{(i-j)!}{(i+j)!}. \quad (5.171)$$

From (5.170), we can see, for example, that $P_0^0 = P_0 = 1$, $P_3^0 = P_3 = (5\mu^3 - 3\mu)/2$, $P_3^1 = (3/2)(\mu^2 - 1)\sin\theta_o$, and $P_3^3 = 15\sin^3\theta_o$. Also, the *orthogonality property* is

$$\int_{-1}^{1} P_i(\mu) P_j(\mu)\, d\mu = \begin{cases} 0 & i \neq j \\ \dfrac{2}{2i+1} & i = j, \end{cases} \quad (5.172)$$

and the *recurrence* of the Legendre polynomial is of the form

$$\mu P_i(\mu) = \frac{i P_{i-1}(\mu) + (i+1) P_{i+1}(\mu)}{2i+1}. \quad (5.173)$$

The expansion in (5.169) is carried only over a finite number of terms. In the P-3 *approximation* $i \leq 3$ and in the P-1 *approximation* $i \leq 1$. The even values are *not* used because of the difficulty in obtaining the *approximate* boundary conditions.

(B) Moments for Plane-Parallel Geometry and P-3 Approximation

The equation of radiative transfer with I expanded is multiplied by directional cosines. For the *one-dimensional* case, we have

$$\int_{-1}^{1} \mu^i \left[\mu \frac{dI}{d\tau} + I - (1 - \omega_a) I_b - \frac{\omega_a}{4\pi} I_0 \right] d\mu = 0 \quad \text{for} \quad i = 0, 1, 2, \ldots. \quad (5.174)$$

Now I is expanded as in (5.169) and for the P-3 approximation. For plane-parallel (azimuthally symmetric) geometry, I is given in terms of the moments (instead of η_i^j) as

$$I(\tau, \theta) = \frac{1}{4\pi} \Big[I_0 + 3 I_1 \cos\theta + \frac{5}{4}(3 I_2 - I_0)(3\cos^2\theta - 1) + \frac{7}{4}(5 I_3 - 3 I_1)(5\cos^3\theta - 3\cos\theta) \Big], \quad (5.175)$$

where

$$I_i(\tau) = \frac{1}{2\pi} \int_{-1}^{1} \mu^i I(\tau, \mu)\, d\mu. \quad (5.176)$$

When the azimuthal symmetry along with identities and the closure conditions are used (Ratzel and Howell, 1983; Ratzel, 1981) and (5.175) is inserted in (5.174), the emerging *moment differential equations* are

$$\frac{dI_1}{d\tau} = (1 - \omega_a)\left(4\sigma T^4 - I_0\right), \quad (5.177)$$

$$\frac{dI_2}{d\tau} = -I_1, \quad (5.178)$$

$$\frac{dI_3}{d\tau} = -I_2 + \frac{(1-\omega_a)}{3}\left(4\sigma T^4 - I_0\right) + \frac{I_0}{3}, \quad (5.179)$$

$$\frac{dI_4}{d\tau} = -\frac{1}{10}\frac{dI_0}{d\tau} - \frac{7}{6} I_3, \quad (5.180)$$

i.e., four equations for the four moments. The *energy equation* provides the relationship between T and I_1.

(C) BOUNDARY CONDITIONS FOR P-3 APPROXIMATION

Four boundary conditions are needed for the solution of the four differential moment equations. The *radiosity* (outgoing intensity) of the bounding surface 1 at $\tau = 0$ is

$$I(0,\mu) = \varepsilon_1 I_b(T_1) + \rho_1^s I(0,-\mu)$$
$$+ 2\int_0^1 I(0,-\mu')\mu'\,d\mu' = f_1(\mu) \quad \text{for } \mu > 0, \quad (5.181)$$

where $\varepsilon_1 + \rho_1^s + \rho_1^d = 1$. This boundary condition is a combination of the boundary conditions given in Sections 5.6.2 and 5.6.3. The *Marshak* representation of the boundary conditions is (Ozisik, 1985, p. 341)

$$\int_0^1 I(0,\mu)\mu^{2i-1}\,d\mu = \int_0^1 f_1(\mu)\mu^{2i-1}\,d\mu \quad \text{for } \mu > 0, \quad (5.182)$$

where $i = 1, 2, 3, \ldots, N+1$ and N is the order of approximation.

For the P-3 approximation, the boundary conditions found by using (5.175) in (5.182) are

$$3I_0 \pm 16(1+2E_j)I_1 + 15I_2 - 32\sigma T^4 = 0, \quad j = 1, 2, \quad (5.183)$$

$$-(2+5R_j)I_0 \pm 16R_j I_1 + 15(2+R_j)I_2$$
$$\pm 32(1+2E_j - R_j)I_3 - 32\sigma T^4 = 0 \quad j = 1, 2, \quad (5.184)$$

where

$$E_j = (\rho_j^s + \rho_j^d)(1 - \rho_j^s - \rho_j^d)^{-1}, \quad (5.185)$$
$$R_j = \rho_j^d(1 - \rho_j^s - \rho_j^d)^{-1}, \quad (5.186)$$

and the *positive signs* are for surface 1 at $\tau = 0$. Extension to convective boundary conditions are given by Kaviany (1985).

(D) MOMENTS AND BOUNDARY CONDITIONS FOR P-1 APPROXIMATION

For the P-1 approximation only the *zeroth* and *first* moment are used. For *isotropic* scattering the differential equations for moments are

$$\frac{dI_1}{d\tau} = (1-\omega_a)(4\sigma T^4 - I_0), \quad (5.187)$$

$$\frac{dI_0}{d\tau} = -3I_1. \quad (5.188)$$

When account is made of a *forward* peak in the scattering, the last equation becomes

$$\frac{dI_0}{d\tau} = -3\left(1 - \omega_a \frac{A_1}{3}\right) I_1. \tag{5.189}$$

The boundary conditions are

$$I_0 \pm 2\left(1 + 2\frac{1-\varepsilon_j}{\varepsilon_j}\right) I_1 - 4\sigma T_j^4 = 0 \quad j = 1, 2. \tag{5.190}$$

For *isotropic* scattering when (5.188) is used, (5.190) becomes

$$\pm \left.\frac{dI_0}{d\tau}\right|_j = \frac{3}{2\left(1 + 2\frac{1-\varepsilon_j}{\varepsilon_j}\right)} \left(I_0 - 4\sigma T_j^4\right). \tag{5.191}$$

5.7.4 DISCRETE-ORDINATES (S-N) APPROXIMATION

Following the approximation of Schuster-Schwarzchild in division of the radiation field into an inward and an outward stream (two-flux approximation), Chandrasekhar (1960, Chapter 2) *increased* the number of these *discrete-streams* or *discrete-ordinates*. The integral in the equation of radiative transfer is approximated by division into increments, e.g., the integrals -1 and $+1$ for μ are divided into $2N$ increments. Since the process of determining the area (integration) is called *quadrature*, the various numerical approximations of the integrals have been referred to as quadratures, e.g., Gaussian quadrature. The result of the approximation is to reduce the integro-differential equation to a set of *coupled ordinary linear* differential equations, which is then either reduced to *algebraic* equations by direct integration (e.g., Hottel et al., 1968; Rish and Roux, 1987) or solved *numerically* (e.g., Carlson and Lathrop, 1968; Truelove, 1987; Fiveland, 1988; Jamaluddin and Smith, 1988; Kumar et al., 1990). The in-scattering term (the integral) is *approximated* by a quadrature, where μ_i is the *quadrature points* between -1 and $+1$ corresponding to a $2N$-order quadrature, and $\Delta\mu_i$ (solid angle increment) is the corresponding quadrature *weight*. Then the one-dimensional radiative transfer equation for intensity at x and in the direction μ_i becomes

$$\mu_i \frac{dI_{\lambda i}(x)}{dx} = -\langle\sigma_{\lambda\,\text{ex}}\rangle I_{\lambda i}(x) + \langle\sigma_{\lambda a}\rangle I_{\lambda b}(x)$$

$$+ \frac{\langle\sigma_{\lambda s}\rangle}{2} \sum_{j=-M, j\neq 0}^{M} \Delta\mu_j I_{\lambda j}(x) \langle\Phi_\lambda\rangle (\theta_i \to \theta_j)$$

$$\text{for } i = -M, -M+1, \ldots, M, \ i \neq 0 \tag{5.192}$$

$$\sum_{j=-M, j\neq 0}^{M} \Delta\mu_j = 2, \tag{5.193}$$

where

$$\langle \Phi_\lambda \rangle (\theta_i \to \theta_j) = \frac{1}{\pi} \int_0^\pi \langle \Phi_\lambda \rangle [\theta_o(\theta_i, \phi_i \to \theta_j, \phi_j)] \, d(\phi_j - \phi_i). \quad (5.194)$$

The boundary conditions are

at $x = 0$, $\quad I_{\lambda i} = \varepsilon_\lambda I_{\lambda b} + \rho_{\lambda s} I_{\lambda -i}$

$$+ 2\rho_{\lambda d} \sum_{j=-1}^{-M} \Delta\mu_j I_{\lambda j} \mu_j \quad i = 1, \ldots, M \quad (5.195)$$

at $x = L$, $\quad I_{\lambda i} = \varepsilon_\lambda I_{\lambda b} + \rho_{\lambda s} I_{\lambda -i}$

$$+ 2\rho_{\lambda d} \sum_{j=1}^{M} \Delta\mu_j I_{\lambda j} \mu_j \quad i = -1, \ldots, -M, \quad (5.196)$$

where $I_i(x) = I(x, \mu_i)$ and $i = 0$ (corresponding to the lateral boundaries) has been avoided because of the one-dimensional geometry assumed (for two- and three-dimensional geometries, $i = 0$ is included). Equation (5.192) is called the *discrete-ordinates equation*.

(A) SELECTION OF QUADRATURE

The interval $(-1, 1)$ is divided according to zeros of the Legendre polynomial $P_M(\mu)$, and the integral is represented as (Chandrasekhar, 1960)

$$\int_{-1}^1 f(\mu) \, d\mu \simeq \sum_{j=-M, j\neq 0}^{M} \Delta\mu_j f(\mu_j). \quad (5.197)$$

The weights $\Delta\mu_j$ are given by

$$\Delta\mu_j = \frac{1}{P'_M(\mu_j)} \int_{-1}^1 \frac{P_M(\mu)}{\mu - \mu_j} \, d\mu, \quad (5.198)$$

where $P'_M(\mu_j) = (dP_M/d\mu)_{\mu=\mu_j}$.

A quadrature of accuracy comparable to the Gauss quadrature is based on the zeros of the *Laguerre polynomials*.

$$\int_0^\infty e^{-x} f(x) \, dx = \sum_{j=-M, j\neq 0}^{M} \Delta\mu_j f(x_j), \quad (5.199)$$

where the x_j's ($j = -M, \ldots, -1, 1, \ldots, M$) are the zeros of $L_M(x)$ and

$$\Delta\mu_j = \frac{1}{L'_M(x_j)} \int_0^\infty \frac{e^{-x} L_M(x)}{x - x_j} \, dx = \frac{1}{x_j} [L'_M(x_j)]^{-2}. \quad (5.200)$$

A simple scheme has been suggested by Fiveland (1987), which consistently gives more accurate heat transfer results than solutions found with the Gauss quadrature. The scheme is based on satisfying the half range moments of intensity. Considering symmetric ordinate points, the following relationship must be satisfied

$$\int_0^1 \mu^i \, d\mu = \sum_{j=1}^{M} \mu_j^i \, \Delta\mu_j \quad \text{for } \mu > 0 \qquad (5.201)$$

and similarly for $\mu < 0$, where $i = 0, 1, 2, \ldots, M$.

Choosing the special case of equal weights,

$$i = 0 \quad \text{gives} \quad \sum_{j=1}^{M} \Delta\mu_j = 1, \qquad (5.202)$$

$$i = 1 \quad \text{gives} \quad \sum_{j=1}^{M} \mu_j = \frac{M}{2}. \qquad (5.203)$$

In general

$$\sum_{j=1}^{M} \mu_j^i = \frac{M}{i+1}, \quad i = 1, \ldots, M. \qquad (5.204)$$

This gives M equations that can be solved for $\mu_j (j = 1, \ldots, M)$. This method is *preferable* to Gauss quadrature because it satisfies the half-range moment,

$$\int_0^1 I_\lambda(\mu) \, d\mu \quad \text{and} \quad \int_{-1}^0 I_\lambda(\mu) \, d\mu, \qquad (5.205)$$

which is important for better prediction of half-range heat fluxes. Gauss quadrature does *not* satisfy half-range moments specially for lower-order quadratures. However, it places *more* points in a *forward* direction so that highly forward scattering phase functions are more accurately represented.

(B) Three-Dimensional Formulation

The formulation described above can be extended to three dimensions (Fiveland, 1988). The *three-dimensional* Cartesian coordinate radiative transfer can be written as

$$\mu_i \frac{\partial I_{\lambda i}}{\partial x} + \xi_i \frac{\partial I_{\lambda i}}{\partial y} + \gamma_i \frac{\partial I_{\lambda i}}{\partial z}$$

$$= -\langle \sigma_{\lambda \, ex} \rangle I_{\lambda i} + \langle \sigma_{\lambda \, a} \rangle I_{\lambda b} + \frac{\langle \sigma_{\lambda \, s} \rangle}{4\pi} \sum_j \Delta\Omega_j I_{\lambda j} \langle \Phi_\lambda \rangle (\theta_o), \qquad (5.206)$$

where μ_i, ξ_i, and γ_i are the *direction cosines* of the ith direction. Using spherical geometry, θ_o can be written in terms of μ, ξ, and γ as

$$\cos\theta_o = \mu_i\mu_j + \xi_i\xi_j + \gamma_i\gamma_j. \qquad (5.207)$$

The choice of *quadrature directions* and *weights* in three dimensions is discussed by Carlson and Lathrop (1968). The basic condition to be satisfied is

$$\sum_j \Delta\Omega_j = 4\pi, \qquad (5.208)$$

where $\Delta\Omega_j$ represents the solid angle associated with the jth direction.

To *preserve* computational invariance, the direction mesh must be made invariant under geometric transformation. Thus, μ_i, ξ_i, and γ_i have the same set of values and are symmetrically located with respect to the origin. The exact directions can be calculated by moment matching, following a procedure similar to that for the one-dimensional formulation.

(C) SOLUTION OF DISCRETE-ORDINATE EQUATIONS

The *finite-difference* solution of the discrete-ordinate equations has been discussed by Carlson and Lathrop and Fiveland (1988). The general scheme consists of evaluating the *intensity* at the *cell center* by relating it to the *intensities* at the *cell faces*. The *source* term comprising in-scattering from other directions and the emission is calculated using this intensity. The intensity at the opposite face is calculated *iteratively*.

For one-dimensional (with no emission) radiation, the finite-difference approximations of the equation of radiative transfer leads to the following equation

$$I_{\lambda j+1}^i = \frac{\left[\left(\dfrac{\mu_i}{\Delta x} - \dfrac{\langle\sigma_{\lambda\,\text{ex}}\rangle}{2}\right) I_{\lambda j+1}^i + \dfrac{\langle\sigma_{\lambda s}\rangle}{2} \sum_{k=-M, k\neq 0}^{M} \Delta\mu_k \langle\Phi_\lambda\rangle (\theta_i \to \theta_j) I_{\lambda j+\frac{1}{2}}^k\right]}{\left(\dfrac{\mu_i}{\Delta x} + \dfrac{\langle\sigma_{\lambda\,\text{ex}}\rangle}{2}\right)}. \qquad (5.209)$$

In cases where emission is important, the condition of *radiative equilibrium* gives (Kumar et al., 1988)

$$I_{b\,j+\frac{1}{2}} = \frac{1}{2}\int_{-1}^{1} I_{j+\frac{1}{2}}^k\, d\mu = \frac{1}{2}\sum_{k=-M, k\neq 0}^{M}\Delta\mu_k I_{j+\frac{1}{2}}^k. \qquad (5.210)$$

The difference equation becomes

$$I^i_{\lambda,j+1} = \frac{\left[\left(\dfrac{\mu_i}{\Delta x} - \dfrac{\langle \sigma_{\lambda\,ex}\rangle}{2}\right)I^i_{\lambda\,j+1} + \dfrac{\langle \sigma_{\lambda\,a}\rangle}{2}\sum\limits_{k=-M,k\neq 0}^{M}\Delta\mu_k I^k_{\lambda\,j+\frac{1}{2}} + \dfrac{\langle \sigma_{\lambda\,s}\rangle}{2}\sum\limits_{k=-M,k\neq 0}^{M}\Delta\mu_k \langle\Phi_\lambda\rangle(\theta_i \to \theta_j)I^k_{\lambda\,j+\frac{1}{2}}\right]}{\left(\dfrac{\mu_i}{\Delta x} + \dfrac{\langle \sigma_{\lambda\,ex}\rangle}{2}\right)}.$$

(5.211)

Iteration begins from $x = 0$ ($j = 1$) for $\mu_i > 0$ and proceeds in the direction of actual irradiation. For $\mu_i < 0$, iteration starts from $x = L$ ($j = n$) and proceeds toward $x = 0$.

For the *three-dimensional* finite-difference equations, we have (Fiveland, 1988)

$$\mu_i A(I^i_{\lambda\,e} - I^i_{\lambda\,w}) + \xi_i B(I^i_{\lambda\,n} - I^i_{\lambda\,s}) + \gamma_i C(I^i_{\lambda\,f} - I^i_{\lambda\,r})$$
$$= -\langle\sigma_{\lambda\,ex}\rangle I_{\lambda\,p} + \langle\sigma_{\lambda\,a}\rangle V_b I_{\lambda\,bp}$$
$$+ V_p \frac{\langle\sigma_{\lambda\,s}\rangle}{4\pi}\sum_j \Delta\mu_j I^j_{\lambda\,p}\langle\Phi_\lambda\rangle(\theta_o), \qquad (5.212)$$

where p is located at the *center* of the cell (control volume); e, w, n, s, f, and r stands for *east*, *west*, *north*, *south*, *front*, and *rear*, respectively; and V_p is the cell volume. The unknowns at the east, north, and front cell *faces* can be *eliminated* by using the relationship

$$I^i_{\lambda\,p} = \alpha I^i_{\lambda\,e} + (1-\alpha)I^i_{\lambda\,w} = \alpha I^i_{\lambda\,n} + (1-\alpha)I^i_{\lambda\,s} = \alpha I^i_{\lambda\,f} + (1-\alpha)I^i_{\lambda\,r}, \quad (5.213)$$

where $\alpha = 1/2$ gives the Carlson and Lathrop *diamond difference* scheme. Using (5.212) and (5.213),

$$I^i_{\lambda\,p} = \frac{\mu_i A I^i_{\lambda\,w} + \xi_i B I^i_{\lambda\,s} + \gamma_i C I^i_{\lambda\,r} + \alpha(S_1 + S_2)V_p}{\mu_i A + \xi_i B + \gamma_i C + \alpha\langle\sigma_{\lambda\,ex}\rangle V_p}, \qquad (5.214)$$

where $A = \Delta y\,\Delta z$, $B = \Delta x\,\Delta y$, and $C = \Delta x\,\Delta y$ and

$$S_1 = \langle\sigma_{\lambda\,a}\rangle I_{\lambda\,bp}, \qquad (5.215)$$

$$S_2 = \frac{\langle\sigma_{\lambda\,s}\rangle}{4\pi}\sum_j \Delta\Omega_j I^j_{\lambda\,p}\langle\Phi_\lambda\rangle(\theta_o). \qquad (5.216)$$

These equations are solved by *point-by-point* iteration with iteration proceeding in the direction of irradiation.

Application of this method, in determination of the effective radiative properties (i.e., the inverse solution), has been discussed by Hendricks (1993).

5.7.5 FINITE-VOLUME METHOD

The finite-volume method is introduced by Raithby and Chui (1990) and has been integrated into computations involving other modes by Chai et al. (1994). In the finite-volume method, the space is subdivided into discrete nonoverlapping volume with a node p located centrally within it. Since direction of intensity is also an independent variable, the solid angle 4π is divided into N_ω discrete, nonoverlapping solid angle segments. The directional nodal intensity I_p^i is obtained by integrating (5.11) over a finite volume and introducing approximations. The intensity is assumed *constant* over the finite volume and the solid angle. The directional intensities arriving into the finite volume are determined by back integration up to a location where the directional intensity is found reasonably accurately by interpolation between nodes. This method is accurate and can also apply to *collimated irradiation* and finite volumes with *nonorthogonal boundaries*.

5.8 Scaling (Similarity) in Radiative Heat Transfer

5.8.1 SIMILARITY BETWEEN PHASE FUNCTIONS

Existence of *similarity* requires *invariance* of the equation of radiative transfer under *simultaneous* variation of a set of variables. For the similarity or scaling law to exist, the variables must be related through a *similarity transformation function* or *scaling group*. In this section, these transformations between the *optical thickness* and the *phase function* are examined. The objective is to *reduce* the *complexity* of the phase function (and the associated representation through series expansion) by *scaling* the *optical thickness*. A strongly anisotropic phase function has to be represented by a large number of terms of a Legendre polynomial. This significantly increases the computation required to solve the equation of transfer. If forward diffraction peak of the phase function is represented by a *delta function*, the rest of the phase function is *weakly anisotropic* and can be represented by a *smaller* number of terms. Obviously, the scaling relations are most useful for particles of *intermediate size*. The phase function for a small particle is not strongly anisotropic and does not require a large number of terms in its polynomial representation. On the other extreme, the *diffraction* for a *very large* particle is so focused in the forward direction that it can be considered to be a *part* of *transmitted* energy and *omitted* from the phase function.

The historical development in this field is given by McKellar and Box (1981); the following is the review of their formulation of the scaling laws.

Considering radiation in a plane-parallel geometry with *no emission* integrated over all wavelengths, we have

$$\mu\frac{\partial I(\tau,\Omega)}{\partial \tau} = -I(\tau,\Omega) + \omega_a \frac{1}{4\pi}\int_{4\pi} \langle\Phi\rangle(\cos\theta_o)I(\tau,\Omega')\,d\Omega'$$

$$= \int_{4\pi} G(\Omega,\Omega')I(\tau,\Omega)\,d\Omega', \qquad (5.217)$$

where $\omega_a = \sigma_s/(\sigma_s + \sigma_a)$ is the *scattering albedo*, the kernel is

$$G = \omega_a \frac{1}{4\pi}\langle\Phi\rangle - \delta(\Omega,\Omega'), \qquad (5.218)$$

and δ is the *Dirac delta* function. Now introduce a new (or *transformed*) *optical thickness* $\hat{\tau}$ and a *transformed kernel* $\hat{G}$ with *linear* transformations (using a constant β), and we have

$$\hat{\tau} = \frac{1}{\beta}\tau, \qquad (5.219)$$

$$\hat{G} = \beta G. \qquad (5.220)$$

Then by substituting for τ and G in (5.217), we find that $I(\tau,\Omega)$ is also the solution to

$$\mu\frac{\partial I(\hat{\tau},\Omega)}{\partial \hat{\tau}} = \int_{4\pi}\hat{G}(\Omega,\Omega')I(\hat{\tau},\Omega')\,d\Omega$$

$$= -I(\hat{\tau},\Omega) + \hat{\omega}_a\frac{1}{4\pi}\int_{4\pi}\langle\Phi\rangle(\cos\theta_o)I(\hat{\tau},\Omega')\,d\Omega', \quad (5.221)$$

where the *transformed albedo* $\hat{\omega}_a$ is obtained by assuming that in transforming τ the *absorption coefficient* is kept *constant*, i.e.,

$$(1-\omega_a) = \frac{1}{\beta}(1-\hat{\omega}_a) \quad \text{or} \quad \hat{\omega}_a = \beta\omega_a + 1 - \beta. \qquad (5.222)$$

For the *transformed phase function*, we have

$$\hat{\omega}\langle\hat{\Phi}\rangle = \beta\omega_a\langle\Phi\rangle + 4\pi(1-\beta)\delta(\Omega-\Omega'). \qquad (5.223)$$

These relations show how a *change* in the *optical thickness* through β (5.219) must accompany *changes* in the *albedo* (5.222) and in the *phase function* (5.223). For $\omega_a = 1$ (*conserved scattering*), we have $\hat{\omega}_a = 1$, as expected. Note that for $\beta = 1$, the phase function transformation results in *isolation* of the forward peak, which also appears in many approximations. Since these approximations use the Legendre polynomial P_i for expansion of the phase function (Chu and Churchill, 1955) and for the solution, we can also expand $\langle\Phi\rangle$, $\langle\hat{\Phi}\rangle$, and δ, i.e.,

$$\langle\Phi\rangle(\cos\theta_o) = \sum_{i=0}^{\infty}(2i+1)A_i P_i(\cos\theta_o), \qquad (5.224)$$

5.8 Scaling (Similarity) in Radiative Heat Transfer

$$\langle \widehat{\Phi} \rangle (\cos\theta_o) = \sum_{i=0}^{\infty}(2i+1)\widehat{A}_i P_i(\cos\theta_o), \tag{5.225}$$

$$\begin{aligned}\delta(\Omega,\Omega') &\equiv \delta(\mu-\mu')\delta(\phi-\phi') \\ &= \frac{1}{2\pi}\delta(1-\cos\theta_o) \\ &= \frac{1}{4\pi}\sum_{i=0}^{\infty}(2i+1)P_i(\cos\theta_o), \end{aligned} \tag{5.226}$$

where

$$A_i = \frac{1}{2}\int_{-1}^{1}\langle \Phi \rangle (\cos\theta_o) P_i(\cos\theta_o)\, d\cos\theta_o. \tag{5.227}$$

The relation between A_i and $\widehat{A}_i$ is found by substituting the expansions (5.224)–(5.226) into (5.221), giving

$$1 - \widehat{\omega}_a \widehat{A}_i = \beta(1-\omega_a A_i) \tag{5.228}$$

or

$$\widehat{\omega}_a \widehat{A}_i = \beta\omega_a A_i + 1 - \beta. \tag{5.229}$$

For *isotropic* scattering where $A_o = 1$ and $A_i = 0$, we have $\widehat{A}_1 = 1$.

The expansion of the phase function in the Legendre polynomial has been *truncated* for convenience of calculation to various extents. These truncations are *justified* as long as the proper *adjustment* is made to the *optical thickness*, according to the preceding *similarity* arguments. These approximations are also reviewed by McKellar and Box and discussed later.

(A) Delta-M Approximation

The *delta-M* (or δ-M, where M is related to the number of terms in the series expansion) is an approximation of the highest order compared to the rest and is recommended by Wiscombe (1977). M is the *order of approximation*. The phase function is written as

$$\langle \Phi_{\delta-M} \rangle (\cos\theta_o) = 2f\delta(1-\cos\theta_o)$$
$$+ (1-f)\sum_{i=0}^{2M-1}(2i+1)\widehat{A}_{\delta-M_i}P_i(\cos\theta_o), \tag{5.230}$$

where

$$A_i = f + (1-f)A_{\delta-M_i}, \tag{5.231}$$

and as $M \to \infty$, then the δ-M representation and the original expansion (5.224) are identical. However, when a small number of terms are used, the coefficients for $i > 2M-1$ are set to zero ($\widehat{A}_{\delta-M_i} = 0$ for $i > 2M-1$), and deviations from the original expansion occur. Using the approximate phase function in the radiative transfer equation, we find

$$\widehat{\tau} = (1 - \omega_a f)\tau, \qquad (5.232)$$

$$\widehat{\omega}_a = \frac{\omega_a(1-f)}{1-\omega_a f}, \qquad (5.233)$$

or

$$\beta = \frac{1}{1-\omega_a f} \quad \text{or} \quad f = \frac{\beta - 1}{\beta \omega_a}. \qquad (5.234)$$

Then, in using the δ-M approximation, the optical thickness and the albedo must be modified according to (5.232)–(5.233).

(B) Delta-Eddington Approximation

The *standard Eddington* approximation is

$$\langle \widehat{\Phi}_E \rangle = 1 + 3A_1 \cos\theta_o, \qquad (5.235)$$

which is an approximation for slightly anisotropic scattering. The δ-Eddington (δ-E) approximation (Joseph et al., 1976) is

$$\langle \widehat{\Phi}_{\delta-E} \rangle (\cos\theta_o) = 2f\delta(1 - \cos\theta_o) + (1-f)(1 + 3\widehat{A}_1 \cos\theta_o), \qquad (5.236)$$

where, in order to satisfy symmetry,

$$\widehat{A}_1 = \frac{A_1 - f}{1 - f} \qquad (5.237)$$

and

$$f = A_2. \qquad (5.238)$$

Also in a δ-E approximation it is implied that $A_i = A_2$ for $i \geq 2$. This makes the approximation suitable for strongly forward scattering and usable along with the two-flux approximation. Then τ and ω_a are scaled according to (5.232)–(5.233)

(C) Transport Approximation

This consists of the sum of a forward delta and an anisotropic scattering (Davison, 1957), i.e.,

$$\langle \Phi_T \rangle (\cos\theta_o) = 2f\delta(1 - \cos\theta_o) + 1 - f, \qquad (5.239)$$

for correct symmetry

$$f = A_1 = g, \qquad (5.240)$$

where g is the *symmetry factor of the scattering phase function*. Note that the result is $A_i = A_1$ for $i \geq 1$. This approximation is *less* accurate than δ-E but more accurate than the standard Eddington approximation. Then τ and ω_a are scaled according to (5.232)–(5.233).

(D) DELTA-HENYEY-GREENSTEIN APPROXIMATION

The original Henyey-Greenstein approximation has only one free parameter g and is given as

$$\Phi_{H-G}(\cos\theta_o) = \frac{1-g^2}{(1+g^2-2\cos\theta_o)^{3/2}}. \tag{5.241}$$

The δ-(H-G) approximation is

$$\langle \Phi_{\delta-(H-G)} \rangle (\cos\theta_o) = 2f\delta(1-\cos\theta_o) + (1-f)\Phi_{H-G}(\cos\theta_o). \tag{5.242}$$

When f and g are chosen to match the first two moments, we have

$$f + (1-f)g = A_1, \tag{5.243}$$

$$f + (1-f)g^2 = A_2, \tag{5.244}$$

or

$$g = \frac{A_1 - A_2}{1 - A_1}, \tag{5.245}$$

and

$$f = \frac{A_2 - A_1^2}{1 - 2A_1 + A_2}. \tag{5.246}$$

Then τ and ω_a are scaled according to (5.232)–(5.233).

These approximate phase functions can be used on the spectral level, and when used as such, the similarity (or scale) conditions should also be used on the spectral bases. Table 5.4 summarizes some of the results. For large-size particles, the approximations have been examined by Crosbie and Davidson (1985) and compared with the exact results of the Mie theory. For ceramic foams, a discussion is given by Hendricks (1993).

5.8.2 SIMILARITY BETWEEN ANISOTROPIC AND ISOTROPIC SCATTERING

When the main objective is to obtain the solution to the distribution of the radiative heat flux and the average incident intensity, scaling of *anisotropic* scattering to *isotropic* scattering can be employed. Lee and Buckius (1982, 1983, 1986) have *normalized* the equation of radiative transfer and obtained the *dimensionless parameters*, which then lead to the similarity or scaling relations. Further, they use the first-order spherical harmonics approximation (P-1) and the two-flux approximation in order to *adapt* the scaling laws to these approximation methods of solution. Consider the *total directional intensity* (the scaling can also be applied at the spectral level) and use $\tau^* = \tau/\tau_L$, $I^* = I/I_0$, $I_b^* = I_b/I_0$, and the *normalized* equation of radiation transfer becomes

$$\frac{\partial I^*}{\partial \tau} = \left[\frac{\tau_L}{\mu}\right] I^* + \left[\frac{(1-w_a)\tau_L}{\mu}\right] I_b^* + \left[\frac{w_a \tau_L}{2\mu}\right] \int_{-1}^{1} I^* \langle \Phi \rangle (\mu, \mu') \, d\mu'. \quad (5.247)$$

Of the parameters shown in (5.247), only two are *independent*, and if the first two are chosen, the transformation to *isotropic* scattering requires

$$\frac{(1-\widehat{w}_a)\widehat{\tau}_L}{\widehat{\mu}} = \frac{(1-w_a)\tau_L}{\mu} \quad (5.248)$$

and

$$\frac{\widehat{\tau}_L}{\widehat{\mu}} = \frac{\tau_L}{\mu} \quad (5.249)$$

with

$$\langle \widehat{\Phi} \rangle (\mu, \mu') = 1. \quad (5.250)$$

Because of the angular dependency, these scaling laws are *not* useful. However, useful *approximate* scaling can be found if the equation of radiation transfer is *integrated* over the solid angles and when the solution is further approximated by keeping only the *leading* terms.

(A) P-1 Approximation

As was discussed, the method of *spherical harmonics* provides high order approximation to the equation of radiative transfer. The P-1 approximation equations, (5.187) and (5.189) can be written as

$$\frac{dq}{d\tau} + (1-w_a)I = 4\pi(1-w_a)I_b, \quad (5.251)$$

$$\frac{dI}{d\tau} + 3\left(1 - w_a \frac{A_1}{3}\right)q = 0, \quad (5.252)$$

where A_1 is the coefficient in expansion (5.224), I is the *average incident intensity*, and q is the *radiation heat flux*. These equations can be normalized such that $q^* = q/q_o$, $\tau^* = \tau/\tau_L$, and $I_b^* = \pi I_b/q_o$. Then when these equations are *normalized* and *combined*, we have

$$\frac{d^2 q^*}{d\tau^{*2}} - 3(1-w_a)\left(1 - w_a \frac{A_1}{3}\right)\tau_L^2 q^* = 4(1-w_a)\tau_L \frac{dI_b^*}{d\tau^*}, \quad (5.253)$$

which suggests scaling of

$$(1 - \widehat{w}_a)\widehat{\tau}_L = (1 - w_a)\tau_L \quad (5.254)$$

and

$$\left(1 - \widehat{w}_a \frac{A_1}{3}\right)\widehat{\tau}_L = \left(1 - w_a \frac{A_1}{3}\right)\tau_L. \quad (5.255)$$

The constant $A_1/3$ is equivalent to the asymmetry factor g. Then for isotropic scattering $g = 0$, we have the transformation for *anisotropic* to *isotropic* scattering given by

$$(1 - \widehat{\omega}_a)\widehat{\tau}_L = (1 - \omega_a)\tau_L, \quad \text{and} \quad \widehat{\omega}_a = \frac{\omega_a\left(1 - \dfrac{A_1}{3}\right)}{1 - \omega_a \dfrac{A_1}{3}}, \qquad (5.256)$$

$$\widehat{\tau}_L = \left(1 - \omega_a \frac{A_1}{3}\right)\tau_L. \qquad (5.257)$$

(B) Two-Flux Approximation

This approximation is given by the two equations (5.140)–(5.141). Lee and Buckius (1982) consider the optically thick limit of this approximation. The two dependent equations can be made independent by first differentiation and then intersubstitution. The resulting *second-order* equations for q^+ and q^- lead to the following *scaling laws*

$$(1 - \widehat{\omega}_a\widehat{\tau}_L) = (1 - \omega_a)\tau_L \quad \text{or} \quad \widehat{\omega}_a = \frac{2\omega_a(1 - f)}{1 - \omega_a(2f - 1)}, \qquad (5.258)$$

$$\widehat{\tau}_L = [1 - \omega_a(2f - 1)]\tau_L, \qquad (5.259)$$

where

$$f = \frac{1}{2}\int_0^1 \langle\Phi\rangle(\mu, \mu')\,d\mu'. \qquad (5.260)$$

Lee and Buckius (1982) examine the *three conditions* of *isothermal emission, boundary incident,* and *radiative equilibrium* and find that the P-1 approximation *scales* lead to *accurate* results (when compared to the non-scaled-more rigorous solutions), *while* the two-flux approximation scales are accurate *only* for the emission problems. The adequacy of the P-1 approximation and the similarity relations (5.256)–(5.257) for optically thick media has been confirmed by Glicksman et al. (1987). However, Glicksman et al. scaled the scattering coefficient *only* (which holds *only* if $\omega_a = 1$).

5.9 Noncontinuum Treatment: Monte Carlo Simulation

Chan and Tien (1974) use *ray-optics* in a *simple* cubic cell and assume *specular* reflection. Then they apply the results of the simple cubic cell to obtain the multicell transmission using the *layer theory* used in the analysis of multilayer coated surfaces. Their model *underpredicts* the *transmittance* significantly. Yang et al. (1983) used *random* arrangement of spheres ($\varepsilon =$

0.42) and ray-optics along with a *Monte Carlo technique*. They show that the entering ray is most likely to have its *first interaction* at a distance of half the radius and a *mean penetration* of 4/3 of the radius. They compute the *probability distribution function* and use this information for computing the transmission through a packed bed. They also find that almost *all* the rays *hit* a sphere surface after traveling a distance of three diameters. Ray-optics *combined* with a *modified* Monte Carlo method is used by Kudo et al. (1987) on a unit cell basis for prediction of the *transmittance* through a packed bed of spheres with gray and diffuse reflection. The simulation of thermal radiation adjacent to bounding solid surfaces has been performed by Tseng et al. (1992). All of these simulations use *diffuse irradiation (not collimated)* as the boundary conditions.

Using an *adjustable* parameter ℓ/d (with ℓ being the cell size and an ℓ/d of 10) Tien and Drolen (1987) show that the prediction of Kudo et al. gives satisfactory agreement with the experimental results of Chen and Churchill (1963) *assuming* that the incident radiation in this experiment is *diffuse*. However, for this, $\varepsilon = 0.906$ where the experiments are for $\varepsilon = 0.4$. The results of Yang et al. are in good agreement with the experiment of Chen and Churchill (assuming diffuse irradiation) for packed beds that are only a few particles deep (i.e., $L/d < 3$). Examination of the experiment of Chen and Churchill shows that their beds are irradiated with a *nearly collimated* beam. This *nondiffuse* boundary condition when used in the Monte Carlo simulations is expected to *change* the *transmittance* significantly. All of these ray-tracing techniques *neglect diffraction*, which is justifiable for large particles.

Singh and Kaviany (1991) extended the Monte Carlo technique to accommodate semitransparent particles as well as emitting particles. They examine *randomly* packed beds as well as arrangements of *variable porosity* based on a *simple cubic packing*. Their method is reviewed below.

The first is a bed of randomly packed spheres. The bed was generated by the computer program PACKS (Jodrey and Tory, 1979) and was previously used by Yang et al. The bed of randomly packed spheres generated by this method has a porosity of 0.42.

The second bed is based on a simple cubic packing. The *layers* are, however, *staggered* with respect to each other. This can be significant when considering a packed bed of particles with *large* absorption. The regular simple cubic structure would result in *some* rays being transmitted *directly* through the voids in this regular structure. Also, from a practical standpoint, irregular arrangements are *more* relevant.

The domain of interest consists of a box with a square cross section bounded by $x = 0$, $x = 1$, $z = 0$, and $z = 1$ with depth equal to the depth of the bed. The irregular arrangement is achieved by generating sphere centers at *four corners* of the square [(0,0), (0,1), (1,0), (1,1)] in the x-z planes at $y = 0.5, 1.5, \ldots$. The centers are then staggered by applying the

5.9 Noncontinuum Treatment: Monte Carlo Simulation

following *transformation* to all four spheres in the layer.

$$x_c = x_c + 0.5(2\xi_x - 1), \tag{5.261}$$

$$z_c = z_c + 0.5(2\xi_z - 1), \tag{5.262}$$

where ξ_x and ξ_z are *random* numbers between 0 and 1. This process is carried out for each layer using newly generated ξ_x and ξ_z for each layer. After tracing a *small* number of rays (say 100), the process is *repeated* on the original center locations using freshly generated random numbers. Spheres of unit diameter result in a porosity of 0.476. To get a higher porosity, the sphere size can be reduced. In this case, the spheres will *no longer* touch each other. For a lower ε, the distance between the layer centers for unit-diameter spheres must be $y_n - y_{n-1} = 0.524/(1-\varepsilon)$, where y_n refers to the the y coordinate of the nth layer. Alternate layers have a sphere at the square center. The layers are staggered by an amount limited by the physical constraint that *no overlap* is allowed. Thus the maximum distance by which a layer can be of stagger varies from 0.5 for a porosity of 0.476 to 0 for a porosity of 0.26. Equations (5.261) and (5.262) must now be changed to

$$x_c = x_c + \gamma(2\xi - 1), \tag{5.263}$$

$$z_c = z_c + \gamma(2\xi - 1), \tag{5.264}$$

where γ is a *function of porosity alone* and represents the extent to which sphere centers can be displaced without overlap. Both of these models are used in conjunction with the *periodic* boundary condition in the x- and z-directions.

5.9.1 OPAQUE PARTICLES

A ray is defined by the coordinates of its starting point $P_o(x_o, y_o, z_o)$ and its direction cosines (ℓ, m, n). The ray enters the bed at a random point in the x-z plane [forming the lower surface $(y = 0)$], i.e.,

$$(x_o, y_o, z_o) = (W\xi_x, 0, W\xi_z), \tag{5.265}$$

where W is the *lateral* dimension of the box being used and ξ_x and ξ_z are *random* numbers between 0 and 1. The angles ϕ and θ are given by

$$\phi = 2\pi\xi, \tag{5.266}$$

$$\theta = \cos^{-1}[1 - \xi(1 - \cos\theta_{\max})], \tag{5.267}$$

where $\theta_{\max}$ is the *maximum* angle that the incident radiation makes with the normal. For diffuse incident flux, $\cos\theta_{\max} = 0$. The direction cosines of the ray are

$$\ell = \sin\theta \cos\phi, \tag{5.268}$$
$$m = \cos\theta, \tag{5.269}$$
$$n = \sin\theta \sin\phi. \tag{5.270}$$

The coordinates of a ray after traveling a length S are given by

$$x = x_0 + \ell S, \tag{5.271}$$
$$y = y_0 + m S, \tag{5.272}$$
$$z = z_0 + n S. \tag{5.273}$$

Substituting in the equation of the sphere, we have

$$(x - x_c)^2 + (y - y_c)^2 + (z - z_c)^2 = R^2, \tag{5.274}$$

i.e., a quadratic equation in S is obtained. A positive *discriminant* indicates that the ray intersects the sphere. Equation (5.274) is solved for all spheres for which it has a positive discriminant. The smaller of the two solutions obtained gives the *actual* point of intersection with a sphere. Also, the distance the ray travels before it intersects a bounding surface is determined. The *minimum* distance a ray travels before it is intercepted by a sphere or a bounding surface is then determined. The sphere or bounding surface corresponding to this solution is the surface that actually intercepts the ray.

If the ray is intercepted by the side walls, the periodic boundary condition is applied. In case it passes through the upper or lower face, the energy associated with the ray is registered as transmission or reflection. If it is intercepted by a sphere, the point of intersection is determined and the direction cosines of the reflected ray for a specularly scattering sphere are found using the laws of reflection which are stated below.

- The incident ray, the reflected ray, and the normal to the surface all lie in the *same plane*.

- The angle of incidence is *equal* to the angle of reflection.

If the sphere is assumed to be *diffusely* scattering, then the ray is scattered in a *random* direction from the point of interception under the restriction that the ray does *not penetrate* the sphere. After reflection, the energy of the reflected ray is given by $E_r = \rho E_i$.

This process is repeated until the ray passes through either the *upper* or *lower* surface. The number of rays used for each simulation ranged from 100,000 to 1,000,000. Packed beds with lower transmittance need more rays for the same accuracy.

5.9.2 SEMITRANSPARENT PARTICLES

Transmitting particles are dealt with by ray tracing inside the sphere, following the laws of reflection and refraction. The angle that the incident radiation makes with the tangent to the surface, i.e., θ_i, is calculated. Then the angle of *refraction* is given by

$$\cos\theta_i = n\cos\theta_r, \quad n = \frac{n_s}{n_f}. \tag{5.275}$$

Next, the *Fresnel coefficients* and the *reflectivity* are calculated in terms of the angles θ_i and θ_r (Siegel and Howell, 1981, p.95, p.101), i.e.,

$$\rho'_{\parallel\lambda} = \left[\frac{\tan(\theta_i - \theta_r)}{\tan(\theta_i + \theta_r)}\right]^2 \quad \text{and} \quad \rho'_{\perp\lambda} = \left[\frac{\sin(\theta_i - \theta_r)}{\sin(\theta_i + \theta_r)}\right]^2 \quad \text{for } \kappa \to 0. \tag{5.276}$$

Thus, the reflected parts of energy are $\rho'_{\parallel\lambda}$ and $\rho'_{\perp\lambda}$. The refracted parts are $1 - \rho'_{\parallel\lambda}$ and $1 - \rho'_{\perp\lambda}$. Then the energy carried by the various rays is (van de Hulst, 1981, p. 204)

$$E_{\parallel,p} = \rho'_{\parallel\lambda} \quad \text{for } P = 0, \tag{5.277}$$

$$E_{\parallel,p} = (1 - \rho'_{\parallel\lambda})^2(\rho'_{\parallel\lambda})^{P-1} \quad \text{for } P = 1, 2, 3, \ldots \quad \text{if } \kappa = 0. \tag{5.278}$$

For the other *polarization*, replace $\parallel$ with $\perp$. However if κ_s is not small, then (5.276) should not be used to calculate the reflectivity. Instead, an exact analysis should be followed (Siegel and Howell, 1981, p.100), although ray tracing beyond $(p = 0)$ will *not* be required because even moderate values of κ_s (for a large particle in the geometrical optics range) make the particle *virtually opaque*.

For *nonpolarized* irradiation, the total energy carried by a ray is given by

$$E = \frac{1}{2}(E_{\perp,p} + E_{\parallel,p}). \tag{5.279}$$

When a ray strikes a sphere, it is either *reflected* $(p = 0)$ or *transmitted* $(p = 1, 2, \ldots)$ with a *reduction* in the energy due to *absorption*. The outcome is decided by generating a random number. Let us define

$$\beta_i = \sum_{j=0}^{i} E_j. \tag{5.280}$$

Then the ray is reflected $(P = 0)$ if

$$\xi < \beta_0 \tag{5.281}$$

and is transmitted with $P = i$ if

$$\beta_i < \xi \leq \beta_{i+1}. \tag{5.282}$$

Generally, tracing up to $P = 2$ or 3 is sufficient. Figure 5.10 shows a sketch of a ray traced up to $P = 2$. The ray incident on the point P_0 can either be reflected or transmitted.

- If the ray is *reflected*, its direction cosines are calculated as in the case of opaque particles. However, the energy carried by the ray remains unchanged.

- If the ray is *transmitted*, the direction cosines of the ray $r_{0,1}$ are found using the laws of refraction.

    - The *incident* ray, the *refracted* ray, and the *normal* to the surface all lie in the same plane.

    - The angle of *refraction* is related to the angle of *incidence* by the *Snell law*.

The coordinates of point P_1 are found by using $P_0 P_1 = 2R \sin \theta_r$. The direction cosines of the rays $r_{1,2}$ and r_1 are found by applying the laws of reflection and refraction at point P_1, respectively. The coordinates of point P_2 and the direction cosines of ray r_2 are found by repeating the preceding steps.

In the case of semitransparent particles ($\kappa \neq 0$), the energy of the rays is reduced by an *attenuating factor* given by

$$F_P = \exp(-4P \kappa \alpha_R \sin \theta_r) \quad \text{for} \quad P = 1, 2, 3, \ldots . \quad (5.283)$$

Therefore, the energy carried by a transmitted ray is given by

$$E_{r\,P} = F_P E_i \quad \text{for} \quad P = 1, 2, \ldots . \quad (5.284)$$

5.9.3 Emitting Particles

The spheres are assumed to have a high enough thermal conductivity so that a sphere can be assumed to be isothermal. The case simulated here is of a bed of absorbing, emitting, and scattering spheres. If the sphere has a reflectivity ρ, the ray undergoes the following.

- Is reflected if $\rho > \xi$ (either *diffuse* or *specularly* reflecting particles may be considered), or

- is absorbed and emitted if $\rho \leq \xi$.

The emission can take place from any *randomly* selected point on the surface. Also, the direction of the emitted ray is determined according to the *Lambert cosine* law as in the case of diffusely reflected rays.

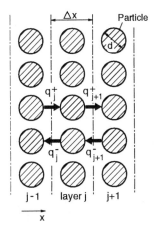

Figure 5.18 Radiative heat flux designation in geometric, layered model.

5.10 Geometric, Layered Model

Using *geometric optics* (radiation size parameter α_R, larger than about five) and the concept of *view factor*, the *emission*, *transmission*, and *reflection* of periodically arranged, *diffuse, opaque* particles has been modeled by Mazza et al. (1991). For a one-dimensional radiative transfer through a porous medium, as rendered in Figure 5.18, the radiative heat flux (across an area A) is given by

$$q_j^- A = \langle T_r \rangle q_{j+1}^- A + \langle \rho_r \rangle q_j^+ A + \langle \epsilon_r \rangle \sigma T_j^4 A, \quad (5.285)$$

$$q_{j+1}^+ A = \langle T_r \rangle q_j^+ A + \langle \rho_r \rangle q_{j+1}^- A + \langle \epsilon_r \rangle \sigma T_j^4 A \quad (5.286)$$

The *effective* radiative properties, i.e., *effective transmissitivity* $\langle T_r \rangle$, *effective reflectivity* $\langle \rho_r \rangle$, and *effective emissivity* (assumed equal to absorptivity) $\langle \epsilon_r \rangle$ are determined for various *two-dimensional* arrangements of *spherical particles*. Emerging correlations, relating these effective properties to the particle surface emissivity ϵ_r and medium porosity ϵ, do *not* appear to depend significantly on the arrangement. These correlations obtained by Mazza et al. are

$$\langle T_r \rangle = 1 - \langle \rho_r \rangle - \langle \epsilon_r \rangle, \quad (5.287)$$

$$\langle \rho_r \rangle = \frac{\pi(1 - \epsilon_r) a_1 N_s d^2}{2}, \quad (5.288)$$

$$\langle \epsilon_r \rangle = \frac{\pi \epsilon_r a_2 N_s d^2}{2}, \quad (5.289)$$

$$a_1 = \frac{2}{3\left[1 + a_3(N_s d^2)^{1.41}\right]^{1/2}}, \tag{5.290}$$

$$a_2 = \frac{1}{\left[1 + a_4(N_s d^2)^{1.74}\right]^{1/2}}, \tag{5.291}$$

$$a_3 = \frac{1.46\epsilon_r + 0.484}{1 + 0.16\epsilon_r}, \tag{5.292}$$

$$a_4 = \frac{1.967\epsilon_r + 0.00330}{1 + 0.07\epsilon_r}, \tag{5.293}$$

where N_s is the number of scatterers per unit *area*.

The correlation applies to $0.630 < N_s d^2 < 1.155$, where the upper limit corresponds to the closest two-dimensional packing of spheres.

The variational upper and lower bounds for the effective emissivity of randomly arranged particles has been obtained by Xia and Strieder (1994a, b).

5.11 Radiant Conductivity Model

The radiative heat transfer for a one-dimensional, plane geometry with emitting particles under the steady-state condition is given by (Vortmeyer, 1978)

$$q_r = \frac{F\sigma}{\frac{1 + \rho_w}{1 - \rho_w} + \frac{L}{d}}(T_1^4 - T_2^4), \tag{5.294}$$

where F is called the *radiative exchange factor* and the properties are assumed to be *wavelength-independent*. If $\rho_w = 0$ and the bed is several particles deep, then the first term of the denominator can be *neglected*. Then, for $T_1 - T_2 < 200$ K, a *radiant conductivity* is defined (Tien and Drolen, 1987)

$$k_r = 4\,F\,d\,\sigma\,T_m^3. \tag{5.295}$$

The approach has many *limitations*, but the single most important limitation is that the value of F *cannot* be *easily* calculated. Of all the methods discussed in the next section, only the Monte Carlo method can be used for calculating F for semitransparent particles. The value of F also depends upon the value of the *conductivity* of the *solid phase*. In the Kasparek experiment (Vortmeyer, 1978) and the Monte Carlo method discussed in Section 5.9, *infinite conductivity* is assumed, which is justified for metals. Similarly, the case of zero conductivity can be easily treated by considering the rays to be emitted from the same point at which they were absorbed. However, the intermediate case, i.e., when the conductivity is comparable to the radiant conductivity shows a *strong dependence* of radiant conductivity on the solid conductivity. The extent of this dependence may be seen

5.11 Radiant Conductivity Model

TABLE 5.5 RADIATION EXCHANGE FACTOR F ($\varepsilon = 0.4$)

MODEL	EMISSIVITY				
	0.2	0.35	0.60	0.85	1.0
Two-flux (diffuse)	0.88	0.91	1.02	1.06	1.11
Two-flux (specular)	1.11	1.11	1.11	1.11	1.11
Discrete ordinate (diffuse)	1.09	1.15	1.25	1.38	1.48
Discrete ordinate (specular)	1.48	1.48	1.48	1.48	1.48
Argo and Smith	0.11	0.21	0.43	0.74	1.00
Vortmeyer	0.25	0.33	0.54	0.85	1.12
Kasparek (experiment)	-	0.54	-	1.02	-
Monte Carlo (diffuse)	0.32	0.45	0.68	0.94	1.10
Monte Carlo (specular)	0.34	0.47	0.69	0.95	1.10

by comparing the difference in the values of F in Table 5.5 corresponding to low and high emissivities. If the conductivity were small, all the F values would be close to those obtained for the $\varepsilon = 0$ case. Thus, a simple tabulation of F as in Table 5.5 is of limited use. On the other hand, this approach is simple. The dependence of F on k_s will be discussed in Section 5.11.2.

5.11.1 Calculation of F

Many different models are available for prediction of F, and these are reviewed by Vortmeyer. Here, the main *emphasis* will be on examining the validity of the radiant conductivity approach by comparing the results of some of these models with the Monte Carlo simulations and with the available experimental results.

A solution to this problem based on the two-flux model is given by Tien and Drolen (1987)

$$F = \frac{2}{d(\overline{\sigma}_{\lambda a} + 2\overline{\sigma}_{\lambda s})}, \qquad (5.296)$$

which can be written as

$$F = \frac{2}{3(1 - \varepsilon)(\eta_{\lambda a} + 2B\eta_{\lambda s})}. \qquad (5.297)$$

For *isotropic* scattering, $B = 0.5$ and (5.297) becomes *independent* of the particle *emissivity* (for *large* particles).

In the Kasparek experiment (described by Vortmeyer), measurements of radiation heat transfer through a number of planar series of *welded steel spheres* were made. Conduction and convection were eliminated by placing the layers a small distance apart and performing the experiment in a vacuum. The high thermal conductivity of the material ensured that the heat resistance of the material was negligible and that the spheres were isothermal. Measurements were made using *polished steel* spheres ($\varepsilon_r = 0.35$) and *chromium oxide-coated* spheres ($\varepsilon_r = 0.85$). The arrangements considered were a cubic ($\varepsilon \simeq 0.5$) and a porosity of 0.4.

The Monte Carlo method, as described in Section 5.9.3, was used to predict the heat transfer through the packed bed used in the Kasparek experiment (unlike the previous sections, we encounter emitting particles). The results confirm the validity of the exchange factor approach as long as the emissivity is not close to zero. The change in the value of the exchange factor resulting from *increasing* the number of layers from 8 to 16 was *less* than 0.01 for $\varepsilon_r \geq 0.20$. The simulation was performed for both *specularly* scattering and *diffusely* scattering spheres. The polished steel spheres can be considered *specularly* scattering, while the chromium oxide spheres would scatter *diffusely*. Table 5.5 shows the results of the Monte Carlo simulation as well as the results obtained from the two-flux model and the models of Argo and Smith (described by Vortmeyer) and Vortmeyer for a porosity of 0.4. The predictions by the Monte Carlo method match the experimental results fairly well, considering that some uncertainty is always present in the emissivity values. Diffuse spheres result in slightly smaller values of F although, as expected, the difference decreases with increasing emissivity and vanishes for $\varepsilon_r = 1$. The change in value of F with porosity also matched the experimental results. For a cubic packing, the value of F increases from 0.47 to 0.51 for $\varepsilon_r = 0.35$, while the experimental value increases from 0.54 to 0.60. For $\varepsilon_r = 0.85$, the value of F increases from 0.94 to 0.97, while the experimental value increases from 1.02 to 1.06.

The predictions based on the two-flux model for diffuse spheres show a very small sensitivity to the emissivity, while those for specularly scattering spheres show no change at all. Using the method of discrete ordinates, *higher* values of F are obtained than those predicted by the two-flux method. Also, for specularly scattering spheres, the heat transfer remains independent of the emissivity. This can be seen from the equation of transfer by applying the condition of radiative equilibrium. Physically, when the spheres are treated as point scatterers, there is clearly no difference between isotropic scattering from a point and emission from it. The mechanism that results in an increase in the radiative heat transfer with increase in the emissivity is that of *transportation* of the absorbed energy through each particle (by *conduction*), i.e., particles *absorb* radiation at one face and *emit* a part of it from the other face. For a dilute medium consisting of *isotropically* scattering *small* particles separated by large distances, the heat transfer is again expected to be *independent* of the particle emissivity.

The effect of solid conductivity on the radiant conductivity k_r (or its dimensionless form F) has been examined by Singh and Kaviany (1994) and their results are reviewed below.

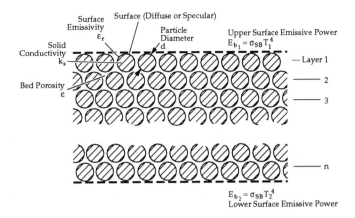

Figure 5.19 The plane-parallel geometry showing the bounding, emitting surfaces and the layers of spherical particles.

5.11.2 Effect of Solid Conductivity

For the one-dimensional, plane-parallel porous medium made of *opaque* spherical particles emitting in the range of blackbody incident radiation from the upper boundary held at T_1 and lower boundary at T_2, Singh and Kaviany (1994) consider the effect of solid conductivity k_s on the radiative heat transfer through the medium. A schematic of the problem considered is given in Figure 5.19.

The two blackbody emissive powers are to be related by the relation

$$E_{b_2} = \gamma E_{b_1}, \tag{5.298}$$

where γ varies between 0 and 1. The spheres have solid conductivity k_s and the matrix has a porosity ϵ. The objective is to find the radiative heat transfer through the medium.

The low and high conductivity limits of this problem have been explored experimentally (Vortmeyer, 1978) and by the Monte Carlo method (Singh and Kaviany, 1991). In the low conductivity asymptote, the rays are considered to be emitted from the same point on the sphere at which they were absorbed. In the high conductivity asymptote, an individual sphere is assumed to be isothermal and a ray absorbed by the sphere is given an equal probability of being emitted from anywhere on the sphere surface. This results in an increase in the radiant conductivity, because the rays absorbed on one side can be emitted from the other side thus by-passing

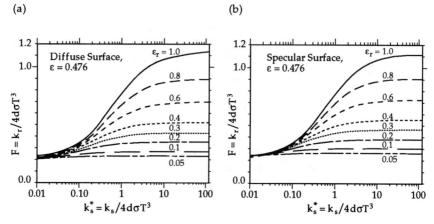

Figure 5.20 Effect of dimensionless solid conductivity on the dimensionless radiant conductivity for (a) diffuse particle surface and (b) specular particle surface.

the radiative resistance. In the general problem the solid and the radiant conductivities can have arbitrary magnitudes. Then, the radiative heat flux q_r for this one-dimensional, plane geometry is given by (5.294). The radiant conductivity k_r is given by (5.295), where

$$F = F(k_s^*, \epsilon_r, \epsilon) \tag{5.299}$$

and T_m is the mean temperature. The dimensionless solid conductivity k_s^* is defined as

$$k_s^* = \frac{k_s}{4d\sigma T_m^3}. \tag{5.300}$$

The mean bed temperature T_m is given in terms of the emissive powers of the bounding black surfaces as

$$T_m^3 = \frac{E_{b_1} - E_{b_2}}{4\sigma(T_1 - T_2)}. \tag{5.301}$$

In this problem $\rho_w = 0$, so the value of the radiant conductivity can be determined from the heat flux through the bed as

$$k_r = \frac{q_r}{(1 + \frac{L}{d})(T_1 - T_2)}. \tag{5.302}$$

The factor 1 in the denominator is often neglected, because $L \gg d$. Here, in the determination of the radiative exchange factor F, the factor 1 is retained, and once the value of k_r is known, the exchange factor F is calculated from (5.295).

TABLE 5.6 CONSTANTS IN THE EXCHANGE FACTOR CORRELATION ($\varepsilon = 0.476$)

	SPECULAR	DIFFUSE
a_1	0.5711	0.5756
a_2	1.4704	1.5353
a_3	0.8237	0.8011
a_4	0.2079	0.1843

Within the bed, the radiation is treated by combining the ray tracing with the Monte Carlo method. The conduction through the spheres is allowed by solving for the temperature distribution in a representative sphere for each particle layer in the bed. The finite-volume approximation is used to solve the heat conduction equation

$$k_s \nabla^2 T = \dot{s}, \tag{5.303}$$

where $\dot{s}$ is the source term which is nonzero for the boundary nodes on the spheres (i.e., surface nodes) and is found from the radiation part of the problem (i.e., the energy absorbed at each location on the surface of the sphere).

The results for $\varepsilon = 0.476$ and various values of ϵ_r and k_s^* have been obtained for both *diffusive* and *specular* surfaces. The results are shown in Figures 5.20(a) and (b). The results for both surfaces are nearly the same. Both low and high k_s^* asymptotes are present. The low k_s^* asymptotes are reached for $k_s^* < 0.10$ and the high k_s^* asymptote is approached for $k_s^* > 10$. There is a monotonic increase with ϵ_r, i.e., as absorption increases, the radiant conductivity increases for high k_s^*.

The results of Figures 5.20(a) and (b) have been correlated using

$$F = a_1 \epsilon_r \tan^{-1}\left(a_2 \frac{k_s^{*a_3}}{\epsilon_r}\right) + a_4, \qquad \text{for given } \varepsilon. \tag{5.304}$$

The best-fit values of the constants are given in Table 5.6.

The computer-intensive nature of the problem prevented a thorough sweep of the porosity range as an independent variable. However, the effect of the porosity in the high conductivity limit has been discussed by Singh and Kaviany (1991) and are given in part in Table 5.5. For example, by decreasing the porosity from 0.6 to 0.5, the magnitude of F changes from 0.47 to 0.51 for $\epsilon_r = 0.35$ (specular surfaces) and from 0.94 to 0.97 for $\epsilon_r = 0.85$ (diffuse surfaces). In practical packed beds, the porosity ranges between 0.3 to 0.6 with a value of 0.4 for randomly arranged loosely packed monosized spheres. Therefore, the sensitivity of the radiant conductivity with respect to the porosity (as compared to other parameters) is not expected to be very significant.

346   5. Radiation Heat Transfer

The variational upper bound on the radiant conductivity, including the conduction through the particle, has been predicted by Wolf et al. (1990).

5.12 Modeling Dependent Scattering

In order to account for the interparticle interactions, the decrease in the efficiencies has been correlated to the porosity. The correlations are discussed by Tien and Drolen (1987) and Tien (1988). For $\alpha_R \to 0$, the *Percus-Yevick model* is used gives

$$\frac{(\eta_{\lambda s})_{\text{dep}}}{(\eta_{\lambda s})_{\text{Mie}}} = \frac{\varepsilon^4}{(3 - 2\varepsilon)^2} \quad \text{for} \quad \frac{C}{\lambda} < 0.5, \tag{5.305}$$

which is *not* accurate for large α_R. Another correlation is that of Hottel et al. (1971), where they recommend the following

$$\log\log \frac{(\eta_{\lambda s})_{\text{Mie}}}{(\eta_{\lambda s})_{\text{dep}}} = 0.25 - \frac{3.83 C}{\lambda}$$

$$= 0.25 - \frac{3.83}{\pi}\alpha_R\left[\frac{0.905}{(1-\varepsilon)^{1/3}} - 1\right] \quad \text{for} \quad \frac{C}{\lambda} > 0.069. \tag{5.306}$$

These relationships, along with the experimental results of Ishimaru and Kuga (1982) and the *liquid model* of Drolen and Tien (1987) are plotted in Figure 5.21. The results are for $\alpha_R = 0.529$ and $n = 1.19$. The Percus-Yevick model (which is for $\alpha_R \to 0$) and the correlation of Hottel et al. (1971) predict the behavior well. The model of Drolen and Tien also predicts the behavior relatively well for this small α_R and $\varepsilon > 0.85$. However, at higher values of α_R, the predictions of this model deviate substantially from that of Hottel et al. Kumar and Tien (1990) model dependent scattering in a cloud of *Rayleigh*-sized particles. Chern et al. (1994) consider dependent scattering from fibers and show that the effect is most pronounced at an intermediate porosity. Singh and Kaviany (1992) model dependent interactions in the geometric range. The model is presented in detail in the next section.

5.12.1 MODELING DEPENDENT SCATTERING FOR LARGE PARTICLES

One approach is to *scale* the independent properties so that dependent computations can be carried out using the equation of radiative transfer with these scaled properties. However, since the *deviations* from the independent theory are a function of the *porosity* and the *complex index of refraction*, we will show that a *simple* scaling of the extent of dependence is *not feasible*. This will be done by examining the probability density functions for

5.12 Modeling Dependent Scattering

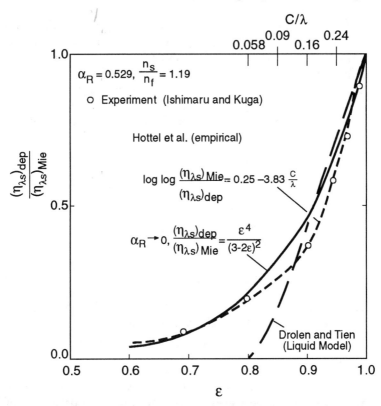

Figure 5.21 Effect of interparticle spacing (or porosity) on the normalized spectral scattering coefficient for spherical particles with optical properties shown.

independent and dependent scattering from both opaque and transparent particles.

Singh and Kaviany (1992) use an approach that *separately* accounts for *multiple scattering* in the representative elementary volume and the *transportation* of radiation through a particle (across a substantial optical thickness). Multiple scattering depends on the *porosity alone* and is accounted for by scaling the *optical thickness* using the porosity. The transmission through semitransparent particles is modeled by *allowing* for the transportation effect while *describing* the intensity field by the method of discrete ordinates. This is done by taking into consideration the *spatial difference* between the point where a ray first interacts with a sphere and the point from which it finally leaves the sphere. This spatial difference corresponds to an *optical thickness* (for a given porosity) across which the ray is transported while undergoing scattering by a particle.

The results of the application of this *dependence-included discrete ordinates method* are shown to be in good agreement with those obtained from the Monte Carlo method. The correct modeling of the physics results in the applicability over the *full range* of *porosity* and *optical properties* and *obviates* the need for calculating and presenting scaling factors in a three-dimensional array.

Kamiuto (1990) has proposed a *heuristic* correlated *dependent scattering theory* that attempts to calculate the dependent properties of large particles from the independent properties. The extinction coefficient and the albedo are scaled as

$$\langle \sigma_{ex} \rangle = \gamma \, \langle \sigma_{ex} \rangle_{ind} \qquad (5.307)$$

and

$$\langle \omega \rangle = 1 - (1 - \langle \omega \rangle_{ind})/\gamma, \qquad (5.308)$$

where

$$\gamma = 1 + \frac{3}{2}(1 - \varepsilon) - \frac{3}{4}(1 - \varepsilon)^2 \quad \text{for} \quad \varepsilon < 0.921. \qquad (5.309)$$

The phase function is left *unchanged*. The results of this theory will be compared to those of the Monte Carlo simulation.

(A) SCALING

In this section, we attempt to find scaling factors so that the independent radiative properties can be scaled to give the dependent properties of the particulate media. The *scaling factor* S_r is assumed to be scalar and scales the optical thickness leaving the *phase function* and *scattering albedo* *unchanged*.

(i) S_r for Opaque Spheres

Consider a plane-parallel particulate medium subject to *diffuse* incident radiation at one boundary. The medium contains particles that are non-emitting in the wavelength range of interest. For opaque particles with nonzero emissivity, the slopes of the transmission curve on a logarithmic scale approach a constant value away from the boundary. The scaling factor S_r is calculated by finding the ratio of the slopes calculated by the Monte Carlo method described in Section 5.9 and by the independent theory. Calculation of slopes should ideally be carried out away from the boundary in order to obtain the *bulk* properties of the bed. The calculations under the assumption of independent scattering were done by the method of discrete ordinates. A distance of about *six* optical thicknesses from the boundary was found to be sufficient to obtain a *constant* slope. For the Monte Carlo method, the case of low porosities and high emissivities presents *some* problems. The intensity can be attenuated by as much as an *order* of magnitude

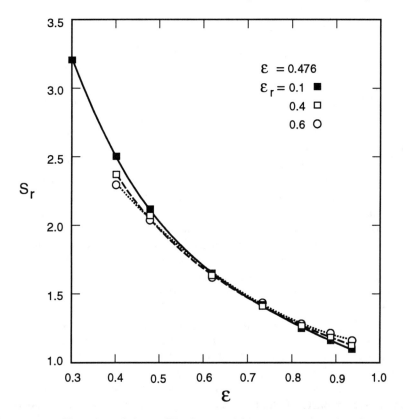

Figure 5.22 Variation of the scaling factor with respect to porosity, for several emissivities.

for every layer of particles. Thus, calculation at large depths becomes difficult because of the very small overall transmission. We are forced, therefore, to *determine* the transmission *close* to the boundary. Also, because of the low overall transmission, the transmission from individual rays becomes important. Therefore, apart from the difficulty in determining the transmission, the bed may actually give different values of the scaling factor at different depths. This difficulty may be overcome by noting that transmissions of the order of 10^{-5} are so small that any change in the scaling factor at large distances from the source, due to transmission resulting from a small number of rays, is of no significance.

Figure 5.22 shows the scaling factor for *opaque* spheres as a function of porosity for different emissivities. The values of S_r can be curve-fitted as

$$S_r = 1 + 1.84(1-\varepsilon) - 3.15(1-\varepsilon)^2 + 7.20(1-\varepsilon)^3 \quad \text{for} \quad \varepsilon > 0.3. \quad (5.310)$$

Since the *effect* of *emissivity* on S_r is small, (5.310) can be used to obtain the value of S_r for *other* emissivities.

(ii) The Basis of Scaling

Figure 5.23 shows the *probability density function* (PDF) for a bed made of *specularly* reflecting *opaque* spheres with a porosity of 0.476. The PDF was obtained by a *direct* Monte Carlo simulation for packed beds of spheres as discussed by Singh and Kaviany (1992). Also plotted are the PDFs calculated from the theory of *independent* scattering and the PDF for the *scaled* properties. The effect of increasing the emissivity is to *increase* the relative importance of the right-hand side of the curve. This is because *multiple reflections* attenuate the energy of a ray undergoing a number of interactions, as a result of short path lengths. Thus, the net contribution to transmittance will come from the rays that include a greater number of longer paths and thus are transmitted with a lesser number of interactions. If the scaled PDF and the Monte Carlo PDF have different shapes, the scaling factor will *change* greatly with the particle emissivity. However, since the scaled PDF is found to *conform* closely to the PDF from the Monte Carlo simulation, the effect of the emissivity on S_r is *small*, as seen in Figure 5.22. Therefore, the scaling can be carried out, treating the scaling factor as a function of porosity *alone*.

Figure 5.24 shows a schematic of the *interaction* of a ray with a transparent sphere. The ray is intercepted by the first sphere at point P_0. Part of the energy is *transmitted* through the sphere and interacts with a second sphere at point Q_0. The distance $P_0 Q_0$ is the distance that this energy travels after interaction at point P_0 and before its interaction with the next sphere. Other parts of the incident energy at P_0 travel different paths, as explained in detail in the next section. Figure 5.25 shows the PDF for a bed of *transparent* particles ($n = 1.5$). The PDF for *semitransparent* particles is similar except that the fraction of rays passing through the sphere has to be *modified* to account for the energy *attenuated* on passing through the sphere. It is clear that the PDF for *nonopaque* particles and that obtained from the independent theory are basically *dissimilar*. Even though scaling factors can still be found for a prescribed *set* of n, κ, and ε, a change in any one of the *three* parameters will change the PDF and thus affect S_r. Therefore, a scaling approach *necessitates* calculation and presentation of scaling factors in a three-dimensional array and is *not* found to be *suitable*.

(B) Dependence-Included Discrete Ordinates Method (DIDOM)

DIDOM models radiation heat transfer in a packed bed of semitransparent spheres. The deviation from the independent theory takes place because of the following two distinct effects.

- *Multiple scattering* within a small elemental volume.

5.12 Modeling Dependent Scattering 351

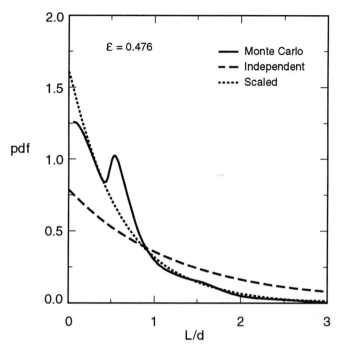

Figure 5.23 Variation of the probability density function as a function of distance, for a bed of opaque particles ($\varepsilon = 0.476$).

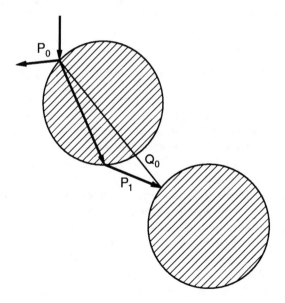

Figure 5.24 A schematic of a ray transmission through a particle and its arrival at an adjacent particle.

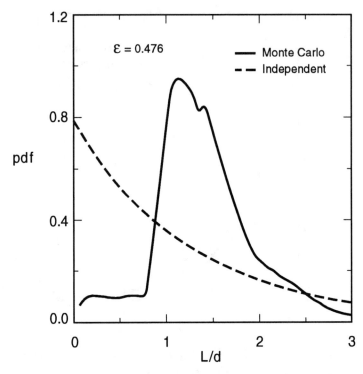

Figure 5.25 Same as Figure 5.19, except for transparent particles ($\varepsilon = 0.476$, $n = 1.5$).

- *Transportation* across a substantial optical thickness.

Multiple scattering is a function of porosity alone and is accounted for by scaling, as shown in the previous section. The transportation effect is modeled by allowing for *transmission* through a sphere while solving the equation of radiative transfer. For this, the method of discrete ordinates has been found to be most suitable. The *key* to understanding and modeling the transportation effect is that a ray may be *scattered* by a particle from a point that is *different* from the point at which the ray first interacts with the particle. This is because of transmission through a particle. In highly porous media ($\varepsilon \to 1$), this effect is of no consequence because the particle size is *small* as compared to the interparticle distance. However, in packed beds, the ray may be transported through a distance that corresponds to a *substantial optical thickness*. Thus, not only is it important to know the direction in which a particle scatters, it is also essential to know the displacement undergone by the ray as it passes through the particle. In this section, we first examine the properties of a single particle. Then the properties of beds are discussed. Finally, the DIDOM is presented.

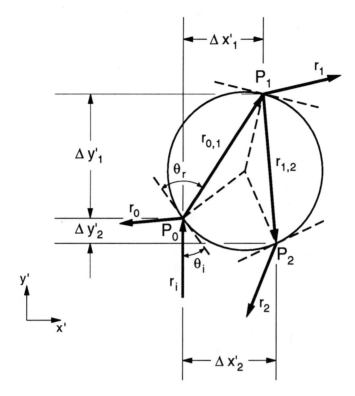

Figure 5.26 A ray tracing through a single particle.

(i) Properties of a Single Particle

The procedure for obtaining the properties of a *single* particle by geometric scattering has already been discussed in Section 5.2. Here we extend this method to also calculate the points from where the rays are scattered. Figure 5.26 shows a sketch of an incident ray being scattered.

For *independent* scattering, the sum of the *gains* in a particular direction resulting from *various* values of P gives the phase function Φ for a nonabsorbing sphere. For an absorbing sphere, the resulting values of the phase function must be divided by the scattering efficiency.

However, for cases where the transportation effect is important, the addition of gains for different values of P is not permissible. This is because rays scattered in the same direction from different points (P_0 for $P = 0$, P_1 for $P = 1$) will *not* have the same effect on transmission in a packed bed. Therefore, along with the gain, information regarding the point from which the ray leaves the sphere must be mentioned. Thus, the phase function will be reported as a three-column array, i.e., $[\Phi(\theta_o, P), \Delta x', \Delta y']$. Here $\Phi(\theta_o, P) = G(\theta_o, P)/\eta_s$, $\Delta x' = x'_P - x'_0$, and $\Delta y' = y'_P - y'_0$ represent the displacement undergone by the ray in a direction *perpendicular* and *parallel* to the incident ray, respectively. Thus, $A_a = \eta_a \pi R^2$, $A_s = \eta_s \pi R^2$, and $[\Phi(\theta_o, P), \Delta x', \Delta y']$ are determined. $\Delta x'$ and $\Delta y'$ are given by

$$\Delta x' = 0, \quad \Delta y' = 0 \quad \text{for} \quad P = 0 \tag{5.311}$$

and

$$\Delta x' = d \sin\theta_r \sum_{P'=1}^{P} \sin\left[(2P'-1)\theta_r - \theta_i\right],$$

$$\Delta y' = d \sin\theta_r \sum_{P'=1}^{P} \cos\left[(2P'-1)\theta_r - \theta_i\right] \quad \text{for} \quad P = 1, 2, \ldots . \tag{5.312}$$

(ii) Properties of Beds

In this section, we will relate the *radiative properties* of a single particle determined in the previous section to the radiative properties of the particulate medium. We assume a one-dimensional plane-parallel slab geometry. The required properties are $\langle \sigma_a \rangle$, $\langle \sigma_s \rangle$, and $[\langle \Phi \rangle (\mu_j \to \mu_i), \Delta k]$. The last one represents the phase function from a direction μ_j to a direction μ_i, and Δk represents the *number of grids* through which it is transported in the direction perpendicular to the slab boundaries. For *monosized* scatterers of porosity ε, we have

$$\langle \sigma_s \rangle = N_s A_{\lambda s} S_r; \tag{5.313}$$

similarly, $\langle \sigma_a \rangle = N_s A_a S_r$.

The procedure for computing $[\langle \Phi \rangle (\mu_j \to \mu_i), \Delta k]$ is outlined here.

(a) To find the phase function for scattering into a direction μ_i from a direction μ_j, we must integrate over the azimuthal angle ϕ. For this purpose we employ a Gaussian quadrature and find discrete values of $\phi_i - \phi_j$ between 0 and π at 24 points.

(b) At every point, we find $\theta_o = \cos^{-1}[\mu_i \mu_j + \sqrt{1 - \mu_i^2}\sqrt{1 - \mu_j^2}\cos(\phi_i - \phi_j)]$.

(c) Up to this point, the treatment is similar to that used when employing a standard DOM with the phase function available at discrete values of θ_o except that each value of P has its own phase function for every θ_o. However, here the numerical integration over ϕ to evaluate

$$\langle \Phi \rangle (\mu_j \to \mu_i) = \frac{1}{\pi} \int_0^\pi \sum_P \langle \Phi \rangle \left[\theta_o(\mu_j, \phi_j \to \mu_i, \phi_i), P\right] \mathrm{d}(\phi_j - \phi_i) \tag{5.314}$$

is not performed. This is because we cannot add $\langle \Phi \rangle [\theta_o(\mu_j, \phi_j \to \mu_i, \phi_i), P]$ terms *unless* they have the same Δk.

5.12 Modeling Dependent Scattering 355

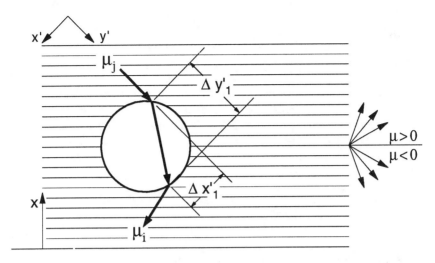

Figure 5.27 A demonstration of the transportation effect for semitransparent particles.

(d) For every $\langle \Phi \rangle \, [\theta_o(\mu_j, \phi_j \to \mu_i, \phi_i), P]$, we find $\Delta k = \text{Integer}\{\Delta k_{x'} + \Delta k_{y'}\}$, where $\Delta k_{x'}$ and $\Delta k_{y'}$ are the contributions of $\Delta x'$ and $\Delta y'$ to Δk. Δk is rounded off to the nearest integer. Figure 5.24 shows how $\Delta x'$ and $\Delta y'$ contribute to Δk. Equations for $\Delta k_{x'}$ and $\Delta k_{y'}$ are written as

$$\Delta k_{x'} = \Delta x' \, \zeta_k \sqrt{1 - \mu_j^2} \, \frac{\mu_j}{|\mu_j|},$$

$$\Delta k_{y'} = \Delta y' \, \zeta_k \, \mu_j, \qquad (5.315)$$

where

$$\zeta_k = \frac{\text{radius of sphere}}{\text{distance between two grids}} = \frac{0.75(1-\varepsilon)S_r}{\Delta \tau}. \qquad (5.316)$$

(e) We calculate $\langle \Phi \rangle (\mu_j \to \mu_i, \Delta k)$, i.e., the phase function from μ_j to μ_i that is transported by Δk number of grid points. Note that

$$\langle \Phi \rangle (\mu_j \to \mu_i) = \sum_{\Delta k} \langle \Phi \rangle (\mu_j \to \mu_i, \Delta k). \qquad (5.317)$$

(iii) DIDOM

The one-dimensional radiative transfer equation at x and in direction μ_i can be written as

$$\mu_i \frac{\mathrm{d}I_i(x)}{\mathrm{d}x} = -\langle\sigma_{\mathrm{ex}}\rangle I_i(x) + \langle\sigma_a\rangle I_b(x) + \frac{\langle\sigma_s\rangle}{2}\Gamma_i(x)$$

$$\text{for } i = -M, -M+1, \ldots, M, \ i \neq 0, \qquad (5.318)$$

where the Γ_i term represents the in-scattering term and accounts for in-scattering into the direction μ_i at location x from all directions at x as well as from all directions at other x locations. After discretization, by evaluating (5.318) at the midpoint between two nodes (k and $k+1$) as in Fiveland (1987), the in-scattering term can be written as

$$\Gamma_{i,k+1/2} = \sum_{k'=1}^{n} \sum_{j=-M, j \neq 0}^{M} \Delta\mu_j I_{j,k'+1/2} \langle\Phi\rangle (\mu_j \to \mu_i, \Delta k)$$

$$\text{for } \Delta k = k - k', \qquad (5.319)$$

where $\Delta\mu_j$ are the quadrature weights corresponding to the direction μ_j and

$$\sum_{j=-M, j\neq 0}^{M} \Delta\mu_j = 2. \qquad (5.320)$$

The boundary conditions are

$$I_i = \varepsilon_r I_b + \rho_s I_{-i} + 2\rho_d \sum_{j=-1}^{-M} \Delta\mu_j I_j \mu_j$$

$$i = 1, \ldots, M \quad \text{at} \quad x = 0, \qquad (5.321)$$

$$I_i = \varepsilon_r I_b + \rho_s I_{-i} + 2\rho_d \sum_{j=1}^{M} \Delta\mu_j I_j \mu_j$$

$$i = -1, \ldots, -M \quad \text{at} \quad x = L. \qquad (5.322)$$

In the case of incident radiation on a transparent boundary, this equation is used with $\varepsilon_r = 1$, $\rho_s = 0$, and $\rho_d = 0$. The intensity at the boundary, in a direction μ_i, is equal to the intensity of the incident radiation in that direction.

As a first step, the values of $\langle\Phi\rangle [\theta_o(\mu_j, \phi_j \to \mu_i, \phi_i), P]$ are calculated at discrete values of $(\phi_j - \phi_i)$ for all combinations of μ_i and μ_j, and the corresponding Δk_s are calculated to obtain $\langle\Phi\rangle (\mu_j \to \mu_i, \Delta k)$. For $\mu_i > 0$, the intensities at $x = 0$ ($k = 1$) are known from the boundary conditions. $I_i (\mu_i > 0)$ is evaluated at $k = 1, \ldots, n$. Similarly, $I_i (\mu_i < 0)$ is evaluated at $k = n, \ldots, 1$.

The in-scattering term $\Gamma_{i,k+1/2}$ is stored in a two-dimensional array, which is updated at every point, e.g., while calculating scattering phase function from direction j to direction i at point $k+1/2$ for $\mu_j > 0$

$$\Gamma_{i,k+1/2+\Delta k} = \Gamma_{i,k+1/2+\Delta k} + \Delta\mu_j I_{j,k+1/2} \langle\Phi\rangle (\mu_j \to \mu_i, \Delta k). \qquad (5.323)$$

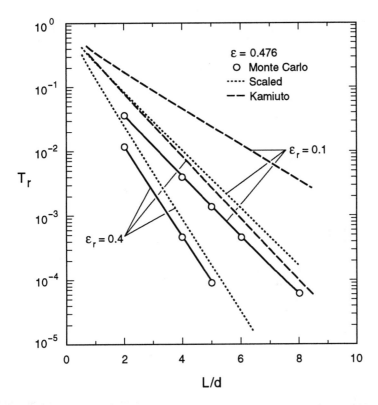

Figure 5.28 Transmittance through a bed of specularly reflecting spherical particles ($\varepsilon = 0.476$).

This calculation is carried out for scattering into other directions, i.e., for different values of i. It is then repeated for all positive values of μ_j. Then the $I_{j,k+1}(\mu_j > 0)$ are calculated and updated and $\Gamma_{j,k+1/2}(\mu_j > 0)$ [used to calculate $I_{j,k+1}(\mu_j > 0)$] is set to zero. This procedure is carried out for $k = 1, \ldots, n$. After the sweep for $\mu_j > 0$ is complete, the calculation for I_i ($\mu_j < 0$) is carried out at $k = n, \ldots, 1$ in a similar manner.

Figure 5.28 shows the *transmittance* through a bed of *specularly* reflecting *opaque* spheres ($\varepsilon = 0.476$) as a function of the bed thickness. The particles are assumed to have a *constant reflectivity*. As expected, the scaled results show the same bulk behavior as the Monte Carlo results. However, the results are offset by a difference that occurs at the boundaries where the bulk properties are no longer valid. The difference is more pronounced for $\varepsilon_r = 0.1$ than for $\varepsilon_r = 0.4$. This is because for $\varepsilon_r = 0.1$ a large amount of energy is reflected at the surface of the bed (before the continuum treatment becomes applicable). The results obtained from the Kamiuto correlated theory are found to overpredict the transmission.

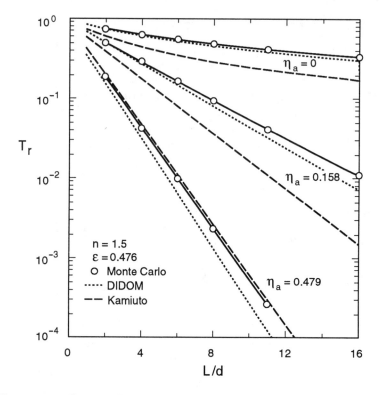

Figure 5.29 Same as Figure 5.28, except for transparent and semitransparent particles ($\varepsilon = 0.476$, $n = 1.5$).

Figure 5.29 illustrates the change in the *transmittance* as a function of bed thickness for *transparent* and *semitransparent* particles ($\varepsilon = 0.476$). The spheres have a refractive index, $n = 1.5$. Three different absorptivities are considered, i.e., $\kappa\alpha_R = 0$, $\kappa\alpha_R = 0.05$, and $\kappa\alpha_R = 0.2$, giving $\eta_a = 0$, $\eta_a = 0.158$, and $\eta_a = 0.479$, respectively. The results of the DIDOM are in good agreement with those of the Monte Carlo method for all these cases. The results from the Kamiuto correlated theory *underpredict* the transmission for transparent particles. As the absorption of the particle is increased, the results become closer to the Monte Carlo results. In the limiting case of opaque particles, the correlated theory overpredicts the transmittance.

Thus, the dependent properties for *opaque* particles are obtained by scaling the optical thickness obtained from the independent theory. Radiative transfer through *semitransparent* particles is modeled by allowing for the *transmission* through the particle *while* solving the *equation of radiative transfer*, thus resulting in the dependence-included discrete ordinates method.

5.13 Summary

In conclusion, we offer some suggestions on how to model the problem of radiative heat transfer in porous media. First, we must choose between a *direct simulation* and a *continuum treatment*. Wherever possible, continuum treatment should be used because of the lower cost of computation. However, the *volume-averaged* radiative properties may not be available in which case continuum treatment cannot be used. Except for Monte Carlo techniques for large particles, direct simulation techniques have not been developed to solve but the simplest of problems. However, direct simulation techniques should be used in case the number of particles is *too* small to justify the use of a continuum treatment and as a tool to verify dependent scattering models.

If the continuum treatment (Section 5.1) is to be employed, we must first identify the elements that make up the system. The choice of elements might be obvious (as in the case of a packed bed of spheres) or some simplifying assumptions might have to be made. Common simplifying assumptions are assuming the system to be made up of cylinders of infinite length (for fibrous media) or assuming arbitrary *convex-surfaced* particles to be spheres of equivalent cross section or volume. Then the properties of an individual particle can be determined (Section 5.2). If the system cannot be broken down into elements, then we have no choice but to determine its radiative properties experimentally.

On the other hand, if we can treat the system as being made up of elements, then we must identify the system as *independent* or *dependent* (Section 5.3). In theory, *all systems* are dependent, but if the deviation from the independent theory is not large, the assumption of independent scattering should be made. The range of validity of this assumption can be *approximately* set at $C/\lambda > 0.5$ and $\varepsilon > 0.95$. If the problem lies in the independent range, then the properties of the bed can be readily calculated (Section 5.4), and the equation of transfer can be solved using one of the methods discussed in Section 5.7.

However, if the system is in the dependent range, some *modeling* of the extent of dependence is necessary to get the properties of the packed bed. Models for particles in the *Rayleigh* range and the *geometric* range (Section 5.13) are available. However, no approach is yet available for particles of *arbitrary* size, and experimental determination of properties is again necessary.

Finally we note that the thermal conductivity of the solid phase influences the radiation properties. When using the radiant conductivity, the results show that k_r can increase by five fold for $k_s \to \infty$ as compared to that for $k_s \to 0$ (for $\varepsilon_r = 1$ and typical porosities).

5.14 References

Azad, F. H., 1985, "Differential Approximation to Radiative Transfer in Semi-Transparent Media," *ASME J. Heat Transfer*, 107, 478–481.

Bayazitoğlu, Y. and Higenyi, J., 1979, "Higher Order Differential Equations of Radiative Transfer," *AIAA J.*, 17, 424–431.

Bohren, G. F. and Huffman, D. R., 1983, *Absorption and Scattering Light by Small Particles*, John Wiley.

Born, M. and Wolf, E., 1988, *Principles of Optics*, Pergamon (Oxford).

Brewster, M. Q., 1983, "Radiative Heat Transfer in Fluidized Bed Combustors," *ASME* paper no. 83-WA/HT-82.

Brewster, M. Q., 1992, *Thermal Radiative Transfer and Properties*, Wiley and Sons, New York.

Brewster, M. Q. and Tien, C.-L., 1982a, "Radiative Transfer in Packed and Fluidized Beds: Dependent versus Independent Scattering," *ASME J. Heat Transfer*, 104, 573–579.

Brewster, M. Q. and Tien, C.-L., 1982b, "Examination of the Two-Flux Model for Radiative Transfer in Particular Systems," *Int. J. Heat Mass Transfer*, 25, 1905–1907.

Carlson, B. G. and Lathrop, K. D., 1968, "Transport Theory—The Method of Discrete Ordinates," in *Computing Methods in Reactor Physics*, Gordon and Breach Science Pub., 171–266.

Cartigny, J. D., Yamada, Y., and Tien, C.-L., 1986, "Radiative Heat Transfer with Dependent Scattering by Particles: Part 1—Theoretical Investigation," *ASME J. Heat Transfer*, 108, 608–613.

Chai, J. C., Lee, L. S., and Patankar, S. V., 1994, "Finite Volume Method of Radiative Heat Transfer," *J. Thermophys. Heat Transfer*, 8, 419–425.

Chan, C. K. and Tien, C.-L., 1974, "Radiative Transfer in Packed Spheres," *ASME J. Heat Transfer*, 96, 52–58.

Chandrasekhar, S., 1960, *Radiation Transfer*, Dover.

Chang, S. L. and Rhee, K. T., 1984, "Blackbody Radiation Functions," *Int. J. Comm. Heat Mass Transfer*, 11, 451–455.

Chen, J. C. and Churchill, S. W., 1963, "Radiant Heat Transfer in Packed Beds," *AIChE J.*, 9, 35–41.

Cheng, P., 1969, "Two-Dimensional Radiating Gas Flow by a Moment Method," *AIAA J.*, 2, 1662–1664.

Chern, B.-C., Howell, J. R., and Moon, T. J., 1994, "Dependent Scattering Effects on Wave Propagation through Filament-Wound Composites," in *Radiation Heat Transfer: Current Research*, Bayazitoglu, Y., et al., Editors, ASME HTD-Vol. 276, American Society of Mechanical Engineers.

Chu, C. M. and Churchill, S. W., 1955, "Representation of Angular Distribution of Radiation Scattered by a Spherical Particle," *J. Opt. Soc. Amer.*, 45, 958–962.

Condiff, D. W., 1987, "Anisotropic Scattering in Three-Dimensional Differential Approximation for Radiation Heat Transfer," *Int. J. Heat Mass Transfer*, 30, 1371–1380.

Crosbie, A. L. and Davidson, G. W., 1985, "Dirac-Delta Function Approximations to the Scattering Phase Function," *J. Quant. Spectrosc. Radiat. Transfer*, 33, 391–409.

Davison, B., 1957, *Neutron Transport Theory*, Oxford.
Deissler, R. G., 1964, "Diffusion Approximation for Thermal Radiation in Gases with Jump Boundary Condition," *ASME J. Heat Transfer*, 86, 240–246.
Driscoll, W. G. and Vaughan, W., eds., 1978, *Handbook of Optics*, McGraw–Hill.
Drolen, B. L. and Tien, C.-L., 1987, "Independent and Dependent Scattering in Packed Spheres Systems," *J. Thermophys. Heat Transfer*, 1, 63–68.
Fiveland, W. A., 1987, "Discrete Ordinate Methods for Radiative Heat Transfer in Isotropically and Anisotropically Scattering Media," *ASME J. Heat Transfer*, 109, 809–812.
Fiveland, W. A., 1988, "Three–Dimensional Radiative Heat Transfer Solutions by the Discrete–Ordinates Method," *J. Thermophys. Heat Transfer*, 2, 309–316.
Glicksman, L., Schuetz, M., and Sinofsky, M., 1987, "Radiative Heat Transfer in Foam Insulation," *Int. J. Heat Mass Transfer*, 30, 187–197.
Grosshandler, W. H. and Monterio, S. L. P., 1982, "Attenuation of Thermal Radiation by Pulverized Coal and Char," *ASME J. Heat Transfer*, 104, 587–593.
Hendricks, T. J., 1993, *Thermal Radiative Properties and Modelling of Reticulated Porous Ceramics*, Ph. D. Thesis, The University of Texas at Austin.
Hendricks, T. J. and Howell, J. R., 1994, "Absorption/Scattering Coefficients and Scattering Phase Functions in Reticulated Porous Ceramics," in *Radiation Heat Transfer: Current Research*, Bayazitoglu, Y., et al., Editors, ASME HTD–Vol. 276, American Society of Mechanical Engineers.
Hottel, H. C., Sarofim, A. F., Dalzell, W. H., and Vasalos, I. A., 1971, "Optical Properties of Coatings, Effect of Pigment Concentration," *AIAA J.*, 9, 1895–1898.
Hottel, H. C., Sarofim, A. F., Evans, L. B., and Vasalos, I. A, 1968, "Radiative Transfer in Anisotropically Scattering Media: Allowance for Fresnel Reflection at the Boundaries," *ASME J. Heat Transfer*, 90, 56–62.
Hottel, H. C., Sarofim, A. F., Vasalos, I. A., and Dalzell, W. H., 1970, "Multiple Scatter: Comparison of Theory with Experiment," *ASME J. Heat Transfer*, 92, 285–291.
Hsieh, C. K. and Su, K. C., 1979, "Thermal Radiative Properties of Glass from 0.32 to 206 μm," *Solar Energy*, 22, 37–43.
Ishimaru, A. and Kuga, Y., 1982, "Attenuation Constant of a Coherent Field in a Dense Distribution of Particles," *J. Opt. Soc. Amer.*, 72, 1317–1320.
Jamaluddin, A. S. and Smith, P. J., 1988, "Predicting Radiative Transfer in Axisymmetric Cylindrical Enclosure Using the Discrete Ordinates Method," *Combust. Sci. Technol.*, 62, 173–186.
Jeans, J. H., 1917, "The Equation of Radiative Transfer of Energy," *Monthly Notices of Royal Astronomical Society*, 78, 445–461.
Jodrey, W. S. and Tory, E. M., 1979, "Simulation of Random Packing of Spheres," *Simulation*, Jan., 1–12.
Joseph, J. H., Wiscombe, W. J., and Weinman, J. A., 1976, "The Delta–Eddington Approximations for Radiative Heat Transfer," *J. Atm. Sci.*, 34, 1408–1422.
Kamiuto, K., 1990, "Correlated Radiative Transfer in Packed Bed-Sphere Systems," *J. Quant. Spectrosc. Radiat. Transfer*, 43, 39–43.
Kaviany, M., 1985, "One–Dimensional Conduction–Radiation Heat Transfer Between Parallel Surfaces Subject to Convective Boundary Conditions," *Int. J. Heat Mass Transfer*, 28, 497–499.

Kerker, M., 1969, *The Scattering of Light and Other Electromagnetic Radiation*, Academic.
Kerker, M., Scheiner, P., and Cooke, D. D., 1978, "The Range of Validity of Rayleigh and Mie Limits for Lorentz–Mie Scattering," *J. Opt. Soc. Amer.*, 68, 135–137.
Ku, J. C. and Felske, J. D., 1984, "The Range of Validity of the Rayleigh Limit for Computing Mie Scattering and Extinction Efficiencies," *J. Quant. Spectrosc. Radiat. Transfer*, 31, 569–574.
Kudo, K., Yang W., Tanaguchi, H., and Hayasaka, H., 1987, "Radiative Heat Transfer in Packed Spheres by Monte Carlo Method," *Heat Transfer in High Technology and Power Engineering Proceedings*, Hemisphere, 529–540.
Kumar, S. and Tien, C.-L., 1990, "Dependent Scattering and Absorption of Radiation by Small Particles," *ASME J. Heat Transfer*, 112, 178–185.
Kumar, S., Majumdar, A., and Tien, C.-L., 1990, "The Differential–Discrete Ordinate Method for Solution of the Equation of Radiative Transfer," *ASME J. Heat Transfer*, 112, 424–429.
Lee, H. and Buckius, R. O., 1982, "Scaling Anisotropic Scattering in Radiation Heat Transfer for a Planar Medium," *ASME J. Heat Transfer*, 104, 68–75.
Lee, H. and Buckius, R. O., 1983, "Reducing Scattering to Non–Scattering Problems in Radiation Heat Transfer," *Int. J. Heat Mass Transfer*, 26, 1055–1062.
Lee, H. and Buckius, R. O., 1986, "Combined Mode Heat Transfer Analysis Utilizing Radiation Scaling," *ASME J. Heat Transfer*, 108, 626–632.
Lee, S. C., 1990, "Scattering Phase Function for Fibrous Media," *Int. J. Heat Mass Transfer*, 33, 2183–2190.
Lee, S. C., White, S., and Grzesik, J. A., 1994, "Effective Radiative Properties of Fibrous Composites Containing Spherical Particles," *J. Thermoph. Heat Transfer*, 8, 400–405.
Liou, K.-N. and Hansen, J. E., 1971, "Intensity and Polarization for Single Scattering Polydisperse Spheres: A Comparison of Ray–Optics and Mie Scattering," *J. Atmospheric Sci.*, 28, 995–1004.
Lorrain, P. and Corson, D. R., 1970, *Electromagnetic Fields and Wave*, Second Edition, Freeman and Co., 422–551.
Mazza, G. D., Berto, C. A., and Barreto, G. F., 1991, "Evaluation of Radiative Heat Transfer Properties in Dense Particulate Media," *Powder Tech.*, 67, 137–144.
McKellar, B. H. J. and Box, M. A., 1981, "The Scaling Group of the Radiative Transfer Equation," *J. Atmospheric Sci.*, 38, 1063–1068.
Mengüc, M. P. and Viskanta, R., 1982, "Comparison of Radiative Heat Transfer Approximations for Highly Forward Scattering Planar Medium," *ASME* paper no. 82–HT–20.
Mengüc, M. P. and Viskanta, R., 1985a, "On the Radiative Properties of Polydispersions: A Simplified Approach," *Combust. Sci. Technol.*, 44, 143–149.
Mengüc, M. P. and Viskanta, R., 1985b, "Radiative Transfer in Three–Dimensional Rectangulated Enclosures Containing Inhomogeneous, Anisotropically Scattering Media," *J. Quant. Spectrosc. Radiat. Transfer*, 33, 533–549.
Mengüc, M. P. and Viskanta, R., 1986, "An Assessment of Spectral Radiative Heat Transfer Predictions for a Pulverized Coal–Fired Furnace," in *Proceedings of 8th International Heat and Mass Conference (San Francisco)*, 2, 815–820.
Modest, M. F., 1993, *Radiative Heat Transfer*, McGraw–Hill, New York.

Modest, M. F. and Azad, F. H., 1980, "The Differential Approximation for Radiative Transfer in an Emitting, Absorbing and Anisotropically Scattering Medium," *J. Quant. Spectrosc. Radiat. Transfer*, 23, 117–120.

Nelson, H. F., Look, D. C., and Crosbie, A. L., 1986, "Two–Dimensional Radiative Back–Scattering from Optically Thick Media," *ASME J. Heat Transfer*, 108, 619–625.

Ozisik, M. N., 1985, *Radiative Transfer and Interaction with Conduction and Convection*, Werbel and Peck.

Palik, E. D., ed., 1985, *Handbook of Optical Constants of Solids*, Academic Press.

Papini, M., 1989, "Study of the Relationship Between Materials and Their Radiative Properties: Application to Natural Fibers for Spectral Wavelength Range 0.25–2.5 μm," *Infrared Phys.*, 29, 35–41.

Penndorf, R. B., 1962, "Scattering and Extinction for Small Absorbing and Nonabsorbing Aerosols," *J. Opt. Soc. Amer.*, 8, 896–904.

Raithby, G. D. and Chui, E. H., 1990, "A Finite–Volume Method for Predicting a Radiant Heat Transfer in Enclosures with Participating Media," *ASME J. Heat Transfer*, 112, 415–423.

Ratzel, A. C., 1981, "P–N Differential Approximation for Solution of One– and Two–Dimensional Radiation and Conduction Energy Transfer in Gray Participating Media," Ph.D. thesis, University of Texas at Austin.

Ratzel, A. C. and Howell, J. R., 1983, "Two–Dimensional Radiation in Absorbing–Emitting Media Using the P–N Approximation," *ASME J. Heat Transfer*, 105, 333–340.

Rish, J. W. and Roux, J. A., 1987, "Heat Transfer Analysis of Fiberglass Insulations with and without Foil Radiant Barriers," *J. Thermophys. Heat Transfer*, 1, 43–49.

Selamet, A., 1989, *Radiation Affected Laminar Flame Propagation*, Ph.D. thesis, University of Michigan.

Selamet, A. and Arpaci, V. S., 1989, "Rayleigh Limit Penndorf Extension," *Int. J. Heat Mass Transfer*, 32, 1809–1820.

Siegel, R. and Howell, J. R., 1981, *Thermal Radiation Heat Transfer*, Second Edition, McGraw-Hill.

Siegel, R., and Howell, J. R., 1992, *Thermal Radiation Heat Transfer*, Third Edition, Hemisphere, Washington.

Singh, B. P. and Kaviany, M., 1991, "Independent Theory Versus Direct Simulation of Radiative Heat Transfer in Packed Beds," *Int. J. Heat Mass Transfer*, 34, 2869–2881.

Singh, B. P. and Kaviany, M., 1992, "Modeling Radiative Heat Transfer in Packed Beds," *Int. J. Heat Mass Transfer*, 35, 1397–1405.

Singh, B. P. and Kaviany, M., 1994, "Effect of Particle Conductivity on Radiative Heat Transfer in Packed Beds," *Int. J. Heat Mass Transfer*, 37, 2579–2583.

Sparrow, E. M. and Cess, R. D., 1978, *Radiative Heat Transfer*, McGraw-Hill.

Tien, C.-L., 1988, "Thermal Radiation in Packed and Fluidized Beds," *ASME J. Heat Transfer*, 110, 1230–1242.

Tien, C.-L. and Drolen, B. L., 1987, "Thermal Radiation in Particulate Media with Dependent and Independent Scattering," *Annual Review of Numerical Fluid Mechanics and Heat Transfer*, 1, 1–32.

Tong, T. W. and Swathi, P. S., 1987, "Radiative Heat Transfer in Emitting–Absorbing–Scattering Spherical Media," *J. Thermophys. Heat Transfer*, 1, 162–170.

Tong, T. W. and Tien, C.-L., 1983, "Radiative Heat Transfer in Fibrous Insulations—Part 1: Analytical Study," *ASME J. Heat Transfer*, 105, 70–75.

Tong, T. W., Yang, Q. S., and Tien, C.-L., 1983, "Radiative Heat Transfer in Fibrous Insulations—Part 2: Experimental Study," *ASME J. Heat Transfer*, 105, 76–81.

Truelove, J. S., 1987, "Discrete Ordinates Solutions of the Radiative Transport Equation," *ASME J. Heat Transfer*, 109, 1048–1051.

Tseng, J. W. C., Xia, Y., and Strieder, W., 1992, "Monte Carlo Calculations of Wall-to-Random-Bed View Factors: Impermeable Spheres and Fibers," *AIChE J.*, 38, 955–958.

van de Hulst, H. C., 1981, *Light Scattering by Small Particles*, Dover.

Vortmeyer, D., 1978, "Radiation in Packed Solids," in *Proceedings of 6th International Heat Transfer Conference* (Toronto), 6, 525–539.

Wang, K. Y. and Tien, C.-L., 1983, "Thermal Insulation in Flow Systems: Combined Radiation and Convection Through a Porous Segment," *ASME*, paper no. 83-WA/HT-81.

Weast, R. C., ed., 1987, *Handbook of Chemistry and Physics*, 68th ed., C.R.C. Press.

Weaver, J. H., ed., 1981, *Optical Properties of Metals*, Fachuvnfarmationszentrum Energie, Physik, Mathematik Gmbh.

Wiscombe, W. J., 1977, "The Delta–M Method: Rapid Yet Accurate Flux Calculations for Strongly Asymmetric Phase Functions," *J. Atm. Sci.*, 34, 1408–1422.

Wolf, J. R., Tseng, J. W. C., and Strieder, W., 1990, "Radiative Conductivity for a Random Void–Solid Medium with Diffusely Reflecting Surfaces," *Int. J. Heat Mass Transfer*, 33, 725–734.

Xia, Y. and Strieder, W., 1994a, "Complementary Upper and Lower Truncated Sum, Multiple Scattering Bounds on the Effective Emissivity," *Int. J. Heat Mass Transfer*, 37, 443–450.

Xia, Y. and Strieder, W., 1994b, "Variational Calculation of the Effective Emissivity for a Random Bed," *Int. J. Heat Mass Transfer*, 37, 451–460.

Yamada, Y., Cartigny, J. D., and Tien, C.-L., 1986, "Radiative Transfer with Dependent Scattering by Particles: Part 2—Experimental Investigation," *ASME J. Heat Transfer*, 108, 614–618.

Yang, Y. S., Howell, J. R., and Klein, D. E., 1983, "Radiative Heat Transfer Through a Randomly Packed Bed of Spheres by the Monte Carlo Method," *ASME J. Heat Transfer*, 105, 325–332.

6
Mass Transfer in Gases

Although throughout the text we consider both *liquids* and *gases*, in this chapter we consider mass transfer in gases *only*. We examine the observed *deviations* from the viscous flow behavior at *low pressures* (or for very small pore sizes) and consider *chemical reactions*. The subject of gas diffusion in porous media has been extensively treated by Jackson (1977), Cunningham and Williams (1980), and Wakao and Kaguei (1982). Jackson gives special attention to the *multicomponent* gas mixtures (more than *two* components), even though for more than three components and at low gas pressures where the *molecular slip* occurs, the treatment becomes very difficult because of the *interdependence* of the individual *mass fluxes*.

In this chapter, we begin with the experimental observation of *Knudsen* on the *slip* flow in *capillary* tubes at *low* pressures. The *high*-pressure (*viscous*), *low*-pressure (*Knudsen*), and *intermediate regime* of gas flow through small tubes and pores will be discussed. Then we will review the various molecular diffusions and the *predictions* based on the *kinetic theory*. This includes a prediction based on the inclusion of the solid matrix as *heavy particles*. Then we will develop the species conservation equation for the flow and mass transfer including chemical reaction in porous media. The local volume-averaging technique, along with the available results for the *total mass diffusivity tensor*, is discussed.

Some of the fundamentals used in this chapter in the development of the volume-averaged *mass* conservation (including components and overall) equation are given here.

- The gas velocity is based on the *mass-averaged velocity*, i.e.,

$$\mathbf{u} = \frac{\sum_{i=1}^{n_r} \rho_i \mathbf{u}_i}{\sum_{i=1}^{n_r} \rho_i}, \qquad (6.1)$$

where $\mathbf{u}_i$ is the *Eulerian* velocity of component i and $\sum_{i=1}^{n_r} \rho_i = \rho$.

- *The ideal gas behavior* is assumed, i.e.,

$$n_i = \frac{p_i}{k_B T}, \quad \rho = \frac{pM}{R_g T}, \quad \rho_i = \frac{p_i M_i}{R_g T}, \quad p = \sum_{i=1}^{n_r} p_i, \qquad (6.2)$$

where R_g is the *universal gas constant* (8.3144 kJ/kg·mol-K), and M

is the *average molecular weight* (kg/kg·mol). The average molecular weight is given as $M = \sum_{i=1}^{n_r} p_i M_i / p$. The universal gas constant is given in terms of the fundamental *Boltzmann* k_B and *Avogadro* N_A *constants* as

$$R_g = N_A k_B, \qquad (6.3)$$

where $k_B = 1.381 \times 10^{-23}$ (J/K) and $N_A = 6.0225 \times 10^{23}$ (molecules/g·mol).

- The *mean free path of molecules*, the average distance the molecules travel between two successive collisions, is given as

$$\lambda = \frac{k_B T}{2^{5/2} \pi R_m^2 p}, \qquad (6.4)$$

where πR_m^2 is the *collision cross section* of the molecules and R_m is the radius and is of the order of 1 Å. This radius does not vary significantly, for example, $2R_m \simeq 2.7$ Å for hydrogen, 3.681 Å for nitrogen, 3.433 Å for oxygen, and 3.61 Å for air. For water vapor $2R_m$ is 4.6 Å, and for carbon dioxide it is 3.996 Å (Tien and Lienhard, 1971, 331–332; Kanury, 1988, p. 53).

- The *Knudsen number*, which is a measure of the *probability* of the *molecule-molecule collision* compared to that for the *molecule-matrix surface collision*, is

$$Kn = \frac{\lambda \text{ (mean free path of gas molecules)}}{C \text{ (average pore size or interparticle clearance)}}. \qquad (6.5)$$

The *interparticle spacing* for spherical particles is estimated by (5.114). When $Kn \gg 1$, the flow is called the *Knudsen flow*, and when $Kn \ll 1$ the flow is called the *viscous flow*. The regime between these two is called the *transition flow* regime.

- *Stoichiometric reaction* of the form

$$\sum_{i=1}^{n_r} \nu_{ri} \chi_i \rightarrow \sum_{i=1}^{n_r} \nu_{pi} \chi_i \qquad (6.6)$$

is assumed, where ν_{ri} is the *stoichiometric coefficient* (number of moles) for the *reactant* component i and ν_{pi} is for the *products*. For components that are not reactants, $\nu_{ri} = 0$ and for those that are not products, $\nu_{pi} = 0$. n_r is the *number of reaction components* (reactants and products) and χ_i represents the *chemical symbol* for the component i.

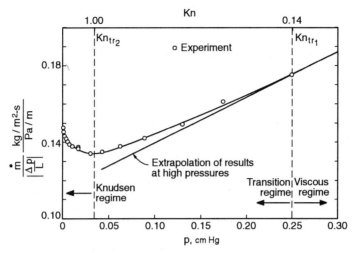

Figure 6.1 The results of Knudsen experiment showing the mass flow rate through a bundle of capillary tubes at various absolute pressures. (From Cunningham and Williams, reproduced by permission ©1980 Plenum.)

6.1 Knudsen Flows

Consider flow of an ideal gas in a small-diameter straight tube. For the fully developed flows and when $Kn \ll 1$, we have the pressure given by the Poiseuille-Hagen relation. This relation given in terms of the mass flow rate $\dot{m}$ is

$$\dot{m} = \rho u_p = -\frac{R^2}{8\mu} \frac{pM}{R_g T} \frac{dp}{dx}, \qquad (6.7)$$

where u_p is the average pore velocity. In 1909, Knudsen performed an experiment with gas flow through capillary tubes. He changed the pressure gradient (pressure drop along the tubes) as well as the absolute pressure. In one of his experiments, he uses carbon dioxide at 25°C flowing through 24 capillary tubes of $d = 3.33 \times 10^{-2}$ mm and $L = 2$ cm. The measured mass flow rate as a function of the average absolute pressure is given in Figure 6.1. For the viscous flow regime ($Kn \ll 1$), we expect from (6.7)

$$\frac{\dot{m}}{\left|\dfrac{dp}{dx}\right|} = \frac{R^2}{8\mu} \frac{pM}{R_g T}. \qquad (6.8)$$

For *isothermal* flows from this, we note that $\dot{m}/|dp/dx|$ is proportional to the *average pressure*.

However, his results for $Kn > 0.14$ *deviate* from this viscous flow behavior with the mass flow being *larger* than that expected for purely viscous flows. The increase in $\dot{m}$ is associated with the *molecular slip* at the fluid-solid interface A_{fs}.

Now we assume a *slip-boundary condition* of the type

$$\tau_R = -\mu \frac{du}{dr} = \alpha u_i \quad \text{on} \quad A_{fs}, \tag{6.9}$$

where τ_R is the *wall shear stress* and α is the *slip friction coefficient*. The no-slip asymptote is reached for $\alpha \to \infty$, where we have $u_i \to 0$ (since τ is finite). Using this boundary condition, we have

$$\frac{\dot{m}}{\left|\frac{dp}{dx}\right|} = \left(\frac{R^2}{8\mu} + \frac{R}{2\alpha}\right) \frac{pM}{R_g T}. \tag{6.10}$$

This accounts for the *extra* mass flow rate; α can be determined from the experimental results.

This revision of the viscous flow regime, which allows for the slip, does *not* predict the flow for $Kn \gg 1$. For that, Knudsen gives a *semiempirical* relation of the form

$$\frac{\dot{m}}{\left|\frac{dp}{dx}\right|} = \left(a_1 p + D_K \frac{1 + a_2 p}{1 + a_3 p}\right) \frac{M}{R_g T}, \tag{6.11}$$

where D_K is the *Knudsen diffusivity* (m^2/s) and is defined such that

$$\frac{\dot{m}}{\left|\frac{dp}{dx}\right|} = D_K \frac{M}{R_g T} \quad \text{or} \quad \dot{m} = -D_K \frac{dp}{dx} \quad \text{for} \quad p \to 0. \tag{6.12}$$

The constants a_1, a_2, a_3, and D_K depend on the gas and tube material properties and are determined experimentally. By comparing (6.10) and (6.11), we note that α is proportional to p, which is expected since $\alpha \to \infty$ as the pressure increases and the viscous flow regime is reached.

Another semiempirical relation is given by Weber (Cunningham and Williams, 1980, p. 51) as

$$\dot{m} = \underbrace{-\frac{K}{\mu} \frac{pM}{R_g T} \frac{dp}{dx}}_{\text{viscous flow}} \underbrace{- \frac{4}{3}\left(1 - \frac{Kn}{1 + Kn}\right) D_{\text{slip}} \frac{M}{R_g T} \frac{dp}{dx}}_{\text{slip flux}}$$

$$\underbrace{- \frac{Kn}{1 + Kn} D_K \frac{M}{R_g T} \frac{dp}{dx}}_{\text{Knudsen flux}}, \tag{6.13}$$

where D_{slip} [which for this problem is $pR/(2\alpha)$] and D_K are determined using the *kinetic theory* (this will be discussed in Section 6.5). This predicts the *minimum* in $\dot{m}/|dp/dx|$ observed in Figure 6.1.

The presence of the minimum in $\dot{m}$ is the result of the *extension* of the viscosity effect (no-slip). This effect is *weakened* by the increase in λ at low pressures. This minimum in $\dot{m}(p)$ is *not* found in most *porous media*. Exceptions are the very *high* porosity media such as steel wools. The reason for this lack of a minimum is that for *long* tubes (or large L/d) and at low pressures, the molecules can *travel* a long distance between any two successive wall or molecule collisions. However, in other porous media, this length is *limited* by the pore length (where usually the average linear dimensions in all directions are approximately equal).

We now turn to the general treatment of mass diffusion in porous media. However, before embarking on the *single-medium* treatment of mass diffusion in porous media, we first identify the various *mass diffusions* and examine the predictions of the kinetic theory for these diffusions. An extension of the kinetic theory of gases, which treats the elements of the solid matrix as particles with large mass, is also reviewed.

6.2 Fick Diffusion

Observations and *phenomenological* descriptions of mass diffusion at relatively *high* pressures show that the diffusion rate (mass flux in the absence of a total pressure gradient) is proportional to the *concentration* (or component density) *gradient*. This means that the *diffusion flux* for component i is

$$\rho_i \mathbf{u}_i = \dot{\mathbf{m}}_i = -D_{mi} \nabla \rho_i \quad \text{for} \quad \nabla p = 0, \tag{6.14}$$

where D_{mi} is the *mass diffusion coefficient* (m^2/s) of component i diffusing into the other components. The coefficient D_{mi} is a function of component density ρ_i as well as the density of the other components. The relationship between $\rho_i \mathbf{u}_i$ and $\nabla \rho_i$ is called the *Fick law*. When dealing with mass diffusion in porous media, the Fick law and the mass diffusion coefficient have to be modified to allow for the presence of the matrix. This is dealt with in a manner similar to that used for heat transfer in Chapters 3 and 4, i.e., through a single-continuum treatment of the gas diffusion/convection. This local volume-averaged treatment is given in Section 6.7.

When dealing with very low pressure, i.e., $Kn \gg 1$ (*Knudsen diffusion*), the presence of the other molecules do not influence the rate of diffusion. Instead the molecule-matrix surface collision limits the flow. In this case, the Fick law and the mass diffusion coefficient are no longer applicable (although a similar relationship based on the molecule-surface collision is used).

6.3 Knudsen Diffusion

When the Knudsen number $Kn = \lambda/C$ is *larger* than 10 (a general criterion that applies to porous media), the resistance to the flow of the component i of the gas mixture is dominated by the molecule-particle collisions. This leads to a diffusive flux given by

$$\rho_i \mathbf{u}_i = \dot{\mathbf{m}}_i = -D_{Ki} \nabla \rho_i \quad i = 1, 2, 3, \ldots, n_r, \tag{6.15}$$

$$\rho \mathbf{u} = \dot{\mathbf{m}} = -\sum_{i=1}^{n_r} D_{Ki} \nabla \rho_i, \tag{6.16}$$

where D_{Ki} is the *Knudsen diffusion coefficient* for component i and is independent of the presence of other components.

6.4 Crossed Diffusion

In principle, the existence of a *gradient* of a given intensive thermodynamic property can produce a *flux* of another property. For example, for a gas mixture, a temperature gradient can cause a flux of the enthalpy as well as a mass diffusion flux (this is called *thermal mass diffusion*). This phenomena is called the *crossed diffusion*. For the i component of the gas mixture, and by considering n different *driving forces* $\nabla \phi$ and assuming *linear* relationships with the driving forces (this is expected to be valid if the *deviations* from the equilibrium state are *not* very large), the diffusion mass flux is given as

$$\dot{\mathbf{m}}_i = -\sum_{j=1}^{n} D_{ik}^j \nabla \phi_j. \tag{6.17}$$

The analysis of *Onsager* is given in Cunningham and Williams. Some of the driving forces that can contribute to *mass transfer* are

- gradient of *concentration* ($D_{ik}^j = D_{mi}$),
- gradient of *total pressure*,
- gradient of *potential energy* (body force), and
- gradient of *temperature* (called the *Soret* effect).

The coefficients D_{ik}^j, other than D_{mi}, are generally very *difficult* to measure, and except in a few cases, their accurate magnitudes are *not* available. In principle, the kinetic theory allows for the prediction of these coefficients. The Soret diffusion for low molecular weight species (H, H_2, and He) is treated by Kee et al. (1988). Based on this treatment, the species diffuse *from* the *low-* to the *high*-temperature regions. We now review the prediction of the various diffusivities using the kinetic theory.

6.5 Prediction of Transport Coefficients from Kinetic Theory

An extensive review of the experimental and theoretical treatments of the transport properties of gases and liquids is given by Reid et al. (1987). The kinetic theory of gases is given by Vincenti and Kruger (1967) and Tien and Lienhard (1971), among others. In relation to flow of gases and inclusion of higher-order approximations, a development is given by Cunningham and Williams (1980, Chapters 2 and 3). The following *brief* introduction deals with the simple treatment of molecular collisions and is based on the results given by Cunningham and Williams. A few important results of the kinetic theory that will be used in the treatment of the diffusivities are given here.

- The *mean molecular* speed for component i is given as

$$\overline{u}_{m\,i} = \left(\frac{8R_g T}{\pi M_i}\right)^{1/2}. \qquad (6.18)$$

- The *average* value of the *velocity* component in the x-direction is

$$\overline{u}_{m\,xi} = \frac{1}{4}\overline{u}_{m\,i}. \qquad (6.19)$$

- The *molecular mass flux* through the y-z plane is given by

$$\dot{n}_{m\,i} = \frac{1}{4} n\,\overline{u}_{m\,i}, \qquad (6.20)$$

where $n = p/(k_B T)$ is *the number of molecules per unit volume*.

- The *net mass transport* of component i in the x-direction is

$$\dot{m}_i = -\frac{1}{2}\overline{u}_{m\,i}\lambda_i\frac{\partial \rho_i}{\partial x}. \qquad (6.21)$$

With these results, we now consider the predictions of the diffusivities.

6.5.1 Fick Diffusivity in Plain Media

For a *binary* gas mixture containing components 1 and 2 and under *isobaric* condition (i.e., the molecular density, including both components, is uniform), the *molecular fluxes* of the two components are equal and opposite in direction. For component 1 diffusing *into* component 2, we have (Cunningham and Williams, 1980, p. 82)

$$D_{21} = D_{12} = D_m = \frac{n_1 \overline{u}_1 \lambda_1 + n_2 \overline{u}_2 \lambda_2}{2(n_1 + n_2)}. \qquad (6.22)$$

The *modified* (because of the presence of the other component) *mean free path* for 1 is

$$\lambda_1^{-1} = 4n_2\pi \left(\frac{R_{m_1} + R_{m_2}}{2}\right)^2 \left(1 + \frac{M_1}{M_2}\right)^{1/2}. \qquad (6.23)$$

These lead to the *Chapman-Enskog* mass diffusivity

$$D_m = \frac{1}{\pi(R_{m_1} + R_{m_2})^2(n_1 + n_2)} \left[\frac{2R_g T}{\pi}\left(\frac{1}{M_1} + \frac{1}{M_2}\right)\right]^{1/2} \qquad (6.24)$$

or

$$D_m = \frac{2^{1/2}(R_g T)^{3/2}}{\pi^{3/2}(R_{m_1} + R_{m_2})^2 p} \left(\frac{1}{M_1} + \frac{1}{M_2}\right)^{1/2}. \qquad (6.25)$$

The predicted pressure dependency *agrees* with the experimental results, while the power for the temperature dependency is *higher* in the experimental results. As useful relations, the *gas dynamic viscosity* and *molecular thermal conductivity* for a *single*-component gas are also given here. From Tien and Lienhard (1971, p. 320), we have

$$\mu = \frac{1}{4\pi^{3/2} R_m^2}(R_g T M)^{1/2} \qquad (6.26)$$

and

$$k = \frac{5}{8}\frac{c_v}{\pi^2 R_m^2}(\pi M R_g T)^{1/2}. \qquad (6.27)$$

More extensive treatments, including that of gas mixtures with more than two components, are given by Reid et al. (1987). Accurate and efficient evaluations of the transport properties (viscosity, thermal conductivity, and mass diffusivity), based on more rigorous treatments of the gas kinetics, are given by Kee et al. (1988).

6.5.2 KNUDSEN DIFFUSIVITY FOR TUBE FLOWS

Since the *surface roughness* determines whether the reflection of the molecules colliding with wall is *specular* or *diffusive*, the general treatment of the *Knudsen* diffusion includes a parameter describing the fraction of the reflection, which is *diffuse*. Here we assume that *all reflections are diffuse*. The Knudsen diffusivity is found from consideration of the momentum conservation (applied to flow through a tube) and leads to

$$\tau_R = \frac{3}{4}nk_B T \frac{\overline{u}_{mx}}{\overline{u}_m} = -\frac{R}{2}\frac{dp}{dx}, \qquad (6.28)$$

where as before τ_R is the shear stress at $r = R$. The *Knudsen* flux is given as

$$\dot{m}_K = \rho \overline{u}_{mx} = \frac{pM}{R_g T}\overline{u}_{mx}. \qquad (6.29)$$

6.5 Prediction of Transport Coefficients from Kinetic Theory

Then, from the preceding two equations we have

$$\dot{m}_K = \frac{4}{3}\frac{\bar{u}M}{R_gT}\tau_R = -\frac{2}{3}\frac{R\bar{u}_mM}{R_gT}\frac{dp}{dx}, \qquad (6.30)$$

which gives the *Knudsen diffusivity* as

$$D_K = \frac{2}{3}R\bar{u}_m = \frac{2}{3}R\left(\frac{8R_gT}{\pi M}\right)^{1/2}. \qquad (6.31)$$

For *porous media*, the Knudsen diffusion coefficient is given as

$$D_K = C_K\left(\frac{R_gT}{M}\right)^{1/2}, \qquad (6.32)$$

where the constant C_K depends on the solid matrix structure and is determined experimentally (this is discussed in Section 6.6).

6.5.3 SLIP SELF-DIFFUSIVITY FOR TUBE FLOWS

The slip friction coefficient α is found by a momentum balance between the wall shear stress and the momentum flux, and this leads to

$$\alpha = \left(\frac{2M}{\pi R_gT}\right)^{1/2}p. \qquad (6.33)$$

Now, since the *viscous flux* is given by the Darcy law and the *slip flux* is given by

$$\dot{m}_{\text{slip}} = -\frac{R}{2\alpha}\frac{pM}{R_gT}\frac{dp}{dx}, \qquad (6.34)$$

the *ratio* of $\dot{m}_{\text{slip}}$ to $\dot{m}_{\text{visc}}$ is found by taking the interparticle spacing C as $C = 2R$, resulting in

$$\frac{\dot{m}_{\text{slip}}}{\dot{m}_{\text{visc}}} = 8Kn = 8\frac{\mu}{\rho\bar{u}_mR}, \qquad (6.35)$$

which can be used to *estimate* this ratio. Finally the *slip self-diffusivity*, which from (6.10) is equal to $Rp/(2\alpha)$, is found as

$$D_{\text{slip}} = \frac{\pi R\bar{u}_m}{8}, \qquad (6.36)$$

which is *independent* of the pressure.

6.5.4 ADSORPTION AND SURFACE FLUX

The gas molecules can be *absorbed to the surface* of the matrix (the surface absorption is called *adsorption*, or specifically, *enrichment* of one or more components in an *interfacial layer*) due to the interaction between the gas and solid *surface molecular forces*. The gain in the gas energy is called the *heat of adsorption* ΔE_a (also called *activation energy*). When the *adsorbed* molecules gain enough energy, they can leave the surface by overcoming this activation energy. Such a gained energy can be due to an increase in the temperature. The *molar heat of absorption* $\Delta E_a M$ varies from 0.1 to 10^2 kcal/mol. The adsorbed molecules can also *drift* along the surface (*surface mobility of adsorbed gases*). The development of the field of *surface flux* is given in Jackson (1977, Chapter 7), Cunningham and Williams (1980, 52–60), and Chappuis (1982).

The adsorbed molecules *oscillate* with *frequency* of the order 10^{-13} s, and at equilibrium, molecules *adhere* and *leave* the surface at the *same* rate. The adsorption time is of the order of 10^{-6} s (Chappuis, 1982, p. 423), which is much *larger* than the *elastic specular* reflection contact time, which is equal to the oscillation period mentioned earlier.

The types of gas-solid bounds are *van der Waals* (*polar* and *nonpolar* gases and *electrical conducting* or *insulating* solids), *ionic*, and *covalent*. A different *classification* of forces is given by Israelachvili (1989, p. 221), where he defines (loosely) the following three categories.

- *Purely electrostatic* forces originating from the Coulomb force between charges. The interaction between charges, permanent dipoles, quadrupoles, etc., fall into this class.

- *Polarization* forces arising from the dipole moments induced in atoms and molecules by the electric fields of nearby charges and permanent dipoles. All interactions in a solvent medium involve polarization effects.

- *Quantum-mechanical* forces, which give rise to covalent bounding and repulsive exchange interactions that balance all reaction forces at very short distances.

The *adsorbed* molecules exert a *two-dimensional pressure* ϕ on the solid surface (similar to the surface tension, ϕ has the unit of force/length), which is related to the three-dimensional pressure by

$$\rho_{\text{ad}} = \frac{A_o p M}{R_g T} \frac{\mathrm{d}\phi}{\mathrm{d}p}, \tag{6.37}$$

where A_o is the specific surface area A_{fs}/V and ρ_{ad} is the *density of the adsorbed gas layer*.

The *adsorption isotherm* gives the relationship between ρ_{ad} and p through *sorption coefficient* H_{ad}, i.e.,

$$\left.\frac{\partial \rho_{\text{ad}}}{\partial p}\right|_T = H_{\text{ad}}. \tag{6.38}$$

6.5 Prediction of Transport Coefficients from Kinetic Theory

For the model of $\rho_{\rm ad}(p)$ given earlier (due to Gibbs) and assuming *linear isotherm for adsorption*, i.e.,

$$\rho_{\rm ad} = H_{\rm ad}\, p, \tag{6.39}$$

we have

$$H_{\rm ad} = \frac{A_o M}{R_g T}\frac{d\phi}{dp}. \tag{6.40}$$

Since $\phi = 0$ for $p = 0$, the preceding can be integrated to give

$$\phi M = m_{\rm ad} R_g T, \tag{6.41}$$

where $m_{\rm ad}$ is the *mass of the adsorbed gas per unit area* (the counterpart of density). This relation is called the *ideal gas law for two-dimensional gases*, which is based on a linear isotherm.

The *surface flux* driven by the pressure gradient ∇p, when the local thermal equilibrium exists between the surface adsorbed and the bulk gas, is of the form

$$\dot{m}_{\rm ad} = \frac{R_g T^{1/2}}{R_{\rm ad} A_o M}\frac{\rho_{\rm ad}^2}{p}\exp\left(\frac{-\Delta E_a M}{R_g T}\right)\nabla p, \tag{6.42}$$

where $R_{\rm ad}$ is the *adsorption resistance* and is a constant that depends on *gas-solid* material properties.

The surface flux can also be given in terms of the Fick diffusion as

$$\dot{m}_{\rm ad} = -D_{\rm ad}\, H_{\rm ad}\, M \exp\left(\frac{-\Delta E_a M}{R_g T}\right)\nabla p, \tag{6.43}$$

which, when compared to (6.42) gives

$$D_{\rm ad} = \frac{R_g T^{1/2}\rho_{\rm ad}^2}{A_o R_{\rm ad} H_{\rm ad}\, pM^2}. \tag{6.44}$$

In general, *empirical relations* are used for $D_{\rm ad}$. One of these is (Cunningham and Williams, 1980, p. 59)

$$D_{\rm ad}\,({\rm m^2/s}) = 1.6\times 10^{-6}\exp\left(-\frac{\Delta E_a M}{R_g T}\right). \tag{6.45}$$

Examination of this shows that as $T \to \infty$, the asymptotic value of $D_{\rm ad}$ is much *smaller* than the bulk mass diffusivity. Therefore, at high temperatures, the surface flux is negligible. Finally, note that the *surface* and *volumetric* fluxes are additive. Surface and Knudsen diffusion in *silica gel* (a desiccant) has been examined by Pesaran and Mills (1987).

6.6 Dusty-Gas Model for Transition Flows

The discussion of the Knudsen flow given in Section 6.1 was limited to flow through tubes. At low pressures, the inclusion of the effect of the solid matrix on the *mass* diffusion/convection requires a treatment different than that presented for the heat diffusion/convection. This is because at *low* pressures the gas/matrix interaction is different than that at *high* pressures. Therefore, in applying the local volume averaging, a special treatment of the transition flows (intermediate pressures) is required. One method of arriving at the *effective mass diffusion tensor* $\mathbf{D}_{me}$ and the *mass dispersion tensor* $\mathbf{D}_m^d$ is by the so-called *dusty-gas model*. In this model the dispersed phase (the solid matrix) is represented as *impermeable* spheres uniformly distributed in space with a *velocity* of zero (with respect to a stationary reference). The spheres are treated as a component of the gas and are assumed to be *larger* and *heavier* than the gas molecules. The drag force on the sphere is balanced by an external force. The *modified* kinetic theory is then applied to the combination particles-molecules ensemble. Note that when dealing with *low* pressures, where the dusty gas model is expected to be valuable because of its direct inclusion of molecule-solid surface interactions, the momentum equation for the fluid must also be *modified* to include the *Knudsen transition* flows. Therefore, the treatment of *dilute* gas flows requires a special treatment (even for the energy equation).

In the presence of both a *total pressure gradient* and a *concentration gradient*, Gunn and King (1969) [a derivation is also given by Jackson (1977, p. 42)] give the mass flow equation (both *viscous* and *diffusion*) for component 1 in a *binary mixture* as

$$\mathbf{m}_1 = -\frac{D_{me}D_{K_1}}{D_{me} + \dfrac{\rho_1 D_{K_1} + \rho_2 D_{K_2}}{\rho}}\frac{pM_1}{R_g T}\nabla\rho_1$$

$$-\left[\frac{D_{K_1}(D_{me} + D_{K_2})}{D_{me} + \dfrac{\rho_1 D_{K_1} + \rho_2 D_{K_2}}{\rho}} + \frac{Kp}{\mu_m}\right]\frac{\rho_1 M_1}{\rho R_g T}\nabla p, \quad (6.46)$$

with

$$p = p_1 + p_2, \quad (6.47)$$

$$p_1 = \frac{\rho_1}{\rho}p, \quad (6.48)$$

$$\rho_1 + \rho_2 = \rho, \quad (6.49)$$

where μ_m is the *viscosity* of the *mixture* and as before $D_{K_1} = C_K(R_g T/M_1)^{1/2}$.

The two constants C_K (unit of m) and D_{me}/D_m depend on the structure of the porous medium and are determined experimentally.

We note that using the Darcy law for the gas mixture flows is allowed when (Gunn and King)

$$M^{1/2} = \frac{\rho_1}{\rho} M_1^{1/2} + \frac{\rho_2}{\rho} M_2^{1/2} \qquad (6.50)$$

is used in determining the flow *rate* of a pure gas equivalent of the binary mixture.

The constant C_K, which gives the *relative* Knudsen flow, was measured by Gunn and King for a *fritted-glass filter* and found to be 7×10^{-7} m. The ratio of the *effective binary mass diffusivity* $D_{m\,e}$ to the binary mass diffusivity D_m is found simultaneously with C_K by using the least-square approximation of the experimental data. Since $D_{m\,e}/D_m$ is *not* expected to depend on the pressure, we expect the results for a given solid matrix (e.g., Section 6.9.2) to be of general applicability (i.e., independent of gas, pressure, and temperature).

The dusty gas model for more than two components is discussed by Jackson (1977, Chapters 3 and 4).

6.7 Local Volume-Averaged Mass Conservation Equation

There are *two* features of the mass transfer in porous media that make for a treatment different than that applied to the heat transfer in Chapters 3 and 4. The first is the *low-pressure* behavior of gases discussed before; the second is the *inclusion* of *chemical reaction* in the pore *volume* and on the *surface* of the matrix (where this surface reaction is *stronger* in the case of catalytic packed-bed reactors).

The local volume-averaged treatment of endothermic/exothermic chemical reactions in porous media requires the simultaneous treatment of the fluid flow, and mass and heat transfer. In the heat transfer treatment, when valid, local thermal equilibrium between phases is imposed, otherwise local thermal nonequilibrium will be assumed (Chapter 7). In the mass transfer treatment, assumptions about local chemical equilibrium or nonequilibrium will be made. Figure 6.2 shows the theoretical treatment of reaction in porous media and the various aspects relating to transport, reaction, and phase change.

The approach taken here is to *arrive* at the local volume-averaged mass conservation equations. This will be done along the lines of the analysis of Slattery (1981, 565–578) using some of the steps already reviewed in Chapters 3 and 4. We assume that both the Fick diffusion law and the *no-slip* condition on the solid surface *hold*. Therefore, we limit the inclusion of the low pressure effects to the *modification* of the *effective* mass diffusivity only. In addition, we will consider the cases where the solid surface is

378   6. Mass Transfer in Gases

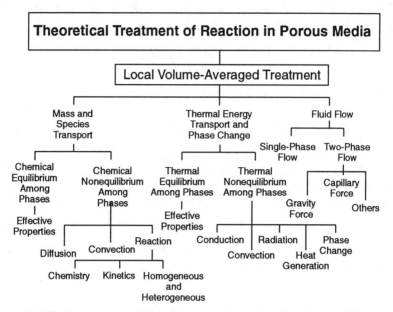

Figure 6.2 Various aspects of the treatment of reaction in porous media.

impermeable. The differential equation of *continuity of component i* is (Bird et al., 1960, p.502)

$$\frac{\partial \rho_i}{\partial t} + \nabla \cdot (\rho_i \mathbf{u}_i) = \dot{n}_i, \qquad (6.51)$$

where the *diffusion equation* is

$$\rho_i \mathbf{u}_i = \rho_i \mathbf{u} - D_{mi} \rho \nabla \frac{\rho_i}{\rho} \qquad (6.52)$$

or

$$\rho_i (\mathbf{u}_i - \mathbf{u}) = -D_{mi} \rho \nabla \frac{\rho_i}{\rho}. \qquad (6.53)$$

By combining the continuity and diffusion equations, we have

$$\frac{\partial \rho_i}{\partial t} + \nabla \cdot (\rho_i \mathbf{u}) = \nabla \cdot D_{mi} \rho \nabla \frac{\rho_i}{\rho} + \dot{n}_i, \qquad (6.54)$$

where $\dot{n}_i$ is the *volumetric rate of production* of component i and $\mathbf{u}$ is the mass averaged velocity given by (6.1). We also note that $\sum_{i=1}^{n_r} \dot{n}_i = 0$.

The differential equation of the *overall continuity* is

$$\frac{\partial \rho}{\partial t} + \nabla \cdot (\rho \mathbf{u}) = 0. \qquad (6.55)$$

6.7 Local Volume-Averaged Mass Conservation Equation

For the total fluid phase, the local volume average of this equation is given in Section 2.7.3, where we used $\mathbf{u}_i = 0$ on A_{fs}. The result is

$$\frac{\partial \langle \rho \rangle^f}{\partial t} + \nabla \cdot \langle \rho \mathbf{u} \rangle^f = 0. \tag{6.56}$$

Now introducing

$$\mathbf{u} = \langle \mathbf{u} \rangle^f + \mathbf{u}' \tag{6.57}$$

and

$$\rho_i = \langle \rho_i \rangle^f + \rho_i', \tag{6.58}$$

and by assuming that $\langle \rho \rangle^f = \rho$, the fluid phase, volume average of (6.54) is taken as before, and we have

$$\frac{\partial \langle \rho_i \rangle^f}{\partial t} + \nabla \cdot \langle \rho_i \mathbf{u} \rangle^f = \nabla \cdot \left[D_{mi} \langle \rho \rangle^f \nabla \langle \frac{\rho_i}{\rho} \rangle^f + \frac{D_{mi}}{V_f} \langle \rho \rangle^f \int_{A_{fs}} \mathbf{n} \frac{\rho_i}{\rho} \, dA \right]$$
$$+ \frac{D_{mi}}{V_f} \langle \rho \rangle^f \int_{A_{fs}} \mathbf{n} \cdot \nabla \frac{\rho_i}{\rho} \, dA + \langle \dot{n}_i \rangle^f. \tag{6.59}$$

Because we have assumed that the matrix is impermeable, the *surface (heterogeneous) reaction* is given by

$$\frac{D_{mi}}{V_f} \langle \rho \rangle^f \int_{A_{fs}} \mathbf{n} \cdot \nabla \frac{\rho_i}{\langle \rho \rangle^f} \, dA \equiv \langle \dot{n}_i \rangle^{fs}. \tag{6.60}$$

This is due to *surface catalytic reaction* or *physical absorption/desorption* and is designated as $\langle \dot{n}_i \rangle^{fs}$. Then we have

$$\frac{\partial \langle \rho_i \rangle^f}{\partial t} + \nabla \cdot \langle \rho_i \mathbf{u} \rangle^f = \nabla \cdot D_{mi} \langle \rho \rangle^f \left[\nabla \frac{\langle \rho_i \rangle^f}{\langle \rho \rangle^f} + \frac{1}{V_f} \int_{A_{fs}} \mathbf{n} \frac{\rho_i}{\rho} \, dA \right]$$
$$+ \langle \dot{n}_i \rangle^{fs} + \langle \dot{n}_i \rangle^f. \tag{6.61}$$

Now decomposing the velocity, we have

$$\frac{\partial \langle \rho_i \rangle^f}{\partial t} + \nabla \cdot \langle \mathbf{u} \rangle^f \langle \rho_i \rangle^f$$
$$= \nabla \cdot \left[D_{mi} \langle \rho \rangle^f \nabla \frac{\langle \rho_i' \rangle^f}{\langle \rho \rangle^f} + D_{mi} \frac{\langle \rho \rangle^f}{V_f} \int_{A_{fs}} \mathbf{n} \frac{\rho_i}{\rho} \, dA - \langle \mathbf{u}' \rho_i' \rangle^f \right]$$
$$+ \langle \dot{n}_i \rangle^{fs} + \langle \dot{n}_i \rangle^f. \tag{6.62}$$

This equation is different than that used for the convection heat transfer by Carbonell and Whitaker (1984) and given in Section 4.3.1, because it includes the *surface* and *volumetric* sources. The volumetric source is *not* expected to be uniform in the pore volume because the *density disturbance* for the i component is not uniform within it.

We write (6.62) as

$$\frac{\partial \langle \rho_i \rangle^f}{\partial t} + \nabla \cdot \langle \mathbf{u} \rangle^f \langle \rho_i \rangle^f = \nabla \cdot \mathbf{D}_{mi} \langle \rho \rangle^f \cdot \nabla \frac{\langle \rho_i \rangle^f}{\langle \rho \rangle^f} + \langle \dot{n}_i \rangle^{fs} + \langle \dot{n}_i \rangle^f, \quad (6.63)$$

where the *total effective mass diffusivity tensor* is

$$\mathbf{D}_{mi} = \mathbf{D}_{mi\,e} + \mathbf{D}_{mi}^d. \quad (6.64)$$

In general, the volume-averaged quantities, e.g., $\mathbf{D}_{mi\,e}$ the *effective mass diffusivity tensor*, $\mathbf{D}_{mi}^d$ the *mass dispersion tensor*, $\langle \dot{n}_i \rangle^f$ the *volume-averaged homogeneous production rate*, and $\langle \dot{n}_i \rangle^{fs}$ the *volume-averaged heterogeneous production rate*, all depend on the chemical kinetics (volumetric and surface).

For gases, the volume-averaged density of component i is related to its partial pressure as

$$\langle \rho_i \rangle^f = \frac{\langle p_i \rangle^f M_i}{R_g T}, \quad (6.65)$$

where $\langle p \rangle^f = \sum_{i=1}^{n_r} \langle p_i \rangle^f$, and we have

$$\langle \rho \rangle^f = \frac{\langle p \rangle^f M}{R_g T}. \quad (6.66)$$

The *extra constraint* on the production is

$$\sum_{i=1}^{n_r} \left[\langle \dot{n}_i \rangle^{fs} + \langle \dot{n}_i \rangle^f \right] = 0. \quad (6.67)$$

6.8 Chemical Reactions

A variety of chemical reactions, of engineering and geological interests, occurs in porous media. Figure 6.3 shows a classification of the various aspects related to the porous medium, fluid, reaction, flow, and phase change. A pore-level rendering of transport, reaction and phase change is shown in Figures 6.4(a) and (b), for single- and two-phase flows, respectively. The reaction in the fluid phase may involve species which are released in the fluid phase by a solid-fluid phase change (melting or devolatization) and the product of reaction occuring in the fluid phase may be in solid (precipitation or condensation).

6.8 Chemical Reactions 381

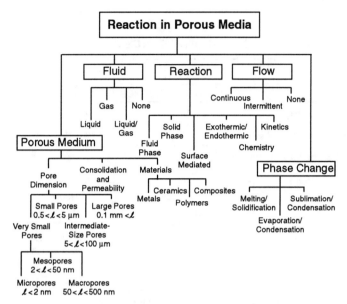

Figure 6.3 Various aspects of the porous media, fluid, reaction, flow and phase change influencing reaction in porous media.

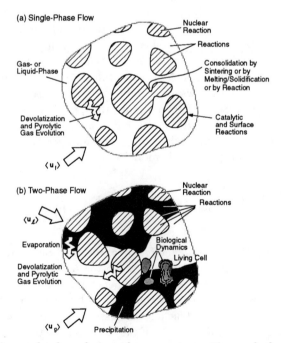

Figure 6.4 A pore-level rendering of transport, reaction and phase change for (a) single-phase, and (b) two-phase flows.

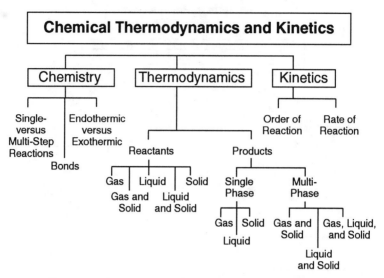

Figure 6.5 Chemistry, thermodynamic, and kinetic aspect of chemical reactions.

The description of the chemical reaction requires *thermodynamic* and *chemical kinetic* data which are generally obtained experimentally. Figure 6.5 shows various aspects of chemistry, *physical* and *chemical* thermodynamics and *chemical* kinetics as related to reaction in porous media.

The *rate* of production of component i depends on the chemical kinetics. For homogeneous reactions, one of the available *phenomenological chemical kinetic expressions* that allows for a more accurate *temperature dependence* is given later (Kuo, 1986, p. 209; Williams, 1988, p. 3). This rate expression is called the *modified Arrhenius model*. For m *number of chemical reactions* occurring, according to (6.6) where n_r is the number of components, the *production rate* of component i is modeled as

$$\dot{n}_i = M_i \sum_{k=1}^{m} (\nu_{pik} - \nu_{rik}) B_k T^{\alpha_k} \exp\left(-\frac{\Delta E_{ak}}{R_g T}\right) \prod_{j=1}^{n_r} \left(\frac{\rho_j}{M_j}\right)^{\nu_{rjk}}, \quad (6.68)$$

where B_k is the *collision-frequency factor* of the reaction k and is called the *preexponential factor* and where α_k is called the *temperature exponent*. B_k, α_k, and the *activation energy* ΔE_{ak} (on a per-molar basis) all depend on the specific chemical reaction. In some reactions, a third component is required for the reaction to proceed (such as in dissociation and recombination). Discussion of this and other aspects of the chemical reaction rate expression are given by Kee et al. (1989).

For heterogeneous reactions, where the reaction occurs on A_{fs}, the reaction rate is given as (Carbonell and Whitaker, 1984)

$$-D_{mi}\rho\mathbf{n}\cdot\nabla\frac{\rho_i}{\rho} = \kappa_i\rho_i \quad \text{on } A_{fs}, \tag{6.69}$$

where κ_i is the *heterogeneous reaction rate coefficient* for component i. In general, the reaction rate coefficient is given in terms of the activation energy and the concentration of the species involved in the reaction. As with the homogeneous reactions, the heterogeneous reactions can be of *any order*. In *combustion*, the *modeling* of heterogeneous reactions has been discussed by Harrison and Ernst (1978), Bruno et al. (1983), Fakheri and Buckius (1983), and Marteney and Kesten (1981). Here we assume the earlier simple relation holds where κ_i is constant and is given *a priori*. A review of catalysis in *combustion* is given by Pfefferle and Pfefferle (1987).

6.9 Evaluation of Total Mass Diffusivity Tensor

As we discussed, the mass diffusion (without convection) in porous media has been studied *extensively* and the effect of low pressures (Knudsen diffusion), surface diffusion, etc., has been addressed. The mass dispersion in porous media has also been studied *extensively*, and the treatment is similar to that for heat transfer; however, these treatments do *not* address the Knudsen and surface diffusions. We now summarize the available predictions of $\mathbf{D}_{m\,e}$ and $\mathbf{D}_m^d$. Here, for simplicity we shall assume that $\mathbf{D}_{mi\,e}$ and $\mathbf{D}_{mi}^d$ are *independent* of the reactions. A discussion of the effect of chemical reaction on $\mathbf{D}_{m\,e}$ is given by Ryan et al. (1980), where they summarize the available experimental results and conclude that this effect is *small*. Then, as for heat transfer, we have

$$\frac{\mathbf{D}_{mi}^d}{D_{mi}} = \frac{\mathbf{D}_{mi}^d}{D_{mi}}(Pe_i, Sc_i, \varepsilon, \text{structure}, \text{etc.})$$

and

$$\frac{\mathbf{D}_{mi\,e}}{D_{mi}} = \frac{\mathbf{D}_{mi\,e}}{D_{mi}}(\varepsilon, \text{etc.}). \tag{6.70}$$

6.9.1 EFFECTIVE MASS DIFFUSIVITY

Assuming that $\mathbf{D}_{m\,e}$ is *not* influenced by the occurrence of chemical reactions, the determination of $\mathbf{D}_{m\,e}$ follows the steps taken in Chapter 3 for $\mathbf{D}_e$. For $\mathbf{u} = 0$ and by taking $\langle n_i\rangle^f = \langle n_i\rangle^{fs} = 0$, the heat diffusion equation (3.19) applies equally to mass diffusion. The results given in Sections 3.2.3 and 3.6 which can be written in terms of $D_{m\,e}/D_{m\,f} = D_{m\,e}/D_m(\varepsilon, D_{m\,s}/D_{m\,f})$ and are plotted in Figures 3.1 and 3.9 can be used

384   6. Mass Transfer in Gases

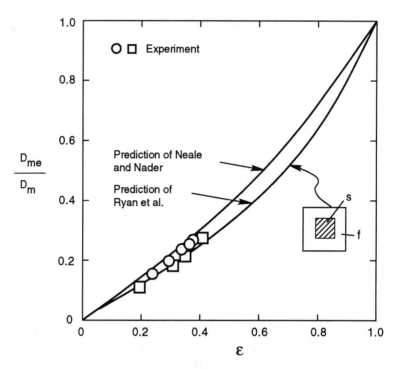

Figure 6.6 Variation of effective mass diffusivity with respect to porosity, for $D_{ms} = 0$. The prediction of Neale and Nader and Ryan et al., along with the experimental results (Ryan et al.), are shown.

for $D_{ms}/D_{mf} > 1$. However, for *most* applications $D_{ms}/D_{mf} < 1$, and $D_{ms}/D_{mf} \to 0$ is *generally* the applicable case. For $D_{ms}/D_{mf} = 0$, which is used in (6.62), $k_s = 0$. Ryan et al. (1980) have solved the equation for $\mathbf{b}_f$ along the lines discussed in Sections 3.2.1 and 3.2.2. Their results are given for *isotropic* media as

$$\frac{D_{me}}{D_{mf}} = \frac{D_{me}}{D_{mf}}(\varepsilon), \quad D_{ms} = 0. \tag{6.71}$$

They use a *two-dimensional periodic structure* with the unit cell being a *square* and the solid phase also being a *square* and located at the *center* of the cell. This unit cell, along with the predictions, is shown in Figure 6.6. Ryan et al. also examine some *anisotropic* media. As evident in Figure 6.6, their predictions, based on their isotropic unit cell, *match* the experimental results for isotropic media fairly well.

For the case of *packed beds of impermeable spheres*, the effective mass diffusion coefficient D_{me} for *binary gas* mixtures has also been determined by Neale and Nader (1973) through a geometric model and verified experimentally. Their model is similar to the one they use for the determination of permeability [Section 2.4.3 (B)], i.e., two concentric spheres placed in a porous medium with the *inner* sphere being *impermeable* and the interstitial volume between the spheres being plain. They find for *isotropic* media

$$\frac{D_{m\,e}}{D_{m\,f}} = \frac{2\varepsilon}{3-\varepsilon}, \qquad D_{m\,s} = 0. \tag{6.72}$$

The ratio $D_{m\,e}/D_{m\,f}$ has *traditionally* been given (Epstein, 1989) in terms of the *tortuosity factor* $(1+L_t^*)^2$ for porous media of arbitrary structure as $\varepsilon/(1+L_t^*)^2$. Note that the tortuosity L_t^* in *modern* usage is that defined in Section 4.3.1 by (4.38) and (4.43). In modern usage, $D_{m\,e}/D_{m\,f} = \varepsilon(1+L_t^*)$ for $D_{m\,s} = 0$, as discussed by Ryan et al.

6.9.2 MASS DISPERSION TENSOR

Theoretical treatment of convection-diffusion mass transfer in *ordered* porous media (spherical particles) is given by Koch et al. (1989) and for disordered porous media (spherical particles) by Koch and Brady (1985). In their analysis for *periodic* structures made of spherical particles, Koch et al. use $D_{m\,f} = D_{m\,s}$ and their unit-cell-based results are discussed in Section 4.4 for heat transfer (with $\alpha_f = \alpha_s$). Their analysis leads to

$$D^d_{m\,\|} = D^d_{m\,\|}\left(Pe_\ell, \varepsilon, \frac{D_{m\,f}}{D_{m\,s}} = 1\right), \tag{6.73}$$

$$D^d_{m\,\perp} = D^d_{m\,\perp}\left(Pe_\ell, \varepsilon, \frac{D_{m\,f}}{D_{m\,s}} = 1\right), \tag{6.74}$$

with $Pe_\ell = u_D\ell/D_m$, where ℓ is the unit cell. As shown in Figure 4.5, their predicted results are in relatively good agreement with the measurements.

For random arrangement of spherical particles, the results of the analysis of Koch and Brady, summarized in Table 4.1, are also applicable for mass transfer. Their results are in the form of

$$D^d_{m\,\|} = D^d_{m\,\|}\left(Pe_R, \varepsilon, \frac{D_{m\,f}}{D_{m\,s}}\right), \tag{6.75}$$

$$D^d_{m\,\perp} = D^d_{m\,\perp}\left(Pe_R, \varepsilon, \frac{D_{m\,f}}{D_{m\,s}}\right), \tag{6.76}$$

where $Pe_R = u_D R/D_m$. Their predictions are in very good agreement with the experimental results.

6.10 Evaluation of Local Volume-Averaged Source Terms

In some applications, including combustion and porous media, the variations of ρ_i in V_f and on A_{fs} are *noticeable* compared to $\langle \rho_i \rangle$. Special attention should be given to the local volume-averaged terms $\langle n_i\rangle^f$ and $\langle n_i\rangle^{fs}$. This area has *not* yet been examined. In the following, we assume that the variations of ρ_i are *negligibly* small compared to $\langle\rho_i\rangle$.

6.10.1 Homogeneous Reaction

The local phase volume-averaged rate of homogeneous production of component i, $\langle n_i\rangle^f$, is obtained by averaging (6.68) over the representative elementary volume. For the local density given by

$$\rho_i = \langle\rho_i\rangle^f + \rho_i', \qquad (6.77)$$

and by *not* including the effect of the T' on $\langle n_i\rangle$, we have

$$\langle n_i\rangle^f = M_i \sum_{k=1}^m (\nu_{pik} - \nu_{rik}) B_k \langle T\rangle^{\alpha_k}$$

$$\times \exp\left(-\frac{\Delta E_{ak}}{R_g \langle T\rangle}\right) \prod_{j=1}^{n_r} \frac{1}{V_f} \int_{V_f} \left(\frac{\langle\rho_i\rangle^f + \rho_i'}{M_i}\right)^{\nu_{rjk}} dV. \qquad (6.78)$$

Now, since ρ_i' occurs over length scale ℓ and significant changes in $\langle\rho_i\rangle^f$ are only allowed over L (a requirement for the local volume-averaging formulation), then $\rho_i' \ll \rho_i$. In some applications $\rho_i' \simeq O(\langle\rho_i\rangle^f)$, and for these a special treatment is needed. The most accurate approach is that of a *direct simulation* (for some periodic structures). For $\ell \ll L$, we can neglect the density disturbance compared to the local phase-averaged density, and we have

$$\langle n_i\rangle^f = M_i \sum_{k=1}^n (\nu_{pik} - \nu_{rik}) B_k \langle T\rangle^{\alpha_k} \exp\left(-\frac{\Delta E_{ak}}{R_g \langle T\rangle}\right) \prod_{j=1}^{n_r} \left(\frac{\langle\rho_i\rangle^f}{M_i}\right)^{\nu_{rjk}}. \qquad (6.79)$$

The local volume-average treatment of reaction in porous media is also given by Nguyen et al. (1982).

6.10.2 Heterogeneous Reaction

The rate of diffusion of component i toward (away from) the surface, when averaged over the internal surface area of the representative elementary volume, is equal to the rate of surface destruction (production) of that component. This is used in (6.60) for the definition of the *heterogeneous production rate* $\langle \dot{n}_i\rangle^{fs}$. We now use the reaction model given by (6.71) to relate $\langle n_i\rangle^{fs}$ to κ_i and $\langle\rho_i\rangle$. We decompose ρ_i as in (6.77). Then we have

$$\frac{D_{mi}}{V_f}\langle\rho\rangle^f \int_{A_{fs}} \mathbf{n}\cdot\nabla\frac{\rho_i}{\langle\rho\rangle^f} dA = -\frac{1}{V_f}\int_{A_{fs}} \kappa_i \rho_i \, dA$$

$$= -\kappa_i \frac{A_{fs}}{V_f}\left(\langle\rho_i\rangle^f + \langle\rho'_i\rangle^{fs}\right), \quad (6.80)$$

where we have assumed that κ_i is *uniform* over A_{fs} and, as before, we have defined

$$\langle\rho'_i\rangle^{fs} = \frac{1}{A_{fs}}\int_{A_{fs}} \rho'_i \, dA. \quad (6.81)$$

Then, from (6.60) and (6.69), we have

$$\langle \dot{n}_i\rangle^{fs} = -\kappa_i \frac{A_{fs}}{V_f}(\langle\rho_i\rangle^f + \langle\rho'_i\rangle^{fs}), \quad (6.82)$$

where A_{fs}/V_f is the fluid-specific surface area.

There *are* applications where $\langle\rho'_i\rangle^{fs}$ is of the order of magnitude of $\langle\rho_i\rangle^f$, for example, in *flames* occurring in porous media with the flame thickness being of the order of the pore size. An example will be discussed in Section 7.4. However, the requirement for the application of the local volume-averaging over a representative elementary volume is that the disturbance ρ'_i be much smaller than $\langle\rho_i\rangle^f$. This leads to

$$\langle \dot{n}_i\rangle^{fs} \simeq -\kappa_i \frac{A_{fs}}{V_f}\langle\rho_i\rangle^f. \quad (6.83)$$

For spherical particles, we have $A_{fs}/V_f = 6(1-\varepsilon)/(\varepsilon d)$.

6.11 Local Chemical Nonequilibrium

For the local volume-averaged solid surface reaction rate given by (6.83), it is assumed that the reaction is kinetically controlled, i.e., the concentration of species i within the representative elementary volume is nearly uniform leading to negligible *diffusional resistance* to species i reaching the solid surface. However, when the pore-velocity is low or when the particles are large, significant diffusional resistance may occur. Then the local concentration of species i at the surface $\langle\rho_i\rangle^{fs}$ may be substantially different than the local pore-volume averaged concentration $\langle\rho_i\rangle^s$. Then in *modeling* this local chemical nonequilibrium, a parallel resistance path for the surface reaction is assumed using both the *kinetic* and the *pore-level diffusion* resistances. This is discussed by Glassman (1987) and Fogler (1992). The model gives

$$\langle \dot{n}_i\rangle^{fs} = \frac{1}{\dfrac{1}{\langle \dot{n}\rangle^{fs}_D} + \dfrac{1}{\langle \dot{n}\rangle^{fs}_{\kappa_i}}}. \quad (6.84)$$

The kinetic component $\langle \dot{n}\rangle^{fs}_{\kappa_i}$ is given by (6.83) and the diffusion component $\langle \dot{n}\rangle^{fs}_D$ is given in terms of the mass transfer resistance (or inverse of the

Sherwood number). This mass transfer resistance is similar to the heat transfer resistance used in the local thermal nonequilibrium treatment (or two-medium treatment) which will be reviewed in Chapter 7. This gives

$$\langle \dot{n} \rangle_D^{fs} = -\frac{Sh_d D_{mi}}{d} \frac{A_{fs}}{V_f}(\langle \rho_i \rangle^f - \langle \rho_i \rangle^{fs}). \tag{6.85}$$

In general, it is assumed that $\langle \rho_i \rangle^{fs} \ll \langle \rho_i \rangle^f$, i.e., that species i is *entirely consumed* in the surface. Then both $\langle \dot{n} \rangle_D^{fs}$ and $\langle \dot{n} \rangle_{\kappa_i}^{fs}$ are *directly proportional* to $\langle \rho_i \rangle^f$. The relationship between the Sherwood number and the pore velocity, pore size, etc., will be discussed in Section 7.2.2. An application of parallel resistance for surface reactions in packed beds is given by Fatehi and Kaviany (1994).

6.12 Modifications to Energy Equation

As a result of exothermic reactions, the temperature of the solid and the fluid increases and the local values of $\langle T \rangle^s$ and $\langle T \rangle^g$ may not be equal (i.e., local thermal nonequilibrium). Phase changes can also occur. These will be discussed in Section 7.4. Here local thermal equilibrium is assumed.

Since the velocity used in the convective term of the energy equation is the *mass-averaged* velocity **u** and because the components may *not* have the same *volumetric enthalpy* $\rho_i i_i$, allowance must be made for the difference in this quantity between components. The component i travels with respect to **u** at a velocity $\mathbf{u}_i - \mathbf{u}$. The enthalpy flux of this component *across* the plane moving with the velocity **u** is $\rho_i i_i(\mathbf{u}_i - \mathbf{u})$ and for the entire component, this is $\sum_{i=1}^{n_r} \rho_i i_i(\mathbf{u}_i - \mathbf{u})$. Then, the *total heat flux* is

$$\mathbf{q} = \rho i \mathbf{u} + \sum_{i=1}^{n_r} \rho_i i_i(\mathbf{u}_i - \mathbf{u}) - k \nabla T, \tag{6.86}$$

where

$$i = \sum_{i=1}^{n_r} \frac{\rho_i}{\rho} i_i. \tag{6.87}$$

As the mass velocity increases, the contribution of the enthalpy flux terms decrease. For a *binary mixture* in *isobaric flow*, the *sum* of the individual enthalpy fluxes crossing the plane moving with the velocity **u** is found from (6.53) and is $D_m(i_2 - i_1)\partial \rho_1/\partial x$. When this is scaled with respect to $\rho i \mathbf{u}$, we have the ratio as $Pe_L^{-1}[(\Delta \rho_{1L})/\rho](|i_2 - i_1|/i)$. This can be much *smaller* than unity when $Pe_L = uL/D_m$ is *large*, i.e., diffusion across the plane moving with velocity **u** can be *neglected* when Pe_L is large. However, note that the local gradient of $\partial \rho_i/\partial x$ can be *fairly* large (for example, in flames).

Also, the *source* term in the energy equation (for *combined f-* and *s*-phase) due to the *volumetric chemical reaction* is

$$\langle \dot{s} \rangle = -\varepsilon \sum_{i=1}^{n_r} \Delta i_{oi} \langle \dot{n}_i \rangle^f, \qquad (6.88)$$

where Δi_o is the *heat of formation* and the chemical reaction in the pore alone is considered. If the surface reaction also occurs, $-\varepsilon \sum_{i=1}^{n_r} \Delta i_{oi} \langle \dot{n}_i \rangle^{fs}$ must be *added* to this.

6.13 References

Bird, R. B., Stewart, W. E., and Lightfoot, E. N., 1960, *Transport Phenomena*, J. Wiley.
Bruno, C., Walsh, P. M., Santa Vicca, D. A., Sinha, N., Yaw, Y., and Bracco, F. V., 1983, "Catalytic Combustion of Propane/Air Mixtures on Platinum," *Combust. Sci. Technol.*, 31, 43–74.
Carbonell, R. G. and Whitaker, S., 1984, "Heat and Mass Transfer in Porous Media," in *Fundamentals of Transport in Porous Media*, Bear and Corapcioglu, eds., Martinus Nijhoff, 123–198.
Chappuis, J., 1982, "Contact Angles," *Multiphase Sci. and Tech.*, 1, 387–505.
Cunningham, R. E. and Williams, R. J. J., 1980, *Diffusion in Gases and Porous Media*, Plenum.
Epstein, N., 1989, "On Tortuosity and the Tortuosity Factor in Flow and Diffusion Through Porous Media," *Chem. Engng. Sci.*, 44, 777–779.
Fakheri, A. and Buckius, R. O., 1983, "Transient Catalytic Combustion on a Flat Plate," *Comb. Flame*, 52, 169–184.
Fatehi, M. and Kaviany, M., 1994, "Adiabatic Reverse Combustion in a Packed Bed," *Combust. Flame*, 99, 1–17.
Fogler, H. S., 1992, *Elements of Chemical Reaction Engineering*, Second Edition, Printice Hall, Englewood Cliffs, New Jersey.
Glassman, I., 1987, *Combustion*, Academic, New York.
Gunn, R. D. and King, C. J., 1969, "Mass Transport in Porous Materials Under Combined Gradients of Composition and Pressure," *AIChE J.*, 15, 507–514.
Harrison, B. K. and Ernst, W. R., 1978, "Catalytic Combustion in Cylindrical Channels: A Homogeneous–Heterogeneous Model," *Combust. Sci. Technol.*, 19, 31–38.
Israelachvili, J. N., 1989, *Intermolecular and Surface Forces*, Academic.
Jackson, R., 1977, *Transport in Porous Catalysts*, Elsevier.
Kanury, A. M., 1988, *Introduction to Combustion Phenomena*, Gordon and Breach Science.
Kee, R. J., Rupley, F. M., and Miller, J. A., 1989, "CHEMKIN-II: A Fortran Chemical Kinetics Package for the Analysis of Gas–Phase Chemical Kinetics," SAND 89–8009.UC–401, Sandia National Laboratories.
Kee, R. J., Warnatz, J., and Miller, J. A., 1988, "A Fortran Computer Code Package for the Evaluation of Gas–Phase Viscosities Conductivities, and Diffusion Coefficient", SAND 83–8209.UC–32, Sandia National Laboratories.

Koch, D. L. and Brady, J., 1985, "Dispersion in Fixed Beds," *J. Fluid Mech.*, 154, 399–427.

Koch, D. L., Cox, R. G., Brenner, H., and Brady, J., 1989, "The Effect of Order on Dispersion in Porous Media," *J. Fluid Mech.*, 200, 173–188.

Kuo, K. K. Y., 1986, *Principles of Combustion*, J. Wiley.

Marteney, P. J. and Kesten, A. S., 1981, "Kinetics of Surface Reactions in Catalytic Combustion," *18th Symposium (International) on Combustion*, Combustion Institute, 1899–1908.

Neale, G. H. and Nader, W. K., 1973, "Prediction of Transport Processes Within Porous Media: Diffusive Flow Processes Within Homogeneous Swarm of Spherical Particles," *AIChE J.*, 19, 112–119.

Nguyen, V. V., Gray, W. G., Pinder, G. F., Botha, J. F., and Crerar, D. A., 1982, "A Theoretical Investigation on the Transport of Chemicals in Reactive Porous Media," *Water Resour.*, 18, 1149–1156.

Pesaran, A. A. and Mills, A. F., 1987, "Moisture Transport in Silica Gel Packed Beds—I. Theoretical Study," *Int. J. Heat Mass Transfer*, 30, 1037–1049.

Pfefferle, L. D. and Pfefferle, W. C., 1987, "Catalysis in Combustion," *Catal. Rev.-Sci. Eng.*, 29, 219–267.

Reid, R. C., Prausnitz, J. M., and Poling, B. E., 1987, *The Properties of Gases and Liquids*, 4th ed., McGraw–Hill.

Ryan, D., Carbonell, R. G., and Whitaker, S., 1980, "Effective Diffusivities for Catalyst Pellets under Reactive Conditions," *Chem. Engng. Sci.*, 35, 10–16.

Slattery, J. C., 1981, *Momentum, Energy, and Mass Transfer in Continua*, R. E. Krieger.

Tien, C.-L. and Lienhard, J. H., 1971, *Statistical Thermodynamics*, Holt, Rinehart and Winston Inc.

Vincenti, W. G. and Kruger, E. H., 1967, *Introduction to Gas Dynamics*, J. Wiley.

Wakao, N. and Kaguei, S., 1982, *Heat and Mass Transfer in Packed Beds*, Gordon and Breach Science.

Williams, F. A., 1988, *Combustion Theory*, Addison–Wesley.

7
Two-Medium Treatment

In this chapter we examine the *single-phase* flow through solid matrices where the assumption of the local thermal equilibrium between the phases is *not* valid. When there is a significant heat *generation* occurring in any one of the phases (solid or fluid), i.e., when the primary heat transfer is by heat generation in a phase and the heat transfer through surfaces bounding the porous medium is less significant, then the local (finite and small) volumes of the solid and fluid phases will be far from the local thermal equilibrium. Also, when the temperature at the *bounding surface changes* significantly with respect to *time*, then in the presence of an interstitial flow and when solid and fluid phases have significantly different *heat capacities* and *thermal conductivities*, the local rate of change of temperature for the two phases will not be equal.

In the *two-medium* treatment of the single-phase flow and heat transfer through porous media, no local thermal equilibrium is assumed between the fluid and solid phases, but it is assumed that each phase is *continuous* and represented with an appropriate *effective total thermal conductivity*. Then the *thermal* coupling between the phases is approached either by the examination of the *microstructure* (for simple geometries) or by *empiricism*. When empiricism is applied, simple *two-equation* (or two-medium) *models* that contain a modeling parameter, h_{sf} (called the *interfacial convective heat transfer coefficient*), are used. As will be shown, only those empirical treatments that contain not only h_{sf} but also the appropriate effective thermal conductivity tensors (for both phases) and the *dispersion tensor* (in the fluid-phase equation) are expected to give *reasonably accurate* predictions.

We begin with the *phase volume averaging* of the energy equations, which shows how the fluid phase dispersion as well as the other convective and conductive effects appear as the coupling coefficients in the energy equations. Then these coefficients, including the interfacial heat convection coefficient, are evaluated for a *simple* porous medium, i.e., capillary tubes. Then we examine the existing heuristic two-medium treatments and show that most of them are *inconsistent* with the results of the local phase volume averaging. Also, in order to examine the cases where the assumptions made in the phase-averaged treatments do *not* hold, we examine *point-wise solutions* to a periodic flow. Finally, the chemical reaction in the fluid phase and departure from local thermal equilibrium, will be examined in an example of premixed combustion in a two-dimensional porous media. For this problem, the results of point-wise (i.e., direct simulation), single- and

392   7. Two-Medium Treatment

two-medium treatments will be compared for the flame speed and flame structure.

7.1 Local Phase Volume Averaging for Steady Flows

The local volume-averaging treatment leading to the coupling between the energy equation for each phase is formulated by Carbonell and Whitaker (1984) and is given in Zanotti and Carbonell (1984) Levec and Carbonell (1985) and Quintard and Whitaker (1993). Their development for the transient heat transfer with a steady flow is reviewed here. Some of the features of their treatment are discussed first.

- For the *transient* behavior, it is assumed that the *penetration depth* (in the fluid and solid phases) is *larger* than the linear dimension of the representative elementary volume. This is required in order to volume-average over the representative elementary volume while satisfying that ΔT_ℓ over this volume is much *smaller* than that over the system ΔT_L, i.e., not all the temperature drop occurs within the representative elementary volume. If ΔT_ℓ is nearly equal to ΔT_L, then the *direct simulation* of the heat transfer over length ℓ has to be performed. Except for very fast transients, the time for the penetration over ℓ, i.e., ℓ^2/α, is much smaller than the time scales associated with the system transients of interest.

- Each phase is treated as a *continuum*. The phase volume-averaged *total thermal diffusivity* tensor will be determined for each phase.

- *Closure constitutive* equations are developed similar to those used when the existence of the local thermal equilibrium was assumed. This requires relating the disturbances in the temperature fields to the gradients of the volume-averaged temperatures and to the difference between the phase volume-averaged temperatures.

7.1.1 Allowing for Difference in Average Local Temperatures

For steady flows with *no heat generation*, the transient energy equation for the fluid and solid phases and the boundary conditions on A_{fs} are

$$(\rho c_p)_f \left(\frac{\partial T_f}{\partial t} + \mathbf{u}_f \cdot \nabla T_f \right) = \nabla \cdot k_f \nabla T_f \quad \text{in} \quad V_f, \tag{7.1}$$

$$(\rho c_p)_s \frac{\partial T_s}{\partial t} = \nabla \cdot k_s \nabla T_s \quad \text{in} \quad V_s, \tag{7.2}$$

7.1 Local Phase Volume Averaging for Steady Flows

$$T_f = T_s, \quad \mathbf{n}_{fs} \cdot k_f \nabla T_f = \mathbf{n}_{fs} \cdot k_s \nabla T_s \quad \text{on} \quad A_{fs}. \tag{7.3}$$

The temperatures and velocity are *decomposed* in terms of the local (representative elementary volume) *disturbance* and *mean* values as

$$T_f = \langle T \rangle^f + T'_f, \tag{7.4}$$

$$T_s = \langle T \rangle^s + T'_s, \tag{7.5}$$

$$\mathbf{u}_f = \langle \mathbf{u} \rangle^f + \mathbf{u}'_f = \mathbf{u}_p + \mathbf{u}'_f. \tag{7.6}$$

Next, the *intrinsic* phase volume average of the energy equation and the boundary conditions are taken, and theorem (2.85) is used. The equation for the phase-averaged temperatures for the fluid phase is

$$(\rho c_p)_f \left(\frac{\partial \langle T \rangle^f}{\partial t} + \langle \mathbf{u} \rangle^f \cdot \nabla \langle T \rangle^f \right)$$

$$= \nabla \cdot \left(k_f \nabla \langle T \rangle^f + \frac{k_f}{V_f} \int_{A_{fs}} \mathbf{n}_{fs} T'_f \, dA \right)$$

$$- (\rho c_p)_f \nabla \cdot \langle \mathbf{u}' T' \rangle^f + \frac{1}{V_f} \int_{A_{fs}} \mathbf{n}_{fs} \cdot k_f \nabla T'_f \, dA, \tag{7.7}$$

where it is assumed that the gradient of the temperature deviation is much larger than the gradient of phase-averaged temperature. Also for the solid phase, we have

$$(\rho c_p)_s \frac{\partial \langle T \rangle^s}{\partial t} = \nabla \cdot \left(k_s \nabla \langle T \rangle^s + \frac{k_s}{V_s} \int_{A_{fs}} \mathbf{n}_{sf} T'_s \, dA \right)$$

$$+ \frac{1}{V_s} \int_{A_{fs}} \mathbf{n}_{sf} \cdot k_s \nabla T'_s \, dA. \tag{7.8}$$

The equations for the disturbance temperatures are

$$(\rho c_p)_f \left(\frac{\partial T'_f}{\partial t} - \nabla \cdot \langle \mathbf{u}' T' \rangle^f \right) + \nabla \cdot \frac{k_f}{V_f} \int_{A_{fs}} \mathbf{n}_{fs} T'_f \, dA$$

$$= \nabla \cdot k_f \nabla T'_f - \frac{1}{V_f} \int_{A_{fs}} \mathbf{n}_{fs} \cdot k_f \nabla T'_f \, dA$$

$$- (\rho c_p)_f \left(\mathbf{u}'_f \cdot \nabla \langle T \rangle^f + \mathbf{u}_f \cdot \nabla T'_f \right), \tag{7.9}$$

$$(\rho c_p)_s \frac{\partial T'_s}{\partial t} + \nabla \cdot \frac{k_s}{V_s} \int_{A_{fs}} \mathbf{n}_{sf} T'_s \, dA$$

$$= \nabla \cdot k_s \nabla T'_s - \frac{1}{V_s} \int_{A_{fs}} \mathbf{n}_{sf} \cdot k_s \nabla T'_s \, dA. \tag{7.10}$$

Next, the boundary conditions for the preceding are obtained from (7.3) and are given as

$$T'_f = T'_s + \left(\langle T \rangle^s - \langle T \rangle^f\right), \tag{7.11}$$

$$\mathbf{n}_{fs} \cdot k_f \nabla T'_f = \mathbf{n}_{fs} \cdot k_s \nabla T'_s + \mathbf{n}_{fs} \cdot \left(k_s \nabla \langle T \rangle^s - k_f \nabla \langle T \rangle^f\right). \tag{7.12}$$

Now by assuming that

$$\frac{\alpha_f t}{\ell_f^2} \gg 1, \quad \frac{\alpha_s t}{\ell_s^2} \gg 1 \tag{7.13}$$

and by making the same order-of-magnitude analysis as those given by (4.47) and (4.48), it can be shown that the left-hand sides of (7.9) and (7.10) are negligibly small.

Since T'_f and T'_s are coupled through the boundary conditions (7.12), the *closure constitutive equation* introduced in (3.23)–(3.24) and (4.35) must be *modified* to include the influence of the gradients of the phase-averaged temperature as well as the differences in the phase-averaged temperatures. These closure equations are constructed as

$$T'_f = \mathbf{b}_{ff} \cdot \nabla \langle T \rangle^f + \mathbf{b}_{fs} \cdot \nabla \langle T \rangle^s + \psi_f \left(\langle T \rangle^s - \langle T \rangle^f\right) + \phi_f, \tag{7.14}$$

$$T'_s = \mathbf{b}_{ss} \cdot \nabla \langle T \rangle^s + \mathbf{b}_{sf} \cdot \nabla \langle T \rangle^f + \psi_s \left(\langle T \rangle^s - \langle T \rangle^f\right) + \phi_s, \tag{7.15}$$

where ϕ_f and ϕ_s are identically *zero* for periodic structures and $\mathbf{b}$ and ψ are functions of the position only. This can be written in the matrix form

$$[T'] = [\mathbf{b}][\nabla \langle T \rangle] + [\psi] \Delta \langle T \rangle, \tag{7.16}$$

where $\Delta \langle T \rangle = \langle T \rangle^s - \langle T \rangle^f$.

7.1.2 EVALUATION OF [b] AND [ψ]

The equations for the transformation vectors [**b**] and scalars [ψ] are found by substituting (7.14)–(7.15) in (7.9)–(7.12) while noting that the left-hand sides of (7.9)–(7.10) are negligibly small. Since the constitutive equations (7.14)–(7.15) are based on *linear* superposition of three different *interactions*, separate equations for each transformation are found. This is equivalent to setting the appropriate coefficients of $[\nabla \langle T \rangle]$ and $\langle T \rangle^s - \langle T \rangle^f$ equal to zero after substitution of (7.14)–(7.15) in (7.9)–(7.12). When this is done, we have in V_f

$$(\rho c_p)_f \left(\mathbf{u}'_f + \mathbf{u}_f \cdot \nabla \mathbf{b}_{ff}\right) = \nabla \cdot k_f \nabla \mathbf{b}_{ff} - \frac{1}{V_f} \int_{A_{fs}} \mathbf{n}_{fs} \cdot k_f \nabla \mathbf{b}_{ff} \, dA, \tag{7.17}$$

$$(\rho c_p)_f \mathbf{u}_f \cdot \nabla \mathbf{b}_{fs} = \nabla \cdot k_f \nabla \mathbf{b}_{fs} - \frac{1}{V_f} \int_{A_{fs}} \mathbf{n}_{fs} \cdot k_f \nabla \mathbf{b}_{fs} \, dA, \quad (7.18)$$

$$(\rho c_p)_f \mathbf{u}_f \cdot \nabla \psi_f = \nabla \cdot k_f \nabla \psi_f - \frac{1}{V_f} \int_{A_{fs}} \mathbf{n}_{fs} \cdot k_f \nabla \psi_f \, dA, \quad (7.19)$$

and in V_s,

$$0 = \nabla \cdot k_s \nabla \mathbf{b}_{ss} - \frac{1}{V_s} \int_{A_{fs}} \mathbf{n}_{sf} \cdot k_s \nabla \mathbf{b}_{ss} \, dA, \quad (7.20)$$

$$0 = \nabla \cdot k_s \nabla \mathbf{b}_{sf} - \frac{1}{V_s} \int_{A_{fs}} \mathbf{n}_{sf} \cdot k_s \nabla \mathbf{b}_{sf} \, dA, \quad (7.21)$$

$$0 = \nabla \cdot k_s \nabla \psi_s - \frac{1}{V_s} \int_{A_{fs}} \mathbf{n}_{sf} \cdot k_s \nabla \psi_s \, dA, \quad (7.22)$$

with the boundary conditions on A_{fs} found as

$$\mathbf{b}_{ff} = \mathbf{b}_{sf}, \quad \mathbf{b}_{fs} = \mathbf{b}_{ss}, \quad \psi_f = \psi_s + 1, \quad (7.23)$$

$$\mathbf{n}_{fs} \cdot k_f \nabla \mathbf{b}_{ff} = \mathbf{n}_{fs} \cdot k_s \nabla \mathbf{b}_{sf} - k_f \mathbf{n}_{fs}, \quad (7.24)$$

$$\mathbf{n}_{fs} \cdot k_f \nabla \mathbf{b}_{fs} = \mathbf{n}_{fs} \cdot k_s \nabla \mathbf{b}_{ss} + k_s \mathbf{n}_{fs}, \quad (7.25)$$

$$\mathbf{n}_{fs} \cdot k_f \nabla \psi_f = \mathbf{n}_{fs} \cdot k_s \nabla \psi_s. \quad (7.26)$$

Furthermore, *periodic* boundary conditions for unit cells with spatial periods ℓ_i are assumed, i.e.,

$$[\mathbf{b}(\mathbf{x} + \ell_i)] = [\mathbf{b}(\mathbf{x})], \quad (7.27)$$

$$[\psi(\mathbf{x} + \ell_i)] = [\psi(\mathbf{x})]. \quad (7.28)$$

The equations for the closure functions, subject to the boundary conditions on A_{fs} and the unit cell boundary, constitute a well-posed problem that can in principle be solved. As a first attempt, the numerical solution for two-dimensional periodic structures can be sought. This has not yet been done. However, the solution for the travel of a thermal pulse through a capillary tube has been obtained by Zanotti and Carbonell.

7.1.3 ENERGY EQUATION FOR EACH PHASE

Assuming that the closure functions can be determined, the equations for the volume-averaged temperatures, (7.7)–(7.8), can be written in terms of the area and volume-averaged quantities involving these functions. These equations are found by substituting (7.14)–(7.15) in (7.7)–(7.8). They are

$$\frac{\partial \langle T \rangle^f}{\partial t}$$
$$+ \left[\langle \mathbf{u} \rangle^f + \alpha_f \frac{A_{fs}}{V_f} \left(2 \langle \mathbf{n}_{fs} \psi_f \rangle^{fs} - \langle \mathbf{n}_{fs} \cdot \nabla \mathbf{b}_{ff} \rangle^{fs} \right) - \langle \mathbf{u}' \psi_f \rangle^f \right] \cdot \nabla \langle T \rangle^f$$
$$+ \left[\alpha_f \frac{A_{fs}}{V_f} \left(-2 \langle \mathbf{n}_{fs} \psi_f \rangle^{fs} - \langle \mathbf{n}_{fs} \cdot \nabla \mathbf{b}_{fs} \rangle^{fs} \right) + \langle \mathbf{u}' \psi_f \rangle^f \right] \cdot \nabla \langle T \rangle^s$$
$$= \nabla \cdot \left[\alpha_f \left(\mathbf{I} + 2 \frac{A_{fs}}{V_f} \langle \mathbf{n}_{fs} \mathbf{b}_{ff} \rangle^{fs} \right) - \langle \mathbf{u}' \mathbf{b}_{ff} \rangle^f \right] \cdot \nabla \langle T \rangle^f$$
$$+ \nabla \cdot \left(2\alpha_f \frac{A_{fs}}{V_f} \langle \mathbf{n}_{fs} \mathbf{b}_{fs} \rangle^{fs} - \langle \mathbf{u}' \mathbf{b}_{fs} \rangle^f \right) \cdot \nabla \langle T \rangle^s$$
$$+ \alpha_f \frac{A_{fs}}{V_f} \langle \mathbf{n}_{fs} \cdot \nabla \psi_f \rangle^{fs} \left(\langle T \rangle^s - \langle T \rangle^f \right), \qquad (7.29)$$

$$\frac{\partial \langle T \rangle^s}{\partial t} + \left[\alpha_s \frac{A_{fs}}{V_s} \left(2 \langle \mathbf{n}_{sf} \psi_s \rangle^{fs} - \langle \mathbf{n}_{sf} \cdot \nabla \mathbf{b}_{sf} \rangle^{fs} \right) \right] \cdot \nabla \langle T \rangle^f$$
$$+ \left[\alpha_s \frac{A_{fs}}{V_s} \left(-2 \langle \mathbf{n}_{sf} \psi_s \rangle^{fs} - \langle \mathbf{n}_{sf} \cdot \nabla \mathbf{b}_{ss} \rangle^{fs} \right) \right] \cdot \nabla \langle T \rangle^s$$
$$= \nabla \cdot \left[\alpha_s \left(\mathbf{I} + 2 \frac{A_{fs}}{V_s} \langle \mathbf{n}_{sf} \mathbf{b}_{ss} \rangle^{fs} \right) \right] \cdot \nabla \langle T \rangle^s$$
$$+ \nabla \cdot \left(2\alpha_s \frac{A_{fs}}{V_s} \langle \mathbf{n}_{sf} \mathbf{b}_{sf} \rangle^{fs} \right) \cdot \nabla \langle T \rangle^f$$
$$+ \alpha_s \frac{A_{fs}}{V_s} \langle \mathbf{n}_{sf} \cdot \nabla \psi_s \rangle^{fs} \left(\langle T \rangle^s - \langle T \rangle^f \right), \qquad (7.30)$$

where $\langle \ \rangle^{fs}$ indicates $(1/A_{fs}) \int_{A_{fs}} (\ ) \, dA$. Since on A_{fs} the temperature disturbances and their derivatives are related through (7.12) and since $\mathbf{u} = 0$, extra relationships result. These are

$$k_f \langle \mathbf{n}_{fs} \cdot \nabla \psi_f \rangle^{fs} = -k_s \langle \mathbf{n}_{sf} \cdot \nabla \psi_s \rangle^{fs}, \qquad (7.31)$$
$$k_f \langle \mathbf{n}_{fs} \cdot \nabla \mathbf{b}_{fs} \rangle^{fs} = -k_s \langle \mathbf{n}_{sf} \cdot \nabla \mathbf{b}_{ss} \rangle^{fs}, \qquad (7.32)$$
$$k_f \langle \mathbf{n}_{fs} \cdot \nabla \mathbf{b}_{ff} \rangle^{fs} = -k_s \langle \mathbf{n}_{sf} \cdot \nabla \mathbf{b}_{sf} \rangle^{fs}, \qquad (7.33)$$
$$\langle \mathbf{n}_{fs} \psi_f \rangle^{fs} = -\langle \mathbf{n}_{sf} \psi_s \rangle^{fs}, \qquad (7.34)$$
$$\langle \mathbf{n}_{fs} \mathbf{b}_{ff} \rangle^{fs} = -\langle \mathbf{n}_{sf} \mathbf{b}_{sf} \rangle^{fs}, \qquad (7.35)$$
$$\langle \mathbf{n}_{fs} \mathbf{b}_{fs} \rangle^{fs} = -\langle \mathbf{n}_{sf} \mathbf{b}_{ss} \rangle^{fs}. \qquad (7.36)$$

The first of these is the equality of the *interfacial conduction heat transfer coefficients*, and as expected these coefficients are coupled to the details of the individual fluid and solid transports. Note that the coefficient 2 appearing in front of the *tortuosity* terms in (7.29) and (7.30) is due to an extra term resulting from the last terms in (7.7) and (7.8). This extra term emerges as $\nabla T'$, is replaced using (7.14) and (7.15), and is then decomposed.

In (7.29)–(7.30), if we take $\langle T \rangle^f = \langle T \rangle^s = \langle T \rangle^f$, i.e., assuming that local thermal equilibrium exists, and multiply (7.29) by ε and (7.30) by $(1-\varepsilon)$, we recover the volume-averaged energy equation for the general case of $k_s \neq 0$ and $\mathbf{u} \neq 0$. When $\mathbf{u} = 0$, we recover the conduction equation (3.22), and when $k_s = 0$, we recover (4.36). Note that when $\langle T \rangle^f = \langle T \rangle^s$, then $T'_f = \mathbf{b}_{ff} \cdot \nabla \langle T \rangle^f$ from (7.14).

The *coupling* between (7.29) and (7.30) is not only due to the heat transfer rate, which is proportional to the difference between the two local-averaged temperatures, $\langle T \rangle^s - \langle T \rangle^f$, but is also due to the difference between the *gradient* of phase-averaged temperatures. In the *first* attempts at modeling the problem by two-medium treatments, [**b**] and $\langle \mathbf{n}\psi \rangle^{fs}$ are taken to be *zero*. Then, *only* the last terms in (7.29) and (7.30) *account* for the interaction between the phases. As will be shown, this is an oversimplification and is *not* justifiable.

Next the energy equation for each phase can be written in a *more compact* form by defining the following coefficients. Note that both the hydrodynamic dispersion, i.e., the influence of the presence of the matrix on the flow (no-slip condition on the solid surface), as well as the interfacial heat transfer are included in (7.29) and (7.30). The *total thermal diffusivity tensors*, $\mathbf{D}_{ff}$, $\mathbf{D}_{ss}$, $\mathbf{D}_{fs}$, and $\mathbf{D}_{sf}$, and the *interfacial convective heat transfer coefficient* h_{sf} are introduced. The total thermal diffusivity tensors include both the effective thermal diffusivity tensor (stagnant) as well as the hydrodynamic dispersion tensor. A *total convective velocity* $\mathbf{v}$ is defined such that we can write (7.29) and (7.30) as

$$\frac{\partial \langle T \rangle^f}{\partial t} + \mathbf{v}_{ff} \cdot \nabla \langle T \rangle^f + \mathbf{v}_{fs} \cdot \nabla \langle T \rangle^s$$

$$= \nabla \cdot \mathbf{D}_{ff} \cdot \nabla \langle T \rangle^f + \nabla \cdot \mathbf{D}_{fs} \cdot \nabla \langle T \rangle^s$$

$$+ \frac{A_{fs}}{V_f (\rho c_p)_f} h_{sf} \left(\langle T \rangle^s - \langle T \rangle^f \right), \qquad (7.37)$$

$$\frac{\partial \langle T \rangle^s}{\partial t} + \mathbf{v}_{sf} \cdot \nabla \langle T \rangle^f + \mathbf{v}_{ss} \cdot \nabla \langle T \rangle^s$$

$$= \nabla \cdot \mathbf{D}_{sf} \cdot \nabla \langle T \rangle^f + \nabla \cdot \mathbf{D}_{ss} \cdot \nabla \langle T \rangle^s$$

$$+ \frac{A_{fs}}{V_s (\rho c_p)_s} h_{sf} \left(\langle T \rangle^f - \langle T \rangle^s \right), \qquad (7.38)$$

where

$$h_{sf} = k_f \langle \mathbf{n}_{fs} \cdot \nabla \psi_f \rangle^{fs} = -k_s \langle \mathbf{n}_{sf} \cdot \nabla \psi_s \rangle^{fs}. \qquad (7.39)$$

As will be discussed, h_{sf} is also used as an *overall convection heat transfer coefficient*. When h_{sf} is determined *experimentally*, it is important to note whether the complete form of equations (7.37)–(7.38) are used for its evaluation. The use of oversimplified versions of (7.37) and (7.38) results in the *inclusion* of the neglected terms into h_{sf}. This simplification results in values for h_{sf} that are valid *only* for those particular experiments. This result will be discussed further in Section 7.2.

7.1.4 EXAMPLE: AXIAL TRAVEL OF THERMAL PULSES

In order to demonstrate the *extent* of the coupling between the local solid- and fluid-phase temperatures, and to show that the coupling cannot only be given by the interfacial heat transfer coupling $h_{sf}[\langle T \rangle^s - \langle T \rangle^f]$, Zanotti and Carbonell considered a *capillary tube model*. Although the model is simple, the implications are far-reaching. The Taylor-Aris dispersion in the fluid phase, which was considered in Chapter 4 in connection with adiabatic solid-fluid interfaces, will now be presented and modified by including the heat transfer through the *interface* and solid phase. Then the *travel* of a thermal pulse through the tube and the developing *phase-lag* between the phases will be determined.

The point energy equations for axisymmetric fully developed flows are

$$\frac{\partial T_f}{\partial t} + u_f \frac{\partial T_f}{\partial x} = \alpha_f \left[\frac{\partial^2 T_f}{\partial x^2} + \frac{1}{r} \frac{\partial}{\partial r} \left(r \frac{\partial T_f}{\partial r} \right) \right] \quad \text{for} \quad 0 \le r \le R_i, \quad (7.40)$$

$$\frac{\partial T_s}{\partial t} = \alpha_s \left[\frac{\partial^2 T_s}{\partial x^2} + \frac{1}{r} \frac{\partial}{\partial r} \left(r \frac{\partial T_s}{\partial r} \right) \right] \quad \text{for} \quad R_i \le r \le R_o. \quad (7.41)$$

The boundary conditions are

$$\frac{\partial T_f}{\partial r} = 0 \quad \text{at} \quad r = 0, \quad (7.42)$$

$$T_f = T_s \quad \text{on} \quad r = R_i, \quad (7.43)$$

$$k_f \frac{\partial T_f}{\partial r} = k_s \frac{\partial T_s}{\partial r} \quad \text{on} \quad r = R_i, \quad (7.44)$$

$$\frac{\partial T_s}{\partial r} = 0 \quad \text{on} \quad r = R_o. \quad (7.45)$$

Then we decompose the temperature and velocity fields as in (7.4)–(7.6) with

$$\langle T \rangle^f = \frac{2}{R_i^2} \int_0^{R_i} T_f \, r \, dr, \quad (7.46)$$

$$\langle T \rangle^s = \frac{2}{R_o^2 - R_i^2} \int_{R_i}^{R_o} T_s \, r \, dr. \quad (7.47)$$

7.1 Local Phase Volume Averaging for Steady Flows

By assuming that the axial conduction in each phase is *negligible* compared to that in the radial direction, Zanotti and Carbonell show that the functions b_{fs} and b_{sf} are zero. Using the same assumptions, the equations for the functions b_{ff}, b_{ss}, ψ_f, and ψ_s become (details leading to these are given by Zanotti and Carbonell)

$$\frac{1}{r}\frac{d}{dr}\left(r\frac{db_{ff}}{dr}\right) = \frac{2}{R_i}\left.\frac{db_{ff}}{dr}\right|_{R_i} + \frac{u'_f}{\alpha_f}, \tag{7.48}$$

$$\frac{1}{r}\frac{d}{dr}\left(r\frac{db_{ss}}{dr}\right) = -\frac{2R_i}{R_o^2 - R_i^2}\left.\frac{db_{ss}}{dr}\right|_{R_i}, \tag{7.49}$$

$$\frac{1}{r}\frac{d}{dr}\left(r\frac{d\psi_f}{dr}\right) = \frac{2}{R_i}\left.\frac{d\psi_f}{dr}\right|_{R_i}, \tag{7.50}$$

$$\frac{1}{r}\frac{d}{dr}\left(r\frac{d\psi_s}{dr}\right) = -\frac{2R_i}{R_o^2 - R_i^2}\left.\frac{d\psi_s}{dr}\right|_{R_i}. \tag{7.51}$$

The velocity distribution is given by the Hagen-Poiseuille relation. The boundary conditions are given by (7.23)–(7.26). In addition, we have

$$\langle b_{ff}\rangle^f = \langle b_{ss}\rangle^s = \langle \psi_f\rangle^f = \langle \psi_f\rangle^s = 0. \tag{7.52}$$

The equations for b and ψ, along with the boundary conditions, are solved in closed form, and the terms in (7.29)–(7.30) are evaluated. The resulting energy equations for the phase-averaged temperatures are

$$\frac{\partial \langle T\rangle^f}{\partial t} + \left(1 + a_1\frac{h_{sf}R_i}{k_f}\right)\langle u\rangle^f \frac{\partial \langle T\rangle^f}{\partial x} - a_1\frac{h_{sf}R_i}{2k_f}\langle u\rangle^f \frac{\partial \langle T\rangle^s}{\partial x}$$

$$= \alpha_f\left[1 + \left(a_2 - a_1^2\frac{h_{sf}R_i}{2k_f}\right)\left(\frac{\langle u\rangle^f R_i}{2\alpha_f}\right)^2\right]\frac{\partial^2 \langle T\rangle^f}{\partial x^2}$$

$$+ \frac{2h_{sf}}{R_o(\rho c_p)_f}\left(\langle T\rangle^s - \langle T\rangle^f\right), \tag{7.53}$$

$$\frac{\partial \langle T\rangle^s}{\partial t} - \frac{(\rho c_p)_f R_i^2}{(\rho c_p)_s(R_o^2 - R_i^2)}a_1\frac{h_{sf}R_i}{2k_f}\langle u\rangle^f \frac{\partial \langle T\rangle^f}{\partial x}$$

$$= \alpha_s\frac{\partial^2 \langle T\rangle^s}{\partial x^2} - \frac{2R_i}{R_o^2 - R_i^2}\frac{h_{sf}}{(\rho c_p)_s}\left(\langle T\rangle^s - \langle T\rangle^f\right), \tag{7.54}$$

where, as before, h_{sf} is the *overall convective heat transfer coefficient* and is determined to be

$$h_{sf}^{-1} = \frac{R_i a_3}{2k_f} + \frac{R_o^2 - R_i^2}{2R_i}\frac{a_4}{k_s}. \tag{7.55}$$

Note that here h_{sf} is *independent* of the velocity. The dimensionless constants a_i are

$$a_1 = \frac{8}{\langle u \rangle^f R_i^4} \int_0^{R_i} r\,dr \int_r^{R_i} \frac{dr}{r} \int_0^r u_f' r\,dr = \frac{1}{6}, \quad (7.56)$$

$$a_2 = \frac{8}{\left[\langle u \rangle^f R_i^2\right]^2} \int_0^{R_i} u_f' r\,dr \int_r^{R_i} \frac{dr}{r} \int_0^r u_f' r\,dr = \frac{1}{12}, \quad (7.57)$$

$$a_3 = \frac{8}{R_i^4} \int_0^{R_i} r\,dr \int_r^{R_i} \frac{dr}{r} \int_0^r r\,dr = \frac{1}{2}, \quad (7.58)$$

$$a_4 = \frac{8 R_i^2}{(R_o^2 - R_i^2)^3} \int_{R_i}^{R_o} r\,dr \int_{R_i}^r \frac{dr}{r} \int_r^{R_o} r\,dr$$

$$= \frac{4 \left(\frac{R_o}{R_i}\right)^4 \ln \frac{R_o}{R_i} - 3 \left(\frac{R_o}{R_i}\right)^4 + 4 \left(\frac{R_o}{R_i}\right)^2 - 1}{2 \left[\left(\frac{R_o}{R_i}\right)^2 - 1\right]^3}. \quad (7.59)$$

Note further that the Taylor-Aris results are obtained by setting $h_{sf} = k_s = 0$, i.e.,

$$\frac{\partial \langle T \rangle^f}{\partial t} + \langle u \rangle^f \frac{\partial T_f}{\partial x} = \alpha_f \left[1 + \frac{\langle u \rangle^f \langle u \rangle^f R_i^2}{48 \alpha_f^2}\right] \frac{\partial^2 \langle T \rangle^f}{\partial x^2}, \quad (7.60)$$

$$\frac{\partial \langle T \rangle^s}{\partial t} = 0. \quad (7.61)$$

Examination of energy equations (7.53) and (7.54) shows that phase volume-averaged equations are coupled not only through the interfacial heat transfer coefficient h_{sf}, but that there *also* is *dispersion* in the fluid phase and a *total convective velocity* for the solid phase. This should be kept in mind in the experimental determination of h_{sf}, where the measured $\langle T \rangle^f$ and $\langle T \rangle^s$ are used to determine h_{sf}. This will be further discussed in Section 7.2.

Next, Zanotti and Carbonell show that in the tube the thermal pulses introduced upstream travel with the speed

$$\frac{\langle u \rangle^f}{1 + \frac{(\rho c_p)_s (R_o^2 - R_i^2)}{(\rho c_p)_f R_i^2}}, \quad (7.62)$$

with a *constant phase-lag* (distance) of

$$\Delta = \frac{(\rho c_p)_f \langle u \rangle^f R_o}{2} \left\{ \frac{1}{h_{sf}\left[1 + \frac{(\rho c_p)_f R_i^2}{(\rho c_p)_s (R_o^2 - R_i^2)}\right]} + \frac{R_i}{2k_f} \right\}, \qquad (7.63)$$

when the elapsed time for the validity of the constant speed and the phase-lag is

$$\frac{t\alpha_f}{R_i^2} \gg \frac{1}{1 + \frac{(\rho c_p)_f R_i^2}{(\rho c_p)_s (R_o^2 - R_i^2)}} \left[\frac{R_i^2}{12(R_o^2 - R_i^2)} + \frac{a_4 k_f}{6k_s}\right] \frac{R_i^2}{6(R_o^2 - R_i^2)}. \qquad (7.64)$$

Note that for a moving fluid from (7.63) the requirements for *zero* phase-lag are *infinite* convective heat transfer coefficient and *infinite* fluid conductivity. However, from (7.64), large k_f requires a very *large* time before (7.63) becomes valid, unless $k_s \gg k_f$. Equations (7.53)–(7.54) show that simplification of the interactions between the two phases by representing the *complex* interactions with a single transfer coefficient multiplied by the temperature difference between the phases is *not* justifiable, at least when *reasonably* accurate *dynamic diagnosis* is expected.

7.2 Interfacial Convective Heat Transfer Coefficient h_{sf}

In the *earlier* treatments of transient heat transfer in packed beds, various *heuristic* models were used instead of the two equations given by (7.37) and (7.38). Wakao and Kaguei (1982) give the history of the development in this area. In the following, some of these models, which all use an *interfacial convection heat transfer coefficient* h_{sf}, will be discussed. The *distinction* should be between h_{sf} found from the energy equations (7.37) and (7.38), and that found from the simplified forms of energy equations. Since these different models are used in the determination of h_{sf}, the literature on the reported value of h_{sf} is rather *incoherent*. Wakao and Kaguei have carefully examined these reported values and classified the modeling efforts.

It should be noted that h_{sf} for a heated *single particle* in an otherwise uniform temperature field is expected to be significantly *different* than that for particles in packed beds. Also, since, in general, the thermal conductivity of the solid is *not* large enough to lead to an *isothermal* surface temperature, the conductivity of the solid also influences the temperature field around it. Therefore, the interstitial convection heat transfer coefficient obtained from a given fluid-solid combination is *not* expected to hold valid for some other combinations. This is evident from the results for h_{sf} given by (7.55) for the simple case of capillary tubes.

The coefficients in (7.37)–(7.38) have been computed for some geometry and range of parameters (Quintard et al., 1995). Simplified h_{sf}-based models can still be used, and we review some of these heuristic models. However, their inadequacy to explain the process and their limitations can *not* be overemphasized.

7.2.1 Models Based on h_{sf}

There are many h_{sf}-based models appearing in the literature. Three such models are given here (Wakao and Kaguei). These are generally for the one-dimensional Darcean flow and heat transfer and for packed beds of spherical particles.

(A) Schumann Model

This is the *simplest* and the *least* accurate of all models. The two equations are given as

$$\frac{\partial \langle T \rangle^f}{\partial t} + \langle u \rangle^f \frac{\partial \langle T \rangle^f}{\partial x} = \frac{h_{sf} A_o}{\varepsilon (\rho c_p)_f} \left(\langle T \rangle^s - \langle T \rangle^f \right), \tag{7.65}$$

$$\frac{\partial \langle T \rangle^s}{\partial t} = -\frac{h_{sf} A_o}{(1-\varepsilon)(\rho c_p)_s} \left(\langle T \rangle^s - \langle T \rangle^f \right), \tag{7.66}$$

where $A_o = A_{fs}/V$ is the specific surface area and $u_p = \langle u \rangle^f$ is the average pore velocity. No account is made of the axial conduction and the dispersion in the solid energy equation. This model is for transient problems *only*.

(B) Continuous-Solid Model

In this model the axial conduction, in both phases, is included through the use of *effective thermal conductivities* k_{fe} and k_{se}. This gives

$$\frac{\partial \langle T \rangle^f}{\partial t} + \langle u \rangle^f \frac{\partial \langle T \rangle^f}{\partial x} = \frac{\langle k \rangle^f}{\varepsilon (\rho c_p)_f} \frac{\partial^2 \langle T \rangle^f}{\partial x^2} + \frac{h_{sf} A_o}{\varepsilon (\rho c_p)_f} \left(\langle T \rangle^s - \langle T \rangle^f \right), \tag{7.67}$$

$$\frac{\partial \langle T \rangle^s}{\partial t} = \frac{\langle k \rangle^s}{(1-\varepsilon)(\rho c_p)_s} \frac{\partial^2 \langle T \rangle^s}{\partial x^2} - \frac{h_{sf} A_o}{(1-\varepsilon)(\rho c_p)_s} \left(\langle T \rangle^s - \langle T \rangle^f \right). \tag{7.68}$$

No account is made of the dispersion and $\langle k \rangle^f$, $\langle k \rangle^s$, and h_{sf} are to be determined experimentally.

(C) DISPERSION-PARTICLE-BASED MODEL

This is an improvement over the continuous-solid model and allows for dispersion. The results are

$$\frac{\partial \langle T \rangle^f}{\partial t} + \langle u \rangle^f \frac{\partial \langle T \rangle^f}{\partial x} = \frac{1}{\varepsilon}\left(\frac{\langle k \rangle}{(\rho c_p)_f} + D^d_{xx}\right)\frac{\partial^2 \langle T \rangle^f}{\partial x^2}$$
$$+ \frac{h_{sf} A_o}{\varepsilon(\rho c_p)_f}\left(T_{sf} - \langle T \rangle^f\right), \quad (7.69)$$

$$\frac{\partial T_s}{\partial t} = \frac{\langle k \rangle^s}{(\rho c_p)_s}\frac{1}{r^2}\frac{\partial}{\partial r}\left(r^2 \frac{\partial T_s}{\partial r}\right), \quad (7.70)$$

$$-k_s \frac{\partial T_s}{\partial r} = h_{sf}\left(T_{sf} - \langle T \rangle^f\right) \quad \text{on} \quad A_{fs}, \quad (7.71)$$

where $T_{sf} = T_s$ on A_{fs}. Wakao and Kaguei suggest $D^d_\parallel / \alpha_f = 0.5 Pe$ with $Pe = \varepsilon u_p d / \alpha_f$. Note that the *bed* effective thermal conductivity $\langle k \rangle$ is included in the fluid-phase equation (Wakao and Kaguei). Also note that the suggested coefficient for Pe in the expression for the dispersion is *smaller* than that given in Chapter 4, where the presence of the local thermal equilibrium was assumed. This particle-based model is the *most* accurate among the three and is widely used. This model is for transient problems *only*.

7.2.2 EXPERIMENTAL DETERMINATION OF h_{sf}

Wakao and Kaguei have critically examined the experimental results on h_{sf} and have selected experiments (*steady-state* and *transient*) which they found to be reliable. They have used (7.69)–(7.71) for the evaluation of h_{sf}. This is a rather indirect method of measuring h_{sf} and as was mentioned the results depend on the model used. They have found the following correlation for h_{sf} for spherical particles (or the dimensionless form of it, the *Nusselt number*)

$$Nu_d = \frac{h_{sf} d}{k_f} = 2 + 1.1\, Re^{0.6}\, Pr^{1/3} \quad \text{for spherical particles}, \quad (7.72)$$

where $Re = \varepsilon u_p d/\nu = u_D d/\nu$. Equation (7.72) gives a $Re \to 0$ asymptote of $h_{sf} d/k_f = 2$, which is more reasonable than $h_{sf} \to 0$ found when models other than (7.69)–(7.71) are used. It should be mentioned that the measurement of h_{sf} becomes *more* difficult and the experimental *uncertainties* become much *higher* as $Re \to 0$. Figure 7.1 shows the experimental

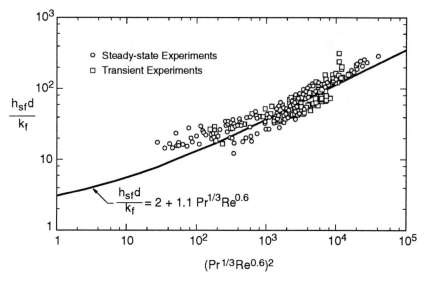

Figure 7.1 Experimental results compiled from many sources by Wakao and Kaguei (for steady-state and transient experiments). Also given is their proposed correlation. (From Wakao and Kaguei, reproduced by permission ©1982 Gordon and Breach Science.)

results compiled by Wakao and Kaguei and their proposed correlation. Note also that at low Re the interfacial convection heat transfer is insignificant compared to the other terms in the energy equations, and, therefore, the suggested $Re \to 0$ asymptote cannot be experimentally verified.

The steady-state results are for the heated spheres (the analogues mass transfer is the sublimation of spherical particles). Note that when heat generation in the solid or fluid phase is involved and the requirement for the existence of the local thermal equilibrium, i.e., (3.1), is satisfied, the volume-averaged treatments given in Chapters 3 and 4 apply.

For ceramic foams, with air as the fluid, Yunis and Viskanta (1993) have *indirectly* measured Nu_d and obtained correlations with Re_d as the variable. They obtain a lower value for the power Re_d. The computed relationship between Nu_d and Re_d, for a periodic structure, will be discussed in Section 4.9.3. The interfacial heat transfer is also discussed in detail by Kaviany (1994).

7.3 Distributed Treatment of Oscillating Flow

For capillary tubes, we can *directly* examine the solution to the *point energy equation* for each phase (instead of using the local phase volume-averaging of the energy equations). This will allow for *inclusion* of any *temporal change* in the fluid flow rate such as in oscillatory flows. When the fluid flow through the matrix changes direction, the transient pore-level fluid dynamics must *also* be considered. The frequency of oscillation determines whether flow *reversal* in the pores leads to formation of *thin boundary layers* (high-frequency *sinusoidal* oscillation) or whenever the flow transients (other than change in direction) are *insignificant* (change in flow directions is separated by a long steady flow period *nonsinusoidal* oscillation). For sinusoidal flow through capillary tubes, the boundary-layer thickness is determined by the *Stokes* or *Womersley number* $R\omega^{1/2}/\nu^{1/2}$, where $\omega = 2\pi f$ is the *angular frequency* (rad/s) and f is the *frequency* (Hz).

Also, using the direct simulation, one of the assumptions made in the phase volume-averaging treatment, namely, that the *penetration* depth is larger than the unit cell, can be avoided. In heat storage/release in capillary tubes, the radial temperature gradient is generally very large, and in some arrangements (Section 4.9), the maximum temperature difference occurs radially and over R. In this and the following section, two-point solutions for transient heat transfer in capillary tubes will be considered. These two examples demonstrate the direct simulation of flow and heat transfer in a class of transient problems where the assumption of the local thermal equilibrium does *not* hold.

The example considered in this section is of a periodic flow (pulsating) in capillary tubes with a step change in the inlet temperature at the end of each half-cycle. In particular, this example considers the following.

- The behavior in the *start-up period,* i.e., when the temporal variation of the temperature field has not reached the *asymptotic* periodic behavior.

- Examination of the *interstitial* convection heat transfer coefficient, i.e., if a prescribed h_{sf} is to be used, which one of the corresponding *constant temperature* or *constant heat flux* surface condition is more appropriate.

- Examination of the thermal entrance effect, i.e., conditions for which the *thermal entrance length* is not short compared to the tube length.

In this section, we consider *nonsinusoidal* oscillations. These have application in heat *regeneration*. Examples are given by Schmidt and Willmott (1981), where they use *constant* values of the interstitial heat transfer coefficient similar to the treatment given in the last section. Flow through a tube of radius R and length L with azimuthal symmetry and subject to flow reversal and inlet temperature change with a period of $1/f$ is considered. Over $1/(2f)$ *elapsed time,* the cold fluid flows in a given direction. Then, over the next $1/(2f)$ elapsed time, the hot fluid flows in the opposite direction. It is assumed that the change in the direction occurs *instantly* at

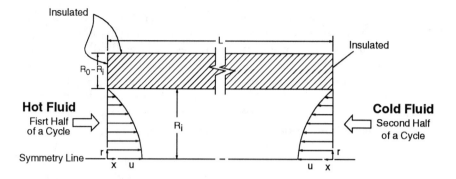

Figure 7.2 A schematic of the nonsinusoidal periodic flow in capillaries. The inlet temperature also changes periodically.

the end of each half. The dimensionless *point* equations for each phase are

$$\frac{fR_i^2}{\alpha_f}\frac{\partial T_f^*}{\partial t^*} + 2Pe\left(1 - \frac{r^{*2}}{R^2}\right)\frac{\partial T_f^*}{\partial x^*}$$

$$= \frac{1}{r^*}\frac{\partial}{\partial r^*}\left(r^*\frac{\partial T_f^*}{\partial r^*}\right) + \frac{\partial^2 T_f^*}{\partial x^{*2}} \quad 0 \leq r^* \leq 1, \quad (7.73)$$

$$\frac{fR_i^2}{\alpha_s}\frac{\partial T_s^*}{\partial t^*} = \frac{1}{r^*}\frac{\partial}{\partial r^*}\left(r^*\frac{\partial T_s^*}{\partial r^*}\right) + \frac{\partial^2 T_s^*}{\partial x^{*2}} \quad 1 \leq r^* \leq \frac{R_o}{R_i}, \quad (7.74)$$

where the *dimensionless* variables are

$$T^* = \frac{T - T_{f\,min}}{T_{f\,max} - T_{f\,min}}, \quad r^* = \frac{r}{R_i}, \quad x^* = \frac{x}{R_i}, \quad Pe = \frac{u_p R_i}{\alpha_f}. \quad (7.75)$$

The boundary and initial conditions state that the solid and fluid temperatures are initially ($t^* = 0$) at $T_{f\,min}$. Then, the inlet temperature is raised to $T_{f\,max}$, and after an elapsed time of $1/(2f)$, the flow changes direction and the inlet temperature becomes $T_{f\,min}$. The cycle is then repeated. Figure 7.2 gives a schematic of the problem considered. By always choosing the origin for the x-axis to be at the entrance of the flow into the tube, we can simplify the mathematical statement. These conditions are given as

$$\frac{\partial T^*}{\partial x^*} = 0, \quad 0 \leq r^* \leq \frac{R_o}{R_i}, \quad x^* = 0, \frac{L}{R_i}, \quad (7.76)$$

$$\frac{\partial T^*}{\partial r^*} = 0, \quad 0 \leq x^* \leq \frac{L}{R_i}, \quad r^* = 0, \frac{R_o}{R_i}, \quad (7.77)$$

$$\frac{k_f}{k_s}\frac{\partial T_f^*}{\partial r^*} = \frac{\partial T_s^*}{\partial r^*}, \quad r^* = 1, \quad (7.78)$$

7.3 Distributed Treatment of Oscillating Flow 407

$$T^* = 0, \quad t^* = 0, \quad \text{for all } r^* \text{ and } x^*, \qquad (7.79)$$

$$T^* = 1, \quad n \leq t^* \leq n + \frac{1}{2}, \quad 0 \leq r^* \leq 1, \; x^* = 0, \qquad (7.80)$$

$$T^* = 0, \quad n + \frac{1}{2} \leq t^* \leq n+1, \quad 0 \leq r^* \leq 1, \; x^* = 0, \qquad (7.81)$$

where $n = 0, 1, 2, \ldots$ is the number of periods.

The *fraction* of heat stored in the solid during one half of the period, when the hot fluid flows through the tube, is

$$\left.\frac{Q_{st}}{Q_{max}}\right|_{t^*=n+1/2} = \left.\frac{2\int_0^1 \int_0^{L/R_i} T_s^* \, r^* \, dr^* \, dx^*}{\left(\frac{R_o^2}{R_i^2} - 1\right)\frac{L}{R_i}}\right|_{t^*=n+1/2}. \qquad (7.82)$$

For *comparison*, the case of a *constant interstitial heat transfer coefficient* will also be considered. For this, the fluid energy equation is written in terms of an averaged fluid temperature $\overline{T}_f^*$ as

$$\frac{fR_i}{\alpha_f}\frac{\partial \overline{T}_f^*}{\partial t^*} + Pe\frac{\partial \overline{T}_f^*}{\partial x^*} + \frac{\partial^2 \overline{T}_f^*}{\partial x^{*2}} = \frac{h_{sf}R_i}{k_f}\left[T_s^*(r^* = 1) - \overline{T}_f^*\right]. \qquad (7.83)$$

For this case, the solid-phase energy equation is given by (7.74), and the initial and boundary conditions are also the same, except that $\overline{T}_f^*$ is used instead of T_f^*.

The finite-difference approximations are applied to (7.73), (7.74), and (7.83) and the results for a *glass-air* combination with $R_i = 1$ mm, $L = 100R_i$, $R_o = 1.3R_i$, $u_p = 2.5$ m/s, and $1/f = 60$ s are given in Figures 7.3 and 7.4. The *dimensionless* parameters are shown below.

$\frac{k_s}{k_f}$	$\frac{(\rho c_p)_s}{(\rho c_p)_f}$	$\frac{\alpha_s}{\alpha_f}$	Pe	$\frac{\alpha_f}{fR_i^2}$	$\frac{x}{LPe} = x^+$
54	1818	0.03	106	1418	0–97

The *Peclet* number is chosen such that the dimensionless *thermal entrance length* x^+ is *not* too large or too small so that the thermal entrance effects can be examined.

For the hydrodynamically and thermally fully developed, the heat transfer coefficient for the case of a constant *surface temperature* $h_{sf,T}$ and that for a *constant heat flux surface condition* $h_{sf,H}$ (Kays and Crawford, 1993) are given as

$$\frac{2h_{sf,H}R}{k_f} = 4.36, \quad x_H^+ = 0.20, \qquad (7.84)$$

408   7. Two-Medium Treatment

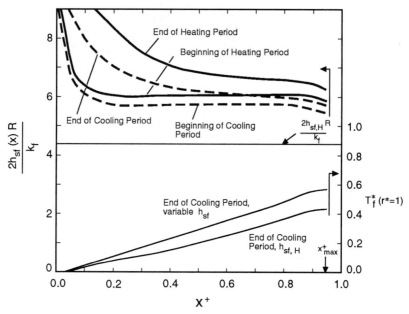

Figure 7.3 Variation of the local heat transfer coefficient and the local surface temperature along the capillary tube. The elapsed times correspond to the beginning and end of the heating and cooling periods.

$$\frac{2h_{sf,T}R}{k_f} = 3.66, \qquad x_T^+ = 0.20, \tag{7.85}$$

where the entrance length x^+ for each case is also given.

Figure 7.3 shows how the *local* convection heat transfer coefficient *changes* along the tube for elapsed times corresponding to the beginning and ends of the heating and cooling periods. The *constant*-heat flux fully developed convective heat transfer is also shown and, as shown, is below the local values. Note that the location $x = 0$ is *always* at the inlet, and, therefore, $h_{sf}(x)$ has its *maximum* always at $x = 0$. The effect of the value of the convective heat transfer on the axial distribution of the surface temperature is also given in Figure 7.3. The results show that constant $h_{sf,H}$ results in *lower* temperatures at the end of the cooling period. However, the temperature is also lower at the end of the heating period (not shown). This results in a smaller heat storage for $h_{sf,H}$. Note that $h_{sf,T}$ (constant surface temperature) results in an *even* larger difference with the results of $h_{sf}(x)$.

Figure 7.4 shows the heat stored at the end of the cooling and heating periods. Note that after about 4 cycles, a quasi-steady behavior is observed. The results show that the amount of heat stored (Q_{st}), as well as the heat exchanged between the heating and cooling periods (ΔQ_{st}), is *larger* when $h_{sf}(x)$ is used instead of $h_{sf,H}$.

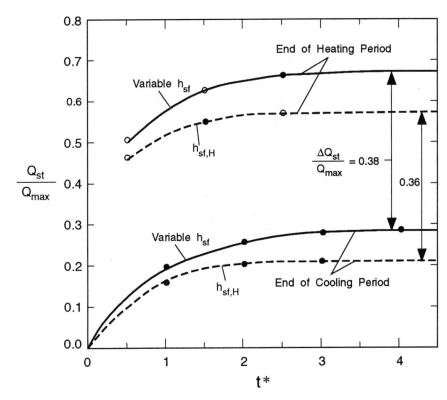

Figure 7.4 Heat stored in the capillary tube at the end of the heating and cooling periods. The results for both a constant and a variable heat transfer coefficient are given.

7.4 Chemical Reaction

One cause of *invalidity* of the local thermal equilibrium assumption, is the occurrence of an *endothermic* or *exothermic* chemical reaction in the fluid or solid (including *surface reactions*) phase. A general discussion of reaction in porous media was given in Section 6.8. A rendering of reaction in porous media is made in Figure 7.5. In a one-dimensional temperature field, the *reaction front* in the porous medium can be moving at the *front velocity* u_F, with respect to a fixed coordinate system x. For example, in combustible porous media and in *diffusion* and *premixed* gaseous reactions where the particular gas velocity does not allow for the immobilization of the reaction front. Then a moving coordinate $x_1 = \text{x} + u_F t$ can be used where u_F can be steady or unsteady. The *upstream* or *nonreacted* conditions are designated with the subscript n and the local thermal equilibrium is assumed.

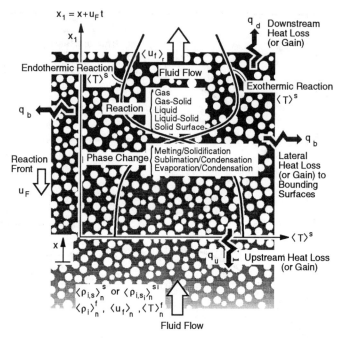

Figure 7.5 A rendering of the reaction front in porous media showing various possible reactions, phase changes, heat losses, and the heat generation/sink. The nonreacted upstream and the reacted downstream conditions are also shown, along with the expected temperature distributions for solid phase.

Heat losses/gains can occur from the front to the upstream and downstream regions and laterally to the bounding lateral surfaces. In general, phase changes, i.e., melting/solidification, sublimation/condensation, or evaporation/condensation can also occur.

As an example, we consider the premixed gaseous reaction in a two-dimensional porous medium. This simple reaction-geometry allows for a close examination of the front structure. Also, using the direct simulations and the volume-averaged treatments (under the assumption of local thermal equilibrium or nonequilibrium), the differences in the predicted front structure and speed can be quantified. The problem has been examined by Sahraoui and Kaviany (1994) and their two-dimensional unit cells are those shown in Figure 3.1 for continuous and discrete solid phases. For the continuous solid phase, fluid flow is allowed through the extended arms of thickness c. These are shown in Figure 7.6. The unit-cell linear dimension is ℓ and a steady flow of gas is assumed with the speed u_F which is equal and opposite to the front speed resulting in a stationary front in the porous media.

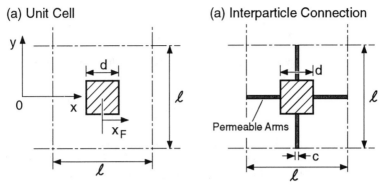

Figure 7.6 Two-dimensional unit-cell models used in the direct simulations.

7.4.1 Two-Dimensional Direct Simulation

The direct simulations of the premixed methane-air combustion is made for the in-line and staggered arrangements of connected or discrete square cylinders. The continuity equation for the gas is given by

$$\frac{\partial \rho_g}{\partial t} + \frac{\partial \rho_g u}{\partial x} + \frac{\partial \rho_g v}{\partial y} = 0. \tag{7.86}$$

The density of the gas phase ρ_g is given by the ideal-gas law at atmospheric pressure (inlet pressure) and for a variable local temperature. In the momentum equation, we assume that the gas velocity is constant and that the buoyancy effect is negligible and write

$$\rho_g \frac{\partial u}{\partial t} + \rho_g u \frac{\partial u}{\partial x} + \rho_g v \frac{\partial u}{\partial y} = -\frac{\partial p}{\partial x} + \mu \left(\frac{\partial^2 u}{\partial x^2} + \frac{\partial^2 u}{\partial y^2} \right), \tag{7.87}$$

$$\rho_g \frac{\partial v}{\partial t} + \rho_g u \frac{\partial v}{\partial x} + \rho_g v \frac{\partial v}{\partial y} = -\frac{\partial p}{\partial y} + \mu \left(\frac{\partial^2 v}{\partial x^2} + \frac{\partial^2 u}{\partial v^2} \right). \tag{7.88}$$

In the energy equations, the specific heat capacity c_p and conductivity k are assumed constant and evaluated at the average temperature (i.e., 1300 K for the equivalence ratio of unity). The gas-phase energy equation is written as

$$(\rho c_p)_g \frac{\partial T_g}{\partial t} + (\rho c_p)_g u \frac{\partial T_g}{\partial x} + (\rho c_p)_g v \frac{\partial T_g}{\partial y}$$
$$= \frac{\partial}{\partial x} k_g \frac{\partial T_g}{\partial x} + \frac{\partial}{\partial y} k_g \frac{\partial T_g}{\partial y} + \dot{n} \Delta i_c \quad \text{in } V_g, \tag{7.89}$$

where $\dot{n}$ is the volumetric rate of generation of product species and Δi_c is the specific heat of combustion. The solid-phase energy equation is

7. Two-Medium Treatment

$$(\rho c_p)_s \frac{\partial T_s}{\partial t} = \frac{\partial}{\partial x} k_s \frac{\partial T_s}{\partial x} + \frac{\partial}{\partial y} k_s \frac{\partial T_s}{\partial y} \quad \text{in} \quad V_s. \tag{7.90}$$

The species conservation equation is written for the product species as

$$\rho_g \frac{\partial Y}{\partial t} + \rho_g u \frac{\partial Y}{\partial x} + \rho_g v \frac{\partial Y}{\partial y} = \frac{\partial}{\partial x} D_m \rho_g \frac{\partial Y}{\partial x} + \frac{\partial}{\partial y} D_m \rho_g \frac{\partial Y}{\partial y} + \dot{n}, \tag{7.91}$$

where $Y = \rho_{p,g}/\rho_g$ is the mass fraction of the product species. For the species diffusivity, we assume that the molecular Lewis number, $Le_m = D/\alpha_g$, is unity. For the species generation term, a first-order, Arrhenius relation is arranged and is given by

$$\dot{n} = A\rho_g(1-Y)e^{-\Delta E_a/R_g T_g}. \tag{7.92}$$

The boundary conditions at the inlet of the computational domain are

$$u = u_F, \quad v = 0, \quad T_g = T_n, \quad \text{and} \quad Y = 0 \quad \text{for} \quad x \to \infty, \tag{7.93}$$

and for the exit

$$\rho_g u = \rho_g(T_n)u_F, \quad v = 0, \quad \frac{\partial T_g}{\partial x} = 0, \quad \text{and} \quad \frac{\partial Y}{\partial x} = 0 \quad \text{for} \quad x \to \infty. \tag{7.94}$$

At the upper and lower boundaries of the domain, symmetry conditions are used for all the quantities. At the solid-gas interface, the no-slip condition is used for velocity, i.e.,

$$u = v = 0 \quad \text{on} \quad A_{gs}. \tag{7.95}$$

The temperature boundary conditions become

$$T_s = T_g, \quad k_s \mathbf{n} \cdot \nabla T_s = k_g \mathbf{n} \cdot \nabla T_g \quad \text{on} \quad A_{gs}. \tag{7.96}$$

For the species, the condition of no mass flux is used, i.e.,

$$\mathbf{n} \cdot \nabla Y = 0 \quad \text{on} \quad A_{gs}. \tag{7.97}$$

The above equations are solved for a variable equivalence ratio Φ which is defined as

$$\Phi = \frac{\left(\dfrac{\nu_F}{\nu_o}\right)_{actual}}{\left(\dfrac{\nu_F}{\nu_o}\right)_{stoich.}}, \tag{7.98}$$

where ν_F and ν_o are the molar coefficients for the fuel and the oxidant in the stoichiometric reaction equation given by (6.6).

In the two-medium treatment, the heat exchange between the two media is modeled using a *prescribed* Nusselt number relation and the temperature difference between the two phases. This Nusselt number for the in-line and staggered arrangements of square cylinders, is seperately computed. The continuity and momentum equations, (7.86) and (7.88), are solved along with the energy equations, (7.89), along with $\dot{n} = 0$ and a prescribed solid surface temperature. In the energy equation, we solve for the dimensionless temperature T^* defined as

$$T^* = \frac{T_g - T_n}{T_s - T_n}. \quad (7.99)$$

In these computations, the gas is assumed to be incompressible. The boundary conditions at the inlet and exit, for velocity and temperature, are those given by (7.93) and (7.94), with $T_n^* = 0$. The surface temperature of all cylinders is constant, i.e., $T_s^* = 1$. The surface area-averaged Nusselt number is found by area averaging the local Nusselt number on a cylinder, i.e.,

$$\langle Nu_\ell \rangle_{A_{gs}} = -\frac{1}{A_{gs}} \int_{A_{gs}} \frac{\ell}{T_s - \langle T \rangle_y^b} \nabla T_g \cdot \mathbf{n} \, dA, \quad (7.100)$$

where $\langle T \rangle_y^b$ is the local, bulk-mixed gas temperature defined as

$$\langle T \rangle_y^b = \frac{\int_{-\ell/2}^{\ell/2} \rho_g u(x,y) T_g(x,y) \, dy}{\int_{-\ell/2}^{\ell/2} \rho_g u(x,y) \, dy}. \quad (7.101)$$

The computed results, showing the variation of the Nusselt number with respect to the particle Reynolds number, particle arrangement, and porosity, are discussed in Section 7.4.3.

7.4.2 VOLUME-AVERAGED MODELS

The local volume-averaged models used in the studies of combustion in porous media are the two-medium treatment which allows for a thermal nonequilibrium between the phases, and the single-medium treatment which assumes a local thermal equilibrium.

(A) TWO-MEDIUM TREATMENT

The two-medium thermal energy and species conservation models used here are

$$\epsilon(\rho c_p)_g \frac{\partial \langle T \rangle^g}{\partial t} + (\rho c_p)_g u_F \frac{\partial \langle T \rangle^g}{\partial x} = \frac{\partial}{\partial x}\left[\epsilon\left(\langle k \rangle^g + (\rho c_p)_g D^d_{xx}\right)\frac{\partial \langle T \rangle^g}{\partial x}\right]$$
$$+ \frac{A_{gs}}{V}\langle Nu_\ell \rangle_{A_{gs}} \frac{k_g}{\ell}(\langle T \rangle^s - \langle T \rangle^g) + \epsilon \langle \dot{n} \rangle^g \Delta i_c. \qquad (7.102)$$

$$(1-\epsilon)(\rho c_p)_s \frac{\partial \langle T \rangle^s}{\partial t} = \frac{\partial}{\partial x}(1-\epsilon)\langle k \rangle^s \frac{\partial \langle T \rangle^s}{\partial x}$$
$$+ \frac{A_{gs}}{V}\langle Nu_\ell \rangle_{A_{gs}} \frac{k_g}{\ell}(\langle T \rangle^g - \langle T \rangle^s). \qquad (7.103)$$

$$\epsilon \rho_g \frac{\partial \langle Y \rangle^g}{\partial t} + \rho_g u_F \frac{\partial \langle Y \rangle^g}{\partial x} = \frac{\partial}{\partial x}\epsilon(\langle D_m \rangle^g + D^d_{m_{xx}})\rho_g \frac{\partial \langle Y \rangle^g}{\partial x} + \epsilon \langle \dot{n} \rangle^g. \qquad (7.104)$$

For the momentum equation, the fluid flow through porous media is governed by the Darcy law. Assuming that the pressure drop in the porous medium is negligible, only the volume-averaged continuity equation is needed which implies that $\rho_g u_F$ is constant for this volume-averaged, one-dimensional flow.

The effective conductivity of the solid and gas phases are determined using the equivalent thermal-circuit model. For the discrete solid phase, the gas phase is assumed to conduct heat through a continuous phase existing between any two, adjacent square cylinders, i.e.,

$$\epsilon \langle k \rangle^g = k_g[1 - (1-\epsilon)^{1/2}]. \qquad (7.105)$$

For the solid, the heat is assumed to be conducted through the remaining volume, i.e.,

$$(1-\epsilon)\langle k \rangle^s = \frac{(1-\epsilon)^{1/2}k_s k_g}{[1-(1-\epsilon)^{1/2}]k_s + (1-\epsilon)^{1/2}k_g}. \qquad (7.106)$$

For the continuous solid phase, we have

$$(1-\epsilon)\langle k \rangle^s = \frac{\left[(1-\epsilon)^{1/2} - \frac{c}{d}(1-\epsilon)^{1/2}\right]k_s k_g}{[1-(1-\epsilon)^{1/2}]k_s + (1-\epsilon)^{1/2}k_g} + \frac{c}{d}(1-\epsilon)^{1/2}k_s, \qquad (7.107)$$

and for the gas phase, the effective conductivity is given by

$$\epsilon \langle k \rangle = \frac{[1-(1-\epsilon)^{1/2}]k_s k_g}{\left[1-\frac{c}{d}(1-\epsilon)^{1/2}\right]k_s + \frac{c}{d}(1-\epsilon)^{1/2}k_g}. \qquad (7.108)$$

For the axial, thermal dispersion coefficient D_{xx}^d, we use the results discussed in Section 4.3.5, and given in Table 4.6 for the in-line arrangement of circular cylinders. This is a good approximation, since the particle shape does not affect the hydrodynamic dispersion significantly (especially for low Peclet numbers). For the effective properties in the species equation, it is assumed that the effective mass diffusivity is the same as the thermal diffusivity of the gas and that the dispersion coefficient for the heat and mass diffusion are the same. Note that since D_{xx}^d depends on k_s/k_g, $D_{m_{xx}} = D_{xx}^d$ only if $k_s/k_g = 0$. Here, the effective, two-medium Lewis number, assumed to be unity, becomes

$$Le_{e_2} = \frac{\frac{\langle k \rangle^g}{(\rho c_p)_g} + D_{xx}^d}{\langle D \rangle^g + D_{m_{xx}}^d} = 1. \qquad (7.109)$$

For the boundary conditions, at the inlet and exit a local thermal equilibrium is assumed. The boundary conditions for the phase-averaged temperature and mass fraction at the inlet are

$$\langle T \rangle^g = \langle T \rangle^s = T_n, \quad \langle Y \rangle^g = 0 \quad \text{for} \quad x \to \infty \qquad (7.110)$$

and at the exit

$$\frac{\partial \langle T \rangle^g}{\partial x} = \frac{\partial \langle T \rangle^s}{\partial x} = \frac{\partial \langle Y \rangle^g}{\partial x} = 0 \quad \text{for} \quad x \to \infty. \qquad (7.111)$$

(B) SINGLE-MEDIUM TREATMENT

For the single-medium thermal energy equation model, i.e., where a local thermal equilibrium (i.e., $\langle T \rangle^s = \langle T \rangle^g = \langle T \rangle$) between the phases is assumed, (7.102) and (7.103) are added to obtain

$$[\epsilon(\rho c_p)_g + (1-\epsilon)(\rho c_p)_s]\frac{\partial \langle T \rangle}{\partial t} + (\rho c_p)_g u_F \frac{\partial \langle T \rangle}{\partial x}$$
$$= \frac{\partial}{\partial x}\left[\langle k \rangle + \epsilon(\rho c_p)_g D_{xx}^d \frac{\partial \langle T \rangle}{\partial x}\right] + \epsilon \langle \dot{n} \rangle^g \Delta i_c. \qquad (7.112)$$

The species conservation equation is the same as that in the two-medium treatment given by (7.104). The species diffusion coefficient is obtained by assuming a unity effective Lewis number for the single-medium, i.e.,

$$Le_{e_1} = \frac{\frac{\langle k \rangle}{(\rho c_p)_g} + D_{xx}^d}{\langle D_m \rangle + D_{m_{xx}}^d} = 1. \qquad (7.113)$$

For the effective conductivity $\langle k \rangle$, we use the exact results of Sahraoui and Kaviany, i.e.,

$$\frac{\langle k \rangle}{k_g} = \frac{f(\epsilon)\frac{k_s}{k_g}(1-\epsilon)^{1/2} + [1-(1-\epsilon)^{1/2}] + [1-f(\epsilon)](1-\epsilon)^{1/2}}{\frac{k_s}{k_g}(1-\epsilon)^{1/2}[1-(1-\epsilon)^{1/2}] + [1-(1-\epsilon)^{1/2}]^2 + (1-\epsilon)^{1/2}}, \tag{7.114}$$

where $f(\epsilon)$ is introduced to account for the two-dimensional effects, and for the square cylinders is given by

$$f(\epsilon) = 0.83 + 0.18(1-\epsilon). \tag{7.115}$$

For the continuous solid phase, the effective conductivity is obtained by adding (7.107) and (7.108). The inlet boundary conditions are given by

$$\langle T \rangle = T_n, \quad \langle Y \rangle^g = 0 \quad \text{for} \quad x \to \infty \tag{7.116}$$

and the exit boundary conditions are given by

$$\frac{\partial \langle T \rangle}{\partial x} = \frac{\partial \langle Y \rangle^g}{\partial x} = 0 \quad \text{for} \quad x \to \infty. \tag{7.117}$$

7.4.3 INTERFACIAL NUSSELT NUMBER

The computed results for the area-averaged Nusselt number $\langle Nu_\ell \rangle_{A_{gs}}$ are shown in Table 7.1 for the in-line and staggered arrangements of square cylinders. The variation of $\langle Nu_\ell \rangle_{A_{gs}}$ between the square particles is not noticeable (except for the first and last cylinders in the computational domain). The results for $\langle Nu_\ell \rangle_{A_{gs}}$ are reported for the range of Reynolds number used in this study. For both particle arrangements, the Nusselt number *decreases* with Re_ℓ, for $Re_\ell < 20$. This is due to the decrease in the local Nusselt number on the *upstream* side of the particle, as shown in the polar plot of the computed, local Nusselt number Nu_ℓ in Figures 7.7(a) and (b). For the in-line arrangement of square cylinders, as Re_ℓ increases the flow becomes more rectilinear and the heat removal from the upstream side *decreases*. At higher Re_ℓ (i.e., $Re_\ell > 20$), Nu_ℓ *increases* slightly at the top surface (along the flow), and therefore, $\langle Nu_\ell \rangle_{A_{gs}}$ is not strongly affected by the variation in Re_ℓ. This is also found by Quintard and Whitaker (1993). For the staggered arrangement, the fluid follows a more tortuous path and the change in $\langle Nu_\ell \rangle_{A_{gs}}$ with respect to Re_ℓ is more pronounced. Figure 7.7(b) shows that as Re_ℓ increases, Nu_ℓ *increases* at the top and upstream surfaces. The increases in the heat transfer on the upstream surface is due to the stagnation point flow behavior as the flow inertia increases.

TABLE 7.1 VARIATION OF AREA-AVERAGE NUSSELT NUMBER WITH RESPECT TO REYNOLDS NUMBER FOR IN-LINE AND STAGGERED ARRANGEMENTS OF SQUARE CYLINDERS, $Pr = 0.67$.

Re_ℓ	$\langle Nu_\ell \rangle_{A_{gs}}$ IN-LINE $\epsilon = 0.9$	$\epsilon = 0.8$	STAGGERED $\epsilon = 0.8$
0.23	7.62	7.45	8.77
2.28	7.12	6.90	8.20
4.55	6.85	6.62	7.97
9.10	6.64	6.41	8.01
22.8	6.53	6.30	8.96
45.5	6.57	6.39	10.6
91.0	6.81	6.52	12.8

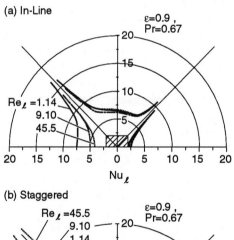
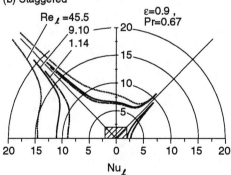

Figure 7.7 Polar presentation of distribution of the local Nusselt number for the two arrangements.

7.4.4 COMPARISON OF RESULTS OF VARIOUS TREATMENTS

Typical steady-state results for the temperature distribution, obtained using the two-dimensional simulation and the one-dimensional volume- averaged models, are shown in Figures 7.8 (a) to (c). The two-dimensional results are for the in-line arrangement of discrete square cylinders with $\epsilon = 0.9$, $k_s/k_g = 100$, $\ell = 1$ mm, and $\Phi = 1$. The two-dimensional result shows a curvature in the flame front due to the presence of the high conductivity solid. The location of maximum rate of reaction, i.e., the flame location x_F, occurs above the solid particle and the particle temperature is uniform because k_s/k_g is rather high. This also causes some upstream heat diffusion, resulting in the enhanced preheat of the reactants. This preheat in addition to increasing the flame speed, causes the local temperature in the gas to increase above the adiabatic temperature. This is shown in Figure 7.8(a), in the region where the local temperatures are higher than 2400 K. The highest temperature occurs at $y/\ell = 0.5$ and is about 2560 K. The *adiabatic flame temperature* T_a for this methane-air system with an equivalence ratio of unity, is 2308 K, and therefore, the maximum temperature difference $T_g - T_a$ is 252 K. In the post-flame region, a redistribution of heat occurs and further downstream a uniform temperature is found. For this case the flame speed is about 5 percent smaller than that of the adiabatic flame speed in plain media u_{F_o}, i.e., $u_F/u_{F_o} = 0.95$. Figure 7.8(b) shows the one-dimensional volume-averaged temperature distributions in the solid and gas phases obtained from the two-medium, local volume-averaged model. The local thermal nonequilibrium between the two phases is apparent. In the reaction zone the gas has a higher temperature, compared to the solid, causing heat transfer from the gas phase (or medium) to the solid phase (or medium). Heat is conducted through the solid medium to the upstream region where the solid medium temperature is higher than the gas medium, and this preheats the gas medium before it arrives in the reaction zone. This enhanced (over the amount occurring through the gas medium) preheat results in an increase in the flame speed and the occurrence of the *excess temperature* (over the adiabtaic temperature). For the two-medium treatment, the maximum temperature reached by the gas medium is 2351 K resulting in an excess temperature of 43 K [this cannot be distinguished in Figure 7.8(b)]. Figure 7.8(c) shows the temperature distribution predicted by the single-medium treatment which assumes a local thermal equilibrium between the gas and solid phases. In the single-medium treatment *no* excess temperature occurs. From Figures 7.8(b) and (c), we notice that the volume-averaged models predict the flame thickness fairly well.

Similarly, typical steady-state results for the temperature distributions for the connected solid phase are shown in Figures 7.9(a) to (c) for the two-dimensional simulation and the volume-averaged models. These solutions

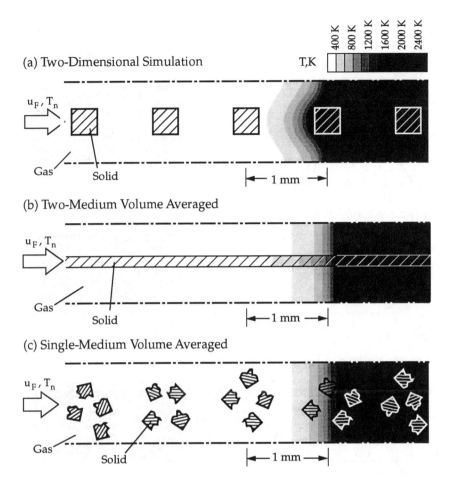

Figure 7.8 Temperature distribution for adiabatic, premixed flame using the in-line arrangement of discrete square cylinders. The results are for $\epsilon = 0.9$, $k_s/k_g = 100$, $\Phi = 1$ and for the different models (a) two-dimensional, (b) one-dimensional, two-medium, and (c) one-dimensional, single-medium.

are also for the in-line arrangement of cylinders with $\epsilon = 0.9$, $k_s/k_g = 100$, $\ell = 1$ mm, $\Phi = 1$, and with $c/d = 0.1$. Figure 7.9(a) shows the temperature distribution for the two-dimensional simulation. The connecting arms between adjacent square cylinders is assumed very permeable (i.e., no resistance to fluid flow) and have the same conductivity as that of the solid. Due to the presence of the connecting arms, the flame thickness and the preheat are further enhanced causing a 60 percent increase in the flame speed compared to the speed in the plain media, i.e., $u_F/u_{F_o} = 1.6$. The maximum

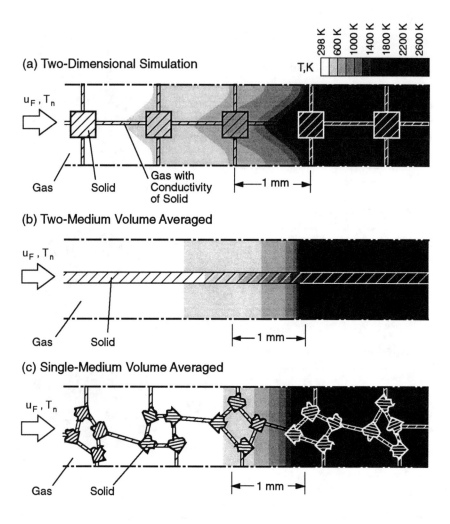

Figure 7.9 Same as Figure 7.8, but for a continuous solid phase ($c/d = 0.1$).

local temperature occurs at $y/\ell = 0.5$ and is about 2680 K, i.e., an excess temperature of 362 K. The results for the two-medium treatment are shown in Figure 7.9(b). As observed in the two-dimensional solution, the two-medium model also shows an increase in the extent of the thermal nonequilibrium, i.e., a higher preheat and an increase in the flame speed. The two-medium treatment predicts the flame thickness fairly well. The results from the single-medium treatment are shown in Figure 7.9(c) and a smaller flame thickness and no excess temperature are predicted.

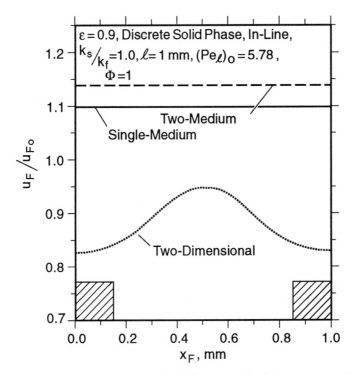

Figure 7.10 Variation of the normalized flame speed with respect to the flame location as obtained from the two-dimensional solutions and the volume-averaged models.

(A) Effect of Flame Location

The effect of flame location is examined for the in-line arrangement using $k_s/k_g = 1$. The variation of the normalized flame speed (again, u_{F_o} is the adiabatic flame speed in plain media) with respect to the location of the flame is shown in Figure 7.10 and the results show that the flame is displaced from the region above the cylinder (i.e., $x_F/\ell = 0$); the flame speed *increases*. This is because when the flame is away from the cylinder, the flow converges in the wake region and causes a uniform temperature in the reaction zone resulting in a higher temperature in the wake region. The gas and cylinder conduct the heat upstream and preheat the nonreacted mixture. The conduction in the wake region is significant because the local Peclet number is small. For the case of $x_F/\ell = 0.5$, the nonuniformity of the temperature in the preheat zone is more pronounced than that for $x_F/\ell = 0$, which results in a larger preheat and consequently a higher flame speed. Figure 7.10 also shows the results of the single- and two-medium treatments. A constant value is shown because these models do not account

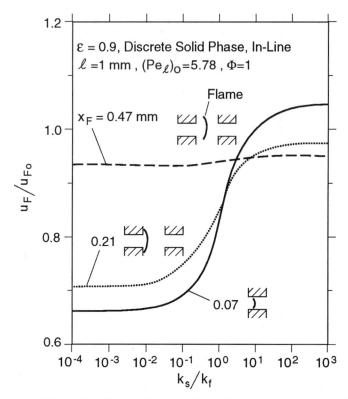

Figure 7.11 Effect of k_s/k_g on the normalized flame speed for three different flame locations within the pore.

for the pore-level variations. The two-medium treatment predicts a flame speed that is higher than the two-dimensional flame speed, for all flame locations.

(B) EFFECT OF k_s/k_g

As was mentioned, one advantage of using a porous medium is the high conductivity of the solid phase which enhances the preheating of the reactant. Figure 7.11 shows that when the flame is located above the cylinder, the flame speed is greatly affected by the solid-phase conductivity. For large k_s/k_g, the preheat is substantial, as shown in Figure 7.12(a) for $x_F/\ell = 0$ and $k_s/k_g = 100$, where the local temperature difference between the solid and gas at two lateral locations $y/\ell = 0$ (passing through the solid) and $y/\ell = 0.5$ (not passing through the solid) is significant due to the conduction through the solid phase. For large k_s/k_g, the flame speed approaches an asymptotic value which depends on the flame location. This is because the cylinders are not connected and the effective conductivity reaches an

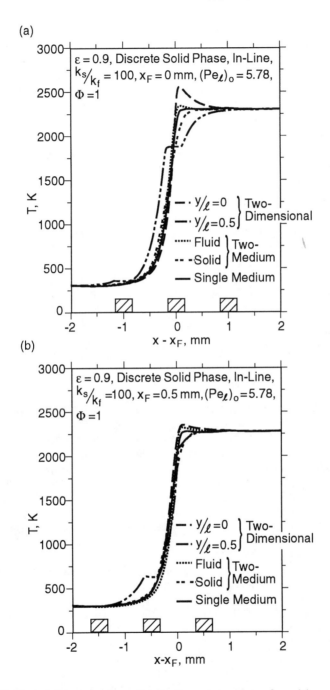

Figure 7.12 Axial temperature distribution at two lateral positions obtained from the two-dimensional solutions and compared with those from the single- and two-medium models, (a) $x_F/\ell = 0$ and (b) $x_F/\ell = 0.5$.

asymptote value as k_s/k_g becomes very large. When the flame is located away from the cylinder (i.e., $x_F/\ell = 0.5$), the effect of the solid conductivity is not significant and the flame velocity is only affected by the local velocity gradient which reduces the flame speed compared to the one-dimensional adiabatic flame speed (in plain media). The local temperature distributions for $x_F/\ell = 0$ at $y/\ell = 0$ and 0.5 are shown in Figure 7.12(b), where it is found that in the y-direction the temperature is more uniform compared to that for $x_F/\ell = 0$. The y-direction nonuniformity existing at the beginning of the preheat zone does not affect the flame speed significantly.

7.5 References

Carbonell, R. G. and Whitaker, S., 1984, "Heat and Mass Transfer in Porous Media," in *Fundamentals of Transport Phenomena in Porous Media*, eds., Bear and Corapcioglu, Martinus Nijhoff, 121–198.

Grassmann, P. and Tuma, M., 1979, "Critical Reynolds Number for Oscillating and Pulsating Tube Flow," (in German), *Thermo-Fluid Dyn.*, 12, 203–209.

Kaviany, M., 1994, *Principles of Convective Heat Transfer*, Springer-Verlag.

Kays, W. M. and Crawford, M. E., 1993, *Convection Heat and Mass Transfer*, Third Edition, McGraw-Hill.

Kurzweg, U. H., Lundgren, E. R., and Lothrop, B., 1989, "Onset of Turbulence in Oscillating Flow at Low Womersley Number," *Phys. Fluids*, A1, 1972–1975.

Levec, J. and Carbonell, R. G., 1985, "Longitudinal and Lateral Thermal Dispersion in Packed Beds, I–II," *AIChE J.*, 31, 581–590, 591–602.

Oliver, F. W. J., 1972, "Bessel Functions of Integer Order," in *Handbook of Mathematical Functions*, Abramowitz and Stegun, eds., NSRDS-NBS, p.379.

Patankar, S. V., 1980, *Numerical Heat Transfer and Fluid Flow*, Hemisphere.

Quintard, M. and Whitaker, S., 1993, "One and Two Equation Models for Transient Diffusion in Two-Phase Systems," *Advan. Heat Transfer*, 23, 269–464.

Quintard, M., Kaviany, M., and Whitaker, S., 1995, "Two-Medium Treatment of Heat Transfer in Porous Media: Numerical Results for Effective Properties," *Adv. Water. Resour.*, submitted.

Sahraoui, M. and Kaviany, M., 1994, "Direct Simulation versus Volume-Averaged Treatment of Adiabatic, Premixed Flame in a Porous Medium," *Int. J. Heat Mass Transfer*, 37, 2817–2834.

Schlichting, H., 1979, *Boundary-Layer Theory*, Seventh Edition, McGraw-Hill.

Schmidt, F. W. and Willmott, A. J., 1981, *Thermal Energy Storage and Regeneration*, McGraw-Hill.

Wakao, N. and Kaguei, S., 1982, *Heat and Mass Transfer in Packed Beds*, Gordon and Breach Science.

Yunis, L. B. and Viskanta, R., 1993, "Experimental Determination of the Volumetric Heat Transfer Coefficient between Stream of Air and Ceramic Foam," *Int. J. Heat Mass Transfer*, 36, 1425–1434.

Zanotti, F. and Carbonell, R. G., 1984, "Development of Transport Equation for Multi-Phase Systems—I—III," *Chem. Engng. Sci.*, 39, 263–278, 279–297, 299–311.

Part II
Two-Phase Flow

8
Fluid Mechanics

The hydrodynamics of the *two-phase* (*liquid-gas*) flow in porous media is addressed in this chapter and very briefly the *fluid-solid* two-phase flow is mentioned at the end of the chapter. Before introducing the volume-averaged momentum and continuity equations for each phase, we review the *elements* of the hydrodynamics of *three-phase systems* (*solid-liquid-gas*). These elements are the *interfacial tensions*, the *static contact angle*, the *moving contact angle*, and the *van der Waals interfacial-layer forces*. We examine the interfacial tension between a liquid and another fluid. For the case of a *static* equilibrium at this interface, we discuss the effect of the *curvature* for the simple problem of *ring formation* between spheres (and cylinders). For dynamic aspects, we examine the combined effect of *capillarity* and *buoyancy* by discussing the *rise* of a *bubble* in a capillary tube. Then we consider more realistic conditions and examine the effects of various factors on the *phase distributions* and the existing results for the phase distributions in flow through packed beds. The *moving contact line* and the effects of *solid surface tension* and the *surface roughness* and *heterogeneities* will then be discussed. For the *perfectly wetting* liquids at equilibrium, a thin *extension* of the liquid is present on the surface. We will examine the thickness and stability of this thin layer. After the phase-volume averaging of the momentum equation, we discuss the various coefficients that appear in the two momentum equations (one for the *wetting phase* and one for the *nonwetting phase*). The coefficients are generally determined *empirically*, because of the complexity of the phase distributions and their strong dependence on the local *saturation*. The *capillary pressure, phase permeabilities, liquid-gas interfacial drag* (due to the difference in the local phase velocities), and the *surface tension gradient-induced shear* at the liquid-gas interface are discussed in detail. The special transient problem of *immiscible* displacement is also examined using simple one-dimensional analyses. In the following paragraphs we review some of definitions used in two-phase flow through porous media and identify the key variables influencing the hydrodynamics.

When compared to the single-phase flows, the two-phase flow in porous media has one significant peculiarity and that is the *wetting* of the surface of the matrix by one of the fluid phases. Although here the attention is basically on a *liquid*-phase wetting the surface, a *gaseous* phase being the *nonwetting* phase, in some applications the two phases can be *two* liquids where one *preferentially* wets the surface. The presence of a curvature at the liquid-gas interface results in a difference between the local gaseous, and liquid-phase pressures (*capillary pressure*). This difference in pressure depends on the fraction of the average pore volume (or porosity of the representative elementary volume) occupied by the wetting phase. This fraction is called the *saturation* and is given as

$$\frac{\varepsilon_\ell}{\varepsilon} = \text{saturation} = s$$

$$= \frac{\text{fraction of the volume occupied by the wetting phase}}{\text{porosity}}. \quad (8.1)$$

As with the single-phase flows, (2.67) and (2.68), the fractions of the representative elementary volume occupied by the liquid and gas phases are

$$\varepsilon_\ell(\mathbf{x}) = \frac{1}{V}\int_V a_\ell(\mathbf{x})\,dV = \frac{V_\ell}{V} = \varepsilon s, \quad (8.2)$$

$$\varepsilon_g(\mathbf{x}) = \frac{1}{V}\int_V a_g(\mathbf{x})\,dV = \frac{V_g}{V} = \varepsilon(1-s), \quad (8.3)$$

where the *distribution functions* for the *wetting* and *nonwetting phases* are given by

$$a_\ell(\mathbf{x}) = \begin{cases} 1 & \text{if } \mathbf{x} \text{ is in the liquid (wetting) phase,} \\ 0 & \text{if } \mathbf{x} \text{ is not in the liquid phase,} \end{cases} \quad (8.4)$$

$$a_g(\mathbf{x}) = \begin{cases} 1 & \text{if } \mathbf{x} \text{ is in the gaseous (nonwetting) phase,} \\ 0 & \text{if } \mathbf{x} \text{ is not in the gaseous phase,} \end{cases} \quad (8.5)$$

$$\varepsilon_s + \varepsilon_\ell + \varepsilon_g = 1, \quad (8.6)$$

$$V_s + V_\ell + V_g = V, \quad (8.7)$$

$$\varepsilon_s = 1 - \varepsilon. \quad (8.8)$$

The subscript ℓ refers to the *liquid* or *wetting phase* and g refers to the *gaseous* or *nonwetting phase*.

As with the fluid dynamics of two-phase flows in plain media, when the two phases do *not* have the same *interstitial velocity* there will be an *interfacial drag* whose determination requires a knowledge of the *interfacial area* $A_{g\ell}$ as well as the local flow field in each phase. This interfacial drag is expected to be important only at *high flow rates*.

In *transient two-phase flows*, one phase *replaces* the other and the dynamics of the *wetting-dewetting* of the surface, which is influenced by the *fluid-fluid interfacial tension*, *solid-fluid interfacial tensions*, and the *solid-surface forces*, must be closely examined. The research on the *dynamics of the contact line* (fluid-fluid-solid contact line) has been advanced in the last decade.

Based on this we expect the following *parameters* (variables) to influence the dynamics of two-phase flow in porous media.

- *Surface tension*: Assuming that a membrane stretches over each interface, the magnitudes of the interfacial tension between each pair of phases are the *fluid-fluid interfacial tension* $\sigma_{g\ell}$, the *wetting fluid-solid interfacial tension* $\sigma_{\ell s}$, and the *nonwetting fluid-solid interfacial tension* σ_{gs}. When in *static equilibrium*, the *vectorial* force balance at the *line of contact* (the *law of Neumann triangle*, Defay and Prigogine, 1966) gives

$$\sigma_{g\ell} + \sigma_{\ell s} + \sigma_{gs} = 0 \quad \text{at contact line.} \quad (8.9)$$

  The tension at $A_{\ell g}$ will be discussed in Section 8.1.1 and that at A_{gs} and $A_{g\ell}$ will be discussed in Section 8.1.4. The *static* mechanical equilibrium of the g-ℓ surface is given by the *Young-Laplace equation*

$$p_c = p_g - p_\ell = \sigma_{g\ell}\left(\frac{1}{r_1} + \frac{1}{r_2}\right)$$

$$\equiv \sigma\left(\frac{1}{r_1} + \frac{1}{r_2}\right) \equiv 2H\sigma \quad \text{on } A_{g\ell}, \quad (8.10)$$

  where p_c is the *capillary pressure* and r_1 and r_2 are the two *principal radii of curvature* of $A_{g\ell}$ and where for simplicity we have used $\sigma_{g\ell} \equiv \sigma$. The *mean curvature of the interface* H is defined as

$$H \equiv \frac{1}{2}\left(\frac{1}{r_1} + \frac{1}{r_2}\right). \quad (8.11)$$

  The *general* dynamic force balance at any point on $A_{\ell g}$ (both normal and tangential components) will be discussed in Section 8.1.1 (D).

- *Wettability*: The extent to which the wetting phase *spreads* over the solid surface. The angle, *measured* in the *wetting phase*, between the solid surface and the g-ℓ interface, is called the *contact angle* θ_c where $\theta_c = 0$ corresponds to *complete wetting*. Presence of surface roughness, adsorbed surface layers, or surfactants influence θ_c *significantly*.

- *Matrix structure*: The size, dimensionality, pore coordinate number, and *topology* of the matrix influences the phase distributions significantly.

- *Viscosity ratio*: μ_g/μ_ℓ influences the relative flow rates *directly* and indirectly through the *interfacial* shear stress. In *fast* transient flows (e.g., immiscible displacement), depending on whether the viscosity of the *displacing* fluid is larger than that of the *displaced* fluid, or vice versa, different displacement *frontal behaviors* are found (Section 8.8).

- *Density ratio*: ρ_g/ρ_ℓ, in *addition* to the *body force*, signifies the relative importance of the *inertial force* for the two phases.
- *Saturation*: This is the extent to which the wetting phase occupies (averaged over the representative elementary volume) the pore space. At very low saturations the wetting phase becomes *disconnected* (or *immobile*). At very high saturations, the nonwetting phase becomes *disconnected*.

In addition, the presence of *temperature* and *concentration gradients* results in *interfacial tension gradients* and influences the phase distributions and flow rates. In dynamic systems, the *history* of the flows and the surface conditions also play a role and lead to the observed *hysteresis* in the phase distributions.

In order to arrive at a local volume-averaged momentum equation for each phase, the effect of the preceding parameters on the *microscopic* hydrodynamics must be examined. In the following, this is done to an extent and then the particular forces that appear in the momentum equations are examined.

8.1 Elements of Pore-Level Flow Structure

In addition to the *imposed pressure gradient* and the *solid surface viscous retarding force* (similar to the Darcy viscous resistance force), the forces affecting the flow of each phase are given in Figure 8.1. For *steady-state* flows, *unlike* the *transient* flows, which involve the local displacement of one of the phases by the other, the *dynamical forces of the moving contact lines* are *not* significant. Other forces, *capillary, gravity, interfacial,* and *inertial* forces can be significant in both steady-state and transient flows.

The flow of the two phases can be arranged as *cocurrent* or *countercurrent*. For each phase the flow can be *laminar* or *turbulent*. Figure 8.2 gives a flow classification for two-phase flow in porous media. For the gas phase at low pressures, the *Knudsen* or *transient* (as compared to *viscous*) regime may be encountered. This would be associated with *very* small vapor pressures in the absence of noncondensibles.

The *phase distributions* depend on the saturations. Figure 8.3 gives a phase distribution classification for consolidated and nonconsolidated (including fluidized) solid particles. In the countercurrent arrangement of the flows, the gaseous phase prevents the flow of the liquid phase whenever the gas flow rate is beyond the *flooding limit*. The *trickle bed* reactors could be arranged with the countercurrent flow with the liquid flowing due to gravity. Prior to flooding, the liquid flow becomes *unstable* due either to the *inertial force* alone or to the interfacial shear and the *inertial forces*. This flow is called the *pulsing flow*. This will be further discussed in Section

8.1 Elements of Pore-Level Flow Structure 431

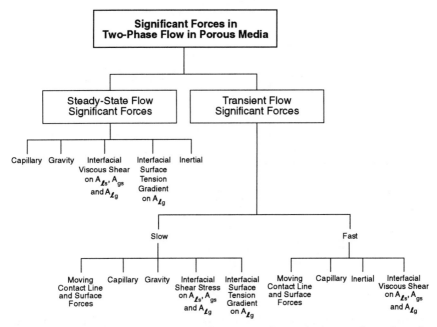

Figure 8.1 Significant forces in steady-state and transient two-phase flows in porous media. The transient flows are further divided into slow and fast transient.

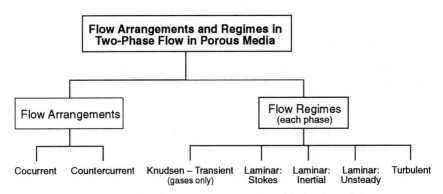

Figure 8.2 Flow arrangements for two-phase flows and the flow regimes for each phase. The molecular flow regime is for the gas phase only.

432   8. Fluid Mechanics

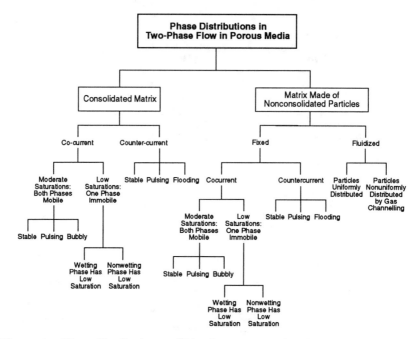

Figure 8.3 Phase distributions, within the representative elementary volume, for solid matrices made of consolidated or nonconsolidated elements.

8.1.2. Before further discussion of the flow regimes, the *interfacial forces* are examined.

8.1.1 SURFACE TENSION

Some aspects of the interfacial tension are addressed below. The solid-liquid-fluid interface (the contact line) is addressed in Section 8.1.4 and some thermodynamic aspects of the surface tension are given in Chapter 9. The mechanical and thermodynamic aspects of the surface tension are reviewed by Defay and Prigogine (1966), Chappuis (1982), and Israelachvili (1989). In general, because the surface forces are of the decaying type, the interface has a finite thickness (except at the critical temperature and pressure where this thickness becomes infinite). The thickness of the interface will be discussed in Chapter 9, and for now we shall treat the interface as having a zero thickness (phase discontinuity at the interface).

(A) INTERFACIAL TENSION BETWEEN A LIQUID AND ITS VAPOR

To increase the interfacial area between a liquid and its vapor, *work* is required. Through an isothermal process, the *stretch* of a planar surface area by dx in the x-direction (while the y-direction dimension L remains the same) requires a *force* f. The associated work required to *move* the liquid molecules to the *surface* is

$$dW = f\,dx = \frac{f}{L}dA_{\ell g} = \sigma\,dA_{\ell g}, \tag{8.12}$$

where $\sigma \equiv f/L = dW/dA_{\ell g}$ is the *surface tension* (force per unit length) or the *work of surface formation per unit of new area formed*, i.e., the work required for forming a *new* unit of interfacial area.

In the thermodynamic treatment of the surface, the *change* in the *excess internal energy* of the surface is given as

$$dU = T\,dS + \sigma\,dA_{\ell g} + \sum_i \mu_i\,dN_i, \tag{8.13}$$

where S is the *entropy*, μ_i is the *chemical potential* of species i, and N_i is the *number* of moles of species i. In (8.13) no allowance is made for the *interfacial volume*. In Chapter 9 we consider a more general treatment of the interface.

The *Helmholtz free energy* F for the surface is

$$F = U - TS, \tag{8.14}$$

where U is the internal energy given earlier. Then we have

$$dF = -S\,dT + \sigma\,dA_{\ell g} + \sum_i \mu_i\,dN_i, \tag{8.15}$$

from which the *definition* of the surface tension in terms of the Helmholtz free energy (or the *thermodynamic definition*) is given as

$$\sigma \equiv \left.\frac{\partial F}{\partial A_{\ell g}}\right|_{T,N_i,\ldots}. \tag{8.16}$$

The interfacial tension is a force *exerted on all points* of the interface and it *acts along* the *tangent* to the interface and in the *direction* that *diminishes* the *interfacial area*.

The surface tension is caused by the forces among the molecules of the liquid phase *on* and *near* the interface. As we discussed in Section 6.5.4, some of these molecular forces are the *ionic* and *metallic bonds* (strong), the *hydrogen bonds* (medium), and the *van der Waals forces* (weak). The thickness of the *disturbed zone* (the *thin* liquid layer over which the surface effects penetrate) is of the order of 10 Å. The van der Waals attraction forces vary as the inverse of *some* power of the separation distance r, e.g., a_1/r^7, where a_1 is a constant that depends on the substance (Chappuis, 1982). This will be further discussed in Sections 8.1.4 and 8.1.5.

434   8. Fluid Mechanics

The surface tension *depends* on the temperature (*decreasing* as the temperature *increases*), and for multicomponent systems it depends on the *concentration* of the solute. A solute that results in the *reduction* of the surface tension is called a *tensio-active solute*. It should be noted that at and near the interface the concentration of the solute, in addition to the diffusion-convection, is governed by the surface forces. Therefore, the surface tension is a *complicated* function of the concentration (at and adjacent to the interface), of the other state variables, and of the imposed fields (e.g., electric field e), i.e.,

$$\sigma = \sigma(T, \frac{\rho_i}{\rho}, \mathrm{e}, \ldots). \tag{8.17}$$

(B) Interfacial Tension Between Two Liquids

When a liquid is in contact with another liquid, the disturbances adjacent to the interface are *less* significant than those of the liquid-vapor interface. Therefore, for a given liquid the surface tension for the case of the liquid-liquid ($\ell_1 - \ell_2$) interface $\sigma_{\ell_1(\ell_2)}$ will be smaller than that for the liquid-vapor ($\ell_1 - g_1$) interface, i.e.,

$$\sigma_{\ell_1(\ell_2)} < \sigma_{(\ell g)_1}. \tag{8.18}$$

Since *both* liquids experience disturbances, the *interfacial tension for the system* $\sigma_{\ell_1\ell_2}$ is the *sum* of the surface tensions, i.e.,

$$\sigma_{\ell_1\ell_2} = \sigma_{\ell_1(\ell_2)} + \sigma_{\ell_2(\ell_1)} < \sigma_{(\ell g)_1} + \sigma_{(\ell g)_2}. \tag{8.19}$$

The *inequality* has been studied and a short review is given by Chappuis (1982). One of the available relations is

$$\sigma_{\ell_1\ell_2} = \sigma_{(\ell g)_1} + \sigma_{(\ell g)_2} - 2a_2(\sigma_{(\ell g)_1}\sigma_{(\ell g)_2})^{1/2}, \tag{8.20}$$

where the constant a_2 depends on both substances, i.e., $a_2 = a_2$ (dipolar moment, polarizability, ionization energy, molecular radius of each liquid).

When the *interaction* forces are of the *same* type, then $a_2 \to 1$, otherwise

$$0.5 \leq a_2 \leq 1.1. \tag{8.21}$$

Table 8.1 gives the surface tension for several liquid-fluid combinations. Note that as expected, for the water-air system the surface tension is much larger than it is for the water–olive oil system.

(C) Static Equilibrium at Liquid-Gas Interface

The history of the analysis of the *static* liquid-fluid interface is given by Myshkis et al. (1987) and by Saez (1983). Saez also discusses the determination of the static *liquid holdup* in packed beds of spheres and cylinders.

TABLE 8.1 SURFACE TENSION FOR SOME LIQUID-GAS AND LIQUID-LIQUID PAIRS (FROM MYSHKIS ET AL.)

SUBSTANCE	MEDIUM IN CONTACT	TEMPERATURE (°C)	SURFACE TENSION σ (dyne/cm)†
Hydrogen	Vapor of the substance	−252.0	2.0
Nitrogen		−195.9	8.3
Nitrogen		−183.0	6.2
Oxygen		−182.7	13.0
Ethyl alcohol		20.0	22.0
Mercury		0.0	513.0
Mercury		20.0	475.0
Mercury	Chloroform	20.0	357.0
Water	Olive oil	20.0	20.0
Water	Air	20.0	72.75
Water		100.0	58.8
Gold		1130.0	1102.0
Gold	Vapor of the substance	1070.0	612.0
Silver		1060.0	750.0

†dyne/cm = 10^3 N/m.

In the mechanical treatment of the interface (*meniscus*), both the *total mechanical energy minimization* (variational formulation) and the straightforward *force balance* can be used. However, the former also allows for a direct *stability analysis*. Both methods result in a relationship among the *surface curvature*, the *hydrostatic pressure*, and the *capillary pressure*. This relationship is called the *Young-Laplace equation*. In addition, the *static contact line* equation is used for the specification of the *contact angle* and this relation is referred to as the *Young equation*.

We begin the analysis of the interfacial location (the meniscus contour) by using a specific example considered by Erle et al. (1971) and by Saez. Consider the liquid bridge between two *spheres*, as shown in Figure 8.4. The separation distance between the spheres is γ. The bridge between *cylinders* (two-dimensional) is also analyzed using the same coordinate axes shown in Figure 8.4. The pressure distribution in the liquid is given by

$$\nabla p_\ell = (\rho_\ell - \rho_g)\mathbf{g}, \tag{8.22}$$

or in the coordinate used

$$p_\ell = p_{\ell_0} + (\rho_\ell - \rho_g)gx, \quad p_{\ell_0} = p_\ell(x=0). \tag{8.23}$$

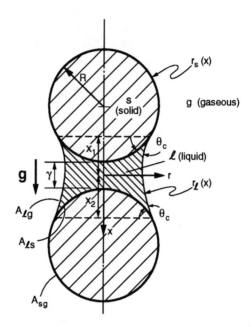

Figure 8.4 Liquid bridge between two spheres separated by a distance γ. Also shown are the coordinate axes used in the analysis, the contact angle θ_c, and the direction of the gravity vector.

The derivation of the Young-Laplace equation, based on the variational formulation for the liquid holdup problem is given by Saez and is reviewed later.

The *total energy* of the system E is

$$E = \sigma_{lg}A_{lg} + \sigma_{ls}A_{ls} + \sigma_{gs}A_{gs} - \int_{V_l} p_l \, dV - \int_{V_g} p_g \, dV + \text{constant}, \quad (8.24)$$

where

$$A_s = A_{ls} + A_{gs}. \quad (8.25)$$

Now define

$$\cos\theta = \frac{\sigma_{gs} - \sigma_{ls}}{\sigma_{lg}}. \quad (8.26)$$

As will be discussed, θ is the contact angle (this equation is the *Young equation*).

By assuming that p_g is uniform in V_g, we have

$$E = \sigma_{lg}(A_{lg} - A_{ls}\cos\theta) - \int_{V_l} p_l \, dV + \int_{V_g} p_g \, dV$$
$$+ (\text{constant} + A_s\sigma_{gs})$$
$$= \sigma(A_{lg} - A_{ls}\cos\theta) - \int_{V_l} p_l \, dV + a_1, \quad (8.27)$$

8.1 Elements of Pore-Level Flow Structure

where we used $\sigma = \sigma_{\ell g}$. Next, we seek an $A_{\ell g}$ such that E is *minimized*, i.e.,

$$\delta E = 0 \quad \text{and} \quad \delta^2 E > 0. \tag{8.28}$$

The *dimensionless* form of (8.27) for the geometry shown in Figure 8.4 and for the scales defined later is given by

$$\begin{aligned}E^*(r_\ell^*) &= \int_{x_1}^{x_2}\left\{2\pi r_\ell^*\left[1+\left(\frac{\mathrm{d}r_\ell^*}{\mathrm{d}x^*}\right)^2\right]^{\frac{1}{2}}(1-\cos\theta)\right.\\ &\qquad\left. -\pi(p_{\ell s}^* - Bo\,x^*)(r_\ell^{*2} - r_s^{*2})\right\}\mathrm{d}x^*\\ &\equiv \int_{x_1}^{x_2} f\left(x^*, r_\ell^*, \frac{\mathrm{d}r_\ell^*}{\mathrm{d}x^*}\right)\mathrm{d}x^*.\end{aligned} \tag{8.29}$$

The function $E^*(r_\ell^*)$ has an *extremum* if

$$\frac{\mathrm{d}f}{\mathrm{d}r_\ell^*} - \frac{\mathrm{d}}{\mathrm{d}x^*}\frac{\mathrm{d}f}{\mathrm{d}\left(\frac{\mathrm{d}r_\ell^*}{\mathrm{d}x^*}\right)} = 0 \quad (\textit{Euler-Lagrange equation}). \tag{8.30}$$

At the *two* endpoints, where f takes on the values f_1 and f_2, we have

$$f_1 + \left(\left.\frac{\mathrm{d}r_s^*}{\mathrm{d}x^*}\right|_1 - \left.\frac{\mathrm{d}r_\ell^*}{\mathrm{d}x^*}\right|_1\right)\left.\frac{\mathrm{d}f}{\mathrm{d}\left(\frac{\mathrm{d}r_\ell^*}{\mathrm{d}x^*}\right)}\right|_1 = 0, \tag{8.31}$$

$$f_2 + \left(\left.\frac{\mathrm{d}r_s^*}{\mathrm{d}x^*}\right|_2 - \left.\frac{\mathrm{d}r_\ell^*}{\mathrm{d}x^*}\right|_2\right)\left.\frac{\mathrm{d}f}{\mathrm{d}\left(\frac{\mathrm{d}r_\ell^*}{\mathrm{d}x^*}\right)}\right|_2 = 0. \tag{8.32}$$

The last two equations are called the *transversality conditions* and must be satisfied by a set of x_1 and x_2 that satisfies the functional f, where x_1 and x_2 are the x-coordinates of the contact line on the spheres.

When (8.30) is applied to (8.29), we have the *Young-Laplace equation*, i.e.,

$$\frac{\mathrm{d}^2 r_\ell^*}{\mathrm{d}x^{*2}} = \frac{1+\left(\frac{\mathrm{d}r_\ell^*}{\mathrm{d}x^*}\right)^2}{r_\ell^*} - (p_{\ell_o}^* + Bo\,x^*)\left[1+\left(\frac{\mathrm{d}r_\ell^*}{\mathrm{d}x^*}\right)^2\right]^{3/2}, \tag{8.33}$$

where

$$x^* = \frac{x}{R}, \quad r^* = \frac{r}{R}, \quad p_{\ell_o}^* = \frac{p_{\ell_o}R}{\sigma}, \quad \Delta\rho = \rho_\ell - \rho_g,$$

$$Bo = \frac{(\Delta\rho)gR^2}{\sigma} \quad (Bond \text{ or } E\ddot{o}tv\ddot{o}s \text{ number}), \tag{8.34}$$

and as before, p_{ℓ_0} is the *reference pressure* at $x = 0$.

For *cylinders*, the Young-Laplace equation is

$$\frac{d^2 r_\ell^*}{dx^{*2}} = -(p_{\ell_0}^* + Bo\,x)\left[1 + \left(\frac{dr_\ell^*}{dx^*}\right)^2\right]^{3/2}. \tag{8.35}$$

The two boundary conditions for the Young-Laplace equation are the prescribed contact angle θ_c at the two ends. The transversality conditions become

$$\theta_c = \tan^{-1}\left.\frac{dr_\ell^*}{dx^*}\right|_1 - \tan^{-1}\left.\frac{dr_s^*}{dx^*}\right|_1 \quad \text{at} \quad x_1, \tag{8.36}$$

$$\theta_c = \tan^{-1}\left.\frac{dr_s^*}{dx^*}\right|_2 - \tan^{-1}\left.\frac{dr_\ell^*}{dx^*}\right|_2 \quad \text{at} \quad x_2, \tag{8.37}$$

where from (8.26) we have

$$\cos\theta_c = \frac{\sigma_{gs} - \sigma_{\ell s}}{\sigma_{\ell g}}, \tag{8.38}$$

which, as was mentioned, is called the *Young equation*.

The stability conditions for the Young-Laplace and Young system of equations are discussed by Saez (using the $\delta^2 E > 0$ condition). Equation (8.33) or (8.35) along with (8.36) and (8.37) are generally solved as initial value equations with θ_c specified and dr_ℓ^*/dx^* guessed at 1. Then, θ at 2 is checked against θ_c and if equality is not found a new dr_ℓ^*/dx^* is guessed at 1. Saez uses this method for the calculation of the liquid holdup and Kaviany (1989) and Rogers and Kaviany (1990) use this method for the evaluation of the surface contour for partially saturated two-dimensional porous surfaces. For three-dimensional menisci and when simplifying symmetries are not present, the computation of the meniscus contour becomes computationally very intensive. When $Bo \to 0$ (or $p_{\ell_0}^* Bo \to 0$), the pressure in the liquid phase is uniform and $-p_{\ell_0} = p_c$.

(D) Liquid Surface Motion Due to Thermo- and Diffuso-Capillarity

The *capillary phenomena*, originating from the existence of the surface tension, influences the dynamics of the liquid-gas flows in two ways. The first one is through the *curvature effect* (which was discussed in the last section). It also affects the flow when the magnitude of the surface tension changes along the liquid-gas interface. This change in the surface tension can be due to a change in the *temperature* (*thermo-capillary*), in the *concentration* (*diffuso-capillary*), or in the *electric potential*.

8.1 Elements of Pore-Level Flow Structure

The surface concentration gradient can be either present on and inside of a *thin* layer adjacent to the surface (*adsorbed surfactant*) or on the surface as well as in the bulk of the liquid (*surface-bulk surfactant*). The gradient in the surface tension, using (8.17), becomes

$$\frac{\partial \sigma}{\partial x_i} = \frac{\partial \sigma}{\partial T}\frac{\partial T}{\partial x_i} + \frac{\partial \sigma}{\partial \frac{\rho_k}{\rho}}\frac{\partial \frac{\rho_k}{\rho}}{\partial x_i} + \cdots . \tag{8.39}$$

In examining the forces that act on a surface and the role of $\partial \sigma/\partial x_i$ (along the surface) on the surface motion, a special treatment of the adsorbed layer is required when surface-active agents are present. Slattery (1990) formulates the role of the adsorbed layer (film) through a *surface viscosity*. We will not address thin surface film behavior. The force balance in the direction *normal* n and *tangent* τ to the interface are given as (Levich and Krylov, 1969)

$$p_\ell - p_g + \sigma\left(\frac{1}{r_1} + \frac{1}{r_2}\right) = 2\mu_\ell \frac{\partial u_{\ell,n}}{\partial n} - 2\mu_g \frac{\partial u_{g,n}}{\partial n} \quad \text{on } A_{\ell g}, \tag{8.40}$$

$$\mu_\ell\left(\frac{\partial u_{\ell,n}}{\partial \tau} + \frac{\partial u_{\ell,\tau}}{\partial n}\right) - \mu_g\left(\frac{\partial u_{g,n}}{\partial \tau} + \frac{\partial u_{g,\tau}}{\partial n}\right) = \frac{\partial \sigma}{\partial \tau} \quad \text{on } A_{\ell g}, \tag{8.41}$$

where, as before, we have chosen ℓ and g to designate the liquid and the fluid (gaseous or liquid) phase, respectively. The interfacial mean curvature $H = (1/r_1 + 1/r_2)/2$ is not known *a priori* and must be determined by simultaneously solving for the $p, \mathbf{u}, T, \rho_i, \ldots$ fields (subject to the boundary conditions, including the interfacial force balance).

For the two-phase flow in porous media, a direct comparison can be made between the viscous and surface tension gradient forces. Assuming that $\partial u_\tau/\partial n \gg \partial u_n/\partial \tau$ and by taking d_n and d_τ as the *length scales* in the normal and tangential directions (associated with a velocity change u_D/ε), we have

$$\frac{\mu_\ell u_{D,\ell}}{d_n \varepsilon_\ell} - \frac{\mu_g u_{D,g}}{d_n \varepsilon_g} = \frac{\partial \sigma}{\partial T}\bigg|_{\rho_i \ldots} \frac{\Delta T}{d_\tau} + \frac{\partial \sigma}{\partial \rho_i}\bigg|_{T \ldots} \frac{\Delta \rho_i}{d_\tau} + \ldots = \frac{\partial \sigma}{\partial \tau}. \tag{8.42}$$

Note that the order of the magnitude of the terms on the left-hand side is rather large (order of the shear stress at the matrix surface). Therefore, for the surface tension gradient to be of the *competing* order, we need

$$\frac{\partial \sigma}{\partial \tau} \sim O\left(\frac{\mu_\ell u_{D,\ell}}{d_n \varepsilon}\right). \tag{8.43}$$

The balance of forces along the normal to the interface indicates that for *practical* cases where the components of the velocities along the normal are *small*, the pressure *difference* between the two phases is equal to the *capillary pressure*.

Note that because of the simultaneous presence of the ℓ and g phases, without the specification of the saturation ($s = \varepsilon_\ell/\varepsilon$) and the phase distributions no general conclusion about the relative magnitudes of the terms in (8.41) can be made. Also, in (8.42), d_n is generally much larger than the unit-cell linear dimension (the magnitude of d_n depends on s and the phase distributions).

(E) Buoyant Motion of a Bubble in a Tube

As a simple example of the contribution of the surface tension to two-phase flows, consider the motion of a bubble in a tube. The problem has been considered by, among others, Reinelt and Saffman (1985), Reinelt (1987), and Martinez and Udell (1990). The case of the rise of a long bubble (due to buoyancy) in a vertical tube with the top end of the tube closed and the drainage of the liquid allowed from the bottom end, is considered by Reinelt. This example allows for *identification* of the significant parameters and examination of the momentum equation and boundary conditions for each phase.

The velocity of the bubble *front* is determined as a function of the Bond number (based on the density difference and the tube radius), i.e., $Bo = (\rho_\ell - \rho_g)gd^2/(4\sigma)$ and the *capillary number* $Ca = \mu_\ell u_b/\sigma$. The analysis is based on the assumption of an *inviscid gas*, i.e., $\mu_g = 0$ and a *negligible inertial force in the liquid phase*, i.e., $Re_\ell \to 0$. Then the governing equations with x chosen in the $-\mathbf{g}$ direction, are

$$\frac{\partial u}{\partial x} + \frac{\partial v}{\partial r} + \frac{v}{r} = 0, \tag{8.44}$$

$$0 = -\frac{\partial p}{\partial x} + \mu_\ell \left(\frac{\partial^2 u}{\partial x^2} + \frac{\partial^2 u}{\partial r^2} + \frac{1}{r}\frac{\partial u}{\partial r}\right) - (\rho_\ell - \rho_g)g, \tag{8.45}$$

$$0 = -\frac{\partial p}{\partial r} + \mu_\ell \left(\frac{\partial^2 v}{\partial x^2} + \frac{\partial^2 v}{\partial r^2} + \frac{1}{r}\frac{\partial v}{\partial r}\right). \tag{8.46}$$

The reference frame is taken with the *finger* (i.e., the advancing bubble front) being *stationary*. The boundary conditions are

on the tube wall ($r = d/2$)

$$u = -u_b, \quad v = 0, \tag{8.47}$$

on $A_{\ell g}$

$$\frac{u - \left(\dfrac{dx_\ell}{dr}\right)v}{\left[1 + \left(\dfrac{dx_\ell}{dr}\right)^2\right]^{1/2}} = 0, \tag{8.48}$$

8.1 Elements of Pore-Level Flow Structure

$$\frac{1-\left(\frac{dx_\ell}{dr}\right)^2}{1+\left(\frac{dx_\ell}{dr}\right)^2}\left(\frac{\partial u}{\partial r}+\frac{\partial v}{\partial x}\right)+\frac{2\left(\frac{dx_\ell}{dr}\right)}{1+\left(\frac{dx_\ell}{dr}\right)^2}\left(\frac{\partial u}{\partial x}-\frac{\partial v}{\partial r}\right)=0, \quad (8.49)$$

$$p_\ell - p_g + \sigma\left(\frac{1}{r_1}+\frac{1}{r_2}\right)$$

$$= 2\mu\frac{\frac{\partial u}{\partial x}-\left(\frac{dx_\ell}{dr}\right)\left(\frac{\partial u}{\partial r}+\frac{\partial v}{\partial x}\right)+\left(\frac{dx_\ell}{dr}\right)^2\frac{\partial v}{\partial r}}{1+\left(\frac{dx_\ell}{dr}\right)^2}, \quad (8.50)$$

where

$$\frac{1}{r_1} = \frac{-\frac{d^2 x_\ell}{dr^2}}{\left[1+\left(\frac{dx_\ell}{dr}\right)^2\right]^{3/2}}, \quad \frac{1}{r_2} = \frac{-\frac{dx_\ell}{dr}}{r\left[1+\left(\frac{dx_\ell}{dr}\right)^2\right]^{1/2}}, \quad (8.51)$$

and p_g is taken as a constant.

For $x \to -\infty$, the velocity distribution in the liquid film in the region $R_b < r < d/2$ (where R_b is the *radius* of the *bubble*) is given by

$$x \to -\infty: \quad u = -\frac{\rho g}{4\mu}\left(\frac{d^2}{4}-r^2+R_b^2\log\frac{2r}{d}\right)-u_b, \quad v=0. \quad (8.52)$$

While for $x \to \infty$, we have $u = -\infty$, and $v = 0$.

The parameters of the problem are $Ca, Bo,$ and $2R_b/d$. Based on the $x \to -\infty$ and $x \to \infty$ conditions, one of the parameters is eliminated. This is done through the following relation,

$$\frac{Ca}{Bo} = \frac{-\frac{16R_b^2}{d^2}+\frac{48R_b^4}{d^4}-\frac{64R_b^4}{d^4}\log\frac{2R_b}{d}}{\frac{32R_b^2}{d^2}}. \quad (8.53)$$

Reinelt solves these equations numerically using the finite-difference approximations and an adaptive composite mesh composed of a curvilinear grid that follows the ℓ-g interface. For $Ca \to 0$, a perturbation solution is available (Bretherton, 1961), which gives

$$Bo - 0.842 = 1.25Ca^{2/9} + 2.24Ca^{1/3}. \quad (8.54)$$

The numerical solution of Reinelt gives $2R_b/d(Ca)$. The preceding analysis shows the role that the interfacial boundary conditions (8.40) and (8.41) play in determining the interfacial contour and the velocity of the interface.

8.1.2 CONTINUOUS PHASE DISTRIBUTION

The phase distributions for two-phase flow through porous media have been reviewed by Dullien (1979, p. 252). In the literature, a *funicular flow regime or state* refers to the *simultaneous* flow of the wetting and nonwetting phases with the wetting phase moving on the outside (wetting the matrix) and the nonwetting phase moving on the inside. However, these phase distributions are *not* generally realized in disordered porous media, where it has been observed that *each* phase moves through its own separate *network* of interconnecting channels. As $\varepsilon_g/\varepsilon$ (nonwetting phase saturation) *increases*, the number of channels carrying this phase *increases* and the number of channels carrying the wetting *decreases*. This flow pattern is called the *channel flow regime*. However, some of the nonwetting phase is observed to be carried by the wetting phase and this is similar to the *funicular flow regime*. The nonwetting phase *preferentially* fills the *larger* pores while the wetting fluid occupies the *smaller* pores. The local pressure *difference* between the two phases is $\langle p_c \rangle \equiv \langle p \rangle^g - \langle p \rangle^\ell$. In general, the networks occupied by each phase are *independent* of the *total* flow rates.

When the two-phase flow is a result of an evaporation phase change in internally heated packed beds made of *nonconsolidated particles*, the vapor (g-phase) can cause *vapor channels* by changing the packing. These are *discrete channels* (or *tunnels*) created in the bed by the inertial and viscous forces of the vapor (Naik and Dhir, 1982). The vapor channels resulting from the displacement of particles have only been found in beds made of relatively small particles ($d < 1.6$ mm). Vapor channels (with no particle displacement) are also found in large particle beds with internal heat generation (Stubos and Buchin, 1988). The available studies of phase distributions and flow regimes in two-phase flow through porous media are mostly limited to isothermal two-phase flows (*not* single-component systems) in packed beds of spheres. The particle size is generally *large* (i.e., $\langle p_c \rangle \to 0$). In a few cases two-phase flow through some *channel-type* passages *etched* in glass plates have been reported.

As the ℓ- or g-phase flow rates (*countercurrent* arrangements) through a matrix are *increased*, a point will be reached where the flow of the ℓ-phase will be *hindered* and the condition of *flooding* is observed. Flooding has also been referred to (Scheidegger, 1974, p. 284) as the state where

$$\left.\frac{\partial \Delta \langle p \rangle}{\partial \langle u \rangle^g}\right|_{\langle u \rangle^\ell} \to \infty, \tag{8.55}$$

i.e., the point at which any further increase in the interstitial g-phase velocity $\langle u \rangle^g$, while keeping $\langle u \rangle^\ell$ constant, results in a large change in the pressure drop across the matrix (along the flow direction) $\Delta \langle p \rangle$.

The flow *transition* from the *trickling* to the *pulsing* flow in two-phase, *cocurrent downflow* through packed beds of spherical particles has been studied by Grosser et al. (1988). The *trickling flow* exists at low flow rates of the ℓ and g phases and in this regime the pressure gradient is time-independent. As the flow rates of the ℓ- or g-phase increases, transition to time-dependent flow (pulsing) occurs. The onset of the pulsation is caused by the *inertial forces* with the *capillary force* tending to oppose them.

For *cocurrent downflow* in packed beds of *large* spheres, models for the flow regime transitions have been proposed by Ng (1986). In his classification, *four* basic regimes are identified, namely, *the trickling, pulsing, spray,* and *bubble/dispersed-bubble* flow regimes. These are briefly reviewed below.

- *Trickling flow regime*: This occurs when the liquid flows over the particles (partially wetting at low $\dot{m}_\ell$ and totally wetting at higher $\dot{m}_\ell$) and the gas flows in the remaining pore space. The transition from partial wetting to complete wetting occurs at the following *transition t-t* (trickle, partial-to-complete wetting) *liquid mass flow rate* $(\dot{m}_\ell)_{t-t}$

$$(\dot{m}_\ell)_{t-t} = \left[\left(\frac{\mu_\ell \rho_\ell}{g}\right)^{1/5} \sigma^{3/5}\right] \frac{\pi}{4} N_c d, \qquad (8.56)$$

where N_c is the *number of particle projections per unit area*. The complete wetting trickle flow is the same as the *annular flow*.

- *Pulsing flow regime*: This occurs when the gas and liquid slugs traverse the column *alternatively*. The flow channels become plugged by the liquid slugs, which are blown off by the gas plugs. The transition *t-p* (trickle-to-pulsing) is initiated just above the constriction in the pore where the gas velocity is at its maximum.

- *Spray flow regime*: The liquid travels down the column in the form of droplets entrained by the continuous gas phase. The transition *t-s* (trickle-to-spray) is marked by the shattering of the liquid films. The gas flow is *turbulent*. The *transition t-s gas mass flow rate* is given by

$$(\dot{m}_g)_{t-s} = \frac{\pi}{4} d_t^2 \rho_g N_c \left(\frac{\sigma}{\rho_g d_t^2}\right)^{1/2}, \qquad (8.57)$$

where $d_t = d \left(2/\pi \sin \pi/3 - 1/2\right)^{1/2}$ is the *equivalent diameter of the minimum throat*.

- *Bubble/dispersed-bubble regimes*: In the bubble flow regime, the gas phase flows as slightly elongated bubbles. As the gas flow rate *increases*, the bubbles become highly *irregular* in shape (the dispersed bubble regime). The transition is marked by the *transitional b-db gas mass flow rate* given by

$$(\dot{m}_g)_{b-db} = \rho_g u_\ell. \qquad (8.58)$$

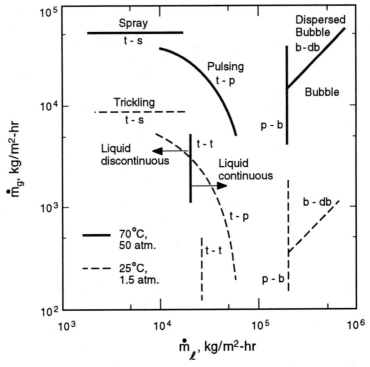

Figure 8.5 Flow regime diagram (from Ng) for cocurrent two-phase flow through a packed bed of spheres of $d = 5$ mm, with water-air as the fluid pair. The effect of temperature and pressure on the transitions is also shown. (From Ng, reproduced by permission ©1986 AIChE.)

The transition from the trickle or pulsing to the bubble flow (p-b) may occur when the turbulent intensity in the liquid flow is large enough to cause the bubble size to be less than the pore size. Figure 8.5 shows the flow regime obtained by Ng for flow of air-water in a randomly arranged bed of spherical particles of $d = 5$ mm and for two different sets of temperature and pressure. The *effects* of pressure and temperature on the transitions are evident. However, note that the *solubility* of air in water increases with increases in temperature and pressure and that the *density* of air also changes substantially with pressure and temperature.

For *countercurrent flows in packed beds* (spherical particles), Tung and Dhir (1988) have developed a flooding liquid flow rate *criterion* that depends *slightly* on the particle diameter. They note that flooding can take place in the annular as well as the bubbly regime. They suggest the following criterion

8.1 Elements of Pore-Level Flow Structure 445

$$\text{Flooding:} \quad |\langle u_\ell \rangle|^{1/2} \left[\frac{6(1-\varepsilon)\rho_\ell}{\varepsilon^3 g(\rho_\ell - \rho_g)d} \right]^{1/4}$$

$$= \begin{cases} 0.83 - 1.24 & \text{for } bubbly \ regime \\ 0 - 0.21 & \text{for } annular \ regime. \end{cases} \quad (8.59)$$

Others have also introduced flooding criteria (e.g., Sherwood et al., 1938).

8.1.3 DISCONTINUOUS PHASE DISTRIBUTIONS

As the volume fraction occupied by one of the fluid phases decreases, a *critical* volume fraction is reached beyond which its phase distribution is no longer continuous. These isolated fluid volumes have curved interfaces, therefore, their state is generally characterized by the capillary pressure $\langle p_c \rangle = \langle p \rangle^\ell - \langle p \rangle^g$. The *smaller* the volume fraction of the *wetting* fluid (i.e., the smaller the saturation of the wetting phase), the *larger* the capillary pressure, i.e., $\partial \langle p_c \rangle / \partial s < 0$, where as noted before, s is used as the saturation of the wetting phase.

Two methods for the experimental determination of $\langle p_c \rangle = \langle p_c \rangle (s)$ are the *gravity drainage* and the *centrifugal drainage* (Dullien, 1979, 29–35). Both of the methods assume that the wetting phase distribution is *continuous*. However, in these experiments at the extreme end (toward the diminishing saturation) of the sample (a porous medium occupied by the two phases) the wetting phase becomes *discontinuous*. Consider the gravity drainage method where an initially fully saturated matrix is allowed to drain by placing the matrix in a container such that the lowest part of the sample is *just* touching the same wetting fluid occupying the container. Then the top portion of the matrix will have the lowest saturation and the lowest part of the matrix, which is in contact with the wetting phase, will have a saturation of unity (and a capillary pressure equal to zero). Now the fluid at a distance h above the $s = 1$ location will have

$$\langle p_c \rangle = (\rho_\ell - \rho_g) gh = \frac{2\sigma}{\langle r_m \rangle} = 2\sigma \cos\theta_c Z(\theta_c) \langle H \rangle, \quad (8.60)$$

where we assume that $\rho_\ell > \rho_g$, and $\langle r_m \rangle$ is the *mean radius of curvature* (averaged over the representative elementary volume). The effect of contact angle θ_c is included through the Melrose (1965) *correlation* and $Z(\theta_c)$ is the *Melrose function* which includes the observed hysteresis to be discussed in Section 8.4.

In the centrifuge method, the capillary pressure for any location $r < r_o$ (where r_o is the location where $s = 1$) is given by

$$\langle p_c \rangle = \frac{1}{2} (\rho_\ell - \rho_g) 4\pi^2 f^2 (r_o^2 - r^2) = 2\sigma \cos\theta_c Z(\theta_c) \langle H \rangle, \quad (8.61)$$

where f is the frequency of rotation.

Both of the methods require an independent measurement of the *local* saturation, since in both methods the saturation distribution is *nonuniform*. Another method of determination of $\langle p_c \rangle = \langle p_c \rangle (s)$ is by the direct measurement of $\langle p \rangle^g - \langle p \rangle^\ell$ in a *uniformly* saturated matrix. We will discuss the behavior of $\langle p_c \rangle = \langle p_c \rangle (s)$ in Section 8.4.

(A) Discontinuous Wetting Phase

The *immobile or irreducible (discontinuous) wetting phase saturation* s_{ir} is referred to as the saturation at which

$$\lim_{s \to s_{ir}} \frac{\partial \langle p_c \rangle}{\partial s} \to -\infty \quad \text{for } decreasing\ s \text{ (i.e., during } drainage\text{)}. \quad (8.62)$$

In practice, depending on the strength of the external force used, a *finite* negative value is used in place of $-\infty$, indicating an *asymptotic approach* to the irreducible wetting phase saturation. Therefore, the measured s_{ir} can *depend* on the method used (this will be discussed in Section 8.4).

The retention of the irreducible saturation by the matrix through the *entrapment* of the wetting phase has been examined by Morrow (1970), where he refers to a volume of the wetting fluid that is hydrodynamically *isolated* and suspended by capillarity as a *pendular element*. The wetting phase is said to be in the *pendular state*. In this state the wetting phase is *not* continuous and, therefore, (8.60) does *not* strictly apply. Morrow performed the gravity drainage method but did *not* allow for a fully saturated lower end by completely isolating the initially fully saturated sample. Then, after allowing for a complete drainage, he measured the irreducible saturation. His results show the following features.

- For matrices that exhibit a distinct irreducible saturation s_{ir} (not *all* matrices have this property, as will be discussed in Section 8.4), all of the wetting fluid is held as *hydraulically isolated* elements that are *not* at a *capillary equilibrium* with each other or with the externally measured capillary pressure. As was mentioned earlier, the externally measured pressure relates only to the interfacial curvature of the surface separating the hydraulically *continuous* segment of the wetting and nonwetting phases.

- In the experiments using packings of clean sand or microspheres, the measured s_{ir} was between 0.06 and 0.10. This was the case for a wide range of particle size distributions and fluid properties.

- When smaller particles are used along with some larger particles, the former cause higher retention of the wetting phase and, therefore, higher overall s_{ir}.

(B) Discontinuous Nonwetting Phase

The *immobile* or *irreducible nonwetting phase saturation* $s_{ir\,g}$ (or the corresponding wetting phase saturation) is referred to as the saturation at which

$$\lim_{s_g \to s_{ir\,g}} \frac{\partial \langle p_c \rangle}{\partial s} \to -\infty,$$

for *increasing* s (i.e., during *imbibition*), (8.63)

where $s_g = \varepsilon_g/\varepsilon$, $s_{ir\,g} = \varepsilon_{ir\,g}/\varepsilon$, and $s + s_g = 1$. During the imbibition, some of the nonwetting phase is trapped, resulting in an irreducible nonwetting phase saturation $s_{ir\,g}$. The magnitude of $s_{ir\,g}$ for different matrices and pairs of wetting and nonwetting fluids has *not* been as rigorously studied. Also, in most cases through sustained inertial or viscous forces of the wetting phase, $s_{ir\,g}$ can be made to *approach* zero.

The trapped body of the nonwetting phase is called a *blob* or *ganglion*. The pore-level phase distributions and their dynamics for the case of one of the fluid phases being discontinuous has been studied by Soo and Slattery (1979) and Slattery (1974). The blob mechanics have also been studied by Ng et al. (1978). The *capillary number*

$$Ca = \frac{\mu_\ell u_{D\ell}}{\sigma} \qquad (8.64)$$

is used as the measure of the ratio of the *dynamic stress* (tending to move the blobs) and the *static stress* (tending to entrap the blobs). Another dimensionless number commonly used is the *Weber number*

$$We = Re\,Ca = \frac{\rho_\ell u_{D\ell}^2 \ell}{\sigma}, \qquad (8.65)$$

which is a measure of the relative strength of the *inertial force* to the *static force*.

Based on their experimental results, Ng et al. find that

$s_{ir\,g}$ is *independent* of Ca,   for $Ca < 2 \times 10^{-5}$,

$s_{ir\,g}$ *decreases* with Ca,   for $2 \times 10^{-5} < Ca < 2 \times 10^{-3}$,   (8.66)

$s_{ir\,g} \to 0$,   for $5 \times 10^{-3} < Ca$.

They note that the *remobilization* of blobs is caused by the Darcy-law pressure difference between a pair of *feet* (extensions of the blobs into the adjacent pores) overcoming the net capillary-pressure difference that the two feet can develop by merely shifting in the pores they occupy. In addition, they have observed that the maximum vertical length of a blob that can remain stationary in a body force field is inversely proportional to the density difference $\rho_\ell - \rho_g$. This dependence of the phase distribution on the direction of the body force is similar to that present in the two-phase flows in plain media.

8.1.4 CONTACT LINE

Our discussion will be limited to the ℓ-g interfaces with *no* mass transfer (immiscible fluids or no phase change). As was discussed, whenever there exist a temperature and concentration gradient around any of the three interfaces, the surface tensions ($\sigma_{\ell s}$, σ_{gs}, and $\sigma_{\ell g}$) are expected to be dependent on these gradients. When no dynamic force is present, the contact line assumes an asymptotic stationary position and a *static contact angle* is observed. In the presence of an ℓ-phase motion, the contact line moves with a *dynamic contact angle*. In the case of a dynamic contact line, the actual contact angle *right* at the *contact line* (three-phase contact line) is different than that slightly *away* from the contact line. In the following, some aspects of the contact line are discussed. A thorough review of the wetting (statics and dynamics) of the solid phase including the wetting transitions, the van der Waals forces, and the fluid dynamics of the moving contact line is given by de Gennes (1985) and de Gennes et al. (1990).

(A) SURFACE TENSION OF SOLID

The *surface energy* of solids has been reviewed by Chappuis (1982) and Israelachvili (1989, Chapter 14), where the treatment is similar to that for liquids presented in Section 8.1.1. Some of the special features of solid surface tension are given here.

The *work of adhesion in vacuum* is the free energy change δW_{12} (or the reversible work done) to separate the *unit* areas of *two* media, 1 and 2, from the contact to infinity. For two identical media, this is called the *work of cohesion per unit area* $\delta W_{11}/\delta A$, where both δW_{12} and δW_{11} are positive.

The *surface energy change* σ (i.e., the surface tension) is the energy required when the surface area of a medium is increased by one unit. This is given by (8.16). Since the process of creating a unit area of surface is equivalent to the separating of two half-unit areas from the contact, we have $\sigma_1 = (1/2)\delta W_{11}/\delta A$.

The *intermolecular* forces that determine the surface energy are the same as those that determine the latent heats and the melting and boiling temperatures. For metals, $\sigma_1 > 1$ N/m, which is much *larger* than that for the low-melting-point solids. The larger magnitude compared to the liquid-gas surface tension is expected because the intermolecular stress in the *bulk* of the solid is orders of magnitude larger than that for liquids.

For the van der Waals interaction between atoms, the *interatomic* (or intermolecular) *potential* can be given as (Israelachvili, 1989, p. 137)

$$\frac{\delta W_{aa}}{\delta A} = -\frac{a_1}{r^6}, \qquad (8.67)$$

where a_1 is a constant coefficient that depends on the atomic properties and r is the interatomic distance. This will be further discussed in Section

8.1.5. When integrated to give the potential between two similar surfaces, we have (Israelachvili, 1989, p. 156)

$$\sigma_1 = \frac{1}{2}\frac{\delta W_{11}}{\delta A} = \frac{a_2}{24\pi d_o^2}, \tag{8.68}$$

where $a_2 = \pi^2 a_1 \rho_n^2$, ρ_n is the *number density* [$\rho_n = 2^{1/2}/(8R_m^3)$, where R_m is the atomic radius] and $d_o = 2R_m/2.5$ is the *interfacial contact separation*. Using these, (8.68) becomes

$$\sigma_1 = \frac{a_2}{2.1 \times 10^{-18}}, \tag{8.69}$$

where the *Hamaker constant* a_2 is given in J and σ_1 is in N/m (or J/m²). The values of a_2 for some substances are given by Israelachvili, and (8.69) predicts the interfacial tension of *non-H-bounding liquids* and *solids* to within 10 to 20 percent of the experimental results.

The surface tension of a solid is larger in vacuum than it is in the presence of any gas or vapor. This is because the adsorbed foreign molecules reduce the intermolecular forces between the molecules of the solid (at and adjacent to the interface). The interfacial tension at the interface of a solid in contact with a fluid is determined from the individual surface tensions according to (8.20), i.e., the same relation that was used for the liquid-fluid interfacial tension.

The solid surface tension for the case where the solid is in *equilibrium* with its vapor is related to that for the solid in vacuum through (Chappuis, 1982, p. 426)

$$\sigma_s - \sigma_{sg} = \frac{R_g T}{M}\int_0^{p^*} m_{ad}(p^*)\,d(\ln p^*), \tag{8.70}$$

where m_{ad} (kg/m²) is the mass adsorbed per unit area (also discussed in Section 6.5.4), and p^* is the gas pressure normalized with respect to the pressure required for the solidification of the gas. This is similar to the adsorption isotherms used for the formation of the liquid phase by the condensation of a gas on solid surfaces (Section 9.5). This relationship shows that the solid surface tension *decreases* as its vapor adheres to the solid surface. The quantity $\sigma_s - \sigma_{sg}$ is called the *spreading film pressure*.

(B) Effect of Surface Heterogeneity, Roughness, and Adsorption

One of the difficulties in the molecular-level treatment of the solid surface is that in practice these surfaces are neither *homogeneous* (in terms of their constituents) nor *smooth*. For example, a machine-finished *metallic* surface typically has a surface layer of the order 100 Å thick made of the *metal* (generally not pure), *oxide fragments, abrasives, lubricants*, etc. Beneath this is a layer of the order of 10 μm thick where the *grains are significantly deformed*. The *plastics* also have surface heterogeneities because both *crystalline* and *amorphous* phases are present on the surface.

The surface *roughnesses* are generally *not isotropic* and the average (root-mean-square) roughness depends on the extent of polishing (1–5 μm for *grinding*, 0.05–0.5 μm for *superpolishing*, and 0.01–0.1 μm for *electrolytic polishing*). Very smooth surfaces are obtained by the *cleavage* of crystals, but only a few materials are easily cleaved. Mica (a mineral) is an example. It has been used for the surface-surface force measurements by using distances of the order of 10^{-3} μm between the surfaces (Horn and Israelachvili, 1981).

The surface tension of the solid is expected to be significantly *altered* in the presence of surface heterogeneity and roughness. As was mentioned, when a solid is exposed to a gas, there will be a multimolecular layer adsorption on the surface (adsorption of gases on solid surfaces will be discussed in Section 9.4). Also, when a solution in contact with a solid surface contains solutes, the molecules of solute and solvent become adsorbed on the solid surface. All of these adsorptions *decrease* the solid surface tension.

(C) Static Contact Line and Contact Angle

In the analysis of the *static equilibrium* at the liquid-fluid interfaces, the *contact line of the three phases* (s, ℓ, g) was characterized by the condition of equilibrium between the components of the surface tensions (σ_{sg}, $\sigma_{s\ell}$, $\sigma_{\ell g}$). This condition is called the *Young equation* and introduces the *static* or *microscopic* or *intrinsic contact angle* θ_c, measured in the ℓ- (wetting-) phase and at the contact line. This is the angle between the local tangent to the solid surface and the tangent to the three-phase contact line. Figure 8.6 shows the *four* forces that act on the contact line (Chappuis, 1982, p. 431). The balance between the *tangential* and the *normal* components leads to [i.e., the decomposition of the law of Neumann triangle (8.9)]

$$\sigma_{gs} - \sigma_{\ell s} = \sigma_{\ell g} \cos\theta_c, \qquad (8.71)$$

$$f_s = \sigma_{\ell g} \sin\theta_c. \qquad (8.72)$$

The *reaction force* f_s that counters the normal component of the liquid-fluid surface tension can cause *deformations* in the solid. Chappuis has observed these deformations by placing a drop of *water* on a *solidifying paraffin* sublayer. Because initially only a very thin layer of solid paraffin is present, this layer easily becomes deformed due to the presence of f_s (i.e., a viscous, incompressible fluid over a *thin* elastically plastically deforming solid substrate).

The difference $\sigma_{gs} - \sigma_{\ell g}$ is also called the *tension of adhesion*. The *work of wetting* (or *dewetting*) of a surface area δA of the solid is $(\sigma_{gs} - \sigma_{\ell g})\delta A$.

8.1 Elements of Pore-Level Flow Structure 451

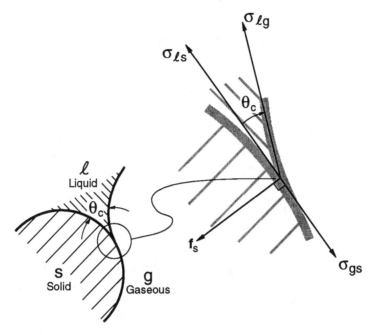

Figure 8.6 The contact angle θ_c, measured in the wetting phase, and the static forces acting on the contact line. The three interfacial tensions and the solid reaction force are shown.

(D) CONTACT ANGLE HYSTERESIS

An ideal smooth homogeneous clean surface in contact with a pair of ℓ and g phases should have a *unique* contact angle. However, real surfaces are heterogeneous contaminated and rough. Therefore, the measured contact angles show multivaluedness. Since any measurement of θ_c requires *introduction* of a solid surface to a pair of g and ℓ phases, these experiments measure a *dynamic* contact angle. Although a large elapsed time is generally allowed after this introduction (or any other such sudden changes), the contact angle approaches different asymptotes depending on the *history* of the motion of the ℓ-phase with respect to the other two phases (s and g phases).

As more liquid is added to a drop that is placed on a horizontal surface by the injection of a liquid through a syringe (Chappuis, 1982, p. 444), the volume of the drop increases, and after each increment, a contact angle θ_a (*advancing contact angle*) is found. Similarly, when the liquid is withdrawn from the drop, a contact angle θ_r (*receding contact angle*) is found. Experiments show that in general $\theta_r < \theta_a$. If a drop is simply placed on a surface and no other disturbance follows (e.g., no addition or withdrawal

TABLE 8.2 STATIC CONTACT ANGLE FOR SOME LIQUID-SOLID PAIRS IN AIR (FROM MYSHKIS ET AL., 1987)

SOLID	LIQUID	CONTACT ANGLE (deg)	SOLID	LIQUID	CONTACT ANGLE (deg)
Glass	Water	0	Steel	Hydrogen	0
Glass	Mercury	128-148	Steel	Nitrogen	0
Glass	Hydrogen	0	Steel	Oxygen	0
Glass	Nitrogen	0	Paraffin	Hydrogen	106
Glass	Oxygen	0	Aluminum	Nitrogen	7
Steel	Water	70-90	Platinum	Oxygen	105

of liquid), then *any* contact angle between θ_a and θ_r can occur and is stable. The difference between θ_a and θ_r is called the *contact angle hysteresis* ($\theta_a - \theta_r$). The hysteresis is caused by the surface *heterogeneity, roughness,* or *adsorption*.

Using the experimental method of *plate immersion* and *emersion* into liquids, Chappuis shows that when the surface heterogeneities are *anisotropic* (i.e., increments of large and small surface tensions placed on the solid surface *parallel* or *perpendicular* to the liquid surface), and when surface tension heterogeneities are *isotropically* distributed, the contact angle hysteresis can be *predicted*. He also shows that the effect of some *regular* and *simple* roughness distributions on the contact angle can be predicted. He critically reviews the methods used for the contact angle measurement. For example, he notes that the *static* contact angle must be measured at the *contact line*, and therefore, its evaluation requires the measurement of the *contact line*, and therefore, its evaluation requires the measurement of the tangent to the meniscus as *close* to the contact line as possible (the name *microscopic* or *intrinsic* refers to this asymptote). Therefore, in principle an asymptote should be found for θ_c using the measured values at *diminishing* distances from the contact line.

The contact angle hysteresis caused by the solid surface tension heterogeneities (defects) has also been analyzed by Joanny and de Gennes (1984). They show that for *weak* heterogeneities no hysteresis is found, but the contact line becomes *wiggly*. In their analyses, both *single* localized defects and *dilute* systems of defects are considered and the advancing and receding contact angles are predicted in terms of the distribution of the defect strength.

Many studies (e.g., Dussan, 1979) have shown that when a drop of liquid is placed on an initially clean surface, a *film* of this liquid is deposited near the contact line. The film thickness is of the order of 10 Å and *spreads* at a speed that depends on the surface roughness and other surface conditions.

The surface spreading is a *surface diffusion* that is gradient-driven, and therefore, the presence of the components of the liquid phase in the g phase influences the spreading rate. This deposited film *decreases* the solid surface tension and in turn changes the contact angle.

Table 8.2 gives some *static* contact angles, which are considered to be *close* to the *intrinsic* values (Myshkis et al., 1987). All the results are for *saturated liquids* near atmospheric pressure. When $\theta_c = 0$, a *complete wetting* occurs.

(E) MOVING CONTACT LINE

When the contact line moves, such as in the spreading of a liquid film or drop, the *observed* (or *macroscopic* or *apparent*) *contact angle* θ_o measured at a short distance (discussed later) from the contact line is *larger* than the *static* (or microscopic or intrinsic) *contact angle* θ_c. An empirical correlation relating θ_o and θ_c shows that the *larger* the speed of the contact line, the larger is θ_c with an asymptotic upper limit of π for θ_o (Dussan, 1979; Slattery, 1991).

Cox (1986), Ngan and Dussan (1989), and de Gennes et al. (1990) have analyzed the quasi-steady meniscus contour (i.e., the variation of the local tangent to the meniscus) for small *capillary* ($Ca_\ell = \mu_\ell u_c/\sigma$, $Ca_g = \mu_g u_c/\sigma$, where u_c is the *velocity* of the *contact line*), *Bond* ($Bo = \Delta\rho g \ell^2/\sigma$, where ℓ is the *large length scale* of the *meniscus*), and *Reynolds* ($Re_\ell = u_c \ell/\nu_\ell$, $Re_g = u_c \ell/\nu_g$) *numbers*. In calculating the velocity field around the moving contact line, a nonintegrable *singularity* is found in the stress at the contact line. This results in a divergent integrand for the *drag* force on the solid boundary. In order to avoid this, a *velocity slip* has been allowed between the fluids and the solid at the contact point (Hocking, 1977; Hocking and Rivers, 1982).

The approach is to allow for the velocity slip over a small distance Δ_s from the contact line (such that $\delta = \Delta_s/\ell \ll 1$). The analysis is based on a *singular* perturbation around δ while using *three* regions of expansion (*inner, intermediate*, and *outer*). When $Ca \ln \delta^{-1} \to 0$, the intermediate region becomes vanishingly *small*. In the intermediate region the length scales are between δ_s and ℓ. The major influence on the contact angle comes from this intermediate region (to the lowest order in Ca). At the next higher order in Ca, the solution depends on the *slip model* used and the overall geometries. One of the slip models used is similar to (2.167) and is

$$\mu_\ell \frac{\partial u_\ell}{\partial n} = \alpha u_\ell \quad \text{on} \quad A_{\ell s} \quad \text{and for} \quad \tau \leq \delta_s, \tag{8.73}$$

where n and τ stand for normal and tangent to the solid surface, τ is measured from the contact line, and α is the *slip coefficient*. The magnitude of δ_s suggested, is between 10^{-7} and 10^{-5} cm.

The preceding studies are based on an *a priori* knowledge of the velocity

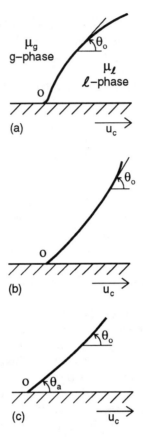

Figure 8.7 (a) The outer region, (b) intermediate region, and (c) inner region around a moving contact line. The contact line velocity is u_c and the microscopic contact angle is taken to be the asymptotic advancing angle θ_a. The macroscopic contact angle θ_o changes depending on the distance from the contact line.

of the contact line, however, this velocity is generally *not* known. Moreover, they assume that a *unique* intrinsic (microscopic) contact angle is known for the system (no hysteresis). One choice for the intrinsic contact angle is the asymptote obtained from the *advancing* drop experiment described in the last section, i.e., taking $\theta_c = \theta_a(u_c \to 0)$. When $\theta_a(u_c \to 0)$ is used as the microscopic contact angle, the experimental results (reviewed by Cox) for the meniscus contour agree with the predictions. The experiments show that θ_o is indeed the same for various values of u_c. In the experiments of Morrow and Nguyen (1982), it was observed that the macroscopic advancing contact angle tends to increase slightly with the contact line velocity while the macroscopic receding contact angle tends to decrease slightly (both to less than 3 degrees).

Figure 8.7 shows the three regions used in the analysis of Cox and the variation of the macroscopic or apparent contact angle θ_o in the three regions. As was mentioned, the inner region has a length scale of the order of δ_s, the intermediate region length scale is between δ_s and ℓ, and the outer region length scale is of the order of ℓ. Here the advancing contact angle at velocity u_c is used as the microscopic angle. The flow around the contact line is discussed by Cox and by Dussan.

The problem of the moving contact line is central in the understanding of the immiscible displacement and other flow transients in porous media and continues to be investigated.

8.1.5 THIN EXTENSION OF MENISCUS

Under steady-state conditions, the meniscus of a liquid *perfectly wetting* a solid surface ($\theta_c = 0$) *extends* as a thin *film liquid* over the solid. In this thin liquid extension, the *capillary force*, *intermolecular forces* (acting between the liquid and the solid), and *gravity force* are in equilibrium (when no concentration and temperature gradients are present). The presence of a surface tension gradient results in an additional surface force.

In the following, the static equilibrium thickness distribution of this layer, including the adsorbed film, is first examined. This is followed by the dynamic equilibrium considerations including evaporation, flow in the liquid phase, and the stability of this thin film when it is heated at $A_{s\ell}$.

(A) STATIC EQUILIBRIUM

According to the Gibbs concept for the interface between phases (in equilibrium), a *transition layer* exists on *each* side of the interface. The pressure in these layers is *anisotropic*, exhibiting a *tensor* behavior (Derjaguin and Churaev, 1978). Now, when any two phases (e.g., a gas and a solid phase) approach each other but are separated by a thin third phase (e.g., a liquid), the third phase grows *thinner*. This is a result of *overlapping transition layers*. In the case of very thin liquid layers, no portion of the liquid will have the same property as the bulk liquid. The pressure in the liquid will be *anisotropic* throughout the liquid and is called the *disjoining pressure*.

Wetting (or *complete wetting*) of a solid surface by a liquid occurs when $\theta_c = 0$. For $\theta_c < \pi/2$, the liquid is said to be *partially wetting*. For $\theta_c > \pi/2$, the liquid is said to be *partially nonwetting*. When $\theta_c = \pi$, it is said that the liquid is *completely nonwetting*. For $\theta_c = 0$, an adsorbed layer extends beyond the meniscus. This is illustrated in Figure 8.8. As shown in the figure, there are *four* regions identified: the *liquid pool*, the *capillary meniscus*, the *transition region* (with a length along the surface of the order of a micron), and the *thin adsorbed film*, which is of the order of

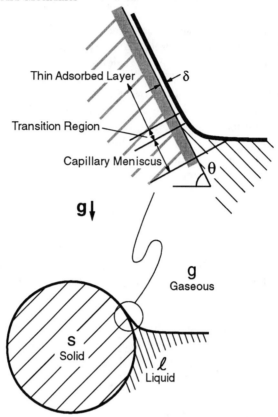

Figure 8.8 Extension of a thin liquid film (including adsorbed layer) over the solid surface for a completely wetting liquid, $\theta_c = 0$. θ is the angle the solid surface makes with the bulk liquid surface. Three of the regions, thin adsorbed layer, transition, and capillary, are labeled.

100 Å thick (Truong and Wayner, 1987). As was mentioned earlier, the chemical potential in the thin film (the thermodynamics of the thin adsorbed liquid layers will be further discussed in Section 9.4) is different than the bulk and is said to have an *excess potential*. The excess is attributed to solid-liquid interactions (the combination of the electrostatic double layers, the van der Waals dispersion forces, etc.). The dominant contribution in the nonpolar liquids is from the van der Waals dispersion forces. These forces can be *predicted* by the *Dzyaloskinskii-Lifshitz-Pitaevskii theory* (Dzyaloskinskii et al., 1961).

The theory is based on the extensions of the Lifshitz treatment of the attractive van der Waals forces between two bodies with characteristic lengths *larger* than the intermolecular distances and with separation distances *larger* than wavelengths of the significant spectral absorption bands.

Truong and Wayner, whose results are reviewed later, have used this theory to calculate the adsorbed film thickness for a few wetting fluids.

The *dispersion force per unit area* f_δ is a function of the *film thickness* δ and of the *dielectric susceptibilities* of the solid, vapor, and liquid (by assuming that δ is larger than the intermolecular distance, *bulk* values are used for the liquid) phases, ε_{ℓ_s}, ε_{ℓ_v}, ε_{ℓ_ℓ}, which are *frequency*-dependent. This force is given as (Truong and Wayner)

$$f_\delta(\delta) = \frac{k_B T}{\pi c_o^3} \sum_{n=0}^{\infty}{}' \varepsilon_{\ell_\ell}^{3/2} \left(\frac{\omega_n}{i}\right)^3 \int_1^\infty \left[\frac{1}{\phi_1(s)\phi_2(v)\exp\left(2p\frac{\omega_n}{i}\varepsilon_{\ell_\ell}^{1/2}\frac{\delta}{c_o}\right) - 1} \right.$$

$$\left. + \frac{1}{\phi_s\left(\frac{\varepsilon_{\ell_s}}{\varepsilon_{\ell_\ell}}\right)\phi_v\left(\frac{\varepsilon_{\ell_v}}{\varepsilon_{\ell_\ell}}\right)\exp\left(2p\frac{\omega_n}{i}\varepsilon_{\ell_\ell}^{1/2}\frac{\delta}{c_o}\right) - 1} \right] p^2\,dp, \qquad (8.74)$$

where

$$\phi_j(x) = \frac{s_j + px}{s_j - px}, \qquad s_j = \left(\frac{\varepsilon_{\ell_j}}{\varepsilon_{\ell_\ell}} - 1 + p^2\right)^{1/2}, \qquad j = s, v, \qquad (8.75)$$

k_B is the Boltzmann constant, $\omega_n = 4n\pi^2 k_B T/(i h_P)$ is the eigen frequency, and h_P is the Planck constant. The prime on the summation sign indicates that $n = 0$ must be given a half weight.

For $\delta \to 0$, i.e., *thin film* asymptote, this dispersion force is approximated by

$$f_\delta(\delta) = \frac{a_1}{6\pi\delta^3}, \qquad \delta < 100 \text{ Å}, \qquad (8.76)$$

where a_1 is called the *Hamaker constant*. For *thick* films the approximation is

$$f_\delta(\delta) = \frac{a_2}{\delta^4}, \qquad \delta > 600 \text{ Å}, \qquad (8.77)$$

where a_2 is called the *retarded dispersion force constant*. For stably wetting films, a_1 and a_2 must be *negative*.

At static equilibrium, the volumetric force balance on the liquid film leads to the balance among the *dispersion*, *capillary*, and *gravity forces*, i.e.,

$$-f_\delta(\delta) + 2\sigma H - \rho g h = 0, \qquad (8.78)$$

where h is measured along $-\mathbf{g}$ (this is further discussed in Section 9.4). For two-dimensional menisci, the *mean curvature* is given as

$$H = \frac{\dfrac{d^2\delta}{dx^2}}{2\left[1+\left(\dfrac{d\delta}{dx}\right)^2\right]^{3/2}}. \qquad (8.79)$$

458   8. Fluid Mechanics

By placing $x = 0$ inside the pool region where $\delta = d$ (d is of the order of a micron) and by assuming the thin or thick film asymptote, we have (e.g., for the thick films)

$$2\sigma H - \frac{a_2}{\delta^4} = \left(2\sigma H - \frac{a_2}{\delta^4}\right)_{x=0} + (\rho_\ell - \rho_g) g x \sin\theta$$
$$+ (\rho_\ell - \rho_g) g (\delta - d) \cos\theta \simeq 2\sigma H(x=0). \tag{8.80}$$

Truong and Wayner have solved this equation for $\delta(x)$, for *hexane* and *octane* liquids on single-crystal *silicon* surfaces. They have also performed some experiments. Note that (8.76) and (8.77) are approximations and both a_1 and a_2 *depend* on δ. For example, using the exact formulation (8.74), they show that $0.32 \times 10^{-29} < -a_2 < 4.26 \times 10^{-29}$ N-m^2 for $10^{-3} < \delta < 1$ μm for the *air/octane/silicon* system. They find that the *average* thickness of the adsorbed layer is 250 Å for the *hexane* system and 195 Å for the *octane* system. They also observe that the transition from 80 percent capillarity to 80 percent dispersion occurs over a distance of 20 μm.

(B) Dynamic Equilibrium

When a temperature gradient exists in the liquid-vapor-solid system, a phase change occurs. In addition, liquid motion driven by buoyancy and surface-tension gradient will also take place. These motions influence the meniscus contour and the liquid film distribution in the transition region. The rate of phase change in the adsorbed layer is zero and is minimal where the van der Waals forces dominates. It increases as the liquid pool is approached.

Truong (1987), Mirzamoghadam and Catton (1988), and Wayner (1989) have examined the motion in the meniscus caused by the heat addition to the solid substrate. In order to prevent bulk or surface ($A_{s\ell}$) nucleation, the solid surface temperature is maintained *slightly above* the saturation temperature. The two-dimensional motion of the liquid is governed by

$$\rho_\ell \left(u \frac{\partial u}{\partial x} + v \frac{\partial u}{\partial y}\right) = -\frac{\partial p_\ell}{\partial x} + \mu \frac{\partial^2 u}{\partial y^2} + \rho_\ell g \sin\theta, \tag{8.81}$$

where x and u are along the solid surface. The pressure gradient in the liquid is given by

$$\frac{\partial p_\ell}{\partial x} = \frac{d}{dx}(2\sigma H) - \rho_\ell g \frac{d\delta}{dx} \cos\theta - \frac{df_\delta(\delta)}{dx}, \tag{8.82}$$

where p_g is assumed to be constant.

The tangential interfacial force balance (8.41) is approximated by

$$\mu_\ell \frac{\partial u}{\partial y} = \frac{d\sigma}{dx} = \frac{d\sigma}{dT}\frac{dT}{dx} \quad \text{on} \quad \delta. \tag{8.83}$$

The solution to the momentum equation requires the knowledge of the temperature field, which in turn is obtained from the liquid and solid substrate energy equations. (This will be further discussed in dealing with the porous/plain media interfacial transport in Chapter 11). When the solid substrate is heated, the surface temperature at the ℓ-s interface is higher where the thin film exists. This results in a smaller value of the surface tension σ, and therefore, the liquid flows *away* from the thin film extension and *toward* the liquid pool (and then returns beneath the surface to complete the *recirculation*).

Burelbach et al. (1988) have examined the occurrence of the *dryout* (the *breakage* of thin films) in the heated thin liquid films. They examine the stability of the flow in thin liquid films. The momentum equation used is that given earlier. For the *two-dimensional* flows and in the *vectorial* form (with a *negligible* gravitational force) the momentum equation becomes

$$\rho\left(\frac{\partial \mathbf{u}}{\partial t} + \mathbf{u}\nabla \cdot \mathbf{u}\right) = -\nabla p + \mathbf{f}_\delta(\delta) + \mu \nabla^2 \mathbf{u}, \tag{8.84}$$

where $\mathbf{f}_\delta(\delta)$ is defined by Burelbach et al. In their analysis, the *base* flow is a flat, thin evaporating (receding ℓ-s interface) liquid film. Whenever the *dynamic* thinning of the film is *followed* by the dominance of the van der Waals forces, rupturing of the film occurs (*rapture instability*).

8.2 Local Volume Averaging

The volume averaging of the momentum and continuity equations for two-phase flow (ℓ and g phases flowing simultaneously) in porous media has been examined by Slattery (1970), Whitaker (1986a,b), Gray (1983), Bear and Bensabat (1989), Hassanizadeh and Gray (1990, 1993), and Gray and Hassanizadeh (1991). In the following, the formulation of Whitaker (1986b) is reviewed, where it is assumed that *both* phases are *continuous*. Creeping flow of the ℓ and g phases is assumed, and the fluids are assumed to be *incompressible*.

Under these assumptions, the *point* momentum and continuity equations for the two phases are

$$0 = -\nabla p_\ell + \rho_\ell \mathbf{g} + \mu_\ell \nabla^2 \mathbf{u}_\ell, \tag{8.85}$$

$$\nabla \cdot \mathbf{u}_\ell = 0, \tag{8.86}$$

$$0 = -\nabla p_g + \rho_g \mathbf{g} + \mu_g \nabla^2 \mathbf{u}_g, \tag{8.87}$$

$$\nabla \cdot \mathbf{u}_g = 0. \tag{8.88}$$

460   8. Fluid Mechanics

The boundary conditions are

$$\mathbf{u}_\ell = 0 \quad \text{on } A_{\ell s}, \tag{8.89}$$

$$\mathbf{u}_g = 0 \quad \text{on } A_{gs}, \tag{8.90}$$

$$\mathbf{u}_\ell = \mathbf{u}_g, \quad -p_\ell \mathbf{n}_{\ell g} + \mathbf{S}_\ell \cdot \mathbf{n}_{\ell g}$$
$$= -p_g \mathbf{n}_{\ell g} + \mathbf{S}_g \cdot \mathbf{n}_{\ell g} + 2\sigma H \mathbf{n}_{\ell g} \quad \text{on } A_{\ell g}, \tag{8.91}$$

where $\mathbf{S}$ is the *shear* stress tensor. The *tangential* component of the interfacial forces are *not* addressed here, and therefore, the effect of the presence of a surface-tension gradient is *not* included. This effect will be addressed in Section 8.2.1. The volume averaging of the ℓ-phase continuity equation gives

$$\langle \nabla \cdot \mathbf{u}_\ell \rangle = \nabla \cdot \langle \mathbf{u}_\ell \rangle + \frac{1}{V}\int_{A_{\ell s}} \mathbf{n}_{\ell s} \cdot \mathbf{u}_\ell \, dA + \frac{1}{V}\int_{A_{\ell g}} \mathbf{n}_{\ell g} \cdot \mathbf{u}_\ell \, dA = 0. \tag{8.92}$$

Since $\mathbf{u}_\ell$ is zero on $A_{\ell s}$, the second term on the right-hand side is zero. Now consider the change in the ℓ-phase content of the volume V. The rate of change of this, i.e., $\partial \varepsilon_\ell / \partial t$, is related to the change of the ℓ-g interfacial position (assuming that $A_{s\ell}$ and A_{sg} are stationary). If the local velocity on the ℓ-g interface is $\mathbf{w}$, then

$$\frac{\partial \varepsilon_\ell}{\partial t} = \varepsilon \frac{\partial s}{\partial t} = \frac{1}{V}\int_{A_{\ell g}} \mathbf{n}_{\ell g} \cdot \mathbf{w} \, dA. \tag{8.93}$$

Now by combining this with (8.92), we have

$$\frac{\partial \varepsilon_\ell}{\partial t} + \nabla \cdot \langle \mathbf{u}_\ell \rangle = \frac{1}{V}\int_{A_{\ell g}} \mathbf{n}_{\ell g} \cdot (\mathbf{u}_\ell - \mathbf{w}) \, dA. \tag{8.94}$$

Similarly, we have

$$\frac{\partial \varepsilon_g}{\partial t} + \nabla \cdot \langle \mathbf{u}_g \rangle = \frac{1}{V}\int_{A_{\ell g}} \mathbf{n}_{g\ell} \cdot (\mathbf{u}_g - \mathbf{w}) \, dA. \tag{8.95}$$

Also from the definitions we have

$$\varepsilon_g + \varepsilon_\ell = \varepsilon \quad \text{or} \quad s_g + s = 1. \tag{8.96}$$

When there is *no interfacial mass transfer*, we have

$$\mathbf{u}_\ell \cdot \mathbf{n}_{\ell g} = \mathbf{u}_g \cdot \mathbf{n}_{\ell g} = \mathbf{w} \cdot \mathbf{n}_{\ell g}, \tag{8.97}$$

and the last terms in (8.94) and (8.95) drop out. The case of interfacial mass transfer (evaporation/condensation) will be discussed in Section 10.1.1.

8.2 Local Volume Averaging

We now turn to the momentum equations. First the dependent variables are *decomposed* into the phase volume-averaged and deviation components, i.e.,

$$p_\ell = \langle p \rangle^\ell + p'_\ell, \qquad \mathbf{u}_\ell = \langle \mathbf{u} \rangle^\ell + \mathbf{u}'_\ell, \qquad (8.98)$$

$$p_g = \langle p \rangle^g + p'_g, \qquad \mathbf{u}_g = \langle \mathbf{u} \rangle^g + \mathbf{u}'_g. \qquad (8.99)$$

Following the substitution of (8.98) and (8.78) into (8.85) and (8.87), equations are obtained for the *intrinsic* phase-averaged quantities and for the deviation components. The deviation component equations, along with closure statements (relating deviations to the phase-averaged quantities through the use of transformation vectors and tensors), are used to arrive at the closure conditions (i.e., differential equations and boundary conditions for the transformation vectors and tensors). Many of the steps are similar to those described in the treatment of the single-phase flows (Section 2.7). These steps are not repeated here, and they can be found in Whitaker (1986a,b). They include order-of-magnitude comparisons among the various terms in the momentum equations and in the interfacial ($A_{\ell g}$) boundary condition (8.91).

In the following, we assume a *periodic* unit cell structure. Note that *not* only should the s-phase distribution be periodic, but the ℓ- and g-phase distributions must also be periodic. This assumption is very *restrictive* because the phase distributions *change* with the local *saturation*. Therefore, the cell size depends on the local saturation. This makes it very difficult to obtain any result that is of general use. However, examination of some *idealized* phase distributions leads to some insight. In the closure conditions, as in Section 2.7.3, relationships are found between the deviation quantities and the *phase-averaged velocities* $\langle \mathbf{u} \rangle^\ell$ and $\langle \mathbf{u} \rangle^g$.

The *transformation tensors* and vectors are introduced through

$$\mathbf{u}'_\ell = \mathbf{B}_{\ell\ell} \cdot \langle \mathbf{u} \rangle^\ell + \mathbf{B}_{\ell g} \cdot \langle \mathbf{u} \rangle^g, \qquad (8.100)$$

$$\mathbf{u}'_g = \mathbf{B}_{g\ell} \cdot \langle \mathbf{u} \rangle^\ell + \mathbf{B}_{gg} \cdot \langle \mathbf{u} \rangle^g, \qquad (8.101)$$

$$p'_\ell = \mu_\ell \mathbf{b}_{\ell\ell} \cdot \langle \mathbf{u} \rangle^\ell + \mu_\ell \mathbf{b}_{\ell g} \cdot \langle \mathbf{u} \rangle^g, \qquad (8.102)$$

$$p'_g = \mu_g \mathbf{b}_{g\ell} \cdot \langle \mathbf{u} \rangle^\ell + \mu_g \mathbf{b}_{gg} \cdot \langle \mathbf{u} \rangle^g, \qquad (8.103)$$

$$H' = \mathbf{h}_\ell \cdot \langle \mathbf{u} \rangle^\ell + \mathbf{h}_g \cdot \langle \mathbf{u} \rangle^g \quad \text{on} \quad A_{\ell g}. \qquad (8.104)$$

Note that the selection of $\langle \mathbf{u} \rangle^\ell$ and $\langle \mathbf{u} \rangle^g$ as the *only* sources for the deviations is rather *arbitrary*. Inclusion of *additional* sources such as $(\langle \mathbf{u} \rangle^\ell - \langle \mathbf{u} \rangle^g)$ and the nonlinear terms such as $(\langle \mathbf{u} \rangle^\ell - \langle \mathbf{u} \rangle^g)^2$ makes the analysis much more difficult.

The [B] tensors and the [b] and [h] vectors are determined through the substitution of (8.100) to (8.104) into (8.85), (8.87), and (8.89) to (8.91).

The momentum equations become (details are given in Whitaker, 1986a,b)

$$0 = -\nabla \langle p \rangle^\ell + \rho_\ell \mathbf{g} - \mu_\ell \mathbf{M}_{\ell\ell} \cdot \langle \mathbf{u} \rangle^\ell - \mu_\ell \mathbf{M}_{\ell g} \cdot \langle \mathbf{u} \rangle^g, \quad (8.105)$$

$$0 = -\nabla \langle p \rangle^g + \rho_g \mathbf{g} - \mu_g \mathbf{M}_{g\ell} \cdot \langle \mathbf{u} \rangle^\ell - \mu_g \mathbf{M}_{gg} \cdot \langle \mathbf{u} \rangle^g, \quad (8.106)$$

where, for example,

$$\mathbf{M}_{\ell\ell} = -\frac{1}{V_\ell} \int_{A_{\ell s}} (-\mathbf{n}_{\ell s} b_{\ell\ell} + \mathbf{n}_{\ell s} \cdot \nabla \mathbf{B}_{\ell\ell}) \, dA$$

$$+ \frac{1}{V_\ell} \int_{A_{\ell g}} (-\mathbf{n}_{\ell g} b_{\ell\ell} + \mathbf{n}_{\ell g} \cdot \nabla \mathbf{B}_{\ell\ell}) \, dA. \quad (8.107)$$

Other equations are found by appropriately changing the indices. By assuming that the inverse of $\mathbf{M}_{\ell\ell}$ and $\mathbf{M}_{gs}$ exists, Whitaker arrives at [from (8.105) and (8.106)]

$$\langle \mathbf{u}_\ell \rangle = -\frac{\mathbf{K}_\ell}{\mu_\ell} \cdot \left(\nabla \langle p \rangle^\ell - \rho_\ell \mathbf{g} \right) + \mathbf{K}_{\ell g} \cdot \langle \mathbf{u}_g \rangle, \quad (8.108)$$

$$\langle \mathbf{u}_g \rangle = -\frac{\mathbf{K}_g}{\mu_g} \cdot \left(\nabla \langle p \rangle^g - \rho_g \mathbf{g} \right) + \mathbf{K}_{g\ell} \cdot \langle \mathbf{u}_\ell \rangle, \quad (8.109)$$

where

$$\mathbf{K}_\ell = \varepsilon_\ell \mathbf{M}_{\ell\ell}^{-1}, \qquad \mathbf{K}_{\ell g} = -\mathbf{M}_{\ell\ell}^{-1} \cdot \mathbf{M}_{\ell g} \frac{\varepsilon_\ell}{\varepsilon_g}, \quad (8.110)$$

$$\mathbf{K}_g = \varepsilon_g \mathbf{M}_{gg}^{-1}, \qquad \mathbf{K}_{g\ell} = -\mathbf{M}_{gg}^{-1} \cdot \mathbf{M}_{g\ell} \frac{\varepsilon_g}{\varepsilon_\ell}. \quad (8.111)$$

The interfacial normal force balance becomes

$$-\langle p \rangle^\ell + \langle p \rangle^g = p_c = 2\sigma \langle H \rangle_{\ell g}, \quad (8.112)$$

where

$$\langle H \rangle_{\ell g} = \frac{1}{A_{\ell g}} \int_{A_{\ell g}} H \, dA \equiv \langle H \rangle. \quad (8.113)$$

The viscous stress terms appearing in (8.91) have been dropped based on the order-of-magnitude comparisons.

The *selection* of the absolute phase velocities $\langle \mathbf{u} \rangle^g$ and $\langle \mathbf{u} \rangle^\ell$ as the sources for the cross influencing has *not* been confirmed experimentally. Schulenberg and Müller (1987) use the difference in the phase velocities to suggest the following *interfacial* drag term for *one-dimensional* flows and when $|\langle u \rangle^g| > |\langle u \rangle^\ell|$ (based on their *experimental* results for packed beds).

Whitaker	Schulenberg and Müller (one-dimensional)	
$\dfrac{\mu_\ell}{K_\ell}\mathbf{K}_{\ell g}\cdot\langle\mathbf{u}_\ell\rangle,$	$\dfrac{W(s)\rho_\ell K}{s\eta\sigma}\left(\dfrac{\langle u_g\rangle}{1-s}-\dfrac{\langle u_\ell\rangle}{s}\right)^2(\rho_\ell-\rho_g)g,$	(8.114)
$\dfrac{\mu_v}{K_v}\mathbf{K}_{g\ell}\cdot\langle\mathbf{u}_g\rangle,$	$-\dfrac{W(s)\rho_g K}{(1-s)\eta\sigma}\left(\dfrac{\langle u_g\rangle}{1-s}-\dfrac{\langle u_\ell\rangle}{s}\right)^2(\rho_\ell-\rho_g)g,$	(8.115)

where η is $K^{1/2}/C_E$, as given in (2.59), and $W(s) = 350s^7(1-s)$ is found from the experimental results.

We expect the drag force to be proportional to the difference in the phase-averaged velocity of the two phases. Therefore, the interfacial drag forms in (8.108) and (8.106), which were originated by the selection of the expansions in (8.100) to (8.104), *cannot* be physically supported. The interfacial shear will be further discussed in Section 8.7.

Returning to Whitaker's formulation the closure problem is given by

$$-\nabla\mathbf{b}_{\ell\ell}+\nabla^2\mathbf{B}_{\ell\ell} = \dfrac{1}{V_\ell}\int_{V_\ell}(-\nabla^2\mathbf{b}_{\ell\ell}+\nabla^2\mathbf{B}_{\ell\ell})\,dV = 0, \qquad (8.116)$$

$$\nabla\cdot\mathbf{B}_{\ell\ell} = 0. \qquad (8.117)$$

The other *three* sets of equations are found by appropriately changing the indices. The boundary conditions are

$$\mathbf{B}_{\ell\ell} = -\mathbf{I}, \quad \mathbf{B}_{g\ell} = 0 \quad \text{on } A_{\ell s}, \qquad (8.118)$$

$$\mathbf{B}_{\ell\ell} = \mathbf{B}_{g\ell}-\mathbf{I}, \quad \mathbf{B}_{\ell g} = \mathbf{B}_{gg}+\mathbf{I} \quad \text{on } A_{\ell g}, \qquad (8.119)$$

$$\mu_\ell\left(-\mathbf{n}_{\ell g}\mathbf{b}_{\ell\ell}+\nabla\mathbf{B}_{\ell\ell}^\dagger\cdot\mathbf{n}_{\ell g}+\mathbf{n}_{\ell g}\cdot\nabla\mathbf{B}_{\ell\ell}\right)$$
$$= \mu_g\left(-\mathbf{n}_{\ell g}\mathbf{b}_{g\ell}+\nabla\mathbf{B}_{g\ell}^\dagger\cdot\mathbf{n}_{\ell g}+\mathbf{n}_{\ell g}\cdot\nabla\mathbf{B}_{g\ell}\right)+2\sigma\mathbf{n}_{\ell g}h_\ell \quad \text{on } A_{\ell g}. \qquad (8.120)$$

Another equation is obtained by using g as the second index in $\mathbf{b}$ and $\mathbf{B}$ earlier and by using $\mathbf{h}_g$.

We also have

$$\mathbf{B}_{g\ell} = 0, \quad \mathbf{B}_{gg} = -\mathbf{I} \quad \text{on } A_{gs}, \qquad (8.121)$$

$$\mathbf{b}_{\ell\ell}(\mathbf{r}+\boldsymbol{\ell}_i) = \mathbf{b}_{\ell\ell}(\mathbf{r}), \quad \mathbf{B}_{\ell\ell}(\mathbf{r}+\boldsymbol{\ell}_i) = \mathbf{B}_{\ell\ell}(\mathbf{r}), \qquad (8.122)$$

$$\langle\mathbf{b}_{\ell\ell}\rangle^\ell = 0, \quad \langle\mathbf{B}_{\ell\ell}\rangle^\ell = 0. \qquad (8.123)$$

Again the *periodicity* condition and the *volume-averaging* condition apply to other components of **b** and **B**, and similar equations are written.

The task of determining **b** and **B** is a difficult one (if not impossible for practical geometries). Even for spherical particles in the *simple-cubic arrangement*, the periodic (but *not* isotropic) ℓ and g phases will be spread over several spheres. An exception will be the idealized trickle-flow packed beds (gravity-driven liquid flow) at low saturations, where we assume that the liquid flows over the spherical particles at flow rates that makes for a stable separated two-phase flow in each cell. Even for this two-phase flow, more than one sphere has been included in order to allow for the *inertial* effect in the liquid (i.e., thinning of the liquid film as it flows downward and accelerates).

The last terms in (8.108) and (8.109) represent the viscous drag of one fluid upon the other. Therefore it is expected that $\mathbf{K}_{\ell g}$ and $\mathbf{K}_{g\ell}$ depend on the viscosity ratio μ_ℓ/μ_g, among other parameters. For example, when $\mu_\ell/\mu_g \to \infty$, the viscous shear exerted on the ℓ phase by the g phase will be *negligible*. Again, note that the interfacial viscous drag terms, appearing in the conventional two-phase momentum equations, appear as a function of the *slip* velocity ($\langle \mathbf{u} \rangle_\ell - \langle \mathbf{u} \rangle_g$) *instead* of the absolute velocity of the other phase.

8.2.1 Effect of Surface Tension Gradient

The tangential interfacial force balance, (8.41), shows that the gradient of the surface tension can influence the interfacial motion. Bear and Bensabat (1989) have included this gradient in their analysis. The result is an extra driving force in the volume-averaged momentum equations. The coefficient of this term, a new conductance, is related to the matrix and fluid properties. When this force is added to (8.108) and (8.109), we have

$$\langle \mathbf{u}_\ell \rangle = -\frac{\mathbf{K}_\ell}{\mu_\ell} \cdot \left(\nabla \langle p \rangle^\ell - \rho_\ell \mathbf{g} \right) + \mathbf{K}_{\ell g} \cdot \langle \mathbf{u}_g \rangle + \mathbf{K}_{\ell \Delta \sigma} \cdot \nabla \langle \sigma \rangle , \quad (8.124)$$

$$\langle \mathbf{u}_g \rangle = -\frac{\mathbf{K}_\ell}{\mu_\ell} \cdot (\nabla \langle p \rangle^g - \rho_g \mathbf{g}) + \mathbf{K}_{g\ell} \cdot \langle \mathbf{u}_\ell \rangle + \mathbf{K}_{g \Delta \sigma} \cdot \nabla \langle \sigma \rangle , \quad (8.125)$$

where $\mathbf{K}_{\ell \Delta \sigma} = \mathbf{K}_{\ell \Delta \sigma} \left(\mu_g, \mu_\ell, s, \text{geometry} \right)$, etc.

Since $\langle \sigma \rangle = \langle \sigma \rangle (\langle \rho_{ig} \rangle, \langle \rho_{i\ell} \rangle, \langle T \rangle, \ldots)$, where ρ_i is the density of the component i that influences the surface tension, we can write

$$\nabla \langle \sigma \rangle = \frac{\partial \langle \sigma \rangle}{\partial \frac{\langle \rho_{ig} \rangle}{\langle \rho_g \rangle}} \nabla \frac{\langle \rho_{ig} \rangle}{\langle \rho_g \rangle} + \frac{\partial \langle \sigma \rangle}{\partial \frac{\langle \rho_{i\ell} \rangle}{\langle \rho_\ell \rangle}} \nabla \frac{\langle \rho_{i\ell} \rangle}{\langle \rho_\ell \rangle} + \frac{\partial \langle \sigma \rangle}{\partial \langle T \rangle} \nabla \langle T \rangle + \ldots . \quad (8.126)$$

Under *some restrictive* assumptions, Bear and Bensabat suggest

$$\mathbf{K}_{\ell \Delta \sigma} = \frac{A_{\ell g}}{V_\ell \mu_\ell (1 + L_{i\ell}^*)} \mathbf{K}_\ell , \quad (8.127)$$

$$\mathbf{K}_{g\Delta\sigma} = \frac{A_{\ell g}}{V_g \mu_g (1 + L_{tg}^*)} \frac{\varepsilon_g \mu_g}{\varepsilon_g \mu_g + \varepsilon_\ell \mu_\ell} \mathbf{K}_g, \qquad (8.128)$$

where $L_{t\ell}^*(s)$ and $L_{tg}^*(s)$ are the *tortuosities* and $A_{\ell g}/V_\ell(s)$ and $A_{\ell g}/V_g(s)$ are the specific area of the interface with respect to the ℓ- and g-phase volumes, respectively.

8.3 A Semiheuristic Momentum Equation

As was the case with the single-phase flows, in two-phase flows we expect to have *deviations* from the *creeping* flow description given by (8.124) and (8.125), whenever the phase velocities are large. We also expect the *interfacial drag* at $A_{\ell g}$ to become important when $\langle \mathbf{u} \rangle^\ell \neq \langle \mathbf{u} \rangle^g$. In the following, we proceed to *include* these terms in a *semiheuristic* manner, as was done in Section 2.9 for the single-phase flows. This is done by the introduction of *empirically* determined coefficients. These coefficients and their dependence on the local saturation and other variables will then be discussed.

8.3.1 INERTIAL REGIME

As was the case for the single-phase flows, when the fluid velocity for one of the phases increases, the inertial regime is reached where an inertial core (region of nearly uniform velocity) is present in this phase. Note that a given phase spreads over many pores, thus, the *definition* of the inertial core and the inertial regime should be interpreted accordingly. Then, in the inertial regimes, pressure drops in *excess* of that given by the Darcy law will be experienced. For single-phase flows, this inertial effect was included through the Ergun correlation (2.59). For two-phase flows in porous media, a similar treatment may be made.

For simplicity we begin by excluding the effects of the interfacial drag (on $A_{\ell g}$) and the surface tension gradient. Then the momentum equations will have the following forms

$$\frac{\partial \langle p \rangle^\ell}{\partial x_j} - \rho_\ell g_j = -\frac{\mu_\ell}{K_\ell} \langle u_{\ell j} \rangle - \frac{\rho_\ell}{K_{\ell i}} |\langle u_{\ell j} \rangle| \langle u_{\ell j} \rangle, \qquad (8.129)$$

$$\frac{\partial \langle p \rangle^g}{\partial x_j} - \rho_g g_j = -\frac{\mu_g}{K_g} \langle u_{gj} \rangle - \frac{\rho_g}{K_{gi}} |\langle u_{gj} \rangle| \langle u_{gj} \rangle, \qquad (8.130)$$

where $K_{\ell i}$ and K_{gi} are to be determined. The *macroscopic inertial coefficients* $K_{\ell i}$ and K_{gi} must satisfy the expected asymptotic behaviors for $s \to 1$ and $s \to 0$. For example, for the ℓ phase, we have

$$\lim_{s \to 1} K_{\ell i} = \frac{K^{1/2}}{C_E} \qquad (8.131)$$

and (2.59) is recovered.

Note that tensors $\mathbf{K}_\ell$ and $\mathbf{K}_g$ are counterparts of $\mathbf{K}$ for the single-phase flows. Therefore, the flow condition under which they are defined is that of the Darcean flow. (This will be further discussed in Section 8.5). This requires that $\mathbf{K}_\ell$ and $\mathbf{K}_g$ be independent of $\langle \mathbf{u}_\ell \rangle$ and $\langle \mathbf{u}_g \rangle$. Then, $\mathbf{K}_\ell$ and $\mathbf{K}_g$ should be determined in the *Darcean* regime and $\mathbf{K}_{\ell i}$ and $\mathbf{K}_{gi}$ are to be determined in the *inertial* regime. In order to obtain simple empirical relations, in some studies $\mathbf{K}_\ell$, $\mathbf{K}_g$, $\mathbf{K}_{\ell i}$, and $\mathbf{K}_{gi}$ are determined *simultaneously* from the best fit to the experimental data. For example, under cocurrent trickling flow conditions in packed beds, Saez and Carbonell (1985) examine the experimental results and suggest that K_ℓ be proportional to $K_{\ell i}$ and K_g be proportional to K_{gi}. This leads to the following *one-dimensional* equations of motion

$$-\frac{\mathrm{d}\langle p \rangle^\ell}{\mathrm{d}x} + \rho_\ell g = \frac{1}{K_\ell}\left(\mu_\ell \langle u_\ell \rangle + \rho_\ell C_E K^{1/2} \langle u_\ell \rangle^2 \right), \qquad (8.132)$$

$$-\frac{\mathrm{d}\langle p \rangle^g}{\mathrm{d}x} + \rho_g g = \frac{1}{K_g}\left(\mu_g \langle u_g \rangle + \rho_g C_E K^{1/2} \langle u_g \rangle^2 \right), \qquad (8.133)$$

where K_ℓ and K_g are in turn given in terms of the Carman-Kozeny permeability (absolute permeability) and the *relative permeabilities*. Therefore, a single permeability (for each phase) is used for both the inertial and viscous regimes. Similar *empirical* efforts by Dhir (1986) do *not* lead to a single permeability relation. It should be mentioned that unlike the single-phase flows, where in the Darcean regime the pressure drop is proportional to the first power of the superficial velocity, in the two-phase flows even in the Darcean regime there are *many transitions*. For example, even if *no* microscopic inertial effects are present, as the liquid flow rate increases the flow *transition* from the *partially* liquid wetted solid surface to the *fully* liquid wetted solid surface occurs in the trickle flow regime in packed beds. Therefore, we have to examine to what extent a Darcean *interpretation* can be given to the two-phase flows and the definitions of $\mathbf{K}_\ell$ and $\mathbf{K}_g$. This will be further discussed in Section 8.5.

8.3.2 Liquid-Gas Interfacial Drag

Now, by including the *microscopic* inertial and *macroscopic* inertial terms (Grosser et al., 1988), by introducing $K_{\ell g1}$, $K_{\ell g2}$, $K_{g\ell 1}$, and $K_{g\ell 2}$ as the coefficients in the liquid-gas interfacial drag forces, and by assuming that this drag is proportional to the difference in the phase velocities and that for *cocurrent* flows $|\langle u_j \rangle^g| > |\langle u_j \rangle^\ell|$, we have the following pair of momentum equations for two-phase flow in porous media

8.3 A Semiheuristic Momentum Equation

$$\frac{\rho_\ell}{\varepsilon s}\left(\frac{\partial \langle u_{\ell j}\rangle}{\partial t}+\langle \mathbf{u}_\ell\rangle\cdot\nabla\langle u_{\ell j}\rangle\right)$$

(ℓ-phase macroscopic inertial force)

$$=-\frac{\partial \langle p\rangle^\ell}{\partial x_j}+\rho_\ell g_j-\frac{\mu_\ell}{K_\ell}\langle u_{\ell j}\rangle-\frac{\rho_\ell}{K_{\ell i}}|\langle u_{\ell j}\rangle|\langle u_{\ell j}\rangle$$

(ℓ-phase pore pressure gradient)   (ℓ-phase body force)   (microscopic interfacial $(A_{\ell s})$ shear stress)   (microscopic inertial force)

$$+\underbrace{\left[K_{\ell g1}|\langle u_j\rangle^g-\langle u_j\rangle^\ell|+K_{\ell g2}(\langle u_j\rangle^g-\langle u_j\rangle^\ell)^2\right]\frac{\langle u_j\rangle^\ell}{|\langle u_j\rangle^\ell|}}_{\text{(microscopic interfacial }(A_{\ell g})\text{ shear stress)}}+\mu_\ell\frac{K_{\ell\Delta\sigma}}{K_\ell}\frac{\partial\sigma}{\partial x_j},$$

(8.134)

(microscopic interfacial $(A_{\ell g})$ surface tension gradient force)

and similarly

$$\frac{\rho_g}{\varepsilon(1-s)}\left(\frac{\partial \langle u_{gj}\rangle}{\partial t}+\langle \mathbf{u}_g\rangle\cdot\nabla\langle u_{gj}\rangle\right)$$

$$=-\frac{\partial \langle p\rangle^g}{\partial x_j}+\rho_g g_j-\frac{\mu_g}{K_g}\langle u_{gj}\rangle-\frac{\rho_g}{K_{gi}}|\langle u_{gj}\rangle|\langle u_{gj}\rangle$$

$$+\left[K_{g\ell 1}|\langle u_j\rangle^g-\langle u_j\rangle^\ell|+K_{\ell g2}(\langle u_j\rangle^g-\langle u_j\rangle^\ell)^2\right]\frac{\langle u_j\rangle^g}{|\langle u_j\rangle^g|}+\mu_g\frac{K_{g\Delta\sigma}}{K_g}\frac{\partial\sigma}{\partial x_j}$$

(8.136)

where we have assumed that all the coefficients are *isotropic*. This assumption simplifies the above equations and is justified because presently *only* the simple isotropic coefficients are available.

Note that from the definition of the phase averaging, i.e., $\langle \mathbf{u}\rangle^\ell = (1/V_\ell)\int_{V_\ell}\mathbf{u}\,dV$, etc., we have

$$\langle \mathbf{u}\rangle^\ell=\frac{\langle \mathbf{u}_\ell\rangle}{\varepsilon s},\qquad \langle \mathbf{u}\rangle^g=\frac{\langle \mathbf{u}_g\rangle}{\varepsilon(1-s)}. \tag{8.137}$$

These momentum equations are solved along with the continuity equations given by (8.94) and (8.95) and the appropriate boundary conditions.

8.3.3 Coefficients in Momentum Equations

Very little is known about the manner in which the coefficients given in (8.134) and (8.135) *depend* on the matrix structure, fluid properties, phase distributions, phase velocities, etc. The most studied coefficients are $\mathbf{K}_\ell$ and $\mathbf{K}_g$, and almost all of the studies are *empirical*. Therefore, no general treatment exists even for these two coefficients. The most important point to be made is that if a *simultaneous* evaluation of these coefficients is *not* made, then by dropping any of them its effect on the flow will be lumped into the rest of the coefficients. Also note that the experiments should be chosen such that each effect can be isolated (e.g., both very low and very high velocities should be included). Therefore, it is important, for example, to simultaneously include the effects of the microscopic inertia and interfacial drag when dealing with isothermal and isoconcentration flows with relatively large velocities and particle sizes. The empirical evaluation of the coefficients should be done systematically where only a few of the forces given in (8.134) and (8.135) are dominant in each stage of the planned experimental investigation. Only for a few very simple cases can theoretical predictions of the coefficient be made.

(A) Phase Permeability Coefficients

These coefficients account for the *drag* force at the solid-fluid interface (e.g., $A_{s\ell}$ for the liquid phase). They have been introduced (Muskat and Meres, 1936) as an extension of the Darcy law to two-phase flows.

In principle, the local phase distribution functions $a_\ell(\mathbf{x})$ and $a_g(\mathbf{x})$ given by (8.1) and (8.2) are the *only* parameters influencing the relative permeabilities. These distributions, in turn, depend on the *matrix structure, surface tension, buoyancy, wettability*, and *local saturation*. Based on the analogy we have made so far with the single-phase flows in the Darcy regime, we expect $\mathbf{K}_\ell$ and $\mathbf{K}_g$ to be *independent* of the magnitude of the velocities. However, this restriction is *not* supported by the experiments, as they show that *several* flow regimes exist in the noninertial (i.e., Darcean) two-phase flow in porous media. These transitions are *flow rate*-dependent, and therefore, the phase permeabilities *should* be velocity-dependent. However, introduction of the velocity dependence in $\mathbf{K}_\ell$ and $\mathbf{K}_g$ is *not* in accord with the analogy proposed by Muskat and Meres between single- and two-phase flow. Also if $\mathbf{K}_\ell = \mathbf{K}_\ell(\mathbf{u}_\ell, \text{etc.})$, then the microscopic inertial terms *cannot* be identified as the sole representation of deviations from the proportionality of the pressure drop to the first power of the phase velocity. Based on these arguments, we *assume* that the phase permeabilities are *independent* of the phase velocities, i.e.,

$$\mathbf{K}_\ell = \mathbf{K}_\ell \left(\text{matrix structures}, s, \sigma, \theta_c, \frac{\rho_\ell}{\rho_g}, \text{history} \right). \tag{8.138}$$

A similar expression can be written for $\mathbf{K}_g$. Note that the matrix structure implies *size, dimensionality, pore coordinate number,* and *topology*. The phase distribution functions a_ℓ and a_g depend on all of the parameters given earlier for $\mathbf{K}_\ell$. We note again that the effects of μ_ℓ/μ_g, $\langle \mathbf{u} \rangle^\ell$, $\langle \mathbf{u} \rangle^g$, and $\nabla \sigma$ *are not* to be included in $\mathbf{K}_\ell$ and $\mathbf{K}_g$. As was mentioned, $\mathbf{K}_\ell$ and $\mathbf{K}_g$ have been studied extensively. The available results will be reviewed in Section 8.5.

(B) Microscopic Inertial Coefficients

These coefficients account for *all* the excess pressure drop due to the relatively large phase velocities (i.e., deviation from the Darcean flow). We expect that

$$\mathbf{K}_{\ell i} = \mathbf{K}_{\ell i}\left(\text{matrix structure}, s, \sigma, \theta_c, \frac{\rho_\ell}{\rho_g}, \text{history}\right). \quad (8.139)$$

Again the effects of μ_ℓ/μ_g, $\langle \mathbf{u} \rangle^\ell$, $\langle \mathbf{u} \rangle^g$, and $\nabla \langle \sigma \rangle$ are *not* to be included in $\mathbf{K}_{\ell i}$. A similar functional dependence is expected for $\mathbf{K}_{gi}$.

Some experimental results are available for $\mathbf{K}_{\ell i}$ and $\mathbf{K}_{gi}$. However, even though they are performed under isothermal and isoconcentration conditions, they are generally done using a *pair* of fluids with $\mu_\ell/\mu_g \neq 1$. Therefore, these coefficients also *include* the effects of the liquid-fluid interfacial drag and the effect of the velocity of the *second* phase. The microscopic inertial coefficients will be further discussed in Section 8.6.

(C) Interfacial Drag Coefficients

These coefficients account for the difference in the viscosities and velocity difference between the ℓ and g phases, which result in an interfacial drag on $A_{\ell g}$. This drag force is *not* necessarily linearly proportional to $\langle \mathbf{u} \rangle^g - \langle \mathbf{u} \rangle^\ell$. When the velocities are rather large, this dependence will become of the type $(\langle \mathbf{u} \rangle^g - \langle \mathbf{u} \rangle^\ell)^n$, where $n > 1$. For $\mathbf{K}_{\ell g1}$, and $\mathbf{K}_{\ell g2}$ to be valid inside and outside the Darcy regime, these coefficients must be independent of $\langle \mathbf{u} \rangle^\ell$ and $\langle \mathbf{u} \rangle^g$. Therefore, we expect

$$\mathbf{K}_{\ell g1} = \mathbf{K}_{\ell g1}\left(\text{matrix structure}, s, \sigma, \theta_c, \frac{\rho_\ell}{\rho_g}, \frac{\mu_\ell}{\mu_g}, \text{history}\right). \quad (8.140)$$

Any effects associated with $\langle \mathbf{u} \rangle^\ell$ or $\langle \mathbf{u} \rangle^g$ or those associated with $\nabla \langle \sigma \rangle$ should *not* be included in this functional dependence. A similar form is found for $\mathbf{K}_{g\ell 1}$ and $\mathbf{K}_{g\ell 2}$. The drag force exerted by the g phase on the ℓ phase is the *negative* of the drag force exerted by the ℓ phase on the g phase. The interfacial drag coefficient will be further discussed in Section 8.7.

(D) Surface-Tension Gradient Coefficients

These account for the interfacial area $A_{\ell g}$ as well as its *projection* along $\nabla \sigma$. This *projected area* depends on the flow direction as well as the phase distributions. In addition, since the contribution of $\nabla \sigma$ as a tangential interfacial force is balanced by the viscous shear forces, (8.41), the viscosities also influence these coefficients. The surface tension gradient coefficients should have the functional form

$$\frac{\mathbf{K}_{\ell \Delta \sigma}}{\mathbf{K}_\ell} = \frac{\mathbf{K}_{\ell \Delta \sigma}}{\mathbf{K}_\ell}\left(\text{matrix structure}, s, \langle \sigma \rangle, \theta_c, \frac{\rho_\ell}{\rho_g}, \frac{\mu_\ell}{\mu_g}, \text{history}, \frac{\boldsymbol{\tau}_{\ell g} \cdot \nabla \sigma}{|\nabla \sigma|}\right). \quad (8.141)$$

A similar relation can be written for $\mathbf{K}_{\ell \Delta \sigma}$. Only *limited* information is available on $\mathbf{K}_{\ell \Delta \sigma}/\mathbf{K}_\ell$ and $\mathbf{K}_{g \Delta \sigma}/\mathbf{K}_g$. The most rigorous study is due to Bear and Bensabat (1989). In their study, *many simplifying* assumptions are made and readily usable relations are developed for *isotropic* phase distributions. These relations are

$$\mu_\ell \frac{K_{\ell \Delta \sigma}}{K_\ell} = \frac{A_{\ell g}}{V_\ell (1 + L^*_{t\ell})}, \quad (8.142)$$

$$\mu_g \frac{K_{g \Delta \sigma}}{K_g} = \frac{A_{\ell g}}{V_g (1 + L^*_{t\ell})} \frac{\varepsilon_g \mu_g}{\varepsilon_g \mu_g + \varepsilon_\ell \mu_\ell}, \quad (8.143)$$

where

$$\frac{A_{\ell g}}{V_g L^*_{t\ell}} = \frac{A_{\ell g}}{V_g L^*_{t\ell}}\left(\text{matrix structure}, s, \langle \sigma \rangle, \theta_c, \frac{\rho_\ell}{\rho_g}, \text{history}\right). \quad (8.144)$$

Note that in *addition* to $\mathbf{K}_{\ell \Delta \sigma} \cdot \nabla \langle \sigma \rangle$ and $\mathbf{K}_{g \Delta \sigma} \cdot \nabla \langle \sigma \rangle$, the gradient of the surface tension appears in the momentum equation indirectly and through the gradient of the *capillary pressure*. The capillary pressure (which relates the two local phase pressures) is generally idealized by $p_c = \langle \sigma \rangle J(s)/(K/\varepsilon)^{1/2}$ and appears in the momentum equation as $\partial p_c / \partial x_i$. This can be decomposed as

$$\frac{\partial p_c}{\partial x_i} = \frac{J(s)}{(K/\varepsilon)^{1/2}} \frac{\partial \langle \sigma \rangle}{\partial x_i} + \frac{\langle \sigma \rangle}{(K/\varepsilon)^{1/2}} \frac{\partial J(s)}{\partial x_i}, \quad (8.145)$$

where again $\langle \sigma \rangle = \langle \sigma \rangle (T, \langle \rho_{ig} \rangle / \langle \rho_g \rangle, \langle \rho_{i\ell} \rangle / \langle \rho_\ell \rangle, \ldots,)$ and $s = s(\mathbf{x})$. We can further expand $\partial \langle \sigma \rangle / \partial x_i$ and $\partial J(s) / \partial x_i$ by using these relations. Therefore, the surface tension gradient *contributes* to the phase momentum equations directly and through the capillary pressure. This has been noted by Scheidegger (1974, p. 261).

8.4 Capillary Pressure

The displacement of the nonwetting phase by the wetting phase is called *imbibition* (or *saturation*), and the reverse is called *drainage* (or *desaturation*). The drainage is done under the influence of gravity or centrifugal force (when the two fluid densities are not the same) or by the imposition of an *externally* applied pressure between the two fluids. The capillary pressure, the pressure *jump* across the nonwetting-wetting phase interface, is generally presented as a function of the saturation, i.e., $\langle p_c \rangle = \langle p_c \rangle (s)$. This function is *multivalued* and this *capillary pressure hysteresis* is such that for a given saturation the capillary pressure is *higher* during the drainage (desaturation). The *hysteresis* is caused by the contact angle hysteresis, which is partly responsible for the hysteresis in the phase distributions (where during a change in the saturation, these distributions depend on the previous phase distributions).

The measurement of the capillary pressure has been reviewed by Dullien (1979, p.20) and one of the methods most often used is the *porous diaphragm tensiometer* (Dullien et al., 1986). In this method the pressure of the wetting phase is measured by allowing for the permeation of this phase by a porous membrane (diaphragm) and into a manometer. The nonwetting phase pressure is generally atmospheric, although it could be measured separately through a membrane, which allows for the *permeation* of this phase *only*.

8.4.1 HYSTERESIS

Under static equilibrium conditions, the capillary pressure, i.e., the jump in pressure across the g-ℓ interface, is averaged over the pore volume (or the representative elementary volume). This average $\langle p \rangle^g - \langle p \rangle^\ell$ is equal to $2\sigma \langle H \rangle$ where $\langle H \rangle$ is the volume-averaged *mean* meniscus *curvature*. Therefore, in principle the $\langle p_c \rangle = \langle p_c \rangle (s)$ relationship is determined by $\langle H \rangle = \langle H \rangle (s)$. As was mentioned, $\langle H \rangle$ depends on the phase distributions. In order to proceed with the determination of $\langle H \rangle = \langle H \rangle (s)$, the matrix structure and the phase distributions should be determined. Note that although $\langle H \rangle = \langle H \rangle (s)$ is expected to describe a *static* equilibrium state, it is generally determined and is of *interest* in processes where the porous medium is *being* saturated or desaturated. In the *experimental* determination of $\langle H \rangle = \langle H \rangle (s)$, a very *large* elapsed time (up to months) is allowed between any Δs. The efforts in modeling hysteresis are reviewed by Dullien (1979) and will be discussed in Section 8.4.2. Here we begin with the experimental results of Dullien et al. (1989), where the results, except for the *lack* of a definite irreducible saturation (for rough surfaces), are similar to those reported by Leverett (1941).

Figure 8.9 is adapted from Dullien et al. (1989) and shows one of the general characteristics of $\langle p_c \rangle = \langle p_c \rangle (s)$, which is that during the *drainage*

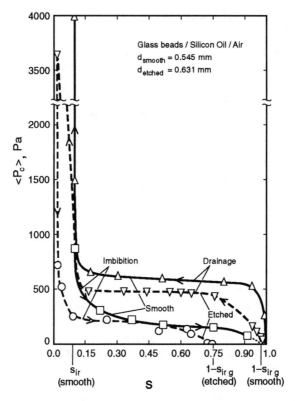

Figure 8.9 The measured capillary pressure (Dullien et al., 1989) versus saturation, for packed beds of smooth and etched spheres. The results show that $s_{ir} \to 0$ for the etched spheres. (From Dullien et al., reproduced by permission ©1989 Academic.)

the measured capillary pressure is *larger*. In the experiment, a silicon oil of dynamic viscosity of 2×10^{-3} kg/m-s, density of 920 kg/m³, and surface tension of 0.025 N/m is used along with air (room temperature, pressure) in a packed bed of glass beads. In one of the experiments, the glass beads where *smooth*, and in the other they were *etched*. The etching was done to examine the concept of *hydraulic continuity/discontinuity*. The estimated precision of the measurements is ±0.005 in s and ±5 Pa in $\langle p_c \rangle = \langle p_c \rangle$. Some of the features of the $\langle p_c \rangle (s)$ function are given here.

- The *irreducible wetting phase saturation* s_{ir} approaches 0.09 for the *smooth* beads as $\langle p_c \rangle \to \infty$ but tends toward zero for the *etched* beads. Note that, as discussed in Section 8.1.3 (A), the variation in parameters other than the surface roughness (e.g., in σ, μ, ρ, θ_c, and d) does *not* change s_{ir}.

- The *irreducible nonwetting phase saturation* $s_{ir\,g}$ depends on the surface roughness and $1 - s_{ir\,g}$ changes from 0.75 to 0.98, i.e., $s_{ir\,g}$ varies between 0.02 and 0.25 and the higher value is for an etched surface.

- Both *saturation* and *desaturation* capillary pressures are higher for the *smaller* spheres (diameters are given in the figure). This is in accordance with the Leverett reduced relationship, i.e., $\langle p_c \rangle = \sigma(K/\varepsilon)^{-1/2} J(s)$, where K is proportional to d^2 (from the Carman-Kozeny relation).

- For a given particle size and saturation, the drainage capillary pressure is larger than the imbibition capillary pressure [i.e., there is a hysteresis in $\langle p_c \rangle = \langle p_c \rangle (s)$].

The first observation mentioned is *explained* in terms of the wetting phase *filling* the eroded (etched) channels. At low wetting phase saturations, these filled microchannels *connect* to the pendular rings, and together they result in a *hydraulic conductivity*. Therefore, as the capillary pressure increases, the wetting phase saturation *continues* to decrease.

The second observation is also related to the filling of the microchannels by the wetting phase. In this case, the *trapping* of the nonwetting phase is made easier (during imbibition) as the microchannels assist in the engulfing of the nonwetting phase.

Since $\langle H \rangle = \langle H \rangle (s)$ depends on the phase distributions and because the interfacial area $A_{\ell g}$ is one of the properties of these distributions, it is instructive to review the treatment of Leverett (1941) for the determination of $A_{\ell g}$ from the $\langle p_c \rangle = \langle p_c \rangle (s)$ curve. His analysis shows that for a given s, $A_{\ell g}$ is *larger* for the drainage branch than it is for the imbibition branch, a result that is evident from the fact that a larger $\langle H \rangle$ (or smaller average mean radius of curvature) requires a larger interfacial area. He uses the thermodynamic definitions of the surface tension (8.16) and the volume (Chapter 9 discusses the thermodynamics of the phases and the interface), i.e.,

$$\left. \frac{\partial F}{\partial p} \right|_{T, N_i, \ldots} \equiv V, \tag{8.146}$$

and assumes a negligible compressibility of the liquid. He then arrives at

$$d \langle F \rangle = - \langle p_c \rangle \, ds. \tag{8.147}$$

Now combining this with (8.16), we have

$$\Delta A_{\ell g} = \frac{\Delta F}{\sigma} = -\frac{1}{\sigma} \int \langle p_c \rangle \, ds, \tag{8.148}$$

i.e., the area under the $\langle p_c \rangle = \langle p_c \rangle (s)$ curve divided by the surface tension gives the *negative* of the ℓ-g interfacial area. As was mentioned, the area under the curve is *larger* for the drainage branch.

8.4.2 MODELS

The function $\langle H \rangle = \langle H \rangle (s)$ including its multivaluedness is expected to be strongly dependent on the matrix structure. In modeling $\langle H \rangle = \langle H \rangle (s)$, a general distinction is made between the imbibition and drainage branches. Further divisions are made based on the matrix structure. The real solid matrices are difficult to directly model because of the three dimensionality and the randomness that exist in these structure. The periodic structures are perhaps the most suitable for the *direct simulation* of the phase distribution. A complete description, even for these structures, is not presently available. However, examination of some of the features of the capillary pressure-saturation curve has revealed some of the mechanics of the phase distributions during a *quasi-steady* saturation or desaturation. An example is the study by Yu and Wardlaw (1986) on the mechanisms of the nonwetting phase trapping during saturation.

The following is a brief review of the capillary pressure modeling efforts as given by Dullien (1979, 42–66). The existing correlations are also given here.

(A) CAPILLARY

Under a *negligible* gravity effect, the capillary pressure in a converging circular tube with a converging angle of ϕ containing a *meniscus* curbing toward the narrower section is given by

$$p_c = p_g - p_\ell = \frac{4\sigma}{d} \cos(\theta_c + \phi), \qquad (8.149)$$

where d is the diameter of the tube at the contact line. When the meniscus curbs towards the diverging section (i.e., when the wetting phase ℓ is filling the portion where the diameter is larger), the sign on the right-hand side changes to negative. We note that a *single* converging (or diverging) tube does *not* model $\langle p_c \rangle = \langle p_c \rangle (s)$ and its hysteresis. Another model for the *drainage* process uses capillary tubes of constant but different radii placed in *parallel* (bundle of capillary tubes). The nonwetting phase *alone* is present on one side of the tubes and it *displaces* the wetting phase, forcing it to the other side. In this model the nonwetting phase penetrates the *largest* capillaries first, filling them. Then the smaller diameter capillary tubes are filled with larger capillary pressures. If the distribution of the diameters are known, the capillary pressure can be found. In the *inverse problem*, from the capillary pressure a corresponding *pore* diameter can be determined. Of course pores are not simple nonconnected capillaries, and this simple model can *only* be viewed as an elementary exercise. Using the probability density function $N(d)$, the change in void volume V of the pores having diameters larger than d but smaller than $d + \Delta d$ is given by $\Delta V = V_t N(d) \Delta d$ with a total void volume of V_t. Then we have $\int_0^\infty N(d)\, d(d) = 1$.

For straight capillary tubes, (8.149) can be written as $p_c d = 4\sigma \cos\theta_c$. By assuming that θ_c is constant, we have $p_c \Delta d + d\Delta p_c = 0$. Now from this, the change in the volume is related to the changes in d and p_c. This is done through

$$\frac{\Delta V}{V_t} = -N(d)\frac{4\sigma\cos\theta}{p_c^2}\Delta p_c = -N(d)\frac{d}{p_c}\Delta p_c \qquad (8.150)$$

or

$$N(d) = -\frac{p_c}{d}\frac{\Delta\dfrac{V}{V_t}}{\Delta p_c}, \qquad (8.151)$$

where V/V_t is the fraction of the pores having diameters less than d. When allowing for the irreducible wetting phase saturation, V_t is taken as the *drainable pore volume*. Then, by measuring $\Delta V/V_t$ and Δp_c, $N(d)$ can be determined.

(B) NETWORK

In a network, pores and their interconnecting capillaries are distributed in a *random* manner. The size and shape of the pores and the diameter and length of the capillaries are prescribed with probability density functions that satisfy certain conservations. The pores are connected to the capillaries according to a probability density function that gives the coordination number (or the number of capillaries connected to a given pore). These elements of the analysis are similar to those applied in the statistical treatment of the thermal conductivity in porous media (Section 3.5.2). Since in two-dimensional networks *only* one phase can be *continuous*, three-dimensional networks are needed for the presentation of the displacement problems (saturation or desaturation). However, Dullien (1979, 44–66) shows that *even* the two-dimensional networks can accurately predict some of the displacement features.

One of earlier *percolation* (penetration through the network) efforts reviewed by Dullien is that of Fatt in 1956. The Fatt two-dimensional model of the desaturation gives an instantaneous saturation (wetting phase) as

$$s = 1 - \frac{\sum_{i=1}^{k} a_{ik}\pi R_i^2 \ell_i}{\sum_{i=1}^{n} N_i \pi R_i^2 \ell_i}, \qquad (8.152)$$

where N_i = number of capillaries of diameter i (i is assigned to signify size and is largest for the smallest size), a_{ik} = number of penetrated capillary tubes of size i when the smallest penetration tube is size k, R_i = capillary tube radius of size i, and ℓ_i = length of the capillary tube of size i.

Dullien considers beds of two- and three-dimensional particles of *simple, body-centered,* and *face-centered cubic arrangements.* For these matrices he has computed the *breakthrough pressure* (the pressure required for the first trace of the intruding phase to reach the opposite side of the network). For a nonwetting phase (mercury) in a previously vacuumed matrix, he has found good agreement with the experiments provided the pore and the channel geometry and the size distributions are realistic. He has also made predictions of the $\langle p_c \rangle = \langle p_c \rangle (s)$ for the mercury intrusion and has found general agreements with the experiments. However, a realistic account of the pore structure is very difficult to implement along with the statistical treatments.

The preceding models attempt to account for the matrix geometry, and because of the extensive computational effort required, they do not address the phase distributions at the pore level. This includes the phase distributions for the two asymptotes $s \to 0$ and $s \to 1$. The case of $s \to 1$ and the associated discontinuity in the distribution of the nonwetting phase has been considered experimentally by Yu and Wardlaw using transparent microstructures. Both concave and convex pores and the connecting capillaries (throats) were etched on glass. They find that as $s \to 1$ most of the discontinuity of the nonwetting phase is caused when the *pore-to-throat-diameter ratio* is larger than 1.75 (for $\theta_c < 70°$). This *critical* ratio depends slightly on the contact angle. During the saturation, both in-pore and in-throat phase distributions are significant. While during the desaturation, the throat phase distributions are the *most* significant. This is the reason for the larger capillary pressure (larger $\langle H \rangle$) during the desaturation. The interface movement during the saturation is influenced by the size, shape, and arrangement of both the pores and the throats in the direction of the advancing displacing phase. Wettability influences the entrapment of the nonwetting phase significantly, and the history of phase distributions is very significant to what follows (hysteresis).

The macroscopic predictions of $\langle p_c \rangle = \langle p_c \rangle (s)$ require the pore-level phase dynamics as well as the ensemble averaging over the system. The trend is toward the *inclusion* of more realistic and complete descriptions of the dynamics of the phase distributions into the statistical models.

(C) EMPIRICAL

Since presently a general theoretical prediction of $\langle p_c \rangle = \langle p_c \rangle (s)$ is not obtainable, correlations are obtained from the experiments. Table 8.3 gives several correlations used in the study of isothermal and isoconcentration two-phase flows in porous media. The correlations are made either to the *imbibition* or the *drainage* experimental data. Therefore, they should be used accordingly. The *Leverett reduced function,* i.e.,

8.4 Capillary Pressure

TABLE 8.3   CORRELATIONS FOR CAPILLARY PRESSURE

CONSTRAINTS	CORRELATION
(a) Water-air-sand	$\langle p_c \rangle = \dfrac{\sigma}{(K/\varepsilon)^{1/2}} \left[0.364(1 - e^{-40(1-s)}) + 0.221(1-s) \right.$ $\left. + \dfrac{0.005}{s - 0.08} \right]$
(b) Water-air-soil and sandstones	$s = s_{\text{ir}} + \dfrac{1 - s_{\text{ir}} - s_{\text{ir }g}}{\left[1 + \left(a_1 \dfrac{\langle p_c \rangle}{\rho_\ell g} \right)^n \right]^{1 - 1/n}}$ where $n > 1$, a_1 is a constant, n and a_1 depend on the matrix and the drainage or imbibition process
(c) Imbibition, non-consolidated sand, from Leverett (1941) data of water-air	$\langle p_c \rangle = \dfrac{\sigma}{(K/\varepsilon)^{1/2}} \left[1.417(1-S) - 2.120(1-S)^2 \right.$ $\left. + 1.263(1-S)^3 \right]$ where $S = \dfrac{s - s_{\text{ir}}}{1 - s_{\text{ir}} - s_{\text{ir }g}}$
(d) Drainage, oil-water in a sandstone	$\langle p_c \rangle = \dfrac{\sigma}{(K/\varepsilon)^{1/2}} \left[a_1 - a_2 \ln(s - s_{\text{ir}}) \right]$ where $a_1 = 0.30$, $a_2 = 0.0633$, $s_{\text{ir}} = 0.15$

(a) Scheidegger (1974) from Leverett (1941) experiment;
(b) van Genuchten (1980); (c) Udell (1985); (d) Pavone (1989).

$$\frac{p_c (K/\varepsilon)^{1/2}}{\sigma \cos \theta_c Z(\theta_c)} = J(s), \qquad (8.153)$$

is generally used. We have included the Melrose (1965) *function* for the effect of contact angle. The effect of wettability on p_c is further discussed by Demond et al. (1994). Note that, $J(s)$, which is also called the *Leverett J-function*, is not universal and depends on the *wettability* and the *surface tension*. Leverett (1941) found this through correlating his experimental results. Then he argued that since $(\langle p_c \rangle / \sigma)(K/\varepsilon)^{1/2}$ is the product of the *mean interfacial curvature* (inverse of the mean radius of curvature of the meniscus in the pore) $\langle p_c \rangle / \sigma$ and the *pore-level length scale* $(K/\varepsilon)^{1/2}$, then it should only depend on s. However, the experimental results of Morrow (using air and a variety of liquids) on the effect of the wettability (contact angle) on $J(s)$ show a *strong* dependence on wettability. [This work is reviewed by Dullien (1979, p. 28)].

As mentioned earlier, the phase distributions during the imbibition and the drainage are different, and the latter depends more strongly on the topology and dimensions of the throats (pore necks). Also, the length scale $(K/\varepsilon)^{1/2}$ is more representative of the *throat* dimension. Therefore, $(K/\varepsilon)^{1/2}$ *cannot* represent the topology and size of both the pores and the throats adequately enough to be used for the drainage and imbibition process. As expected, the J-function represents the *drainage* more *adequately*.

For water-air-solid particle systems, van Genuchten (1980) has correlated the results for the capillary pressure and the relative permeabilities. His correlation is also shown in Table 8.3. He examines the properties of the function given in Table 8.3 and gives a procedure for the *estimation* of a_1 and n.

8.5 Relative Permeability

Muskat and Meres (1936) recommended that the phase permeabilities $\mathbf{K}_\ell$ and $\mathbf{K}_g$ given in (8.134) and (8.135) be treated as *isotropic* and given by

$$K_\ell = K K_{r\ell}, \tag{8.154}$$

$$K_g = K K_{rg}, \tag{8.155}$$

where K is the *absolute permeability* used in the single-phase flows and $K_{r\ell}$ and K_{rg} are the *relative permeability* of the ℓ and g phases, respectively. Then the relative permeability accounts for the phase distributions a_ℓ and a_g, as given in (8.1), i.e.,

$$K_{r\ell} = K_{r\ell}\left(\text{matrix structure}, s, \sigma, \theta_c, \frac{\rho_\ell}{\rho_g}, \text{history}\right). \tag{8.156}$$

A similar dependence is expected for K_{rg}. Furthermore, the approach taken by Muskat and Meres was the simplification of this relation to a *saturation* dependence only, i.e.,

$$K_{r\ell} = K_{r\ell}(s), \tag{8.157}$$

$$K_{rg} = K_{rg}(s), \tag{8.158}$$

which, although it is still in popular usage, has proven to be generally *inadequate*. However, for a given liquid-fluid-solid combination, solid matrix structure, and a range of saturation, *limited success* has been obtained using (8.157) and (8.158). Before beginning a discussion of the general behavior of $K_{r\ell}$ (and K_{rg}) as given by (8.157) and (8.158), it must be emphasized that the relative permeabilities are defined, through (8.154) and (8.155), to represent *only* the drag force on $A_{\ell s}$ (or A_{gs}). This is further elaborated later.

8.5.1 Constraint on Applicability

The tensors $\mathbf{K}_\ell$ and $\mathbf{K}_g$, and the scalars $K_{r\ell}$ and K_{rg} in (8.154) and (8.155) are defined for the Darcy (Stokes) flow with a negligible interfacial drag at $A_{\ell g}$ and a negligible *Marangoni effect* ($\nabla \langle \sigma \rangle \to 0$). Therefore, the concept of the relative permeability, as introduced by Muskat and Meres, should be interpreted to represent the phase distributions and their influence on the viscous drag on $A_{\ell s}$ and A_{gs}. From what is known about flow transitions occurring at low velocities (Section 8.1.2), the determination of the relative permeability for flows that show no effect of the flow rates on these coefficients and for $\mu_\ell/\mu_g = 1$ is rather impossible. It should be noted that in some studies, the effect of the microscopic (and in some cases the macroscopic) inertial, the interfacial drag at $A_{\ell g}$ ($\mu_\ell/\mu_g \neq 1$), and the Marangoni effect have all been included in $K_{r\ell}$ and K_{rg}. This can be considered an enlargement of the degree of empiricism commonly practiced in the treatment of transport through porous media. However, by lumping all the effects into one single coefficient for accurate prediction, this coefficient must be a very *complex* function of a large number of variables. Therefore, the concept of the phase permeability (and the relative permeability) should be distinguished from the liquid-fluid interfacial drag, the inertial, and the surface tension gradient effects.

8.5.2 Influencing Factors

Since the simultaneous flow of the two phases has been made analogous to that of the Darcean single-phase flow, the flow of each phase at a given local saturation depends on how the two phases are distributed over the representative elementary volume. The factors that can be influencing the phase distributions are the *local saturation, matrix (or pore) structure, history, contact angle, surface tension,* and *density ratio.* These effects are generally examined experimentally. The various techniques that are used to measure the relative permeabilities and their advantages and disadvantages are reviewed by Dullien (1979, 269–273). Osoba et al. (1951) have compared the results from the various techniques and have found that in general good agreement exists among them. We now examine these factors.

(A) Saturation

The local saturation, i.e., the extent to which the liquid fills the pores in the representative elementary volume, influences the phase distributions in this volume and in turn influences the resistance to the flow of each phase. The irreducible wetting phase saturation s_{ir} and the irreducible nonwetting phase saturation $s_{\text{ir}\,g}$ form the bounds for the local saturation, i.e., $s_{\text{ir}} < s < 1 - s_{\text{ir}\,g}$. As $s \to s_{\text{ir}}$, the concentration of the wetting phase around the corners will occur, and as $s \to 1 - s_{\text{ir}\,g}$ a more uniform distribution will result. Observations on phase distributions over the entire range of s are limited. Considering the complexity of the phase distribution functions a_ℓ and a_g, the local *flow channels* through which each phase flows is a complex function of the local saturation. The *excess resistance*, i.e., $1/K_{r\ell}$ also becomes a complex function of s. Exponent n ranging from unity to as large as seven has been proposed for $K_{r\ell} = s^n$. One of the important features of the excess resistances $1/K_{r\ell}$ and $1/K_{rg}$ is that the *simultaneous flow* of the phases causes excess resistances for *both* phases. This means that in general for a given saturation we have

$$K_{r\ell} + K_{rg} < 1, \quad \text{given } s. \tag{8.159}$$

Suggestions that $K_{r\ell} + K_{rg} = 1$, as made by Wyllie (1962), *cannot* be supported from the fluid mechanics considerations. Experimental results show that $K_{r\ell} + K_{rg}$ can be *even* less than 0.1 for the *intermediate* values of s, while in some cases in the two extreme values of s, the sum of the two conductances asymptotically approaches unity. The intermediate saturations correspond to a *mutual* and *significant* resistance *offered* by each phase to the flow of the other phase. The wetting phase tends to cover the solid surface area in a manner that results in the *smallest* liquid-fluid interfacial area and this results in a large resistance to the flow of this phase. Since the minimization of A_{fg} requires spreading over many pores, the flow channels through which the wetting phase flows become constricted, and the wetting phase has to travel through tortuous paths.

We note that the *local saturation* can be a function of the phase velocities. This influence is through the interfacial and volumetric forces that govern the distribution of phases (the Navier-Stokes and continuity equations and the interfacial force and mass balances). This influence of the phase velocity on the relative permeability through s is not in accord with the proposed analogy between Darcean single-phase and two-phase flows. This dependence will be demonstrated through a simple model in Section 8.5.3 (A).

(B) MATRIX STRUCTURE

The solid phase distribution influences the distributions of the ℓ and g phases and, in turn, the relative permeabilities. Since the ℓ and g phases flow through the passages that are nonuniform in cross section, the distributions of the phases (and their point velocities) are significantly influenced by the topology of these passages. The absolute permeability is a measure of the average throat size, and we expected that in general $K_{r\ell}$ and K_{rg} would also depend on K. However, note that K does not give all the information about the microstructure and, therefore, cannot be used as the only parameter describing the appropriate small length scale for the two-phase flow.

Figure 8.10(a) is adapted from Dullien (1979, p. 266) and is based on the results of Morgan and Gordon (1970). The figure shows the variation of $K_{r\ell}$ and K_{rg} with respect to s for *three* different sandstones with the *water-oil* as the ℓ-g phase pair. The sample with $K = 1.3 \times 10^{-12}$ m^2 contains large well-connected pores. For the sample with $K = 2.0 \times 10^{-14}$ m^2, the pores are small and well connected. For the sample with $K = 3.6 \times 10^{-14}$, large pores are few and are connected by small pores. The results given in Figure 8.10(a) show that the large pores allow for a smaller irreducible saturation s_{ir} and a smaller resistance to flow of the wetting phase. When both large and small pores are present, the relative permeability is *controlled* by the *small* pores, as shown by the similarity between the results for the last two samples. As expected, both the absolute and wetting phase relative permeability decreases when the average pore size is smaller. The dependence of the phase distributions on the matrix structure in two-phase flows has been qualitatively studied for packed beds of spherical particles (Tung and Dhir, 1988; Levec et al., 1986; Ng, 1986). A quantitative description of the two-phase flow through matrices, including the prediction of the interfacial area, is not available (this lack of information also exists to an extent for the two-phase flow in plain media).

(C) History

As was discussed in connection with the capillary pressure (obtained through quasi-static process), phase distributions experience hysteresis depending on whether a given saturation is reached through the drainage or imbibition branch. Figure 8.10(b) is also adapted from Dullien (1979, p. 256) and is based on the experimental results of Osoba et al. (1951). The results show multivaluedness for the relative permeability-saturation relation. The hysteresis is more pronounced for the nonwetting phase (gas) than it is for the wetting phase (oil). The porous media is a sandstone.

For a given s, the assumed Stokes flows (both phases) can result in phase distributions that are different from the static distributions. This difference has not yet been systematically studied. Therefore, the similarities and differences between the hysteresis of the relative permeability (of the wetting phase), which is obtained under a *fluid motion*, and the capillary pressure, which is obtained under a *quasi-static* condition, *cannot* be presently quantified.

(D) Contact Angle

As expected, the wettability influences both the *dynamic* and the *quasi-static* phase distributions. No microscopic observations are available, but the measured relative permeabilities show a pronounced dependence on wettability. Dullien (1979, p. 258) reviews the available data on this effect. For example in an oil-water-sandstone system with an s (for water) equal

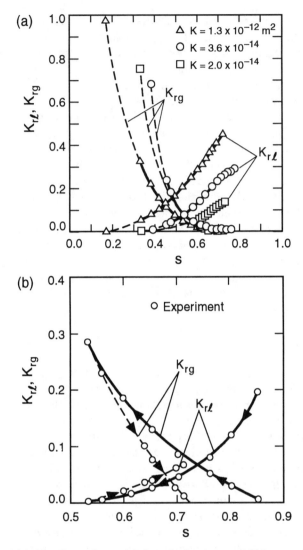

Figure 8.10 (a) The dependence of the relative permeabilities on the absolute permeability and saturation. Figure is adapted from Dullien and the experiment is for water-oil-sandstone systems. (b) Saturation hysteresis in the relative permeabilities. The irreducible nonwetting phase saturation also shows hysteresis. (From Dullien, reproduced by permission ©1979 Academic.)

to 0.2, the relative permeability of the oil (surfactants are added to the oil in order to alter its contact angle) is *decreased* by 16 percent when the contact angle is *changed* from 0 to 47 degrees. When the contact angle is changed from 0 to 180 degrees, this decrease became 34 percent. The results presented by Dullien show that this dependency on the contact angle persists over the entire reachable saturation ($s_{ir} \leq s \leq 1 - s_{ir\,g}$). Also, the relative permeability hysteresis persists for all degrees of wettability.

(E) SURFACE TENSION

For a given local saturation, when the pore length scale is small, the surface tension can significantly influence the meniscus contour. Since the resistance to the flow of each phase depends on the topology of the conduits they pass through, the relative permeability (inverse of the excess resistance) is dependent on the surface tension. In *large* pores the capillarity becomes less significant (but buoyancy becomes important). Note that by assuming a Darcean behavior for the two-phase flows, the Stokes flow field (noninertial) is not expected to influence the relative permeability. Therefore, in determining the capillary effect, the surface tension cannot be scaled with respect to the viscous forces. Therefore, the capillary number $\mu u_D/\sigma$, as recommended by Lefebvre du Prey (1973), cannot in principle be used. However, the viscous force appearing in the balance of the normal and the tangential forces at the liquid-fluid interface, (8.40) and (8.41), are expected to influence the phase distributions. Saez et al. (1986) have analyzed the pore-level fluid mechanics for a *periodically constricting tube* model and have shown that the phase velocities *influence* the local saturation. Therefore, the viscous forces indirectly (through s) influence the relative permeability.

As the surface tension increases, the relative permeabilities are expected to decrease. Therefore, addition of the surface active agents increases both of the relative permeabilities. The *upper bounds* (i.e., $\sigma \to 0$) for the relative permeabilities depend on the matrix structure.

(F) DENSITY RATIO

Density ratio accounts for the effects of the buoyancy on the phase distributions (in light of the Darcean flow). When scaled with respect to the surface tension and the pore length scale, we obtain the Bond (or Eötvös) number as

$$Bo = \frac{\rho_g \left(\dfrac{\rho_\ell}{\rho_g} - 1\right) g d^2}{\sigma}. \tag{8.160}$$

For a negligible motion in both phases and for a *Bo* smaller than unity, the buoyancy force is less important than the static surface tension force. Then the meniscus contours are controlled by the surface tension and the contact angle (within the assumption of a static or a quasi-static contact line). For Bond numbers larger than unity, the gravity force influences the phase distributions and the direction of flow with respect to the gravity vector becomes important.

Finally, we note again that the phase velocities (magnitude and direction) influence the relative permeability through occurrence of the transitions which, although in the noninertial regime, are phase-velocity-dependent. Moreover, the flow arrangements, i.e., cocurrent or countercurrent, as well as the direction of the liquid flow with respect to the gravity vector (in each of the two arrangements), can influence $\mathbf{K}_\ell$ and $\mathbf{K}_g$. However, we have assumed that dependence on the phase velocities is weak, so that we can arrive at Darcean-equivalent phase permeabilities for the two-phase flows.

8.5.3 MODELS

Models based on concepts similar to those used for the prediction of the capillary pressure have been used for the prediction of the relative permeabilities. The modeling of the relative permeability is even more difficult because in addition to the hysteresis, which is common between $\langle p_c \rangle$ and $K_{r\ell}$ (and K_{rg}), the models should predict $K_{r\ell}$ and K_{rg}, simultaneously. The existing capillary tube, statistical, and network models have been reviewed by Scheidegger (1974, 274–283) and Dullien (1979, 276–284). A network model which shows the hysteresis and is based on pore-bodies interconnected by pore-throats is given by Jerauld and Salter (1990). In the following, the periodically constricted tube model of Saez et al. (1986), used for the trickling flow in packed beds of spherical particles, and some of the existing empirical correlations are examined.

(A) PERIODICALLY CONSTRICTED TUBE

Presently, the *direct* (*point* solution) simulation of the two-phase flow in porous media is limited to an axisymmetric fully liquid-covered, *periodically constricted tube model* by Saez et al. (1986), where a *cocurrent* flow *along the direction of gravity* is considered. Figure 8.11 gives a schematic of the unit cell they consider. The geometry is that of a variable-area tube, where the variation in the cross-sectional area (perpendicular to the x-axis) is chosen such that it corresponds to that in a unit cell of spheres packed in the *simple cubic arrangement* (porosity is 0.48 for this packing). This representation of the simple cubic structure by a periodically constricted tube results in a computed absolute permeability that is 20 percent lower (Stokes flow solutions) than that obtained by the Carman-Kozeny relation.

The thickness of the liquid at $x = 0$ is prescribed and is assumed to be that at $x = \ell$, where ℓ is the unit-cell length along x. This assumes no acceleration, and therefore, Stokes flow (with gravity) occurs in both the liquid and gaseous phases. This assumption *cannot* be strongly supported because the flow of a liquid film over curved surfaces involves thinning (when flow

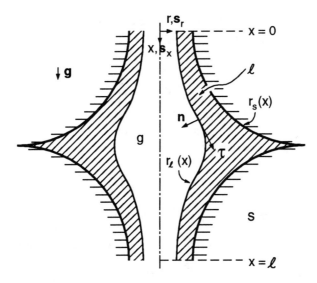

Figure 8.11 The unit cell used by Saez et al. for a periodically constricting tube model of two-phase flow. The direction of the gravity vector is also shown and both phases flow (Darcean) along this vector.

is along **g**) and thickening (when the velocity has a component along $-\mathbf{g}$ or when the flow is decelerating). However, without this assumption, the computation becomes more *intensive* as more than one period in the tube has to be included in the computational domain.

The principal radii are

$$r_1 = -\ell \frac{\left[1+\left(\frac{dr_\ell}{dx}\right)^2\right]^{3/2}}{\frac{d^2 r_\ell}{dx^2}} \quad \text{and} \quad r_2 = r_\ell \left[1+\left(\frac{dr_\ell}{dx}\right)^2\right]. \qquad (8.161)$$

The normal and the tangent unit vectors on A_{fs} are

$$\mathbf{n}_{fs} = \frac{-\mathbf{s}_r + \left(\frac{dr_\ell}{dx}\right)\mathbf{s}_x}{\left[1+\left(\frac{dr_\ell}{dx}\right)^2\right]^{1/2}} \quad \text{and} \quad \boldsymbol{\tau}_{fs} = \frac{\mathbf{s}_x + \left(\frac{dr_\ell}{dx}\right)\mathbf{s}_r}{\left[1+\left(\frac{dr_\ell}{dx}\right)^2\right]^{1/2}}, \qquad (8.162)$$

where the parameters are defined in Figure 8.11.

The interfacial force balances, (8.40)–(8.41), are simplified by assuming that

$$\left(\frac{r_s}{r_\ell}-1\right)\frac{\mu_g}{\mu_\ell} \ll 1. \qquad (8.163)$$

This leads to the *elimination* of the gas-phase shear stress terms. The surface tension gradient is also *neglected*, leading to the simplification of (8.91) to the following vectorial equation for the normal component of interfacial forces

$$\left[p_\ell - p_g + \sigma\left(\frac{1}{r_1} + \frac{1}{r_2}\right)\right]\mathbf{n} = \mathbf{S}_\ell \cdot \mathbf{n}. \qquad (8.164)$$

The pressure in each phase is decomposed into a *linear area-averaged* component and a *periodic* deviation component, i.e.,

$$p_\ell = p'_\ell - \frac{\Delta \bar{p}\, x}{\ell}, \qquad (8.165)$$

$$\bar{p} = \frac{1}{A}\int_A p'\, dA, \qquad (8.166)$$

where $\Delta \bar{p}$ is the pressure drop across a unit cell and p'_ℓ is periodic, i.e.,

$$p'_\ell(x, r) = p'_\ell(x + \ell, r), \qquad (8.167)$$

$$\bar{p}(x) = \bar{p}(x + \ell). \qquad (8.168)$$

In addition, the velocity components are also periodic.

The solutions to the Stokes flows (in ℓ and g phases), subject to the continuity of the velocity at the interface and the *kinematic condition* ($\mathbf{u}_\ell \cdot \mathbf{n}_\ell$) = 0, is found by the finite element approximation method. For the case of a *constant area tube*, where r_ℓ and r_g are constant, $r_1 \to \infty, r_2 \to \infty$, and $s = r_\ell^2/r_s$, their *closed-form* solutions for the *relative* permeabilities for the case where the effect of the gas-side shear stress on the liquid flow is *not neglected*, are

$$K_{r\ell} = s^2 + \frac{2\left(1 - \frac{\rho_g}{\rho_\ell}\right)(1-s)^2\left[s + (1-s)\ln(1-s)\right]\left[1 - s\left(1 - 2\frac{\mu_g}{\mu_\ell}\right)\right]}{\left(1 - \frac{\rho_g}{\rho_\ell}\right)(1-s)^2\left[2\frac{\mu_g}{\mu_\ell}\ln(1-s) - 1\right] - 72\frac{\mu_g \langle u_g\rangle}{\rho_g g}\frac{(1-\varepsilon)^2}{d^2\varepsilon^3}} \qquad (8.169)$$

and

$$K_{rg} = \frac{(1-s)\left\{A\left[(1-s) + 2\frac{\mu_g}{\mu_\ell}(1-s)\right] + \left(1 - \frac{\rho_g}{\rho_\ell}\right)(1-s)s^2\left[1 - 2\frac{\mu_g}{\mu_\ell}\ln(1-s)\right]\right\}}{A + \left(1 - \frac{\rho_g}{\rho_\ell}\right)s^2}, \qquad (8.170)$$

where

$$A = -72\frac{\mu_\ell \langle u_\ell\rangle}{\rho_g g}\frac{(1-\varepsilon)^2}{d^2\varepsilon^3} - 2(1-s)\left[s + (1-s)\ln(1-s)\right] \qquad (8.171)$$

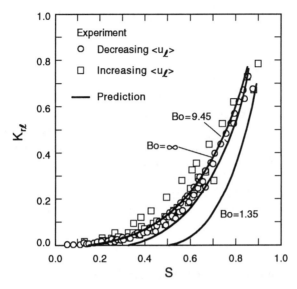

Figure 8.12 Comparison between the predictions of Saez et al. using the periodically constricting tube model, with the experiments of Levec et al., for the liquid-phase relative permeability as a function of the liquid saturation. (From Levec et al., reproduced by permission ©1986 AIChE.)

and d is the *equivalent* particle diameter, which is taken as $6\,V_s/A_s$. These equations show that $K_{r\ell}$ *depends* on $\langle u_g \rangle$, although for most practical cases this dependence is very *weak*. Also, K_{rg} depends on $\langle u_\ell \rangle$, and this is a much *stronger* influence compared to that of $\langle u_g \rangle$ on $K_{r\ell}$.

When the liquid flow is *decoupled* from the gas flow, the periodically constricted tubes give $s = s(Bo, \langle u_\ell \rangle)$, where $Bo = \rho_\ell g \ell^2 / \sigma$. The influence of Bo on the unit-cell saturation originates from the interfacial force balance given by (8.164). For a given s, Saez et al. obtain $K_{r\ell} = K_{r\ell}(s, Bo)$, i.e., $K_{r\ell}$ depends *indirectly* on $\langle u_\ell \rangle$ through s. Figure 8.12 gives the numerical results for the periodically constricted tube and those of the experiments done by Levec et al. (1986) for packed beds of spheres. The Bond number dependence predicted by the model is *not* found experimentally for the $0.44 \leq Bo \leq 1.90$ used in the experiments. Also, the model does *not* include any hysteresis-related features, such as the dependence of the phase distributions on the decreasing or increasing liquid flow rate, but the experimental results *show* a history-dependence.

As is evident from this modeling effort, the *parameters* influencing $K_{r\ell}$ and K_{rg} are *too* large to allow for a comprehensive *direct simulation* (point solution) of the two-phase flow in porous media. However, the model of Saez et al. *clarifies* the forces and interactions in the Darcy regime and shows that the concept of $K_{r\ell} = K_{r\ell}(s)$ is an *approximation* that is only justified when placed in the context of the general empirical treatments that presently exist.

(B) EMPIRICAL

Table 8.4 gives some of the available *empirical* correlations and the specific *conditions* for which they were *intended*. The soil-related correlations are given by Corey (1986). A review of the two- and three-phase relative permeability correlations is given by Delshad and Pope (1989). As was discussed, $K_{r\ell}$ and K_{rg} are *complicated* functions of many more parameters in addition to s (or S), and any two-phase flow analysis that uses a correlation for $K_{r\ell}(s)$ and $K_{rg}(s)$ should perform an examination of the *sensitivity* of the predicted results with respect to the *variations* in the $K_{r\ell}$ and K_{rg} models.

8.6 Microscopic Inertial Coefficient

The momentum equations (8.134) and (8.135) are rather heuristic and do *not* include all the features of the two-phase flow in porous media. For example, by choosing to make $\mathbf{K}_\ell$ and $\mathbf{K}_g$ independent of $\langle \mathbf{u}_g \rangle$ and $\langle \mathbf{u}_\ell \rangle$, all the velocity effects, other than the proportionality of the pressure gradient to the velocity to the first power, are included in the other terms. This was done in order to develop two-phase flow momentum equations that are similar to these for the single-phase flows and to ensure that the asymptotic behaviors for $s \to 1$ and $s \to 0$ are *readily* recovered. Here, the microscopic inertia term $(\rho_\ell / K_{\ell i}) \langle u_{\ell j} \rangle^2$ and its g-phase counterpart account only for that part of the pressure drop that is due to the fluid-solid drag force and is proportional to $\langle u_{\ell j} \rangle^2$ and $\langle u_{gj} \rangle^2$.

As with $\mathbf{K}_\ell$ and $\mathbf{K}_g$, we assume that $\mathbf{K}_{\ell i}$ and $\mathbf{K}_{gi}$ are independent of the phase velocities and that they have functional forms as given by (8.139). Then the factors influencing the microscopic inertial coefficients are the same as those that influenced the phase permeabilities, i.e., ultimately only the phase distributions influence $\mathbf{K}_{\ell i}$ and $\mathbf{K}_{gi}$. The inertial coefficients have gathered *less* attention than the phase permeabilities. This is because *unless* the pore size is *large*, the inertial regime is *not* encountered. Therefore, only in relatively *coarse packed beds* and high-porosity matrices is this effect important. In connection with the packed beds, Tutu et al. (1983), Saez and Carbonell (1985), Schulenberg and Müller (1987), and Tung and Dhir (1988) have examined these coefficients by using the experimental data for packed beds of spheres. No *detail* pore-level hydrodynamic analysis leading to the determination of these coefficients is presently available.

All of the empirical treatments deal with isothermal steady-state one-dimensional two-phase flows. While Tutu et al. and Schulenberg and Müller *simultaneously* include the effect of the interfacial (liquid-gas) drag, the

TABLE 8.4 CORRELATIONS FOR RELATIVE PERMEABILITIES

CONSTRAINTS	CORRELATION
(a) Sandstones and limestones, oil-water	$K_{r\ell} = S^4, \quad K_{rg} = (1-S)^2(1-S^2)$
(b) Nonconsolidated sand, well sorted	$K_{r\ell} = S^3, \quad K_{rg} = (1-S)^3$
(c) Nonconsolidated sand, poorly sorted	$K_{r\ell} = S^{3.5}, \quad K_{rg} = (1-S)^2(1-S^{1.5})$
(d) Connected sandstone, limestone, rocks	$K_{r\ell} = S^4, \quad K_{rg} = (1-S)^2(1-S^2)$
(e) Sandstone-oil-water	$K_{r\ell} = S^3, \quad K_{rg} = 1 - 1.11S$
(f) Soil-water-gas	$K_{rg} = (1 - s_{\text{ir}} - s_{\text{ir }g} - s)^{1/2} \left\{ (1 - s^{1/m})^m \right.$ $\left. - \left[1 - (1 - s_{\text{ir}} - s_{\text{ir }g})^{1/m}\right]^m \right\}^2$ where m is found from experiments
(g) Glass spheres-water (water vapor)	$K_{r\ell} = S^3$ $K_{rg} = 1.2984 - 1.9832S + 0.7432S^2$
(h) Trickling flow in packed bed	$K_{r\ell} = S^{2.0}$ for increasing liquid flow rate $K_{r\ell} = \begin{cases} S^{2.9} & S \geq 0.2 \\ 0.25 S^{2.0} & S < 0.2 \end{cases}$ for decreasing liquid flow rate $K_{rg} = (1-s)^n$ where $n = n(Re_g, \text{increase or decrease in } \langle u \rangle^\ell)$ $n = 4.8$ has been suggested by Saez and Carbonell (1985)

(a) Corey given by Wyllie (1962); (b)-(d), Wyllie (1962); (e) Scheidegger (1974, p. 255); (f) Mualem (1976) given in Delshad and Pope (1989); (g) Verma et al. (1984); (h) Levec et al. (1986).

other two investigations (although they use fluid pairs for which $\mu_g/\mu_\ell \neq 1$) do *not* account for this effect while determining $\mathbf{K}_{\ell i}$ and $\mathbf{K}_{gi}$.

A simple model by Tung and Dhir accounts for the *excess* pressure gradient due to the presence of the second phase by *modifying* the flow cross-sectional area for each of the phases. This model also allows for *simple* phase distributions by considering the various two-phase flow regimes such as *bubbly, bubbly-slug, pure slug, slug-annular*, and *pure annular*, which are assumed to occur in this order as s is progressively *decreased*. Because a large particle diameter is assumed, the effect of the *surface tension* and the *contact angle*, as well any *hysteresis*, is not included.

According to existing models, a *relative inertial permeability* is defined (assuming isotropic behavior) as

$$\frac{\rho_\ell}{K_{\ell i}}|\langle u_{\ell j}\rangle|\langle u_{\ell j}\rangle = \rho_\ell \frac{C_E}{K^{1/2}}\frac{1}{K_{r\ell i}}|\langle u_{\ell j}\rangle|\langle u_{\ell j}\rangle, \qquad (8.172)$$

$$\frac{\rho_g}{K_{gi}}|\langle u_{gj}\rangle|\langle u_{gj}\rangle = \rho_g \frac{C_E}{K^{1/2}}\frac{1}{K_{rgi}}|\langle u_{gj}\rangle|\langle u_{gj}\rangle \qquad (8.173)$$

where in general

$$K_{r\ell i} = K_{r\ell i}\left(\text{matrix structure}, s, \sigma, \frac{\rho_\ell}{\rho_g}, \text{history}\right), \qquad (8.174)$$

$$K_{rgi} = K_{rgi}\left(\text{matrix structure}, s, \sigma, \frac{\rho_\ell}{\rho_g}, \text{history}\right). \qquad (8.175)$$

However, most of the efforts to determine $K_{r\ell i}$ and K_{rgi}, which have all been empirical, have assumed

$$K_{r\ell i} = K_{r\ell i}(s), \qquad (8.176)$$

$$K_{rgi} = K_{rgi}(s). \qquad (8.177)$$

The correlations recommended for packed beds of spheres are given in Table 8.5. A further simplification is generally made by assuming that

$$K_{r\ell i} = K_{r\ell}, \qquad (8.178)$$

$$K_{rgi} = K_{rg}. \qquad (8.179)$$

This assumes that the phase distributions are *not* influenced by the dominance of the inertial force, an assumption that is *not* expected to hold, especially for the *coarse* particle packed beds. As was mentioned, no direct simulation of the two-phase flows is presently available.

8.7 Liquid-Gas Interfacial Drag

We assume that the interfacial area $A_{\ell g}$ and the other properties of the phase distributions are determined by the local saturation. When $\mu_\ell \neq \mu_g$, the interfacial drag on $A_{\ell g}$ can become significant when $\langle u \rangle^g \neq \langle u \rangle^\ell$. For the Stokes flow in each phase, an interfacial drag similar to the Darcean resistance is expected. This drag force will be proportional to $|\langle u \rangle^g - \langle u \rangle^\ell|$, i.e., the absolute value of the *relative velocity* and the interfacial area

TABLE 8.5 CORRELATIONS FOR MICROSCOPIC INERTIAL COEFFICIENTS

CONSTRAINTS	CORRELATION
(a) Packed beds made of large spheres, air-water flow, no net liquid flow	$K_{rgi} = K_{rg} = \left(\dfrac{1-s}{0.83}\right)^3, 0.17 \leq s \leq 1$ $K_{rgi} = K_{rg} = 1, s \leq 0.17$
(b) Cocurrent trickle flow in packed beds	$K_{r\ell i} = K_{r\ell}, \quad K_{rgi} = K_{rg}$ where $K_{r\ell}$ and K_{rg} are given in Table 8.4
(c) Packed beds made of large particles	$K_{r\ell i} = s^5, \quad 0 \leq s \leq 0.7$ $K_{rgi} = (1-s)^6, \quad 0 \leq s \leq 0.7$ $K_{rgi} = 0.1(1-s)^4, \quad 0.7 \leq s \leq 1$
(d) Cocurrent and countercurrent flow in packed beds	$K_{r\ell i} = K_{r\ell} = s^3$ Bubbly and slug flow, $0.6 \leq s \leq 1$: $K_{rgi} = \left[\dfrac{1-\varepsilon}{1-\varepsilon(1-s)}\right]^{2/3}(1-s)^3$ Annular flow, $0 \leq s \leq 0.26$: $K_{rgi} = \left[\dfrac{1-\varepsilon}{1-\varepsilon(1-s)}\right]^{2/3}(1-s)^2$ Transition flow, $0.26 \leq s \leq 0.6$: K_{rgi} is a smooth function with these two asymptotes

(a) Tutu et al. (1983); (b) Saez and Carbonell (1985);
(c) Schulenberg and Müller (1987); (d) Tung and Dhir (1988).

$A_{\ell g}$. As the phase velocities increase, the inertial effects become more important and the drag force will also have $(\langle u \rangle^g - \langle u \rangle^\ell)^2$ dependence.

The estimation of the interfacial area $A_{\ell g}$ (or the *specific interfacial area* $A_{\ell g}/V$) for two-phase flows in plain media has been reviewed by Kocamustafaogullari and Ishii (1983). Their focus is on the more analytically amenable *dispersed (bubble or droplet) flow regimes*. The extension of their findings to the two-phase flow in porous media is *not* expected to be very fruitful because the bubbly regime makes up a small region of the practical interest and the *funicular* flow does not occur in relatively small particle size practical beds. However, in packed beds made of large monosized particles, where the interfacial drag is expected to be significant, an approach based on the *drag theory* may be useful. By assuming the shape of the bubble or droplet (generally spheres or thin ellipsoids) and the phase saturations, the most difficult variable to be determined is the *number density*. This is found empirically based on algebraic or differential models. In determining the drag coefficients, interactions among the elements of the dispersed phase are generally neglected, i.e., isolated elements in infinite domains are assumed. The application of the drag theory to two-phase flows in high-permeability media is limited to the simple analysis of Tung and Dhir.

Tutu et al., Schulenberg and Müller, and Tung and Dhir have obtained correlations for the interfacial drag for packed beds of large spheres. The correlations are given in Table 8.6. The experimental results of Tutu et al. show that the drag force is independent of the gas flow rate (in their experiments the net liquid velocity is zero, i.e., gas is flown into a bed containing a liquid) and depends only on the saturation. This has not been confirmed by the other studies. In Table 8.6, the drag model of Tutu et al., based on a drag coefficient $c_D = c_D(s)$, is given. However, as was mentioned, their experimental results actually show that $c_D = c_D(s, |\langle u \rangle^g - \langle u \rangle^\ell|)$. The correlations in Table 8.6 are of the form given in (8.134) and (8.135), but *isotropy* is assumed, i.e.,

$$K_{g\ell 1}|\langle u \rangle^g - \langle u \rangle^\ell|\frac{\langle u \rangle^\ell}{|\langle u \rangle^\ell|} + K_{g\ell 2}(\langle u \rangle^g - \langle u \rangle^\ell)^2 \frac{\langle u \rangle^\ell}{|\langle u \rangle^\ell|}$$

$$= \frac{K_{g\ell 1}}{\varepsilon}\left(\frac{\langle u_g \rangle}{1-s} - \frac{\langle u_\ell \rangle}{s}\right)\frac{\langle u \rangle^\ell}{|\langle u \rangle^\ell|} + \frac{K_{g\ell 2}}{\varepsilon^2}\left(\frac{\langle u_g \rangle}{1-s} - \frac{\langle u_\ell \rangle}{s}\right)^2 \frac{\langle u \rangle^\ell}{|\langle u \rangle^\ell|}. \quad (8.180)$$

A similar equation is written for the ℓ phase.

Caution should be taken in using any of the correlations given in Table 8.6, as these empirical relations are derived based on flows that are almost always in the inertial regime, and therefore, the viscous and inertial parts of the solid-fluid and fluid-fluid interfacial drags have *not been evaluated separately*.

8.8 Immiscible Displacement

The immiscible displacement refers to a *transient* two-phase flow where no mass transfer occurs across $A_{\ell g}$ (i.e., no mass diffusion or phase change) and where the saturation of one of the phases *reduces* from a high value to a low value during this process. In particular, it refers to the displacements where a *front* develops with the phase being *displaced* having a large saturation downstream of this front and the displacing phase having a large saturation behind this front. The pore-level physics of the dynamic displacement

TABLE 8.6 CORRELATIONS FOR LIQUID-FLUID INTERFACIAL DRAG COEFFICIENTS

CONSTRAINTS	CORRELATIONS
(a) No net liquid flow, packed bed of *large* spheres, bubbly flow, $s \to 1$	$K_{g\ell 1} = 0, \quad K_{g\ell 2} = \dfrac{3\rho_\ell C_D}{4d\varepsilon^2(1-s)},$ where $C_D = C_D(s)$
(b) Packed bed of *large* spheres, glass (ethanol water solutions) air, large $\langle u \rangle^\ell$, $\langle u \rangle^g$, and $s > 0.5$	$K_{g\ell 1} = 0,$ $K_{g\ell 2} = -\dfrac{\rho_\ell(\rho_\ell - \rho_g)gK\varepsilon^2}{s\sigma}\dfrac{C_E}{K^{1/2}}W(s),$ $K_{\ell g1} = 0, \quad K_{g\ell 2} = -K_{\ell g2}\dfrac{s}{(1-s)}\dfrac{\rho_\ell}{\rho_g},$ for $|\langle u \rangle^g| > |\langle u \rangle^\ell|,$ where $\dfrac{C_E}{K^{1/2}} = \dfrac{1.75(1-\varepsilon)}{d\varepsilon^3}, \quad W(s) = 350s^7(1-s)$
(c) Packed bed of *large* spheres, for $s < 0.7$, see reference	$1 \le s \le 0.7$ $K_{\ell g 1} = -\dfrac{a_1 \nu_\ell \varepsilon(\rho_\ell - \rho_g)}{d_b^2\left(1 - \dfrac{\rho_g}{\rho_\ell}\right)},$ $K_{\ell g 2} = -\dfrac{a_2\left[s + \dfrac{\rho_g}{\rho_\ell}(1-s)\right]s\varepsilon(\rho_\ell - \rho_g)}{d_b\left(1 - \dfrac{\rho_g}{\rho_\ell}\right)},$ where $d_b = 1.35\left[\dfrac{\sigma}{g(\rho_\ell - \rho_g)}\right]^{1/2}$ and $a_1 = a_1(s), \quad a_2 = a_2(s)$ $K_{g\ell 1} = -K_{\ell g1}\dfrac{s}{1-s}, \quad K_{g\ell 2} = -K_{\ell g2}\dfrac{s}{1-s}$

(a) Tutu et al. (1983); (b) Schulenberg and Müller (1987); (c) Tung and Dhir (1988).

of one phase (wetting or nonwetting) by another (wetting or nonwetting), is rather complicated. These are similar to the dynamic aspects of the drainage (displaced fluid is wetting) and the imbibition (displaced phase is nonwetting), except in the drainage and imbibition the change in local saturation is slow enough such that a quasi-static state exists. The *factors* that influence the phase distributions during the invasion are given here.

- *Matrix structure*: Pore *connectivity* and *topology* are significant. The degree of the randomness within the matrix determines the complexity of the sequential *branching* during the invasion.

- *Relative permeabilities*: Generally both phases are present and their mobilities depend on the *local saturation*.

- *Viscosity ratio and viscosity of displacing fluid*: When the viscosity of the invading phase is larger, the front is more stable. Since both phases move, the relative viscosities (and the relative permeabilities) determine the relative motion.

- *Density ratio and density of displaced and displacing fluids*: The buoyancy and inertial forces influence the stability and topology of the interface.

- *Surface tension*: As the surface tension increases and when the invading phase is wetting, it spreads further into the bed.

- *Wettability of displaced fluid*: When the invading phase is nonwetting, it flows in channels (or *fingers*) with widths of the order of the pore size and independent of the flow rate. For the invasion of the wetting phase, this width is flow-rate- and surface-tension-dependent.

- *Velocity of the displacing fluid*: At lower flow rates, the invading wetting phase *spreads* over many pores, while with the increasing flow rate this width *decreases*.

- *Interfacial mass transfer (phase change)*: *Condensation* of the invading gaseous phase at the front tends to *stabilize* the front.

- *Length scales d and ℓ*: Significant variations in saturation are found over distances of the order of d. This makes the use of the local volume-averaged equations *questionable* in describing the saturation distribution. The noncontinuum approach is through analyses of *fingers* (a *three-dimensional moving interface*) or through the *percolation* analyses. The fluid mechanics of the displacement, which addresses the pore-level interfacial tracking and the attempts to include most of the significant factors, is presently being investigated. The existing more heuristic one-dimensional models, which are of practical use, have not been rigorously examined. In the following, both of these are briefly reviewed.

8.8.1 INTERFACIAL INSTABILITIES

Given the preceding factors, the following three groups have been identified as dominant in determining the type of displacement that occurs.

- *Mobility ratio,* $m = (\mu_g K_{r\ell})/(\mu_\ell K_{rg})$. For $m > 1$ the mobility is said to be *favorable* for *stable invasion*; for $m < 1$, it is said to be *unfavorable*.

- *Capillary number,* $Ca = \langle u \rangle^i \mu_i / \sigma$, where subscript i stands for invading phase.

- *Wettability of the invading phase.*

Figure 8.13 is a classification given by Lenormand and Zarcone (1985) for the case where a *wetting* fluid is displaced by a nonwetting fluid. The *diffusion-limited aggregation* is a diffusion-like fingering where, unlike the *invasion percolation*, the capillarity is not significant. The transition from the invasion percolation to the diffusion limited aggregation takes place near $Ca = 10^{-7}$ and the transition between the invasion percolation and *continuum* description takes place near $Ca = 10^{-2}$. This should be considered as a *rough classification* due to the limited observations on which it is based. Reviews of the phase displacement dynamics are given by Payatakes (1982), Homsy (1987), and Adler and Brenner (1988). We begin with the experimental observations of Stokes et al. (1986), where the case of unfavorable mobility ratio is considered. In order to visualize the displacement, their packed bed of spheres is *thin* and in some cases only several particle diameters thick. They use oil-water (solutions) glass beads where the injection and withdrawal *sites* are uniformly distributed along the injection and withdrawal walls. The effect of buoyancy is *not* examined. They note that the *invasion* by the *wetting fluid* (imbibition) results in *fingers* with widths that *decrease* as $\langle u \rangle^\ell$ increases. They find that this width w correlates with

ℓ-phase invading: $\quad w/(KCa)^{1/2} = a_1$, for a given μ_ℓ/μ_g,

where a_1 is a constant, $Ca = \langle u \rangle^\ell \mu/\sigma$, and μ is the *larger* of the two viscosities. As $\langle u \rangle^\ell$ decreases, the finger nearly *covers* the entire cross section. The *drainage* is different in that the *invading nonwetting phase* appears as *fingers* with widths of the order of the particle size, i.e.,

g-phase invading: $\quad w/d \simeq a_2$, independent of $Ca = \langle u \rangle^g \mu/\sigma$.

Stability of the fingers has been studied by Scheidegger (1969), and the pore-scale fingering in porous media for the *nonwetting* phase invading has been examined by Chen and Wilkinson (1985). The latter uses a Monte Carlo simulation (square lattice of connected tubes with randomly chosen

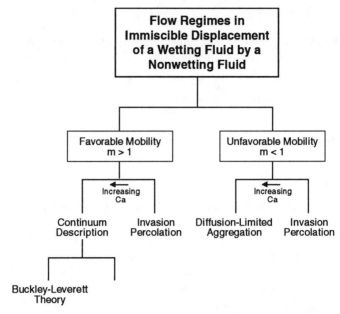

Figure 8.13 A classification of the flow regimes in invasion by the nonwetting phase, according to the limited observations reported by Lenormand and Zarcone.

radii) and performs experiments on etched glass networks. They find out that for narrow size distributions of tubes, the fingers form an *ordered* pattern, while for a wide size distribution the structure of this pattern is *chaotic*. In steam injection into wetted media, the condensation of the steam and the resulting *volume change* tend to *stabilize* the advancing front (Miller, 1975). The specific problem of steam injection will be discussed in Chapter 12 as part of the two-phase flow with phase change.

For simplicity, in the analysis and experiments (e.g., for visualization), the Darcean two-phase flow in porous media is sometimes compared with that of the Stokes flow in a *thin fluid layer* confined between two parallel *plates* (a problem studied by Hele-Shaw in 1898). This analog is called the *Hele-Shaw model* and was originally introduced as an analog by Saffman and Taylor (1958). However, this model does not include the significant features such as the *wettability* and *small* radius of curvature associated with the presence of both of the phases in a pore. Also, in two-phase flow through media, the saturation does *not* undergo discontinuity. Therefore, the interface in the two-phase flow in a Hele-Shaw cell is *never* realized in porous media. A review of the Hele-Shaw flows is given by Homsy (1987). The Hele-Shaw theory and experiments describe the two-dimensional features of the flow (averaged over the small layer thickness). For single-phase flows, the two-dimensional momentum equation for small layer thickness b

$(b/L \to 0)$ and for $Re = uL/\nu \to 0$ is given by

$$\nabla \overline{p} = -\frac{12\mu}{b^2}\overline{\mathbf{u}} + \rho \mathbf{g}, \tag{8.181}$$

where $12/b^2$ is the *equivalent permeability*, L is the characteristic length of the system, and the bar stands for the averaging carried over the thickness. The two-phase flow momentum equations and the interfacial force balance for the Hele-Shaw are discussed by Homsy, along with some solutions.

One method for the analysis of the two-phase flows in porous media is based on the *percolation concepts* and a review is given by Alder and Brenner. A porous medium is viewed as consisting of a statistically distributed network of pores. Generally *only* one phase is allowed in a pore, and *only* one phase is continuous. Some of the features are similar to those discussed in connection with the modeling of the capillary pressure and relative permeabilities.

8.8.2 BUCKLEY-LEVERETT FRONT

Assuming that the continuum treatment is applicable and that a stable advancing front exists, Buckley and Leverett (1942) *analytically* treated the one-dimensional quasi-steady two-phase flow problem. Reviews of their work and the extensions of it are given by Morel-Seytoux (1969), Dullien (1992), Wooding and Morel-Seytoux (1976), and Homsy (1987). The inclusion of *gravity* and *capillarity* is also addressed by Okusu and Udell (1989).

The one-dimensional momentum equations for each phase in the *absence* of significant local acceleration, inertial ℓ-g interfacial drag, and surface tension gradient are the simplified form of (8.134) and (8.135) and are given by

$$\langle u_\ell \rangle = -\frac{KK_{r\ell}}{\mu_\ell}\left(\frac{\partial p_\ell}{\partial x} + \rho_\ell g \sin\theta\right), \tag{8.182}$$

$$\langle u_g \rangle = -\frac{KK_{rg}}{\mu_g}\left(\frac{\partial p_g}{\partial x} + \rho_\ell g \sin\theta\right), \tag{8.183}$$

where θ is measured from the horizontal plane with $\theta = \pi/2$ corresponding to positive velocities pointing *against* the gravity vector.

The continuity equations (8.94)-(8.95) can be written in terms of the local *saturation* as

$$\varepsilon\frac{\partial s}{\partial t} + \frac{\partial \langle u_\ell \rangle}{\partial x} = 0, \tag{8.184}$$

$$-\varepsilon\frac{\partial s}{\partial t} + \frac{\partial \langle u_g \rangle}{\partial x} = 0, \tag{8.185}$$

where we have used $s = \varepsilon_\ell/\varepsilon$ and $s_g = \varepsilon_g/\varepsilon = 1 - s$. Combining (8.184) and (8.185) we have

$$\frac{\partial(\langle u_\ell\rangle + \langle u_g\rangle)}{\partial x} = \frac{\partial \langle u_t\rangle}{\partial x} = 0, \qquad (8.186)$$

where $u_t = \langle u_\ell\rangle + \langle u_g\rangle$ is the *total local superficial* (Darcy) velocity.

The *fractions of the total volumetric flow rate* are defined as $f_\ell = \langle u_\ell\rangle/\langle u_t\rangle$ and $f_g = \langle u_g\rangle/\langle u_t\rangle = 1 - f_\ell$. By subtracting the ℓ-momentum equation from the g-momentum equation and by *introducing* the capillary pressure $p_c = p_g - p_\ell$ and the density difference $\rho_\ell - \rho_g$, we have

$$f_g = \frac{1 - \dfrac{KK_{rg}}{\mu_\ell \langle u_t\rangle}\left[\dfrac{\partial p_c}{\partial x} - (\rho_\ell - \rho_g)g\sin\theta\right]}{1 + \dfrac{\mu_g}{\mu_\ell}\dfrac{K_{r\ell}}{K_{rg}}}$$

$$= \frac{1 - \dfrac{K_{rg}(s)}{Ca}\left[\dfrac{\partial J(s)}{\partial x^*} - Bo\right]}{1 + m}, \qquad (8.187)$$

where $p_c = \sigma J(s)/(K/\varepsilon)^{1/2}$ is assumed, $x^* = x/(K\varepsilon)^{1/2}$, $Ca = \langle u_t\rangle\mu_\ell/\sigma$, and $Bo = (\rho_\ell - \rho_g)gK\sin\theta/\sigma$.

Since $\langle u_\ell\rangle = f_\ell\langle u_t\rangle$ and $\partial\langle u_t\rangle/\partial x = 0$, the continuity for the ℓ-phase (8.184), can be written as

$$\varepsilon\frac{\partial s}{\partial t} + \langle u_t\rangle\frac{\partial f_\ell}{\partial x} = 0. \qquad (8.188)$$

Now, assuming that $f_\ell = f_\ell(s)$, which requires that $K_{r\ell} = K_{r\ell}(s)$ and $K_{rg} = K_{rg}(s)$ and that $s = s(x,t)$, we have

$$\frac{\partial f_\ell}{\partial x} = \frac{df_\ell}{ds}\frac{\partial s}{\partial x}. \qquad (8.189)$$

Then

$$\varepsilon\frac{\partial s}{\partial t} + \langle u_t\rangle\frac{df_\ell}{ds}\frac{\partial s}{\partial x} = 0, \qquad (8.190)$$

which is the *Buckley-Leverett equation*. On the constant s lines we have

$$\frac{\partial s}{\partial t}dt + \frac{\partial s}{\partial x}dx = 0, \quad \text{iso-}s\text{ lines}. \qquad (8.191)$$

Then the Buckley-Leverett equation becomes

$$\left.\frac{dx}{dt}\right|_s = u_F = \frac{\langle u_t\rangle}{\varepsilon}\frac{df_\ell}{ds}, \qquad (8.192)$$

where u_F is the velocity of the *point* (or *front*) having saturation s. In terms of the location, we have

$$x(s) = x_o + \frac{1}{\varepsilon}\frac{df_\ell}{ds}\int_0^t \langle u_t\rangle\, dt. \qquad (8.193)$$

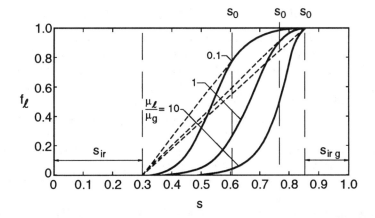

Figure 8.14 Variation of the wetting-phase fraction of the volumetric flow with respect to saturation, for three different viscosity ratios. s_o is the point of tangent to the $f_\ell(s)$ curve. (From Wooding and Morel-Seytoux, reproduced by permission ©1976 Ann. Rev. Fluid Mech.)

Since u_F is proportional to df_ℓ/ds, we need to examine how df_ℓ/ds changes with s.

Figure 8.14 is adapted from Wooding and Morel-Seytoux (1976) and gives the variation of $f_\ell(s)$ for $J(s) = 0$, $Bo = 0$, and for typical $K_{r\ell}(s)$, $K_{rg}(s)$, such as those given in Table 8.4. The results are given for several viscosity ratios. The *wetting phase* is the displacing fluid, and, therefore, the *imbibition* branch of $K_{r\ell}$ and K_{rg} should be used. Note that it is also assumed that the flow is stable one-dimensional and quasi-steady and that the local volume-averaging treatment is justifiable. This treatment is assumed to hold even if $\mu_\ell/\mu_g < 1$. As will be shown, the *front* will only be *stable* for saturation larger than a *critical value*.

Now, since in a stable flow, s decreases *monotonically* with x ($x = 0$ is the location of injection of the ℓ phase), then we expect u_s to decrease as s increases such that the higher saturations are always behind the lower saturations. For this to occur, we need $d^2 f_\ell/ds^2 < 0$, so that through (8.192) we have a *monotonic* decrease in u_s (and df_ℓ/ds) with respect to s. Therefore, in Figure 8.14 for $s_o \leq s \leq s_{ir}$ where $d^2 f_\ell/ds^2 > 0$, no stable flow exists. Only for $1 - s_{ir\, g} \leq s \leq s_o$ are stable flows encountered. The *front* of the displacement will have a *front saturation* s_o and a *front velocity* u_{F_o}, and behind it, s increases up to the value $1 - s_{ir\, g}$ at the injection point.

Extension of these results to cases where $J(s) \neq 0$ and $Bo \neq 0$ has been done by Okusu and Udell, among others. They examine the effect of Ca and Bo on the speed of the front u_{F_o} and on the saturation distribution in the *variable velocity zone*, $1 - s_{ir\, g} \leq s \leq s_o$.

8.8.3 STABILITY OF BUCKLEY-LEVERETT FRONT

The *stability* of the planar saturation front has been studied, among others, by Yortsos and Huang (1984), Jerauld et al. (1984a,b), Chikhliwala and Yortsos (1985), and Chikhliwala et al. (1988). Here we consider only the *marginal states* as determined by the application of the *linear stability theory* and corresponding to *zero growth rate* for the introduced disturbances. The following are the results obtained by Yortsos and Huang and their treatment of the *jump in saturation* across the front. They also consider the stabilizing effect of the capillarity, but we consider only the jump problem discussed in the last section (Figure 8.14).

Behind the front (upstream), the saturation is $s_{-\infty}$; downstream the saturation is s_∞. The front saturation s_o and velocity u_{F_o} are determined as outlined earlier, once the fluids and matrix are identified. We now consider the stability of this flow. For no gravity and capillarity effects, the momentum equations (8.182) and (8.183) are simplified ($p_c = 0$, $p_g = p_\ell = p$) and (8.184)–(8.186) give the mass conservation for the phases. The *base flow* is the front just discussed in Section 8.8.2, and the *perturbed field* is found by introduction of perturbations of the type

$$s = \bar{s} + s', \quad p = \bar{p} + p', \tag{8.194}$$

$$s' = A_s(x_1)e^{i(a_y y + a_z z) + \beta t}, \quad p' = A_p(x_1)e^{i(a_y y + a_z z) + \beta t}, \tag{8.195}$$

where $x_1 = x - u_{s_o}t$ is the *moving coordinate axis* and the *wave numbers* a_y and a_z are related to *wavelengths* λ_y and λ_z through

$$a_y = \frac{2\pi}{\lambda_y}, \quad a_z = \frac{2\pi}{\lambda_z}. \tag{8.196}$$

The jump conditions across the front and the *characteristic equation* that relates the *growth rate* β to the wave numbers a, $a^2 = a_y^2 + a_z^2$, and the system variables is given by Yortsos and Huang. Before stating the characteristic equation, we introduce the following variables, which are also used in the equation.

The *total mobility* is defined as

$$m_t = m_1 \frac{K_{r\ell}(s)}{K_{r\ell}(s_{-\infty})} + \frac{K_{rg}(s)}{K_{rg}(s_\infty)}, \tag{8.197}$$

with m_1 as the ratio of the *maximum mobility* of the two phases, i.e.,

$$m_1 = \frac{\mu_g K_{r\ell}(s_{-\infty})}{\mu_\ell K_{rg}(s_\infty)}. \tag{8.198}$$

The wetting phase and fractional volumetric flow rates at *far downstream* and *far upstream* locations are given as

$$f_{\ell,\infty} = \cfrac{1}{1 + \cfrac{K_{r\ell}^2(s_{-\infty})\mu_g}{K_{rg}(s_\infty)K_{r\ell}(s_\infty)\mu_\ell}} = \cfrac{1}{1 + m_1\cfrac{K_{r\ell}(s_{-\infty})}{K_{r\ell}(s_\infty)}}, \qquad (8.199)$$

$$f_{\ell,-\infty} = \cfrac{1}{1 + \cfrac{K_{r\ell}(s_{-\infty})K_{rg}(s_{-\infty})\mu_g}{K_{rg}^2(s_\infty)\mu_\ell}} = \cfrac{1}{1 + m_1\cfrac{K_{rg}(s_{-\infty})}{K_{rg}(s_\infty)}}. \qquad (8.200)$$

Now the characteristic equation is

$$\beta^* = a^* \frac{(f_{\ell,-\infty} - f_{\ell,\infty})(m_{t,-\infty} - m_{t,\infty})}{m_{t,-\infty} + m_{t,\infty}}, \qquad (8.201)$$

where β^* and a^* are the *dimensionless* growth rate and wave number, respectively. The *marginal states* (zero growth) are found from the preceding by setting $\beta^* = 0$. Now we have the *condition* for the *marginal stability* as

$$(m_{t,\infty})_c = (m_{t,-\infty})_c, \quad \text{for } Bo = 0, \ J(s) = 0, \qquad (8.202)$$

where subscript c stands for the *critical* value. This can be written as

$$m_{1c}\frac{K_{r\ell}(s_\infty)}{K_{r\ell}(s_{-\infty})} + 1 = m_{1c} + \frac{K_{rg}(s_{-\infty})}{K_{rg}(s_\infty)}. \qquad (8.203)$$

For the Buckley-Leverett front, we have an *immobile* (for displacing phase) saturation far downstream, i.e., $s_\infty = s_{\text{ir}}$, therefore, $k_{r\ell}(s_\infty) = 0$, and from (8.199), we have $f_{\ell,\infty} = 0$. For this case we have the *marginal stability criterion* given by

$$m_{1c} = 1 - \frac{K_{rg}(s_{-\infty})}{K_{rg}(s_\infty)}. \qquad (8.204)$$

For $m_1 > m_{1c}$, the *front is stable*. When $K_{rg}(s_{-\infty}) = 0$, the *critical mobility* will be unity. This corresponds to a *single* wetting phase displacing a single nonwetting phase (*Saffman-Taylor stability*). We will examine this problem again in Section 12.5.1 (C), where the saturation front discussed here will also be a *condensation* front. We will also discuss the *effect* of the capillarity and gravity on the stability of the front.

8.9 Fluid-Solid Two-Phase Flow

As in applications such as infiltration, in granular precipitations resulting from fluid phase reaction, and in ablation of the solid matrix, the two-phase flow of *fluid-particulates* in porous media also has been studied. The filtration treatments for both *aerosols* (particles in gas phase) and *hydrosols* (particles in liquid phase) has been reviewed by Tien (1989). The particles of order of nanometer suspended or settled in gas are called aerogels (discussed in Section 3.4.2). The particles travelling through a porous medium can be *deposited* on the solid matrix. The *mechanisms* for deposition are *inertial impaction* (which is the major deposition mechanism for aerosols with particles larger than 1 μm), *interception* (a geometric phenomena), *sedimentation* (due to density difference between fluid and particles), *electrostatic forces* (for aerosols), *Brownian diffusion* (for submicron particles), and *sieving* (for particles larger than the pore-throat dimension).

The motion and deposition of particles in a two-dimension, periodic arrangement of cylinders is discussed by Marshall et al. (1994).

8.10 References

Adler, P. M. and Brenner, H., 1988, "Multiphase Flow in Porous Media," *Ann. Rev. Fluid Mech.*, 20, 35–59.
Bear, J. and Bensabat, J., 1989, "Advective Fluxes in Multiphase Porous Media Under Non-Isothermal Conditions," *Transp. Porous Media*, 4, 423–448.
Bretherton, F. P., 1961, "The Motion of Long Bubbles in Tubes," *J. Fluid Mech.*, 10, 160–188.
Buckley, S. E. and Leverett, M. C., 1942, "Mechanism of Fluid Displacement in Sands," *Trans. AIME*, 146, 107–116.
Burelbach, J. P., Bankoff, S. G., and Davis, S. H., 1988, "Non-Linear Stability of Evaporating/Condensing Liquid Films," *J. Fluid Mech.* 195, 463–494.
Chappuis, J., 1982, "Contact Angle," in *Multiphase Science and Technology*, 1, 387–505.
Chen, J.-D. and Wilkinson, D., 1985, "Pore-Scale Viscous Fingering in Porous Media," *Phys. Rev. Letters*, 55, 1892–1895.
Chikhliwala, E. D., Huang, A. B., and Yortsos, Y. C., 1988, "Numerical Study of the Linear Stability of Immiscible Displacement in Porous Media," *Transp. Porous Media*, 3, 257–276.
Chikhliwala, E. D. and Yortsos, Y. C., 1985, "Theoretical Investigations of Finger Growth by Linear and Weakly Nonlinear Stability Analysis," SPE 14367, Las Vegas, Nevada.
Corey, A. T., 1986, *Mechanics of Immiscible Fluids in Porous Media*, Water Resources Publications.
Cox, R. G., 1986, "The Dynamic of the Spreading of Liquids on a Solid Surface Part 1. Viscous Flow, Part 2. Surfactants," *J. Fluid Mech.* 168, 169–200.
de Gennes, P. G., 1985, "Wetting: Statics and Dynamics," *Rev. Mod. Phys.*, 57, 827–863.
de Gennes, P. G., Hua, X., and Levinson, P., 1990, "Dynamic of Wetting: Local Contact Angles," *J. Fluid Mech.*, 212, 55–63.
Defay, R. and Prigogine, I., 1966, *Surface Tension and Adsorption* (English edition), J. Wiley.
Delshad, M. and Pope, G. A., 1989, "Comparison of the Three-Phase Oil Relative Permeability Models," *Transp. Porous Media*, 4, 59–83.

Demond, A. H., Desai, F. N., and Hayes, K. F., 1994, "Effect of Cationic Surfactants on Organic Liquid-Water Capillary Pressure-Saturation Relationships," *Wat. Resourc. Res.*, 30, 333–342.

Derjaguin, B. V. and Churaev, N. V., 1978, "On the Question of Determining the Concept of Disjoining Pressure and Its Role in Equilibrium and Flow of Thin Films," *J. Colloid Interface Sci.*, 66, 389–398.

Dhir, V. K., 1986, "Some Aspects of Two-Phase Flow Through Porous Media," *Nuc. Engng. Design*, 95, 275–283.

Dullien, F. A. L., 1979, *Porous Media: Fluid Transport and Pore Structure*, Academic.

Dullien, F. A. L., Lai, F. S. T., and Macdonald, I. F., 1986, "Hydraulic Continuity of Residual Wetting Phase in Porous Media," *J. Colloid Interface Sci.*, 109, 201–218.

Dullien, F. A. L., 1992, *Porous Media: Fluid Transport and Pore Structure*, Second Edition, Academic.

Dullien, F. A. L., Zarcone, C., Macdonald, I. F., Collins, A., and Bochard, R. D. E., 1989, "The Effects of Surface Roughness on Capillary Pressure Curves and the Heights of Capillary Rise in Glass Bead Packs," *J. Colloid Interface Sci.*, 127, 362–372.

Dussan, E. B., 1979, "On the Spreading of Liquids on Solid Surfaces: Static and Dynamic Contact Lines," *Ann. Rev. Fluid Mech.*, 11, 371–400.

Dzyaloskinskii, I. E., Lifshitz, E. M., and Pitaevskii, L. P., 1961, "The General Theory of van der Waals Forces," *Adv. Phys.*, 165–209.

Erle, M. A., Dyson, D. C., and Morrow, N. R., 1971, "Liquid Bridge Between Cylinders, in a Torus, and Between Spheres," *AIChE J.*, 17, 115–121.

Gray, W. G., 1983, "General Conservation Equations for Multi-Phase Systems: 4. Constitutive Theory Including Phase Change," *Adv. Water Resour.*, 6, 130–140.

Gray, W. G. and Hassanizadeh, S. M., 1991, "Unsaturated Flow Theory Including Interfacial Phenomena," *Wat. Resourc. Res.*, 27, 1855–1863.

Grosser, K., Carbonell, R. G., and Sundaresan, S., 1988, "Onset of Pulsing in Two-Phase Co-current Down Flow Through a Packed Bed," *AIChE J.*, 34, 1850–1860.

Hassanizadeh, S. M. and Gray, W. G., 1990, "Mechanisms and Thermodynamics of Multiphase Flow in Porous Media Including Interface Boundaries," *Adv. Water Resources*, 13, 169–186.

Hassanizadeh, S. M. and Gray, W. G., 1993, "Thermodynamic Basis of Capillary Pressure in Porous Media," *Wat. Resourc. Res.*, 29, 3389–3405.

Hocking, L. M., 1977, "A Moving Fluid interface Part 2. The Removal of the Trace Singularity by a Slip Flow," *J. Fluid Mech.* 79, 209–229.

Hocking, L. M. and Rivers, A. D., 1982, "Spreading of a Drop by Capillary Action," *J. Fluid Mech.* 121, 425–442.

Homsy, G. M., 1987, "Viscous Fingering in Porous Media," in *Ann. Rev. Fluid Mech.*, 19, 271–311.

Horn, R. G. and Israelachvili, J. N., 1981, "Direct Measurements of Structural Forces Between Two Surfaces in Non-Polar Liquid," *J. Chem. Phys.*, 75, 1400–1411.

Israelachvili, J. N., 1989, *Intermolecular and Surface Forces*, Academic.

Jerauld, G. R., Davis, H. T., and Scriven, L. E., 1984a, "Frontal Structure and Stability in Immiscible Displacement," SPE/DOE 12691, Tulsa, Oklahoma.

Jerauld, G. R., Davis, H. T., and Scriven, L. E., 1984b, "Stability Fronts of Permanent Form in Immiscible Displacement," SPE 13164, presented in Houston, Texas.

Jerauld, G. R. and Salter, S. J., 1990, "The Effect of Pore-Structure on Hysteresis in Relative Permeability and Capillary Pressure: Pore-Level Modeling," *Transp. Porous Media*, 5, 103–151.

Joanny, J. F. and de Gennes, P. G., 1984, "A Model for Contact Angle Hysteresis," *J. Chem. Phys.*, 81, 552–562.

Kaviany, M., 1989, "Forced Convection Heat and Mass Transfer from a Partially Liquid Covered Surface," *Num. Heat Transfer*, 15A, 445–469.

Kocamustafaogullari, G. and Ishii, M., 1983, "Interfacial Area and Nucleation Site Density in Boiling Systems," *Int. J. Heat Mass Transfer*, 26, 1377–1387.

Lefebvre du Prey, E. J., 1973, "Factors Affecting Liquid-Liquid Relative Permeabilities of a Consolidated Porous Media," *Soc. Pet. Eng. J.*, 13, 39–47.

Lenormand, R. and Zarcone, C., 1985, "Two-Phase Flow Experiments in a Two-Dimensional Permeable Medium," *Phys. Chem. Hydro.*, 6, 497–506.

Levec, J., Saez, A. E., and Carbonell, R. G., 1986, "The Hydrodynamics of Trickling Flow in Packed Beds, Part II: Experimental Observations," *AIChE J.*, 32, 369–380.

Leverett, M. C., 1941, "Capillary Behavior in Porous Solids," *Trans. AIME*, 142, 152–169.

Levich, V. G. and Krylov, V. S., 1969, "Surface-Tension-Driven Phenomena," *Ann. Rev. Fluid Mech.*, 293–316.

Marshall, H., Sahraoui, M., and Kaviany, M., 1994, "An Improved Analytic Solution for Analysis of Particle Trajectories in Fibrous, Two-Dimensional Filters," *Phys. Fluids*, 6, 507–520.

Martinez, M. J. and Udell, K. S., 1990, "Axisymmetric Creeping Motion of Drops Through Circular Tubes," *J. Fluid Mech.*, 210, 565–591.

Melrose, J. C., 1965, "Wettability as Related to Capillary Action in Porous Media," *SPE J.*, 5, 259–271.

Miller, C. A., 1975, "Stability of Moving Surfaces in Fluid Systems with Heat and Mass Transfer Transport, III. Stability of Displacement Fronts in Porous Media," *AIChE J.*, 21, 474–479.

Mirzamoghadam, A. and Catton, I., 1988, "A Physical Model of the Evaporating Meniscus," *ASME J. Heat Transfer*, 110, 201–207.

Morel-Seytoux, H. J., 1969, "Introduction to Flow of Immiscible Liquids in Porous Media," in *Flow Through Porous Media*, Academic, 455–516.

Morgan, J. T. and Gordon, D. T., 1970, "Influence of Pore Geometry on Water-Oil Relative Permeability" *J. Pet. Tech.*, 22, 1199–1208.

Morrow, N., 1970, "Irreducible Wetting-Phase Saturations in Porous Media," *Chem. Engng. Sci.*, 25, 1799–1815.

Morrow, N. R. and Nguyen, M. D., 1982, "Effect of Interface Velocity on Dynamic Contact Angle at Rough Surface," *J. Colloid Interface Sci.*, 89, 523–531.

Mualem, D., 1976, "A New Model for Predicting the Hydraulic Conductivity of Saturated Porous Media," *Water Resour. Res.*, 12, 513–522.

Muskat, M. and Meres, A. W., 1936, "The Flow of Heterogeneous Fluid Through Porous Media," *Physics*, 7, 346–363.

Myshkis, A. D., Babskii, V. G., Kopachevskii, N. D., Slobozhanin, L. A., and Tyuptsov, A. D., 1987, *Low-Gravity Fluid Mech.: Mathematical Theory of Capillary Phenomena*, Springer-Verlag.

Naik, A. S. and Dhir, V. K., 1982, "Forced Flow Evaporation Cooling of a Volumetrically Heated Porous Layer," *Int. J. Heat Mass Transfer*, 25, 541–552.

Ng, K. M., 1986, "A Model for Flow Regime Transitions in Co-current Down-Flow Trickle-Bed Reactors," *AIChE J.*, 32, 115–122.

Ng, K. M., Davis, H. T., and Scriven, L. E., 1978, "Visualization of Blob Mechanics in Flow Through Porous Media," *Chem. Engng. Sci.*, 33, 1009–1017.

Ngan, C. G. and Dussan, E. B., 1989, "On the Dynamics of Liquid Spreading on Solid Surfaces," *J. Fluid Mech.* 209, 191–226.

Okusu, N. M. and Udell, K. S., 1989, "Immiscible Displacement in Porous Media Including Gravity and Capillary Forces," in *Multiphase Transport in Porous Media–1989*, ASME FED-Vol. 82 (HTD-Vol. 127), 13–21.

Osoba, J. S., Richardson, J. G., Kerver, J. K., Hafford, J. A., and Blair, P. M., 1951, "Laboratory Measures of Relative Permeability," *Trans. Amer. Inst. Min. Eng.*, 192, 47–56.

Pavone, D., 1989, "Explicit Solution for Free-Fall Gravity Drainage Including Capillary Pressure," in *Multiphase Transport in Porous Media–1989*, ASME FED-Vol. 92 (or HTD-Vol. 127), 55–62.

Payatakes, A. C., 1982, "Dynamics of Oil Ganglia During Immiscible Displacement in Water-Wet Porous Media," *Ann. Rev. Fluid Mech.*, 14, 365–393.

Reinelt, D. A., 1987, "The Rate at Which a Long Bubble Rises in a Vertical Tube," *J. Fluid Mech.*, 175, 557–565.

Reinelt, D. A. and Saffman, P. G., 1985, "The Penetration of a Finger into a Viscous Fluid in a Channel and Tube," *SIAM J. Sci. Stat. Comput.*, 6, 542–561.

Rogers, J. and Kaviany, M., 1990, "Variation of Heat and Mass Transfer Coefficients During Drying of Granular Beds," *ASME J. Heat Transfer*, 112, 668–674.

Saez, A. E., 1983, *Hydrodynamics and Lateral Thermal Dispersion for Gas-Liquid Cocurrent Flow in Packed Beds*, Ph.D. thesis, University of California-Davis.

Saez, A. E. and Carbonell, R. G., 1985, "Hydrodynamic Parameters for Gas-Liquid Co-current Flows in Packed Beds," *AIChE J.*, 31, 52–62.

Saez, A. E., Carbonell, R. G., and Levec, J., 1986, "The Hydrodynamics of Trickling Flow in Packed Beds, Part I: Conduit Models," *AIChE J.*, 32, 353–368.

Saffman, P. G. and Taylor, G. I., 1958, "The Penetration of a Fluid into a Porous Medium or Hele Shaw Cell Containing a More Viscous Liquid," *Proc. Roy. Soc.* (London), A245, 312–329.

Scheidegger, A. E., 1969, "Stability Conditions for Displacement Processes in Porous Media," *Can. J. Phys.*, 47, 209–214.

Scheidegger, A. E., 1974, *The Physics of Flow Through Porous Media*, Third Edition, University of Toronto Press.

Schulenberg, T. and Müller, U., 1987, "An Improved Model for Two-Phase Flow Through Beds of Coarse Particles, " *Int. J. Multiphase Flow*, 13, 87–97.

Sherwood, T. K., Shipley, G. H., and Holloway, F. A. L., 1938, "Flooding Velocities in Packed Columns," *Ind. Engng. Chem.*, 30, 765–769.

Slattery, J. C., 1970, "Two-Phase Flow Through Porous Media," *AIChE J.*, 16, 345–352.

Slattery, J. C., 1974, "Interfacial Effects in the Entrapment and Displacement of Residual Oil," *AIChE J.*, 20, 1145–1154.

Slattery, J. C., 1990, *Interfacial Transport Phenomena*, Springer-Verlag.

Soo, G. O. and Slattery, J. C., 1979, "Interfacial Tension Required for Significant Displacement of Residual Oil," *Soc. Pet. Eng. J.*, 83–90.

Stokes, J. P., Weitz, D. A., Golbub, J. P., Dougerty, A., Robbins, M. O., Chaikin, P. M., and Lindsay, H. M., 1986, *Phys. Rev. Lett.*, 57, 1718–1721.

Stubos, A. K. and Buchin, J.-M., 1988, "Modeling of Vapor Channeling Behavior in Liquid Saturated Debris Bed," *ASME J. Heat Transfer*, 110, 968–975.

Tien, C., 1989, *Granular Filtration of Aerosols and Hydrosols*, Butterworths, Boston.

Truong, J. G., 1987, *Ellipsometric and Interferometric Studies of Thin Liquid Films Wetting on Isothermal and Non-Isothermal Solid Surfaces*, Ph.D. thesis, Rensselear Polytechnic Institute.

Truong, J. G. and Wayner, P. C., 1987, "Effect of Capillary and Liquid: Theory and Experiment," *J. Chem. Phys.*, 87, 4180–4188.

Tung, V. X. and Dhir, V. K., 1988, "A Hydrodynamic Model for Two-Phase Flow Through Porous Media," *Int. J. Multiphase Flow*, 14, 47–64.

Tutu, N. K., Ginsberg, T., and Chen, J. C., 1983, "Interfacial Drag for Two-Phase Flow Through High Permeability Porous Beds," in *Interfacial Transport Phenomena*, ASME, 37–44.

Udell, K. S., 1985, "Heat Transfer in Porous Media Considering Phase Change and Capillarity—The Heat Pipe Effect," *Int. J. Heat Mass Transfer*, 28, 485–494.

van Genuchten, M. Th., 1980, "A Closed-Form Equation for Predicting the Hydraulic Conductivity of Unsaturated Soils," *Soil Sci. Soc. Amer. J.*, 44, 892–898.

Verma, A. K., Pruess, K., Tsang, C. F., and Withespoon, P. A., 1984, "A Study of Two-Phase Concurrent Flow of Steam and Water in an Unconsolidated Porous Medium," in *Heat Transfer in Porous Media and Particulate Flows*, ASME HTD-Vol. 46, 135–143.

Wayner, P. C., 1989, "A Dimensionless Number for Contact Line Evaporative Heat Sink," *ASME J. Heat Transfer*, 111, 813–815.

Whitaker, S., 1986a, "Flow in Porous Media I: A Theoretical Derivation of Darcy's Law," *Transp. Porous Media*, 1, 3–25.

Whitaker, S., 1986b, "Flow in Porous Media II: The Governing Equations for Immiscible, Two-Phase Flow," *Transp. Porous Media*, 1, 105–125.

Wooding, R. A. and Mortel-Seytoux, H. J., 1976, "Multi-Phase Fluid Flow Through Porous Media," *Ann. Rev. Fluid Mech.*, 8, 233–274.

Wyllie, M. R. J., 1962, "Relative Permeabilities", in *Petroleum Production Handbook*, McGraw-Hill, Volume 2, Chapter 25, (25-1)–(25-14).

Yortsos, Y. C. and Huang, A. B., 1984, "Linear Stability Analysis of Immiscible Displacement Including Continuously Changing Mobility and Capillary Effects: Part I—Simple Basic Flow Profiles," SPE/DOE 12692, Tulsa, Oklahoma.

Yu, L. and Wardlaw, N. C., 1986, "Mechanisms of Non-Wetting Phase Trapping During Imbibition at Slow Rates," *J. Colloid Interface Sci.*, 109, 473–486.

9

Thermodynamics

In this chapter we examine the thermodynamics of the liquid-gas systems in porous media. The treatment centers around the examination of two effects, namely, the effect of the *liquid-gas interfacial tension* and *curvature* (when the *radius of curvature* becomes very *small*), and the effect of the *solid-fluid interfacial forces* which results in a significant surface adsorption (when *specific interfacial area* becomes very *large* for small pores). First the *classical* treatments, which are centered around the *effect* of the meniscus *curvature* on the *thermodynamic state*, are discussed. Both *single-* and *multicomponent* systems are considered. Then we examine the thermodynamics of *thin liquid-film* extensions of perfectly wetting liquids and the role of the van der Waals forces on the equilibrium state. Next, the *capillary condensation* (adsorption) and evaporation (desorption) in small pores is discussed. The *hysteresis* in the adsorption isotherm, as well as the other features of adsorption and desorption are discussed. The curvature arguments made for the existence of the hysteresis are discussed before the introduction of the more *modern* descriptions based on the *molecular interaction theories*. Then two of these modern theories and their predictions of the *phase change* and the *stability* of the thin liquid films in small pores are discussed. Some thermodynamic aspects of solid-liquid phase change are discussed in Section 12.6.

9.1 Thermodynamics of Single-Component Capillary Systems

In this section we review the classical thermodynamic definitions of the surface tension while treating the interface as a *transitional layer*. We will also arrive at the Gibbs-Duhem relation, which will be used in the following sections. Then we refer to the molecular theory of the fluid interfaces, which is capable of determining the interfacial-layer *thickness*.

9.1.1 WORK OF SURFACE FORMATION

Here we consider simple compressible single-component systems. The system is made of two distinct phases and a thin *interfacial layer* (*transitional layer*) separating them. The *work received* by a system made of the two phases and the *transition* layer with volumes V_ℓ, V_g, and V_t, pressures p_ℓ, p_g, and p_t, and an interface of surface area $A_{\ell g}$ is (Defay and Prigogine, 1966)

$$dW = -p_\ell \, dV_\ell - p_g \, dV_g - p_t \, dV_t + \sigma \, dA_{\ell g}, \tag{9.1}$$

where ℓ and g stand for the liquid and gaseous phases and t stands for the *transition* layer. We can write this work as the sum of the work received by the two *homogeneous* phases and the transition layer, i.e.,

$$dW = dW_f + dW_g + dW_t, \tag{9.2}$$

with

$$dW_t = -p_t \, dV_t + \sigma \, dA_{\ell g}. \tag{9.3}$$

The first term on the right-hand side of (9.3) is the *compressibility work* and the second term is the *work of surface formation*.

The interfacial (or surface) layer has been assigned a pressure and volume, and we now add to its variables. These variables are the *internal energy* U_t, the *entropy* S_t, the *Helmholtz free energy* (or *Helmholtz function*) F_t, and the *Gibbs free energy* G_t. Under *thermal equilibrium* (i.e., $T_\ell = T_g = T_t = T$), we have

$$F_t = U_t - TS_t \quad \text{and} \quad G_t = F_t + p_t V_t - \sigma A_{\ell g}. \tag{9.4}$$

9.1.2 First and Second Laws of Thermodynamics

For the entire system (the two phases and the interfacial layer), the conservation of energy states that

$$dU = dQ + dW, \tag{9.5}$$

or using (9.1), we have

$$dQ = dU + p_\ell \, dV_\ell + p_g \, dV_g + p_t \, dV_t - \sigma \, dA_{\ell g}, \tag{9.6}$$

where the *change* in the entropy for an *infinitesimal* heat addition from the surroundings dQ, is

$$dS = \frac{dQ_{\text{int}}}{T} + \frac{dQ}{T}. \tag{9.7}$$

In this equation dQ_{int} is due to the *irreversible* internal processes and

$$dQ_{\text{int}} \geq 0. \tag{9.8}$$

Using the definition of the Helmholtz free energy, we have

$$F = U - TS, \tag{9.9}$$

or when differentiated,

$$dF = dU - T \, dS - S \, dT. \tag{9.10}$$

9.1 Thermodynamics of Single-Component Capillary Systems

Now, substituting for $T\,dS$ from (9.7) and (9.10), we have

$$dU = T\,dS - p_\ell\,dV_\ell - p_g\,dV_g - p_t\,dV_t + \sigma\,dA_{\ell g} - dQ_{\text{int}} \qquad (9.11)$$

or

$$dF = -S\,dT - p_\ell\,dV_\ell - p_g\,dV_g - p_t\,dV_t + \sigma\,dA_{\ell g} - dQ_{\text{int}}. \qquad (9.12)$$

From the last equation, we *define*

$$\begin{aligned}
\left.\frac{\partial F}{\partial T}\right|_{V_i, A_{\ell g}} &= -S, \\
\left.\frac{\partial F}{\partial V_\ell}\right|_{T, V_i(i \neq \ell), A_{\ell g}} &= -p_\ell, \\
\left.\frac{\partial F}{\partial V_g}\right|_{T, V_i(i \neq g), A_{\ell g}} &= -p_g, \qquad (9.13) \\
\left.\frac{\partial F}{\partial V_t}\right|_{T, V_i(i \neq t), A_{\ell g}} &= -p_t, \\
\left.\frac{\partial F}{\partial A}\right|_{T, V_i} &= \sigma,
\end{aligned}$$

where $i = g, \ell$, and t. When the irreversible heat is due to the *change* in the *chemical variables*, we have

$$-dQ_{\text{int}} = \mu'_\ell\,dN_\ell + \mu'_g\,dN_g + \mu'_t\,dN_t. \qquad (9.14)$$

Then, from this and (9.12), we have

$$dF = -S\,dT - p_\ell\,dV_\ell - p_g\,dV_g - p_t\,dV_t + \sigma\,dA_{\ell g}$$
$$+ \mu'_\ell\,dN_\ell + \mu'_g\,dN_g + \mu'_t\,dN_t. \qquad (9.15)$$

Now by considering $F = F(T, V_i, A_{\ell g}, N_i)$ and $i = \ell, g$, and t, we have

$$\frac{\partial F}{\partial N_\ell} = \mu'_\ell, \qquad (9.16)$$

$$\frac{\partial F}{\partial N_g} = \mu'_g, \qquad (9.17)$$

$$\frac{\partial F}{\partial N_t} = \mu'_t, \qquad (9.18)$$

where μ'_i is the *complete chemical potential* (Defay and Prigogine, 1966, p. 53), and N_i is the *number of moles* in the volume occupied by i. The complete chemical potentials are related to *ordinary chemical potentials* (shown without primes) through

$$\mu'_\ell = \mu_\ell + \frac{\partial F_t}{\partial N_\ell}, \tag{9.19}$$

$$\mu'_g = \mu_g + \frac{\partial F_t}{\partial N_g}, \tag{9.20}$$

$$\mu'_t = \mu_t. \tag{9.21}$$

Therefore, when these derivatives of the free energy of the interface are *negligible* (compared to μ_ℓ and μ_g), the complete and ordinary chemical potentials become equal.

For the interfacial layer, we have

$$dF_t = -S_t\,dT + V_t\,dp_t - \sigma\,dA_{\ell g} + \mu'_t\,dN_t. \tag{9.22}$$

Also, from the state variables we have

$$dF_t = -p_t\,dV_t - V_t\,dp_t + \sigma\,dA_{\ell g} + A_{\ell g}\,d\sigma + \mu'_t\,dN_t + N_t\,d\mu'_t. \tag{9.23}$$

The last two give

$$S_t\,dT - V_t\,dp_t + A_{\ell g}\,d\sigma + N_t\,d\mu'_t = 0, \tag{9.24}$$

which is called the *Gibbs-Duhem equation* for the interfacial layer.

The *heat of extension of the surface* is defined as (Defay and Prigogine)

$$\text{heat of extension} = -T\frac{\partial \sigma}{\partial T}. \tag{9.25}$$

The *total energy of extension of the surface* is

$$\left.\frac{\partial U}{\partial A_{\ell g}}\right|_{T,V_i,N_i} = \sigma - T\frac{\partial \sigma}{\partial T}. \tag{9.26}$$

9.1.3 THICKNESS OF INTERFACIAL LAYER

The molecular theory of fluid interfaces deals with the *short range* intermolecular forces acting on the interface. In the molecular treatment, a *continuous* variation in the density is allowed. This theory has been studied by (among others) Cahn and Hilliard (1958) and Bongiorno and Scriven (1976). The thickness of the interface *increases* with the *increase* in the temperature (saturated state) and becomes *infinite* as the *critical temperature* T_c is reached. The *surface tension decreases* with the increase in the temperature and vanishes as the critical temperature is reached.

Cahn and Hilliard construct a model that gives the *interfacial free energy* as the integral over the interfacial volume of the sum of *two* contributions;

one is a function of the *local* value of the properties, the other is proportional to the square of the *local gradient* of the *density*. They show that the ratio of the *interfacial layer thickness* to the *mean interaction distance* of the intermolecular forces has an *asymptotic* value between *one* and *two* for $T/T_c \to 0$ and tends to infinity as $T/T_c \to 1$. Since the mean interaction distance is of the order of 10 Å, the thickness of the interfacial layer is not expected to be more than $o(100$ Å$)$ for $T/T_c < 0.9$. They show that the temperature dependency of both the thickness of the interfacial layer and the surface tension are *not* expressible in simple single-term correlations. Near T_c the variation of the thickness follows $T_c^{1/2}(T_c - T)^{-1/2}$ and the variation of the surface tension follows $T_c^{3/2}(T_c - T)^{-3/2}$.

One of the *simple* correlations for the variation of the surface tension with the temperature is (Hsieh, 1975)

$$\sigma = \sigma_o \left(1 - \frac{T - 273.16}{T_c - \Delta T_m - 273.16}\right)^n, \qquad (9.27)$$

where ΔT_m indicates that by using this simple form the interfacial tension diminishes *before* the critical state is reached. The range of the constants are $\Delta T_m \simeq 6-8$ K and $1 \leq n \leq 2$. For water $\sigma_o = 0.0755$ N/m, $T_c = 647.3$ K, $\Delta T_m = 6.14$ K, and $n = 1.2$. Equations of the type given earlier were *first* suggested by van der Waals.

9.2 Effect of Curvature in Single-Component Systems

The equilibrium properties of single-component capillary systems has been studied by Defay and Prigogine (1966, Chapter 15). Some of their developments are given here. Their treatment is based on the following assumptions.

- The interfacial volume is negligible, i.e., $V_t \to 0$, then $F_t \to 0$, $V_t \to 0$, and $N_t \to 0$, and the interfacial layer is *not* included in the analysis.

- The meniscus *mean radius of curvature* is *larger* than the *interfacial thickness*, i.e., $H^{-1} > 100$ Å.

- The *thermal equilibrium* exists, i.e.,

$$T_\ell = T_g. \qquad (9.28)$$

- The *mechanical equilibrium* exists (*Laplace equation*), i.e.,

$$p_g - p_\ell = 2H\sigma. \qquad (9.29)$$

- The *chemical equilibrium* exists, i.e.,

$$\mu_g = \mu_\ell = \mu. \tag{9.30}$$

- The *single-component* capillary system is completely specified by any *two* variables, e.g., T and H or p_g and H.

- The *perfect gas behavior* is assumed for the gaseous phase, i.e.,

$$p_g = \frac{\rho_g R_g T}{M} \quad \text{or} \quad \frac{\rho_g}{M} = \frac{p_g}{R_g T} \equiv \frac{1}{v_g}, \tag{9.31}$$

where R_g is the universal gas constant, M is the molecular weight, ρ_g/M is the *molar density* of the gaseous phase, and v_g is the *molar volume*.

- The first-order *liquid compressibility* is assumed, i.e.,

$$\kappa_\ell = \frac{1}{\rho_\ell} \left.\frac{\partial \rho_\ell}{\partial p_\ell}\right|_H = -\frac{1}{v_\ell} \left.\frac{\partial v_\ell}{\partial p_\ell}\right|_H, \tag{9.32}$$

where κ_ℓ is the *compressibility* of the liquid.

- Although the treatment is general, we focus on $p_g > p_\ell$, i.e., where the meniscus is *curbed toward* the gaseous phase.

The *effect* of the curvature on the *surface tension* is negligibly small, except when the mean radius of curvature is less than 100 Å. This effect is not considered in the following, but the details are available in Defay and Prigogine (1966, 256–258).

The *molar-based* Gibbs-Duhem equation (9.24) for the ℓ and g phases becomes

$$s_\ell \, dT - v_\ell \, dp_\ell + d\mu_\ell = 0, \tag{9.33}$$

$$s_g \, dT - v_g \, dp_g + d\mu_g = 0, \tag{9.34}$$

where from (9.29) and (9.30), we have

$$d\mu_\ell = d\mu_g = d\mu, \tag{9.35}$$

$$dp_g - dp_\ell = d(2H\sigma). \tag{9.36}$$

Central to the analysis of two-phase flow in porous media is the capillary pressure. In the following analysis, the effect of the meniscus curvature also appears through $2H\sigma$, which is the same as the capillary pressure. Therefore, knowing the capillary pressure (which is a function of the *saturation* and the other variables, as reviewed in Section 8.4), the effect of the curvature on the thermodynamic properties and on the states can be found. Equation (9.36) is called the *Laplace equation*.

9.2.1 Vapor Pressure Reduction

For a *constant temperature* process, we can combine (9.33), (9.34), and (9.35) to arrive at

$$v_\ell \, dp_\ell = v_g \, dp_g, \qquad T = \text{constant}. \tag{9.37}$$

Now, using (9.36) to eliminate p_ℓ, we have

$$d(2H\sigma) = \frac{v_\ell - v_g}{v_\ell} \, dp_g = dp_g - \frac{v_g}{v_\ell} \, dp_g, \tag{9.38}$$

where as we mentioned σ is a *weak* function of H. By assuming a perfect gas behavior, (9.31), we have

$$d(2H\sigma) = dp_g - \frac{R_g T}{v_\ell p_g} \, dp_g, \qquad T = \text{constant}, \tag{9.39}$$

which can be integrated from $p_g = p_{g_0}$ for $H_o = 0$, to p_g and H. By assuming that v_ℓ remains *constant* (i.e., $\kappa_\ell \to 0$), we have

$$2H\sigma = (p_g - p_{g_0}) - \frac{R_g T}{v_\ell} \ln \frac{p_g}{p_{g_0}}, \qquad T \text{ and } v_\ell \text{ are kept constant.} \tag{9.40}$$

This is the *Kelvin equation* and shows that for the case considered here ($p_g > p_\ell$), the *vapor pressure decreases* as H increases (or the mean radius of curvature decreases). The first term on the right-hand side of (9.40) is generally *negligible*. For water at 18°C and a mean radius of curvature of 1 μm, we have $p_g/p_{g_0} = 0.9891$, and for a mean radius of curvature of 100 Å, this becomes 0.897.

If we allow for the change of v_ℓ with the pressure, a *correction* can be added to (9.40). However, this correction is small. To show this, we integrate (9.32) to arrive at

$$\ln \frac{v_\ell}{v_{\ell_0}} = -\kappa_\ell (p_\ell - p_{\ell_0}). \tag{9.41}$$

Now we note that for water $\kappa_\ell = 5 \times 10^{-11}$ dyne/cm^2, and therefore the change in v_ℓ is generally negligible.

Note that in general

$$p_g - p_{g_0} \ll p_g - p_\ell = 2H\sigma. \tag{9.42}$$

Therefore, for single-component systems and for low p_g, the liquid pressure can become *negative* when H is large.

9.2.2 Reduction of Chemical Potential

When considering a *constant temperature* process, (9.33) and (9.34) along with (9.35) give

$$d\mu = v_g \, dp_g = v_\ell \, dp_\ell, \qquad T = \text{constant}. \tag{9.43}$$

Also from (9.36) and (9.43), we have

$$d(2H\sigma) = dp_g - \frac{v_g}{v_\ell}dp_g = dp_g - \frac{d\mu}{v_\ell}. \qquad (9.44)$$

Now by assuming that v_ℓ remains *constant* and by integrating this with $H = 0$ for μ_o, p_{g_o} to H for μ, p_g, we have

$$2H\sigma = (p_g - p_{g_o}) - \frac{\mu - \mu_o}{v_\ell} \qquad (9.45)$$

or

$$\mu = \mu_o - (2H\sigma)v_\ell + (p_g - p_{g_o})v_\ell, \qquad (9.46)$$

where again the *second term* on the right-hand side is generally *negligible*. This shows that the presence of a curvature *reduces the chemical potential*.

9.2.3 Increase in Heat of Evaporation

Assuming that the *interfacial area* remains constant during evaporation, the amount of the *reversible heat* required for steady evaporation is the *heat of evaporation* $\Delta i_{\ell g}$ and is

$$\Delta i_{\ell g} = T(s_g - s_\ell), \quad \text{constant } T \text{ and } H. \qquad (9.47)$$

Since at constant T we have $s_g = s_g(p_g)$ and $s_\ell = s_\ell(p_\ell)$, then we have

$$di_{\ell g} = T\left(\frac{\partial s_g}{\partial p_g}dp_g - \frac{\partial s_\ell}{\partial p_\ell}dp_\ell\right). \qquad (9.48)$$

From the Gibbs-Duhem equations, (9.33) and (9.34), at equilibrium, we have

$$\left.\frac{\partial s_g}{\partial p_g}\right|_T = -\left.\frac{\partial v_g}{\partial T}\right|_{p_g} \quad \text{and} \quad \left.\frac{\partial s_\ell}{\partial p_\ell}\right|_T = -\left.\frac{\partial v_\ell}{\partial T}\right|_{p_\ell}. \qquad (9.49)$$

Then, (9.48), (9.37), and (9.49) can be combined, and we have

$$di_{\ell g} = -T\left(\left.\frac{\partial v_g}{\partial T}\right|_{p_g}\frac{v_\ell}{v_g - v_\ell} - \left.\frac{\partial v_\ell}{\partial T}\right|_{p_\ell}\frac{v_g}{v_g - v_\ell}\right)d(2H\sigma). \qquad (9.50)$$

For a perfect gas behavior and for $v_\ell \ll v_g$, we have

$$di_{\ell g} = -\left(v_\ell - T\left.\frac{\partial v_\ell}{\partial T}\right|_{p_\ell}\right)d(2H\sigma). \qquad (9.51)$$

The integration between $H_o = 0$, $i_{\ell g_o}$ and H, $\Delta i_{\ell g}$ (assuming that v_ℓ, T, and $\partial v_\ell/\partial T|_{p_\ell}$ remain constant) gives

$$\Delta i_{\ell g} = \Delta i_{\ell g_o} + \left(v_\ell - T\left.\frac{\partial v_\ell}{\partial T}\right|_{p_\ell}\right)(2H\sigma), \qquad (9.52)$$

9.2 Effect of Curvature in Single-Component Systems

where in general $v_\ell > T(\partial v_\ell/\partial T)_{p_\ell}$ and therefore, as H increases the *heat of evaporation increases*. However, this increase is not substantial and for water at 4°C and a mean radius of curvature of 1 μm, we have $i_{\ell g}/i_{\ell g_0} = 1.00006$.

Note that in determining the *total* heat needed for a change in H and for the phase change, the heat of the extension of the surface given by (9.25), i.e., $-T\partial\sigma/\partial T(2Hv_\ell)$ must also be *supplied* to account for the change in H. Again, even this term is *negligibly* small, except for very large values of H.

When *evaporation* by *desorption* is considered, as is the case in porous media with small pores (Section 9.5), the *heat of desorption* is also added to the heat of evaporation. Estimation of the heat of desorption is based on the *Hückel formulation* (Defay and Prigogine, 1966, 236–239). The analysis assumes that the relationship between the volume of the adsorbed liquid $V_{\ell\,\text{ad}}$ and H_{des} is known as $H_{\text{des}} = H_{\text{des}}(V_{\ell\,\text{ad}})$ where H_{des} is the *desorption mean curvature*. A closed system with a total of $N = N_g + N_\ell$ liquid and vapor molecules is assumed. This system is given by T, p_g, and N_ℓ (knowing N). The *Clausius-Clapeyron equation* is

$$\Delta i_{\ell g_0} = R_g T^2 \frac{d(\ln p_{g_0})}{dT}. \tag{9.53}$$

A similar equation for $i_{\ell g}$ is developed. The derivation is given in Defay and Prigogine, and the result is

$$\Delta i_{\ell g} = R_g T^2 \left.\frac{d(\ln p_g)}{dT}\right|_{N_\ell}. \tag{9.54}$$

The Kelvin equation (9.40) is approximated as

$$\ln \frac{p_g}{p_{g_0}} = -\frac{v_\ell}{R_g T}(2H_{\text{des}}\sigma). \tag{9.55}$$

By assuming that $v_\ell = v_\ell(T)$, by differentiating the Kelvin equation with respect to T, and by using (9.54) and (9.55) we have

$$\Delta i_{\ell g} = \Delta i_{\ell g_0} + (2H_{\text{des}}v_\ell)\left(\sigma - T\frac{d\sigma}{dT}\right) - (2H_{\text{des}}\sigma)T\frac{\partial v_\ell}{\partial T}$$

$$- (2\sigma v_\ell)T\left.\frac{\partial H_{\text{des}}}{\partial T}\right|_{N_\ell}, \tag{9.56}$$

where

$$\frac{\partial H_{\text{des}}}{\partial T} = \frac{\partial H_{\text{des}}}{\partial V_{\ell\,\text{ad}}}\frac{\partial v_\ell}{\partial T}N_\ell. \tag{9.57}$$

For very large H_{des}, the second and the last terms in (9.56) can be *large*. Hückel has found that in evaporation-desorption of sulfur dioxide in active carbon the change is $i_{\ell g}/i_{\ell g_0} = 1.38$ for a $H_{\text{des}}^{-1} = 11$ Å. However, for most two-phase flow applications this increase in $i_{\ell g}$ is *not* noticeable.

Note that for the desorption we have $H_{\text{des}} = H_{\text{des}}(V_{\ell\,\text{ad}})$, which is generally different than $H_{\text{ad}} = H_{\text{ad}}(V_{\ell\,\text{ad}})$ obtained from the *adsorption* results. This *hysteresis* should be kept in mind in calculating the *heat of adsorption* (or the *capillary condensation*). This will be discussed in Section 9.5.

9.2.4 LIQUID SUPERHEAT

In two-phase single-component systems, we know that for a small vapor bubble to exist in a liquid, it is necessary to *raise* the liquid temperature above $T_o(p_\ell)$. This amount of the liquid *superheat* $T - T_o$ is determined later.

From (9.33), (9.34), and (9.35) and by assuming that the *liquid pressure* does *not* change ($dp_\ell = 0$), we have

$$dT(s_g - s_\ell) - v_g\,dp_g = 0. \tag{9.58}$$

We also have $\Delta i_{\ell g} = T(s_g - s_\ell)$, which, when combined with the preceding, gives

$$\frac{\Delta i_{\ell g}}{T}\,dT - v_g\,dp_g = 0. \tag{9.59}$$

Now using

$$d(p_g - p_\ell) = dp_g = d(2H\sigma) \tag{9.60}$$

and treating the vapor as an ideal gas, we have

$$\frac{\Delta i_{\ell g}}{T^2}\,dT = R_g\frac{d(2H\sigma)}{2H\sigma + p_\ell} = R_g\,d[\ln(2H\sigma + p_\ell)]. \tag{9.61}$$

Next, by assuming that $\Delta i_{\ell g}$ remains constant as H and T change (we can use an *average* value for $\Delta i_{\ell g}$), and by integrating the preceding between T_o, $H = 0$ and T, H, we have

$$\Delta i_{\ell g}\left(\frac{1}{T_o} - \frac{1}{T}\right) = R_g \ln\frac{2H\sigma + p_\ell}{p_\ell}. \tag{9.62}$$

This equation shows that $T > T_o$, i.e., the liquid becomes *superheated* in the presence of a meniscus curbing toward the vapor.

If the variation of $\Delta i_{\ell g}$ with respect to T is *also* taken into account using a *linear* relation such as

$$\Delta i_{\ell g} = a_1 - a_2 T, \tag{9.63}$$

we will have

$$a_1\left(\frac{1}{T_o} - \frac{1}{T}\right) - b\ln\frac{T}{T_o} = R_g \ln\frac{2H\sigma + p_\ell}{p_\ell}. \tag{9.64}$$

Using this equation for *water* with $p_\ell = 1$ atm, Defay and Prigogine (1966, p. 242) show that for $H^{-1} = 10$ μm the superheat is $T - T_o = 3.3$ K; the corresponding change in the surface tension is $\sigma - \sigma_o = -0.71$ dyne/cm, where $\sigma_o = 55.46$ dyne/cm, and the change in the heat of evaporation is $\Delta i_{\ell g} = -0.17$ kJ/mol, where $\Delta i_{\ell g_o} = 40.40$ kJ/mol. For $H^{-1} = 1$ μm, the superheat becomes $T - T_o = 21.4$ K and for $H^{-1} = 0.1$ μm, it becomes $T - T_o = 76$ K. Therefore, a *significant* superheat can exist if the bulk and the wall nucleation sites are *absent*. Note that the *homogeneous nucleation limit* for bulk boiling requires much larger superheat (Lienhard, 1982).

By knowing the liquid superheat, the mean curvature H can be determined from this equation. The mean curvature can also be obtained from the Kelvin equation (9.40) and the Laplace equation (9.36). The two equations can be written as

$$p_{g_o}(T) - p_\ell = -\frac{R_g T}{v_\ell} \ln \frac{p_g}{p_{g_o}(T)}, \qquad (9.65)$$

$$p_g - p_\ell = 2H\sigma, \qquad (9.66)$$

where in (9.65) the $p_g - p_{g_o}$ terms has been neglected compared to the logarithmic term. The vapor pressure p_{g_o} is that of the saturated state at T, which is the *temperature of superheated state*.

Udell (1983) reports measured superheats (for a packed bed of sand particles of $150 < d < 208$ μm, with water as the fluid) that agree with these predictions. He also uses the *van der Waals equation of state*

$$p = \frac{R_g T}{v - b} - \frac{a}{v^2} \qquad (9.67)$$

for a *graphical* presentation of the liquid superheat state and the lowering of the chemical potential. From the Gibbs-Duhem relation (9.33) and (9.34) at a constant temperature, we have

$$\mu = \int v \, dp + a_1(T) \quad \text{at constant } T, \qquad (9.68)$$

where a_1 is a constant. This equation, combined with (9.67), allows for the graphical presentation of μ.

9.2.5 CHANGE IN FREEZING TEMPERATURE

We can consider the freezing (or melting) of a liquid (solid) in a capillary tube as a liquid-solid phase change in porous media. Now, depending on the *filling history* and the specific volumes of the two phases, the presence of a curved solid-liquid interface can result in either $p_s > p_\ell$ or $p_\ell > p_s$. The Laplace equation is written as

$$p_s - p_\ell = \pm(2H_{s\ell}\sigma_{s\ell}), \qquad (9.69)$$

where $\sigma_{s\ell}$ is the surface tension of the s-ℓ interface; see Section 8.1.4 (A). We can write (9.59) for the s-ℓ system as

$$\frac{\Delta i_{s\ell}}{T}\,dT - \sigma_s\,dp_s = 0. \tag{9.70}$$

Again, by using $dp_\ell = 0$, we have

$$\frac{dT}{T} = \mp \frac{v_s}{\Delta i_{s\ell}}\,d(2H_{s\ell}\sigma_{s\ell}). \tag{9.71}$$

When this is integrated with T_0, $H = 0$ and T, H, and by taking $i_{s\ell}$ as a constant, we have

$$\ln \frac{T}{T_0} = \mp \frac{v_s(2H_{s\ell}\sigma_{s\ell})}{\Delta i_{s\ell}}. \tag{9.72}$$

Then the *freezing point can decrease* if $p_s - p_\ell > 0$. For example if $v_s > v_\ell$, then $p_s > p_\ell$, $T < T_0$ and the excess pressure can cause significant *stress* in the solid matrix (in building material this stress causes damage).

Antoniou (1964) has performed experiments on freezing of water in porous glass slabs where the effect of the curvature and the solid-liquid surface forces have resulted in a *decrease* of the freezing point to $-22°C$. A significant elongation of the slab is also observed as a result of the freezing. The liquid pressure and freezing temperature increase with the curvature when the meniscus *curbs toward* the liquid. Also, both the *heat of fusion* $\Delta i_{s\ell}$ and the *heat of sublimation* Δi_{sg} depend on the curvature and their values can be found in a manner similar to that followed in Section 9.2.3.

9.2.6 CHANGE IN TRIPLE-POINT TEMPERATURE

When the *three* phases of a pure substance are in equilibrium and a curvature exists at the interface between the various phase combinations, we have (for a *given selection* of the signs of the radii of curvature) the Laplace equations

$$p_g - p_\ell = -2H_{g\ell}\sigma_{g\ell}, \tag{9.73}$$

$$p_g - p_s = -2H_{gs}\sigma_{gs}, \tag{9.74}$$

$$p_s - p_\ell = -2H_{s\ell}\sigma_{s\ell}, \tag{9.75}$$

which lead to

$$2H_{gs}\sigma_{gs} + 2H_{s\ell}\sigma_{s\ell} = 2H_{g\ell}\sigma_{g\ell}. \tag{9.76}$$

Therefore, the radii of curvature are *related* through (9.76). We can write the Gibbs-Duhem equation for the *solid phase* as

$$\sigma_s\,dT - v_s\,dp_s + d\mu_s = 0. \tag{9.77}$$

Since by using (9.76) the specification of only *two pressure jumps* (curvatures) is needed, we consider the s-ℓ and the ℓ-g interfaces. From the first two Laplace equations given earlier and the *three* Gibbs-Duhem equations, we can eliminate $d\mu$, dp_g, dp_ℓ, and dp_s. The result is

$$\left(\frac{s_g - s_s}{v_g - v_s} - \frac{s_g - s_\ell}{v_g - v_\ell}\right) dT = \frac{v_\ell}{v_g - v_\ell} d(2H_{g\ell}\sigma_{g\ell}) + \frac{v_s}{v_g - v_s} d(2H_{gs}\sigma_{gs}). \tag{9.78}$$

Next, we use $\Delta i_{s\ell} = T(s_\ell - s_s)$, $v_\ell \ll v_g$, and $v_s \ll v_g$ to arrive at

$$\Delta i_{s\ell} \frac{dT}{T} = v_\ell\, d(2H_{g\ell}\sigma_{g\ell}) - v_s\, d(2H_{gs}\sigma_{gs}). \tag{9.79}$$

Now by integrating this equation with T_o, $H_{g\ell} = 0$, $H_{gs} = 0$, and T, $H_{g\ell}$, and H_{gs}, we have

$$\ln \frac{T}{T_o} = \frac{2}{\Delta i_{s\ell}}(2H_{g\ell}\sigma_{g\ell} - 2H_{gs}\sigma_{gs}). \tag{9.80}$$

Note that the *sign* convention used for the curvatures is that given in (9.73) to (9.75) and is chosen for convenience only. These signs should be changed according to the specific phase changes that occur in the physical system. For example, if

$$p_g > p_\ell \quad \text{and} \quad p_g > p_s, \tag{9.81}$$

we have

$$\ln \frac{T}{T_o} = \frac{2}{\Delta i_{s\ell}}(-2H_{g\ell}\sigma_{g\ell} + 2H_{gs}\sigma_{gs}). \tag{9.82}$$

Then, if $2H_{gs}\sigma_{gs} < 2H_{g\ell}\sigma_{g\ell}$, the *triple-point temperature is reduced* because of the curvature.

9.3 Multicomponent Systems

When more than one species is present, even only in one of the phases, the thermodynamic treatment must include the effects of the variation in the composition on the rest of the state variables. When an ideal behavior (similar to that adapted for gases) is assumed for the *liquid phase*, then certain *simple* relationships among the state variables can be found. Defay and Prigogine have covered this subject and in this section a few of their developments that can be applied to transport in porous media are given.

9.3.1 SURFACE TENSION OF SOLUTION

An *ideal solution* is that one for which the component chemical potential is expressible in the form

$$\mu_i = \mu_{i_o}(T, p) + R_g T \ln x_i, \tag{9.83}$$

where x_i is the *mole fraction of component* i. All solutions *behave* ideally as $x_i \to 0$. Those that follow this even for $x_i \to 1$ are called the *perfect solutions*. Perfect behavior is approached when the molecules of each species are similar in size. This *size similarity* means that they also occupy the same area on the *interface* between the ℓ and g phases. For a perfect *binary* solution (i.e., $x_{1\ell} + x_{2\ell} = 1$) and for a planar interface, we have for the surface tension of the solution (Defay and Prigogine, 1966, p.167)

$$\exp\left(-\frac{\sigma a}{R_g T}\right) = x_{1\ell} \exp\left(-\frac{\sigma_1 a}{R_g T}\right) + x_{2\ell} \exp\left(-\frac{\sigma_2 a}{R_g T}\right), \qquad (9.84)$$

where σ_1 and σ_2 are the surface tension of the pure substance 1 and 2, respectively.

When σ_1 and σ_2 are nearly the same, then we have

$$\sigma = \sigma_1 x_{1\ell} + \sigma_2 x_{2\ell} \quad \text{for} \quad \left|\frac{(\sigma_1 - \sigma_2)a}{R_g T}\right| \ll 1. \qquad (9.85)$$

Then the *partial molar surface area* a, which is given as

$$a = \frac{\partial A_{g\ell}}{\partial N_{1t}}, \qquad (9.86)$$

is the *same* for all the species, because of their assumed similar size. N_{1t} is the *number of moles of species* 1 *in the interfacial layer*.

For a *monolayer surface*, we have

$$A_{\ell g} = (N_{1t} + N_{2t})a, \qquad (9.87)$$

where a is approximately the area occupied by one mole of either of the species spread as a *monolayer* with a close-packed arrangement. In *real* interfacial layers more than one molecular layer exists and the molecular arrangement depends on the mole fractions in the ℓ and g phases, as well as on σ, T, and p (for planar interfaces). For a curved surface, the relationship between a and the variables is *even* more complex.

9.3.2 Vapor Pressure Reduction

The effect of the curvature on the *partial pressures* of the components present in the liquid phase is similar to that for the single-component liquid (the Kelvin equation). This can be shown by noting that the Laplace equation is not altered except that σ depends on the compositions in the bulk fluids and at the interface. Also at the equilibrium state, we need to extend (9.30) to apply to each component, i.e.,

$$\mu_{i\ell} = \mu_{ig}, \qquad (9.88)$$

where i stands for the ith component. As before, we have

$$\frac{d\mu_{i\ell}}{dp_\ell} = v_{i\ell}, \quad \frac{d\mu_{ig}}{dp_g} = v_{ig}, \quad \text{or} \quad dp_\ell = \frac{d\mu_{i\ell}}{v_{i\ell}}. \tag{9.89}$$

For a perfect gas behavior, the chemical potential is given by

$$\mu_{ig} = \mu_i(T) + R_g T \ln \frac{p_{ig}}{p_g}, \tag{9.90}$$

where p_{ig} is the *partial pressure* of the species i. Then, from (9.89) and (9.90), we have

$$dp_\ell = \frac{R_g T}{v_{i\ell}} d\left(\ln \frac{p_{ig}}{p_g}\right). \tag{9.91}$$

Now using the Laplace equation, we obtain

$$d(2H\sigma) = dp_g - \frac{R_g T}{v_{i\ell}} d\left(\ln \frac{p_{ig}}{p_g}\right). \tag{9.92}$$

Next by integrating this equation between p_{g_0}, p_{ig_0}, and $H = 0$ and p_g, p_{ig}, and H, we arrive at

$$2H\sigma = (p_g - p_{g_0}) - \frac{R_g T}{v_{i\ell}} \ln \frac{p_{ig}}{p_{ig_0}}, \tag{9.93}$$

where again the *first term* on the right-hand side is *negligible* compared to the second term. This equation is similar to (9.40) and allows for the calculation of the *species vapor pressure reduction* in the mixture (i.e., the Kelvin equation for multicomponent systems).

9.4 Interfacial Thermodynamics of Meniscus Extension

For *perfectly wetting* liquids, we now consider the thin liquid film spreading over the solid surface and we consider the condition where the liquid is in a thermodynamic equilibrium state. This subject was discussed in Section 8.1.5 in connection with the van der Waals dispersive forces. These forces depend on the liquid film thickness δ, as given by (8.74) with the two approximate asymptotes given by (8.76) and (8.77) for the small and large δ limits. In the *transition region*, these forces as well as the capillarity and the gravity can be important. We now follow the analysis of Truong and Wayner (1987) concerning the thermodynamics of the thin liquid film in the transition region. A *single-component* system is assumed. The chemical potential of the *liquid film* $\mu_{\ell f}$ is taken as the *sum* of the chemical potential of the *bulk liquid* μ_{ℓ_o}, the contribution made by the *van der Waals interactions* μ_{ℓ_δ}, and the contribution due to the *curvature* μ_{ℓ_H}, i.e.,

$$\mu_{\ell f}(p_\ell, T) = \mu_{\ell_o}(T) + \mu_{\ell_\delta}(\delta, T) + \mu_{\ell_H}(H, T), \tag{9.94}$$

where it has been assumed that the *liquid* is *incompressible*.

For the vapor phase we assume an ideal behavior, and then we will have

$$\mu_g(p_g, T) = \mu_{g_0}(p_{g_0}, T) + R_g T \ln \frac{p_g}{p_{g_0}}, \qquad (9.95)$$

where p_{g_0} is the saturation vapor pressure of the bulk (away from the interface) liquid.

We begin by taking $H = 0$ and determining $\mu_{\ell_\delta}(\delta, T)$. At equilibrium we have

$$\mu_g = \mu_{\ell f}. \qquad (9.96)$$

When this is used along with (9.95), we arrive at

$$\mu_{\ell_\delta} = R_g T \ln \frac{p_g}{p_{g_0}}. \qquad (9.97)$$

The case of $\mu_{\ell_\delta} < 0$ corresponds to a vapor pressure reduction caused by the notable existence of the van der Waals dispersive forces in the liquid phase. This is similar to the Kelvin equation for the curvature effect.

Now we include the effect of gravity. The change in the chemical potential of the liquid (along $-\mathbf{g}$), while keeping the other variables constant, is given by

$$\frac{d\mu_\ell}{dz} = -g. \qquad (9.98)$$

The integration of this equation with $z = 0$, μ_{ℓ_0} and $z = h$, μ_ℓ, where h is positive and is measured along $-\mathbf{g}$, gives

$$\mu_{\ell f} = \mu_{\ell_0} - gh. \qquad (9.99)$$

Then, from (9.97) and (9.99), we have

$$\mu_{\ell_\delta} = -gh = R_g T \ln \frac{p_g}{p_{g_0}} = f(\delta) v_\ell. \qquad (9.100)$$

For the case of $H = 0$, this equation gives

$$f(\delta) = -\frac{gh}{v_\ell} < 0, \qquad (9.101)$$

which when used in (9.100) leads to $\mu_{\ell_\delta} < 0$.

Next we consider $H \neq 0$ and use the Kelvin equation (9.40) (where, as we discussed, the first term on the right-hand side is negligible). From this equation, by using the equilibrium condition (9.96) and by taking the condition where $f(\delta) = 0$, we have

$$\mu_{\ell_H} = -(2H\sigma)\sigma_\ell < 0. \qquad (9.102)$$

Now by using

$$\mu_g = \mu_{\ell f} = \mu_{\ell_0} + \mu_{\ell\delta} + \mu_{\ell_H}, \qquad (9.103)$$

along with (9.97), (9.99), (9.100), and (9.102), we arrive at

$$f(\delta) - 2H\sigma = \frac{R_g T}{v_\ell} \ln \frac{p_g}{p_{g_0}} \qquad (9.104)$$

or

$$f(\delta) - 2H\sigma = -\frac{gh}{v_\ell}. \qquad (9.105)$$

Then, the vapor pressure reduction can be determined from the *modified Kelvin equation* (9.104), which accounts for both the *curvature effect* and the *effect of the thin liquid film dispersion forces*. The force balance (9.105) describes the thermodynamic and mechanical equilibrium along the thin film and can be used to determine the variations of δ and H along the film in the transition region. Truong and Wayner give some *numerical* and *experimental* results for the variation of δ in this transition region.

The thermodynamic stability condition for the thin liquid film is given by the *requirement for the thermodynamic stability*, i.e.,

$$\left.\frac{\partial \mu_{\ell\delta}}{\partial \delta}\right|_T \geq 0. \qquad (9.106)$$

Note that $\mu_{\ell\delta} < 0$ for the stable *planar* thin liquid films. Then it is further required by (9.106) that $\mu_{\ell\delta}$ *increase* as δ *increases*. If (9.106) is *not* satisfied, then no *wetting* film will be formed. Dzyaloskinskii et al. (1961) have examined this stability condition for several possible variations of $\mu_{\ell\delta}$ with respect to δ. We will consider the stability of the thin films in small pores in Section 9.6.2.

9.5 Capillary Condensation

Adsorption of a superheated gas by a solid surface was discussed briefly in Chapter 6. In the following we consider the conditions (temperature, pressure, and pore size) where the *adsorbed* vapor can condense to a liquid (or a condensed phase resembling a liquid state) in the pores of the solid. Due to the *lowering* of the vapor pressure, we will have $p_g(T)/p_{g_0}(T) < 1$. As was mentioned, $p_{g_0}(T)$ is the saturation pressure where $H = 0$ and δ (film thickness) is large, i.e., where *no* capillary or dispersive force effects exist. The subject of adsorption has been reviewed by Gregg and Sing (1982), where the gas-solid molecular interactions as well as the classifications of various physical adsorptions are given. In general, since the *desorption* is as significant as the adsorption, the desorption branch of the *adsorption isotherm* is also evaluated. For small pore sizes, a *hysteresis loop*, similar to that found for the capillary pressure, is found in the adsorption isotherm.

Some features of the adsorption of gases by solids, as well as the capillary condensation, are discussed later. The material is that given by Gregg and Sing, where more details about the adsorption and desorption in porous media can be found. The adsorption technology is discussed in a volume edited by Slejko (1985).

9.5.1 Adsorption by Solid Surface

As was mentioned in Chapter 6, the enrichment of one or more of the components in an interfacial layer is called *adsorption*. We consider interfaces where *one* of the phases is solid and the other gaseous. Also we consider only single-component systems. In the *physical adsorption*, which is considered here, *no* chemical bounding occurs between the adsorbed gas and the solid surface.

The adsorption forces were discussed in Section 6.5.4. Various classifications of the surface and molecular forces are available. Yet another classification is given by Gregg and Sing, where according to this classification the *solid* can contain *no ions* or *positive charges, concentrated positive charges,* or *concentrated negative charges* and the *gas* can have *spherically symmetric shells, π-bounds* or *lone pairs of electrons, positive charges concentrated on peripheries of the molecules,* or *both electron density* and *positive charges*.

If we begin with a surface with no adsorbed molecules and then expose the surface to a gas at a *low* pressure and then *increase* the gas pressure, we will first observe the *formation* of a *monolayer*. As more molecules are adsorbed to the surface, the gas *fractional coverage* increases until a *densely occupied monolayer* is formed. A densely occupied monolayer to an extent acts as an *extension* of the solid surface and, therefore, attracts more gas molecules (but the interaction forces are weaker). This leads to the formation of the *multilayers*.

(A) Adsorption Isotherm

For a given matrix structure the amount of the gas adsorbed on the internal surface of the matrix is proportional to the volume of the matrix as long as the sampling volume is large (many times the representative elementary volume). If the density of the matrix does *not* change noticeably, then we can express the *adsorption capacity of the matrix* by

$$\frac{\text{g of gas adsorbed}}{\text{g of solid}} \equiv m^*_{\text{ad}}, \tag{9.107}$$

where we expect

$$m^*_{\text{ad}} = m^*_{\text{ad}}(p_g, T, \text{molecular properties of gas and solid},$$
$$\text{matrix structure, history}). \tag{9.108}$$

For a given gas-solid system (including matrix structure) and temperature, where *no hysteresis* is assumed, we can write

$$m^*_{\text{ad}} = m^*_{\text{ad}}(p_g) \quad \text{at constant } T, \tag{9.109}$$

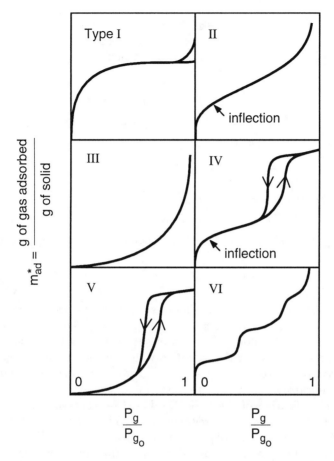

Figure 9.1 Classification of the adsorption isotherms into six separate groups. (From Gregg and Sing, reproduced by permission ©1982 Academic.)

where m_{ad}^* is also the called the *adsorption isotherm*. Note that in Chapter 6, for convenience, we expressed the adsorption isotherms through the sorption coefficient, (6.38).

Experimental results show that $m^*(p_g)_{T,...}$ is not necessarily single-valued and after the monolayers are formed and if the pore size is in a specific range, then $m^*(p_g)_{T,...}$ would be *history*-dependent due to the capillary condensation. In general, $m_{ad}^*(p_g)_{T,...} < m_{des}^*(p_g)_{T,...}$ for the same p_g, T, and gas-solid system, i.e., for a *given* p_g, T, etc., the amount of mass adsorbed is *larger* in the *desorption* branch.

In the classification of the adsorption isotherms, the adsorption isotherms have been divided into six *groups* or *types* (Gregg and Sing, 1982, p. 3).

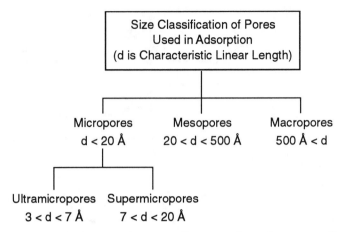

Figure 9.2 Size classification of very small pores, where the average linear size of the pore is d. (From Gregg and Sing, 1982.)

These are shown in Figure 9.1. The classification has been *originated* by Brunauer et al. Note that types IV and V possess a hysteresis loop.

For small pores, the pore size determines the extent to which the adjacent solid surfaces *interact* (or the extent to which the adsorbed molecules are *simultaneously* affected by these surfaces). For larger pores, this interaction is *less* significant, but the pore size determines the extent of the capillary effect. A classification of the pore size, based on the average pore width (or diameter) d, is given by Gregg and Sing. They also suggest a further classification of the very small pores (micropores) as given in Figure 9.2. For the pore sizes that are *only* a few times *larger* than the molecular collision cross section $2R_m$ (examples of R_m are given in the introduction to Chapter 6), *only* the monolayer adsorption occurs. The porous media possessing this range of pore size is called the *ultramicropores*. For micropores *no* hysteresis is expected.

Note that in practice a *range* of pore sizes is present, i.e., the micro-, meso-, and macropores *coexist*. Then, the adsorption isotherm demonstrates an *averaged* behavior, when the solid surface area is *nearly* uniformly divided among these different pores. The different adsorption isotherms are discussed now.

The *type I isotherm* occurs in the micropores where the pore size is only a few molecular diameters large. Then the potential fields from the neighboring walls overlap, resulting in the enhancement of the net interaction between the gas molecules and solid surfaces. This results in the nearly complete filling of the pores at low relative pressures p_g/p_{g_0}. The adsorbed mass can be modeled by

$$m_{ad} = \frac{c\dfrac{p_g}{p_{g_0}}}{1+c\dfrac{p_g}{p_{g_0}}} N_m M. \tag{9.110}$$

In this equation the *monolayer capacity* (moles of gas adsorbed per gram of solid) N_m is given by $N_m = A_o/(A_m N_A)$, $A_m = \pi R_m^2$, where R_m is the molecular radius, N_A is the Avogadro number, A_o is the *mass* specific surface area (surface area of *one gram* of solid), and c is a constant determined experimentally. The theoretical bases for (9.110) were originated by Langmuir in 1916 using the kinetic theory. At equilibrium and for a monolayer adsorption, he assumes

$$\underbrace{a_1 \frac{N_A}{2(MR_gT)^{1/2}} p_g x_o}_{\text{rate of condensation}} = \underbrace{n_m(1-x_o)f \exp\left(\frac{\Delta i_a'}{R_gT}\right)}_{\text{rate of evaporation}}, \tag{9.111}$$

where a_1 is the fraction of incident molecules that condense, x_o is the fraction of the bare sites (for a monolayer adsorption), n_m is the number of the adsorption sites per unit area, f is the frequency of oscillation of the molecules perpendicular to the surface, and $\Delta i_a'$ is the *isosteric molar heat of adsorption*. This can be solved for $1-x_o$, which is the fraction of sites occupied, and the result is

$$1 - x_o = m_{ad}^* = \frac{m_{ad}}{N_m M} = \frac{bp_g}{1+bp_g}, \tag{9.112}$$

which is the *Langmuir equation*. In this equation the simple theory gives $b = a_1 N_A/[2(MR_gT)^{1/2} n_m f]\exp(-\Delta i_a'/R_gT)$, however, as was mentioned earlier, in practice b (or c) is determined experimentally.

The *type II isotherm* occurs in adsorption of gases by the nonporous solids. First a monolayer is formed and then the condensation continues by the adsorption of the molecules to the first layer. The process then continues, but the potential field weakens as more layers are formed. The adsorption isotherm is modeled by

$$m_{ad} = \frac{c\dfrac{p_g}{p_{g_0}}}{\left(1-\dfrac{p_g}{p_{g_0}}\right)\left[1+(c-1)\dfrac{p_g}{p_{g_0}}\right]} N_m M, \tag{9.113}$$

which is called the *Brunauer-Emmett-Teller equation* where again c is determined experimentally.

The *type III isotherm* is similar to the type II except that the multilayer is developed *simultaneously* with the completion of the monolayer. So the point of inflection does *not* appear. This is characteristic of weak gas-solid interactions. In addition to nonporous solids, type III can also occur in the macropores.

The *type IV* and *V isotherms* occur in the mesopores or micropore solids. For type V isotherms, the lack of the inflection point at low p_g/p_{g_0} is *again* due to the building of the higher layers before the completion of the monolayer. The asymptote shown for $p_g/p_{g_0} \to 1$ does *not* always exist and a final increase in the amount adsorbed may be found as $p_g/p_{g_0} \to 1$. The steep increase in m^*_{ad} during the increase in p_g/p_{g_0}, i.e., during the adsorption branch, has been associated with the *coalescing* of the clusters of molecules. The steep decrease in m^*_{des} during the decrease in p_g/p_{g_0} has been associated with the evaporation of the capillary-condensed molecules. We will discuss other physical reasons for this *jump* in $\dot{m}^*$ in Section 9.6.

The *type VI isotherm* is also called the *stepped isotherm* and is created by the *rapid* fall of the interaction potential with the distance from the surface. For uniform surfaces, each layer becomes complete at a relative pressure p_g/p_{g_0} because of the large difference in the interaction potential between the successive layers. Then each layer gives rise to a step corresponding to a buildup of that layer. This is followed by an increase in p_g/p_{g_0} before the pressure is high enough for the adsorption of the next layer.

(B) Isosteric Heat of Adsorption

In Chapter 6 we dealt with the adsorption in terms of the adsorption energy added to the adsorbed gas molecules. In principle this energy is *shared* between the solid and adsorbed gas molecules, but for convenience the *molar energy of adsorption* Δu_a is *assigned* to the adsorbed gas as

$$\Delta u_a = u_{fa} - u_g, \qquad (9.114)$$

where u_g is the *molar internal energy* of the *bulk gas* and u_{fa} is the *molar internal energy* of the *adsorbed gas* (which can be in a condensed state). Note that Δu_a varies as more and more molecules are adsorbed (first partial coverage of a monolayer, then a complete monolayer, and so on for the second and higher layers). Therefore, we should add the qualifier *integral* to emphasize that Δu_a depends on the *previous* state of the *coverage*.

The *molar integral enthalpy of adsorption* Δi_a is the same as the *molar heat of adsorption*. The *differential molar enthalpy of adsorption* $\Delta i'_a$ is defined as (Gregg and Sing, 1982, p. 14)

$$\Delta i'_a \equiv \left.\frac{\partial u_{fa}}{\partial N_a}\right|_{T,p} - i_g, \qquad (9.115)$$

where N_a is the number of the gas moles adsorbed.

Gregg and Sing (1982, 16–17) show that

$$-\frac{\Delta i'_a}{R_g T^2} = \left.\frac{\partial \ln p_g}{\partial T}\right|_{N_a}, \qquad (9.116)$$

where $\Delta i'_a$ is called the *molar isosteric enthalpy of adsorption* (or *isosteric heat of adsorption*). This can be written as

$$(\ln p_g)_{N_a} = \Delta i'_a \frac{1}{R_g T} + a_1, \qquad (9.117)$$

where a_1 is a constant. Then, by plotting $(p_g)_{N_a}$ versus T, $\Delta i'_a$ can be determined, where, as noted, N_a is kept constant. Note that $\Delta i'_a$ is also a function of N_a. If the molar heat of condensation $i_{g\ell}$ was taken out of $\Delta i'_a$, the *net isosteric heat of adsorption* is found as $\Delta i'_a - \Delta i_{\ell g}$, where $\Delta i'_a < 0$ and $\Delta i_{g\ell} < 0$, i.e., energy is *evolved* in adsorption and condensation.

9.5.2 Condensation in a Mesoporous Solid

The mesoporous solids can offer significantly large specific areas, and therefore, can be used in the adsorption processes. Also, *when* the Kelvin relation for the vapor pressure lowering (due to meniscus curvature) is valid in this range of pore sizes, the adsorption isotherm is used for the *pore size prediction*. As will be shown for mesoporous solid, it is not clear that the Kelvin equation is accurate for the *description* of the adsorption isotherm, especially the *hysteresis* loop. A bed made of nonconsolidated *small* particles shows a type II isotherm. If this bed is *compacted* using relatively high pressures, the isotherms show a type IV behavior with an *increase* in the amount of gas adsorbed in the *capillary condensation/evaporation* regime. Figure 9.3 is adapted from Gregg and Sing (1982, p. 112) and shows this regime as well as the *multilayer adsorption* regime. Some of the features of the type IV adsorption isotherm are given here.

- There is a *transitional relative pressure* $(p_g/p_{g_0})_t$ where above this transitional value the adsorption by a mesoporous solid becomes distinct from that for the nonporous solids. This transition is shown in Figure 9.3.

- For $p_g/p_{g_0} < (p_g/p_{g_0})_t$, the adsorption is restricted to a *thin* layer on the solid surfaces. Therefore, *no* distinction appears in the general behavior of $m^*_{\text{ad}}(p_g/p_{g_0})$ in this regime.

- For $p_g/p_{g_0} > (p_g/p_{g_0})_t$, the capillary condensation commences starting from the *smallest* pores [as suggested by the Kelvin equation (9.40)].

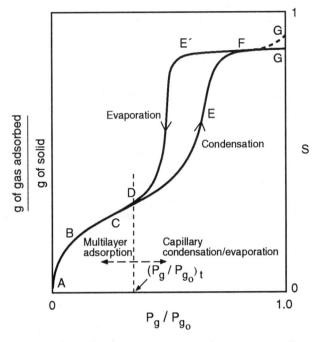

Figure 9.3 Type IV adsorption isotherm with the multilayer adsorption and the capillary condensation/evaporation regimes specified. The condensation/evaporation regime is marked by a hysteresis loop. (From Gregg and Sing, reproduced by permission ©1982 Academic.)

- When $p_g/p_{g_o} = 1$, almost all the pores are filled with the liquid phase, and the *saturation* s (Chapter 8) tends to unity (called *Gurvitsch rule*).

- The hysteresis loop in some gas-solid systems can appear with point F moving toward the right. Also the DEF and DE'F branches can become nearly *vertical*.

- The amount adsorbed (or remaining adsorbed) is always *larger* along the *desorption* branch.

- The increase in p_g/p_{g_o} with increase in m^*_{ad} can be *tentatively* explained by the Kelvin relation. This states that as the condensation proceeds and the film thickness and radius of curvature of the liquid in the pores increase, the vapor pressure approaches the saturation pressure. This is further discussed by Defay and Prigogine (1966, 222–227). The hysteresis in the capillary condensation/evaporation is discussed later.

- Traditionally the Kelvin relation has been used to explain the various features of the type IV isotherms. As will be shown this classical thermodynamic approach has limited applicability and the more recent molecular theories (Section 9.6) are more successful in predicting these features. Some of the curvature-based arguments and their limitations are discussed later.

- Due to the curvature alone (i.e., the Kelvin equation), the vapor pressure for nitrogen at 77.35 K and $H^{-1} = 500$ Å is only reduced by about two percent, i.e., $(p_g/p_{g_0})(H^{-1} = 500$ Å$) = 0.9810$. For water at $T = 18°$C and $H^{-1} = 1000$ Å, this is 0.9891. Therefore, using the Kelvin relation the condensation in pores larger than 500 Å will *not* be *measurably* different than that in nonporous solids.

- For pore sizes (and meniscus curvatures) less than 20 Å the Kelvin equation is not expected to hold, because the interfacial layer thickness is not substantially smaller than the pore size.

- Even in the mesopore range, as was discussed in Sections 8.1.5 and 9.4, the *van der Waals* dispersive forces and the other *charge-polarity*-based forces influence the liquid film thickness. We also expect the contact angle to be different from that for the bulk liquid when the liquid film is very thin or the pore is very small. Therefore, the application of the Kelvin equation for the determination of the pore size can only be considered *tentative*. The capillary condensation influenced by van der Waals forces will be discussed in Section 9.6.

Since the observed $(p_g/p_{g_0})_{\text{ad}} > (p_g/p_{g_0})_{\text{des}}$ for a given m^*_{ad} and T is generally explained in terms of the Kelvin equation and by using the observation that condensation begins at the solid surface, we now consider this explanation. Based on these arguments, the condensation begins after a thin adsorbed film is created over the solid, which is the nucleation surface for the condensation. In order to use the Kelvin equation, geometrically specific arguments are used in which it is shown that the curvature during the adsorption and desorption branches is *different*. The simple form of the Kelvin equation (9.40) can be written as

$$\frac{p_g}{p_{g_0}} = \exp\left(-\frac{2\sigma v_\ell}{R_g T} H\right). \tag{9.118}$$

Therefore, we need to show that

$$H_{\text{ad}} = \frac{1}{2}\left(\frac{1}{r_1} + \frac{1}{r_2}\right)_{\text{ad}} < H_{\text{des}} = \frac{1}{2}\left(\frac{1}{r_1} + \frac{1}{r_2}\right)_{\text{des}}$$

for constant m^*_{ad} and T, (9.119)

i.e., the mean radius of curvature is *larger* in the *adsorption branch*. One of the examples used is that of the condensation inside a cylinder open on both sides. Condensation *begins* at the walls and then proceeds inward (no gravity effect) and if $(r_1)_{ad}$ is the radius of the cylindrical meniscus, then $(r_2)_{ad} = \infty$. When $p_g/p_{g_0} \to 1$ and the cylinder is filled, the desorption will begin and evaporation takes place from the open ends with hemispherical menisci where $(r_1)_{des} = (r_1)_{des}$. Now if we assume that $(1/r_1)_{ad} < (2/r_1)_{des}$, then we have $H_{ad} < H_{des}$. However, this assumption is not *rigorously* justified. Other geometric arguments, including the *ink bottle, tubes with periodically varying cross section, two- and three-dimensional wedges, slit-shaped pores*, and *packed bed of spheres*, are also given in Gregg and Sing (1982, 126–131). The proof of (9.119) for any saturation (or m^*_{ad}), without directly solving the dynamic capillary equations subject to prescribed geometry and contact angle, will not be rigorous. Therefore, most arguments merely support one's intuition that for a given s and T during desorption (reduction of saturation by evaporation) the mean radius of curvature is *smaller*. This follows a trend similar to that of the non-evaporative drainage branch of the capillary pressure, where the radius of curvature was found to be smaller during drainage (for a given saturation s). Note that the fluid behavior in narrow pores is much more complex than that depicted by the Kelvin relation. This will be further discussed in the next section, where it is suggested that the occurrence of hysteresis is due to the existence of a *metastable state* in the *phase transition* or the existence of *metastable films*.

9.6 Prediction of Fluid Behavior in Small Pores

The hysteresis appearing in the adsorption isotherm for the mesopore porous media can be in part explained by the *fluid models* that include the van der Waals forces. One of the pore geometries that has been studied is that of a *long cylinder* of radius R. In the existing treatments, the solid-fluid interaction potential is modeled with a different extent of rigor. Then the stability of the fluid is studied. In the *phase-change transition theory*, the *marginal state*, where the *gaseous phase* changes into the *liquid phase* (or vice versa), is analyzed. In the *thin liquid film stability theory*, the transition from the *layered (film) coverage* to the *partial liquid coverage* of the entire *pore space* is examined. Both approaches treat the fluid as a continuum assuming that the fluid domain under the study is several molecules thick and that the density distribution is continuous. The understanding of the physics of the surface adsorption is still maturing. In the following, both of these theories and their *predictions* of the *hysteresis* in small pores are discussed.

9.6.1 PHASE TRANSITION IN SMALL PORES: HYSTERESIS

As the pore size decreases, i.e., as the wall spacing decreases, the overlapping of the surface potentials results in a larger influence on the fluid *physioadsorption* and the phase transition. When the pore size is of the order of the molecular diameter, the interaction of the fluid-solid and the fluid-fluid potentials can result in a critical point in the gas-liquid coexistence curve. We consider the recent advances made in the area of gas-liquid transition in pores under the influence of the van der Waals forces. One of the theoretical treatments of adsorption and capillary coexistence is based on the *density functional*, which uses the *simple mean-field* approach. One of the analyses is that of Peterson et al. (1986). In this treatment, the determination of the *inhomogeneous* potential and the density fields in a cylindrical pore is made using *prescribed* solid-fluid and fluid-fluid *potentials*.

The *normalized number density distribution* $\rho(\mathbf{x})$ for the fluid is to be determined. The density can be normalized using the *hard sphere diameter* d. The *grand potential functional* is given as

$$\Omega(\rho) = E_h + E_{ff} + E_{sf}. \tag{9.120}$$

The formulation, as well as some of the details of the components of the grand potential, is also given by Evans et al. (1986). The first of the *three* potentials on the right-hand side is

$$E_h = \int F_h(\rho)\,d\mathbf{x}_1, \tag{9.121}$$

where F_h is Helmholtz free energy density of a uniform *hard sphere fluid* of density ρ. The second one is

$$E_{ff} = \frac{1}{2}\int u_{ff}(\mathbf{x}_{12})\rho(\mathbf{x}_1)\rho(\mathbf{x}_2)\,d\mathbf{x}_2\,d\mathbf{x}_1, \tag{9.122}$$

where u_{ff} is the *attractive part of the fluid-fluid potential*. The third one is

$$E_{sf} = \int (\mu_{sf} - \mu)\rho(\mathbf{x}_1)\,d\mathbf{x}_1, \tag{9.123}$$

where μ_{sf} is the *external (solid-fluid) potential*, μ is the chemical potential, $\mathbf{x}_1$ is located in the fluid, and $\mathbf{x}_2$ is located in the solid. The *minimization* of the grand potential leads to the expression for *equilibrium density* $\rho(\mathbf{x})$. This expression is

$$\mu_h(\rho) = \frac{\partial F_h}{\partial \rho} = \mu - \mu_{sf}(\mathbf{x}_1) - \mu_{ff}(\mathbf{x}_1), \tag{9.124}$$

where

$$\mu_{ff}(\mathbf{x}_1) = \int u(\mathbf{x}_{12})\rho(\mathbf{x}_2)\,d\mathbf{x}_2 \tag{9.125}$$

is the potential at $\mathbf{x}_1$ due to fluid-fluid interactions.

The hard sphere model uses the density ρ and porosity ε (this is for the fluid phase), and the model used is (*Percus-Yevick approximation*)

$$\frac{\mu_h}{k_B T} = \ln \frac{\rho}{\varepsilon} + \frac{(1-\varepsilon)\left[14 - 13(1-\varepsilon) + 4(1-\varepsilon)^2\right]}{2\varepsilon^3}, \qquad (9.126)$$

where $\varepsilon = 1 - \pi\rho/6$, ρ is normalized using d^3 and the hard sphere diameter d is prescribed as a function of temperature.

The solid-fluid potential for a constant density solid ρ_s is given as

$$\mu_{sf} = \rho_s \int v(\mathbf{x}_{12})\,d\mathbf{x}_2, \qquad (9.127)$$

where $v(\mathbf{x}_{12})$ is the *interaction potential* between a fluid particle at $\mathbf{x}_1$ and a volume element in the solid at $\mathbf{x}_2$. The potentials u and v are taken to be of the *Lennard-Jones* type *potential* (Barber and Loudun, 1989), i.e.,

solid-fluid:
$$v = 4v_0 \left[\left(\frac{2R_{mv}}{r}\right)^{12} - \left(\frac{2R_{mv}}{r}\right)^6\right], \qquad (9.128)$$

fluid-fluid:
$$u = \begin{cases} -u_0 - 4u_0\left[\left(\frac{2R_{mu}}{r}\right)^{12} - \left(\frac{2R_{mu}}{r}\right)^6\right] & r < 2^{1/6}(2R_{mu}), \\ 4u_0\left[\left(\frac{2R_{mu}}{r}\right)^{12} - \left(\frac{2R_{mu}}{r_c}\right)^{12} - \left(\frac{2R_{mu}}{r}\right)^6 + \left(\frac{2R_{mu}}{r_c}\right)^6\right] & \\ & 2^{1/2}(2R_{mu}) < r < r_c, \\ 0 & r > r_c, \end{cases}$$
$$(9.129)$$

where v_0 and u_0 are the *well depths* and $2R_{mv}$ and $2R_{mu}$ are the collision diameters.

The cut-off of the fluid-fluid potential at r_c is needed in order to make the fluid-fluid interactions more realistic. The parameters are taken to be those *characteristic* of argon in a pore of solid carbon dioxide (i.e., Ar–CO_2 system). The numerical values for this system are

$$\frac{v_0}{k_B} = 119.8 \text{ K}, \quad 2R_{mu} = 3.405 \text{ Å}, \quad r_c = 2.5(2R_{mu}), \qquad (9.130)$$

$$\frac{v_0}{u_0} = 1.277, \quad \text{and} \quad \frac{R_{mv}}{R_{mu}} = 1.094. \qquad (9.131)$$

9.6 Prediction of Fluid Behavior in Small Pores

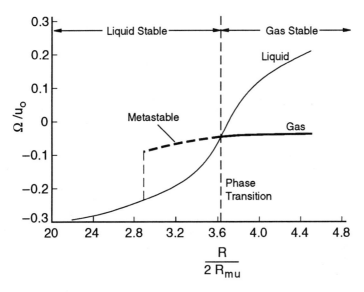

Figure 9.4 Phase stability diagram for the Ar–CO$_2$ system predicted by Peterson et al. (From Peterson et al., reproduced by permission ©1986 Roy. Soc. Chem.)

For cylindrical pores with $r = 0$ on the cylinder axis, we can write (9.124) as

$$\mu_h[\rho(r_1)] = \mu - \mu_{sf}(r_1) - \mu_{ff}(r_1), \tag{9.132}$$

$$r = (x^2 + y^2)^{1/2}, \quad \theta = \tan^{-1}\frac{y}{x}, \quad z = z. \tag{9.133}$$

When $\varepsilon \to 1$ (small fluid density), then (9.132) gives

$$\frac{\mu_h}{k_B T} = \ln \rho(r_1), \quad \text{for } \varepsilon \to 1. \tag{9.134}$$

Therefore, in a pore of radius R and for $r_1 \to R$ (where $\mu_{sf} > \mu_{ff}$), this equation becomes

$$\rho(r_1) = \exp\left(\frac{\mu - \mu_{sf}}{k_B T}\right) \quad \text{for } \varepsilon \to 1, \ r_1 \to R. \tag{9.135}$$

Some of the calculation details are given by Peterson et al. Their results for the normalized grand potential are given in Figure 9.4. The results are for a given temperature and pressure and for the Ar–CO$_2$ system. When two solutions are found (i.e., the liquid *or* the gas phase can be present), then the one with the *lower* Ω is the stable one. Also shown in the figure is a *metastable* gas phase, which occurs when the pore is *initially* filled with

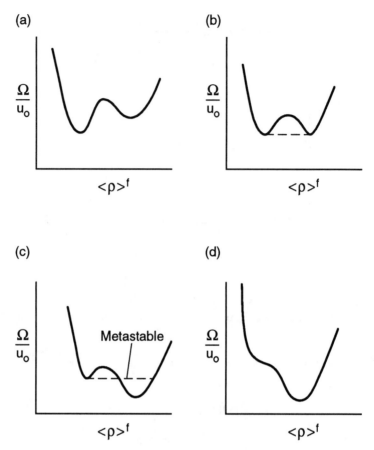

Figure 9.5 Phase stability as determined from the minimum in the grand free energy Ω. (a) and (d) are single-phase (gas and liquid) stable states, (b) is for the critical point, and (c) is for the metastable gas phase. (From Peterson et al., reproduced by permission ©1986 Roy. Soc. Chem.)

the gas phase and the radius R of the cylinder is *reduced*. In this case, the *transition to a liquid-filled* pore occurs at $R/(2R_{mu}) = 2.85$ instead of the point of the *thermodynamic phase equilibrium* $R/(2R_{mu}) \simeq 3.6$.

The existence of the liquid or gas phase is determined from the variation of the grand free energy with respect to the mean (pore volume-averaged) density. This is depicted in Figure 9.5 which is also adapted from Peterson et al. In this figure, in case (a) the gas is *stable* because the minimum in the grand free energy occurs at the lower density. In case (b), the gas and liquid *coexist*. This corresponds to a *critical point* on the gas-liquid coexistence curve. In case (c), the liquid is the *stable* phase, and in case (d), the liquid is the stable phase and *no* local minimum in the gas phase is present. The *metastable gas* phase shown in Figure 9.4 corresponds to

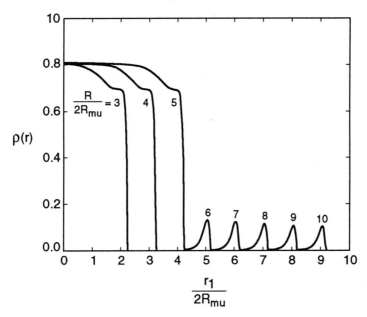

Figure 9.6 Distribution of the normalized density in a capillary tube for various tube radii. The prediction is by Peterson et al. The gas-liquid transition for this system occurs at $R/(2R_{mu}) = 5.07$. (From Peterson et al., reproduced by permission ©1986 Roy. Soc. Chem.)

case (c). The *existence* of the metastable state can lead to the occurrence of *hysteresis*, i.e., depending on whether the transition state is *reached* from the liquid or the gaseous phase side, *different potential* paths are taken.

The results presented earlier show that as $R/(2R_m)$ becomes very *small*, *only* the liquid phase is present in the pores. Note that, in principle, the present analysis is limited to the relatively *large* pores where the *molecular fluctuations* are not significant. Also, we have included this analysis *only* to show the present *direction* in the modeling of the capillary condensation and phase change in small pores. Therefore, we only point out the *qualitative* trends in the available results.

For the preceding system, the distribution of the *normalized density* in the pore as a function of the normalized pore size is also given by Peterson et al. These results are shown in Figure 9.6. The results are for $k_B T/u_o = 0.6$, and $p_g/p_{go} = 0.6$. For $R/(2R_{mu}) = 5.07$ the *gas-liquid transition* occurs. For $R/(2R_{mu}) < 5.07$ *only* the liquid phase (larger density) is found and this phase covers the *entire* pore with a *very sharp* decrease in the density near the *confining wall*. Therefore, the liquid is *not* wetting the solid. For $R/(2R_{mu}) > 5.07$ *only* the gaseous phase is present and the central core

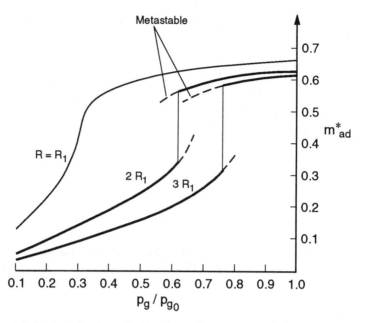

Figure 9.7 Predicted adsorption isotherm by Evans et al. for various capillary tube radii. $R_1 \simeq 10 R_{mu}$ and $R_c = 20 R_{mu}$. (From Evans et al., reproduced by permission ©1986 Roy. Soc. Chem.)

of the cylinder is much *less* populated than a small annulus zone that exists near the wall. Again the population *decreases* as the wall is approached.

As p_g/p_{g_0} increases [for a given $R/(2R_{mu})$], a *transition* from the gaseous to the liquid phase occurs. This transition appears as a *jump* in the adsorption isotherm where the amount of the gas adsorbed increases very rapidly.

Evans et al. (1986) have analyzed the capillary system in a manner similar to Peterson et al., except for the *models* used for the potentials. Their results for an adsorption isotherm are given in Figure 9.7. The adsorption is determined from

$$m^*_{ad} = \frac{2d^3}{R^2} \int_0^R \rho r_1 \, dr_1. \tag{9.136}$$

Their results are for a wetting liquid phase $\theta_c = 0$ and $T/T_c = 0.8$, where T_c is the *bulk critical temperature*. For a small tube radius ($R_1 \simeq 10 R_{mu}$), the amount of mass adsorbed increases monotonically with p_g/p_{g_0}. However, as R is increased a *jump appears* and the *magnitude* of this jump in $\dot{m}^*$ increases as R increases. The metastable portion of the isotherm also decreases as R decreases such that for a very small R (i.e., $R_1 < R_c$, where R_c is the *critical radius* and for this example it is about $20 R_{mu}$) the metastable branch and the jump *disappear*, i.e., *no hysteresis* occurs. Since for $R_1 < R_c$ no *first-order transition* occurs, the isotherm for $R_1 < R_c$ is called *supercritical*. The Kelvin equation can be written as

$$R_g T \ln \frac{p_g}{p_{g_0}} \simeq -\frac{2\sigma \cos\theta_c}{R\left(\dfrac{1}{v_\ell} - \dfrac{1}{v_g}\right)}, \qquad (9.137)$$

where θ_c is the contact angle. Evans et al. show that this equation *does not* predict the isotherm accurately for completely wetting liquids ($\theta_c = 0$). For the completely wetting case, a *thick* liquid-like film is present between the wall and the gas (Evans et al.).

9.6.2 STABILITY OF LIQUID FILM IN SMALL PORES: HYSTERESIS

For the case of low temperatures, an alternative approach to the description of the fluid behavior in small pores is the application of the *quantum hydrodynamic theory* of *long-wavelength excitations* to thin liquid films. Cole and Saam (1974) used this approach as well as the thermodynamic properties (and stability) of thin liquid films. The expanded version of their developments are given by Saam and Cole (1975), and the related refinement of the parameters used to express the van der Waals forces are given by Vidali and Cole (1981), and Rauber et al. (1982). *Experimental* verifications using extremely uniform pore diameters have been made by Awschalom et al. (1986) for oxygen and krypton at 89.50 K and 118.0 K, respectively.

Here we do not review the quantum hydrodynamic treatment of Saam and Cole and the simplifying assumptions made about the liquid behavior at *low temperatures*. It should be mentioned that part of their treatment is specific to liquid helium, but their predictions apply to other cryogenic fluids that do not have the superfluid behavior of helium. Their development of thermodynamics of stability is along the lines of those we have discussed before. Their analysis shows that *during* the filling (*adsorption*), as the layer of adsorbed film thickens [Figure 9.8(a)] in a cylindrical pore of radius R, normal mode excitations become *sensitive* to the geometry and *instability* occurs. This instability gives rise to the *partial* filling of the pores, which corresponds to the *minimization* of the grand potential via the *surface area* minimization [Figure 9.8(b)]. Thermodynamic considerations also give this *critical film thickness* $R - R_{ca}$. Also, *during* the emptying (*desorption*) of the pores, they show that there exists a *metastable* film thickness $R - R_{cd}$, where it becomes *favorable* (minimum grand potential) for the system to form a *layered configuration* (i.e., liquid film). The van der Waals forces are included in both the hydrodynamic and the thermodynamic treatments. In the liquid-film stability approach discussed here and the phase-change transition treatment given in the last section, the underlying molecular approach is the same. However, no *direct* comparison between the two predictions is available.

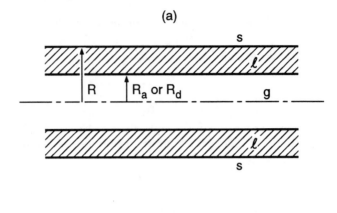

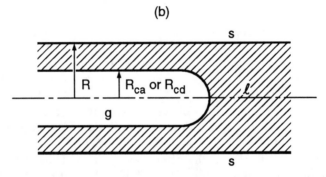

Figure 9.8 (a) A layer of adsorbed liquid film in a cylinder and (b) a partially liquid-filled cylinder. R_{ca} and R_{cd} are the critical radii for transition between the film to the partial filling and the partial filling to film, respectively. (From Awschalom et al., reproduced by permission ©1986 Amer. Phys. Soc.)

From Section 9.4, we have

$$\mu = \mu_o(p_{go}, T) + \mu_\delta - v_\ell(2H\sigma). \tag{9.138}$$

For a liquid film inside a cylinder, we use $R_a = R - \delta$ and ∞ for the two principal radii of curvature, and then we have

$$\mu = \mu_o(p_{go}, T) + v_\ell f(\delta) - \frac{v_\ell \sigma}{R_a}, \tag{9.139}$$

with the requirement for the stability given by (9.106), i.e.,

$$\frac{\partial \mu}{\partial \delta} \geq 0. \tag{9.140}$$

This gives

$$\frac{\partial \mu}{\partial \delta} = -\frac{\partial \mu}{\partial R_a} = -v_\ell \frac{\partial f}{\partial R_a} + \frac{v_\ell \sigma}{R_a^2} \geq 0 \tag{9.141}$$

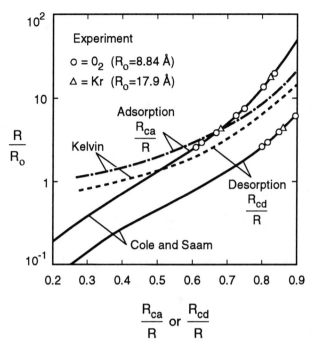

Figure 9.9 The predicted critical liquid film thickness $(R - R_{ca})$ during adsorption and the metastable liquid film thickness $(R - R_{cd})$ during desorption, and the corresponding measured values. (From Awschalom et al., reproduced by permission ©1986 Amer. Phys. Soc.)

or

$$\sigma \geq R_a^2 \frac{\partial f}{\partial R_a}, \quad \text{for } stable \text{ liquid film.} \tag{9.142}$$

At the *marginal state*

$$\sigma = R_{ca}^2 \left.\frac{\partial f}{\partial R_a}\right|_{R_{ca}}, \tag{9.143}$$

where f has been developed for the cylindrical pores by Saam and Cole and is given in terms of a hypergeometric function. The final expression for R_{ca} is

$$\frac{R^2}{R_o^2} = \frac{R_{ca}^2}{R^2}\left(1 - \frac{R_{ca}^2}{R^2}\right)^{-5/2} P_{3/2}^1\left[\frac{\left(1 + \dfrac{R_{ca}^2}{R^2}\right)}{\left(1 - \dfrac{R_{ca}^2}{R^2}\right)}\right], \tag{9.144}$$

with

$$R_a = \left(\frac{3\pi v_o}{v_\ell \sigma}\right)^{1/2}, \tag{9.145}$$

where v_o is the *van der Waals solid-fluid coefficient* and $P^1_{3/2}$ is the associated Legendre function, which is given by

$$P^1_{3/2}(z) = \frac{5}{2\pi} \int_0^\pi \left[z + (z^2-1)^{1/2} \cos\theta\right]^{3/2} \cos\theta \, d\theta. \tag{9.146}$$

For the *thermodynamic* analysis of the emptying branch, we start from (9.104), which is rewritten as

$$f(r) - \frac{2\sigma}{R_d} = \frac{R_g T}{v_\ell} \ln \frac{p_g}{p_{go}}. \tag{9.147}$$

The *asymptote* for $r \to R_{cd}$ is given as

$$\frac{R}{R_d} = \frac{R_o^2}{R^2}\left[-\frac{R^3 v_\ell f(R_{cd})}{3\pi v_o} + \frac{R^3 v_\ell f(r)}{3\pi v_o}\right] + \frac{R}{R_{cd}}. \tag{9.148}$$

The result of the *minimization* of the grand potential is

$$2\frac{R^2}{R_{cd}^2} \int_0^{R_{cd}} f(r) r \, dr = f(R_{cd}) + \frac{\sigma}{R_{cd}}. \tag{9.149}$$

Figure 9.9 shows the predictions of Cole and Saam, as well as the predictions using the Brunauer-Emmett-Teller equation (9.113), which is based on the Kelvin reasoning of the curvature effects. The experimental results of Awschalom et al. are also shown. The Kelvin predictions underestimate R_{ca} and R_{cd} for small values of R/R_o. The results show that the van der Waals forces play a *significant* role in determining the jump in the adsorption isotherm. The experimental results are not satisfactorily represented by the correlations based on the layering theory.

Note that in attempting to describe the hysteresis in the adsorption isotherm using the curvature arguments, the aim is to show that $H_{\text{des}} > H_{\text{ad}}$ for a given m^*_{ad}. As we discussed, these arguments are not successful. The arguments based on the film thickness stability aim at showing that $(m^*_{\text{ad}})_{\text{des}} > (m^*_{\text{ad}})_{\text{ad}}$ by arriving at $R_{cd} > R_{ca}$. These arguments can be more rigorously supported.

9.7 References

Antoniou, A. A., 1964, "Phase Transformation of Water in Porous Glass," *J. Phys. Chem.*, 68, 2754–2764.

Awschalom, D. D., Warnock, J., and Shafer, M. W., 1986, "Liquid-Film Instabilities in Confined Geometries," *Phys. Rev. Lett.*, 57, 1607–1610.

Barber, D. J. and Loudun, R., 1989, *An Introduction to the Properties of Condensed Matter*, Cambridge University.

Bongiorno, V. and Scriven, L. E., 1976, "Molecular Theory of Fluid Interfaces," *J. Colloid Interface Sci.*, 57, 462–475.

9.7 References

Cahn, J. W. and Hilliard, J. E., 1958, "Free Energy of a Nonuniform System. I. Interfacial Free Energy," *J. Chem. Phys.*, 28, 258–267.

Cole, M. W. and Saam, W. F., 1974, "Excitation Spectrum and Thermodynamic Properties of Liquid Films in Cylindrical Pores," *Phys. Rev. Lett.*, 32, 985–988.

Defay, R. and Prigogine, I., 1966 (English edition), *Surface Tension and Adsorption*, with collaboration of A. Bellemans, and translated from French by D. H. Everett, J. Wiley.

Dzyaloskinskii, I. E., Lifshitz, E. M., and Pitaevskii, L. P., 1961, "The General Theory of van der Waals Forces," *Adv. Phys.*, 165–209.

Evans, R., Marini Bettolo Marconi, U., and Tarazona, P., 1986, "Capillary Condensation and Adsorption in Cylindrical and Slit-Like Pores," *J. Chem. Soc., Faraday Trans.*, 82, 1763–1787.

Gregg, S. J. and Sing, K. S. W., 1982, *Adsorption, Surface Area and Porosity*, Academic.

Hsieh, J.-S., 1975, *Principles of Thermodynamic*, McGraw-Hill, Chapter 6.

Lienhard, J. H., 1982, "Corresponding States Correlations of the Spinodal and Homogeneous Nucleation Limits," *ASME J. Heat Transfer*, 104, 379–381.

Peterson, B. K., Walton, J. P. R. B., and Gubbins, K. E., 1986, "Fluid Behavior in Narrow Pores," *J. Chem. Soc., Faraday Trans.*, 82, 1789–1800.

Rauber, S., Klein, J. R., Cole, M. W., and Bruch, L. W., 1982, "Substrate-Mediated Dispersion Interaction between Adsorbed Atoms and Molecules," *Surf. Sci.*, 123, 173–178.

Saam, W. F. and Cole, M. W., 1975, "Excitation and Thermodynamics for Liquid-Helium Films," *Phys. Rev. B*, 11, 1086–1105.

Slejko, F. L., ed., 1985, *Adsorption Technology*, Marcel Dekker.

Truong, J. G. and Wayner, P. C., 1987, "Effect of Capillary and van der Waals Dispersion Forces on the Equilibrium Profile of a Wetting Liquid: Theory and Experiment," *J. Chem. Phys.*, 87, 4180–4188.

Udell, K. S., 1983, "Heat Transfer in Porous Media Heated from above with Evaporation, Condensation and Capillary Effects," *ASME J. Heat Transfer*, 105, 485–492.

Vidali, G. and Cole, M. W., 1981, "The Interaction between an Atom and a Surface at Large Separation," *Surf. Sci.*, 110, 10–18.

10
Conduction and Convection

As we discussed in Chapter 8, our knowledge of the *pore-level* fluid mechanics in two-phase flow through porous media is rather incomplete. In this chapter, we discuss thermal dispersion, i.e., convective heat transfer at the pore level, using the available knowledge about the subject. This knowledge is *even* more inconclusive. We begin with the local volume averaging of the energy equation, and then we arrive at the *effective thermal conductivity tensor* and the *thermal dispersion tensor* for the *three-phase system* (liquid-gas-solid). The same closure conditions used in the single-phase flow treatments are used. Then we examine the various features of these *tensors* such as their *anisotropy*, and we discuss some of the available models and empirical relations for the various elements of these tensors. We conclude by noting that near the *bounding surfaces*, the phase distribution nonuniformities lead to substantial variations in the magnitude of the components of the effective thermal conductivity and the dispersion tensors.

10.1 Local Volume Averaging of Energy Equation

The principles of the local volume averaging as applied to the *conduction* equation (Section 3.2), the single-phase flow *convection* equation (Section 4.3), and the two-phase flow *momentum* equation (Section 8.2), will now be applied to the two-phase flow energy equation. The concept is that developed by Whitaker (1977), where the extensive derivations are given. We expect to arrive at a local volume-averaged energy equation in which the *effective thermal conductivity* is the combined contribution of the three phases (s, ℓ, and g) to the molecular conduction, and the *thermal dispersion* is the combined dispersion in the ℓ and g phases. We consider the general case of transient temperature fields with a local heat generation and the ℓ-g phase change. The closure conditions lead to equations for the transformation vectors. For simplicity, we consider the simple case of a unit cell in a *periodic* fluid-solid structure.

10.1.1 Averaging

The point differential energy equation for the three phases are

$$(\rho c_p)_s \frac{\partial T_s}{\partial t} = k_s \nabla^2 T_s + \dot{s}_s, \qquad (10.1)$$

$$(\rho c_p)_\ell \left(\frac{\partial T_\ell}{\partial t} + \mathbf{u}_\ell \cdot \nabla T_\ell\right) = k_\ell \nabla^2 T_\ell + \dot{s}_\ell, \tag{10.2}$$

$$(\rho c_p)_g \left(\frac{\partial T_g}{\partial t} + \mathbf{u}_g \cdot \nabla T_g\right) = k_g \nabla^2 T_g - \nabla \cdot \sum_{i=1}^{n_r} \rho_{ig} i_{ig} \mathbf{u}_{ig} + \dot{s}_g, \tag{10.3}$$

where we have allowed for a *multicomponent* gas phase, while the liquid and solid phases are assumed to have *constant* and *uniform* concentrations. The gas-phase energy equation is that discussed in Chapter 6, and for a reacting gas the source term $\dot{s}_g$ contains the *enthalpies* of formation as given in Section 6.11. The gas phase heat capacity is defined similar to (6.87) and is

$$(c_p)_g = \sum_{i=1}^{n_r} \frac{\rho_{ig}(c_p)_{ig}}{\rho_g}. \tag{10.4}$$

The *boundary conditions* on $A_{\ell g}$ are obtained using $\mathbf{w}$ as the interfacial velocity and $\mathbf{n}$ and $\boldsymbol{\tau}$ as the interfacial normal and tangent unit vectors (Chapter 8), and are

mass, tangential: $\quad \mathbf{u}_\ell \cdot \boldsymbol{\tau}_{\ell g} + \mathbf{u}_g \cdot \boldsymbol{\tau}_{g\ell} = 0, \tag{10.5}$

mass, normal: $\quad \rho_\ell (\mathbf{u}_\ell - \mathbf{w}) \cdot \mathbf{n}_{\ell g}$
$\qquad\qquad + \rho_g (\mathbf{u}_g - \mathbf{w}) \cdot \mathbf{n}_{g\ell} = 0, \tag{10.6}$

species, condensable: $\quad \rho_\ell (\mathbf{u}_\ell - \mathbf{w}) \cdot \mathbf{n}_{\ell g}$
$\qquad\qquad + \rho_{cg} (\mathbf{u}_{cg} - \mathbf{w}) \cdot \mathbf{n}_{g\ell} = 0, \tag{10.7}$

species, noncondensable: $\quad \rho_{ng} (\mathbf{u}_{ng} - \mathbf{w}) \cdot \mathbf{n}_{g\ell} = 0, \tag{10.8}$

energy: $\quad \rho_\ell i_\ell (\mathbf{u}_\ell - \mathbf{w}) \cdot \mathbf{n}_{\ell g} + \rho_g i_g (\mathbf{u}_g - \mathbf{w}) \cdot \mathbf{n}_{g\ell}$
$\qquad = k_\ell \nabla T_\ell \cdot \mathbf{n}_{\ell g} + k_g \nabla T_g \cdot \mathbf{n}_{g\ell}$
$\qquad - \sum_{i=1}^{n_r} \rho_i \mathbf{u}_i i_i \cdot \mathbf{n}_{g\ell}, \tag{10.9}$

temperature: $\quad T_\ell = T_g, \tag{10.10}$

where we have used the subscripts c for the *condensable* and n for the *noncondensable* gas components.

We also have

$\mathbf{u}_\ell = 0, \quad k_\ell \nabla T_\ell \cdot \mathbf{n}_{\ell s} + k_s \nabla T_s \cdot \mathbf{n}_{s\ell} = 0, \quad T_\ell = T_s \quad \text{on } A_{\ell s}, \tag{10.11}$

$\mathbf{u}_g = 0, \quad k_g \nabla T_g \cdot \mathbf{n}_{gs} + k_s \nabla T_s \cdot \mathbf{n}_{sg} = 0, \quad T_g = T_s \quad \text{on } A_{gs}. \tag{10.12}$

10.1 Local Volume Averaging of Energy Equation

The continuity and momentum equations for two-phase flow have been given in Chapter 8, where (8.94) states the mass continuity for phase ℓ and (8.95) for phase g. The mass conservation equation for each component of the gas is given by (6.63), where the effective mass diffusivity and mass dispersion are also included in this equation.

Next the temperature (and the velocity and density) in each phase is *decomposed* into a phase volume-averaged component $\langle T \rangle^s$ and a deviation component T'_s, as was done in *all* the previous treatments of the local volume averaging given in Chapters 2, 3, 4, 6, 7, and 8.

We now proceed with the local volume-averaging treatment. The details of the derivation are given by Whitaker. The local volume averaging of the *solid-phase* energy equation is already given by (3.14), except for the heat generation and storage terms. We will restate (3.14), assuming a constant k_s, as

$$(1-\varepsilon)(\rho c_p)_s \frac{\partial \langle T \rangle^s}{\partial t}$$

$$= k_s \nabla \cdot \left[\nabla (1-\varepsilon) \langle T \rangle^s + \frac{1}{V} \int_{A_{s\ell}} T_s \mathbf{n}_{s\ell} \, dA + \frac{1}{V} \int_{A_{sg}} T_s \mathbf{n}_{sg} \, dA \right]$$

$$+ \frac{k_s}{V} \left(\int_{A_{s\ell}} \nabla \cdot T_s \mathbf{n}_{s\ell} \, dA + \int_{A_{sg}} \nabla \cdot T_s \mathbf{n}_{sg} \, dA \right) + \langle \dot{s}_s \rangle. \qquad (10.13)$$

In developing the local volume averaging of the *liquid phase* energy equation we note that due to *evaporation* or *condensation* at $A_{\ell g}$ there is a *jump* in the normal component of the velocity across $A_{\ell g}$. The averaged equation is similar to (4.34) except that this contribution at $A_{\ell g}$ must also be added. The result is

$$\varepsilon s (\rho c_p)_\ell \frac{\partial \langle T \rangle^\ell}{\partial t} + (\rho c_p)_\ell \langle \mathbf{u}_\ell \rangle \cdot \nabla \langle T \rangle^\ell + (\rho c_p)_\ell \nabla \cdot \langle \mathbf{u}'_\ell T'_\ell \rangle$$

$$+ \frac{(\rho c_p)_\ell}{V} \int_{A_{\ell g}} T'_\ell (\mathbf{u}_\ell - \mathbf{w}) \cdot \mathbf{n}_{\ell g} \, dA$$

$$= k_\ell \nabla \cdot \left(\nabla \varepsilon s \langle T \rangle^\ell + \frac{1}{V} \int_{A_{\ell s}} T_\ell \mathbf{n}_{\ell s} \, dA + \frac{1}{V} \int_{A_{\ell g}} T_\ell \mathbf{n}_{\ell g} \, dA \right)$$

$$+ \frac{k_\ell}{V} \left(\int_{A_{\ell s}} \nabla T_\ell \cdot \mathbf{n}_{\ell s} \, dA + \int_{A_{\ell g}} \nabla T_\ell \cdot \mathbf{n}_{\ell g} \, dA \right) + \langle \dot{s}_\ell \rangle. \qquad (10.14)$$

For the *gaseous phase* volume-averaged energy equation we have

$$\sum_{i=1}^{n_r} \langle \rho_{ig} \rangle (c_p)_{ig} \frac{\partial \langle T \rangle^g}{\partial t} + \sum_{i=1}^{n_r} (c_p)_{ig} \langle \rho_{ig} \mathbf{u}_{ig} \rangle \cdot \nabla \langle T \rangle^g + \nabla \cdot \sum_{i=1}^{n_r} (c_p)_{ig} \langle \rho'_{ig} \mathbf{u}'_{ig} T'_g \rangle$$

$$+ \frac{1}{V} \int_{A_{\ell g}} \sum_{i=1}^{n_r} (\rho c_p)_{ig} T'_g (\mathbf{u}_{ig} - \mathbf{w}) \cdot \mathbf{n}_{g\ell} \, dA + \frac{\partial}{\partial t} \sum_{i=1}^{n_r} (c_p)_{ig} \langle \rho'_{ig} T'_g \rangle$$

$$= k_g \nabla \cdot \left[\nabla \varepsilon (1-s) \langle T \rangle^g + \frac{1}{V} \int_{A_{gs}} T_g \mathbf{n}_{gs} \, dA + \frac{1}{V} \int_{A_{g\ell}} T_g \mathbf{n}_{g\ell} \, dA \right]$$

$$+ \frac{k_g}{V} \left(\int_{A_{gs}} \nabla T_g \cdot \mathbf{n}_{gs} \, dA + \int_{A_{g\ell}} \nabla T_g \cdot \mathbf{n}_{g\ell} \, dA \right) + \langle \dot{s}_g \rangle. \tag{10.15}$$

Next, we assume that the *local thermal equilibrium* exists among the three phases, i.e.,

$$\langle T \rangle^\ell = \langle T \rangle^g = \langle T \rangle^s = \langle T \rangle, \tag{10.16}$$

and add the energy equation for the three phases. The result is

$$\left[(1-\varepsilon)(\rho c_p)_s + \varepsilon s (\rho c_p)_\ell + \varepsilon (1-s) \sum_{i=1}^{n_r} \langle \rho_i \rangle^g (c_p)_{ig} \right] \frac{\partial \langle T \rangle}{\partial t}$$

$$+ \left[(\rho c_p)_\ell \langle \mathbf{u}_\ell \rangle + \sum_{i=1}^{n_r} (c_p)_{ig} \langle \rho_{ig} \mathbf{u}_{ig} \rangle \right] \cdot \nabla \langle T \rangle$$

$$+ \frac{1}{V} \left[\int_{A_{\ell g}} (\rho c_p)_\ell T'_\ell (\mathbf{u}_\ell - \mathbf{w}) \cdot \mathbf{n}_{\ell g} \, dA + \int_{A_{g\ell}} \sum_{i=1}^{n_r} (\rho c_p)_{ig} T'_g (\mathbf{u}_{ig} - \mathbf{w}) \, dA \right]$$

$$= \nabla \cdot \left\{ \nabla [k_s(1-\varepsilon) + k_\ell \varepsilon s + k_g \varepsilon (1-s)] \langle T \rangle + \frac{k_s - k_\ell}{V} \int_{A_{s\ell}} T_s \mathbf{n}_{s\ell} \, dA \right.$$

$$\left. + \frac{k_\ell - k_g}{V} \int_{A_{\ell g}} T_\ell \mathbf{n}_{\ell g} \, dA + \frac{k_g - k_s}{V} \int_{A_{gs}} T_g \mathbf{n}_{gs} \, dA \right\}$$

$$+ \frac{1}{V} \left[\int_{A_{s\ell}} (k_s \nabla T_s - k_\ell \nabla T_\ell) \cdot \mathbf{n}_{s\ell} \, dA + \int_{A_{\ell g}} (k_\ell \nabla T_\ell - k_g \nabla T_g) \cdot \mathbf{n}_{\ell g} \, dA \right.$$

$$\left. + \int_{A_{gs}} (k_g \nabla T_g - k_s \nabla T_s) \cdot \mathbf{n}_{gs} \, dA \right] - (\rho c_p)_\ell \nabla \cdot \langle \mathbf{u}'_\ell T'_\ell \rangle$$

$$- \nabla \cdot \sum_{i=1}^{n_r} \int_{A_{g\ell}} (c_p)_{ig} \rho'_{ig} (\mathbf{u}'_{ig} - \mathbf{w}) T'_g \cdot \mathbf{n}_{\ell g} \, dA + \langle \dot{s} \rangle, \tag{10.17}$$

where

$$\langle \dot{s} \rangle = \langle \dot{s}_\ell \rangle + \langle \dot{s}_g \rangle + \langle \dot{s}_s \rangle + \frac{\partial}{\partial t} \sum_{i=1}^{n_r} (c_p)_{ig} \langle \rho'_{ig} T'_g \rangle. \tag{10.18}$$

From boundary conditions (10.11) and (10.12), two of the interfacial area integrals in (10.17) are zero. We also have

$$\frac{1}{V}\int_{A_{\ell g}} (k_\ell \nabla T_\ell - k_g \nabla T_g) \cdot \mathbf{n}_{\ell g}\, dA$$

$$= \frac{1}{V}\int_{A_{\ell g}} \left[\rho_\ell i_\ell (\mathbf{u}_\ell - \mathbf{w}) \cdot \mathbf{n}_{\ell g} + \sum_{i=1}^{n_r} \rho_{ig} i_{ig}(\mathbf{u}_{ig} - \mathbf{w}) \cdot \mathbf{n}_{g\ell}\right] dA. \qquad (10.19)$$

Next, assuming that $i = i(T)$, and by allowing for the heat of evaporation $i_{\ell g}$ for the c component (i.e., *condensible* component) of the gas, we have

$$i_\ell = i_{\ell_o} + (c_p)_\ell (T_\ell - T_{cg,s}), \qquad (10.20)$$

$$i_{ig} = i_{ig_o} + (c_p)_{ig}(T_g - T_{cg,s}), \qquad (10.21)$$

$$\Delta i_{\ell g} = i_{cg_o} - i_{\ell_o} + (c_p)_{cg}(\langle T \rangle - T_{cg,s}) - (c_p)_\ell (\langle T \rangle - T_{cg,s}), \qquad (10.22)$$

where $T_{cg,s}$ is the *saturation temperature* of the condensible component at one atmosphere. Using these we have

$$\rho_\ell \left[i_\ell - (c_p)_\ell T'_\ell\right](\mathbf{u}_\ell - \mathbf{w}) \cdot \mathbf{n}_{\ell g}$$

$$+ \sum_{i=1}^{n_r} \rho_{ig} [i_{ig} - (c_p)_{ig} T'_g](\mathbf{u}_{ig} - \mathbf{w}) \cdot \mathbf{n}_{g\ell} = \Delta i_{\ell g} \rho_\ell (\mathbf{u}_\ell - \mathbf{w}) \cdot \mathbf{n}_{\ell g}. \qquad (10.23)$$

We also define the *volumetric evaporation rate* $\langle \dot{n} \rangle$ as

$$\langle \dot{n} \rangle \equiv \frac{1}{V}\int_{A_{\ell g}} \rho_\ell (\mathbf{u}_\ell - \mathbf{w}) \cdot \mathbf{n}_{\ell g}\, dA. \qquad (10.24)$$

Now using (10.19), (10.23), and (10.24), the energy equation (10.17) becomes

$$\left[(1-\varepsilon)(\rho c_p)_s + \varepsilon s(\rho c_p)_\ell + \varepsilon(1-s)\sum_{i=1}^{n_r} \langle \rho_i \rangle^g (c_p)_{ig}\right] \frac{\partial \langle T \rangle}{\partial t}$$

$$+ \left[(\rho c_p)_\ell \langle \mathbf{u}_\ell \rangle + \sum_{i=1}^{n_r} (c_p)_{ig} \langle \rho_{ig} \mathbf{u}_{ig} \rangle\right] \cdot \nabla \langle T \rangle + \Delta i_{\ell g} \langle \dot{n} \rangle$$

$$= \nabla \cdot \left\{\nabla [k_\ell \varepsilon s + k_g \varepsilon (1-s) + k_s (1-\varepsilon)] \langle T \rangle + \frac{k_s - k_\ell}{V} \int_{A_{s\ell}} T_s \mathbf{n}_{s\ell}\, dA \right.$$

$$\left. + \frac{k_\ell - k_g}{V} \int_{A_{\ell g}} T_\ell \mathbf{n}_{\ell g}\, dA + \frac{k_g - k_s}{V}\int_{A_{gs}} T_g \mathbf{n}_{gs}\, dA\right\}$$

$$- (\rho c_p)_\ell \nabla \cdot \langle \mathbf{u}'_\ell T'_\ell \rangle - \nabla \cdot \sum_{i=1}^{n_r} (c_p)_{ig} \langle \rho'_{ig} \mathbf{u}'_{ig} T'_g \rangle + \langle \dot{s} \rangle. \qquad (10.25)$$

Whitaker (1977, p. 158) shows that by using the *Gray theorem*, i.e.,

$$\langle T \rangle^s \nabla (1-\varepsilon)$$
$$= \frac{1}{V} \left(\int_{A_{s\ell}} (T'_s - T_s) \mathbf{n}_{s\ell} \, dA + \int_{A_{sg}} (T'_s - T_s) \mathbf{n}_{sg} \, dA \right), \quad (10.26)$$

the integrands in the conduction terms are changed such that T_s, T_ℓ, and T_g are replaced with T'_s, T'_ℓ, and T'_g. The resulting energy equation is similar to that for the case of the single-phase flow and *for* $k_s = 0$ this equation becomes (4.34).

10.1.2 Effective Thermal Conductivity and Dispersion Tensors

The energy equation given in the last section contains terms that are similar to those that were previously labeled as the effective thermal conductivity and thermal dispersion. For simplicity we consider a *single-component gas*, and then as before we use the transformation (closure equations), i.e.,

$$T'_s = \mathbf{b}_s \cdot \nabla \langle T \rangle, \quad T'_\ell = \mathbf{b}_\ell \cdot \nabla \langle T \rangle, \quad T'_g = \mathbf{b}_g \cdot \nabla \langle T \rangle. \quad (10.27)$$

We also define the *effective conductivity tensor* for two-phase flow in porous media as

$$\mathbf{K}_e \equiv [k_\ell \varepsilon s + k_g \varepsilon (1-s) + k_s (1-\varepsilon)] \mathbf{I} + \frac{k_s - k_\ell}{V} \int_{A_{s\ell}} \mathbf{n}_{s\ell} \mathbf{b}_s \, dA$$
$$+ \frac{k_\ell - k_g}{V} \int_{A_{\ell g}} \mathbf{n}_{\ell g} \mathbf{b}_\ell \, dA + \frac{k_g - k_s}{V} \int_{A_{gs}} \mathbf{n}_{gs} \mathbf{b}_g \, dA. \quad (10.28)$$

Also, define the *dispersion tensor* for two-phase flow in porous media as

$$(\rho c_p)_\ell \mathbf{D}^d = -(\rho c_p)_\ell \langle \mathbf{u}'_\ell \mathbf{b}_\ell \rangle - (\rho c_p)_g \langle \mathbf{u}'_g \mathbf{b}_g \rangle. \quad (10.29)$$

The *total conductivity tensor* $\mathbf{D}$ is defined as

$$\mathbf{D} = \frac{\mathbf{K}_e}{(\rho c_p)_\ell} + \mathbf{D}^d. \quad (10.30)$$

By using these in the thermal energy equation we have

$$[(1-\varepsilon)(\rho c_p)_s + \varepsilon s (\rho c_p)_\ell + \varepsilon (1-s) \langle \rho \rangle^g c_{p_g}] \frac{\partial \langle T \rangle}{\partial t}$$
$$+ [(\rho c_p)_\ell \langle \mathbf{u}_\ell \rangle + (\rho c_p)_g \langle \mathbf{u}_g \rangle] \cdot \nabla \langle T \rangle + \Delta i_{\ell g} \langle \dot{n} \rangle$$
$$= \nabla \cdot [\mathbf{K}_e + (\rho c_p)_\ell \mathbf{D}^d] \cdot \nabla \langle T \rangle + \langle \dot{s} \rangle. \quad (10.31)$$

For the *multicomponent* gas phase, the gas dispersion term becomes complicated due to inclusion of the deviation component of the density for each species. If we assume that we can arrive at **D** for a *multicomponent* gas phase, we can then write the energy equation as

$$\left[(1-\varepsilon)(\rho c_p)_s + \varepsilon s(\rho c_p)_\ell + \varepsilon(1-s)\sum_{i=1}^{n_r}\langle\rho_i\rangle^g (c_p)_{ig}\right]\frac{\partial\langle T\rangle}{\partial t}$$

$$+ \left[(\rho c_p)_\ell \langle \mathbf{u}_\ell\rangle + \sum_{i=1}^{n_r}\langle\rho_{ig}\rangle^g (c_p)_{ig}\langle\mathbf{u}_g\rangle\right]\cdot\nabla\langle T\rangle + \Delta i_{\ell g}\langle\dot{n}\rangle$$

$$= \nabla\cdot\left[\mathbf{K}_e + (\rho c_p)_\ell \mathbf{D}^d\right]\cdot\nabla\langle T\rangle + \langle\dot{s}\rangle, \tag{10.32}$$

where we have made the simplifying approximation that

$$\langle\rho_{ig}\mathbf{u}_{ig}\rangle \simeq \langle\rho_{ig}\rangle^g \langle\mathbf{u}_g\rangle. \tag{10.33}$$

For the simple case of single-component systems, we can arrive at the *closure equations* and their boundary conditions. This can be done by assuming *periodic* three-phase structures (phase distributions) with simple unit cells [such as the one used for the relative permeability in Section 8.5.3 (A)]. Then, $\mathbf{u}'_\ell$ and $\mathbf{u}'_g$ as found in that section will be used to solve for [**b**] in the [**b**] equation. The [**b**] equation is the simplified [T'] equation and the procedure for its derivation is similar to that given in Sections 3.2 and 4.3. In Section 10.3.2 we will consider the solution to the [**b**] equation for the case of a single-component system and where $\langle\dot{n}\rangle = 0$.

10.2 Effective Thermal Conductivity

In the *absence* of any motion, heat generation, and phase change, the energy equation (10.32) becomes

$$\left[(1-\varepsilon)(\rho c_p)_s + \varepsilon s(\rho c_p)_\ell + \varepsilon(1-s)\sum_{i=1}^{n_r}\langle\rho_i\rangle^g (c_p)_{ig}\right]\frac{\partial\langle T\rangle}{\partial t}$$

$$= \nabla\cdot\mathbf{K}_e\cdot\nabla T \quad \text{for} \quad \mathbf{u} = 0. \tag{10.34}$$

Then, for given phase distributions we can determine $\mathbf{K}_e$ for the representative elementary volume V, as it was done for the single-fluid phase f-s systems. We note that the local saturation (wetting phase), which is defined as

$$s = \frac{1}{\varepsilon V}\int_V a_\ell(\mathbf{x})\,dV, \tag{10.35}$$

is an integrated quantity and does not contain the *all* necessary information for the determination of $\mathbf{K}_e$.

Then we expect a functional relationship for the effective thermal conductivity tensor, of the form

$$\mathbf{K}_e = \mathbf{K}_e[k_s, k_\ell, k_g, a_\ell(\mathbf{x}), a_g(\mathbf{x})]. \tag{10.36}$$

However, $\mathbf{K}_e$ can be given in terms of the more *readily* measurable quantities such as

$$\mathbf{K}_e = \mathbf{K}_e(k_s, k_\ell, k_g, \mathbf{u}_g, \mathbf{u}_g, \sigma, \frac{\mu_g}{\mu_\ell}, \frac{\rho_g}{\rho_\ell}, s, \theta_c, \varepsilon,$$

$$\text{solid structure, history}). \tag{10.37}$$

This replacement of the variables is done *noting* that the phase distributions *depend* on the velocity field, etc. We also expect *two* asymptotic behaviors for $\mathbf{K}_e$, which for *isotropic* phase distributions are given

$$\text{for } s \to 1 \qquad \mathbf{K}_e = k_{e(s-\ell)}\mathbf{I} = k_e(s=1)\mathbf{I},$$

$$\text{for } s \to 0 \qquad \mathbf{K}_e = k_{e(s-g)}\mathbf{I} = k_e(s=0)\mathbf{I},$$

$$k_e = k_e(k_s, k_f, \varepsilon, \text{solid structure}). \tag{10.38}$$

Figures 10.1 (a) and (b) show a rendering of a two-dimensional phase distributions in a representative elementary volume. The condition envisioned is that of the two-phase flow through a packed bed of randomly arranged spherical particles. The *flow* is along the x-axis, where in Figure 10.1(a) each phase is *continuous* in that direction. In the y- and z-directions the two phases are taken to be *discontinuous*. We have *envisioned* that each phase flows in channels that spread over a few particles. Again, we note that because the phase distributions depend on $\mathbf{u}$, i.e.,

$$a_\ell(\mathbf{x}) = a_\ell(\mathbf{x})[\mathbf{u}_g(\mathbf{x}), \mathbf{u}_\ell(\mathbf{x}), \text{etc.}], \tag{10.39}$$

$$a_g(\mathbf{x}) = a_g(\mathbf{x})[\mathbf{u}_g(\mathbf{x}), \mathbf{u}_\ell(\mathbf{x}), \text{etc.}], \tag{10.40}$$

even if we are considering the effective thermal conductivity (also called the *stagnant effective thermal conductivity*) we must acknowledge that $\mathbf{K}_e = \mathbf{K}_e[\mathbf{u}_g(\mathbf{x}), \mathbf{u}_\ell(\mathbf{x}), \text{etc.}]$ through the influence of $\mathbf{u}_g$ and $\mathbf{u}_\ell$ on a_ℓ and a_g. One of the significant consequences of this dependence is that $\mathbf{K}_e$ is noticeably *anisotropic*. The various components of $\mathbf{K}_e$ are shown in Figure 10.1. We note that in the *existing* treatments the simple relation $\mathbf{K}_e = K_e(s)$ is assumed.

Since in two-phase flow and heat transfer in porous media for any direction the bulk effective thermal conductivity is generally much smaller than the bulk thermal dispersion, the available studies on $\mathbf{K}_e$ are limited. In the following we briefly discuss the *anisotropy* of $\mathbf{K}_e$ and then review the available treatments.

10.2 Effective Thermal Conductivity 553

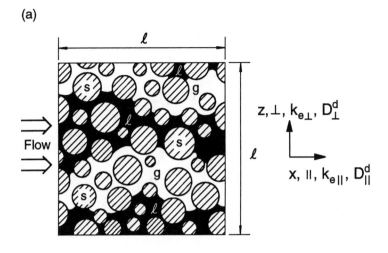

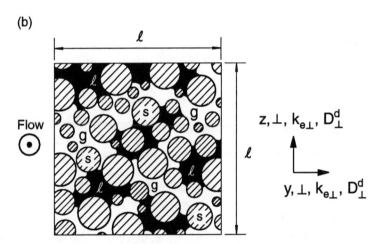

Figure 10.1 A rendering of solid-, liquid-, and gas-phase distributions for flow through a partially saturated representative elementary volume. (a) The flow examined is along the x-axis and (b) the flow is perpendicular to this plane.

10.2.1 ANISOTROPY

In order to point out the extent of anisotropy in $\mathbf{K}_e$ we use the following simple arguments. These are based on the *assumption* that $\mathbf{K}_e = \mathbf{K}_e(s)$. We begin by introducing $k_{e\parallel}$ and $k_{e\perp}$ with the flow being along the x-direction, as in Figure 10.1. We assume symmetry around the x-axis and write

$$\mathbf{K}_e = \begin{pmatrix} k_{exx} & 0 & 0 \\ 0 & k_{eyy} & 0 \\ 0 & 0 & k_{ezz} \end{pmatrix} = \begin{pmatrix} k_{e\parallel} & 0 & 0 \\ 0 & k_{e\perp} & 0 \\ 0 & 0 & k_{e\perp} \end{pmatrix}. \tag{10.41}$$

In Figures 10.1 (a) and (b) the *ensemble*-averaged phase distributions in the y-z plane is *isotropic*, while those in the x-y and x-z planes are *anisotropic*. The effective thermal conductivity in the x-direction, $k_{e\parallel}$, can be approximated as $k_e(s=0)$ and $k_e(s=1)$ arranged in *parallel*, i.e.,

$$\frac{k_{e\parallel}}{k_\ell} = s\frac{k_e(s=1)}{k_\ell} + (1-s)\frac{k_e(s=0)}{k_\ell}. \tag{10.42}$$

For the components perpendicular to the flow, perhaps a *geometric mean* of $k_e(s=1)$ and $k_e(s=0)$ may be appropriate. This gives

$$\frac{k_{e\perp}}{k_\ell} = \left[\frac{k_\ell(s=1)}{k_\ell}\right]^s \left[\frac{k_\ell(s=0)}{k_\ell}\right]^{1-s}. \tag{10.43}$$

However, we note that the extent of the *wettability* can significantly alter $a_\ell(\mathbf{x})$ and $a_g(\mathbf{x})$, and therefore, the presence or the lack of *liquid films* on the solid surface can significantly influence $k_{e\parallel}$ and $k_{e\perp}$. We note that the geometric mean and the parallel arrangement given above are not accurate. However, the anisotropy of $\mathbf{K}_e$ is evident.

10.2.2 CORRELATIONS

Presently *no* rigorous solutions for $k_{e\parallel}$ and $k_{e\perp}$ are available. Although not attempted, one of the readily solvable problems would be that of the *periodically* constricted tube introduced by Saez et al. (1986). This model was used for the determination of the relative permeability in Section 8.5.3 (A). Examination of this model (Figure 8.11) reveals the *anisotropy* of $\mathbf{K}_e$ and the *failure* of (10.42) and (10.43) in accurately predicting $k_{e\parallel}$ and $k_{e\perp}$. For this simple unit-cell phase distribution (which is an approximation to the simple-cubic arrangement of monosize spheres), we expect

$$k_{e\parallel} \gg k_{e\perp}, \tag{10.44}$$

because of the *smaller* thermal conductivity of the gas phase.

Most of the reported experimental results for $\mathbf{K}_e(s)$ are obtained for $\mathbf{u} \neq 0$, and therefore, *are* the results of the *simultaneous* evaluation of $\mathbf{K}_e(s)$ and $\mathbf{D}^d(s)$. In these experiments, generally $\mathbf{K}_e \ll (\rho c_p)_\ell \mathbf{D}^d$. An exception is the experiment of Somerton et al. (1974) where only $\mathbf{K}_e = \mathbf{I}k_e(s)$ was determined, but with no examination of the anisotropy. Another one is that of Matsuura et al. (1979a), where the velocity was *reduced* sufficiently to allow for the determination of $k_e(s)$ with some accuracy, again no *directional* dependence was considered.

Table 10.1 summarizes some of the available empirical correlations for $k_{e\parallel}(s)$ and $k_{e\perp}(s)$, for packed beds of spherical particles. The experiments of Specchia and Baldi (1979), Hashimoto et al. (1976), and Weekman and Myers (1965) are *not* of high enough accuracies at low velocities, and therefore, their values for a_1 are *not* expected to be accurate (Matsuura et al., 1979a). Also, Matsuura et al. (1979b) could not resolve the saturation-dependence of k_e in their experiments. Somerton et al. (1974) *do* find a saturation dependence as given in Table 10.1. Also shown in the table are the *estimates* based on the parallel arrangement and geometric mean.

Examination of Table 10.1 shows that further study of the parameters influencing $\mathbf{K}_e$ is needed. This includes the effects of the wettability and the *significance* of the expected *hysteresis* in $\mathbf{K}_e(s)$.

10.3 Thermal Dispersion

We now examine the contribution of each flowing phase (ℓ and g) to the thermal dispersion tensor. In the single-phase flows we found *different* Peclet number-dependence of $\mathbf{D}^d$ for ordered and disordered solid phase distributions. Here we have to include in the evaluation of the dispersion tensor in each phase the phase distribution of the other fluid phase in addition to the solid phase distribution. The available experimental results show that the dispersion within the liquid phase makes a *larger* contribution to $\mathbf{D}^d$ than that for the gaseous phase. But we expect that for *low* saturations the gas phase dispersion contribution will become significant compared to the effective molecular conduction.

For a single-component system, with $\partial/\partial t \to 0$, $\langle \dot{n} \rangle \to 0$, $\langle \dot{s} \rangle = 0$, (10.31) can be rewritten as

$$\left[\langle \mathbf{u}_\ell \rangle + \frac{(\rho c_p)_g}{(\rho c_p)_\ell} \langle \mathbf{u}_g \rangle \right] \cdot \nabla T = \nabla \cdot \left[\frac{\mathbf{K}_e}{(\rho c_p)_\ell} + \mathbf{D}_\ell^d + \frac{(\rho c_p)_g}{(\rho c_p)_\ell} \mathbf{D}_g^d \right] \cdot \nabla T, \quad (10.45)$$

where we have *defined* the liquid- and gas-phase dispersion coefficients as

$$\mathbf{D}_\ell^d \equiv -\varepsilon s \langle \mathbf{u}_\ell' \mathbf{b}_\ell \rangle^\ell, \quad \mathbf{D}_g^d \equiv -\varepsilon(1-s) \langle \mathbf{u}_g \mathbf{b}_g \rangle^g. \quad (10.46)$$

In (10.46), $\mathbf{D}_\ell^d$ and $\mathbf{D}_g^d$ depend on the pore-level velocity and phase distributions (in their respective phases). These phase velocity distributions are coupled at $A_{\ell g}$, and therefore, the motion of one phase influences the other. The general relations for $\mathbf{D}_\ell^d$ and $\mathbf{D}_g^d$ will be of the form

$$\mathbf{D}_\ell^d = \mathbf{D}_\ell^d [k_s, k_\ell, k_g, a_\ell(\mathbf{x}), a_g(\mathbf{x}), (\rho c_p)_\ell, (\rho c_p)_g,$$
$$\mathbf{u}_\ell, \mathbf{u}_g, \mu_\ell, \mu_g, \sigma, \theta_c, \text{history}]. \quad (10.47)$$

A similar relationship can be given for $\mathbf{D}_g^d$. Note that the various flow regimes (such as the trickle, pulsing, and bubble regimes in the cocurrent

TABLE 10.1 CORRELATIONS FOR TWO-PHASE EFFECTIVE THERMAL CONDUCTIVITY

CONSTRAINTS	CORRELATION
(a) Alumina spheres-water-air	$\dfrac{k_{e\perp}}{k_\ell} = a_1$ where a_1 is determined experimentally
(b) Glass spheres-water-air ($\varepsilon = 0.375$)	$\dfrac{k_{e\perp}}{k_\ell} = a_1, \quad 0.13 < s < 0.6$ where a_1 is determined experimentally
(c) Glass spheres-water-air ($\varepsilon = 0.4$)	$\dfrac{k_{e\perp}}{k_\ell} = 1.5, \quad 0.08 < s < 0.21$
(d) Glass and ceramic spheres-water-air	$\dfrac{k_{e\perp}}{k_\ell} = \dfrac{k_e(s=0)}{k_\ell}$
(e) Nonconsolidated sands-brine-air (moist), no directional dependence investigated	$\dfrac{k_{e\parallel}}{k_\ell} = \dfrac{k_{e\perp}}{k_\ell}$ $= \dfrac{k_e(s=0)}{k_\ell} + s^{1/2}\left[\dfrac{k_e(s=1) - k_e(s=0)}{k_\ell}\right]$
(f) Disordered porous media, not experimentally verified	$\dfrac{k_{e\parallel}}{k_\ell} = s\dfrac{k_e(s=1)}{k_\ell} + (1-s)\dfrac{k_e(s=0)}{k_\ell}$
(g) Disordered porous media, not experimentally verified	$\dfrac{k_{e\perp}}{k_\ell} = \left[\dfrac{k_e(s=1)}{k_\ell}\right]^s \left[\dfrac{k_e(s=0)}{k_\ell}\right]^{1-s}$

(a) Weekman and Myers (1965); (b) Hashimoto et al. (1976); (c) Matsuura et al. (1979a); (d) Specchia and Baldi (1979); (e) Somerton et al. (1974) and Udell and Fitch (1985); (f) *parallel arrangement*; (g) *geometric mean*.

and countercurrent downflow in packed beds of spheres described in Section 8.12) are *all* represented by a_ℓ and a_g. We expect to recover the two asymptotic behaviors, i.e.,

$$\text{for } s \to 1: \quad \mathbf{D}_\ell^d = \mathbf{D}_\ell^d(s=1) = \varepsilon \mathbf{D}^d, \quad \mathbf{D}_g^d = 0,$$

$$\text{for } s \to 0: \quad \mathbf{D}_\ell^d = 0, \quad \mathbf{D}_g^d = \mathbf{D}_g^d(s=0) = \varepsilon \mathbf{D}^d, \quad (10.48)$$

where $\mathbf{D}_\ell^d(s=1)$ and $\mathbf{D}_g^d(s=0)$ are the single-phase flow thermal dispersion tensor (multiplied by ε) given in Chapter 4. From the results of

Chapter 4 we know that

$$\mathbf{D}^d(s=0 \text{ or } s=1) = \mathbf{D}^d\left(Pe, \frac{k_s}{k_f}, \varepsilon, \text{solid structure}\right) \qquad (10.49)$$

with a *weak* dependence on Pr.

Further examination of Figures 10.1 (a) and (b) shows that due to *anisotropy* of the phase distributions $a_\ell(\mathbf{x})$ and $a_g(\mathbf{x})$, the dispersion coefficients $\mathbf{D}_\ell^d$ and $\mathbf{D}_g^d$ are also *anisotropic*.

Given later are a discussion of this anisotropy in $\mathbf{D}^d$, a simple model for the *axial* dispersion, a simple model for the *transverse* dispersion, and the existing *empirical* correlations for the transverse dispersion.

10.3.1 ANISOTROPY

As before, we assume that the *off-diagonal* components of $\mathbf{D}^d$ are *zero*, i.e.,

$$\mathbf{D}^d = \begin{pmatrix} D_{xx}^d & 0 & 0 \\ 0 & D_{yy}^d & 0 \\ 0 & 0 & D_{zz}^d \end{pmatrix} = \begin{pmatrix} D_\parallel^d & 0 & 0 \\ 0 & D_\perp^d & 0 \\ 0 & 0 & D_\perp^d \end{pmatrix}$$

$$= \begin{pmatrix} D_{\ell\parallel}^d + D_{g\parallel}^d & 0 & 0 \\ 0 & D_{\ell\perp}^d + D_{g\perp}^d & 0 \\ 0 & 0 & D_{\ell\perp}^d + D_{g\perp}^d \end{pmatrix}. \qquad (10.50)$$

We now examine Figures 10.1 (a) and (b) where $a_\ell(\mathbf{x})$ and $a_g(\mathbf{x})$ are for a flow along the x-direction. We expect a strong anisotropy in D^d. Note that for the *single-phase* flows we had

$$D_\perp^d < D_\parallel^d \quad \text{for } s=0 \text{ or } s=1. \qquad (10.51)$$

In addition, the discontinuity in both the ℓ and g phases in the y-z plane further magnifies this difference. Therefore, we expect

$$D_{g\perp}^d \ll D_{g\parallel}^d \quad 0 < s < 1,$$
$$D_{\ell\perp}^d \ll D_{\ell\parallel}^d \quad 0 < s < 1. \qquad (10.52)$$

For *large* saturations (wetting phase) we expect the wetting phase to be spread over many pores such that the presence of the ℓ-g interface would *not* significantly alter D_ℓ^d within the ℓ phase. Furthermore, D_ℓ^d is expected to be much *larger* than D_g^d for *large* saturations. Then we can *suggest*

$$D_\parallel^d \sim D_{\ell\parallel}^d \sim \varepsilon s D_\parallel^d(s=1) \quad \text{for large } s,$$
$$D_\perp^d \sim D_{\ell\perp}^d \sim \varepsilon s D_\perp^d(s=1) \quad \text{for large } s. \qquad (10.53)$$

For the *intermediate* and *small* saturations (but $s > s_{\mathrm{ir}}$), the phase distributions can significantly influence the magnitude of $D_\parallel^d$ and $D_\perp^d$ and a simple linear saturation dependence *cannot* be used.

10.3.2 Models

Because of the geometrical complexities and their strong dependence on the saturation, so far only a *few* models have been devised for examination of the dependence of $\mathbf{D}^d$ on the various parameters identified in (10.47). We begin with a simple model (deterministic) based on the *closure* of the local volume averaging (i.e., evaluation of [**b**]) and a *stochastic*-circuit model based on the *percolation* concept.

(A) Axial Dispersion for a Periodic Structure

For the two-dimensional case of a liquid film falling (liquid thickness δ_ℓ) over a *vertical* solid slab (solid thickness δ_s) and *separating* this solid phase from a gaseous phase film (gas thickness δ_g), Saez (1983) and Saez et al. (1986) solves the [**b**] equation to arrive at D_{xx}^d. This is similar to the Taylor dispersion in tubes discussed in Chapter 4. The *three* vertical layers make up a *unit cell* in a three-phase system, and these unit cells are placed vertically *next* to each other (*periodic* boundary conditions can be used along the horizontal axis in the *y*-direction). Steady-state conditions are assumed and the *convection* in the *gaseous* phase is *neglected*. The two-dimensional (*x-y* plane) energy and momentum equations for the three phases are given as

$$(\rho c_p)_\ell u_\ell \frac{\partial T_\ell}{\partial x} = k_\ell \left(\frac{\partial^2 T_\ell}{\partial x^2} + \frac{\partial^2 T_\ell}{\partial y^2} \right), \quad 0 \leq y \leq \delta_\ell, \quad (10.54)$$

$$0 = \mu_\ell \frac{\partial^2 u_\ell}{\partial y^2} + \rho_\ell g, \quad 0 \leq y \leq \delta_\ell, \quad (10.55)$$

$$0 = \frac{\partial^2 T_s}{\partial x^2} + \frac{\partial^2 T_s}{\partial y^2}, \quad -\delta_s \leq y \leq 0, \quad (10.56)$$

$$0 = \frac{\partial^2 T_g}{\partial x^2} + \frac{\partial^2 T_g}{\partial y^2}, \quad \delta_\ell \leq y \leq \delta_g + \delta_\ell. \quad (10.57)$$

The boundary conditions are

$$\frac{\partial T_s}{\partial y} = 0 \quad \text{at} \quad y = -\delta_s, \quad (10.58)$$

$$u_\ell = 0, \quad T_\ell = T_s, \quad k_\ell \frac{\partial T_\ell}{\partial y} = k_s \frac{\partial T_s}{\partial y}, \quad \text{at} \quad y = 0, \quad (10.59)$$

$$\frac{\partial u_\ell}{\partial y} = 0, \quad T_\ell = T_g, \quad k_\ell \frac{\partial T_\ell}{\partial y} = k_g \frac{\partial T_g}{\partial y}, \quad \text{at} \quad y = \delta_\ell, \quad (10.60)$$

$$\frac{\partial T_g}{\partial y} = 0, \quad \text{at} \quad y = \delta_\ell + \delta_g. \quad (10.61)$$

The $y = -\delta_s$ and $y = \delta_\ell + \delta_g$ boundary conditions are based on the assumed *symmetry* of the unit cell at these boundaries.

The *area*-averaged equations (averaged over the y-z plane), after the *introduction* of the phase average and the deviation components of the temperature and velocity fields are

$$(\rho c_p)_\ell \langle u \rangle^\ell \frac{\partial \langle T \rangle^\ell}{\partial x} = -(\rho c_p)_\ell \frac{\partial}{\partial x} \langle u'_\ell T'_\ell \rangle + k_\ell \frac{\partial^2 T'}{\partial y^2}$$

$$+ \frac{k_\ell}{\delta_\ell} \left(\left.\frac{\partial T'}{\partial y}\right|_{\delta_\ell} - \left.\frac{\partial T'}{\partial y}\right|_0 \right), \quad (10.62)$$

$$0 = \frac{\partial^2 \langle T \rangle^s}{\partial x^2} + \frac{1}{\delta_s} \frac{k_\ell}{k_s} \left.\frac{\partial T'_\ell}{\partial y}\right|_0, \quad (10.63)$$

$$0 = \frac{\partial^2 \langle T \rangle^g}{\partial x^2} + \frac{1}{\delta_g} \frac{k_\ell}{k_g} \left.\frac{\partial T'_\ell}{\partial y}\right|_{\delta_g}, \quad (10.64)$$

where, for example,

$$\langle T \rangle^\ell = \frac{1}{\delta_\ell} \int_0^y T_\ell \, dy. \quad (10.65)$$

The [b] equations are found from the T' equations. (The derivations of the T' equations are not repeated here but are similar to those given in Sections 3.2 and 4.3.) The final forms of the [b] equations are

$$(\rho c_p)_\ell u'_\ell = k_\ell \frac{d^2 b_\ell}{dy^2} - \frac{k_\ell}{\delta_\ell} \left(\left.\frac{db_\ell}{dy}\right|_{\delta_\ell} - \left.\frac{db_\ell}{dy}\right|_0 \right), \quad (10.66)$$

$$0 = \frac{d^2 b_s}{dy^2} - \frac{1}{\delta_s} \frac{k_\ell}{k_s} \left.\frac{db_\ell}{dy}\right|_0, \quad (10.67)$$

$$0 = \frac{d^2 b_g}{dy^2} + \frac{1}{\delta_g} \frac{k_\ell}{k_g} \left.\frac{db_\ell}{dy}\right|_{\delta_\ell}, \quad (10.68)$$

with

$$\frac{db_s}{dy} = 0 \quad \text{at } y = -\delta_s, \quad (10.69)$$

$$b_\ell = b_s \quad \text{at } y = 0, \quad (10.70)$$

$$b_\ell = b_g \quad \text{at } y = \delta_\ell, \quad (10.71)$$

$$\frac{db_g}{dy} = 0 \quad \text{at } y = \delta_\ell + \delta_g, \quad (10.72)$$

and

$$\langle b_s \rangle^s = \langle b_\ell \rangle^\ell = \langle b_g \rangle^g = 0. \quad (10.73)$$

After the imposition of local thermal equilibrium, the local area-averaged energy equation becomes

560   10. Conduction and Convection

$$(\rho c_p)_\ell \langle u_\ell \rangle \frac{\partial T}{\partial x} = \left[\frac{\delta_s}{\delta_s + \delta_\ell + \delta_g} k_s + \frac{\delta_\ell}{\delta_s + \delta_\ell + \delta_g} k_\ell + \frac{\delta_g}{\delta_s + \delta_\ell + \delta_g} k_g \right.$$
$$\left. - \frac{(\rho c_p)_\ell}{\delta_s + \delta_\ell + \delta_g} \int_0^{\delta_\ell} (b_\ell u'_\ell)\, dy \right] \frac{\partial^2 \langle T \rangle}{\partial x^2} = D_{xx} \frac{\partial^2 \langle T \rangle}{\partial x^2}. \quad (10.74)$$

The solution to (10.66)–(10.73) is found, noting that

$$u_\ell = \frac{\rho_\ell g}{2\mu_\ell} y(2\delta_\ell - y), \quad \langle u \rangle^\ell = \frac{\rho_\ell g \delta_\ell^2}{3\mu_\ell}, \quad u'_\ell = u_\ell - \langle u \rangle^\ell. \quad (10.75)$$

Also the *saturation* (wetting phase) and porosity are

$$s = \frac{\delta_\ell}{\delta_s + \delta_\ell + \delta_g}, \quad \varepsilon = \frac{\delta_\ell + \delta_g}{\delta_s + \delta_\ell + \delta_g}. \quad (10.76)$$

The *axial dispersion coefficient* becomes

$$D_{xx}^d = D_\parallel^d = D_{\ell\parallel}^d = \frac{Pe_\ell^2}{24} \varepsilon s \left[\frac{37}{180} + \frac{5\varepsilon s}{12} + \frac{a_1 - a_2}{a_3} \right], \quad (10.77)$$

when the Peclet number based on the *hydraulic diameter* is

$$Pe_\ell = \frac{\langle u \rangle^\ell}{\alpha_\ell} (6 s \delta_\ell) \quad (10.78)$$

and

$$a_1 = -\frac{3}{20}\varepsilon s - \frac{14}{15}\frac{k_\ell}{k_g}\varepsilon(1-s),$$

$$a_2 = -\frac{9}{10}\varepsilon s - \frac{7}{30}\frac{k_\ell}{k_s}\varepsilon(1-s),$$

$$a_3 = -9\varepsilon^2 s^2 - 4\frac{k_\ell}{k_s}\varepsilon s(1-\varepsilon) - 4\frac{k_\ell}{k_g}\varepsilon^2 s(1-s)$$
$$-4\frac{k_\ell^2}{k_s k_g}(1-\varepsilon)\varepsilon(1-s). \quad (10.79)$$

For a given saturation, (10.77) shows D_{xx}^d is proportional to Pe_ℓ^2. This is similar to the trend found for the single-phase flows in ordered matrices (Chapter 4), i.e., the Taylor dispersion for laminar flow in tubes. However, $s = s(\langle u \rangle^\ell, \text{etc.})$; therefore, the dependence on Pe_ℓ is in general *more complex*. Saez et al. *relate* the saturation to Pe_ℓ and a *normalized gravity* term $(\rho_\ell^2 g / \mu_\ell^2)(6\delta_\ell)^3$ for the *trickle flow* in packed beds (gas phase stationary; liquid phase flowing along g). They use the available and their own experimental results to find the correlation constants. The assumed *ideal* hydrodynamics of the fully developed (no inertial effect) vertical film flow in the preceding example is *far* from that of the actual flow around *touching* spheres (which are generally randomly arranged). Therefore, we do *not* expect this analysis to lead to an *accurate* prediction of $D^d_{\ell\parallel}$ in the trickle flow regime in packed beds. *However*, we note that the familiar Taylor-Aris dispersion behavior is obtained. As will be noted for $D^d_{\ell\parallel}$, no experimental results are available for comparison.

(B) A Model for Lateral Dispersion in Disordered Media

The lateral phase distributions in the trickle flow in packed beds has been simulated by Crine et al. (1980), and an approximation to these distributions has been used by Crine (1982) for the determination of the lateral dispersion coefficient. The phase distributions are obtained by dividing the packed bed into a set of *elementary transport cells* (pores). A stochastic model of the *connectivity* between the pores, based on the *configurational entropy* of this clustering phenomenon, is developed. The model predicts the *most likely* liquid phase and velocity distributions within the bed. Figure 10.2 shows a portion of the *lateral* (perpendicular to flows) *cross section* of the three-dimensional network used by Crine. The liquid phase distribution in the network is shown. Note the *lack* of the continuity of the liquid phase, as was also depicted in Figure 10.1(b).

No rigorous treatments of **D** for this phase distribution are available. Crine (1982) suggests a *parallel* arrangement of the *resistances* as shown in Figure 10.3. The lateral distribution of the phases is simplified as shown in the figure. Note that the liquid phase distribution in the y-direction ($\perp_1$) is *not* the same as that in the z-direction ($\perp_2$), i.e., *anisotropy* exists *even* in the phase distributions in the *plane* perpendicular to the flow. Crine only considers $\perp_1$. In $\perp_1$, the three parallel *resistances* are the *fully irrigated* path having a resistance f^2, the *fully dry* path having a resistance $(1-f)^2$, and the resistance of the path of *dry-irrigated* (in series) given by $2f(1-f)$. These resistances, as well as the definition of f, are also shown in Figure 10.3.

Based on these resistances, Crine arrives at the total lateral ($\perp_1$) effective thermal conductivity as

$$D_\perp = \frac{k_{e\perp}}{(\rho c_p)_\ell} + D^d_{e\perp}, \tag{10.80}$$

$$\frac{k_{e\perp_1}}{(\rho c_p)_\ell} = (1-f)\frac{k_{e\perp}(s=0)}{(\rho c_p)_\ell}, \tag{10.81}$$

$$\varepsilon D^d_{\perp_1} = f^2 D^d_\perp(s=1) + (1-f)\frac{1}{\dfrac{1}{\dfrac{k_{e\perp}(s=0)}{(\rho c_p)_\ell}} + \dfrac{1}{D^d_\perp(s=1)}}, \tag{10.82}$$

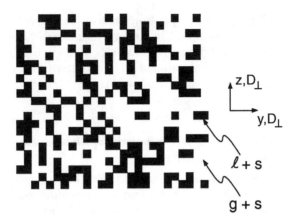

Figure 10.2 Liquid phase distribution in the plane perpendicular to the flow, obtained by a stochastical model by Crine. (From Crine, reproduced by permission ©1982 Gordon and Breach Science.)

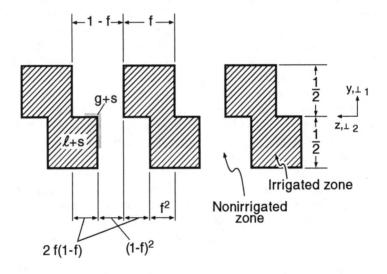

Figure 10.3 The model of Crine for calculation of the lateral dispersion coefficient in partially saturated porous media. Note that the phase distributions are anisotropic. (From Crine, reproduced by permission ©1982 Gordon and Breach Science.)

where, based on the single-phase flow results for *disordered* structures (Chapter 4), he uses

$$\frac{D_\perp^d(s=1)}{\alpha_\ell} = a_1 Pe_\ell, \quad Pe_\ell = \frac{\overline{u}_\ell d}{\alpha_\ell}. \tag{10.83}$$

The parameters f and $\bar{u}_\ell$ are defined as

$$f = \frac{\langle u_\ell \rangle}{\langle u_\ell \rangle + u_{\ell_o}}, \quad \langle u_\ell \rangle \equiv f\bar{u}_\ell = f(\langle u_\ell \rangle + u_{\ell_o}), \tag{10.84}$$

where f is the *fraction of particles irrigated* and u_{ℓ_o} is the *local stable liquid velocity* in an *isolated rivulet* and is determined from the experimental results. For large $\langle u_\ell \rangle$, only a small fraction of the particles are irrigated. $\bar{u}_\ell$ is the particle-level velocity and is assumed to be uniform. Using this, we have in terms of $k_{e\perp}(s=0)$, a_1, and f

$$D_{\perp_1} = (1-f)\frac{k_{e\perp}(s=0)}{(\rho c_p)_\ell} + a_1 f \langle u_\ell \rangle d + \frac{1}{\dfrac{1}{\dfrac{k_{e\perp}(s=0)}{(\rho c_p)_\ell}} + \dfrac{1}{\dfrac{a_1}{f}\langle u_\ell \rangle d}}. \tag{10.85}$$

The *three* parameters, $k_{e\perp}(s=0)$, a_1, and f, are determined using the *experimental* result. Crine finds that the Pe_ℓ and the saturation dependence of $D_{\perp_1}$ can be *predicted* satisfactorily by (10.85) for the *trickle flow* in packed beds of large particles.

10.3.3 CORRELATIONS FOR LATERAL DISPERSION COEFFICIENT

The existing experimental results for the *total* effective thermal conductivity are for the *lateral* component *only*, i.e., $D_\perp$. In the experiments, the measurement of the temperature distribution is made in the *plane* perpendicular to the flow. Then a two-dimensional or axisymmetric steady-state temperature field and a *single-component* gas phase are assumed. For these conditions and with *no* phase change or internal energy generation, the energy equation becomes

$$[(\rho c_p)_\ell \langle u_\ell \rangle + (\rho c_p)_g \langle u_g \rangle] \frac{\partial T}{\partial x} = (\rho c_p)_\ell D_\perp \left(\frac{\partial^2 T}{\partial y^2} + \frac{1}{y}\frac{\partial T}{\partial y} \right), \tag{10.86}$$

where it is assumed that

$$D_\perp \left(\frac{\partial^2 T}{\partial y^2} + \frac{1}{y}\frac{\partial T}{\partial y} \right) \gg D_\| \frac{\partial^2 T}{\partial x^2}. \tag{10.87}$$

Note that we expect $D_\| > D_\perp$; therefore, for the preceding to be valid, the axial temperature variations should be made as small as possible. The measured temperature T is generally measured in the *pores* and is not volume-averaged. However, if the local temperature deviations are small compared to the temperature drop across the system, this measured T can be reasonably taken as $\langle T \rangle$.

For large Pe_ℓ and Pe_g, the contribution of $k_{e\perp}$ becomes small and $D_\perp \sim D_\perp^d$. Also, for sufficiently large saturations (or liquid flow rates) $D_{\ell\perp}^d \gg D_{g\perp}^d$. Therefore, most of the studies have reported *very weak* dependence of $D_\perp^d$ on Pe_g. No noticeable *hysteresis* has been reported in $D_\perp^d(s)$; therefore, we expect that (10.47) would reduce to

$$D_{\ell\perp}^d = D_{\ell\perp}^d(k_s, k_\ell, s, u_\ell, \sigma, \theta_c, \varepsilon, \text{solid structure}). \tag{10.88}$$

No systematic examination of dependence on the wettability or the surface tension is yet available. For large Pe_ℓ, we expect k_s/k_ℓ *not* to influence $D_{\ell\perp}^d$.

The experiments discussed later are generally for *spherical particles* and for *water-air* as the fluid pair. Most of them are concerned with the *trickle flow* in such beds, where $s = s(Pe_\ell, \text{etc.})$. Therefore, the experiments are generally designed to determine

$$D_{\ell\perp}^d = D_{\ell\perp}^d(Pe_\ell), \text{ given flow regime, } \varepsilon, \text{ solid structure,}$$
$$\sigma, \theta_c, \text{ and large } Pe_\ell. \tag{10.89}$$

Note that this simplification only holds for a given *flow regime*. In the flow regimes that occur for the various combinations of $\langle u_\ell \rangle$ and $\langle u_g \rangle$, a separate correlation in the form of (10.89) must be introduced for each regime.

The lateral temperature nonuniformity is caused by the *controlled* cooling or heating occurring at the *bounding surface* (generally a cylindrical shell) or by the *controlled* nonuniform inlet temperature. We will discuss both methods and give the correlations of the type given in (10.89), which have been found by using these experimental results.

(A) Heat Supply/Removal at Bounding Surfaces

This boundary condition has been used by Weekman and Myers (1965), Hashimoto et al. (1976), Matsuura et al. (1979b), and Specchia and Baldi (1979). The axisymmetric temperature field given by (10.86) is solved subject to

$$T = T_i, \quad x = 0, \ 0 \le y \le R, \tag{10.90}$$

$$\frac{\partial T}{\partial y} = 0, \quad x \ge 0, \ y = 0, \tag{10.91}$$

$$h_{bw}(T_w - T) = \frac{D_\perp}{(\rho c_p)_\ell} \frac{\partial T}{\partial y}, \quad x \ge 0, \ y = R. \tag{10.92}$$

In this formulation, the packed bed-wall *heat transfer coefficient* (or a *temperature slip coefficient*) h_{bw} applied at the bed-wall interface is introduced in place of the *actual* boundary condition

$$T = T_w, \quad x \ge 0, \ y = R. \tag{10.93}$$

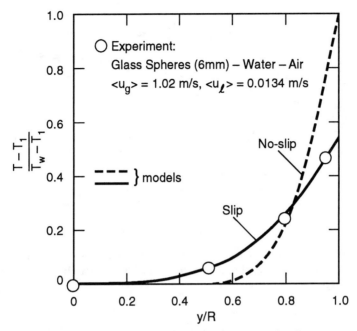

Figure 10.4 The results of Specchia and Baldi showing the failure of the uniform $D_{e\perp}$ analysis. The slip model matches the experimental results in the bulk and near the bounding surface. (From Specchia and Baldi, reproduced by permission ©1979 Gordon and Breach Science.)

This is done to *implicitly* allow for the *variation* of the total effective thermal conductivity near the bounding surface, i.e., to account for

$$D_\perp = D_\perp(y) \quad \text{and} \quad D_\perp(y/R \to 1) \ll D_\perp(y/R \sim 0). \quad (10.94)$$

This type of slip boundary condition was examined in dealing with conduction (Section 3.7) and the single-phase flows (Section 4.10), and we will again refer to this nonuniformity in Section 10.3.4. Specchia and Baldi also discuss the *errors* associated with the assumption of a *constant* $D_\perp$ across the bed (up to the bounding surface) and the role of the *correction parameter* h_{bw}.

As with the single-phase flows, the magnitude of $D_\perp$ is *smaller* near the bounding surface causing a larger temperature drop in this region. As was mentioned, the same phenomenon occurs in two-phase flows and this sharp drop is then modeled by the slip coefficient h_{bw}. Figure 10.4 shows the temperature distribution obtained by Specchia and Baldi using the *slip* and *no-slip* boundary conditions. In the no-slip model, a *uniform* $D_{\ell\perp}$ is used causing a smaller temperature gradient; therefore, the *disagreement* with the experimental results *occur*. The slip coefficient h_{bw} is obtained by *matching* the experimental results with the model prediction.

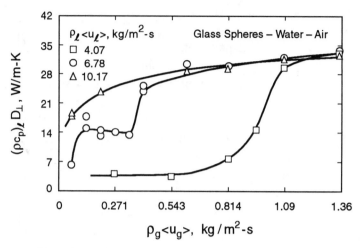

Figure 10.5 Experimental results of Weekman and Myers showing the effect of the flow transition on $D_\perp$. (From Weekman and Myers, reproduced by permission ©1965 AIChE.)

Since *evaporation* (in the *heated* wall experiments) or *condensation* (in the *cooled* wall experiments) takes place along the flow, in practice the *heat capacity* of the gaseous phase is adjusted to allow for this *change* in the vapor concentration. However, an *averaged* value along the bed is taken and $(\rho c_p)_g$ is treated as *constant*. The solution to (10.86) subject to (10.90)–(10.92) is

$$\frac{T - T_1}{T_w - T_1} = 1 - 2 \sum_{i=1}^{\infty} \frac{J_0(b_i y/R) e^{-a_1 b_i^2 x}}{b_i J_1(b_i) \left(1 + \frac{b_i^2}{Bi^2}\right)}, \tag{10.95}$$

where

$$Bi = \frac{h_{bw}(\rho c_p)_\ell R}{D_{e\perp}}, \quad a_1 = \frac{D_\perp}{(\rho c_p)_\ell R^2 \left[(\rho c_p)_\ell \langle u_\ell \rangle + (\rho c_p)_g \langle u_g \rangle\right]}, \tag{10.96}$$

and

$$b_i J_1(b_i) = Bi J_0(b_i). \tag{10.97}$$

From (10.95) and by using the measured $T(y,x)$, the unknowns h_{bw} and $D_\perp$ are found *simultaneously* using the least-square approximation.

Using the *no-slip* boundary condition (10.93), Weekman and Myers determined the *uniform* $D_\perp$ and examined the effect of the gas and liquid flow rates on $D_\perp$. They found *noticeable* change in $D_\perp$ as the flow *transition* from the *trickle* to the *pulsing* flow (Section 8.1.2) took place. Figure 10.5 shows some of their results. For a given *liquid mass flow* rate $\rho_\ell \langle u_\ell \rangle$, as the *gas mass flow* rate $\rho_g \langle u_g \rangle$ is *increased* the transition occurs and is marked by a *sudden* increase in $D_\perp$. The flow transition occurs at lower $\rho_g \langle u_g \rangle$ as the liquid mass flow rate increases. Here we note that the results given in Figure 10.5 are based on the assumption of a *uniform* $D_\perp$ and any flow *channeling* or other flow *maldistributions* are *not* accounted for. Therefore, we have included Figure 10.5 only to point out the existence of various flow regimes and the dependence of $D_\perp$ on the corresponding phase distributions $a_\ell(\mathbf{x})$ and $a_g(\mathbf{x})$ associated with these regimes.

All of the experimental results show that the liquid lateral dispersion coefficient becomes *larger* in two-phase flow, i.e.,

$$\frac{D_{\ell\perp}(s<1)}{D_{\ell\perp}(s=1)} > 1, \quad \text{given } \langle u_\ell \rangle, \tag{10.98}$$

i.e., the thermal dispersion in the liquid phase is *enhanced* due to the presence of the ℓ-g interface and the wetting of the solid. This enhancement diminishes as $s \to 1$. At higher gas flow rates, the *pulsations* also cause enhancement of the liquid phase dispersion.

In Table 10.2 we summarize some of the correlations for $D_\perp^d$. While some of these show a dependence on Pe_ℓ to the *first* power, the correlation of Saez et al. shows a power of 0.8. We note again that because of the presence of the various regimes, *no* one correlation is expected to describe the entire range of practical flow rates.

We have tabulated these correlations for completeness and in order to point out the unanswered discrepancies among them. The f-factor used by Crine lacks universality, while the correlation of Saez et al. *overpredicts* the results of Weekman and Myers and Specchia and Baldi. Both Saez et al. and Crine avoid h_{bw} by using a nonuniform inlet temperature (in their experiments) for establishing a lateral temperature nonuniformity. We discuss their approach next.

(B) Nonuniform Inlet Temperature

For a *two-dimensional* geometry with the boundary conditions corresponding to a *nonuniform* inlet temperature (lateral step change), i.e.,

$$T = T_1, \quad x = 0, \ y \geq 0,$$
$$T = T_2, \quad x = 0, \ y \leq 0,$$
$$T = T_1, \quad x \geq 0, \ y \to \infty,$$
$$T = T_2, \quad x \geq 0, \ y \to -\infty, \tag{10.99}$$

the solution to (10.86) is (Carslaw and Jaeger, 1986)

$$\frac{T-T_1}{T_2-T_1} = \frac{1}{2}\text{erfc}\left\{\frac{y[(\rho c_p)_\ell \langle u_\ell \rangle + (\rho c_p)_g \langle u_g \rangle]^{1/2}}{2\left[\frac{D_\perp}{(\rho c_p)_\ell}x\right]^{1/2}}\right\}. \tag{10.100}$$

TABLE 10.2 CORRELATIONS FOR TWO-PHASE FLOW LATERAL DISPERSION COEFFICIENT

CONSTRAINTS	CORRELATION
(a) Cocurrent, downward flow in packed bed of spheres, air-water-glass, wall heating	$\dfrac{D_\perp^d}{\alpha_\ell} = 0.00174 Pe_\ell + 0.172 Pe_g$ $Pe_\ell = \dfrac{\langle u_\ell \rangle d}{\alpha_\ell}, \quad Pe_g = \dfrac{\langle u_g \rangle d}{\alpha_g}$
(b) Glass and ceramic spheres and ceramic rings, with water-air, wall heating	$D_\perp^d = D_{g\perp}^d + D_{\ell\perp}^d$ $D_{g\perp}^d = D_{g\perp}^d(s=0), \quad D_{g\perp}^d \ll D_{\ell\perp}^d$ $\dfrac{\langle u_\ell \rangle d}{D_{\ell\perp}^d} = 338 Re_\ell^{0.67} H^{0.29} \left(\dfrac{dA_o}{\varepsilon} \right)^{-2.7}$ where $H(Re_\ell, \text{etc})$ is the *liquid holdup*, A_o is the specific surface area, $Re_\ell = \langle u_\ell \rangle d / \nu_\ell$, and H is correlated by Saez and Carbonell (1985) and Crine (1982)
(c) Alumina spheres-water-air, nonuniform inlet temperature	Given by (10.85), and for large Pe_ℓ $\dfrac{D_\perp}{\alpha_\ell} = a_1 f Pe_\ell, \quad Pe_\ell = \dfrac{\langle u_\ell \rangle d}{\alpha_\ell}$
(d) Glass spheres-air-water, $\varepsilon = 0.4$, nonuniform inlet temperature, trickle flow	$\dfrac{D_\perp^d}{\alpha_\ell} = 0.35 Pe_\ell^{0.8}, \quad 50 \leq Pe_\ell \leq 3000$ $Pe_\ell = \dfrac{\langle u \rangle^\ell}{\alpha_\ell} \dfrac{d\varepsilon s}{1-\varepsilon}$
(e) Glass and alumina spheres-water (or water glycerin) air, cocurrent, downflow, wall heating	$\dfrac{D_\perp}{\alpha_\ell} = a_1 Pe_\ell + 0.095 Pe_g \dfrac{k_g}{k_\ell}$ $Pe_\ell = \dfrac{\langle u_\ell \rangle d}{\alpha_\ell}, \quad Pe_g = \dfrac{\langle u_g \rangle d}{\alpha_g}$ α_g is based on vapor-air properties $a_1 = a_1(Re_\ell, d_t/d, \text{etc.})$; d_t is tube diameter

(a) Weekman and Myers (1965); (b) Specchia and Baldi (1979); (c) Crine (1982); (d) Saez (1983); (e) Hashimoto et al. (1976).

Also, for an *axisymmetric* temperature field with

$$T = T_1, \quad x = 0, \ 0 \leq y \leq R_1,$$
$$T = T_2, \quad x = 0, \ R_1 \leq y \leq R,$$

$$\frac{\partial T}{\partial y} = 0, \quad x \geq 0, \quad y = 0,$$

$$\frac{\partial T}{\partial y} = 0, \quad x \geq 0, \quad y = R, \tag{10.101}$$

the solution to (10.86) is given by Crine (1982). These two-dimensional and axisymmetric temperature fields are measured at discrete x and y locations. Then, using these measurements, $D_\perp$ is found.

Since no temperature slip coefficient h_{bw} is needed, these *adiabatic wall methods* can give a *more accurate* result for $D_\perp$. We have already discussed the relationship used by Crine (1982), i.e., (10.85), where he determines (assuming $D_\perp = D_{\perp_1}$) the *three* constants in that correlation using the values found for $D_{e\perp}$. Saez et al. (1986) correlates $D_\perp$ directly with Pe_ℓ. Their correlations are also shown in Table 10.2. Since no theoretical treatment of $D_\perp$ is available, the validity of the correlations given in Table 10.2 cannot be verified. We also note that most of these correlations do *not* satisfy the asymptotes given by (10.48). Again this table only represents existing correlations. These correlations lack universality and should only be used where conditions are very similar to those in the original experiments.

10.3.4 Dispersion near Bounding Surfaces

At and near the bounding surface the phase distributions $a_\ell(\mathbf{x})$ and $a_g(\mathbf{x})$ are expected to be substantially different than in the bulk. This, in addition to velocity nonuniformity near this surface, causes significant variation in the local values of $\mathbf{K}_e$ and $\mathbf{D}^d$. We have already made a reference to the single-phase counterpart of this problem (Sections 3.7 and 4.10). For the two-phase flows in Section 10.3.3 (A), we noticed how a temperature slip was devised to account for this nonuniformity in $\mathbf{D}_\ell$. Since no rigorous treatment of the anisotropy and nonuniformity of $\mathbf{K}_e$ and $\mathbf{D}^d$ near bounding surfaces (fluid or solid) are available, we only point out to the importance of these deviations from the bulk behavior.

10.4 References

Carslaw, H. S. and Jaeger, J. C., 1986, *Conduction of Heat in Solids*, Clarendon.
Crine, M., 1982, "Heat Transfer Phenomena in Trickle-Bed Reactors," *Chem. Eng. Commun.*, 19, 99–114.
Crine, M., Marchot, P., and L'Homme, G., 1980, "Liquid Flow Maldistributions in Trickle-Bed Reactors," *Chem. Eng. Commun.*, 7, 377–388.
Hashimoto, K., Muroyama, K., Fujiyoshi, K., and Nagata, S., 1976, "Effective Radial Thermal Conductivity in Co-current Flow of a Gas and Liquid Through a Packed Bed," *Int. Chem. Eng.*, 16, 720–727.

Matsuura, A., Hitcka, Y., Akehata, T., and Shirai, T., 1979a, "Effective Radial Thermal Conductivity in Packed Beds with Gas-Liquid Down Flow," *Heat Transfer—Japanese Research*, 8, 44–52.

Matsuura, A, Hitake, Y., Akehata, T., and Shirai, T., 1979b, "Apparent Wall Heat Transfer Coefficient in Packed Beds with Downward Co-Current Gas-Liquid Flow," *Heat Transfer—Japanese Research*, 8, 53–60.

Saez, A. E., 1983, *Hydrodynamics and Lateral Thermal Dispersion for Gas-Liquid Cocurrent Flow in Packed Beds*, Ph.D. thesis, University of California-Davis.

Saez, A. E. and Carbonell, R. G., 1985, "Hydrodynamic Parameters for Gas-Liquid Co-Current Flow in Packed Beds," *AIChE J.*, 31, 52–62.

Saez, A. E., Carbonell, R. G., and Levec, J., 1986, "The Hydrodynamics of Trickling Flow in Packed Beds, Part I: Conduit Models," *AIChE J.*, 32, 353–368.

Somerton, W. H., Keese, J. A., and Chu, S. C., 1974, "Thermal Behavior of Unconsolidated Oil Sands," *SPE J.*, 14, 513–521.

Specchia, V. and Baldi, G., 1979, "Heat Transfer in Trickle-Bed Reactors," *Chem. Eng. Commun.*, 3, 483–499.

Udell, K. S. and Fitch, J. S., 1985, "Heat and Mass Transfer in Capillary Porous Media Considering Evaporation, Condensation and Noncondensible Gas Effects," in *Heat Transfer in Porous Media and Particulate Flows*, ASME HTD 46, 103–110.

Weekman, V. W. and Myers, J. E., 1965, "Heat Transfer Characteristics of Cocurrent Gas-Liquid Flow in Packed Beds," *AIChE J.*, 11, 13–17.

Whitaker, S., 1977, "Simultaneous Heat, Mass, and Momentum Transfer in Porous Media: A Theory of Drying," *Adv. Heat Transfer*, 13, 119–203.

11

Transport Through Bounding Surfaces

When a *gaseous plain* medium bounds a partially saturated porous medium, then interfacial heat and mass transfer can occur across this interface. This interfacial *convective heat and mass transfer* requires temperature and concentration gradients and fluid motions. Figure 11.1 depicts this interface A_{pa} between a *porous* and a *plain* (ambient) medium. A discussion of this interfacial transport is given by van Brakel (1982). This interface is *not* planar and the distribution of the phases on this interface is very *complex* and depends on the *surface saturation* (defined for a *thin* interfacial layer), the solid topology, the wettability, and the history. There is, in general, some *motion* within the interfacial *liquid* phase. In addition to this and the motion in the gaseous plain medium, the liquid and gaseous phases in the *bulk* of the porous medium experience motion. The porous medium gaseous phase moves due to the concentration and the total gas-phase pressure gradients. The liquid phase, in addition to the saturation gradient, can undergo motion due to the gradient of the surface-tension, buoyancy, wind-shear, solid-fluid surface forces (as discussed in Chapter 9), or combinations of these forces. In this chapter, we review the elements and existing treatments of the interfacial transport across *partially saturated* surfaces. Since the problem involves the *simultaneous* heat and mass transfer, a complete treatment is very involved. However, as with other phenomena we examine some simple systems in detail and evaluate the experimental results for the more complex systems.

11.1 Evaporation from Heated Liquid Film

In general, the rate of heat and mass transfer from a partially saturated porous medium to a convective ambient depends on the phase distributions on and adjacent to the surface. The phase distributions $a_\ell(\mathbf{x})$ and $a_g(\mathbf{x})$ are governed by the various forces influencing the flow of the liquid and the gaseous phase. As shown in Figure 11.1, unlike the liquid phase, the gaseous phase is continuous across the *nominal* interface. To a great extent, the van der Waals forces influence the phase distributions and the heat transfer for the case of a *perfectly wetting* liquid. In Section 8.1.5, we studied some of the hydrodynamic aspects of these extended menisci. Here we consider the heat transfer aspects. The presence of these solid-fluid forces enhance

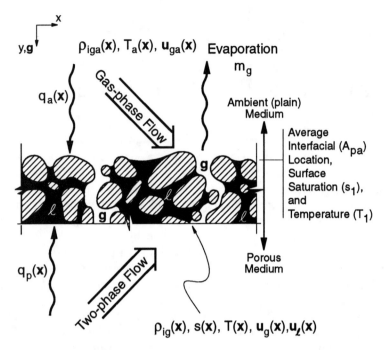

Figure 11.1 Phase distributions near a porous plain medium interface. The nominal (average) interfacial location and variables used in the analysis (for plain and porous media) are also shown.

the evaporation rate from these menisci. Also this simple problem offers a means for approaching the problem of interfacial heat and mass transfer. Figure 11.2 gives a schematic of the problem and the variables. This figure is an adaptation of Figure 8.8. In dealing with this problem, we make the following simplifying assumptions.

- $\theta_c = 0$, i.e., a *complete wetting* of the solid surface is assumed.
- A *single-component* system, i.e., the vapor pressure, is equal to the total gas phase pressure.
- *No* motion in the *gaseous* phase is allowed.
- There are two-dimensional steady-state temperature and velocity fields in the *liquid*.
- There is a *smooth* and *homogeneous* solid surface [Section 8.1.4 (B)].
- The meniscus is *heated* at $A_{\ell s}$ with $T_{\ell s}$ *slightly* above the *saturation temperature* T_s such that *no boiling* occurs.

11.1 Evaporation from Heated Liquid Film

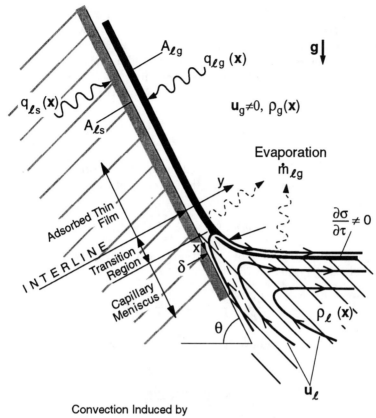

Figure 11.2 A schematic of the problem of evaporation around the interline, showing the variables, liquid motions, and forces involved this motion.

This problem has been studied by Bressler and Wyatt (1970), Wayner et al. (1976), Wayner (1978), Holm and Goplen (1979), Cook et al. (1981), Truong (1987), Truong and Wayner (1987), Mirzamoghadam and Catton (1988), and Wayner (1989). We begin by examining the heat transfer rate in the transition region (Figure 11.2) and we include the capillary meniscus and liquid motion.

11.1.1 SIMPLE MODEL FOR TRANSITION REGION

In the transition region shown in Figure 11.2, the liquid evaporates at the liquid surface $A_{\ell g}$ while *flowing toward* the adsorbed layer such that the liquid mass flow rate into the transition region is equal to the evaporation rate from the surface. Note that there is no evaporation from the adsorbed layer. Wayner (1989) has examined the heat transfer rate in the transition region using some *experimentally determined* film thickness distributions. In this region, the *inertial* and *body forces* are neglected in (8.81). Then, subject to

$$u_\ell = 0 \quad \text{at } y = 0, \tag{11.1}$$

$$\frac{\partial u_\ell}{\partial y} = 0 \quad \text{at } y = \delta, \tag{11.2}$$

the simplified two-dimensional momentum equation is solved. The resulting *liquid mass flow rate* (per unit depth) in terms of the average velocity $\overline{u}_\ell$ is given by

$$\rho_\ell \overline{u}_\ell \delta(x) = \frac{\delta^3(x)}{3\nu_\ell} \frac{dp_\ell}{dx}. \tag{11.3}$$

By neglecting the capillary and gravity terms in (8.82), dp_ℓ/dx for the small δ asymptote of f_δ, i.e., (8.76), becomes

$$\frac{dp_\ell}{dx} = -\frac{df_\delta}{dx} = \frac{d}{dx} \frac{a_1}{6\pi\delta^3(x)}. \tag{11.4}$$

Using this in (11.3), we have (positive u_ℓ is along x)

$$\rho_\ell \overline{u}_\ell \delta(x) = \frac{a_1}{\pi} \frac{\left(\dfrac{d\delta}{dx}\right)}{\nu_\ell \delta}, \tag{11.5}$$

where $d\delta/dx$ can be determined from the energy conservation consideration (this will be done in the next section), but Wayner uses the available experimental results for the *average* film thickness and thickness gradient.

For a steady-state flow and evaporation, the *heat* flow rate required for the evaporation of this mass flow rate is

$$-\Delta i_{\ell g} \rho_\ell \overline{u}_\ell \delta(x) = -\frac{\Delta i_{\ell g} a_1 \left(\dfrac{d\delta}{dx}\right)}{6\pi\nu_\ell \delta}. \tag{11.6}$$

In order to obtain a heat *flux*, consider an average film thickness $\overline{\delta}$, then the average length of the transition regime is $\overline{\delta}/(d\delta/dx)$, and we have

$$\overline{q} = -\frac{\Delta i_{\ell g} a_1 \left(\dfrac{d\delta}{dx}\right)^2}{6\pi\nu_\ell \overline{\delta}^2}. \tag{11.7}$$

Wayner gives $d\delta/dx = 0.017$ and $\overline{\delta} = 100$ Å as *typical* results for *octane-gold* as the fluid-solid system, where the gold surface is heated slightly above the boiling point of octane. Then, using the approximate numerical values for the Hamaker constant for this system he finds $\bar{q} = 8 \times 10^3$ W/m². This is *relatively* high for *nonboiling* liquid films and is the result of the *small film thickness* and the *steady flow* of liquid into this film. Note that the *length* of the transition region is rather *small* [of the order of 10 μm, Section 8.1.5 (A)].

A more complete treatment of the hydrodynamics and the heat transfer in the meniscus will be given later. However, this simplified analysis shows how the liquid is *pulled* toward the interline and evaporated. It also points to the significance of parameters $d\delta/dx$ and $\bar{\delta}$. Note that *large* $d\delta/dx$ and *small* $\bar{\delta}$ are needed, i.e., a *short length* is required for the transition region for it be an effective heat dissipation region.

11.1.2 Inclusion of Capillary Meniscus

In order to determine the heat transfer in the capillary meniscus and to include the effect of the *motion* in the *capillary meniscus* on the transition region, Moosman and Homsy (1980) and later Mirzamoghadam and Catton (1988) combine the transition and the capillary region and include the relevant forces (Figure 11.2). They assume that $\mathbf{u}_g = 0$, and therefore, *no* gas-phase flow-induced motion is *allowed* in the liquid phase. Their analysis leads to the determination of the following features.

- The *film thickness* and the *temperature* of the adsorbed layer, i.e., $\delta(x=0) = \delta_o$ and $T(x=0) = T_o$ are obtained.

- The *velocity* and *temperature fields* in the transition and the capillary regions, including the temperature distribution along the ℓ-g interface $T_{\ell g}(x)$ and the meniscus thickness $\delta(x)$ are determined.

- The *temperature distribution* along the s-ℓ interface $T_{s\ell}(x)$ is obtained.

The liquid meniscus is bounded on the left-hand side by $x = 0$ where the *interline* is located, on the right-hand side by $x = \ell$ where the meniscus contour reaches an *asymptote* (i.e., the liquid surface becomes horizontal, $\boldsymbol{\tau} \cdot \mathbf{g} = 0$), at the bottom by the *solid surface* $y = 0$, and on the top by the *free surface* $y = \delta(x)$.

Since δ is *small*, the temperature distribution along y is nearly *linear*, and the local heat flux, which is

$$q_{\ell s}(x) = k_\ell \frac{T_{s\ell}(x) - T_{\ell g}(x)}{\delta(x)} = \dot{m}_{\ell g} \Delta i_{\ell g}, \qquad (11.8)$$

can be readily evaluated knowing the quantities on the right-hand side. These quantities are determined from the following physical and mathematical descriptions.

(A) Adsorbed Layer

Assuming that the small δ asymptote for the dispersion force, i.e., (8.76) holds, then the *effect* of this force on the *saturation temperature* T_s is approximated by using the Clausius-Clapeyron equation, i.e.,

$$\frac{dp}{dT} = \frac{\Delta i_{\ell g}}{T_s \left(\frac{1}{\rho_g} - \frac{1}{\rho_\ell} \right)} \tag{11.9}$$

or

$$\frac{T_s \, dp_\ell}{\rho_\ell} = -\Delta i_{\ell g} \, dT_s \quad \text{constant } p_g. \tag{11.10}$$

This effect (i.e., the *rise in the temperature*) is found by the integration of this equation and by using $p_\ell = f_\delta$. The result is

$$T = T_s \left[1 - \frac{f_\delta(\delta_0)}{\rho_\ell \Delta i_{\ell g}} \right] = T_0, \quad f_\delta(\delta_0) = \frac{a_1}{6\pi\delta_0^3}, \quad \text{at } x = 0. \tag{11.11}$$

The *temperature* of the *adsorbed layer* is uniform at T_0. By including the gravity effect, the volumetric force balance for the adsorbed layer is the simplified form of (8.81), i.e.,

$$-\frac{df_\delta}{dx} + \rho_\ell g \sin\theta = \frac{a_1}{2\pi\delta^4} \frac{d\delta}{dx} + \rho_\ell g \sin\theta = 0, \tag{11.12}$$

from which the slope of δ at $x = 0$ is found as

$$\left. \frac{d\delta}{dx} \right|_0 = \frac{2\rho_\ell g \sin\theta \delta_0^4}{a_1} \quad \text{at } x = 0. \tag{11.13}$$

(B) Liquid Motion

Liquid motion is caused by a combination of the volumetric and surface forces, namely, the *hydrostatic force*, the *van der Waals dispersion forces*, the *viscous force*, and the *surface-tension gradient*. The *inertial force* is expected to be *small*. The momentum equation is that given by (8.81), except for the negligible inertial force. The liquid pressure gradient is that given by (8.82) where the *capillarity* is also written as a *volumetric force* (based on the *one-dimensional* assumption made for the pressure distribution). Therefore, we have

$$0 = -\frac{dp_\ell}{dx} + \mu_\ell \frac{\partial^2 u_\ell}{\partial y^2} + \rho_\ell g \sin\theta, \tag{11.14}$$

$$\frac{\partial u}{\partial x} + \frac{\partial v}{\partial y} = 0, \tag{11.15}$$

11.1 Evaporation from Heated Liquid Film

$$-\frac{dp_\ell}{dx} = \frac{d}{dx}(2\sigma H) - \rho_\ell g \frac{d\delta}{dx} + \frac{a_1}{2\pi\delta^4}\frac{d\delta}{dx}. \qquad (11.16)$$

The *mean curvature* for the two-dimensional surface is given by (8.79). The boundary conditions are

$$u_\ell = 0 \qquad \text{at } y = 0, \qquad (11.17)$$

$$u_\ell = u_{\ell g} \qquad \text{at } y = \delta, \qquad (11.18)$$

$$\mu_\ell \left.\frac{\partial u_\ell}{\partial y}\right|_\delta = \frac{d\sigma}{dx} \qquad \text{at } y = \delta. \qquad (11.19)$$

The last boundary condition is obtained from (8.41) with $\mathbf{u}_g = 0$, and by assuming that $(\partial/\partial\tau)\mathbf{u}_\ell \cdot \mathbf{n} \ll (\partial/\partial n)\mathbf{u}_\ell \cdot \boldsymbol{\tau}$, i.e., a negligible tangential gradient of the normal velocity.

The *recirculating* flow in the meniscus, including the liquid surface velocity $u_{\ell g}$, is determined using (11.14)–(11.19) along with (11.11) and (11.13), and the condition of no curvature at $x = 0$ and $x = \ell$, i.e.,

$$\frac{d^2\delta}{dx^2} = 0, \qquad \text{at } x = 0, \ell. \qquad (11.20)$$

Also needed is the surface tension gradient. This is obtained by using (8.39) and by assuming that this gradient is solely due to the temperature gradient, i.e.,

$$\frac{\partial\sigma}{\partial x} = \frac{\partial\sigma}{\partial T}\frac{\partial T}{\partial x}. \qquad (11.21)$$

A linear approximation is made to (9.27) for $T_{\ell g} \ll T_c$. This gives

$$\sigma = \sigma(T_s)\left[1 - \frac{a_1}{T_c}(T - T_s)\right], \qquad (11.22)$$

$$\frac{\partial\sigma}{\partial T} = -\frac{a_1\sigma(T_s)}{T_c}. \qquad (11.23)$$

(C) TEMPERATURE FIELD

The energy equation for the liquid with a negligible axial conduction is

$$(\rho c_p u)_\ell \frac{\partial T_\ell}{\partial x} + (\rho c_p v)_\ell \frac{\partial T_\ell}{\partial y} = k_\ell \frac{\partial^2 T}{\partial y^2}. \qquad (11.24)$$

The temperature distribution along the y-axis is assumed to be linear, i.e.,

$$T_\ell = T_\ell(x, y) = T_{\ell s}(x) - [T_{\ell s}(x) - T_{\ell g}(x)]\frac{y}{\delta(x)}. \qquad (11.25)$$

As the meniscus becomes thicker, an asymptotic behavior is assumed. For the temperature this asymptote is taken as

$$\frac{dT}{dx} = 0 \quad \text{at} \quad x = \ell. \tag{11.26}$$

A simple solid surface temperature distribution is used based on the assumptions that $T_{\ell s}(x = 0) = T_{\ell s}(x = \ell)$ and that a uniform heat flux q_o is applied through a solid slab with a thickness Δ and a thermal conductivity k_s. This temperature distribution is

$$T_{\ell s} = \frac{q_o}{k_s \Delta}\left(\frac{x^2}{2} - \frac{x\ell}{2}\right) + T_o. \tag{11.27}$$

This formulation (thermodynamics, fluid dynamics, and heat transfer) allows for the determination of $T_{\ell g}(x)$ and $\delta(x)$. Mirzamoghadam and Catton solve (11.14)–(11.16) and (11.24) using the *integral* method. They also performed experiments with *water* and *freon* on a *copper* surface. Some of their typical results are given in Figures 11.3 (a) to (c). The results are for $\delta_o = 45$ Å and $\theta = 30°$. The film thickness increases monotonically with x while the temperature difference $T_{\ell s}(x) - T_{\ell g}(x)$ and the mean velocity $\overline{u}_\ell$ undergo *maxima*. The locations of these maxima are within the transition region. By using the solution to $\delta(x)$, the heat flux can be determined from (11.8). Their results are in general agreement with the simplified analysis of Wayner given in the previous section, i.e., the heat transfer rate is large and the average flux is of the *order* of 10^3 W/m^2 in the transition region. Outside the transition region (i.e., in the capillary meniscus region), the heat transfer rate drops to that associated with the *natural* convection. Their results also show that for $20° < \theta < 30°$ the heat transfer rate is *maximized*.

For porous media with a small pore (or particle) size, i.e., $d < 10^3 \delta_o$ and for $\theta_c \to 0$, we expect the heat transfer from the partially saturated surfaces to be greatly influenced by the contribution from the extended menisci. Note that these analyses are for the heat supplied through $A_{\ell s}$. However, we expect a similar behavior when the heat is supplied through $A_{\ell g}$ (ambient convection). For liquid-solid systems with $\theta_c \neq 0$, this formulation of the extended liquid film does *not* apply, and the liquid phase distribution will be *strongly* history-dependent.

11.2 Mass Diffusion Adjacent to a Partially Saturated Surface

Assuming that the surface roughness resulting from the *partial* exposure of the solid matrix does *not* noticeably influence the ambient gaseous flow field near the porous-plain interface, the process of mass transfer from the

11.2 Mass Diffusion Adjacent to a Partially Saturated Surface

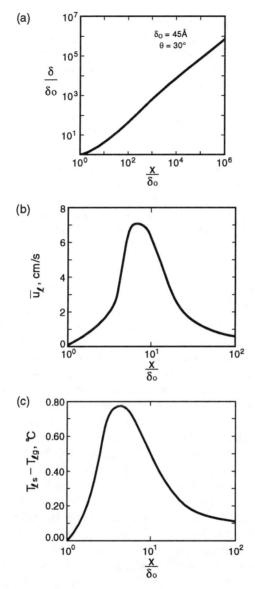

Figure 11.3 Distributions of (a) liquid film thickness, (b) average velocity in the film, and (c) temperature difference across the film, as predicted by Mirzamoghadam and Catton for $\delta_o = 45$ Å and $\theta = 30°$. (From Mirzamoghadam and Catton, reproduced by permission ©1988 ASME.)

surface can be examined with an emphasis on the *mass diffusion* at and around this surface. This would require the *mean surface roughness* Δ to be much smaller than the momentum boundary-layer thickness δ. In convection heat and mass transfer from the surface of a partially saturated porous medium, the vapor molecules diffuse *laterally* and encounter a resistance as they leave the liquid surface and travel through *surface passages* to arrive in the plain medium. Figure 11.1 gives a two-dimensional rendering of these passages with differing widths and lengths. In the following, we consider this mass diffusion at and near the porous-plain interface by first examining the case where the average pore size $d = 2R$ is small enough such that the vapor molecules can *drift* along the *dry patches* and make the surface *appear* as if it were *fully* covered by the vapor molecules (with the vapor density corresponding to the saturation temperature of the surface liquid). Then, for this large *Knudsen number* $Kn = \lambda/R$ case, we examine an available model for the mass transfer rate. Next, we consider the case of *small* Knudsen numbers where the vapor molecules have to *diffuse* through inert molecules as they pass through the passages. A simple diffusion-convection model will be examined for this condition.

11.2.1 LARGE KNUDSEN NUMBER MODEL

The mass diffusion of interest here is that of the *condensible component c* of the gas phase. The density of this component at the liquid surface is $\rho_{cg\,o}$ and is different than that in the ambient $\rho_{cg\,a}$. This concentration difference is generally due to the heat addition (heat removal) to partially liquid saturated porous media which results in the surface evaporation (condensation). Schlünder (1988a,b) proposes a model for partially liquid-covered surfaces based on the diffusion of the evaporated molecules adjacent to the surface. Although the model is *not* rigorous, it shows the significant role played by the drifting of the vapor molecules for porous media with a small pore size. When the *mean free path* of the vapor molecules is of the order of the pore size, these molecules can *drift* near the surface and therefore cover the entire surface area (both the wet and dry patches). Figure 11.4 shows this model developed by Schlünder. A model based on a different distribution of the liquid phase is given by Maneval and Whitaker (1990). In their models the liquid patches spread over many pores with the dry patches (which also cover many pores) covering the remainder of the surface. In the Schlüder model the surface is covered by spherical liquid patches of radius R with their centers a distance ℓ apart. This is a periodic surface structure with a planar unit-cell length 2ℓ. The *areal surface saturation* s_1 based on the *projected area* is

$$s_1 = \frac{\pi R^2}{4\ell^2}. \tag{11.28}$$

The surface of the liquid patch is at the saturated state with a density $\rho_{cg\,o}$. At a *mean* depth of penetration λ, based on the molecular drift, this density drops to $\rho_{cg\lambda}$. It is assumed that within this penetration depth *no* resistance is offered by the *inert* gas components, i.e., the *Knudsen*

11.2 Mass Diffusion Adjacent to a Partially Saturated Surface

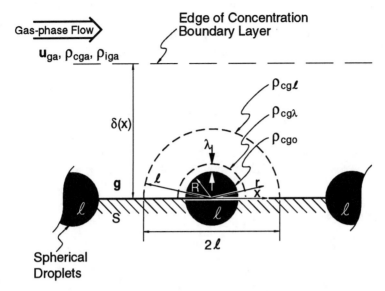

Figure 11.4 The surface model of Schlünder (1988a) for a partially liquid-covered surface with a large Knudsen number. The parameters used in the model are also shown.

diffusion occurs in this region. The unit cell is extended perpendicular to the surface a distance ℓ (a pseudo-hemispherical unit cell), where from the edge of the penetration depth $r = R + \lambda$ to the edge of the unit cell $r = \ell$, the *Fick* diffusion occurs. From the outside of the unit cell to the edge of the concentration boundary layer, i.e., from $r = \ell$ to $r = \delta$, again the Fick diffusion is assumed in the Schlünder model. Since the Schmidt number of near unity is normally encountered, this boundary-layer thickness can be taken to be the momentum boundary-layer thickness. It is assumed that the surface *roughness* will not influence the hydrodynamics significantly because $R/\delta \to 0$.

By assuming a *negligible* resistance due to the presence of the inert molecules in the region $R < r < R + \lambda$, the drifting vapor molecules can cover the entire porous-plain interface A_{pa}. As a result, the mass transfer rate from this interface can reach the same magnitude as that of the *totally* liquid-covered surface. This is the $\ell/R \to 1$ or $s_1 \to 1$ asymptote. The deviation from this asymptote can be given by

$$\frac{\dot{m}_g(s_1)}{\dot{m}_g(s_1 = 1)} = \frac{\rho_{cg\,\ell} - \rho_{cg\,\infty}}{\rho_{cg\,o} - \rho_{cg\,\infty}}. \tag{11.29}$$

A one-dimensional and steady *lateral* transfer of the vapor from the surface toward the boundary layer is assumed. Next, the three diffusion resistances described earlier, i.e., the Knudsen and two Fick diffusions, are assumed to be resisting the mass flow in a *serial* arrangement. Some of the assumptions and details are given in Schlünder (1988a,b) and his result for an *accommodation* coefficient of unity (in the Knudsen diffusion) is

$$\frac{\dot{m}_g(s_1)}{\dot{m}_g(s_1 = 1)} = \frac{1}{1 + \frac{R}{2\delta s_1}\left[\frac{2\lambda}{R} + \frac{1}{1 + \lambda/R} - \left(\frac{4s_1}{\pi}\right)^{1/2}\right]} = f\left(\frac{R}{\delta}, \frac{\lambda}{R}, s_1\right),$$

(11.30)

where, as was mentioned, λ/R is the *Knudsen number* and the definition of Kn is given by (6.5). The two asymptotes for (11.30) are

$$\frac{\dot{m}_g(s_1)}{\dot{m}_g(s_1 = 1)} = \frac{1}{1 + \frac{2R}{\pi\delta}\left(\frac{\pi}{4s_1}\right)^{1/2}\left[\left(\frac{\pi}{4s_1}\right)^{1/2} - 1\right]} \quad \text{for } R \gg \lambda,$$

(11.31)

$$\frac{\dot{m}_g(s_1)}{\dot{m}_g(s_1 = 1)} = \frac{1}{1 + \frac{\lambda}{\delta s_1}} \quad \text{for } R \ll \lambda. \quad (11.32)$$

The boundary-layer thickness is a function of the gas Reynolds number based on the system length scale ($Re_L = u_g L/\nu_g$), and a Schmidt number of unity is assumed. The mean free path can be calculated using (6.4). For example, for ethanol at the saturation condition (one atmosphere), we have $\lambda = 0.07$ μm, while δ is generally of the order of mm. Schlünder has plotted (11.30) for $\lambda/\delta = 10^{-4}$, and his results are shown in Figure 11.5 for the normalized mass flow rate (i.e., the evaporation rate) as a function of the surface saturation s_1 for various values R/λ. The results show that even for $R/\lambda = 10^3$ a surface saturation as low as 0.6 can result in *nearly* the same mass transfer rates as that for $s_1 = 1$. The case of $R/\lambda = 10^3$ corresponds to $R \simeq o(10$ μm), which is of the order of the particle (or pore) size encountered for sandstones. For yet smaller R, the surface saturation can become even smaller before any *noticeable* deviation from the fully saturated surface is found.

It should be pointed out that this analysis is not rigorous and the resistance within each region and the arrangement of the resistances are not as simple as assumed. However, the *constant drying rate*, i.e., the independence of $\dot{m}_g$ from s_1, encountered in the drying of porous media having *very small* pores, can be explained using the Schlünder model. (This will be further discussed in Section 11.5.)

11.2.2 SMALL KNUDSEN NUMBER MODEL

When the pore size is large enough such that the Knudsen diffusion does not result in the nearly uniform distribution of the vapor near the surface,

11.2 Mass Diffusion Adjacent to a Partially Saturated Surface

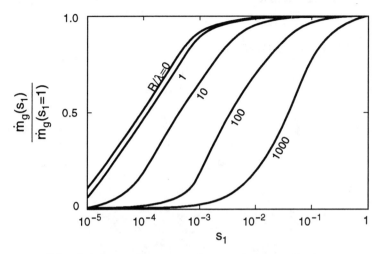

Figure 11.5 Effect of the surface saturation on the normalized mass transfer rate for various values of $1/Kn$. The prediction is by Schlünder (1988a) and for $\lambda/\delta = 10^{-4}$. (From Schlünder, reproduced by permission ©1988 Pergamon.)

a one-dimensional model based on the combined effects of the *Fick diffusion* and *convection*, such as the one developed by Tao and Kaviany (1991), may be used to examine the effect of the surface saturation on the evaporation rate. Figure 11.6 shows this surface model and its associated features. This model is based on the following simplifying assumptions.

- A one-dimensional *lateral* (i.e., perpendicular to the envisioned planar surface A_{pa}) *diffusion* occurs in the *indented* regions, i.e., through the $0 \leq y \leq \Delta$ region.

- A one-dimensional *lateral diffusion-convection* occurs in the $\Delta \leq y \leq \Delta + \delta$ region.

- A *simple cubic* arrangement of solid spherical particles is assumed on the surface.

- A *planar* liquid surface (no capillarity) is assumed for the liquid meniscus.

- A Schmidt number of near *unity* is assumed.

The *surface saturation* for this model is defined as the fraction of the void space (in the top most layer of the particles) that is occupied by the liquid phase, i.e.,

$$s_1(\Delta/R) = 1 - \frac{\Delta/2R}{1 - \pi/6}\left(\frac{\pi \Delta^2}{12R^2} - \frac{\pi \Delta}{4R} + 1\right). \qquad (11.33)$$

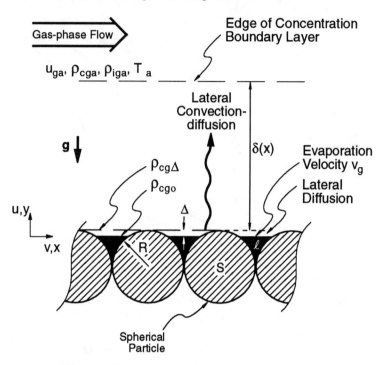

Figure 11.6 A surface model for a partially liquid-covered surface. The model is for small Knudsen numbers.

The analysis of each of the regions, as well as the determination of the concentration at the boundary between these two regions, are given here.

(A) DIFFUSION LAYER

In this layer, $0 \leq y \leq \Delta$, the vapor molecules *diffuse* through a variable area A_g channel. This diffusion is given by

$$\dot{m}_g(s_1) = -A_g N D_m \frac{d\rho_{cg}}{dy} = \text{constant}. \tag{11.34}$$

The boundary condition is

$$\rho_{cg} = \rho_{cg\,o} \quad \text{at } y = 0. \tag{11.35}$$

We also designate the vapor density at the boundary between the two regions as

$$\rho_{cg} = \rho_{cg\,\Delta} \quad \text{at } y = \Delta. \tag{11.36}$$

In (11.34) D_m is the *binary* mass diffusion coefficient for component c diffusing through the inert component and N is the number of particles per unit area.

11.2 Mass Diffusion Adjacent to a Partially Saturated Surface

The unit-cell-based variable cross-sectional area for the diffusion of the vapor molecules is given by

$$A_g\left(\frac{y}{R},\frac{\Delta}{R}\right) = 4\pi R^2\left[\frac{y^2}{4R^2} + \left(1-\frac{\Delta}{R}\right)\frac{y}{2R} + \frac{1}{4\pi} - \frac{\Delta}{2R} + \frac{\Delta^2}{4R^2}\right]. \quad (11.37)$$

By integrating (11.34) and by using this variable area, we have

$$\dot{m}_g = 2RND_m\frac{\rho_{cg\,o} - \rho_{cg\,\Delta}}{a_1}, \quad (11.38)$$

where

$$a_1 = \frac{2}{\pi a_2}\left[\tan^{-1}\left(\frac{1}{a_2}\right) - \tan^{-1}\left(\frac{1-\Delta/R}{a_2}\right)\right] \quad (11.39)$$

and

$$a_2 = \left(\frac{4}{\pi} - 1\right)^{1/2}. \quad (11.40)$$

(B) CONVECTION-DIFFUSION LAYER

The one-dimensional (lateral) treatment of the boundary-layer convection is made similar to that done for the *stagnant film* This approximate treatment is known to give satisfactory results. Glassman (1977) discusses this treatment, which uses the local boundary-layer thickness and a mass (also energy) transport equation given by

$$v_g\frac{d\rho_{cg}}{dy} = D_m\frac{d^2\rho_{cg}}{dy^2}, \quad (11.41)$$

with the boundary conditions

$$-D_m\frac{d\rho_{cg}}{dy} = \dot{m}_g = \rho_{cg\,\Delta}v_g \quad \text{at} \quad y = \Delta,\ \eta = 0, \quad (11.42)$$

$$\rho_{cg} = \rho_{cg\,\Delta} \quad \text{at} \quad y = \Delta,\ \eta = 0, \quad (11.43)$$

$$\rho_{cg} = \rho_{cg\,o} \quad \text{at} \quad y = \Delta+\delta,\ \eta = \delta, \quad (11.44)$$

where $\eta = y - \Delta$. By solving (11.41), we obtain the mass transfer rate as

$$\dot{m}_g = \frac{\rho_g D_m}{\delta}\ln\left(\frac{\rho_g - \rho_{cg\,a}}{\rho_g - \rho_{cg\,\Delta}}\right). \quad (11.45)$$

For $s_1 = 1$, we have

$$\dot{m}_g(s_1 = 1) = \frac{\rho_g D_m}{\delta}\ln\left(\frac{\rho_g - \rho_{cg\,a}}{\rho_g - \rho_{cg\,o}}\right). \quad (11.46)$$

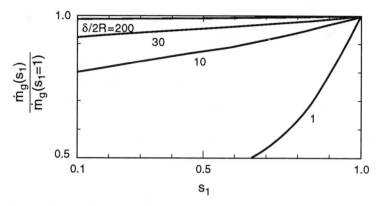

Figure 11.7 Effect of the surface saturation on the normalized mass transfer rate for evaporation of ethanol into air. Based on the small Knudsen number surface model of Tao and Kaviany (1991a).

(C) DETERMINATION OF $\rho_{cg\,\Delta}$

By using (11.38), (11.45), and (11.46) with $N = 1/(4R^2)$, we have

$$\frac{\dot{m}_g(s_1)}{\dot{m}_g(s_1=1)} = \frac{\delta(\rho_{cg\,o} - \rho_{cg\,\Delta})}{2R\rho_g a_1 \ln\left(\dfrac{\rho_g - \rho_{cg\,a}}{\rho_g - \rho_{cg\,o}}\right)}$$

$$= f\left(s_1, \frac{\delta}{2R}\right) = \frac{\ln\left(\dfrac{\rho_g - \rho_{cg\,\Delta}}{\rho_g - \rho_{cg\,o}}\right)}{\ln\left(\dfrac{\rho_g - \rho_{cg\,a}}{\rho_g - \rho_{cg\,o}}\right)}. \qquad (11.47)$$

Also, by using (11.38) and (11.45), we have

$$\ln\left(\frac{\rho_g - \rho_{cg\,a}}{\rho_g - \rho_{cg\,\Delta}}\right) = \frac{\rho_{cg\,o} - \rho_{cg\,\Delta}}{2Ra_1}, \qquad (11.48)$$

which can be used to solve for $\rho_{cg\,\Delta}$. As expected, (11.47) shows that the extra resistance associated with the diffusion of the vapor through the variable area region of thickness Δ *results* in a smaller mass flow rate (note that $\rho_{cg\,\Delta} \leq \rho_{cg\,o}$).

Tao and Kaviany (1991) solve (11.47) for the evaporation of ethanol into air (with $\rho_{cg\,a} = 0$), and their results are given in Figure 11.7. The results are given as the variation of the normalized mass flow rate with respect to the surface saturation for *various* values of $\delta/(2R)$. The results show that for $\delta/(2R) > 200$ *no* significant change in the mass transfer rate is found when the surface saturation becomes *less* than unity. For a *transition* value of $\delta/(2R) = 30$, the effect of s_1 on $\dot{m}_g$ begins to be significant. Based on this, for a boundary-layer thickness of the order of mm, the pore (or particle) size should be of the order of 100 μm (or more) for the effect of the surface saturation to become noticeable.

11.3 Convection from Heterogeneous Planar Surfaces

As a prelude to the *inclusion* of the *hydrodynamics* into the surface models, we begin with the planar surfaces. The effect of the surface *roughness* caused by the exposure of the solid matrix will be addressed in Section 11.4. We begin by considering a simple *planar* porous, plain media interface. We examine the effect of the extent to which this interface is occupied by the liquid phase, on the interfacial mass transfer rate. We first consider the mass transfer from a single active strip and then consider a surface with a uniform distribution of the active sources.

11.3.1 Mass Transfer from a Single Strip

Here we consider parallel flow over planar surfaces. When examining the mass transfer (or heat transfer) from a planar surface with a discrete source distribution, we note that at low ambient velocities the boundary-layer treatment *fails* and the *axial* diffusion becomes important. For example, for a two-dimensional flow and a *discrete* source of length w located at x_1, the *leading* and *trailing* edge effects result in an *increased* transfer rate compared to the boundary layer-based predictions. This problem has been studied by Ling (1963) and by Ackerberg et al. (1978), among others. The boundary-layer treatment, using the integral method, is given by Kays and Crawford (1980, p. 147). The boundary-layer behavior is approached at high Peclet numbers. In Figure 11.8 the experimental results of Ackerberg et al. are shown for the average dimensionless transfer rate as a function of the Peclet number. The ambient velocity field is given in terms of the assumed linear velocity distribution near the surface. The experimental results are in good agreement with the numerical treatment based on elliptic equations. The high Pe asymptote is also shown.

Note that the results are for *small* enough w such that over the w the concentration (or temperature) boundary layer is inside the momentum boundary layer. The results given in Figure 11.8 show that for the partially liquid-covered surfaces (in this case a discrete liquid patch on a planar surface) the *sidewise* (along the flow) *diffusion* would increase the mass transfer rate at *low* Pe [$Pe = (\mathrm{d}u_g/\mathrm{d}y)w^2/(4\alpha_g)$]. This sidewise or axial diffusion also occurs on the nonplanar surfaces, but for the nonplanar surfaces a secondary flow in the surface indentation occurs that adds a Reynolds number dependency (this will be discussed in Section 11.4).

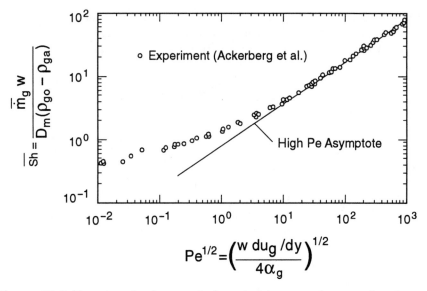

Figure 11.8 Mass transfer from a single wet strip on a planar surface to a convective ambient with linear velocity distribution near the surface. The high-velocity asymptote is also shown. (From Ackerberg et al., reproduced by permission ©1978 Cambridge University.)

11.3.2 SIMULTANEOUS HEAT AND MASS TRANSFER FROM MULTIPLE SURFACE SOURCES

We now examine the evaporation from liquid patches on a planar surface caused by the heat transfer from the moving ambient gaseous phase. The heat is transferred to the liquid *directly* by the gas and *indirectly* from the gas to the solid and then to the liquid. Therefore, the thermal conductivity of the solid *also* influences the evaporation rate. Then, unlike the last example, in the case of the simultaneous heat and mass transfer, the heat transfer in the substrate should be *included* in the analysis. For a *low* conductivity solid and with the liquid patches having a circular cross section (i.e., holes made in a plate and completely filled with liquid), Plumb and Wang (1982) perform experiments in which heated air flows over this surface. The patches of radius R are *uniformly* distributed with the distance between the centers of these patches being ℓ. A water-plexiglas system is used with the air velocities mostly in the turbulent range. The surface saturation is defined as

$$s_1 = \frac{\pi R^2}{\ell^2}. \tag{11.49}$$

In their experiment, s_1 is varied between 0.049 to 0.442. Note that for the circular patches the case of $s_1 = 1$, i.e., the fully liquid-covered surface, cannot be obtained by simply increasing R. The shape of the wet patches must change after the circles touch one another. They correlate their experimental results using the fully saturated surface as the datum and find that the parameters s_1 and $Pe = 2u_g R/\alpha_g$ can correlate the results for $Pe < 1000$. This correlation is

$$\frac{\overline{Sh}(s_1)}{\overline{Sh}(s_1 = 1)} = a_1 s_1 Pe^{-1/2}, \qquad (11.50)$$

where for their experiment $a_1 = 27$. The Sherwood number averaged over the surface is defined as $\overline{Sh} = \dot{m}_g R/(\Delta \rho_{cg} D_m)$.

For $Pe > 1000$, i.e., at high velocities, there is a *significant scatter* in the experimental data. The *increase* in $\overline{Sh}$ (for $s_1 < 1$) with a *decrease* in Pe can be due to the sidewise diffusion we discussed in the previous section. Since R as large as 6.35 mm was used in the experiments, a significant *liquid motion* in the cavities and possibly the *spill-over* of the liquid can occur. The liquid motion can be due to the wind shear, surface-tension gradient, and buoyancy.

This example shows how the *surface saturation* influences the *simultaneous* heat and mass transfer from a planar surface. For nonplanar surfaces and when the solid thermal conductivity is *not* small, other parameters must be included. We will discuss these parameters in the following sections. First we consider two-dimensional surfaces and then we examine the surface of packed beds of spheres.

11.4 Convection from Heterogeneous Two-Dimensional Surfaces

In order to further examine the dependence of the interfacial heat and mass transfer rates for the porous plain medium interface depicted in Figure 11.1, we consider the case of the steady-state convective transfer from a simple periodic two-dimensional surface. The *simplest* geometry that can still have some *realistic* features is that of the in-line arrangement of cylinders. In the following, we examine a numerical simulation of the flow, concentration, and temperature fields. This example does *not* include the heat transfer through the substrate, i.e., it uses *prescribed, uniform*, and *identical* temperatures for the liquid and solid surfaces. Then we review an experimental study in which the surface temperature variations on the solid and liquid surfaces are measured. In both of these studies, we examine the variation of the *mass transfer rate* with respect to the surface saturation. We expect to find a functional dependence of

$$\frac{\overline{\dot{m}}(s_1)}{\overline{\dot{m}}(s_1 = 1)} = f\left(s_1, Pe, \theta_c, Bo, \text{solid structure}, Ma, Ra, Re, \text{etc.}\right), \quad (11.51)$$

where the undefined parameters will be defined later.

11.4.1 A Simple Surface Model

A simple two-dimensional surface with the gaseous phase flowing parallel to it is used by Kaviany (1989) for the study of the dependence of the mass transfer rate on the surface saturation. The surface is made of cylinders with each cylinder *just* touching the two neighboring cylinders, where all of these points, as well as the centers of the cylinders, are on an axis that is parallel to the flow. The volume above this axis and between the cylinders is partially filled with a liquid that evaporates to the ambient gas because of the difference in the vapor density between the surface of the liquid and the free stream gas. The liquid motion is neglected by assuming that the *Marangoni* and *Rayleigh* numbers defined as

$$Ma = \frac{\partial \sigma}{\partial T} \frac{(\Delta T) R}{\alpha_\ell \mu_\ell} \tag{11.52}$$

and

$$Ra = \frac{g \beta_\ell (\Delta T) R^3}{\nu_\ell \alpha_\ell}, \tag{11.53}$$

are smaller than the *critical* values that mark the *transition* to *significant* liquid motions. These critical values have not yet been determined for the geometry under consideration, i.e., the space between cylinders. The existing results for square cavities *suggest* that for large cylinders ($R > 2$ mm) and for water as the fluid, the surface-tension gradient and the buoyancy-induced motions can become significant. Also, for large enough u_g, the *wind shear* induced motion can be significant and the liquid surface will not be symmetric with respect to the axis perpendicular to the flow and passing through the contact point of the two cylinders. It is also assumed that the surface *evaporation velocity* v_o does not perturb the flow field significantly, i.e.,

$$\frac{v_o}{u_{ga}} Re^{1/2} \tag{11.54}$$

is *small*, where $Re = u_{ga} R / \nu_g$ and u_{ga} is the free stream velocity.

Constant properties are assumed, therefore, the concentration and temperature fields are computed independently using a body-fitted coordinate system and the finite difference approximations, along with the stream function-vorticity formulation of the velocity field. For the computational economy, instead of *periodic* boundary conditions, which are more realistic, the flow arriving at the unit cell (the unit cell is made of two one-quarter cylinders, i.e., $\ell = 2R$) is assumed to be *unperturbed*. However, the *qualitative* trends predicted by this model are expected not to be significantly dependent on this treatment of the upstream boundary condition. The meniscus contours are determined as described in Section 8.1.1 (C), and it was found that for the assumed complete wetting ($\theta_c = 0$) and for the prescribed $Bo = \rho_g g R^2 / \sigma = 0.14$, the *radius of curvature* was *nearly* uniform over the meniscus. The surface saturation is given in terms of the void

11.4 Convection from Heterogeneous Two-Dimensional Surfaces

space above the point of contact, i.e.,

$$s_1(\text{volume}) = \frac{\text{liquid volume}}{\left(2 - \frac{\pi}{2}\right)R^2}. \tag{11.55}$$

For comparison, two other surface saturations are also defined. These are

$$s_1(\text{area, planar}) = \frac{\text{stream-wise projected length of the meniscus}}{2R}, \tag{11.56}$$

$$s_1(\text{area, nonplanar}) = \frac{\text{meniscus arc length}}{2R}. \tag{11.57}$$

The temperature field is also computed by assuming that the entire surface is maintained at a temperature different than the ambient temperature. This does *not* occur in practice, as will be discussed in the next section. However, the introduction of a temperature difference between the solid and liquid surfaces along with a different temperature for the ambient gas results in an extra parameter, which has to be determined by the inclusion of the heat transfer in the substrate (the solid-liquid composite substrate). This will also be discussed in the next section.

Typical velocity, concentration, and temperature fields are shown in Figures 11.9 (a)-(c), for $s_1(\text{volume}) = 0.32$ and $Re = 100$. The *vortex* formed in the indentation *moves* to the *left* as the Reynolds number *increases*. The local Sherwood number is defined as

$$Sh = \frac{\dot{m}_g R}{(\rho_{cg\,o} - \rho_{cg\,a})D_m}. \tag{11.58}$$

The distribution of Sh for this saturation and at several Reynolds numbers is shown in Figure 11.10. Note that the sidewise diffusion results in the *enhancement* of the mass transfer rate at the two ends of the meniscus (as was discussed in Section 11.3.1).

The effect of the surface saturation on the normalized average mass transfer rate is shown in Figure 11.11. The computed results are for three values of s_1, namely, $s_1(\text{volume}) = 0.20, 0.32$, and 0.50. The results show a strong Reynolds number dependence. All the numerical results are for $Sc = 0.7$, and any dependence on Sc is *not* explored. Therefore, we *cannot* choose the Peclet number as the only parameter. The results for the normalized average mass transfer rate are given as

$$\frac{\overline{\dot{m}}_g(s_1)}{\overline{\dot{m}}_g(s_1 = 1)} = f(s_1, Re, Sc). \tag{11.59}$$

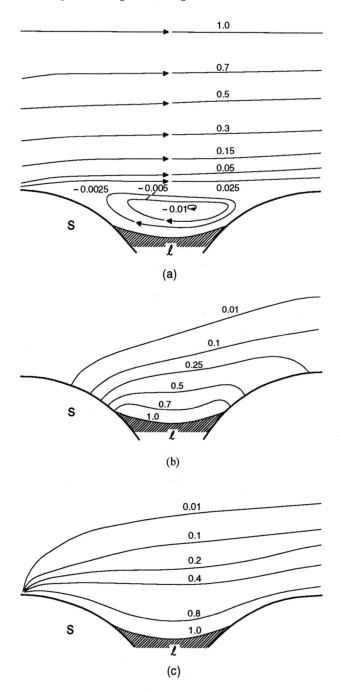

Figure 11.9 Lines of constant values of (a) stream function, (b) vapor mass concentration, and (c) temperature for s_1(volume) $= 0.32$ and $Re = 100$.

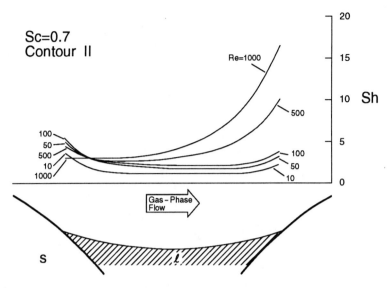

Figure 11.10 Distribution of the normalized local mass transfer rate for various Reynolds numbers and $s_1(\text{volume}) = 0.32$.

Also, Figure 11.11 shows that for a given saturation, $\overline{m}_g$ increases as Re decreases. This is in agreement with the measurements of Plumb and Wang (1982), but the saturation dependence is more *complex* than the linear relationship in their correlation, i.e., (11.50).

Also shown in Figure 11.11 is the *presumed* relation

$$\frac{\overline{m}_g[s_1(\text{area, nonplanar})]}{\overline{m}_g(s_1 = 1)} = s_1(\text{area, nonplanar}). \tag{11.60}$$

As is evident, this simplification is *not* acceptable. Another simple relationship shown in Figure 11.11 is based on the assumptions boundary-layer behavior and a planar (projected) area treatment. In this treatment, for the liquid starting at x_1 away from the starting point of the momentum boundary layer (leading edge) the local Sherwood number is (Kays and Crawford, 1980)

$$Sh_x = \frac{0.332 Sc^{1/2} Re_x^{1/2}}{\left[1 - \left(\frac{x_1}{x}\right)^{3/4}\right]^{1/4}}. \tag{11.61}$$

When the Sherwood and Reynolds numbers are *redefined* and are based on the *particle radius*. The average Sherwood number is found for a liquid strip between x_1 and x_2 located on the x-axis, and this is

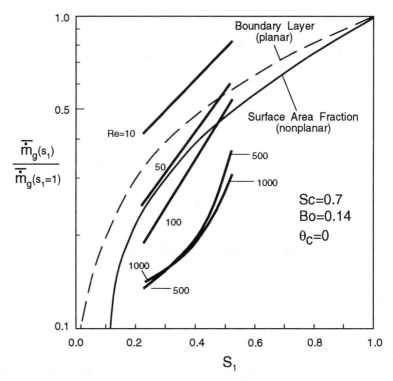

Figure 11.11 Effect of surface saturation on normalized average mass transfer rates for several Reynolds numbers. A simple linear relation and a boundary-layer treatment based on the projected area are also shown.

$$\overline{Sh} = 0.332 Sc^{1/2} Re^{1/2} \frac{1}{x_2 - x_1} \int_{x_1}^{x_2} \frac{dx}{x^{1/2}\left[1 - \left(\frac{x_1}{x}\right)^{3/4}\right]^{1/2}} \quad (11.62)$$

with

$$\overline{Sh}(s_1 = 1) = 0.471 Sc^{1/3} Re^{1/2}. \quad (11.63)$$

Also, note that

$$\frac{\overline{\dot{m}_g}(s_1)}{\overline{\dot{m}_g}(s_1 = 1)} = \frac{(x_2 - x_1)\overline{Sh}(s_1)}{2\overline{Sh}(s_1 = 1)}. \quad (11.64)$$

As was mentioned, this boundary-layer treatment, which also does *not* include the curvature effect, gives *inaccurate* predictions. This is because the *sidewise* diffusion and the *vortex* formed in the indentation both alter the local and averaged transfer rates significantly.

11.4 Convection from Heterogeneous Two-Dimensional Surfaces

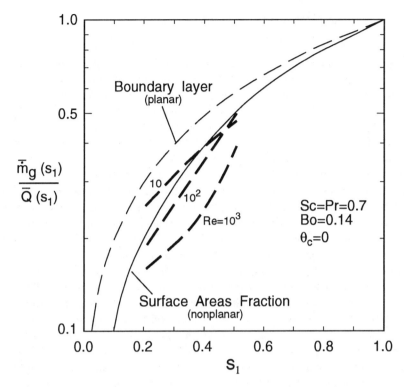

Figure 11.12 Variation of ratio of dimensionless average mass transfer rate to dimensionless heat transfer rate as a function of the surface saturation for various Reynolds numbers. A simple linear relation and a prediction based on boundary-layer treatment of the project area are also shown.

For $s_1 < 1$ the *analogy* between the heat and mass transfer will no longer hold. This is because the *concentration* and *temperature* fields are significantly *different* [Figure 11.9 (b) and (c)]. For the uniform surface temperature assumed, the ratio of the dimensionless *average* mass transfer rate to the dimensionless *average* heat transfer rate, as computed for this model, is given in Figure 11.12, as a function of the surface saturation. We note that unlike the area fraction and the boundary-layer (planar) models, this ratio is Reynolds number dependent. However, for $Re = 10$ and 10^2, the computed results *nearly* follow the prediction based on the area fraction. We note again that in practice the solid and liquid surfaces will have *different* temperatures and the surface temperature of each of these phases will *not* be uniform because of the conduction in the solid phase and the conduction-convection in the liquid phase. Therefore, this *simplified* surface model can only offer suggestive trends for $\overline{m}_g(s_1)/\overline{m}_g(s_1 = 1)$ and $\overline{m}_g(s_1)/\overline{Q}(s_1)$.

11.4.2 Experimental Observation on Simultaneous Heat and Mass Transfer

The effect of the surface saturation on the evaporation rate is investigated experimentally by Tao and Kaviany (1991), where glass cylinders *spaced* a distance $\ell/(2R) - 1 = 0.1835$ (simulating the simple cubic arrangement of spheres) are placed with the r-axis perpendicular to the flow. The single row of cylinders makes up the upper surface of an ethanol bath where ethanol can cover the cylinders to various extents by manipulation of the hydrostatic pressure. The *distance* from the top of the cylinder to the lowest location on the meniscus Δ is measured using a video camera. This method is *not* very accurate but is *noninvasive*. Then, using the measured Δ and the Young-Laplace description of the meniscus, the surface saturation is computed. Since by specifying Δ and the solid surface topology both s_1 and p_c are determined, from this analysis, Tao and Kaviany (1991) also compare the predicted p_c *against* the one measured using a manometer. They find that the measured p_c is much *lower* than that predicted. The smaller p_c, indicating a *larger* radius of curvature, *cannot* be completely explained. However, lack of the ideal conditions assumed in the *static* treatment of the meniscus, is the reason for the discrepancies. The capillary pressure and Δ are measured during evaporation. Since during the evaporation, the liquid surface temperature is *not* uniform (as measured by an infrared imager) and the wind-shear induced motion is *present*, the Young-Laplace static description of the meniscus is *not* expected to hold. However, no *dynamic* analysis leading to prediction of the surface saturation has been attempted. We should point out that *even* for a static system, in general, the measured p_c (or s_1) does *not* follow the predicted results. An example is the measurements of the *static holdup* $[s(\mathbf{u} = 0)$ for packed beds] by Saez and Carbonell (1985). They point to the *imperfect* wettability as the source of the discrepancies. The effect of the heating of the solid on the wettability is shown in the experiments of Orell and Bankoff (1971).

We now discuss the experimental results of Tao and Kaviany (1991). The surface saturation is based on a unit cell of length ℓ (along the flow) and a depth of $\ell - 1/2[(\ell/2R) - 1]$. Both the length and the depth are taken in the *plane* of the flow. Therefore, the cylinders are completely exposed before the surface saturation of zero is achieved. Some of the experimental results of Tao and Kaviany are shown in Figure 11.13 for three different $Bo = \rho_g g R^2/\sigma$, corresponding to particle diameters of 0.501, 1.499, and 3.277 mm. The Reynolds number $Re = u_{ga} R/\nu_g$ ranges from 8 to 150. The air is heated to 70°C while the supplied ethanol is at room temperature (about 25°C). The results show that for small Reynolds numbers ($u_{ga} = 0.56$ m/s and $Bo = 0.02$), the *sidewise* diffusion causes normalized mass transfer rate $\overline{m}_g(s_1)/\overline{m}_g(s_1 = 1)$ to become slightly *larger* than unity for a large range of s_1. The $Bo \to 0$ asymptote is the *diffusion* limit considered in Section 11.2.1. As was mentioned, the accuracy in the predicted s_1 (using Δ) is *not*

11.4 Convection from Heterogeneous Two-Dimensional Surfaces 597

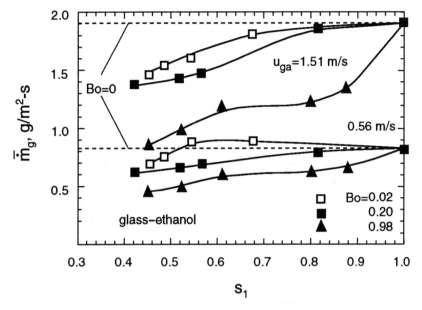

Figure 11.13 The measured mass transfer rate as a function of the surface saturation for a surface made of glass cylinders with ethanol as the liquid. The results are for three Bond numbers and two ambient air velocities.

very high. However, the results show a strong *Bo dependence* as evident from the magnitude of $\partial \dot{m}_g(s_1)/\partial s_1$, which is nearly zero for low *Bo* and becomes much larger for larger *Bo*.

Since in the model discussed in Section 11.4.1, the cylinders were in *contact* with each other (lowest surface porosity) and due to the many simplifying assumptions made there (including decoupling of the heat and mass transfer), *no direct* comparison can be made between the predictions given in Figure 11.11 and the measured results given in Figure 11.13. However, for the case of $Bo = 0.2$ (which is close to that used in Section 11.4.1) from the experimental results given in Figure 11.13, we note a rather *weak* s_1 dependence for large values of s_1 and a much *stronger* dependence for the smaller values of s_1. This trend is also found in Figure 11.11.

Using an infrared imager, Tao and Kaviany report the averaged surface temperature distributions. A typical distribution is shown in Figure 11.14. Note that the liquid surface temperature is *not* uniform, i.e., except for the two *cores*, one on the *top* of the cylinder and the other in the *middle* of the meniscus, there is a *large* axial temperature gradient. Using the measured surface temperature for the solid and the liquid and by using an adiabatic boundary condition for the lower surface, i.e., at $\ell - 1/2[(\ell/2R) - 1])$, and the symmetry boundary condition on the two lateral surfaces, the local heat flux arriving at the solid surface is *computed* (Tao, 1989). This is done

598   11. Transport Through Bounding Surfaces

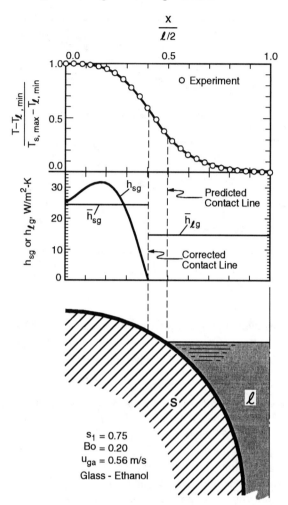

Figure 11.14 The measured surface temperature distribution in half of a unit cell. The inferred local and average heat transfer coefficients on the solid surface and the average heat transfer coefficient on the liquid surface are also shown.

by using a composite mesh for the solid-liquid system and by applying the finite difference approximations. *No* motion in the liquid is included. Based on this, the local and averaged heat transfer coefficient (h_{sg} and $\overline{h}_{sg}$) at A_{sg} can be determined. The *remainder* of the heat flow needed for the evaporation (determined from $\overline{\dot{m}}A_{\ell g}i_{\ell g}$, where $\dot{m}$ is measured) arrives from the liquid surface. This is used to determine $\overline{h}_{\ell g}$. The value of these heat transfer coefficients are also shown in Figure 11.14. The predicted contact line (from the Young-Laplace formulation) leads to a nonphysical heat flow distribution and has to be *corrected* as shown in this figure. The average mass transfer coefficient for the liquid surface is also determined using the average liquid surface temperature. The results show that as s_1 decreases at the liquid surface the *analogy* between the heat and mass transfer deteriorates.

This example shows how the surface transfer rates (heat and mass) are *influenced* by the substrate composition and properties. The dependence of the transfer rates on the surface saturation can, in principle, be predicted by the inclusion of the *radiative* heat transfer and the *motion* of the liquid. However, such analysis even for the simple steady-state two-dimensional case examined will be computationally very extensive.

11.5 Simultaneous Heat and Mass Transfer from Packed Beds

So far we have examined the heat and mass transfer from partially liquid-covered planar and two-dimensional nonplanar surfaces. For these surfaces some *rigorous* theoretical treatments can be made without any special efforts, e.g., the *two-dimensional* meniscus contour was readily estimated by the static formulation. The determination of the *three-dimensional* meniscus contour requires a substantial effort, especially if the liquid motion has to be included (as was discussed in the last section, this motion may be significant for Bo as low as 0.2). We now examine some of the experimental results on the *transient simultaneous* heat and mass transfer from packed beds. The study is that of Rogers and Kaviany (1990) where the beds are initially *fully* saturated and then exposed to a convective ambient. The experimental conditions are for $0.00135 \leq Bo \leq 21.72$ and $Re = u_{ga} R/\nu_g = 21$ to 2666, where glass spheres water, and plexiglas spheres water are used and $\rho_{cg\,a} \neq 0$. The ambient, convecting air is kept at around 60°C, while the sample is initially at about 22°C. For $Bo \leq 0.3$ ($d \leq 1.5$ mm) the packing is random, and for $Bo \geq 0.3$ the packing is the simple-cubic arrangement. Figure 11.15 shows the variation of the normalized mass transfer rate (the bed length along the flow L is 15 cm) with respect to time. The vapor density on the liquid surface $\rho_{cg\,o}$ is evaluated at the wet-bulb temperature. This causes an initial increase in $\overline{Sh}$. After the initial bed heatup period for $Bo \leq 0.00135$, a nearly constant mass transfer rate period is reached. This behavior is that modeled by Schlünder, i.e., the large Knudsen number model discussed in Section 11.2.1. For this regime the capillarity and sidewise diffusion dominate, resulting in no effect of the surface saturation on the mass transfer rate. We can extrapolate between the results for $Bo = 0.00135$ and $Bo = 0.0337$ and set the following limit:

$$Bo \leq 0.01 \quad \text{capillarity and streamwise diffusion dominate.} \quad (11.65)$$

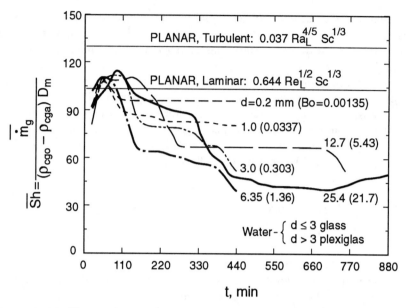

Figure 11.15 Measured variation of the averaged dimensionless mass transfer rate with respect to time for beds made of spherical particles and initially fully saturated with water. The predictions based on a planar fully liquid-covered surface are also given.

The results for the beds of a simple-cubic arrangement show that as the liquid evaporates the liquid front *recedes*, i.e., the gravity dominates over the capillarity. For these beds, *significant* liquid surface area changes are encountered as the liquid surface recedes. This is manifested in the variations in the mass transfer rate. For the *largest* Bo considered ($d = 25.4$ mm and $Bo = 21.7$), the effect of this liquid surface area change is the *most* pronounced. By further examining Figure 11.15, we note that for $Bo \geq 0.30$ the gravity dominates, i.e.,

$$Bo \geq 0.30 \quad \text{gravity dominates.} \qquad (11.66)$$

Although we speak of the receding liquid surface, we note that liquid *rings* remain around the contact points, thus creating *extra* surface area for the mass transfer.

In the analysis of Rogers and Kaviany, uniform surface temperature is assumed for the liquid and solid surfaces and these temperatures are measured with thermocouples. These average temperatures are used to estimate $\overline{h}_{sg}$ and $\overline{h}_{\ell g}$ by considering the conduction heat transfer within the solid and by estimating the meniscus contour. The meniscus computation uses the measured water remaining in the bed and the two-dimensional static

Young-Laplace formulation. They also find that the *analogy* between heat and mass transfer *deteriorates* as the surface saturation decreases and the solid becomes exposed.

In reviewing the materials presented in this chapter on the *dependence* of the *surface mass transfer* rate on the *surface saturation*, we point out the following.

- For *Bond numbers* below a *critical* value Bo_c, which depends on the gas-liquid-solid system (say $Bo_c = \rho_\ell g R_c^2/\sigma \simeq 0.01$) and as long as the surface saturation is not below a critical value s_{1c}, which also depends on the gas-liquid-solid system (say, $s_{1c} \simeq 0.3$), we expect the mass transfer rate to be the *same* as that for the fully liquid-covered surface. The *model* based on the mass diffusion also predicts this phenomena when the Knudsen number is large.

- For *larger* Bond numbers, we expect the gravity and the exposed solid to *influence* the heat and mass transfer rates. The influence *cannot* be lumped into a dependence on the surface saturation *alone*, because the *thermal conductivity* of the phases, the gas-phase Reynolds, Prandtl, and Schmidt numbers, wettability, and convection within the liquid phase all play significant roles and influence the heat and mass transfer rates. The *analogy* between the heat and mass transfer also deteriorates significantly. The mass diffusion model for small Knudsen numbers discussed in Section 11.2.2 is expected to be valid only for Bo slightly above Bo_c.

11.6 References

Ackerberg, R. C., Patel, R. D., and Gupta, S. K., 1978, "The Heat/Mass Transfer to a Finite Strip at Small Péclet Numbers," *J. Fluid Mech.*, 86, 49–65.

Bressler, R. G. and Wyatt, P. W., 1970, "Surface Wetting Through Capillary Grooves," *ASME J. Heat Transfer*, 92, 126–132.

Cook, R., Tung, C. Y., and Wayner, P. C., 1981, "Use of Scanning Microphotometer to Determine the Evaporative Heat Transfer Characteristics of the Contact Line Regions," *ASME J. Heat Transfer*, 103, 325–330.

Glassman, I., 1977, *Combustion*, Academic, 185–187.

Holm, F. W. and Goplen, S. P., 1979, "Heat Transfer in the Meniscus Thin-Film Transition Regime," *ASME J. Heat Transfer*, 101, 543–547.

Kaviany, M., 1989, "Forced Convection Heat and Mass Transfer from a Partially Liquid-Covered Surface," *Num. Heat Transfer*, 15A, 445–469.

Kays, W. M. and Crawford, M. E., 1980, *Convective Heat and Mass Transfer*, McGraw-Hill.

Ling, S. C., 1963, "Heat Transfer from a Small Isothermal Spanwise Strip on an Insulated Boundary," *ASME J. Heat Transfer*, 85, 230–236.

Maneval, J. E. and Whitaker, S., 1990, "Effects of Saturation Heterogeneities on the Interfacial Mass Transfer Relation," in *Drying '89*, 238–245, Hemisphere.

Mirzamoghadam, A. and Catton, I., 1988, "A Physical Model of the Evaporating Meniscus," and "Holographic Interferometry Investigation of Enhanced Tube Meniscus Behavior," *ASME J. Heat Transfer*, 201–213.

Moosman, S. and Homsy, G. M., 1980, "Evaporating Menisci of Wetting Fluids," *J. Colloid Interface Sci.*, 73, 212–223.

Orell, A. and Bankoff, G., 1971, "Formation of Dry Spots in Horizontal Liquid Film Heated from Below," *Int. J. Heat Mass Transfer*, 14, 1835–1842.

Plumb, O. A. and Wang, C.-C., 1982, "Convective Mass Transfer from Partially Wetted Surfaces," ASME paper no. 82-HT-59.

Rogers, J. A. and Kaviany, M., 1990, "Variation of Heat and Mass Transfer Coefficient During Drying of Granular Beds," *ASME J. Heat Transfer*, 112, 668–674.

Saez, A. E. and Carbonell, R. G., 1985, "Hydrodynamic Parameters for Gas-Liquid Co-Current Flow in Packed-Beds," *AIChE J.*, 31, 52–62.

Schlünder, E. U., 1988a, "On Mechanisms of the Constant Drying Rate Period and its Relevance to Diffusion Controlled Catalytic Gas Phase Reactions," *Chem. Engng. Sci.*, 43, 2685–2688.

Schlünder, E. U., 1988b, "Über den Mechanismus des ersten Trochnungsabschnittes und seine mögliche Bendentung für diffusions-Kontrollierte Katalytische Gasphasen-Reaktionen," *Chem.-Ing.-Tech.*, 60, 117–120.

Tao, Y.-X., 1989, *Effect of Surface Saturation on Evaporation Rate*, Ph.D. thesis, University of Michigan, Ann Arbor.

Tao, Y.-X. and Kaviany, M., 1991, "Burning Rate of Liquid Supplied Through Wick," *Combust. Flame*, 86, 47–61.

Tao, Y.-X. and Kaviany, M., 1991, "Simultaneous Heat and Mass Transfer From a Two-Dimensional, Partially Liquid Covered Surfaces," *ASME J. Heat Transfer*, 113, 875–882.

Truong, J. G., 1987, *Ellipsometric and Interferometric Studies of Thin Liquid Films Wetting on Isothermal and Non-Isothermal Solid Surfaces*, Ph.D. thesis, Rensselear Polytechnic Institute.

Truong, J. G. and Wayner, P. C., 1987, "Effects of Capillary and van der Waals Dispersion Forces on the Equilibrium Profile of a Wetting Liquid: Theory and Experiment," *J. Chem. Phys.*, 87, 4180–4188.

van Brakel, J., 1982, "Mass Transfer in Convective Drying," in *Advances in Drying*, 1, 217–267.

Wayner, P. C., 1978, "The Effect of the London-van der Waals Dispersion Force on Interline Heat Transfer," *ASME J. Heat Transfer*, 100, 155–159.

Wayner, P. C., 1989, "A Dimensionless Number for the Contact Line Evaporative Heat Sink," *ASME J. Heat Transfer*, 111, 813–815.

Wayner, P. C., Kao, Y. K., and La Croix, L. V., 1976, "The Interline Heat-Transfer Coefficient of an Evaporating Wetting Film," *Int. J. Heat Mass Transfer*, 19, 487–492.

12
Phase Change

In this chapter we examine *evaporation* and *condensation* in porous media in detail and briefly review *melting* and *solidification* in porous media. The heat supply or removal causing these to occur is generally through the *bounding surfaces* and these surfaces can be *impermeable* or *permeable*. We begin by considering condensation and evaporation adjacent to *vertical* impermeable surfaces. These are the counterparts of the film condensation and evaporation in plain media. The presence of the solid matrix results in the occurrence of a *two-phase flow region* governed by gravity and capillarity. The study of this two-phase flow and its effect on the condensation or evaporation rate (i.e., the heat transfer rate) has begun recently. Evaporation from *horizontal* impermeable surfaces is considered next. Because the evaporation is mostly from *thin-liquid films* forming on the solid matrix (in the *evaporation zone*), the evaporation does not require a significant *superheat*. The onset of dryout, i.e., the *failure* of the gravity and capillarity to keep the surface wet, occurs at a *critical* heat flux but only *small* superheat is required. We examine the predictions of the critical heat flux and the treatment of the vapor-film and the two-phase regions. We also examine the case of *thin* porous-layer *coating* of *horizontal* surfaces and review the limited data on the porous-layer thickness dependence of the heat flow rate versus the superheat curve. Then we turn to *permeable* bounding surfaces and examine the *moving condensation front* occurring when a vapor is injected in a liquid-filled solid matrix. Finally we examine the heating at a permeable bounding surface where the surface temperature is *below* the saturation temperature and the resulting *surface* and *internal* evaporations result in the *gradual* drying of the surface. The melting and solidification of single and multicomponent systems are discussed as part of phase change in *condensed phase*.

12.1 Condensation at Vertical Impermeable Bounding Surfaces

When a *vertical impermeable* surface bounding a semi-infinite vapor-phase domain is cooled *below the saturation temperature* and when nucleation sites are present, condensation begins, and if the condensed phase (liquid) wets the surface perfectly, the *film-condensation* flow occurs. For *phase-density buoyant* flows (i.e., flows caused by the density difference between phases and in presence of the gravity field), if the vapor is in the *pores* of a

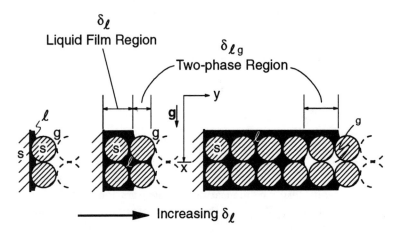

Figure 12.1 Condensation at an impermeable vertical surface for cases with $\delta_\ell/d < 1$, $\delta_\ell/d \simeq 1$, and $\delta_\ell/d > 1$.

solid matrix bounded by this vertical surface, the liquid flows through the matrix both *along* the gravity vector (due to buoyancy) and *perpendicular* to it (due to capillarity). A *single-phase region* (*liquid-film region*) must be present adjacent to the wall where the liquid is subcooled up to the edge of this film (a *distance* δ_ℓ from the surface). Beyond δ_ℓ the capillarity results in a *two-phase region* that is *nearly* isothermal. For the phase-density buoyant, film condensation flow in plain media, the film thickness is generally *small*. δ_ℓ increases as $x^{1/4}$ with x taken along the gravity vector and measured from the leading edge of the film. This is characteristic of natural convection in plain media. For condensation in porous media the local film thickness can be much *larger* than the pore (or particle) size, which will allow for a local volume-averaged treatment of this fluid-solid heterogeneous system. However, when the local film thickness is *smaller* or *comparable* in size to the pore size, the local volume-averaged treatments will no longer be applicable. Figure 12.1 depicts the *local* flow of liquid for *cases* where $\delta_\ell/d < 1$, $\delta_\ell/d \simeq 1$, and $\delta_\ell/d > 1$. As was mentioned, for cases with a *large* δ_ℓ/d, a two-phase region *also* exists *extending* from $y = \delta_\ell$ to $y = \delta_{\ell g}$.

The case of *combined* buoyant-forced (i.e., applied external-pressure gradient, film condensation flow in porous media) has been examined by Renken et al. (1994) and here we examine the case when there is no external pressure gradient.

In the following, we only examine in sufficient detail the fluid flow and heat transfer for thick liquid films, i.e., $\delta_\ell/d > 1$. When $\delta_\ell \leq d$, direct (point-wise) simulations (or simplified two-medium models) must be applied and these are not directly addressed.

12.1.1 THICK LIQUID-FILM REGION ($\delta_\ell/d \gg 1$)

In order to study two-phase flow and heat transfer for this phase change problem, we *assume* that the local volume-averaged conservation equations (including the assumption of local thermal equilibrium) are *applicable*. For this problem, we note the following.

- The *liquid-film* and *two-phase* regions each must contain many pores, i.e., for *gravity-capillarity*-dominated ($Bo \simeq 1$) and *capillarity*-dominated ($Bo < 1$) flows, we require

$$\frac{\delta_\ell}{d} \gg 1 \quad \text{and} \quad \frac{\delta_{\ell g}}{d} \gg 1 \quad \text{for} \quad Bo = \frac{g(\rho_\ell - \rho_g)K/\varepsilon}{\sigma} \ll 1, \quad (12.1)$$

$$\frac{\delta_\ell}{d} \gg 1 \quad \text{for} \quad Bo \lesssim 1. \quad (12.2)$$

For the *gravity*-dominated flows, the two-phase region is *absent*. This large Bond number asymptote will be discussed in Section 12.1.1 (C).

- The liquid *viscosity* varies with temperature $\mu_\ell = \mu_\ell(T)$ and can be included in the analysis. However, when μ_ℓ is evaluated at the *average film temperature*, this variation can be represented sufficiently accurately.

- The solid structure and the *hydrodynamic* nonuniformities can cause large variations of ε, K, $k_{\ell\perp}$, and $D_\perp^d$ near the bounding surface. Special attention should be given to these nonuniformities. For the *continuum* treatment, we require that

$$(\nabla \varepsilon)d \ll 1, \quad (12.3)$$

i.e., a *large* porosity variation near the bounding surface requires a special treatment.

- The *boundary-layer* treatment is assumed to be justifiable, i.e., $\delta_\ell/L \ll 1$.

- At and near the bounding surface the lateral effective thermal conductivity $k_{\ell\perp}(y = 0)$ depends on the thermal conductivity of the *bounding surface* (in addition to the other parameters affecting the bulk values of $k_{\ell\perp}$, Section 3.7). Therefore, in using

$$\left(k_{\ell\perp}\frac{\partial T}{\partial y}\right)_{y=0} \simeq \dot{m}_{g\ell} i_{\ell g}, \quad (12.4)$$

a special attention should be given to evaluation of $k_{\ell\perp}(y=0)$, whenever the experimental results for $T(x,y)$ are used for the evaluation of $(\partial T/\partial y)_{y=0}$. Generally, the local condensation rate $\dot{m}_{g\ell}$ is measured *instead* of $T(x,y)$. This is because δ_ℓ is generally small, and, therefore, the accuracy of $T(x, 0 < y < \delta_\ell)$ is not high *enough* to result in acceptable gradients.

- The gaseous phase is assumed to be a simple single-component system. The presence of *noncondensible* gases, which results in a significant reduction in the condensation rate (because of their accumulation near $y = \delta_\ell$ and their resistance to the flow of vapor toward this interface), can be addressed by the inclusion of the species conservation equation (i.e., the mass diffusion equation). For the $y \geq \delta_\ell + \delta_{\ell g}$ domain, the single-phase mass diffusion given by (6.63) can be used. For the two-phase region, a mass diffusion equation for the gaseous phase can be written by noting the *anisotropy* of the effective mass diffusion and the dispersion tensors (as discussed in Section 10.1.2 for the energy equation). As will be discussed, the determination of the saturation distribution for this condensation problem is not yet satisfactorily resolved, and therefore, the analysis of the effect of the noncondensibles has not yet been carried out with any reasonable accuracy. Since the temperature gradient in the two-phase region is negligibly small (this gradient is caused by the vapor pressure reduction, Section 9.2.1), there is a discontinuity in the gradient of the temperature at δ_ℓ (Figure 12.2). Then, following the standard procedures, the analytical treatment is based on the separation of the domains. Here there are *three* domains, namely, $0 \leq y \leq \delta_\ell(x)$, $\delta_\ell(x) \leq y \leq \delta_\ell + \delta_{\ell g}(x)$, and $\delta_\ell + \delta_{\ell g}(x) \leq y \leq \infty$. In the following, we treat the first two domains, the third domain is *assumed* to have uniform fields.

(A) Liquid-Film Region

The single-phase flow and heat transfer in this region can be described by the continuity equation (2.88), the momentum equation (2.143), and the energy equation (4.221). We will deal only with the volume-averaged velocities, e.g., $\langle u_\ell \rangle = u_\ell$, therefore, we drop the averaging symbol from the superficial (or Darcean) velocities. For the two-dimensional steady-state boundary-layer flow and heat transfer, we have (the coordinates are those shown in Figure 12.2)

$$\frac{\partial u_\ell}{\partial y} + \frac{\partial v_\ell}{\partial x} = 0, \tag{12.5}$$

$$\frac{\rho_\ell}{\varepsilon}\left(u_\ell \frac{\partial u_\ell}{\partial x} + v_\ell \frac{\partial u_\ell}{\partial y}\right) = \frac{\mu_\ell(x,y)}{\varepsilon}\frac{\partial^2 u_\ell}{\partial y^2} - \frac{\mu_\ell(x,y)}{K}u_\ell + g(\rho_\ell - \rho_g), \tag{12.6}$$

$$u_\ell \frac{\partial T}{\partial x} + v_\ell \frac{\partial T}{\partial y} = \frac{\partial}{\partial y}\left[\frac{k_{\ell\perp}(y)}{(\rho c_p)_\ell} + \varepsilon D_\perp^d(x,y)\right]\frac{\partial T}{\partial y}. \tag{12.7}$$

The boundary conditions are

$$u_\ell = v_\ell = 0, \quad T = T_0 \quad \text{at } y = 0,$$

$$u_\ell = u_{\ell i}, \quad T = T_s \quad \text{at } y = \delta_\ell, \tag{12.8}$$

where T_s is the *saturation temperature*.

12.1 Condensation at Vertical Impermeable Bounding Surfaces

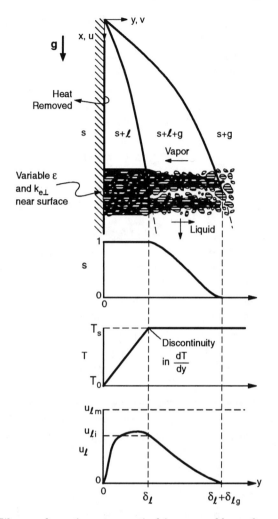

Figure 12.2 Film condensation at a vertical impermeable surface with $\delta_\ell/d > 1$ and $\delta_{\ell g}/d > 1$. Distributions of saturation, temperature, and liquid phase velocity are also depicted.

Since the liquid velocity at δ_ℓ, $u_{\ell i}$ is *not* known, an *extra* boundary condition is needed. For single-phase flows using the Brinkman treatment, we have

$$(\mu'_\ell)_{\delta_\ell^-} \left.\frac{\partial u_\ell}{\partial y}\right|_{\delta_\ell^-} = (\mu'_\ell)_{\delta_\ell^+} \left.\frac{\partial u_\ell}{\partial y}\right|_{\delta_\ell^+} \quad \text{at} \quad y = \delta_\ell. \quad (12.9)$$

This allowed us to make the transition at $y = \delta_\ell$ due to the solid matrix structural change (e.g., *discontinuity* in permeability), where μ'_ℓ depends on the matrix structure (Section 2.11.6). Presently, we do not have much knowledge about μ'_ℓ for two-phase flows (even though we have assumed that the vapor shear is *not* significant because at $y = \delta_\ell$ we have $s = 1$). The simplest, but not necessarily an accurate assumption, is that of $(\mu'_\ell)_{\delta_\ell^-} = (\mu'_\ell)_{\delta_\ell^+}$. We will use this in Section 12.1.1 (D).

The *maximum* velocity possible in the liquid phase is found by neglecting the macroscopic inertial and viscous terms. The result is

$$u_{\ell m} = \frac{K}{\mu_\ell} g(\rho_\ell - \rho_g), \qquad (12.10)$$

where, since this *idealized* flow is one-dimensional, as μ_ℓ decreases with increase in y, $u_{\ell m}$ increases with a maximum at $y = \delta_\ell$. However, in practice u_ℓ does not reach $u_{\ell m}$ because of the *lateral* flow toward the two-phase region. Also, in most cases of practical interest $\delta_\ell/d = o(1)$, and, therefore, the velocity no-slip condition at $y = 0$ causes a significant flow retardation throughout the liquid-film region (Section 2.10.1). The velocity reaches its maximum value at $y = \delta_\ell$ if the shear stress $\mu_\ell \partial u_\ell/\partial y$ at δ_ℓ is zero, otherwise it peaks at $y < \delta_\ell$. This is also depicted in Figure 12.2.

The convection heat transfer in the liquid-film region is generally negligible (Rose, 1988), therefore, (12.7) can be written as

$$D_\perp \frac{\partial T}{\partial y} = \left[\frac{k_{\ell \perp}}{(\rho c_p)_\ell} + \varepsilon D_\perp^d\right] \frac{\partial T}{\partial y} = \text{constant}. \qquad (12.11)$$

For $k_\ell < k_s$, $D_\perp$ is *generally* smallest near the bounding surface (A_f/A_t is largest) resulting in an expected *significant* deviation from the *linear* temperature distribution found in the film condensation in *plain media*.

(B) Two-Phase Region

The *two-phase* flow and heat transfer are given by the continuity equations for the ℓ and g phases (8.94) and (8.95), the momentum equations (8.135) and (8.137), and the energy equation (10.31). The two-phase region is assumed to be *isothermal* by neglecting the effect of the curvature (i.e., saturation) on the thermodynamic equilibrium state. This is justifiable, except for the very small pores (large p_c). For the steady-state flow considered here, we have (for the assumed *isotropic* phase permeabilities)

$$\frac{\partial u_\ell}{\partial x} + \frac{\partial v_\ell}{\partial y} = 0, \qquad (12.12)$$

$$\frac{\partial u_g}{\partial x} + \frac{\partial u_g}{\partial y} = 0, \qquad (12.13)$$

$$\frac{\rho_\ell}{\varepsilon s}\left(u_\ell \frac{\partial u_\ell}{\partial x} + v_\ell \frac{\partial u_\ell}{\partial y}\right) = -\frac{\partial p_\ell}{\partial x} + \rho_\ell g - \frac{\mu_\ell}{K K_{r\ell}} u_\ell, \qquad (12.14)$$

12.1 Condensation at Vertical Impermeable Bounding Surfaces 609

$$\frac{\rho_\ell}{\varepsilon s}\left(u_\ell\frac{\partial v_\ell}{\partial x}+v_\ell\frac{\partial v_\ell}{\partial y}\right)=-\frac{\partial p_\ell}{\partial y}-\frac{\mu_\ell}{KK_{r\ell}}v_\ell, \qquad (12.15)$$

$$\frac{\rho_g}{\varepsilon(1-s)}\left(u_g\frac{\partial u_g}{\partial x}+v_g\frac{\partial u_g}{\partial y}\right)=-\frac{\partial p_g}{\partial x}+\rho_g g-\frac{\mu_g}{KK_{rg}}u_g, \qquad (12.16)$$

$$\frac{\rho_g}{\varepsilon(1-s)}\left(u_g\frac{\partial v_g}{\partial x}+v_g\frac{\partial v_g}{\partial y}\right)=-\frac{\partial p_g}{\partial y}-\frac{\mu_g}{KK_{rg}}v_g, \qquad (12.17)$$

along with
$$p_g - p_\ell = p_c(s, \text{etc.}). \qquad (12.18)$$

When the *Leverett* idealization (8.154) is used, (12.18) reduces to $p_c = p_c(s)$. The convective terms in (12.14)–(12.17) can be significant when the effect of *thickening* of δ_ℓ and $\delta_{\ell g}$ and the effect of g and p_c tend to *redistribute* the phases along the x-axis (flow development effects). If the capillarity is more significant than the gravity, i.e., $\sigma/[g(\rho_\ell - \rho_g)K/\varepsilon]$ is larger than unity, then we expect larger $\delta_{\ell g}$, and vice versa. The overall *energy* balance yields

$$\int_0^x \left(k_\ell \perp \frac{\partial T}{\partial y}\right)_{y=0} dx = \int_0^{\delta_\ell} \rho_\ell i_{\ell g} u_\ell \, dy + \int_{\delta_\ell}^{\delta_{\ell g}} \rho_g i_{\ell g} u_\ell \, dy$$

$$+ \int_0^{\delta_\ell} (\rho c_p)_\ell u_\ell (T_s - T) \, dy, \qquad (12.19)$$

where T_s is the *saturation temperature*. The last term on the right-hand side makes a negligible contribution to the overall heat transfer. The boundary conditions for u_ℓ, etc., at δ_ℓ and $\delta_{\ell g}$ will be discussed in connection with the approximations made to (12.14)–(12.17) in Section 12.1.1 (D).

(C) LARGE BOND NUMBER ASYMPTOTE

Although for the cases where d is small enough to result in $\delta_\ell/d \gg 1$ the *capillarity* will *also* become important, we begin by considering the *simple* case of *negligible* capillarity. As will be shown, the capillary pressure causes *lateral flow* of the liquid, thus *tending to decrease* δ_ℓ. However, the presence of the lateral flow also tends to *decrease* the longitudinal velocity in the liquid-film region, and this *tends to increase* δ_ℓ. The sum of these two effects makes for a δ_ℓ, which may *deviate* significantly from the large Bond number asymptotic behavior. Therefore, the *limitation* of the large Bond number asymptote, especially its *overprediction* of u_ℓ, should be kept in mind.

Assuming that $Bo \to \infty$, we *replace* the boundary condition on u_ℓ at location δ_ℓ with a *zero* shear stress condition (i.e., $\delta_{\ell g} = 0$ and only two regions are present). In addition, we have the *initial* conditions

$$u_\ell = v_\ell = 0, \quad T = T_s \quad \text{at } x = 0. \qquad (12.20)$$

610   12. Phase Change

Next, we examine variations in $\mu_\ell(T)$, $\varepsilon(y)$, and $D_\perp(Pe_\ell, y)$, where $Pe_\ell = \overline{u}_\ell d/\alpha_\ell$. The variation in μ_ℓ can be *nearly* accounted for by using $\mu_\ell[(T_s + T_o)/2]$ in (12.6). For the packed beds of spheres, the variation of ε is significant *only* for $0 < y < 2d$. White and Tien (1987) have *included* the effect of the variable porosity by using the variation in A_f/A_t (Section 2.12.2). Here we assume that $\delta_\ell/d \gg 1$ and therefore, we do *not* expect the *channeling* to be significant. For the case of $\delta_\ell/d \simeq o(1)$, which will be discussed in Section 12.1.2, this porosity variation *must* be considered. We note that Pe_ℓ can be *larger* than unity and that the average liquid velocity $\overline{u}_\ell$ increases with x, and in general the variation of $D_\perp$ with respect to y should be *included*.

A similarity solution is available for (12.5)–(12.7) subject to *negligible* macroscopic inertial and viscous forces, i.e., small permeabilities, and constant $D_\perp$ (Parmentier, 1979; Cheng, 1981). The inertial and viscous forces are included by Kaviany (1986) through the *regular perturbation* of the similarity solution for plain media, i.e., the Nusselt solution (Rose, 1988). The *perturbation parameter* used is

$$\xi_x = 2\left[\frac{\varepsilon g(\rho_\ell - \rho_g)}{\rho_\ell \nu_\ell^2}\right]^{-1/2} \frac{\varepsilon x^{1/2}}{K}, \qquad (12.21)$$

and for large ξ_x, the Darcean flow exists. The other dimensionless parameters are the *subcooling parameter* $c_{p\ell}(T_s - T_o)/i_{\ell g}$ and Prandtl number $Pr_\ell = \alpha_\ell/\nu_\ell$. The results show that, as is the case with the plain media, the film thickness is small. For example, for water with $Pr_\ell = 10$ and $c_{p\ell}(T_s - T_o)/i_{\ell g} = 0.004$ to 0.2 (corresponding to 2 to $100°C$ subcooling), the film thickness δ_ℓ is between 1 and 8 mm for $K = 10^{-10}$ m^2 and between 0.1 and 0.8 mm for $K = 10^{-8}$ m^2. By using the Carman-Kozeny relation and $\varepsilon = 0.4$, we find that the *latter* permeability results in $\delta_\ell/d = 0.03 - 0.25$, which *violates* the local volume-averaging requirement.

The results of Parmentier and Cheng are (for $\delta_\ell/d \gg 1$)

$$\frac{2^{1/2} Nu_x}{Gr_x^{1/4}} = \frac{2}{\pi^{1/2}} \left[\frac{Pr_\ell}{\xi_x \mathrm{erf}\left(\Delta Pr_\ell^{1/2}/\xi_x^{1/2}\right)}\right]^{1/2} \quad \text{for } Bo > 1, \qquad (12.22)$$

$$\frac{i_{\ell g}}{2 c_{p\ell}(T_s - T_o)} + \frac{1}{\pi} = \frac{1}{\pi \left[\mathrm{erf}\left(\Delta Pr_\ell^{1/2}/\xi_x^{1/2}\right)\right]^2} \quad \text{for } Bo > 1, \qquad (12.23)$$

where for a given $i_{\ell g}/[c_{p\ell}(T_s - T_o)]$, Pr_ℓ and ξ_x, Δ and Nu_x are found, and where

$$Nu = \frac{qx}{(T_s - T_o)k_{e\perp}}, \quad \Delta = \frac{\delta_\ell}{x}\left(\frac{Gr_x}{4}\right)^{1/4}, \quad Gr_x = \frac{\varepsilon g(\rho_\ell - \rho_g)x^3}{\rho_\ell \nu_\ell^2}. \qquad (12.24)$$

12.1 Condensation at Vertical Impermeable Bounding Surfaces

Note that for small K, the Bond number $[g(\rho_\ell - \rho_g)K/\varepsilon]/\sigma$ is *also* small, therefore, the capillarity (i.e., the two-phase region) must be included. This is attended to next.

(D) SMALL BOND NUMBER APPROXIMATION

Presently no *rigorous* solution to the combined liquid-film and two-phase regions is available. However, some *approximate* solutions are available (Shekarriz and Plumb, 1986; Majumdar and Tien, 1988; Chung et al., 1990). The available experimental results (Plumb et al., 1990; Chung et al.) are *not* conclusive as they are either for $\delta_\ell/d \simeq o(1)$ or when they contain a significant *scatter*.

By considering capillary-affected flows, we expect that $v_g \gg u_g$, since $\partial s/\partial y \gg \partial s/\partial x$. Also, because the inertial force is *negligible* for the vapor flow, we reduce (12.16) and (12.17) to

$$0 = -\frac{\partial p_g}{\partial y} - \frac{\mu_g}{KK_{rg}}v_g \quad \text{or} \quad v_g = -\frac{1}{\mu_g}KK_{rg}\frac{\partial p_g}{\partial y} = -\frac{\dot{m}_{g\ell}}{\rho_g}. \quad (12.25)$$

Note that from (12.13), we find that v_g is constant along y. For the liquid phase, *both* u_ℓ and v_ℓ are significant, and v_ℓ changes from a relatively large value at δ_ℓ to zero at $\delta_\ell + \delta_{\ell g}$, therefore, $v_\ell \partial v_\ell/\partial y$ will not be negligibly small. Also, $v_g \partial u_\ell/\partial y$ may not be negligible. Then we have

$$\frac{\rho_\ell}{\varepsilon s}v_\ell\frac{\partial u_\ell}{\partial y} = -\frac{\partial p_\ell}{\partial x} + \rho_\ell g + \frac{\mu_\ell}{KK_{r\ell}}u_\ell, \quad (12.26)$$

$$\frac{\rho_\ell}{\varepsilon s}v_\ell\frac{\partial v_\ell}{\partial y} = -\frac{\partial p_\ell}{\partial y} - \frac{\mu_\ell}{KK_{r\ell}}v_\ell. \quad (12.27)$$

Here we assume that the gaseous phase *hydrostatic* pressure is negligible. We note that the approximations made in the evaluations of $K_{r\ell}$ and p_c cause more errors in the determination of u_ℓ and v_ℓ than the exclusion of the inertial terms. Furthermore, we expect $\partial p_\ell/\partial x$ to be small. Then we can write (12.26) and (12.27) as

$$0 = \rho_\ell g - \frac{\mu_\ell}{KK_{r\ell}}u_\ell, \quad \text{or} \quad u_\ell = \frac{g}{\nu_\ell}KK_{r\ell}, \quad (12.28)$$

$$0 = -\frac{\partial p_\ell}{\partial y} - \frac{\mu_\ell}{KK_{r\ell}}v_\ell \quad \text{or} \quad v_\ell = -\frac{1}{\mu_\ell}KK_{r\ell}\frac{\partial p_\ell}{\partial y}. \quad (12.29)$$

The velocity distribution given by (12.28) is that of a *monotonic* decrease from the value of $u_{\ell i}$ at δ_ℓ to zero at $\delta_\ell + \delta_{\ell g}$. The specific distribution depends on the *prescribed* $K_{r\ell}(s, \text{etc.})$. The distribution of v_ℓ given by (12.29) is more complex, because $-\partial p_\ell/\partial y$ increases as s decreases (as $y \to \delta_\ell + \delta_{\ell g}$). Although $\partial p_g/\partial y$ is needed to derive the vapor to δ_ℓ, we note that

$$\frac{\partial p_\ell}{\partial y} = -\frac{\partial p_c}{\partial y} + \frac{\partial p_g}{\partial y} \simeq \frac{\partial p_c}{\partial y}. \quad (12.30)$$

From the experimental results on $p_c(s)$, Section 8.4, we can conclude that in the two-phase region v_ℓ also *decreases* monotonically with y. By using (12.30), we write (12.29) as

$$v_\ell = \frac{1}{\mu_\ell} K K_{r\ell} \frac{\partial p_c}{\partial y}. \tag{12.31}$$

The momentum equations (12.28) and (12.31) can be inserted in (12.12), and when p_c and $K_{r\ell}$ are given in terms of s, the following *saturation equation* is obtained:

$$\frac{\partial u_\ell}{\partial x} + \frac{\partial v_\ell}{\partial y} = \frac{gK}{v_\ell} \frac{\partial K_{r\ell}}{\partial x} + \frac{K}{\mu_\ell} \frac{\partial}{\partial y} \left(K_{r\ell} \frac{\partial p_c}{\partial y} \right), \tag{12.32}$$

with

$$\begin{aligned} s &= 1 \quad \text{at} \quad y = \delta_\ell, \\ s &= 0 \quad \text{at} \quad y = \delta_\ell + \delta_{\ell g}, \\ s &= 0 \quad \text{at} \quad x = 0. \end{aligned} \tag{12.33}$$

The evaluation of $u_{\ell i}$, δ_ℓ, and $\delta_{\ell g}$ requires the analysis of the liquid-film region. For a *negligible* inertial force and with the use of the viscosity evaluated at the average film temperature $\bar{\mu}$ and the definition of $u_{\ell m}$, i.e., (12.10), we can integrate (12.6) and (12.8) to arrive at (Chung et al.)

$$u_\ell = \frac{u_{\ell i} - u_{\ell m} + u_{\ell m} \cosh \dfrac{\delta_\ell}{(K/\varepsilon)^{1/2}}}{\sinh \dfrac{\delta_\ell}{(K/\varepsilon)^{1/2}}} \sinh \frac{y}{(K/\varepsilon)^{1/2}}$$

$$+ u_{\ell m} \left[1 - \cosh \frac{y}{(K/\varepsilon)^{1/2}} \right]. \tag{12.34}$$

Now using $(\mu'_\ell)_{\delta_\ell^-} = (\mu'_\ell)_{\delta_\ell^+}$ in (12.9) and the *overall* mass balance given by

$$\int_0^x \dot{m}_{g\ell} \, dx = \int_0^{\delta_\ell(x)} \rho_\ell u_\ell \, dy + \int_{\delta_\ell(x)}^{\delta_{\ell g}(x)} \rho_\ell u_\ell \, dy, \tag{12.35}$$

Chung et al. solve for the previously mentioned *three* unknowns. Their experimental and predicted results for $d = 0.35$ mm are shown in Figure 12.3, where $Ra_x = Gr_x Pr_\ell$. We note that in their experiment (a closed system) the condensate collects, i.e., the liquid film thickens, at the bottom of the cooled plate. When the plate length is very large, this non-ideal lower portion behavior may be neglected. However, in their experiment with the relatively short plate this can significantly influence the phase distributions

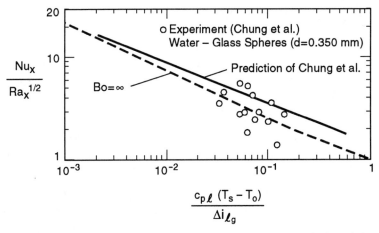

Figure 12.3 Prediction and experimental results of Chung et al. (1990) for condensation (in a packed bed of spheres) at a bounding vertical impermeable surface. The large Bo results of Cheng (1981) are also shown. (From Chung et al., reproduced by permission ©1990 ASME.)

and velocities. The Bond number is 4×10^{-5}. For this size particle and for $c_{p\ell}(T_s - T_o)/i_{\ell g} = 0.1$, they find $u_{\ell i}/u_{\ell m} \simeq 0.06$, i.e., the velocity at δ_ℓ is *much less* than the maximum Darcean velocity given by (12.10). Note that although the velocity in the liquid phase is so *small*, the thickness of the liquid-film region is *not substantially different* than that found for the single-layer ($Bo > 1$) model of Cheng (as evident from the heat transfer rate). The analysis of Chung et al. shows that $\delta_\ell/d \simeq 3.3$. They also use $(A_f/A_t)(y)$ of Section 2.12 in their computation. Note that for this analysis to be meaningful, δ_ℓ/d has to be larger than, say, 10, so that the variation of u_ℓ in $0 \leq y \leq \delta_\ell$ can be predicted with *sufficient* accuracy. Therefore, the applicability of the analysis to their experimental condition is questionable.

12.1.2 THIN LIQUID-FILM REGION ($\delta_\ell/d \simeq 1$)

Presently, no treatment for $\delta_\ell \simeq d$ (Figure 12.1) is available. The single-continuum (local thermal equilibrium) and local volume-averaging treatments used in the last section are not applicable. This is because changes in ε and all the temperature drops occur over a distance of the order of d. These violate the requirements for the selection of a representative elementary volume and for the presence of the local thermal equilibrium. In addition, we encounter the following extra complexities.

- Wettability of the solid matrix θ_c significantly influences the liquid phase distribution $a_\ell(\mathbf{x})$.

- The lateral effective thermal conductivity varies significantly with δ_ℓ. In particular, $k_{e\perp}(y=0)$ will depend on δ_ℓ in addition to other parameters.

- The *bulk* properties such as K, $K_{r\ell}$, K_{rg}, and $D_\perp$ do *not* apply anywhere in the liquid-film or two-phase region. This is because the thickness of these regions *never* approaches the threshold lengths required for achievement of a bulk behavior.

- The liquid flow may be noticeably *unsteady*, and because of the small liquid film thickness, the local Nusselt number also becomes noticeably unsteady (although the frequency of oscillations may be high).

- Since $\partial T/\partial y \simeq o[(T_s - T_o)/d]$ and since condensation occurs on the *liquid film surface*, a large temperature *difference* occurs between the solid and liquid phases. This requires at least a two-medium treatment (Chapter 7) for the liquid film region and a three-medium treatment for the two-phase region.

White and Tien (1987) and Chung et al. (1990) have suggested some *ad hoc* extensions of the continuum treatment ($\delta_\ell/d \gg 1$) to the cases with $\delta_\ell \simeq d$. However, it is expected that for the predictions to be valid when δ_ℓ/d is near unity (or smaller) a *direct simulation* of the flow and heat transfer (including the phase change) is needed. As has been noted in the previous chapters, even a two-dimensional simulation would be fairly involved.

12.2 Evaporation at Vertical Impermeable Bounding Surfaces

For *plain media*, the *film evaporation* adjacent to a *heated vertical surface* is similar to the *film condensation*. In porous media, we also expect *some* similarity between these two processes. For the reasons given in the last section, we will not discuss the cases where $\delta_g/d \simeq 1$, where δ_g is the *vapor-film region thickness*. When $\delta_g/d \gg 1$ and because the liquid flows (due to capillarity) toward the surface located at $y = \delta_g$, we also expect a large two-phase region, i.e., $\delta_{g\ell}/d \gg 1$. Then a local volume-averaged treatment, similar to that applied in Section 12.1.1, can be applied.

The *asymptotic* solution for $Bo > 1$, where $\delta_{g\ell} = 0$ (as was $\delta_{\ell g}$), is the same as that given in Section 12.1.1 (C), and the similarity solution given there holds. Parmentier (1979) has discussed this asymptotic solution. When given in terms of Ra_x, this solution is (given the superheating parameter and Ra_x, then Nu_x and Δ_g are solved *simultaneously*)

$$\frac{Nu_x}{Ra_x^{1/2}} = \frac{1}{\pi^{1/2}\mathrm{erf}(\Delta_g/2)} \quad \text{for } Bo \to \infty, \tag{12.36}$$

$$\frac{c_{pg}(T_o - T_s)}{\Delta i_{\ell g}} = \frac{\pi^{1/2}\Delta_g}{2}\exp(\Delta^2/2)\mathrm{erf}(\Delta_g/2) \quad \text{for } Bo \to \infty, \quad (12.37)$$

with

$$Ra_x = \frac{(\rho_\ell - \rho_g)gKx}{\rho_g \nu_g \alpha_\ell}, \quad \Delta_g = \frac{\delta_g}{x}Ra_x^{1/2}. \quad (12.38)$$

Note that (12.22) and (12.23) are given in terms of the perturbation parameter ξ_x, but otherwise are *identical* to (12.36) and (12.37). As we discussed, for small Bond numbers, which is the case whenever $\delta_g/d \gg 1$, a nearly *isothermal* two-phase region *exists*. Although no treatment of this two-phase region is available, the approach will be that outlined in Section 12.1.1 (D). The vapor will be *rising* due to buoyancy, similar to the falling of the liquid given by (12.28), except here we include the *hydrostatic* pressure of the liquid phase. This gives (defining x to be along $-\mathbf{g}$)

$$0 = (\rho_\ell - \rho_g)g - \frac{\mu_g}{KK_{rg}}u_g \quad \text{or} \quad u_g = \frac{(\rho_\ell - \rho_g)g}{\mu_g}KK_{rg}. \quad (12.39)$$

The *lateral* motion of the vapor is due to the capillarity, and in a manner similar to (12.29), we can write

$$0 = -\frac{\partial p_g}{\partial y} - \frac{\mu_g}{KK_{rg}}v_g \quad \text{or} \quad v_g = -\frac{1}{\mu_g}KK_{rg}\frac{\partial p_g}{\partial y}. \quad (12.40)$$

The *axial* motion of the liquid phase is negligible and the lateral flow is similar to that given by (12.25) and is described as

$$0 = -\frac{\partial p_\ell}{\partial y} - \frac{\mu_\ell}{KK_{r\ell}}v_\ell \quad \text{or}$$

$$v_\ell = \frac{\dot{m}_{\ell g}}{\rho_\ell} = -\frac{1}{\mu_\ell}KK_{r\ell}\frac{\partial p_\ell}{\partial y} \simeq \frac{1}{\mu_\ell}KK_{r\ell}\frac{\partial p_c}{\partial y}. \quad (12.41)$$

The *saturation* will increase monotonically with y with $s = 0$ at $y = \delta_g$, and $s = 1$ at $y = \delta_{g\ell}$. In principle, we then expect the results of Chung et al. (1990) for the condensation to apply to the evaporation. However, since the available experimental results for p_c (*drainage* versus *imbibition*) discussed in Section 8.4, show a hysteresis, and also due to the *lack* of symmetry in $K_{r\ell}(s)/K_{rg}$, discussed in Section 8.5, we do not expect a *complete* analogy.

12.3 Evaporation at Horizontal Impermeable Bounding Surfaces

We now consider heat *addition* to a *horizontal* surface bounding a liquid-filled porous media from below. When the temperature of the bounding surface is at or above the saturation temperature of the liquid occupying the porous media, evaporation occurs. We limit our discussion to matrices that remain unchanged. When nonconsolidated particles make up a bed, the evaporation can cause void channels through which the vapor escapes (Eckert et al. 1985; Reed, 1986; and Stubos and Buchin, 1988). Here we begin by mentioning a phenomenon observed in some experiments (Sondergeld and Turcotte, 1977) where *no significant superheat is required* for the evaporation to start, i.e., evaporation is through *surface evaporation* of thin liquid films covering the solid surface (in the *evaporation zone*). By choosing the surface heating (or external heating), we will not address the volumetrically heated beds (e.g., Naik and Dhir, 1982). For *small* Bond numbers, for *small* heat flux, i.e., when $(T_o - T_s) \simeq 0$, the vapor generated at the bottom surface moves upward, the liquid flows downward to replenish the surface, and the surface remains wetted. When a *critical heat flux* q_{cr} is exceeded, a *vapor film* will be formed adjacent to the heated surface, and a two-phase region will be present above this vapor-film region. The two-phase region will have an *evaporation zone* where the temperature is not uniform and a nearly isothermal region where no evaporation occurs. As *Bo* increases, the role of capillarity diminishes. For $Bo > 1$, the behavior is *nearly* the same as when no rigid matrix is present, i.e., the conventional pool boiling curve will be nearly observed.

In the following, we examine the experimental results (Fukusako et al., 1986) that support these *two* asymptotic behaviors, i.e., $Bo > 1$, the *high-permeability* asymptote, and $Bo \ll 1$, the *low-permeability asymptote*. Then we discuss the *low*-permeability asymptote using a one-dimensional model (Sondergeld and Turcotte, 1977; Bau and Torrance, 1982; Udell, 1985). The one-dimensional model is also capable of predicting q_{cr}, i.e., the *onset of dryout*. We note that the *hysteresis* observed in isothermal two-phase flow in porous media is also found in evaporation-condensation and that the q versus $T_o - T_s$ curve shows a *decreasing q* (or $T_o - T_s$) and an *increasing q* (or $T_o - T_s$) *branch*.

12.3.1 Effect of Bond Number

In the experiments of Fukusako et al. (1986), packed beds of spheres (glass, steel, and aluminum) occupied by fluorocarbon refrigerants were heated from below to temperatures above saturation. Their results for the glass (Freon-11) system with a *bed height* of 80 mm and for four different particle sizes (Bond numbers) are given in Figure 12.4. Also shown are their results for pool boiling (no particles). They have not reported the heat flux corresponding to $(T_o - T_s) < 8°C$ for this solid-fluid system, because all of their results are for $q \geq 10^4$ W/m². For $Bo \to \infty$, i.e., with no solid matrix present, the *conventional* pool boiling curve is obtained, i.e., as $(T_o - T_s)$ increases, *after* the *required* superheat for nucleation is exceeded, the *nucleate* (with a maximum), the *transient* (with a minimum), and the

12.3 Evaporation at Horizontal Impermeable Bounding Surfaces 617

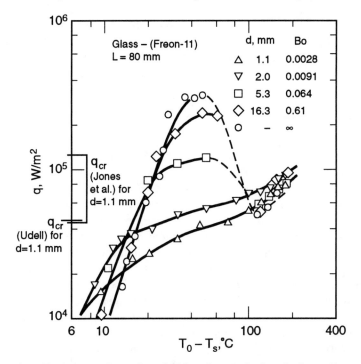

Figure 12.4 Experimental results of Fukusako et al. for the bounding surface heating of the liquid filled beds of spheres to temperatures above saturation. The case of pool boiling ($Bo \to \infty$) is also shown along with the dryout heat flux predicted by Udell (1985) and given in the measurement of Jones et al. (1980). (From Fukusako et al., reproduced by permission ©1986 ASME.)

film boiling regimes are observed. For large particles, this behavior is not significantly altered. However, as the particle size reduces, this maximum and minimum become *less* pronounced, and for very small particles, they *disappear*. For the solid-fluid system used in their experiment, this transition appears to occur at $Bo \leq 0.0028$. Note that we have only been concerned with the heating from *planar horizontal* surfaces. For example, the experimental results of Fand et al. (1987) for heating of a 2-mm-diameter *tube* in a packed bed of glass spheres-water ($d = 3$ mm, $Bo = 0.003$) shows that *unlike* the results of Fukusako et al. *no* monotonic increase in q is found for $(T_o - T_s) > 100°C$. Tsung et al. (1985) use a heated sphere in a bed of *spheres* ($d \geq 2.9$ mm). Their results also show that as d *decreases*, the left-hand portion of the q versus ΔT curve moves *upward*, i.e., the required superheat for a given q is smaller in porous media (compared to plain media).

The results given in Figure 12.4 show that for very *small* values of Bo ($Bo \leq 0.01$), the surface temperature also increases monotonically with the heat flux. This supports the theory of a heat removal mechanism that does not change, unlike that observed in the pool boiling in plain media. For the system shown in Figure 12.4 and for $d = 1.1$ mm the *transition* from the *surface wetted* condition to the *formation* of a vapor film occurs at the critical heat flux [Udell, 1985, this will be discussed in Section 12.3.2 (C)]

$$q_{\text{cr}} = \frac{K\Delta i_{\ell g}(\rho_\ell - \rho_g)g}{\nu_g}\left[1 + \left(\frac{\nu_g}{\nu_\ell}\right)^{1/4}\right]^{-4} = 4.4 \times 10^4 \text{ W/m}^2. \quad (12.42)$$

Jones et al. (1980) measure q_{cr} using various fluids. They obtain a range of q_{cr} with the *lowest* value very close to that predicted by (12.42). These values of q_{cr} are also shown in Figure 12.4. Based on the *theory* of evaporation in porous media given earlier, no vapor film is present until q exceeds q_{cr} (for small Bo). Then, for $q > q_{\text{cr}}$, the surface temperature *begins* to rise, i.e., $T_0 - T_s > 0$. Now by further examination of Figure 12.4, we note that this *low* Bond number asymptote (i.e., the liquid wetting of the surface for $(T_0 - T_s) = 0$ followed by the *simultaneous* presence of the vapor film and the two-phase regions in series for $(T_0 - T_s) > 0$) is *not* distinctly found from the experimental results of Fukusako et al. Although as $Bo \to 0$ the *trend* in their results supports this theory, the Bo *encountered* in their experiment can yet be too *large* for the realization of this asymptotic behavior.

In Figure 12.5, q versus ΔT curves are drawn based on the $Bo \to \infty$ and the $Bo \to 0$ asymptotes and the intermediate Bo results of Fukusako et al. The experimental results of Udell (1985) for $Bo \to 0$ do *not* allow for the verification of the $Bo \to 0$ curve in this figure, i.e., the verification of q versus ΔT for $Bo \to 0$ is not *yet* available. Also, the effect of the particle size (Bond number) on the heat flux for $(T_0 - T_s) > 100°$C is *not* rigorously tested, and the trends shown are based on the *limited* results of Fukusako et al., for this temperature range. In examining the q versus ΔT behavior for porous media, we note the following.

- The *thermal conductivity* of the solid matrix greatly influences the q versus ΔT curve, and the results of Fukusako et al. are for a *non-metallic* solid matrix. Later we will examine some of the results for metallic matrices and show that as k_s *increases*, q *increases* (for a given $T_0 - T_s$).

- In the experiments discussed earlier, no mention of any *hysteresis* has been made in the q versus $(T_0 - T_s)$ curve. However, as will be shown, at least for *thin* porous layer coatings, *hysteresis* has been found, and in the q *decreasing* branch, the corresponding $(T_0 - T_s)$ for a given q is *much larger* than that for an *increasing* q branch.

- In the behavior depicted in Figure 12.5, it is assumed that *only* the particle size is changing, i.e., particle shape, porosity, fluid properties, heated surface, etc., all remain the *same*.

12.3 Evaporation at Horizontal Impermeable Bounding Surfaces

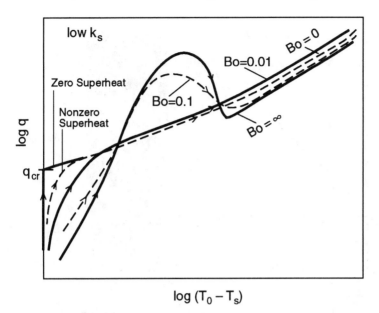

Figure 12.5 Effect of the Bond number on the q versus $T_0 - T_s$ curve is depicted based on the $Bo \to 0$ and ∞ asymptotes and the experimental results of Fukusako et al. The solid phase thermal conductivity is low.

- The theoretical *zero superheat* at the onset of evaporation is not realized, and experiments do show a *finite* $\partial q/\partial T_0$ as $(T_0 - T_s) \to 0$ (see Figure 12.5).

The analysis for large particle sizes is expected to be difficult. For example, when $(T_0 - T_s) > 0$ and a thin vapor film is found on the heated surface, the thickness of this film will be *less* than the particle size, therefore, $(T_0 - T_s)$ occurs over a distance less than d. Since $k_s \neq k_s$, this violates the assumption of the local thermal equilibrium. Also, as the particle size increases, boiling occurs with a *large* range of bubble sizes, i.e., the bubbles may be smaller and larger (elongated) than the particle size.

In the following section, we examine the $Bo \to 0$ asymptotic behavior by using the volume-averaged governing equations. This one-dimensional analysis allows for an estimation of q_{cr} and the length of the isothermal two-phase region for $q > q_{\mathrm{cr}}$.

12.3.2 A One-Dimensional Analysis for $Bo \ll 1$

Figure 12.6 depicts the one-dimensional model for evaporation in porous media with heat addition q from the impermeable lower bounding surface maintained at $T_0 > T_s$, where T_s is the saturation temperature. The vapor-film region has a thickness δ_g, and the two-phase region has a length $\delta_{g\ell}$.

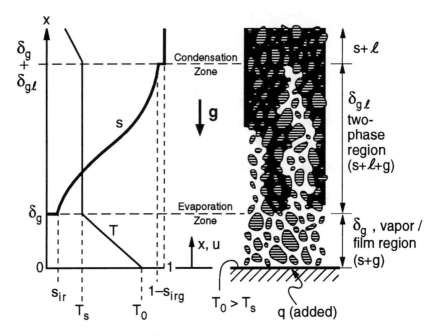

Figure 12.6 Evaporation due to the heat addition from below at temperatures above the saturation. The vapor-film region, the two-phase region, and the liquid region, as well as the evaporation and condensation zones are shown. Also shown are the distributions of temperature and saturation within these regions.

For $\delta_g \leq x \leq \delta_g + \delta_{g\ell}$, the saturation is expected to *increase monotonically* with x. The vapor generated at the evaporation zone (the thickness of this zone is in practice finite but here taken as zero) at $x = \delta_g$, moves upward (buoyancy-driven), *condenses* (condensation occurs in the condensation zone which is taken to have zero thickness) at $x = \delta_g + \delta_{\ell g}$, and returns as liquid (buoyancy- and capillary-driven). By allowing for irreducible saturations s_{ir} and $s_{ir\,g}$, i.e., assuring continuous phase distributions for the two-phase flow, we have to assume an *evaporation zone* just below $x = \delta_g$ in which s undergoes a step change and evaporation occurs. A similar zone is assumed to exist above $x = \delta_g + \delta_{g\ell}$ over which s undergoes another step change and condensation occurs (*condensation zone*). Next, we consider cases with $\delta_g > d$ and $\delta_{g\ell} \gg d$, where we can apply the volume-averaged governing equations based on *bulk* properties. As was discussed in Section 9.2.4, for $s < 1$, the liquid will be in a superheated state depending on the local radius of curvature of the meniscus. Therefore, the two-phase region is *only* approximately *isothermal*. For steady-state conditions, the heat supplied q is removed from the upper single-phase (liquid) region. Since the heat supplied to the liquid region causes an *unstable* stratification, natural convection can occur that can influence the two-phase region (Sondergeld and Turcotte, 1977; Ramesh and Torrance, 1990). In the following one-dimensional analysis, this phenomenon is *not* considered.

(A) VAPOR-FILM REGION

The one-dimensional heat conduction for the stagnant vapor-film region is given by

$$q = -k_e(x)\frac{\mathrm{d}T}{\mathrm{d}x}. \tag{12.43}$$

Since $k_g/k_s < 1$, we expect that for the packed beds near the bounding surface the magnitude of k_e will be *smaller* than the bulk value (see Section 3.7). Therefore, a *nonlinear* temperature distribution is expected near this surface. However, if we assume k_e to be constant within δ_g, then we will have

$$q = k_\ell \frac{T_o - T_s}{\delta_g}, \tag{12.44}$$

where, for a given q, we have $T_o - T_s$ and δ_g as the *unknowns*. Generally, $T_o - T_s$ is also *measured*, which leads to the determination of δ_g. We note again that between the vapor film and the two-phase region an evaporation zone exists in which the saturation and temperature are expected to change continuously. If a *jump* in s was allowed across it, it would be inherently *unstable* and would invade the two adjacent regions *intermittently*. The condensation zone at $x = \delta_g + \delta_{g\ell}$ is expected to have a similar behavior. The present one-dimensional model does *not* address the examination of these zones.

(B) TWO-PHASE REGION

The analysis of the two-phase region is given by Sondergeld and Turcotte (1977), Bau and Torrance (1982), and more completely by Udell (1985) and Jennings and Udell (1985). The vapor that is generated at $x = \delta_g$ and is given by

$$(\rho_g u_g)_{\delta_g} = \frac{q}{\Delta i_{\ell g}}, \tag{12.45}$$

flows *upward* primarily due to buoyancy. By allowing for the variation in p_g, the momentum equation for the gas phase will be (8.135), except that the inertial, drag, and surface-tension gradient terms are *negligible* because of the small Bond number assumption. This gives

$$0 = -\frac{\mathrm{d}p_g}{\mathrm{d}x} + \rho_g g - \frac{\mu_g}{KK_{rg}}u_g, \tag{12.46}$$

where we have used $u_g = \langle u_g \rangle$, $p_g = \langle p \rangle^g$, and $K_g = KK_{rg}$. Since the net flow at any cross section is zero, we have

$$\rho_g u_g + \rho_\ell u_\ell = 0, \tag{12.47}$$

as the continuity equation. The momentum equation for the liquid phase (8.135) becomes

$$0 = -\frac{dp_\ell}{dx} + \rho_\ell g - \frac{\mu_\ell}{KK_{r\ell}} u_\ell, \qquad (12.48)$$

where the local pressure p_g and p_ℓ are related through the capillary pressure (12.18). By using (12.46)–(12.48) and (12.18), we have

$$\frac{dp_c}{dx} = -\frac{q}{Ki_{\ell g}}\left(\frac{\nu_g}{K_{rg}} + \frac{\nu_\ell}{K_{r\ell}}\right) + (\rho_\ell - \rho_g)g. \qquad (12.49)$$

Next, by assuming that the Leverett J-function is applicable and that $K_{r\ell}$ and K_{rg} can be given as functions of s only, (12.49) can be written in terms of the saturation *only*. Udell uses the p_c correlation given in Table 8.3 and the relative permeabilities suggested by Wyllie as given in Table 8.4. By using these, (12.49) becomes

$$\frac{\sigma}{(K/\varepsilon)^{1/2}}\frac{dJ}{dx} = -\frac{q}{Ki_{\ell g}}\left[\frac{\nu_g}{(1-S)^3} + \frac{\nu_\ell}{S^3}\right] + (\rho_\ell - \rho_g)g$$

$$= \frac{\sigma}{(K/\varepsilon)^{1/2}}\frac{dJ}{dS}\frac{dS}{dx}, \qquad (12.50)$$

where, as before,

$$S = \frac{s - s_{\text{ir}}}{1 - s_{\text{ir}} - s_{\text{ir }g}}. \qquad (12.51)$$

Next, we can translate the origin of x to δ_g, and then by integrating over the two-phase zone, we will have

$$\delta_{g\ell} = \int_0^1 \frac{\dfrac{\sigma}{(K/\varepsilon)^{1/2}}\dfrac{dJ}{dS}}{-\dfrac{q}{Ki_{\ell g}}\left[\dfrac{\nu_g}{(1-S)^3} + \dfrac{\nu_\ell}{S^3}\right] + (\rho_g - \rho_\ell)g}\, dS. \qquad (12.52)$$

Whenever q, the liquid and vapor properties, and K are known, $\delta_{g\ell}$ can be determined from (12.52) and the saturation distribution can be found from (12.50) and (12.51).

Note that when in (12.49) the viscous and gravity forces exactly balance, and the capillary pressure gradient, and therefore, the saturation gradient become zero. For this condition, we will have the magnitude $\delta_{g\ell}$ tending to infinity. This is evident in (12.52). The heat flux corresponding to this condition is called the *critical heat flux* q_{cr}. For $q > q_{\text{cr}}$, the thickness of the two-phase region *decreases monotonically* with q. Figure 12.7 shows the prediction of Udell as given by (12.52), along with his experimental results for the normalized $\delta_{g\ell}$ as a function of the normalized q. For large q, an *asymptotic* behavior is observed. The critical heat flux q_{cr} (normalized) is also shown for the specific cases of $\nu_\ell/\nu_g = 0.0146$ and $Bo = 5.5 \times 10^{-7}$.

12.3 Evaporation at Horizontal Impermeable Bounding Surfaces

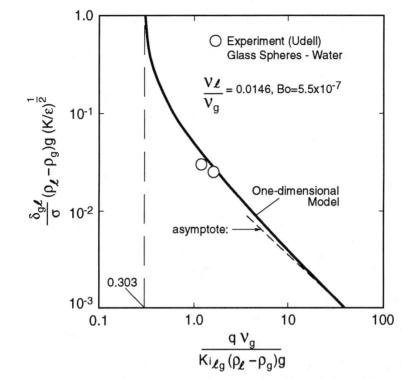

Figure 12.7 Variation of the normalized thickness of the two-phase region as a function of the normalized heat flux for evaporation from the heated horizontal surface. (From Udell, reproduced by permission ©1985 Pergamon.)

(C) Onset of Film Evaporation

The saturation at which the saturation gradient is *zero* (and $\delta_{g\ell} \to \infty$) is found by setting the denominator of (12.52) to zero, i.e.,

$$\frac{q_{\mathrm{cr}}}{K\Delta i_{\ell g}}\left[\frac{\nu_g}{(1-S_{\mathrm{cr}})^3} + \frac{\nu_\ell}{S_{\mathrm{cr}}^3}\right] = (\rho_\ell - \rho_g)g. \qquad (12.53)$$

For this *critical reduced saturation* S_{cr}, the critical heat flux is given by (12.42).

Bau and Torrance (1982) use a different relative permeability-saturation relation and arrive at a slightly different relation. Jones et al. (1980) use a similar treatment and find a relationship for q_{cr} that gives values lower than those predicted by (12.42) by a factor of approximately 2. It should be noted that these predictions of q_{cr} are *estimations* and that the effects of *wettability*, *solid matrix structure* (all of these studies consider spherical particles only), and *surface tension* (all of which influence the phase distributions) are included *only* through the *relative* permeabilities. These permeabilities, in turn, are given as *simple* functions of the saturation *only*. Therefore, the use of realistic and accurate relative permeability relations is *critical* in the prediction of q_{cr}.

12.4 Evaporation at Thin Porous-Layer Coated Surfaces

Evaporation *within* and *over* thin porous layers is of interest in wicked heat pipes and in surface modifications for the purpose of heat transfer enhancement. The case of very thin layers, i.e., $\delta/d \simeq 1$ where δ is the *porous-layer thickness*, has been addressed by Konev et al. (1987), Styrikovich et al. (1987), and Kovalev et al. (1987). Due to the lack of the *local* thermal equilibrium in the two-phase region inside the thin porous layer, we will *not* pursue the analysis for the case of $\delta/d \simeq 1$.

When $\delta/d \gg 1$ but $\delta/(\delta_g + \delta_{g\ell}) < 1$, the two-phase region *extends* to the plain medium surrounding the porous layer. Presently, *no* detailed experimental results exist for horizontal surfaces coated with porous layers with $\delta \neq \delta_g + \delta_{g\ell}$. The experimental results of Afgan et al. (1985) are for heated horizontal *tubes* (diameter D) and as will be shown in their experiments $\delta < \delta_g + \delta_{g\ell}$. Their porous layers are made by the *sintering* of *metallic* particles. The particles are *spherical* (average diameter $d = 81$ μm) and are fused onto the tube in the process of sintering. From the various porous-layer coatings they use we have selected the following *three* cases in order to demonstrate the general trends in their results.

- A layer of thickness $\delta/d \simeq 27$ with $K = 1.4 \times 10^{-10}$ m^2, $\varepsilon = 0.70$, $Bo = 2.6 \times 10^{-5}$, made of stainless-steel particles ($k_s = 14$ W/m-K), and coated over a 16-mm-diameter stainless-steel tube ($D/\delta = 7.3$).
- A layer of thickness $\delta/d \simeq 6.7$ with $K = 3 \times 10^{-11}$ m^2, $\varepsilon = 0.50$, $Bo = 7.2 \times 10^{-6}$, made of titanium particles ($k_s = 21$ W/m-K), and coated over an 18-mm-diameter stainless-steel tube ($D/\delta = 33$).
- A layer of thickness $\delta/d \simeq 5.5$ with $K = 2.0 \times 10^{-12}$ m^2, $\varepsilon = 0.30$, $Bo = 8.9 \times 10^{-7}$, made of stainless-steel particles, and coated over a 3-mm-diameter stainless-steel tube ($D/\delta = 6.7$).

We have used the *mean particle size* d of 81 μm and the Carman-Kozeny equation for the calculation of the permeability. The fluid used is *water*. Their experimental results for these three cases are given in Figure 12.8. In their experimental results q is larger in the *desaturation branch*, while in the experimental results of Bergles and Chyu (1982) q is larger in the *saturation branch*. In order to examine whether the porous-layer thicknesses used in these experiments are *larger* than $\delta_g + \delta_{g\ell}$, we apply the prediction

12.4 Evaporation at Thin Porous-Layer Coated Surfaces

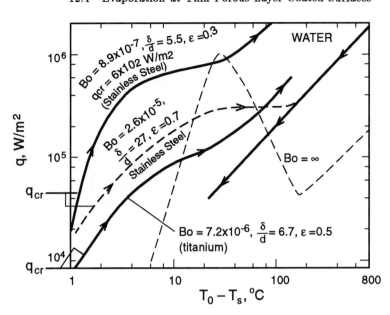

Figure 12.8 Experimental results of Afgan et al. for evaporation from tubes coated with porous layers and submerged in a pool of water. (From Afgan et al., reproduced by permission ©1985 Pergamon.)

of Udell (1985) for the thickness of the two-phase region. His results were shown in Figure 12.7. The asymptote for heat fluxes much *larger* than the critical heat flux is given by

$$\frac{\delta_{g\ell}(\rho_g - \rho_\ell)g(K/\varepsilon)^{1/2}}{\sigma} \frac{q\nu_g}{K\Delta i_{\ell g}(\rho_\ell - \rho_g)g} = 0.0368 \quad q \gg q_{cr} \quad (12.54)$$

or

$$\delta_{g\ell} = 0.0368 \frac{\sigma \Delta i_{\ell g}(K\varepsilon)^{1/2}}{q\nu_g} \quad q \gg q_{cr}. \quad (12.55)$$

For $\delta_{g\ell} = \delta$, we have

$$q(\delta_{g\ell} = \delta) = \frac{0.0368\sigma \Delta i_{\ell g}(K\varepsilon)^{1/2}}{\delta \nu_g}. \quad (12.56)$$

For those cases presented in Figure 12.8, we have calculated the required q for $\delta_{g\ell} = \delta$. The values are

$$q\left(\delta_{g\ell} = \delta, \frac{\delta}{d} = 27, \varepsilon = 0.7 \text{ sample}\right) = 1.10 \times 10^6 \text{ W/m}^2,$$

$$q\left(\delta_{g\ell} = \delta, \frac{\delta}{d} = 6.7, \varepsilon = 0.5 \text{ sample}\right) = 1.73 \times 10^6 \text{ W/m}^2, \quad (12.57)$$

$$q\left(\delta_{g\ell} = \delta, \frac{\delta}{d} = 5.5, \varepsilon = 0.3 \text{ sample}\right) = 4.23 \times 10^5 \text{ W/m}^2.$$

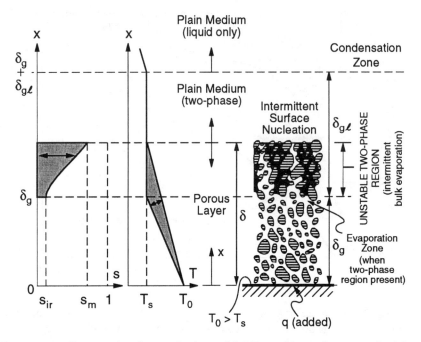

Figure 12.9 Evaporation from a horizontal impermeable surface coated with a porous layer with $\delta < \delta_g + \delta_{g\ell}$. The speculated intermittent drying of the layer and the associated temperature and saturation distributions are shown.

We note that these heat fluxes are lower bounds, because δ is actually occupied by the vapor-film region, evaporation zone, as well as the two-phase region. For the porous layer to contain both of the layers, we need heat fluxes much larger than those given by (12.57), i.e.,

$$\delta \geq \delta_g + \delta_{g\ell} \quad \text{or} \quad q > q(\delta = \delta_{g\ell}). \tag{12.58}$$

Upon examining the experimental results given in Figure 12.8, we note that except for the $\delta/d = 5.5$ layer, we have $q < q(\delta = \delta_{g\ell})$, i.e., the two-phase region extends *beyond* the porous layer and into the plain medium.

No rigorous analysis for the case of $\delta < \delta_g + \delta_{g\ell}$ is available. Assuming that the theory of isothermal two-phase is applicable, we postulate that the portion of the two-phase region that is inside the porous layer will be *unstable*. This instability will be in the form of intermittent drying of this portion, i.e., the entire porous layer becoming intermittently invaded by the vapor phase only. When the porous media is dry there will be a nucleate boiling at the interface of the porous plain medium. Figure 12.9 depicts such an intermittent drying. When the two-phase region extends into the porous layer, the two-phase region will be at the saturation temperature (assuming negligible liquid superheat due to the capillarity). The evaporation takes place in the evaporation zone, just below the two-phase zone.

The saturation at $x = \delta$ will be smaller than $1 - s_{ir\,g}$. This saturation is designated by s_m, which is similar to that for thick porous layers discussed in the previous sections. When the porous layer dries out, the evaporation will be at $x = \delta$. The frequency of this transition (i.e., intermittent drying of the porous layer) decreases as the porous-layer thickness increases and should become zero for $\delta > \delta_g + \delta_{g\ell}$. It should be mentioned that in principle the theory of evaporation-isothermal two-phase region, *cannot* be extended to thin porous-layer coatings. The above given arguments are only speculative. The theory of thin porous layers has not yet been constructed.

We now return to Figure 12.8. For the $\delta/d = 27$ case we estimate the bulk value of k_e for the vapor-film region $(k_s/k_g = 500, \varepsilon = 0.7)$ by using (3.68), and we find k_e to be 0.132 W/m-K. We note that the photomicrographs of Afgan et al. show that the particle distribution near the bounding surface is significantly *different* than that in the bulk. Therefore, this k_e is only an estimate. From (12.44), we have

$$\delta_g = \frac{k_\ell(T_o - T_s)}{q}. \qquad (12.59)$$

For $q = 4 \times 10^4$ W/m^2 and $T_o - T_s = 100°$C, we have $\delta_g/d = 0.41$, i.e., the vapor-film region is *less* than one particle thick. Then, for $\delta/d = 27$ only part of the two-phase region is in the porous layer.

For the $\delta/d = 5.5$ and 6.7 cases, the vapor-film region thickness is also small and nearly a particle diameter thick. However, the remaining space occupied by the two-phase region is also very small. Therefore, both the vapor-film and two-phase regions do not lend themselves to the analyses based on the existence of the local thermal equilibrium and the local volume averaging. The two thin porous layers, $\delta/d = 5.5$ and 6.7, result in different heat transfer rates (for a given $T_o - T_s$), and this difference is also due to the structure of the solid matrix and the value of D/d. For the case of $\delta/d = 5.5$, the one-dimensional analysis predicts that the two-phase region is entirely placed in the porous layer (although the validity of this analysis for such small $\delta_{g\ell}/d$ is seriously *questionable*). This indicates that the liquid supply to the heated surface is *enhanced* when the capillary action can transport the liquid through the entire two-phase region. The *optimum* porous layer thickness, which results in a *small* resistance to vapor and liquid flows, a *large* effective thermal conductivity for the vapor-film region, and possibly some two- and three-dimensional motions, has not yet been rigorously analyzed.

12.5 Moving Evaporation or Condensation Front

So far we have been considering heat transfer *across* an impermeable surface bounding a porous medium and the associated *liquid-vapor* phase change. There are some *nearly* one-dimensional phase-change problems where the bounding surfaces are *permeable* and a stationary or moving phase-change front develops *within* the porous medium. An example of simultaneous heat and mass transfer across a bounding permeable surface is the phase-change *transpiration* cooling where the liquid is supplied at one end of a porous medium and the vapor leaves the other side [Figure 12.10 (a)]. In this example the surface that *receives* the heat has a temperature *above* the saturation temperature $T_s(p_g)$, where p_g is the total gas-phase pressure. This can result in a *steady-state* condition with the interface of the vapor and the two-phase region and the interface of the two phases and the liquid region both becoming stationary. However, these interfaces are generally very unstable, at least when the heat input q is rather *large*. Another example involving a similar phase change is in the injection of a vapor that condenses inside a porous medium and then adds to and displaces the liquid phase [Figure 12.10(b)]. In this case, a *quasi-steady-state* moving condensation front occurs.

Both of these examples involve the simultaneous presence of temperatures equal to or *larger* than the saturation temperature T_s and permeable bounding surfaces. We also encounter problems in which the condensation or evaporation occurs at temperatures *below* T_s. An example of this is the *surface* convective *heating* of partially saturated porous media. This results in both *surface* and *internal* evaporations. In this *transient* problem in the first period where $s > s_{ir}$ everywhere, the liquid *flows* toward the *surface* (called *funicular regime*). This period is followed by a period where $s \leq s_{ir}$ in a region *adjacent* to the heated boundary and an *evaporation front* develops (called *moving-front regime*). Figures 12.11 (a) and (b) show these two regimes in transient *desaturation* by surface heat addition.

In the following one-dimensional analyses of phase change occurring in porous media with permeable bounding surfaces, we first consider the cases involving temperatures equal to or above T_s. Next we examine the cases with $T(x) < T_s$. For the first case of $T(x) \geq T_s$, in the domain a two-phase region *exists* that is nearly at T_s, and an interface separates this region from the generally *subcooled* liquid region. For this class of problems, we only consider the single-component systems. For the case of $T(x) < T_s$, the *funicular regime* ends after a *critical time* where the saturation at and near the surface falls below s_{ir}. After the critical time, the vapor generated at the evaporation front *diffuses* toward the heated surface. As expected, the effect of the *noncondensible* (inert) components of the gas phase on the motion of the vapor is *significant* and this will be discussed. Finally we discuss propagation of a condenstaion front in an otherwise dry porous medium.

12.5 Moving Evaporation or Condensation Front

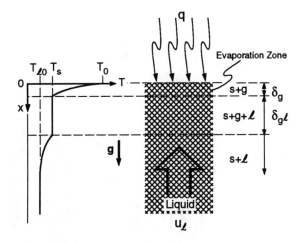

(a) Transpiration cooling using heat of vaporization

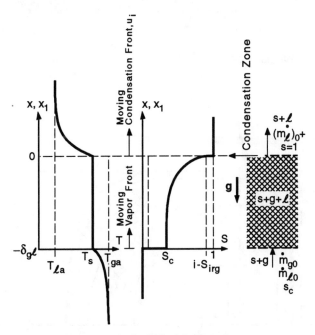

(b) Displacement of a liquid by injection of vapor

Figure 12.10 Examples of evaporation and condensation fronts in porous media with permeable bounding surfaces: (a) transpiration cooling using the heat of vaporization and (b) displacement of a liquid by the injection of its vapor.

630   12. Phase Change

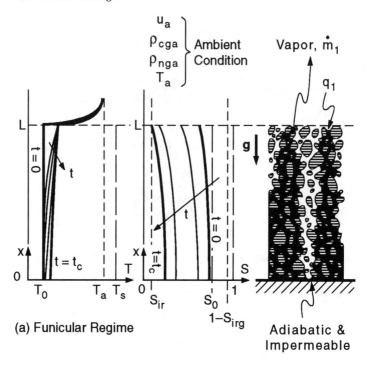

(a) Funicular Regime

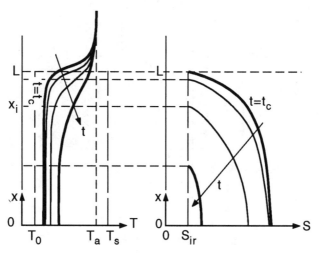

(b) Moving Interface Regime

Figure 12.11 Heat supply at a permeable bounding surface with $T < T_s$. (a) The porous medium loses liquid by surface and internal evaporation in the funicular regime and (b) by internal evaporation in the moving interface regime.

12.5.1 Temperatures Equal to or Larger than Saturation Temperature

There is a class of phase-change problems in which the *single-phase* liquid or vapor flow (saturation unity or zero) enters one of the boundaries of a solid matrix and then leaves the *opposite* boundary after a *complete* phase change (i.e., after vaporization or condensation) as a single-phase flow (saturation zero or unity). Inside the matrix there is a two-phase region *separating* the two single-phase regions. The *interfaces* between the two-phase region and these single-phase regions are generally *idealized* with jumps in the saturation and, as was mentioned, these interfaces are *not* well characterized and may be unstable. In Figure 12.10 (a) and (b) we have idealized as one-dimensional the problems of the transpiration cooling using the heat of vaporization and the displacement of a liquid by the injection of a vapor that condenses inside the matrix. As with the evaporation and condensation adjacent to vertical impermeable surfaces, these two problems have *some* similarities. In Figure 12.10(a) the heat arriving at the permeable surface flows against the flowing vapor until it arrives at the *evaporation zone*. Right after this evaporation zone the two-phase region begins. The problem of a moving condensation front has been studied rather extensively. In Figure 12.10(b) the *subcooled* liquid condenses the vapor at the condensation zone. This problem has been studied by Miller (1975), Yortsos (1982), Menegus and Udell (1985), Stewart et al. (1987), Basel and Udell (1989), and Stewart and Udell (1989). The *velocity* of the *condensation front*, which in part is determined by the extent of the subcooled state of the liquid, and the *stability* of this *interface* (also called the *liquid front*), are examined later.

(A) Two-Phase Region

This region is nearly *isothermal*. The liquid superheat due to the meniscus curvature is large only for very small permeabilities and can be included as discussed in Section 9.2.4. The continuity equations are (8.94) and (8.95) and the momentum equations are (8.135) and (8.137). By assuming that the drag at $A_{\ell g}$ is negligible, we will arrive at

$$\varepsilon \frac{\partial s}{\partial t} + \frac{\partial u_\ell}{\partial x} = 0, \tag{12.60}$$

$$-\varepsilon \frac{\partial s}{\partial t} + \frac{\partial u_g}{\partial x} = 0, \tag{12.61}$$

$$0 = -\frac{\partial p_\ell}{\partial x} - \rho_\ell g - \frac{\mu_\ell}{K_\ell} u_\ell + \frac{\rho_\ell}{K_{\ell i}} u_\ell^2, \tag{12.62}$$

$$0 = -\frac{\partial p_g}{\partial x} - \rho_g g - \frac{\mu_g}{K_g} u_\ell + \frac{\rho_g}{K_{gi}} u_g^2, \tag{12.63}$$

along with the prescribed K_ℓ, $K_{\ell i}$, K_g, and K_{gi}. We have used x as the coordinate axis with $x = 0$ located at the *condensation front* at a time t. At the start of the two-phase region, i.e., $x = -\delta_{g\ell}$, the saturation gradient is zero, therefore, $p_g = p_\ell$. Figure 12.10(b) shows the coordinate system and the *expected* saturation distribution. A pressure gradient $(\partial p/\partial x)_{-\delta_{g\ell}}$ drives the two-phase flow and we designate the saturation at $-\delta_{g\ell}$, corresponding to $(\partial s/\partial x)_{-\delta_{g\ell}} = 0$, as s_c. Because we are dealing with the *funicular regime* for the two-phase region, we require that $s_c > s_{\mathrm{ir}}$. Also as will be shown, for a given vapor flow rate and a set of relative permeabilities and a Bond number, s_c is *uniquely* determined (i.e., *cannot* be arbitrarily assigned). The inlet flow into the two-phase region is given by

$$\dot{m}_t = \rho_\ell u_{\ell_0} + \rho_g u_{g_0} = \dot{m}_{\ell_0} + \dot{m}_{g_0} \quad \text{at } x = -\delta_{g\ell}, \tag{12.64}$$

with

$$\frac{\dot{m}_{g_0}}{\dot{m}_t} < 1.$$

At $x_1 = 0^-$, the saturation is $1 - s_{\mathrm{ir}\,g}$ and both the liquid *velocity* and the *saturation jumps* occur in the condensation zone ($x = 0$). For $x_1 = 0^+$, the liquid velocity is $(u_\ell)_{0+}$ and the saturation is unity.

Following Stewart et al., the coordinate axis is *moved* along with the liquid front and is *scaled* using the *small length scale*. This gives

$$x_1^* = \frac{x_1}{(K/\varepsilon)^{1/2}} \frac{\dot{m}_t \nu_g}{\varepsilon \sigma} = \frac{1}{(K/\varepsilon)^{1/2}} \frac{\dot{m}_t \nu_g}{\varepsilon \sigma}(x - u_i t)$$

$$= \frac{Ca_t}{(K\varepsilon)^{1/2}}(x - u_i t), \tag{12.65}$$

with the *total capillary number* defined as

$$Ca_t = \frac{\dot{m}_t \nu_g}{\sigma}. \tag{12.66}$$

Now by using (12.65) in (12.60) and (12.61), we have

$$\frac{\partial}{\partial x_1^*}(u_\ell - \varepsilon u_i s) = 0, \tag{12.67}$$

$$\frac{\partial}{\partial x_1^*}[u_g - \varepsilon u_i(1-s)] = 0. \tag{12.68}$$

After the integrations, we will have

$$u_\ell = u_{\ell_0} + \varepsilon u_i(s - s_c), \tag{12.69}$$

$$u_g = u_{g_0} - \varepsilon u_i(s - s_c). \tag{12.70}$$

12.5 Moving Evaporation or Condensation Front

As will be shown, u_i depends on u_{ℓ_0}, u_{g_0}, and other variables. Also, at $x_1 = 0^-$, we have $(u_\ell)_{0^-} = u_{\ell_0} + \varepsilon u_i(1 - s_{\text{ir}\,g} - s_c)$. The two momentum equations can be *combined* using the capillary pressure, and when the Leverett reduction and the relation for $K_{\ell i}$ and K_{gi} given by (8.172) and (8.173) are used, we have

$$\frac{dS}{dx_1^*} = \frac{\dfrac{\mu_\ell}{\mu_g}\dfrac{u_\ell}{u_g}\left(\dfrac{1}{K_{r\ell}} + \dfrac{C_E u_\ell K^{1/2}}{\nu_\ell}\right) - \dfrac{1}{K_{rg}} - \dfrac{C_E u_g K^{1/2}}{\nu_g} + \dfrac{\varepsilon Bo}{Ca_g}}{\dfrac{Ca_t}{Ca_g}\dfrac{dJ}{ds}}, \qquad (12.71)$$

where

$$Bo = \frac{(\rho_\ell - \rho_g)g(K/\varepsilon)}{\sigma} \qquad (12.72)$$

and

$$Ca_g = \frac{u_g \mu_g}{\sigma}. \qquad (12.73)$$

For dS/dx_1^* to be zero at $x_1^* = Ca_t(K/\varepsilon)^{-1/2}(-\delta_{g\ell} - u_i t)$, the numerator of (12.71) must *vanish*. This is the condition used for the determination of s_c. For a given $K_{rg}(s)$ and $K_{r\ell}(s)$, we have the *functional* dependence given by

$$s_c = s_c\left(\frac{\varepsilon Bo}{Ca_{g_0}}, \frac{\mu_\ell u_{\ell_0}}{\mu_g u_{g_0}}, \frac{u_{\ell_0} K^{1/2}}{\nu_\ell}, \frac{u_{g_0} K^{1/2}}{\nu_g}\right) \qquad (12.74)$$

and s_c is determined from

$$\frac{\mu_\ell}{\mu_g}\frac{u_{\ell_0}}{u_{g_0}}\left[\frac{1}{K_{r\ell}(s_c)} + \frac{C_E u_{\ell_0} K^{1/2}}{\nu_\ell}\right]$$

$$- \frac{1}{K_{rg}(s_c)} - \frac{C_E u_{g_0} K^{1/2}}{\nu_g} + \frac{\varepsilon Bo}{Ca_{g_0}} = 0. \qquad (12.75)$$

Stewart et al. solve for s_c for the case of $C_E = 0$, and by using the relative permeabilities suggested by Wyllie (Table 8.4), they show that as $\varepsilon Bo/Ca_{g_0}$ and $(\mu_\ell/\mu_g)(u_{\ell_0}/u_{g_0})$ increase, s_c also increases.

(B) Liquid Region

The liquid flow rate is determined from (12.69) by using $s = 1$, i.e.,

$$(\rho_\ell u_\ell)_{0^+} = (\rho_\ell u_\ell + \rho_g u_g)_{0^-} = \dot{m}_t + \varepsilon u_i(\rho_\ell - \rho_g)(1 - s_c). \qquad (12.76)$$

The heat *removal* from the condensation zone is by axial conduction (and against the flow of the liquid) and is given by the energy equation for single-phase flows. For the one-dimensional problem considered, (4.221) becomes

$$[\varepsilon(\rho c_p)_\ell + (1-\varepsilon)(\rho c_p)_s]\frac{\partial T}{\partial t} + (\rho c_p)_\ell u_\ell \frac{\partial T}{\partial x} = (\rho c_p)_\ell \frac{\partial}{\partial x} D_\parallel \frac{\partial T}{\partial x}. \qquad (12.77)$$

Assuming that $D_\parallel$ is constant and using (12.65), this becomes

$$\frac{d^2T}{dx_1^{*2}} = \frac{(K/\varepsilon)^{1/2}}{(\rho c_p)_\ell D_\parallel Ca_t} \{[\varepsilon(\rho c_p)_\ell + (1-\varepsilon)(\rho c_p)_s]\, u_i$$

$$-(\rho_\ell u_\ell)_0 + c_{p\ell}\} \frac{dT}{dx_1^*} \equiv -\gamma^* \frac{dT}{dx_1^*}, \qquad (12.78)$$

with

$$T = T_s \quad \text{at} \quad x_1^* = 0, \qquad (12.79)$$

$$T = T_{\ell a} \quad \text{as} \quad x_1^* \to \infty. \qquad (12.80)$$

The solution to (12.78)–(12.80) is

$$\frac{T - T_{\ell a}}{T_s - T_{\ell a}} = e^{-\gamma^* x_1^*}, \qquad (12.81)$$

where γ^* introduced earlier is given by

$$\gamma^* = \frac{(K\varepsilon)^{1/2}}{Ca_t D_\parallel} \{[\varepsilon + (1-\varepsilon)C]\, u_i - (u_\ell)_{0+}\}. \qquad (12.82)$$

The *heat balance* at the condensation front gives

$$\rho_g u_g i_{\ell g} = -D_\parallel (\rho c_p)_\ell \frac{dT}{dx_1} \quad \text{at} \quad x_1 = 0, \qquad (12.83)$$

which, when used to find u_i, gives (Stewart et al.)

$$\frac{\rho_\ell u_i}{\dot{m}_t} = \frac{1 + \dfrac{\dot{m}_{go}}{\dot{m}_t} \dfrac{1}{Ja}}{(1-\varepsilon)C + \varepsilon + \varepsilon\dfrac{\rho_g}{\rho_\ell} + \dfrac{1-s_c}{Ja} - \varepsilon\dfrac{\rho_\ell - \rho_g}{\rho_{\ell i}}(1-s_c)}, \qquad (12.84)$$

$$C = \frac{(\rho c_p)_s}{(\rho c_p)_\ell}, \qquad Ja = \frac{c_{p\ell}(T_s - T_{\ell a})}{\Delta i_{\ell g}}. \qquad (12.85)$$

The *critical saturation* s_c is found from (12.75), and u_i is evaluated from (12.84). The results of Menegus and Udell show that most of the saturation variation (in the two-phase region) occurs over a *very short* distance, i.e., the two-phase region is nearly *iso-saturation* and *isothermal*. The results of Stewart et al. show that u_i decreases with an *increase* in $\varepsilon Bo/Ca_t$ and with an *asymptotic* behavior for a *large* $\varepsilon Bo/Ca_t$.

(C) Stability of Condensation Front

The stability of the planar condensation front has been studied by Miller (1975), Yortsos (1982), and Stewart and Udell (1989). Miller applies the *linear stability theory* (infinitesimal disturbances) and obtains the *characteristic equation* for the *marginally* stable states. He uses the preceding one-dimensional velocity and temperature fields (with a constant saturation in the two-phase region) as the *base fields*. Yortsos, in addition, allows for the *lateral* heat transfer by using a *transient penetration model* (i.e., a pseudo one-dimensional base temperature field). Stewart and Udell compare the predicted stability map of Miller with their experiments and find good agreements.

The stability analysis begins by decomposing the fields into the *base* and *perturbation* components, i.e.,

$$p_g = \overline{p}_g + p'_g, \tag{12.86}$$

$$p_\ell = \overline{p}_\ell + p'_\ell, \tag{12.87}$$

$$T = \overline{T}_g + T'. \tag{12.88}$$

For the perturbation components, we assume *periodic* behavior in time and in the y-z plane, i.e.,

$$p'_g = A_g(x^*)e^{i(a_y y + a_z z) + \beta t}, \tag{12.89}$$

$$p'_\ell = A_\ell(x^*)e^{i(a_y y + a_z z) + \beta t}, \tag{12.90}$$

$$T' = A_T(x^*)e^{i(a_y y + a_z z) + \beta t}. \tag{12.91}$$

The interface location is also perturbed according to

$$x_1^{*'} = A_x e^{i(a_y y + a_z z) + \beta t}, \tag{12.92}$$

where the wave numbers are related to wavelengths by

$$a_y = \frac{2\pi}{\lambda_y}, \quad a_z = \frac{2\pi}{\lambda_z}, \tag{12.93}$$

and

$$a^2 = a_y^2 + a_z^2, \quad \nabla_1^2 = \frac{\partial^2}{\partial y^2} + \frac{\partial^2}{\partial z^2} = -a^2. \tag{12.94}$$

For a *constant saturation* two-phase region [$s = s_c$ as given by (12.75)], the momentum equations simplify to the Laplace equation for the pressure (note that $p_g = p_\ell$). The liquid region momentum equation is already in this form (because of the uniform permeability). The energy equation for this region is also written in the three-dimensional form. These equations are

$$\nabla \cdot \frac{K_g}{\mu_g} \nabla (p_g - \rho_g g x) = 0 \quad \text{for } x \leq 0, \tag{12.95}$$

12. Phase Change

$$\nabla \cdot \left(\frac{K_g}{\mu_g} + \frac{K_\ell}{\mu_\ell} \right) \nabla p_\ell = 0 \quad \text{for } x \leq 0, \tag{12.96}$$

$$\nabla \cdot \frac{K_\ell}{\mu_\ell} \nabla (p_\ell - \rho_\ell g x) = 0 \quad \text{for } x \geq 0, \tag{12.97}$$

$$[\varepsilon + (1-\varepsilon)C] \frac{\partial T}{\partial t} = D_\| \nabla^2 T + \frac{K_\ell}{\mu_\ell} \nabla (p_\ell - \rho_\ell g x) \cdot \nabla T \quad \text{for } x \geq 0. \tag{12.98}$$

Next, the decompositions given by (12.86)–(12.88) are used in these equations, the resulting equations are *linearized*, and the *mean field* is subtracted from them. The results are

$$\nabla^2 p_g' = \nabla^2 p_\ell' = 0 \quad \text{for} \quad x \leq 0, \tag{12.99}$$

$$\nabla^2 p_\ell' = 0 \quad \text{for} \quad x \geq 0, \tag{12.100}$$

$$[\varepsilon + (1-\varepsilon)C] \frac{\partial T'}{\partial t} = D_\| \nabla^2 T' + \frac{K_\ell}{\mu_\ell} \nabla p_\ell' \cdot \nabla T$$

$$+ \frac{K_\ell}{\mu_\ell} \nabla (\bar{p}_\ell - \rho_\ell g x) \cdot \nabla T' \quad \text{for } x \geq 0. \tag{12.101}$$

We now use the periodic forms given in (12.89) and (12.92) for the perturbation components and use the *moving* coordinate axis x_1. This change of the coordinate axis influences (12.101) only. The pressure perturbation equation for the two-phase region, (12.99), becomes

$$\frac{d^2 A_g}{d x_1^2} - a^2 A_g = 0 \quad \text{for} \quad x_1 \leq 0, \tag{12.102}$$

with the solution (satisfying the decay at large distances from the interface) given by

$$A_g = a_g e^{a x_1}. \tag{12.103}$$

The liquid pressure equations, (12.100), become

$$\frac{d^2 A_\ell}{d x_1^2} - a^2 A_\ell = 0 \quad \text{for} \quad x_1 \geq 0, \tag{12.104}$$

with the solution given by

$$A_\ell = a_\ell e^{-a x_1}. \tag{12.105}$$

The energy equation becomes

$$\frac{d^2 A_T}{d x_1^2} + \gamma \frac{d A_T}{d x_1} - \left\{ a^2 \frac{[\varepsilon + (1-\varepsilon)C]\omega}{D_\|} \right\} A_T$$

$$= \frac{K_\ell}{\mu_\ell} (T_s - T_{\ell a}) \frac{\gamma}{D_\|} e^{-\gamma x_1} \frac{d A_\ell}{d x_1}, \tag{12.106}$$

12.5 Moving Evaporation or Condensation Front

where γ is *dimensional* and related to dimensionless γ^* by

$$\gamma = \gamma^* \frac{Ca_t}{(K\varepsilon)^{1/2}}. \tag{12.107}$$

The solution to (12.13) is (Stewart and Udell)

$$A_T = a_T e^{-\gamma_1 x_1} + a_\ell b e^{-(a+\gamma)x_1}, \tag{12.108}$$

where

$$b = \frac{K_\ell}{\mu_\ell} \frac{\gamma a(T_s - T_{\ell a})}{[\varepsilon + (1-\varepsilon)C]\beta + D_\parallel \gamma a}, \tag{12.109}$$

$$\gamma_1 = \frac{\gamma}{2} + \left\{ \left(\frac{\gamma}{2}\right)^2 + a^2 + \frac{[\varepsilon + (1-\varepsilon)C]\beta}{D_\parallel} \right\}^{1/2}. \tag{12.110}$$

The *marginal states* are found for $\beta = 0$ (no decay or growth). At the interface the pressure and temperature are continuous and the mass (and heat) flowing in and out of the interface is *equal*. These linearized boundary conditions are (Stewart and Udell)

$$\left(A_g + \frac{d\overline{p}_g}{dx_1} A_x \right)_{x_1=0-} = \left(A_\ell + \frac{d\overline{p}_\ell}{dx_1} A_x \right)_{x_1=0+}, \tag{12.111}$$

$$\left(A_T + \frac{d\overline{T}}{dx_1} A_x \right)_{x_1=0-} = 0, \tag{12.112}$$

$$\left(\rho_g \frac{K_g}{\mu_g} \frac{dA_g}{dx_1} A_x \right)_{x_1=0-} = \left(\rho_\ell \frac{K_\ell}{\mu_\ell} \frac{dA_\ell}{dx_1} A_x \right)_{x_1=0+}, \tag{12.113}$$

$$\rho_g \frac{K_g}{\mu_g} i_{\ell g} \frac{dA_g}{dx_1} = (\rho c_p)_\ell D_\parallel \left(\frac{dA_T}{dx_1} + \frac{d^2 \overline{T}}{dx_1^2} A_x \right)_{0+}. \tag{12.114}$$

Using the forms for A_T, A_ℓ, and A_g and the solutions for $\overline{p}_g$, $\overline{p}_\ell$, and $\overline{T}$, these boundary conditions lead to *four* equations for a_T, a_ℓ, a_g, and A_x. This system of equations has a *nontrivial* solution when (Stewart and Udell)

$$\frac{\rho_\ell}{\rho_\ell - \rho_g} \left[\frac{(u_\ell)_{0+} \nu_\ell}{gK_\ell} - \frac{\rho_g}{\rho_\ell} \frac{u_g \nu_g}{gK_g} \right] - 1$$

$$- \frac{\rho_\ell}{\rho_\ell - \rho_g} \left(\frac{\gamma \nu_g D_\parallel}{gK_g} + \frac{\gamma \nu_\ell D_\parallel}{gK_\ell} \right) \frac{\gamma_1 - \gamma}{\frac{a}{Ja} - \gamma_1 + \gamma + a} = 0. \tag{12.115}$$

This expression is *different* than in the Miller solution only in the existence of the term $+a$, which is *not* present in Miller's solution. We repeat that

$$\gamma = \frac{[\varepsilon + (1-\varepsilon)C] u_i - (u_\ell)_{0+}}{D_\parallel}, \quad \gamma_1 = \frac{\gamma}{2} + \left[\left(\frac{\gamma}{2}\right)^2 + a^2 \right]^{1/2}, \tag{12.116}$$

with $(u_\ell)_{0+}$ given by (12.76), u_i by (12.84), and s_c by (12.75).

From (12.115), the *wave number* can be found once the fluid permeabilities and the inlet flow rates $u_{\ell o}$ and u_{go} are specified. We can write (12.115) as

$$\frac{\rho_\ell}{\rho_\ell - \rho_g}(M_\ell - M_g) - 1 - \frac{\rho_\ell}{\rho_\ell - \rho_g}D_\gamma \frac{\frac{\gamma}{2} + \left[\left(\frac{\gamma}{2}\right)^2 + a^2\right]^{1/2} - \gamma}{\frac{a}{Ja} - \frac{\gamma}{2} - \left[\left(\frac{\gamma}{2}\right)^2 + a^2\right]^{1/2} + \gamma + a} = 0, \quad (12.117)$$

with M_ℓ, M_g, and D_γ defined as abbreviations for the expressions appearing in (12.115). Then the critical *wave number* a_c is obtained by finding the minimum in a versus one of the parameters. Using water and the phase permeabilities suggested by Wyllie and $(\rho c_p)_\ell D_\parallel = k_e$, Stewart and Udell find the critical wave number with $\varepsilon Bo/Ca_t$ as the parameter for $u_{\ell_0} = 0$. The experimental results of Stewart and Udell *qualitatively* agree with these predictions.

The condensation at the front tends to *stabilize* this moving front. When the flow is in the direction of gravity, gravity stabilizes the front [in this case the -1 term in (12.117) is replaced by $+1$]. When the *dimensionless mobility of the vapor* M_g is *larger* than that of the liquid, this tends to *destabilize* the front.

12.5.2 TEMPERATURES BELOW SATURATION TEMPERATURE

We now consider a *multicomponent* gaseous system, i.e., $p_{cg} < p_g$, and temperatures *below* the saturation temperature at the total pressure, i.e., $T(x) < T_s(p_g)$. Figure 12.11 (a) and (b) depict the problem considered. In this problem the *surface saturation* s_1 changes with time, i.e., $s_1 = s_1(t)$. The heat is supplied at $x = L$ by an *ambient* gas flowing at free stream conditions ρ_{cga}, ρ_{nga}, p_{ga}, u_{ga}, and T_a. An interfacial *convection heat transfer coefficient* h_{pa} (s_1, u_a, etc.) and an interfacial *convection mass transfer coefficient* h_{mpa} (s_1, u_a, etc.) are *prescribed*. In Chapter 11 we discussed $h_{pa}(s_1)$ and $h_{mpa}(s_1)$ and showed that in *general* the heat and mass transfer *analogy* does *not* hold for partially saturated surfaces undergoing simultaneous heat and mass transfer. For *large* Knudsen numbers (Section 10.2.1), i.e., very small particles, this analogy holds and these coefficients become independent of s_1 (even for relatively small values of s_1). Even when the surface dries out, we expect the heat and mass transfer analogy to continue to hold for these *small* particle packed beds.

The bed is initially at $T_0 < T_a < T_s$, $s_{ir} < s_0 < 1 - s_{ir\,g}$, and at equilibrium, i.e., $p_{cg}(T_0, s_0)$, where we have included the effect of the curvature on the equilibrium state. The heat supplied at $x = L$ produces partial surface

12.5 Moving Evaporation or Condensation Front

evaporation, and once the surface temperature rises above T_o, the rest of the heat penetrates into the bed and produces *internal* evaporation. When $\partial T/\partial x > 0$, we have $\partial p_{cg}/\partial x > 0$, and the vapor moves *inward*. The amount of vapor per unit area leaving the surface is given by $h_{b\,mp}(\rho_{cga} - \rho_{cg1}) > 0$. The surface evaporation causes a decrease in s_1, which results in $\partial s/\partial x < 0$ and the supply of the liquid to the surface (we will discuss the role of gravity later). The saturation throughout the bed decreases with time with s_1 decreasing the fastest. This decrease in s causes an increase in the resistance to liquid flow and consequently requires that $|\partial s/\partial x|$ become increasingly larger. At the surface, where $|\partial s/\partial x|$ is *largest*, the decrease in s_1 and the increase in $|\partial s/\partial x|_{x=L}$ *eventually* cause s_1 to become equal to s_{ir}. The time at which this occurs is called the *critical time*, thereafter, the mobility of the liquid near the surface is rather *intermittent* and intermittent drying of the surface occurs. This intermittent drying continues for a period. After this period, the surface remains dry and an *evaporation front* develops and penetrates into the medium (evaporation front regime). The vapor moves from the front outward toward the surface through the dry matrix and the noncondensible gases. On the other side, the liquid flows to this front.

This description is rather *idealized* in order to allow for the treatment of this problem as a *stable* one-dimensional transient two-phase flow in porous media. Because of the $T_a < T_s$ *limitation* imposed, the porous plain media interfacial heat and mass transfer coefficients dominantly *control* the process during the funicular state. Therefore, this problem does not make for a *critical* evaluation of the theoretical treatment of two-phase flow in porous media (by comparing the predictions with the experimental results). Also, during the evaporation-front period the heat transfer though the dry region (solid, vapor, and inert gas region) determines the evaporation rate, and the liquid flow in the wet region becomes *less* significant. However, because of technological interests, we present the mathematical treatment of this problem later. This problem has been considered by Luikov (1966), Whitaker (1977), Whitaker and Chou (1983), Plumb et al. (1985), Stanish et al. (1986), Kaviany and Mittal (1987), Ilic and Turner (1989), and Rogers and Kaviany (1991), among many others.

(A) Funicular Regime

In this regime, the liquid phase is connected, and the period is realized from $t = 0$ where $s = s_o(x)$ to $t = t_c$ where $s_1 = s(x = L) = s_{\text{ir}}$. The marking of the end of this period, the *critical time*, will depend on *internal* (partially saturated porous media) and *external* (ambient) variables. Here our goal is to *predict* t_c.

We begin from the momentum equations (8.135) and (8.137), i.e.,

$$0 = -\frac{\partial p_\ell}{\partial x} - \rho_\ell g - \frac{\mu_\ell}{K_\ell} u_\ell + \mu_\ell \frac{K_{\ell\Delta\sigma}}{K_\ell} \frac{\partial \sigma}{\partial x}, \qquad (12.118)$$

12. Phase Change

$$0 = -\frac{\partial p_g}{\partial x} - \rho_g g - \frac{\mu_g}{K_g} u_g. \tag{12.119}$$

In the preceding we have neglected the *macroscopic* and *microscopic* inertial terms, as well as the effect of $\partial \sigma/\partial x$ on the gas phase velocity.

We can write (12.118) and (12.119) as

$$u_\ell = -\frac{K_\ell}{\mu_\ell}\left(\frac{\partial p_\ell}{\partial x} - \rho_\ell g\right) + K_{\ell \Delta \sigma}\frac{\partial \sigma}{\partial x}, \tag{12.120}$$

$$u_g = -\frac{K_g}{\mu_g}\left(\frac{\partial p_g}{\partial x} - \rho_g g\right), \tag{12.121}$$

where u_g is the mass velocity of the mixture as defined by (6.1). The conservation of mass is given by (8.94) and (8.95) as

$$\rho_\ell \varepsilon \frac{\partial s}{\partial t} + \rho_\ell \frac{\partial u_\ell}{\partial x} = -\dot{n}, \tag{12.122}$$

$$\varepsilon \frac{\partial \rho_g(1-s)}{\partial t} + \frac{\partial \rho_g u_g}{\partial x} = \dot{n}, \tag{12.123}$$

where $\dot{n}$ is the volumetric evaporation rate (kg/m³-s). The species conservation equation for the *noncondensible* components of the gas phase is given by the adaptation of (6.63) to the two-phase flows. We treat only the *binary* diffusion and write the species conservation equation by using the superficial gas-phase velocity and the longitudinal *total* effective mass diffusion coefficient, i.e.,

$$\frac{\partial \varepsilon \rho_{ng}(1-s)}{\partial t} + \frac{\partial}{\partial x}\rho_{ng} u_g = \frac{\partial}{\partial x}\varepsilon D_{m\,\|}\rho_g \frac{\partial}{\partial x}\frac{\rho_{ng}}{\rho_g}, \tag{12.124}$$

where because of the low $Pe_g = u_g(K/\varepsilon)^{1/2}/\alpha_g$ encountered, we have

$$D_{m\,\|} = D_{m\,e} + D_{m\,\|}^d \simeq D_{m\,e}(\varepsilon, s, D_m), \tag{12.125}$$

where we have implied that $D_{m\,e} > D_{m\,g}^d > D_{m\,\ell}^d$, i.e., the noncondensibles diffuse more readily in the gaseous phase than in the liquid phase. We use (6.72) and then modify it to account for the presence of the liquid phase. This gives

$$\frac{D_{m\,e}}{D_m} = \frac{2\varepsilon}{3-\varepsilon}(1-s). \tag{12.126}$$

The energy equation is that given by (10.32) as

$$\left[(1-\varepsilon)C + \varepsilon s + \varepsilon(1-s)\frac{(\rho c_p)_{cg} + (\rho c_p)_{ng}}{(\rho c_p)_\ell}\right]\frac{\partial T}{\partial t}$$

$$+\left[u_\ell + \frac{(\rho c_p)_{cg} + (\rho c_p)_{ng}}{(\rho c_p)_\ell}u_g\right]\frac{\partial T}{\partial x} + i_{\ell g}\dot{n} = \frac{\partial}{\partial x}D_\|\frac{\partial T}{\partial x} + \langle \dot{s}\rangle. \quad (12.127)$$

Whenever a supplemental *dielectric* heating exists, we use

$$\langle \dot{s}\rangle = 2\pi f \varepsilon_\ell \varepsilon_o e^2, \quad (12.128)$$

where ε_o is the permittivity of free space ($8.8542 \times 10^{-12} A^2 - s^2/N - m^2$), e is the electric field intensity, f is the frequency, and ε_ℓ is the *dielectric loss factor* (or *constant*) and is a property of the liquid and a function of both saturation and temperature (Jones et al., 1974).

The longitudinal total effective thermal conductivity $D_\|$ is determined as given by (10.46). This gives

$$D_\| = \frac{k_{e\|}}{(\rho c_p)_\ell} + \varepsilon s D^d_{\ell\|} + \varepsilon(1-s)D^d_{g\|} \simeq \frac{k_{e\|}}{(\rho c_p)_\ell}, \quad (12.129)$$

where correlations such as those listed in Table 10.1 are used for $k_{e\|}$.

The gas can be treated as behaving ideally given by (6.2). The vapor is in an equilibrium state and the vapor-pressure reduction and the liquid superheat, if significant, are given by (9.40) and (9.62). The capillary pressure $p_c = p_g - p_\ell$ and the relative permeabilities may be presented in the simplified forms given in Tables 8.3 and 8.4. Presently, the only relation available for $K_{\ell\Delta\sigma}$ is that given by (8.143). Note also that the gradient of the capillary pressure for the case where $\nabla\sigma \neq 0$ takes the form given by (8.145). The boundary conditions are

$$\dot{m}_1 = \rho_\ell u_\ell + \rho_{cg}u_g - \rho_g D_{m\ell}\frac{\partial}{\partial x}\frac{\rho_{cg}}{\rho_g}$$

$$= h_{mpa}(\rho_{cg} - \rho_{cg_o}) \quad \text{at } x = L, \quad (12.130)$$

$$p_{cg} + p_{ng} = p_{ga}, \quad (12.131)$$

$$q_1 = \rho_\ell u_\ell i_{\ell g} + k_{e\|}\frac{\partial T}{\partial x} = h_{pa}(T_a - T) \quad \text{at } x = L. \quad (12.132)$$

In principle, the total pressure, temperature, and vapor concentration are continuous across the $x = L$ interface. However, we have used a *jump* boundary condition for the last two by introducing the *film coefficients*. At $x = 0$, we have

$$\frac{\partial \rho_{cg}}{\partial x} = \frac{\partial \rho_{ng}}{\partial x} = \frac{\partial s}{\partial x} = u_\ell = u_g = \frac{\partial T}{\partial x} = 0, \quad \text{at } x = 0. \quad (12.133)$$

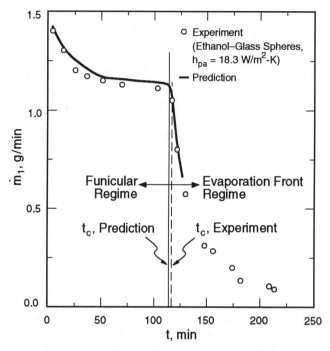

Figure 12.12 Predicted and measured mass transfer rate for a bed of glass spheres initially filled with ethanol and then convectively heated by the ambient air.

The initial conditions are $T = T_o$, and by using $p_g = p_{ga}$, we have the initial saturation distribution $s_o(x)$ given by (12.120) as

$$\frac{dp_c(s_o)}{dx} + (\rho_\ell - \rho_g)g = 0. \tag{12.134}$$

By using the experimental results for h_{mpa} and h_{pa} for a packed bed of 0.10-mm glass spherical particles that are initially saturated with ethanol and then subjected to heated ambient moving air (channel flow), Rogers and Kaviany (1991) measure $\dot{m}(t)$ and $T(x)$. For the funicular regime, the analogy between the interfacial heat and mass transfer remains nearly valid, and the coefficients h_{mpa} and h_{pa} are related through (Bird et al., 1960)

$$h_{mpa} = \frac{h_{pa}}{(\rho c_p)_g L_\ell^{2/3}} \quad \text{for large Knudsen numbers and } s_1 > s_{ir}. \tag{12.135}$$

By treating these coefficients as constants, a nearly constant value of $\dot{m}_1$ is found by solving the equations for T, s, p_g, and ρ_{ng}. This is the so-called *constant drying rate period* and is characteristic of the funicular regime for large Knudsen numbers (Section 11.2.1). Rogers and Kaviany

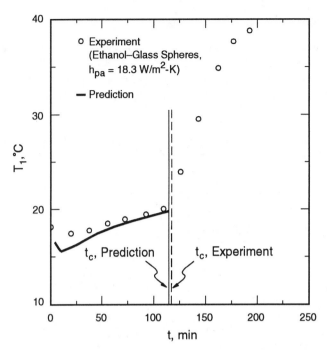

Figure 12.13 Predicted and measured surface temperature variation with respect to time (same conditions as Figure 12.12).

use a finite-volume analysis along with a finite difference approximation of the interfacial conditions and obtain $T(x,t)$, $s(x,t)$, $p_g(x,t)$, $\rho_{ng}(x,t)$, $\dot{m}_1(t)$, and $q_1(t)$.

Their experimental results for $\dot{m}_1(t)$ and $T_1(t)$ are given in Figures 12.12 and 12.13, respectively. Their predicted critical time t_c, i.e., the time of the onset of the surface drying, is also shown in these figures. For $0 \leq t \leq t_c$, their experiment and the analysis show that heat is initially transferred to the surface from the bed as well as the ambient air, resulting in a larger rate of evaporation. Eventually the surface temperature rises and the heat penetrates (against the liquid flow) into the bed and increases the sensible heat of the bed.

The critical time, as predicted by this formulation, is also marked in Figures 12.12 and 12.13 and is in good agreement with the experiments. Note that

$$t = t_c \quad \text{when} \quad s_1 = s_{ir}. \tag{12.136}$$

For $t > t_c$ the surface dries out and the surface temperature rises leading to a smaller heat transfer rate q_1 and a smaller $\dot{m}_1$. This trend is also found in Figures 12.12 and 12.13.

(B) Moving-Interface Regime

We now idealize the events after $t = t_c$ by neglecting the intermittency of the surface wetting (which occurs during a short period after $t = t_c$) and assume that a moving evaporation front is formed right after $t = t_c$. Across this interface, $x = x_i$, we impose the conservation of energy and species, i.e.,

$$k_{e\parallel} \left.\frac{\partial T}{\partial x}\right|_{x_i^+} = k_{e\parallel} \left.\frac{\partial T}{\partial x}\right|_{x_i^-} + \Delta i_{\ell g} \rho_{cg} \left[u_{cg}(x_i^+) - u_{cg}(x_i^-)\right], \qquad (12.137)$$

$$\rho_{cg}\left[u_{cg}(x_i^+) - u_{cg}(x_i^-)\right] = \rho_\ell u_\ell(x_i^-) - \varepsilon s \frac{\partial x_i}{\partial t}, \qquad (12.138)$$

$$\rho_{ng} u_g(x_i^+) - \rho_g D_{m\parallel} \frac{\partial}{\partial x} \frac{\rho_{ng}(x_i^+)}{\rho_g(x_i^+)}$$

$$= \rho_{ng} u_g(x_i^-) - \rho_g D_{m\parallel} \frac{\partial}{\partial x} \frac{\rho_{ng}(x_i^-)}{\rho_g(x_i^-)}, \qquad (12.139)$$

where from (6.35)

$$u_{cg} = u_g - \frac{\rho_g}{\rho_{cg}} D_{m\parallel} \frac{\partial}{\partial x} \frac{\rho_{cg}}{\rho_g}. \qquad (12.140)$$

The saturation at $x = x_i$ is s_{ir}. The interfacial temperature can be obtained from (12.137), the interfacial position from (12.138), and the interfacial value of the density of the noncondensible from (12.140). The vapor is treated as being at equilibrium at $x = x_i$, but superheated for $x > x_i$. The saturation is zero, and the single-phase gaseous flow and heat transfer occur (with both diffusion and convection) for $x > x_i$.

The predictions of Rogers and Kaviany for $\dot{m}_1(t)$, at the early stages of the moving interface regime is also given in Figure 12.12. The rates of heat and mass transfer are controlled by the heat diffusion inward (toward $x = x_i$) and the mass diffusion and convection outward (from $x = x_i$). Note the sharp drop in the mass flow rate shown in Figure 12.12 (and the increase in the surface temperature shown in Figure 12.13). This is due to the existence of the dry region.

12.5.3 Condensation Front Moving into Dry Porous Media

In contast to fronts seperating semi-infinite, single-phase domains discussed in Section 12.5.1, we now consider a condensation front moving into a dry porous media and forming behind it a partially liquid-saturated region. Also, in this class of vapor injection problems considered, the *inlet pressure*, instead of the vapor flow rate, is maintained constant. Because of the strong role of gravity on the saturation distribution in the two-phase region,

12.5 Moving Evaporation or Condensation Front

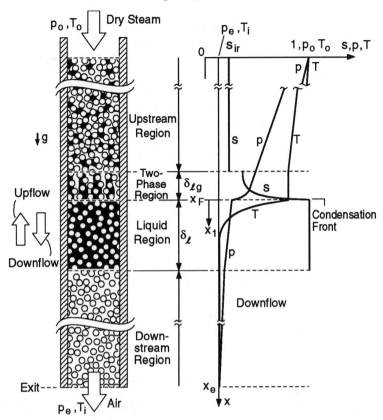

Figure 12.14 Anticipated axial distributions of the temperature, pressure, and liquid saturation, and a rendering of the various regions along the packed column (for the downflow).

interesting differences are found between the *upflow* (against gravity) and *downflow* (along gravity) of the vapor. The problem has been analyzed by Hanamura and Kaviany (1995) for steam injection into initially air filtered packed beds with initial temperature below T_s.

Figure 12.14 shows the model used by Hanamura and Kaviany (1994) for the downflow of steam with the anticipated axial distributions of the temperature, pressure, and liquid saturation within the various defined regions. The porous medium is initially dry and at a temperature $T_i < T_s$. Four different regions are defined, namely, the *upstream*, followed by *two-phase*, *liquid*, and *downstream regions*. The inlet pressure is p_o, and the inlet thermodynamic quality is assumed unity.

Steam condensation is assumed to occur at the condensation zone (of a negligibly small thickness) x_F located between the liquid and two-phase regions. In the liquid region, the liquid saturation is assumed as unity. In the two-phase region, the liquid saturation varies, both phases are mobile, and the capillary pressure is nonzero. No liquid motion is allowed in the upstream region. Although the condensation-front speed is not constant, under the condition of a constant inlet pressure, a quasi-steady state behavior is assumed through all the regions. This is justifiable when the change in front speed is small with respect to time and will be discussed later. The gas, liquid, and solid phases are assumed to be in local thermal equilibrium and the front is assumed to be stable and physical properties are assumed constant.

The dimensionless, two-phase flow conservation equations for the liquid- and gas-phase mass and momentum, and for the thermal energy, obtained from (8.94), (8.95), (8.135), (8.136), and (10.32), are

$$\varepsilon Bo \frac{\partial s}{\partial t^*} + \frac{\partial Ca_\ell}{\partial x^*} = \dot{n}_\ell^*, \qquad (12.141)$$

$$-\varepsilon Bo \frac{\partial s}{\partial t^*} + \mu^* \frac{\partial Ca_g}{\partial x^*} = \dot{n}_g^*, \qquad (12.142)$$

$$\frac{\partial p_\ell^*}{\partial x^*} = \frac{Bo_\ell}{Bo} - \frac{Ca_\ell}{Bo} \left(\frac{1}{K_{r\ell}} + \frac{C_E Re_\ell}{K_{r\ell i}} \right), \qquad (12.143)$$

$$\frac{\partial p_g^*}{\partial x^*} = \frac{Bo_g}{Bo} - \frac{Ca_g}{Bo} \left(\frac{1}{K_{rg}} + \frac{C_E Re_g}{K_{rgi}} \right), \qquad (12.144)$$

$$K_{r\ell} = S^3, \quad K_{rg} = (1-S)^3, \quad S = \frac{s - s_{ir}}{1 - s_{ir}}, \qquad (12.145)$$

$$K_{r\ell i} = s^6, \quad K_{rgi} = (1-s)^6, \qquad (12.146)$$

$$[(1-\varepsilon) + \varepsilon s(\rho c_p)_\ell^* + \varepsilon(1-s)(\rho c_p)_g^*] Bo \frac{\partial T^*}{\partial t^*}$$
$$+ [sCa_\ell(\rho c_p)_\ell^* + (1-s)Ca_g(\rho c_p)_g^* \mu^*] \frac{\partial T^*}{\partial x^*}$$
$$= \left[s + (1-s)(\rho c_p)_{g\ell}^* \frac{D_g}{D_\ell} \right] (\rho c_p)_\ell^* \frac{Bo}{(Pe_\ell)_m} \frac{\partial^2 T^*}{\partial x^{*2}} + \frac{\dot{n}_\ell^*}{Ja}, \qquad (12.147)$$

where the normalized length, time, temperature, and pressure and the dimensionless parameters are

$$x^* = \frac{xBo}{(K\varepsilon)^{1/2}}, \quad t^* = \frac{tBo^2\sigma}{(K\varepsilon)^{1/2}\mu_\ell}, \qquad (12.148)$$

$$T^* = \frac{T - T_i}{T_{F,sat} - T_i} = \frac{T - T_i}{\Delta T}, \qquad (12.149)$$

12.5 Moving Evaporation or Condensation Front

$$p_\ell^* = \frac{p_\ell - p_e}{\sigma}\left(\frac{K}{\varepsilon}\right)^{1/2}, \quad p_g^* = \frac{p_g - p_e}{\sigma}\left(\frac{K}{\varepsilon}\right)^{1/2}, \qquad (12.150)$$

$$Ca_\ell = \frac{\mu_\ell u_\ell}{\sigma}, \quad Ca_g = \frac{\mu_g u_g}{\sigma}, \qquad (12.151)$$

$$Re_\ell = \frac{\rho_\ell u_\ell K^{1/2}}{\mu_\ell}, \quad Re_g = \frac{\rho_g u_g K^{1/2}}{\mu_g}, \qquad (12.152)$$

$$Bo = \frac{(\rho_\ell - \rho_g)gK}{\sigma}, \quad Bo_\ell = \frac{\rho_\ell gK}{\sigma}, \quad Bo_g = \frac{\rho_g gK}{\sigma}, \quad \rho^* = \frac{\rho_g}{\rho_\ell}, \qquad (12.153)$$

$$(\rho c_p)_\ell^* = \frac{(\rho c_p)_\ell}{(\rho c_p)_s}, \quad (\rho c_p)_g^* = \frac{(\rho c_p)_g}{(\rho c_p)_s}, \quad (\rho c_p)_{g\ell}^* = \frac{(\rho c_p)_g}{(\rho c_p)_\ell}, \qquad (12.154)$$

$$\mu^* = \frac{\mu_\ell}{\mu_g}, \quad (Pe_\ell)_m = \frac{(K\varepsilon)^{1/2}\sigma}{D_\ell \mu_\ell}, \quad Ja = \frac{c_{p_\ell}\Delta T}{\Delta i_{\ell g}}\frac{1}{(\rho c_p)_\ell^*}. \qquad (12.155)$$

The modified Péclet number $(Pe_\ell)_m$ is based on the total axial diffusivity and the square root of the absolute permeability, and ΔT is the difference between the initial temperature and the saturation temperature evaluated at front pressure. For the upflows, the sign of the gravity term is changed.

(A) UPSTREAM REGION, $0 \leq< x^* \leq x_F^* - \delta_{\ell g}^*$

In this region the liquid remains in the pores after the passing of the front and is considered immobile; no phase change occurs and the capillary pressure is assumed constant. Then we have

$$\dot{n}_\ell^* = \dot{n}_g^* = 0, \quad u_\ell = 0 \text{ (i.e., } Ca_\ell = 0\text{)}, \quad u_g = u_{g,o} \text{ (i.e., } Ca_g = Ca_{g,o}\text{)},$$

$$p_c^* = p_g^* - p_\ell^* = \text{constant}, \quad T^* = T_{sat}^*(p_g). \qquad (12.156)$$

Since $p_c^* = p_g^* - p_\ell^*$ is constant, subtracting (12.143) from (12.144) will produce

$$\frac{\partial p_c^*}{\partial x^*} = -1 - \frac{Ca_{g,o}}{Bo}\left(\frac{1}{K_{rg}} + \frac{C_E Re_{g,o}}{K_{rgi}}\right) = 0 \quad (s \geq s_{ir}). \qquad (12.157)$$

Equation (12.157) is used to determine the continuous but immobile liquid saturation $s_{im} = s_{im}(x^*)$, and as evident, this distribution depends on the ratio of the Bond and capillary numbers. When there does not exist a real positive solution for s_{im}, then it is assumed that $s_{im} = s_{ir}$. As will be shown, this is the case for the downflow. However, for the upflow s_{im} varies with x^* starting from s_{ir} at the first transition point $x_{tr,1}^*$ and increasing monotonically until the second transition point $x_{tr,2}^*$. Further, the saturation gradient (i.e., the capillary pressure gradient) is negligibly small.

(B) TWO-PHASE REGION, $x_F^* - \delta_{\ell g}^* \leq x^* \leq x_F^*$

In this region, both the liquid and vapor are mobile, and no phase change occurs.

$$\dot{n}_\ell^* = \dot{n}_g^* = 0, \quad T^* = 1. \tag{12.158}$$

The propagation velocity of the condensation front can be regarded as the pore velocity of the liquid at the front. Using a moving coordinate system based on the condensation-front speed, i.e., $x_1^* = x^* - (u_F^*/Bo) t^*$ (where $u_F^* = u_F \mu_\ell/\sigma$), (12.141) and (12.142) are rewritten as

$$\frac{\partial}{\partial x_1^*}(Ca_\ell - \varepsilon s u_F^*) = 0, \tag{12.159}$$

$$\frac{\partial}{\partial x_1^*}(\mu^* Ca_g + \varepsilon s u_F^*) = 0. \tag{12.160}$$

Using $Ca_\ell = \varepsilon u_F^*$ (i.e., $u_\ell = \varepsilon u_F$) and $\mu^* Ca_g = \mu^* Ca_{g,o} - \varepsilon u_F^*$ (i.e., $u_g = u_{g,o} - \varepsilon u_F$) at the front, where $s = 1$, the solutions to the above equations are

$$Ca_\ell = \varepsilon s u_F^*, \tag{12.161}$$

$$\mu^* Ca_g = \mu^* Ca_{g,o} - \varepsilon s u_F^*. \tag{12.162}$$

Writing the capillary pressure in terms of the Leverett J-function, the liquid saturation distribution is determined from the following equation, which is derived by subtracting (12.143) from (12.144).

$$\frac{\partial S}{\partial x_1^*} = -\frac{\frac{Ca_g}{Bo}\left[\left(\frac{1}{K_{rg}} + \frac{C_E Re_g}{K_{rgi}}\right) - \frac{Ca_\ell}{Ca_g}\left(\frac{1}{K_{r\ell}} + \frac{C_E Re_\ell}{K_{r\ell i}}\right) + \frac{Bo}{Ca_g}\right]}{\left(\frac{dJ(S)}{dS}\right)}, \tag{12.163}$$

where

$$p_c^* = \frac{p_c}{\sigma}\left(\frac{K}{\varepsilon}\right)^{1/2} = J(S), \tag{12.164}$$

12.5 Moving Evaporation or Condensation Front

$$J(S) = 1.417(1-S) - 2.120(1-S)^2 + 1.263(1-S)^3. \quad (12.165)$$

The boundary condition for the scaled liquid saturation is $S = 1$ at the condensation front.

In the two-phase region the scaled liquid saturation increases monotonically from S_c at $x_1^* = -\delta_{\ell g}^*$ to $S = 1$ at the front, where S_c is the critical liquid saturation and is obtained by imposing the condition of a zero liquid saturation gradient, i.e., $dS/dx_1^* = 0$. This corresponds to setting the numerator of (12.163) equal to zero. In the second term in the numerator, the relative permeabilities $K_{r\ell}$ and $K_{r\ell i}$ are proportional to the third and sixth powers of liquid saturation, respectively, while the liquid velocity u_ℓ (i.e., Ca_ℓ) is proportional to the first power of the liquid saturation. Since the magnitude of the second term in (12.163) increases with a decrease in the liquid saturation, the magnitude of the critical absolute liquid saturation is always greater than s_{ir}.

(C) CONDENSATION FRONT, $x^* = x_F^*$

At the front, the energy released due to phase change is transferred to the condensed liquid, i.e.,

$$-\frac{Bo}{(Pe_\ell)_m} Ja(\rho c_p)_\ell^* \frac{\partial T^*}{\partial x_1^*} \bigg|_{x_1^* = 0+} = \rho^*(\mu^* Ca_{g,o} - \varepsilon u_F^*). \quad (12.166)$$

Through the integration of (12.172) given below, this heat flux is also given through

$$(1-\varepsilon)u_F^* - \varepsilon(\rho c_p)_\ell^* \Delta Ca_\ell = -\frac{Bo}{(Pe_\ell)_m}(\rho c_p)_\ell^* \frac{\partial T^*}{\partial x_1^*}\bigg|_{x_1^* = 0+}, \quad (12.167)$$

The second term on the left-hand side of equation (12.167) represents the sensible heat of the condensed liquid, ΔCa_ℓ is determined from (12.171) below. Then $Ca_{g,o}$ (i.e., $u_{g,o}$) is determined from the (12.166) and (12.167), which give

$$\rho^*(\mu^* Ca_{g,o} - \varepsilon u_F^*) = Ja\left[(1-\varepsilon)u_F^* - \varepsilon(\rho c_p)_\ell^* \Delta Ca_\ell\right]. \quad (12.168)$$

(D) LIQUID REGION, $x_F^* \leq x^* \leq x_F^* + \delta_\ell^*$

In the liquid region the porous medium is completely saturated with the subcooled liquid, and we have

$$s = 1, \quad \dot{n}_\ell^* = \dot{n}_g^* = 0, \quad \mu_\ell = \mu_{\ell,i}. \quad (12.169)$$

The thickness of the liquid region δ_ℓ^*, which begins at x_F^*, is determined from the total mass balance and is

$$\delta_\ell^* = Ja\, x_F^* \frac{1-\varepsilon}{\varepsilon} - \int_0^{x_F^*} s\, dx^*. \quad (12.170)$$

The pore liquid velocity is the sum of the front velocity and the rate of increase in the thickness of the liquid region. Then, the superficial velocity u_ℓ (i.e., Ca_ℓ) and ΔCa_ℓ are determined from

$$Ca_\ell = \varepsilon(u_F^* + \Delta Ca_\ell) = \varepsilon\left(u_F^* + \frac{\partial \delta_\ell^*}{\partial t^*}\right)$$

$$= \varepsilon u_F^* \left(1 + Ja\frac{1-\varepsilon}{\varepsilon} - \frac{1}{u_F^*}\frac{\partial}{\partial t^*}\int_0^{x_F^*} s\, dx^*\right). \tag{12.171}$$

The energy equation is also transformed to the moving coordinate system and gives

$$\{-[(1-\varepsilon) + \varepsilon(\rho c_p)_\ell^*]\, u_F^* + (\rho c_p)_\ell^* Ca_\ell\}\frac{\partial T^*}{\partial x_1^*}$$

$$= (\rho c_p)_\ell^* \frac{Bo}{(Pe_\ell)_m}\frac{\partial^2 T^*}{\partial x_1^{*2}}. \tag{12.172}$$

Integrating this for $0 \leq x^* \leq \infty$ and using the zero gradient conditions in the far-field, gives (12.167). The solution to equation (12.172), subject to the boundary conditions $T^* = 1$ at $x_1^* = 0$ and $T^* \to 0$ as $x_1^* \to \infty$, is

$$T^* = \exp\left[-(Pe_\ell)_m[(1-\varepsilon) + \varepsilon(\rho c_p)_\ell^*]\right.$$

$$\left. \times \left(Ja\frac{1-\varepsilon}{\varepsilon} - \frac{1}{u_F^*}\frac{\partial}{\partial t^*}\int_0^{x_F^*} s\, dx^*\right)\frac{u_F^* x_1^*}{Bo}\right]. \tag{12.173}$$

The total axial diffusivity D_ℓ is related to the liquid Péclet number Pe_ℓ and the effective conductivity k_e through the empirical relation given in Chapter 4, i.e.,

$$\frac{D_\ell}{\alpha_\ell} = \frac{k_e}{k_\ell} + 0.5 Pe_\ell, \quad Pe_\ell = \frac{u_\ell R}{\alpha_\ell}, \tag{12.174}$$

where α_ℓ is the thermal diffusivity of the liquid and R is the radius of the particle. The effective conductivity used is the empirical relation of Krupiczka given in Chapter 3.

(E) Downstream Region, $x_F^* + \delta_\ell^* \leq x^* \leq x_e^*$

There is no liquid in the downstream region, and at $x^* = x_F^* + \delta_\ell^*$ the gas velocity is assumed to be equal to the liquid velocity. This region is described by

12.5 Moving Evaporation or Condensation Front   651

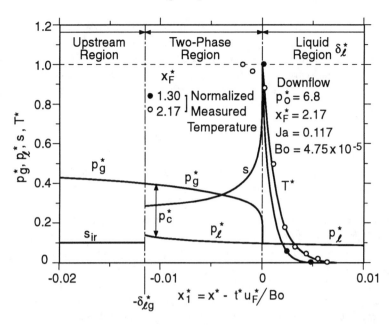

Figure 12.15 Predicted axial (in the moving coordinate x_1^*) distributions of the normalized temperature, liquid saturation, and gas- and liquid-phase pressures. The results are for the downflow and $x_F^* = 2.17$. The results for the temperature at $x_F^* = 1.30$ are also included.

$$\dot{n}_g^* = 0, \quad u_g = u_\ell(x^* = x_F^* + \delta_\ell^*)(\text{i.e.}, Ca_g = Ca_\ell), \quad K_{rg} = K_{rgi} = 1,$$

$$p_g^*(x_F^* + \delta_\ell^*) = p_\ell^*(x_F^* + \delta_\ell^*)(\text{i.e., no meniscus-curvature effect}),$$

$$\mu_g = \mu_a, \quad \rho_g = \rho_a. \tag{12.175}$$

Hanamura and Kaviany solve the above governing equations numerically and also perform some experiments with steam injection into a pack bed of glass particles.

Figure 12.15 shows some typical distributions of the temperature, liquid saturation, and pressure (gas- and liquid-phase) around the condensation front, where the abscissa is the moving coordinate x_1^*. The numerical results for this front structure are for a *downflow*, $x_F^* = 2.17$, and $p_o^* = 6.8$. The temperature distribution for the same condition, but for $x_F^* = 1.30$, is also depicted. The normalized temperature, normalized using the initial temperature and the saturation temperature at the front, drops rapidly within a short distance in the liquid region. Note that the measured temperature is averaged over the width of the thermocouple junction (diameter of about 0.5 mm). A good agreement is found between the experimental and numerical results. The spatial origin in the experimental results is adjusted to follow the numerical results in the liquid region, because, due to the small pressure variations around the front, the exact front location cannot be determined experimentally. A direct comparison between the experimental and numerical results is not made for the case x_F^* approaching zero, because the front speed is too large for an accurate transient measurement of temperature. Note that a quasi-steady propagation is assumed. In the transformation to the moving coordinate system, the variation of u_F^* with time can be included as

$$\frac{\partial x_1^*}{\partial x^*} = 1 - \frac{\partial u_F^*}{\partial x^*}\frac{t^*}{Bo}, \quad \frac{\partial x_1^*}{\partial t^*} = -\frac{u_F^*}{Bo} - \frac{\partial u_F^*}{\partial t^*}\frac{t^*}{Bo}. \qquad (12.176)$$

The difference between the spatial temperature distributions obtained with and without this variation is only about 10 percent (at most). This close agreement between the experimental and predicted results indicates that the assumption of a quasi-steady state behavior at any front location is rather justifiable. For the downflow, (12.157) has no real positive solutions (for any ratio of Bo to $Ca_{g,o}$). In the two-phase region, the absolute liquid saturation increases sharply from that close to the critical liquid saturation to unity. The second term on the right-hand side of (12.163) increases with a decrease in liquid saturation. As a result, the critical liquid saturation is not equal to the irreducible liquid saturation, i.e., there is a saturation jump at the beginning of the two-phase region. As the liquid saturation increases, the gas-phase pressure drastically decreases along the flow direction. The liquid-phase pressure is lower than the gas-phase pressure by the magnitude of the capillary pressure p_c^*.

Figure 12.16 shows the predicted axial distribution of the liquid saturation for various front locations and for the *upflow* with $p_o^* = 1.6$. For the downflow, the liquid saturation in the upstream region is constant, i.e., $s = s_{ir}$, as shown in Figure 12.14. As a result, δ_ℓ^* increases almost linearly with x_F^*. The predicted δ_ℓ^* for $p_o^* = 6.8$ is in good agreement with the experimental results (obtained through visual observations and by measurement of the accumulated amount of the condensate flow). For the upflow, the liquid saturation in the upstream region is maintained at s_{ir} for u_F^*/Bo higher than a critical velocity (equal to 0.177), which occurs at a location referred to as the first (i.e., the discontinuous irreducible to continuous immobile liquid saturation) transition and corresponds to $Bo/Ca_{g,o} = 1.18$. Note that $Bo/Ca_{g,o}$, which is independent of Ja, increases with Bo, because of the increase in the significance of the inertial term. As $Bo \to 0$, the asymptote will be $Bo/Ca_{g,o} = 1$. After the front passes through this first transition point $x_{tr,1}^*$, the liquid saturation in the upstream region begins to increase with increase in x_F^*. This saturation distribution, $s_{im} = s_{im}(x_F^*)$, is determined from the balance between the gravity force and the sum of the viscous and inertial forces in (12.157). Then, the thickness of the liquid region begins to decrease and eventually the liquid region disappears at a location referred to as the second

12.5 Moving Evaporation or Condensation Front

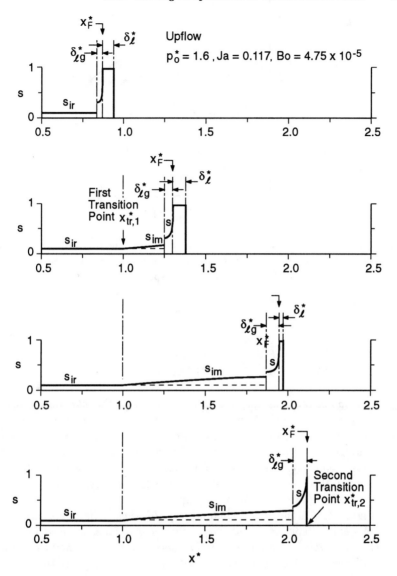

Figure 12.16 Predicted axial distributions of the liquid saturation for the upflow and for various front locations. The results are for $p_o^* = 1.6$.

transition $x_{tr,2}^*$ (where $\delta_\ell^* = 0$). The numerical integration is halted at $x_{tr,2}^*$. Therefore, for the upflow, as observed experimentally, no liquid region passes through the top end of this packed column. The thickness of the two-phase region increases monotonically with increase in x_F^*, since the capillary force becomes dominant as u_F^* decreases.

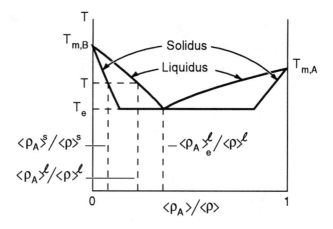

Figure 12.17 The thermodynamic equilbrium phase diagram for a binary solid-liquid system. The eutectic temperature and species A mass fraction, and a dendritic temperature and liquidus and solidus species A mass fractions, are also shown.

12.6 Melting and Solidification

In *single-component systems* (or *pure substances*) the *chemical composition* in all phases are the *same*. In *multicomponent systems*, the chemical composition of a given phase changes in response to *pressure* and *temperature* changes and these compositions are *not* the same in all phases. For *single-component systems*, *first-order phase transitions* occur with a discontinuity in the first derivative of the Gibbs free energy, given by (9.4). In the transitions, T and p remain constant.

In Figure 12.17, the *thermodynamic equilibrium, solid-liquid phase diagram* of a *binary* (species A and B) system is shown for a *nonideal solid* solution (i.e., *miscible* liquid but *immiscible* solid phase). The melting temperatures of pure substances are shown with $T_{m,A}$ and $T_{m,B}$. At the *eutectic-point* mole fraction, designated by the subscript e, both solid and liquid can *coexist* at *equilibrium*. In this diagram the *liquidus* and *solidus* lines are approximated as *straight lines*. A *dendrictic temperature* T and the *dendritic mass fractions* of species $\langle \rho \rangle^s / \langle \rho \rangle^s$ and $\langle \rho \rangle^\ell / \langle \rho \rangle^\ell$, are also shown. The *equilibrium partition ratio* k_p is used to relate the solid- and liquid-phase mass fractions of species $\langle \rho \rangle^s / \langle \rho \rangle^s$ and $\langle \rho \rangle^\ell / \langle \rho \rangle^\ell$ on the liquidus and solidus lines, at a given temperature and pressure, i.e.,

$$k_p = \left. \frac{\langle \rho_A \rangle^s / \langle \rho \rangle^s}{\langle \rho_A \rangle^\ell / \langle \rho \rangle^\ell} \right|_{T,p}. \qquad (12.177)$$

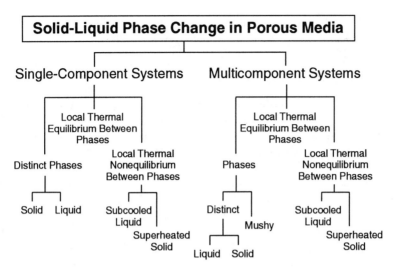

Figure 12.18 A classification of solid-liquid phase change in porous media.

A finite, *two-phase region* (called the *mushy region*) can exist for $k_p < 1$ and corresponds to the case where species A has a *limited solubility* in the solid phase. For $k_p = 1$ a discrete phase change occurs with no mushy region. This mushy region is a solid-liquid mixture where it is generally assumed that the solid phase is *continuous* and therefore treated as a permeable solid, i.e., a *porous medium* with the liquid being capable of motion through the porous media.

Figure 12.18 gives a classification of the solid-liquid phase change in porous media. For single-component systems at a given pressure, melting or solidification as a *distinct phase change* is assumed at a saturation temperature corresponding to the pressure, i.e., $T_{m,A}$ for species A. Although at the interface between the solid and liquid this saturation condition holds, the *local, bulk phases* (i.e., away from the interface) may *not* be at this temperature, and therefore, a *subcooled liquid* (during *solidification*) or a *superheated solid* (during *melting*) may be assumed. For multicomponent systems in addition to the distinct liquid and solid phases, a mushy (two-phase) region also exists. We note that the solid phase may *not* only contain the phase-change substance but can contain an *inert* (not changing phase) solid substance. For example, during solidification of a multicomponent liquid (i.e., molten) when the pore space of a solid matrix is filled with a much higher melting temperature. During this process the solid fraction increases due to the formation of both a solid phase and a mushy region from the liquid phase (e.g., Singh and Behrendt, 1994).

A review of melting and solidification of single-component systems is given below, followed by a discussion of the multicomponent systems. A more extensive treatment is given by Kaviany (1994).

12.6.1 Single-Component Systems

As an example of solid-liquid phase change in porous media, we consider *melting* of the solid matrix by *flow* of a *superheated liquid* through it. The analysis, based on local *thermal nonequilibrium* between solid and liquid phases, has been performed by Plumb (1994) and is reviewed below.

Because of the phase change, the solid-liquid interfacial location changes and this interfacial mass transfer $\langle \dot{n} \rangle$ must be included [as done in the treatment of two-phase flow and heat transfer (with phase change) in Sections 8.3 and 10.1]. The lack of local thermal equilibrium and introduction of local solid and liquid temperature, $\langle T \rangle^s$ and $\langle T \rangle^\ell$, can be addressed similar to the two-medium thermal treatment (without phase change) discussed in Chapter 7. For *constant* solid and liquid *densities* and starting with the continuity equation, as given by (8.94), but allowing for the interfacial mass transfer, we have

$$\frac{\partial \langle 1-\varepsilon \rangle}{\partial t} + \nabla \cdot \langle \mathbf{u}_s \rangle + \frac{1}{\rho_s V} \int_{s\ell} \rho_s (\mathbf{u}_s - \mathbf{w}) \cdot \mathbf{n}_{s\ell} \, dA = 0, \qquad (12.178)$$

where $\mathbf{w}$ is the interfacial velocity. For the liquid-phase mass conservation, we have

$$\frac{\partial \varepsilon}{\partial t} + \nabla \cdot \langle \mathbf{u}_\ell \rangle + \frac{1}{\rho_\ell V} \int_{s\ell} \rho_\ell (\mathbf{u}_\ell - \mathbf{w}) \cdot \mathbf{n}_{\ell s} \, dA = 0. \qquad (12.179)$$

Now as in (10.24), define the *volumetric rate of phase change* (i.e., *melting*) as

$$\langle \dot{n} \rangle \equiv -\frac{1}{V} \int \rho_\ell (\mathbf{u}_\ell - \mathbf{w}) \cdot \mathbf{n}_{\ell s} \, dA. \qquad (12.180)$$

The interfacial mass balance written for the solid-liquid system, is similar to that for the liquid-gas system, i.e., (10.6), and we have

$$\rho_\ell (\mathbf{u}_\ell - \mathbf{w}) \cdot \mathbf{n}_{\ell s} = \rho_s (\mathbf{u}_s - \mathbf{w}) \cdot \mathbf{n}_{s\ell}. \qquad (12.181)$$

Here we assume that the *solid is stationary*, i.e., $\mathbf{u}_s = 0$. Using (12.178) and (12.179) in (12.180) and (12.181), we have the phasic mass conservation equations as

$$\rho_s \frac{\partial (1-\varepsilon)}{\partial t} = -\langle \dot{n}_\ell \rangle, \qquad (12.182)$$

$$\rho_\ell \frac{\partial \varepsilon}{\partial t} + \rho_\ell \nabla \cdot \langle \mathbf{u}_\ell \rangle = \langle \dot{n}_\ell \rangle. \qquad (12.183)$$

12.6 Melting and Solidification

For the phasic energy equations, we start from (10.13) and (10.14), with no gas phase and heat source. The approximations are introduced and as expected the most rigorous treatment involves the local nonequilibrium treatment of Chapter 7 with the phase change treatment of Chapter 10. Plumb suggests *simplified, semiheuristic* phasic energy equations. These are based on the assumption that the solid is locally at the *melting temperature* T_m which makes the solid-phase energy equation *trivial*, and for a *one-dimensional* transport gives

$$\frac{\partial \varepsilon \langle T \rangle^\ell}{\partial t} - \frac{\langle \dot{n}_\ell \rangle T_m}{\rho_\ell} + \frac{\partial}{\partial x} \varepsilon \langle T \rangle^\ell \langle u_\ell \rangle$$

$$= \frac{\partial}{\partial x} \left[\frac{\langle k \rangle}{(\rho c_p)_\ell} + \varepsilon D^d_{xx} \right] \frac{\partial \langle T \rangle^\ell}{\partial x} - \frac{h_{s\ell}}{(\rho c_p)_\ell} \frac{A_{\ell s}}{V} (\langle T \rangle^\ell - T_m), \quad (12.184)$$

$$\langle T \rangle^s = T_m \quad (12.185)$$

$$\langle \dot{n}_\ell \rangle \Delta i_{\ell s} = h_{sf} \frac{A_{\ell s}}{V} (\langle T \rangle^\ell - T_m). \quad (12.186)$$

When (12.184) and (12.186) are combined, we have

$$\frac{\partial \langle T \rangle^\ell}{\partial t} + \langle u_\ell \rangle \frac{\partial \langle T \rangle^\ell}{\partial x} = \frac{\partial}{\partial x} \left[\frac{\langle k \rangle}{(\rho c_p)_\ell} + \varepsilon D^d_{xx} \right] \frac{\partial \langle T \rangle^\ell}{\partial x}$$

$$- \left[\frac{h_{s\ell}}{(\rho c_p)_{\ell \varepsilon}} \frac{A_{\ell s}}{V} + \langle \dot{n}_\ell \rangle c_{p_\ell} \varepsilon \right] (\langle T \rangle^\ell - T_m). \quad (12.187)$$

In (12.187), $\langle k \rangle$ is the effective conductivity given in Section 3.6, D^d_{xx} is the axial dispersion coefficient given in Section 4.8.4, and $h_{s\ell}$ is the interfacial heat transfer coefficient discussed in Section 7.2.2.

Using the above conservation equations and for an adiabatic system (i.e., no heat losses), subject to a *prescribed* inlet liquid velocity and liquid superheat, $\langle T \rangle^\ell_o - T_m$, flowing into a wettable solid matrix with porosity ε_o, Plumb determines the *porosity distribution* in the melting front. The *approximate melt-front speed* is determined from the overall energy balance and by *neglecting the axial conduction* and is

$$u_F = \frac{\rho_\ell c_{p_\ell} (\langle T \rangle^\ell_o - T_m) \langle u_\ell \rangle_o}{\rho_s \Delta i_{\ell s} \left[(1 - \varepsilon)_o + \dfrac{\rho_\ell}{\rho_s} \dfrac{c_{p_\ell} (\langle T \rangle^\ell_o - T_m)}{\Delta i_{\ell s}} \right]}. \quad (12.188)$$

The numerical results show that the thickness of the melting front is proportional to the liquid velocity to a power of 0.4. At low velocities the *melt-front thickness* can become nearly the same as the pore (or particle) size and at very low velocities, diffusion dominates the axial heat transfer.

12.6.2 MULTICOMPONENT SYSTEMS

Melting and solidification in multicomponent systems is of interest in geological and engineering applications. In solidification, *dendritic growth* of crystals has been analyzed under the assumption of *thermal* and *chemical equilibrium* (as a *columnar dendritic* growth) and also under the assumption of *nonequilibrium*, i.e., *liquid subcooling* (as a dispersed *equiaxial dendritic* growth). The process of solidification of multicomponent liquids is discussed by Kurz and Fisher (1992) and a review with the geological applications is given by Hupport (1990) and with engineering applications by Beckermann and Viskanta (1993). A more extensive review is given by Kaviany (1994) and excerpts of this review are given here. The melting of multicomponent solids is reviewed by Woods (1992). Here brief reviews of the equilibrium and nonequilibrium treatments are given below.

(A) EQUILIBRIUM TREATMENT OF SOLIDIFICATION

As an example of liquid-solid phase change in solid-fluid flow systems with the assumption of local thermal equilibrium imposed, consider the formulation of solid-fluid phase change (solidification/melting or sublimation/frosting) of a *binary* mixture. For this problem, the equilibrium condition extends to the local *thermodynamic equilibrium* where the local phasic temperature (*thermal equilibrium*), pressure (*mechanical equilibrium*), and chemical potential (*chemical equilibrium*) are assumed to be equal between the solid and the fluid phases. This is stated as

$$\langle T \rangle^s = \langle T \rangle^f, \tag{12.189}$$

$$\langle p \rangle^s = \langle p \rangle^f, \tag{12.190}$$

$$\langle \mu_A \rangle^s = \langle \mu_A \rangle^f, \quad \langle p \rangle^s = \langle p \rangle^f. \tag{12.191}$$

The local volume averaging of the energy equations, with allowance for a phase change in a binary system, has been discussed by Bennon and Incropera (1987), Rappaz and Voller (1990), Poirier et al. (1991), and Hills et al. (1992). In their *single-medium* treatment, i.e., a local volume-averaged description with the assumption of local thermodynamic equilibrium, a distinct interface between the *region* of *solid phase* and the *region* of *fluid phase* has not been assumed, instead, the fluid-phase volume fraction (i.e., the porosity) is allowed to change *continuously*. This *continuous medium* treatment is in contrast to the *multiple-medium treatment* which allows for separate *solid*, *fluid*, and *mushy* (i.e., solid–fluid) media (e.g., Worster, 1991; Kim and Kaviany, 1992; Vodak et al., 1992). Since the *explicit* tracking of the various distinct interfaces, as defined in the multiple-medium treatment, is *not* needed in the continuous single-medium treatment, it is

12.6 Melting and Solidification

easier to implement. Also, for binary systems the equilibrium temperature depends on the local species concentrations and due to the variation of the concentration within the medium, the phase transition occurs over a range of temperatures, and therefore, the single-medium treatment is even more suitable. In the following, the single-medium treatment of Bennen and Incropera is reviewed. Many simplifications made in the development of this treatment are discussed by Hills et al. Alternative derivations and assumptions are discussed by Rappaz and Voller and Poirier et al., among others.

The energy equation for each phase and for the case of negligible viscous dissipation and radiative heat flux is the simplified form of

$$\frac{\partial \rho_s i_s}{\partial t} + \nabla \cdot \rho_s \mathbf{u}_s i_s = \nabla \cdot k_s \nabla T_f + \dot{s}_s, \tag{12.192}$$

$$\frac{\partial \rho_f i_f}{\partial t} + \nabla \cdot \rho_f \mathbf{u}_f i_f = \nabla \cdot k_f \nabla T_f + \dot{s}_f. \tag{12.193}$$

The specific enthalpies are defined as

$$i_s = \int_0^T c_{p_s} \, dT + i_{s_o}, \tag{12.194}$$

$$i_f = \int_0^T c_{p_f} \, dT + i_{f_o}, \tag{12.195}$$

where c_{p_s} and c_{p_f} are the *mixture*-specific heat of the phases (with each phase made of species A and B). The local phase-volume-averaged energy equations are found by the averaging of (12.192) and (12.193) over their respective phase. When these two equations are added,

$$\frac{\partial}{\partial t}[(1-\varepsilon)\langle\rho\rangle^s \langle i\rangle^s + \varepsilon \langle\rho\rangle^f \langle i\rangle^f]$$

$$+\nabla \cdot [(1-\varepsilon)\langle\rho\rangle^s \langle i\rangle^s \langle\mathbf{u}\rangle^s + \varepsilon \langle\rho\rangle^f \langle i\rangle^f \langle\mathbf{u}\rangle^f]$$

$$= \nabla \cdot [(1-\varepsilon)k_s + \varepsilon k_f]\nabla \langle T\rangle + \frac{k_f - k_s}{V}\int_{A_{fs}} T'_f \mathbf{n}_{fs} \, dA$$

$$+ \nabla \cdot [(1-\varepsilon)\langle\rho\rangle^s \langle\mathbf{u}'i'\rangle^s + \varepsilon \langle\rho\rangle^f \langle\mathbf{u}'i'\rangle^f], \tag{12.196}$$

where the porosity ε was defined by $\varepsilon = V_f/V$ and we have used $\varepsilon \dot{s}_f + (1-\varepsilon)\dot{s}_s = 0$, for no net energy production. Note that in contrast to the porous media with no change, the local porosity can change significantly due to the phase change. In the treatment of Bennen and Incropera, the last three terms on the right-hand side, which influence the effective thermal conductivity and introduce the dispersion tensor, have been neglected. The dispersion contribution has been assumed *negligible*, due to the assumed low velocities. If the dispersion contributions are included, then (12.196) can be written as

$$\frac{\partial}{\partial t}[(1-\varepsilon)\langle\rho\rangle^s \langle i\rangle^s + \varepsilon\langle\rho\rangle^f \langle i\rangle^f]$$
$$+\nabla \cdot [(1-\varepsilon)\langle\rho\rangle^s \langle i\rangle^s \langle \mathbf{u}\rangle^s + \varepsilon\langle\rho\rangle^f \langle i\rangle^f \langle \mathbf{u}\rangle^f]$$
$$= \nabla \cdot [\mathbf{K}_e + (1-\varepsilon)\langle\rho\rangle^f \langle c_p\rangle^f \mathbf{D}_s^d + \varepsilon\langle\rho\rangle^f \langle c_p\rangle^f \mathbf{D}_f^d] \cdot \nabla\langle T\rangle$$
$$\equiv \nabla \cdot \langle\rho\rangle^f \langle c_p\rangle^f \mathbf{D} \cdot \nabla\langle T\rangle, \quad (12.197)$$

where $\mathbf{K}_e$, $\mathbf{D}_f^d$, $\mathbf{D}_s^d$ and $\mathbf{D}$ are defined in a manner similar to those given in Chapters 3 and 4.

Then the solid–fluid, i.e., *phase-mass-averaged*, enthalpy is defined using the volume-averaged density $\langle\rho\rangle = (1-\varepsilon)\langle\rho\rangle^s + \varepsilon\langle\rho\rangle^f$, and we have

$$\langle\rho\rangle\langle i\rangle \equiv (1-\varepsilon)\langle\rho\rangle^s \langle i\rangle^s + \varepsilon\langle\rho\rangle^f \langle i\rangle^f. \quad (12.198)$$

The *mass concentration* of the solid and liquid phases are $\varepsilon\langle\rho\rangle^s/\langle\rho\rangle$ and $(1-\varepsilon)\langle\rho\rangle^f/\langle\rho\rangle$, respectively.

The phase-mass-averaged velocity is

$$\langle\rho\rangle\langle\mathbf{u}\rangle \equiv (1-\varepsilon)\langle\rho\rangle^s \langle\mathbf{u}\rangle^s + \varepsilon\langle\rho\rangle^f \langle\mathbf{u}\rangle^f. \quad (12.199)$$

The convective term can be written as a phase-mixture component and a phase-relative component as (Bennon and Incropera)

$$(1-\varepsilon)\langle\rho\rangle^s \langle i\rangle^s \langle\mathbf{u}\rangle^s + \varepsilon\langle\rho\rangle^f \langle i\rangle^f \langle\mathbf{u}\rangle^f = \langle\rho\rangle\langle\mathbf{u}\rangle\langle i\rangle$$
$$+(1-\varepsilon)\langle\rho\rangle^s (\langle\mathbf{u}\rangle^s - \langle\mathbf{u}\rangle)(\langle i\rangle^s - \langle i\rangle) + \varepsilon\langle\rho\rangle^f (\langle\mathbf{u}\rangle^f - \langle\mathbf{u}\rangle)(\langle i\rangle^f - \langle i\rangle). \quad (12.200)$$

Then (12.197) is written as

$$\frac{\partial}{\partial t}\langle\rho\rangle\langle i\rangle + \nabla \cdot \langle\rho\rangle\langle\mathbf{u}\rangle\langle i\rangle = \nabla \cdot \langle\rho\rangle^f \langle c_p\rangle^f \mathbf{D} \cdot \nabla\langle T\rangle$$
$$-\nabla \cdot [(1-\varepsilon)\langle\rho\rangle^s (\langle\mathbf{u}\rangle^s - \langle\mathbf{u}\rangle)](\langle i\rangle^s - \langle i\rangle)$$
$$+\varepsilon\langle\rho\rangle^f (\langle\mathbf{u}\rangle^f - \langle\mathbf{u}\rangle)(\langle i\rangle^f - \langle i\rangle). \quad (12.201)$$

Now using the definition of the specific enthalpy, the temperature gradient can be replaced with the gradient in the specific enthalpy of the solid as

$$\nabla\langle T\rangle = \frac{1}{\langle c_p\rangle^s}\nabla\langle i\rangle + \frac{1}{\langle c_p\rangle^s}\nabla(\langle i\rangle^s - \langle i\rangle). \quad (12.202)$$

Then the equation of conservation of thermal energy, (12.201), becomes

$$\frac{\partial}{\partial t}\langle\rho\rangle\langle i\rangle + \nabla \cdot \langle\rho\rangle\langle\mathbf{u}\rangle\langle i\rangle$$
$$= \nabla \cdot \frac{\langle\rho\rangle^f \langle c_p\rangle^f \mathbf{D}}{\langle c_p\rangle^s} \cdot \nabla\langle i\rangle + \nabla \cdot \frac{\langle\rho\rangle^f \langle c_p\rangle^f \mathbf{D}}{\langle c_p\rangle^s} \cdot \nabla(\langle i\rangle^s - \langle i\rangle)$$
$$-\nabla \cdot [(1-\varepsilon)\langle\rho\rangle^s (\langle\mathbf{u}\rangle^s - \langle\mathbf{u}\rangle)(\langle i\rangle^s - \langle i\rangle)$$
$$+\varepsilon\langle\rho\rangle^f (\langle\mathbf{u}\rangle^f - \langle\mathbf{u}\rangle)(\langle i\rangle^f - \langle i\rangle)]. \quad (12.203)$$

The last two terms can be written as $-\nabla \cdot (1-\varepsilon) \langle \rho \rangle^s (\langle \mathbf{u} \rangle^s - \langle \mathbf{u} \rangle)(\langle i \rangle^s - \langle i \rangle)$.

In addition, the equations for the conservation of overall *mass*, *momentum*, and *species* are also derived for this two-phase system by Bennen and Incropera, and are

$$\frac{\partial \langle \rho \rangle}{\partial t} + \nabla \langle \rho \rangle \langle \mathbf{u} \rangle = 0, \tag{12.204}$$

$$\frac{\partial \langle \rho \rangle \langle \mathbf{u} \rangle}{\partial t} + \nabla \cdot \langle \rho \rangle \langle \mathbf{u} \rangle \langle \mathbf{u} \rangle = -\nabla \langle p \rangle + \nabla \cdot \langle \mu \rangle^f \frac{\langle \rho \rangle}{\langle \rho \rangle^f} \nabla \langle \mathbf{u} \rangle$$

$$\times \frac{\langle \mu \rangle^f \langle \rho \rangle}{\mathbf{K} \langle \rho \rangle^f}(\langle \mathbf{u} \rangle - \langle \mathbf{u} \rangle^s) + \langle \rho \rangle \mathbf{f}, \tag{12.205}$$

$$\frac{\partial \langle \rho_A \rangle}{\partial t} + \nabla \cdot \langle \rho_A \rangle^f \langle \mathbf{u} \rangle$$

$$= \nabla \cdot \langle \rho \rangle^s \mathbf{D}_m^s \cdot \nabla \frac{\langle \rho_A \rangle^s}{\langle \rho \rangle^s} + \nabla \cdot \langle \rho \rangle^f \mathbf{D}_m^f \cdot \nabla \frac{\langle \rho_A \rangle^f}{\langle \rho \rangle^f}$$

$$- \nabla \cdot [(1-\varepsilon) \langle \rho \rangle^s (\langle \mathbf{u} \rangle^s - \langle \mathbf{u} \rangle)(\frac{\langle \rho_A \rangle^s}{\langle \rho \rangle^s} - \frac{\langle \rho_A \rangle}{\langle \rho \rangle})$$

$$+ \varepsilon \langle \rho \rangle^f (\langle \mathbf{u} \rangle^f - \langle \mathbf{u} \rangle)(\frac{\langle \rho_A \rangle^f}{\langle \rho \rangle^f} - \frac{\langle \rho_A \rangle}{\langle \rho \rangle})], \tag{12.206}$$

where $\langle \rho \rangle \mathbf{f}$ is the volumetric body force and $\mathbf{K}$ is the permeability tensor which changes with *time*, is *nonuniform*, and can be significantly *anisotropic*. The momentum equation (12.205) reduces to the Darcy law when $\mathbf{u} = 0$, and only the fluid properties, i.e., $\langle p \rangle = \langle p \rangle^f$, $\langle \rho \rangle = \langle \rho \rangle^f$, are used and the inertia, boundary, and body forces are negligible. However, it is slightly different from the semi-empirical, extended momentum equation given in Chapter 2. A discussion and the derivation of a more general momentum equation, applicable to higher velocities, is given by Ganesan and Poirier (1990). The *total phase, mass diffusivity tensors* $\mathbf{D}_m^s$ and $\mathbf{D}_m^f$ are defined in a manner similar to $\mathbf{D}$ discussed above. For the case of a stationary solid phase, $\mathbf{u}_s = 0$, and we have

$$\mathbf{u} = \frac{\varepsilon \langle \rho \rangle^f}{\langle \rho \rangle} \langle \mathbf{u} \rangle^f. \tag{12.207}$$

The distribution of *phase mass fractions* $\langle \rho \rangle^s / \langle \rho \rangle$ and $\langle \rho \rangle^f / \langle \rho \rangle$ and the *phasic mass fraction of species A*, $\langle \rho_A \rangle^f / \langle \rho \rangle^f$ and $\langle \rho_A \rangle^s / \langle \rho \rangle^s$, are found using the above governing equations *and* the equilibrium phase diagram. This requires the assumption of local chemical equilibrium, which in turn assumes that the presence of temperature and concentration gradients do *not* alter the phase equilibria.

Note that

$$\frac{\langle \rho_A \rangle}{\langle \rho \rangle} = \frac{(1-\varepsilon)\langle \rho_A \rangle^s}{\langle \rho \rangle} + \frac{\varepsilon \langle \rho_A \rangle^\ell}{\langle \rho \rangle}. \qquad (12.208)$$

Then the *solid-phase mass fraction*, under *phase equilibrium*, can be expressed in terms of k_p as

$$\frac{(1-\varepsilon)\langle \rho \rangle^s}{\langle \rho \rangle} = \frac{\dfrac{\langle \rho_A \rangle^\ell}{\langle \rho \rangle^\ell} - \dfrac{\langle \rho_A \rangle}{\langle \rho \rangle}}{\dfrac{\langle \rho_A \rangle^\ell}{\langle \rho \rangle^\ell} - \dfrac{\langle \rho_A \rangle^s}{\langle \rho \rangle^s}} = \frac{1}{1-k_p}\frac{T-T_f}{T-T_{m,B}}, \qquad (12.209)$$

where referring to Figure 12.17, T is the local temperature, T_f is the liquidus temperature (i.e., on the liquidus line) corresponding to $\langle \rho \rangle^s / \langle \rho \rangle$, and $T_{m,B}$ is the melting temperature for $\langle \rho_A \rangle^s / \langle \rho \rangle \to 0$ (i.e., pure species B). Then k_p is taken as the *ratio* of the slopes of the liquidus and solidus lines. The phase volume averaged mass fractions are related to the total volume-averaged mass fraction through

$$\frac{\langle \rho_A \rangle^s}{\langle \rho \rangle^s} = \frac{k_p}{1+(1-\varepsilon)\dfrac{\langle \rho \rangle^s}{\langle \rho \rangle}(k_p-1)}\frac{\langle \rho_A \rangle}{\langle \rho \rangle}, \qquad (12.210)$$

$$\frac{\langle \rho_A \rangle^\ell}{\langle \rho \rangle^\ell} = \frac{1}{1+(1-\varepsilon)\dfrac{\langle \rho \rangle^s}{\langle \rho \rangle}(k_p-1)}\frac{\langle \rho_A \rangle}{\langle \rho \rangle}. \qquad (12.211)$$

The unknowns ε, $\langle i \rangle^f$, $\langle \mathbf{u} \rangle^f$, $\langle \rho_A \rangle$, $\langle \rho_A \rangle^f$, and $\langle \rho_A \rangle^s$, along with the intermediate variables $\langle i \rangle$, $\langle \rho \rangle$, and $\langle \mathbf{u} \rangle$, are found from (12.204)–(12.206) and (12.207)–(12.211), and the definitions of averages. Bennon and Incropera (1987) apply these equations and obtain numerical solutions for a two-dimensional solidification problem.

(B) Nonequilibrium Treatment of Solidification

In the following, as examples of nonequilibrium treatments of solidification in a binary system, a *kinetic-diffusion* controlled *dendritic* crystal growth and a *buoyancy-influenced dendritic* crystal growth, will be examined.

(i) A Dispersed-Element Model for Kinetic-Diffusion Controlled Growth

Assuming that a *total* number n_s of *spherical* crystals are nucleated per unit volume at a supercooling of $\Delta T_{sc} = T_m - T_\ell$, then these crystals can grow to *final* grain radius of R_c

$$R_c = \left(\frac{3}{4\pi n_s}\right)^{1/3}. \tag{12.212}$$

The initially spherical crystal will have a *dendritic* growth and between the nucleation at time $t = 0$ and the complete growth (end of the solidification) at $t = t_f$, the *grain envelope* radius R will grow from a radius R_{s_o} (i.e., the initial radius of the nuclei) to R_c. The grain envelope is initially around the crystal and during the dendritic growth the envelope contains solid and liquid phases, and finally at the end of the solidification, it will again contain *only* solid. The content of the grain envelope is treated as a *dispersed element* and the *liquid fraction* (or porosity) ε_s of this *dispersed element* is initially zero and then *increases* rapidly followed by an eventual *decrease*, finally becoming zero again. This dispersed element and its surrounding liquid makes a *unit-cell model* and has been introduced and analyzed by Rappaz and Thévoz (1987). A schematic of this unit-cell model is shown in Figure 12.19(a). The growth of the dispersed element subject to the *cooling rate* of the unit cell given by the heat transfer rate $q_c 4\pi R_c^2$ with *no liquid motion* has been analyzed.

The elemental-liquid fraction ε_e and the *cell porosity* ε_c are defined as

$$\varepsilon_e = 1 - \frac{V_s}{V_e}, \qquad \varepsilon_c = 1 - \frac{V_e}{V_c} = 1 - \frac{R^3}{R_c^3}. \tag{12.213}$$

The solid volume is

$$V_s = (1 - \varepsilon_e)V_e = (1 - \varepsilon_e)\frac{4}{3}\pi R^3. \tag{12.214}$$

The quantities ε_e and ε_c are determined from the heat and mass transfer analyses. An *apparent* solid radius can be defined as

$$R_s^3 = \frac{V_s}{\frac{4}{3}\pi} = (1 - \varepsilon_e)R^3 \quad \text{or} \quad R_s = (1 - \varepsilon_e)^{1/3} R. \tag{12.215}$$

For convenience, the *volume fraction* of the *solid* in the *unit cell* f_s and the *volume fraction* of the *dispersed element* in the *unit cell* f_c, which are related to ε_e and ε_c, are used in the analysis. These are

$$f_s = \frac{V_s}{V_c} = \frac{R_s^3}{R_c^3} = (1 - \varepsilon_e)(1 - \varepsilon_c), \quad f_c = \frac{V_e}{V_c} = \frac{R^3}{R_c^3}. \tag{12.216}$$

Assuming an equilibrium phase diagram, such as that shown in Figure 12.19(b), the definition of the equilibrium partition ratio k_p was given by (12.177) and is repeated here.

$$k_p = \frac{\left.\frac{\rho_A}{\rho}\right|_{\ell s}}{\left.\frac{\rho_A}{\rho}\right|_{s\ell}}\bigg|_{T,p}, \tag{12.217}$$

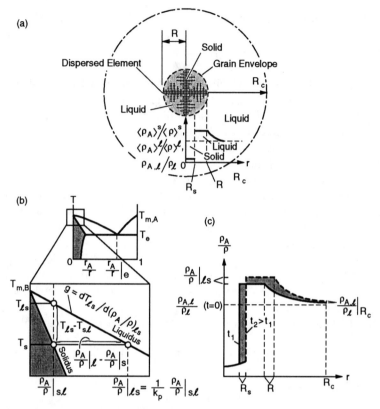

Figure 12.19 (a) Unit-cell model of the equiaxed dendritic growth of a crystal. The liquids within the grain envelope and within the element are shown along with the mass fraction distribution of the species A. (b) The idealized phase diagram. (c) The mass fraction distribution of the species A for two different elapsed times. (From Rappaz and Thévos, reproduced by permission ©1987 Pergamon.)

where the solid-liquid interfacial concentrations are $(\rho_A/\rho)_{s\ell}$ and $(\rho_A/\rho)_{\ell s}$ on the *solidus* and *liquidus* lines, respectively. The liquidus line mass fraction is related to the temperature using

$$T_{\ell s} - T_{s\ell} = -\gamma \left(\left.\frac{\rho_A}{\rho}\right|_{\ell s} - \left.\frac{\rho_A}{\rho}\right|_{s\ell} \right) \text{ or } T_e - T_{m,B} = \gamma \left.\frac{\rho_A}{\rho}\right|_e, \quad (12.218)$$

where

$$\gamma = \left.\frac{dT_{\ell s}}{d(\rho_A/\rho)_{\ell s}}\right|_p = \frac{T_e - T_{m,B}}{(\rho_A/\rho)_e}. \quad (12.219)$$

The *mass fraction difference* can also be written as

$$(\rho_A/\rho)_{\ell s} - (\rho_A/\rho)_{s\ell} = (\rho_A/\rho)_{s\ell}\frac{1 - k_p}{k_p}. \quad (12.220)$$

12.6 Melting and Solidification

As the crystal grows the concentration of species A in the solid increases *slightly*. This corresponds to a negligible mass diffusion in the solid ($D_s \to 0$). Other assumptions about the magnitude of D_s (i.e., $D_s \to \infty$ or a finite D_s) do *not* change the predicted growth rate (Rappaz and Thévoz, 1987). The rejected species A (i.e., *solute*) will result in the *increase* of the concentration of species A in the liquid contained in the envelope and in the remainder of liquid in the unit cell. This is depicted in Figure 12.19(c) where the concentration distributions in the *solid* ($0 \leq r \leq R_s$), in the *interelemental liquid* ($R_s \leq r \leq R$), and in the *cell liquid* ($R < r \leq R_c$) are shown for two elapsed times. In the model, the *liquid* concentration within the element is assumed to be uniform and its magnitude $(\rho_A/\rho)_{\ell s}$ is given by the phase diagram and as a function of T_e. Then in this model the *dendritic tip* concentration and the temperature are $(\rho_A/\rho)_{\ell s}$ and T_e, respectively, and their interrelation is given by (12.218).

The mass fraction distribution in the liquid region is determined for the species conservation equation which for constant ρ_ℓ and D_ℓ gives

$$\frac{\partial}{\partial t}\frac{\rho_{A,\ell}}{\rho_\ell} = D_\ell \left(\frac{\partial}{\partial r^2} + \frac{2}{r}\frac{\partial}{\partial r}\right)\frac{\rho_{A,\ell}}{\rho_\ell} \quad R \leq r \leq R_c. \qquad (12.221)$$

The initial and boundary conditions are

$$\frac{\rho_{A,\ell}}{\rho_\ell}(r,0) = \frac{\rho_{A,\ell}}{\rho_\ell}(t=0) \quad R_s \leq r \leq R_c, \qquad (12.222)$$

$$\frac{\partial}{\partial r}\frac{\rho_{A,\ell}}{\rho_\ell} = 0 \quad r = R_c. \qquad (12.223)$$

The concentration at $r = R$ is found from the species balance made over $0 \leq r \leq R_c$. This is done by the integration of the distributions shown in Figure 12.19(c), i.e.,

$$\int_0^{R_s} k_p \left.\frac{\rho_A}{\rho}\right|_{\ell s} 4\pi r\, dr + \left.\frac{\rho_A}{\rho}\right|_{\ell s} \frac{4}{3}\pi(R^3 - R_s^3)$$
$$+ \int_R^{R_c} \frac{\rho_{A,\ell}}{\rho_\ell} 4\pi r\, dr = \frac{\rho_{A,\ell}}{\rho}(t=0)\frac{4}{3}\pi R_c^3. \qquad (12.224)$$

By differentiating this with respect to time and using (12.221), we have

$$-4\pi D_\ell R^2 \left.\frac{\partial}{\partial r}\frac{\rho_A}{\rho_\ell}\right|_R = -\frac{4}{3}\pi \left.\frac{d}{dt}\frac{\rho_A}{\rho}\right|_{\ell s}(R^3 - R_s^3)$$
$$+ 4\pi R^2 \frac{dR}{dt}(1 - k_p)\left.\frac{\rho_A}{\rho}\right|_{\ell s}. \qquad (12.225)$$

This states that the outward solute flow rate through $r = R$ is determined by the solute *ejected* due to the solidification and the *temperature-caused change* in the mass fraction of the interdendritic liquid.

For *large* solid and liquid conductivities, within the cell the solid, the interdendritic liquid, and the cell liquid can all be in a *near-thermal equilibrium*. For a temperature change occurring on the boundaries of the cell, the assumption of a uniform temperature within the cell requires that the cell Biot number, i.e.,

$$Bi = \frac{Nu_{d_c}^{ext}}{6} \frac{k_\infty}{\langle k \rangle_{V_c}}, \quad Nu_{d_c}^{ext} = \frac{q_c 2R_c}{(T_c - T_\infty)k_\infty} \quad (12.226)$$

be less than 0.1. This condition is assumed to hold (for metals) and a single temperature T_c is used for the unit cell.

The thermal energy balance on the unit cell gives

$$q_c 4\pi R_c^2 = \left[\rho_\ell \Delta i_{\ell s} \frac{df_s}{dt} + \langle \rho c_p \rangle_{V_c} \frac{dT_c}{dt} \right] + \frac{4}{3}\pi R_c^3, \quad (12.227)$$

where $\Delta i_{\ell s}$ is the heat of solidification and

$$\langle \rho c_p \rangle_{V_c} = (1 - \varepsilon_e)(1 - \varepsilon_c)(\rho c_p)_s + [\varepsilon_e(1 - \varepsilon_c) + \varepsilon_c](\rho c_p)_\ell. \quad (12.228)$$

The variation of temperature can be replaced by that of the concentration by assuming that the cell temperature is the equilibrium temperature at the liquidus line. Then using (12.218), (12.227) becomes

$$\frac{3q_c}{R_c} = \Delta i_{\ell s} \frac{df_s}{dt} + \langle \rho c_p \rangle_{V_c} \gamma \frac{d}{dt} \left. \frac{\rho_A}{\rho} \right|_{\ell s}. \quad (12.229)$$

The growth rate of the *dendritic tip* is modeled using the available results for low Péclet number growth in *unbounded* liquid. The model used by Rappaz and Thévoz is

$$\frac{dR}{dt} = \frac{\gamma D_\ell}{\pi^2 (k_p - 1) \frac{\sigma_{\ell s}}{\Delta s_{\ell s}} \frac{\rho_{A,\ell}}{\rho_\ell}(t=0)} \left(\left. \frac{\rho_A}{\rho} \right|_{\ell s} - \left. \frac{\rho_{A,\ell}}{\rho_\ell} \right|_R \right)^2. \quad (12.230)$$

This kinetic condition is discussed by Hills and Roberts (1993). The distribution of $\rho_{A,\ell}/\rho_\ell$ in $R \leq r \leq R_c$ as well as the variations of f_s, f_c, $(\rho_A/\rho)_{\ell s}$, and $(\rho_{A,\ell}/\rho_\ell)_R$ with respect to time are determined by solving (12.215), (12.224), (12.225), (12.229), and (12.230), simultaneously. This is done numerically and some of their results are reviewed below.

The results are for the solidification of Al–Si with an initial silicon concentration $(\rho_{A,\ell}/\rho_\ell)(t=0) = 0.05$, an eutectic concentration of 0.108, $D_\ell = 3 \times 10^{-9}$ m^2/s, $k_p = 0.117$, $\gamma = -7.0$, $\rho_\ell \Delta i_{\ell s} = 9.5 \times 10^8$ J/m^3, $T_{m,B}$(aluminum) $= 660°$C, $T_e = 577°$C, $\sigma_{\ell s}/\Delta s_{s\ell} = 9 \times 10^{-8}$ m-K, and $\langle \rho c_p \rangle_{V_c} = 2.35 \times 10^6$ J/m^3-K. Figure 12.20 (a) shows the results for $\Delta T_{sc} = 0°$C, $R_c = 100$ μm, $R_{s_o} = 1$ μm, and a cooling rate that results in the *total* solidification of the cell in 10 s.

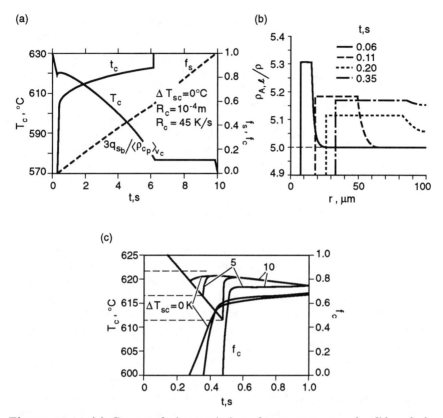

Figure 12.20 (a) Computed time variation of temperatures and solid and element volume fractions. (b) Radial distribution of the species A in the liquid for four different elapsed times. (c) Same as (a) but for three different supercooling conditions. (From Rappaz and Thévoz, reproduced by permission ©1987 Pergamon.)

The variation of temperature T_c and solid and element volume fractions f_s and f_c, respectively, with respect to time, up to the time of the complete solidification, are shown. Since the mass fraction $(\rho_A/\rho)_{\ell s}$ has to *increase* in order for the growth of the dendrite to begin, as stated by (12.230), then no change in dT/dt occurs until the point of *undershoot* is reached, and growth begins with a *large* and sudden increase in f_c. Since the growth rate of the dendritic tip is larger than the rate of the solidification allowed by the heat removal, the volume fraction of the interdendritic liquid ε_s increases substantially. Therefore, f_s increases much slower than f_c. This increase in f_s *decreases* $(\rho_A/\rho)_{\ell s}$ and *increases* T_c, as stated by (12.227) and (12.229). The increase in T_c is called *recalescence*. The concentration distributions in the interdendritic liquid and in the cell liquid are shown in Figure 12.20(b) for several elapsed times. The elapsed times correspond to the time of undershoot, shortly after the rise in T_c begins, when T_c reaches a maximum, and shortly after the decrease in T_c begins. For $t > 0.20$ s, the temperature *decreases* while $(\rho_A/\rho)_{\ell s}$ *increases*. For an elapsed time slightly larger than 6 s, the element grows to the maximum radius R_c and only the solidification of the interdendritic liquid occurs. This ends when all this liquid is solidified. The undershoot temperature predicted for no supercooling, shown in Figure 12.20(a), will correspond to the supercooling ΔT_{sc} when substantial liquid supercooling exists. Figure 12.20(c) shows the results for the same conditions as in Figures 12.20(a) and (b), except 5 and 10°C supercooling are allowed. The results show that the larger the supercooling, the faster the temperature rises after the initial growth (i.e., *accelerated recalescence*).

(ii) Inclusion of Buoyant Liquid and Crystal Motions

The unit-cell-based, diffusion-controlled dendritic growth discussed above has been extended to thermo- and diffusobuoyant *convection* by using local phase-volume-averaged conservation equations and local thermal and chemical *nonequilibrium* among the liquid phase, the solid-particles (equiaxed dendritics) phase, and the confining surfaces of the mold. As before, the crystals are assumed to be formed by the bulk nucleation in a supercooled liquid. The liquid temperature $\langle T \rangle^\ell$, the liquid volume fraction ε and its solid particle counterparts $\langle T \rangle^s$ and $1 - \varepsilon$, the velocities $\langle \mathbf{u} \rangle^\ell$ and $\langle \mathbf{u} \rangle^s$, the volumetric solidification rate $\langle \dot{n} \rangle^s$, and the concentration of species A in each phase $\langle \rho_A \rangle^\ell / \rho_\ell$, $\langle \rho_A \rangle^s / \langle \rho \rangle_s$, are all determined from the solution of the local phase-volume-averaged conservation equations. The *thermodynamic* conditions are applied similar to those in the above diffusion treatment, but the growth rate of the dendritic tip is *not* prescribed. Instead, the interfacial heat and mass transfer is modeled using interfacial Nusselt and Sherwood numbers for the particulate flow and heat transfer. Also, interfacial heat and mass transfer is prescribed as functions of the Reynolds number, and is based on the relative velocity and solid particle diameter d.

The local, phase-volume-averaged treatment of flow and heat and mass transfer has been addressed by Voller et al. (1989), Prakash (1990), Beckermann and Ni (1992), Prescott et al. (1992), and Wang and Beckermann (1992), and a review is given by Beckermann and Viskanta (1993). In the following, the conservations and the thermodynamic conditions are examined, and then some of available results on the growth and the motion of the bulk-nucleated crystals are reviewed.

Laminar flow is assumed and the modeling of $\langle \mathbf{S} \rangle^\ell$, $\langle \mathbf{S} \rangle^s$, and τ_d are pursued similar to the hydrodynamics of the particulate flow. Since the solid particles are *not* spherical, the dendritic arms and other geometric parameters should be included in the models. Ahuja et al. (1992) develop a drag coefficients for equiaxed dendrites.

The effective media properties $\mathbf{D}_m^\ell$, $\mathbf{D}^\ell$, $\mathbf{D}_m^s$, and $\mathbf{D}^s$, which include both

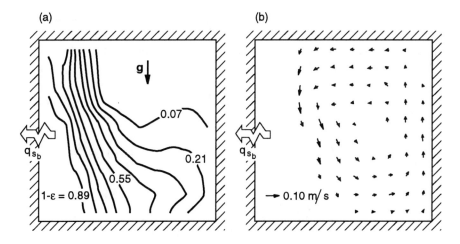

Figure 12.21 Typical solid-volume fraction and motion in a mold cooled on the left wall. (From Beckermann and Ni, reproduced by permission ©1992 Technomic.)

the *molecular* (i.e., *conductive*) and the hydrodynamic dispersion components, are also modeled. Due to the *lack* of any predictive correlations for the *nonequilibrium* transport, local thermal equilibrium conditions are used. For the interfacial convective transport, the local Nusselt and Sherwood numbers are prescribed. The effect of the solid particles geometry must also be addressed (Dash and Gill, 1984).

The numerical results of Beckermann and Ni (1992) for the solidification of a AlOCu mixture are shown in Figures 12.21(a) and (b). The results are for an initial copper concentration of 0.05 cooled in a two-dimensional square mold by extracting heat from the *left* vertical surface (all other surfaces are ideally *insulated*) at a constant and uniform rate q_{s_b}. An initial grain diameter (spherical crystal) of 1 μm is assumed along with a nucleation model. Their results for the solid-volume fraction $1 - \varepsilon$ and the solid-phase velocity $\langle \mathbf{u} \rangle^s$ for an elapsed time of 40 s are shown. The solid concentration is largest near the cooled surface and on the left-lower corner, and the solid velocity vanishes in these areas. Thermo- and diffusobuoyant motion remains intensive in the areas of *low* solid-volume fraction.

As was mentioned, the liquid- and solid-phase flows are assumed to be *laminar*. Then time averaging and the modeling of the turbulent transport are not needed. The interfacial heat and species fluxes are modeled using the *interfacial-averaged* Nusselt and Sherwood numbers. These interfacial convections resulting from the thermal and chemical nonequilibrium are modeled for *both* the liquid- and the solid-side of the interface and are designated by superscripts ℓ and s. These are in dimensionless forms, Nu_d^ℓ,

12. Phase Change

Nu_d^s, Sh_d^ℓ and Sh_d^s. Then conservation equations are found, except now the solid-side interfacial convection is added.

The phasic, local-volume-averaged overall mass, species, momentum, and energy equations for the liquid phase are

$$\frac{\partial}{\partial t}\varepsilon\rho_\ell + \nabla \cdot \varepsilon\rho_\ell \langle \mathbf{u}\rangle^\ell = \langle \dot n\rangle^\ell, \qquad (12.231)$$

$$\frac{\partial}{\partial t}\varepsilon \langle\rho_A\rangle^\ell + \nabla \cdot \varepsilon \langle \mathbf{u}\rangle^\ell \langle\rho_A\rangle^\ell = \nabla \cdot \varepsilon\rho_\ell \mathbf{D}_m^\ell \cdot \nabla \frac{\langle\rho_A\rangle^\ell}{\rho_\ell} + \langle \dot n\rangle^\ell \left.\frac{\langle\rho_A\rangle}{\rho}\right|_{\ell s}$$

$$+\frac{A_{\ell s}}{V}Sh_d^\ell \frac{\rho_\ell D_\ell}{d}\left(\left.\frac{\rho_A}{\rho}\right|_{\ell s} - \frac{\langle\rho_A\rangle^\ell}{\rho_\ell}\right), \qquad (12.232)$$

$$\frac{\partial}{\partial t}\varepsilon\rho_\ell \langle\mathbf{u}\rangle^\ell + \nabla \cdot \varepsilon\rho_\ell \langle\mathbf{u}\rangle^\ell \langle\mathbf{u}\rangle^\ell = -\varepsilon\nabla \langle p\rangle^\ell + \nabla \cdot \varepsilon \langle\mathbf{S}\rangle^\ell + \varepsilon\rho_\ell \mathbf{g}$$

$$+\langle \dot n\rangle^\ell \langle\mathbf{u}_\ell\rangle_{A_{\ell s}} - \frac{\rho_s}{\tau_d}(1-\varepsilon)(\langle\mathbf{u}\rangle^\ell - \langle\mathbf{u}\rangle^s), \qquad (12.233)$$

$$\frac{\partial}{\partial t}\varepsilon\rho_\ell \langle i\rangle^\ell + \nabla \cdot \rho_\ell \langle\mathbf{u}\rangle^\ell \langle i\rangle^\ell = \nabla \cdot \varepsilon(\rho c_p)_\ell \mathbf{D}_\ell \cdot \nabla \langle T\rangle^\ell$$

$$+\langle \dot n\rangle^\ell \langle i_\ell\rangle_{A_{\ell s}} + \frac{A_{\ell s}}{V}Nu_d^\ell \frac{k_\ell}{d}(T_\ell - \langle T\rangle^\ell). \qquad (12.234)$$

Similarly, for the solid phase we have

$$\frac{\partial}{\partial t}\varepsilon\rho_s + \nabla \cdot \varepsilon\rho_s \langle\mathbf{u}\rangle^s = \langle \dot n\rangle^s, \qquad (12.235)$$

$$\frac{\partial}{\partial t}(1-\varepsilon)\langle\rho_A\rangle^s + \nabla \cdot (1-\varepsilon)\langle\mathbf{u}\rangle^s \langle\rho_A\rangle^s = \nabla \cdot (1-\varepsilon)\rho_s \mathbf{D}_m^s \cdot \nabla \frac{\langle\rho_A\rangle^s}{\rho_s}$$

$$+\langle \dot n\rangle^\ell \left.\frac{\langle\rho_A\rangle}{\rho}\right|_{s\ell} + \frac{A_{\ell s}}{V}Sh_d^s \frac{\rho_s D_s}{d}\left(\left.\frac{\rho_A}{\rho}\right|_{s\ell} - \frac{\langle\rho_A\rangle^s}{\rho_s}\right), \qquad (12.236)$$

$$\frac{\partial}{\partial t}(1-\varepsilon)\rho_s \langle\mathbf{u}\rangle^s + \nabla \cdot (1-\varepsilon)\rho_s \langle\mathbf{u}\rangle^s \langle\mathbf{u}\rangle^s = -(1-\varepsilon)\nabla \langle p\rangle^s$$

$$+\nabla\cdot(1-\varepsilon)\langle\mathbf{S}\rangle^s + (1-\varepsilon)\rho_s \mathbf{g} + \langle\dot n\rangle^s \langle\mathbf{u}_s\rangle_{A_{\ell s}} + \frac{\rho_s}{\tau_d}(1-\varepsilon)(\langle\mathbf{u}\rangle^\ell - \langle\mathbf{u}\rangle^s), \qquad (12.237)$$

$$\frac{\partial}{\partial t}(1-\varepsilon)\rho_s \langle i\rangle^s + \nabla\cdot(1-\varepsilon)\rho_s \langle \mathbf{u}\rangle^s \langle i\rangle^s = \nabla\cdot(1-\varepsilon)(\rho c_p)_s \mathbf{D}_s \cdot \nabla \langle T\rangle^s$$

$$+ \langle \dot{n}\rangle^s \langle i_s\rangle_{A_{\ell s}} + \frac{A_{\ell s}}{V} Nu_d^s \frac{k_s}{d}(T_\ell - \langle T\rangle^s). \qquad (12.238)$$

The interfacial balance of the overall mass, species, and energy yields

$$\langle \dot{n}\rangle^\ell = \langle \dot{n}\rangle^s, \qquad (12.239)$$

$$\frac{A_{\ell s}}{V} Sh_d^\ell \frac{\rho_\ell D_\ell}{d}\left(\left.\frac{\rho_A}{\rho}\right|_{\ell s} - \frac{\langle \rho_A\rangle^\ell}{\rho_\ell}\right)$$

$$+ \frac{A_{\ell s}}{V} Sh_d^s \frac{\rho_s D_s}{d}(\left.\frac{\rho_A}{\rho}\right|_{s\ell} - \frac{\langle \rho_A\rangle^s}{\rho_s}) + \langle \dot{n}\rangle^s (\left.\frac{\rho_A}{\rho}\right|_{s\ell} - \left.\frac{\rho_A}{\rho}\right|_{\ell s}) = 0, \qquad (12.240)$$

$$\frac{A_{\ell s}}{V} Nu_d^\ell(T_\ell - \langle T\rangle^\ell) + \frac{A_{\ell s}}{V} Nu_d^s(T_\ell - \langle T\rangle^s)$$

$$+ \langle \dot{n}\rangle^s(\langle i_s\rangle_{A_{\ell s}} - \langle i_\ell\rangle_{A_{\ell s}}) = 0. \qquad (12.241)$$

The specific enthalpies are related to temperature using

$$\langle i_\ell\rangle_{A_{\ell s}} = c_{p_\ell}T_\ell + \Delta i_{\ell s}, \quad \langle i_s\rangle_{A_{\ell s}} = c_{p_s}T_s,$$
$$\langle i\rangle^\ell = c_{p_\ell}\langle T\rangle_\ell \Delta i_{\ell s}, \quad \langle i\rangle^s = c_{p_s}\langle T\rangle_s. \qquad (12.242)$$

The interfacial temperature T_ℓ is related to the interfacial mass fraction $(\rho_A/\rho)_{\ell s}$ through which can be written as

$$\left.\frac{\rho_A}{\rho}\right|_{\ell s} = \left.\frac{\rho_A}{\rho}\right|_e \frac{T_{m,B} - T_\ell}{T_{m,B} - T_e}. \qquad (12.243)$$

A phase diagram similar to Figure 12.19(b) is assumed, and for a given temperature T_ℓ the mass fraction on the solidus line is related to that on the liquidus line using (12.218).

12.7 References

Afgan, N. M., Jovic, L. A., Kovalev, S. A., and Lenykov, V. A., 1985, "Boiling Heat Transfer from Surfaces with Porous Layers," *Int. J. Heat Mass Transfer*, 28, 415–422.

Ahuja, S., Beckermann, C., Zakhem, R., Weidman, P.D., and de Groh III, H.C., 1992, "Drag Coefficient of an Equiaxed Dendrite Settling in an Infinite Medium," Beckermann, C., et al., Editors, ASME HTD-Vol. 218, 85–91, American Society of Mechanical Engineers, New York.

Basel, M. D. and Udell, K. S., 1989, "Two-Dimensional Study of Steam Injection into Porous Media," in *Multiphase Transport in Porous Media*–1989, *ASME HTD-Vol.* 127 (FED-Vol. 82), 39–46.

Bau, H. H. and Torrance, K. E., 1982, "Boiling in Low-Permeability Porous Materials," *Int. J. Heat Mass Transfer*, 25, 45–55.

Beckermann, C. and Ni, J., 1992, "Modeling of Equiaxed Solidification with Convection," in *Proceedings, First International Conference on Transport Phenomena in Processing*, Güceri, S.J., Editor, 308–317, Technomic, Lancaster, PA.

Beckermann, C. and Viskanta, R., 1993, "Mathematical Modeling of Transport Phenomena During Alloy Solidification," *Appl. Mech. Rev.*, 46, 1–27.

Bennen, W. D. and Incropera, F. P., 1987, "A Continuum Model for Momentum, Heat and Species Transport in Binary Solid-Liquid Phase Change Systems-I., and -II.," *Int. J. Heat Mass Transfer*, 30, 2161–2187.

Bergles, A. E. and Chyu, M. C., "Characteristics of Nucleate Pool Boiling from Porous Metallic Coatings," *ASME J. Heat Transfer*, 104, 279–285.

Bird, R. B., Stewart, W. E., and Lightfoot, E. N., 1960, *Transport Phenomena*, Wiley.

Cheng, P., 1981, "Film Condensation Along an Inclined Surface in a Porous Medium," *Int. J. Heat Mass Transfer*, 24, 983–990.

Chung, J. N., Plumb, O. A., and Lee, W. C., 1990, "Condensation in a Porous Region Bounded by a Cold Vertical Surface," in *Heat and Mass Transfer in Frost, Ice, Packed Beds, and Environmental Discharge*, ASME HTD-Vol. 139, 43–50.

Dash, S. K. and Gill, N. M., 1984, "Forced Convection Heat and Momentum Transfer to Dendritic Structures (Parabolic Cylinder and Paraboloids of Revolution)," *Int. J. Heat Mass Transfer*, 27, 1345–1356.

Eckert, E. R. G., Goldstein, R. J., Behbahani, A. I., and Hain, R., 1985, "Boiling in an Unconsolidated Granular Medium," *Int. J. Heat Mass Transfer*, 28, 1187–1196.

Fand, R. M., Zheng, T., and Cheng, P., 1987, "The General Characteristics of Boiling Heat Transfer from a Surface Embedded in a Porous Medium," *Int. J. Heat Mass Transfer*, 30, 1231–1235.

Fukusako, S., Komoriga, T., and Seki, N. 1986, "An Experimental Study of Transition and Film Boiling Heat Transfer in Liquid-Saturated Porous Bed," *ASME J. Heat Transfer*, 108, 117–124.

Ganesan, S. and Poirier, D. R., 1990, "Conservation of Mass and Momentum for the Flow of Interdendritic Liquid During Solidification," *Metall. Trans.*, 21B, 173–181.

Hanamura, K. and Kaviany, M., 1995, "Propagation of Condensation Front in Steam Injection into Dry Porous Media," *Int. J. Heat Mass Transfer*, in press.

Hills, R. N., Loper, D. E., and Roberts, P. H., 1992, "On Continuum Models for Momentum, Heat and Species Transport in Solid-Liquid Phase Change Systems," *Int. Comm. Heat Mass Transfer*, 19, 585–594.

Hills, R. N. and Roberts, P.H., 1993, "A Note on the Kinetic Conditions at a Supercooled Interface," *Int. Comm. Heat Mass Transfer*, 20, 407–416.

Hupport, H. E., 1990, "The Fluid Mechanics of Solidification," *J. Fluid Mech.*, 212, 209–240.

Ilic, M. and Turner, I. W., 1989, "Convective Drying of a Consolidated Slab of Wet Porous Material," *Int. J. Heat Mass Transfer*, 32, 2351–2362.

Jennings, J. D. and Udell, K. S., 1985, "The Heat Pipe Effect in Heterogeneous Porous Media," in *Heat Transfer in Porous Media and Particulate Flows, ASME HTD-Vol.* 46, 93–101.

Jones, P. L., Lawton, J., and Parker, I. M., 1974, "High Frequency Paper Drying: Part I—Paper Drying in Radio and Microwave Frequency Fields," *Trans. Instn. Chem. Engrs.*, 52, 121–131.

Jones, S. W., Epstein, M., Gabor, J. D., Cassulo, J. D., and Bankoff, S. G., 1980, "Investigation of Limiting Boiling Heat Fluxes from Debris Beds," *Trans. Amer. Nucl. Soc.*, 35, 361–363.

Kaviany, M., 1986, "Boundary-Layer Treatment of Film Condensation in the Presence of a Solid Matrix," *Int. J. Heat Mass Transfer*, 29, 951–954.

Kaviany, M. and Mittal, M., 1987, "Funicular State in Drying of a Porous Slab," *Int. J. Heat Mass Transfer*, 30, 1407–1418.

Kaviany, M., 1994, *Principles of Convective Heat Transfer*, Springer-Verlag, New York.

Kim, C.-J. and Kaviany, M., 1992, "A Fully Implicit Method for Diffusion-Controlled Solidification of Binary Alloys," *Int. J. Heat Mass Transfer*, 35, 1143–1154.

Konev, S. K., Plasek, F., and Horvat, L., 1987, "Investigation of Boiling in Capillary Structures," *Heat Transfer—Soviet Res.*, 19, 14–17.

Kovalev, S. A., Solv'yev, S. L., and Ovodkov, O. A., 1987, "Liquid Boiling on Porous Surfaces," *Heat Transfer—Soviet Res.*, 19, 109–120.

Kurz, W. and Fisher, D. J., 1992, *Fundamentals of Solidifcation*, Third Edition, Trans Tech Publications, Switzerland.

Luikov, A. V., 1966, *Heat and Mass Transfer in Capillary Porous Bodies*, Pergamon (original Russian version, 1961).

Majumdar, A. and Tien, C.-L., 1988, "Effects of Surface Tension on Films Condensation in a Porous Medium," *ASME J. Heat Transfer*, 112, 751–757.

Menegus, D. K. and Udell, K. S., 1985, "A Study of Steam Injection into Water Saturated Capillary Porous Media," in *Heat Transfer in Porous Media and Particulate Flows, ASME HTD-Vol.* 46, 151–157.

Miller, C. A., 1975, "Stability of Moving Surfaces in Fluid Systems with Heat and Mass Transport, III: Stability of Displacement Fronts in Porous Media," *AIChE J.*, 21, 474–479.

Naik, A. S. and Dhir, V. K., 1982, "Forced Flow Evaporative Cooling of a Volumetrically Heated Porous Layer," *Int. J. Heat Mass Transfer*, 25, 541–552.

Parmentier, E. M., 1979, "Two-Phase Natural Convection Adjacent to a Vertical Heated Surface in a Permeable Medium," *Int. J. Heat Mass Transfer*, 22, 849–855.

Plumb, O. A., Burnett, D. B., and Shekarriz, A., 1990, "Film Condensation on a Vertical Flat Plate in a Packed Bed," *ASME J. Heat Transfer*, 112, 235–239.

Plumb, O. A., Spolek, G. A., and Olmstead, B. A., 1985, "Heat and Mass Transfer in Wood During Drying," *Int. J. Heat Mass Transfer*, 28, 1669–1678.

Plumb, O. A., 1994, "Convective Melting of Packed Beds," *Int. J. Heat Mass Transfer*, 37, 829–836.

Poirier, D. R., Nandapurkar, P. J., and Ganesan, S., 1991, "The Energy and Solute Conservation Equations for Dendritic Solidification," *Metall. Trans.*, 22B, 889–900.

Prakash, C., 1990, "Two-Phase Model for Binary Solid-Liquid Phase Change, Part I: Governing Equations, Part II: Some Illustration Examples," *Num. Heat Transfer*, 18B, 131–167.

Prescott, P. J., Incropera, F.P., and Gaskell, D.R., 1992, "The Effects of Undercooling, Recalescence and Solid Transport on the Solidification of Binary Metal Alloys," in *Transport Phenomena in Materials Processing and Manufacturing*, ASME HTD-Vol. 196, American Society of Mechanical Engineers, New York.

Ramesh, P. S. and Torrance, K. E., 1990, "Stability of Boiling in Porous Media," *Int. J. Heat Mass Transfer*, 33, 1895–1908.

Rappaz, M. and Thévoz, Ph., 1987, "Solute Model for Equiaxed Dendritic Growth," *Acta Metall.*, 35, 1487–1497.

Rappaz, M. and Voller, V. R., 1990, "Modelling of Micro-Macrosegregation in Solidification Processes," *Metall. Trans.*, 21A, 749–753.

Reed, A. W., 1986, "A Mechanistic Explanation of Channels in Debris Beds," *ASME J. Heat Transfer*, 108, 125–131.

Renken, K. J., Carneiro, M. J., and Meechan, K., 1994, "Analysis of Laminar Forced Convection Condensation within Thin Porous Coating," *J. Thermophy. Heat Transfer*, 8, 303–308.

Rogers, J. A. and Kaviany, M., 1991, "Funicular and Evaporation Front Regimes in Drying of Ceramic Beds," *Int. J. Heat Mass Transfer*, 35, 469–480.

Rose, J. W., 1988, "Fundamentals of Condensation Heat Transfer: Laminar Film Condensation," *JSME Int. J. Series II*, 31, 357–375.

Shekarriz, A. and Plumb, O. A., 1986, "A Theoretical Study of the Enhancement of Filmwise Condensation Using Porous Fins," *ASME paper no. 86-HT-31*.

Singh, M. and Behrendt, D. R., 1994, "Reactive Melt Infiltration of Silicon-Niobium Alloys in Microporous Carbons," *J. Mater. Res.*, 9, 1701–1708.

Sondergeld, C. H. and Turcotte, D. L., 1977, "An Experimental Study of Two-Phase Convection in a Porous Medium with Applications to Geological Problems," *J. Geophys. Res.*, 82, 2045–2053.

Song, M., Choi, J., and Viskanta, R., 1993, "Upward Solidification of a Binary Solution Saturated Porous Medium," *Int. J. Heat Mass Transfer*, 36, 3687–3695.

Stanish, M. A., Schajer, G. S., and Kayihan, F., 1986, "A Mathematical Model of Drying for Hygroscopic Porous Media," *AIChE J.*, 32, 1301–1311.

Stewart, L. D. and Udell, K. S., 1989, "The Effect of Gravity and Multiphase Flow on the Stability of Condensation Fronts in Porous Media," in *Multiphase Transport in Porous Media—1989*, ASME HTD-Vol. 127 (FED-Vol. 82), 29–37.

Stewart, L. D., Basel, M. D., and Udell, K. S., 1987, "The Effect of Gravity on Steam Propagation in Porous Media," in *Multiphase Transport in Porous Media*, ASME HTD-Vol. 91 (FED-Vol. 60), 31–42.

Stubos, A. K. and Buchin, J.-M., 1988, "Modeling of Vapor Channeling Behavior in Liquid-Saturated Debris Beds," *ASME J. Heat Transfer*, 110, 968–975.

Styrikovich, M. A., Malyshenko, S. P., Andrianov, A. B., and Tataev, I. V., 1987, "Investigation of Boiling on Porous Surfaces," *Heat Transfer—Soviet Res.*, 19, 23–29.

Tao, Y.-X. and Gray, D. M., 1995, "Prediction of Snow-Melt Infiltration into Frozen Soils," *Num. Heat Transfer*, to appear.

Tsung, V. X., Dhir, V. K., and Singh, S., 1985, "Experimental Study of Boiling Heat Transfer from a Sphere Embedded in Liquid Saturated Porous Media," in *Heat Transfer in Porous Media and Particulate Flows, ASME HTD-Vol. 46*, 127–134.

Udell, K. S., 1985, "Heat Transfer in Porous Media Considering Phase Change and Capillarity—The Heat Pipe Effect," *Int. J. Heat Mass Transfer*, 28, 485–495.

Vodak, F., Cerny, R., and Prikryl, P., 1992, "A Model of Binary Alloy Solidification with Convection in the Melt," *Int. J. Heat Mass Transfer*, 35, 1787–1791.

Voller, V. R., Brent, A. D., and Prakash, C., 1989, "The Modeling of Heat, Mass and Solute Transport in Solidification Systems," *Int. J. Heat Mass Transfer*, 32, 1719–1731.

Wang, C. Y. and Beckermann, C., 1992, "A Multiphase Micro-Macroscopic Model of Solute Diffusion in Dendritic Alloy Solidification," in *Micro/Macro Scale Phenomena in Solidification*, Beckermann, C., et al., Editors, ASME HTD-Vol. 218, 43–57, American Society of Mechanical Engineers, New York.

Wang, C. Y. and Beckermann, 1993, "A Multiphase Solute Diffusion Model for Dendritic Alloy Solidification," *Metall. Trans.*, 24A, 2787–2802.

Whitaker, S. and Chou, W. T.-H., 1983, "Drying Granular Porous Media–Theory and Experiment," *Drying Tech.*, 1, 3–33.

Whitaker, S., 1977, "Simultaneous Heat, Mass and Momentum Transfer in Porous Media: A Theory of Drying," in *Advances in Heat Transfer*, 13, 119–203.

White, S. M. and Tien, C.-L., 1987, "Analysis of Laminar Film Condensation in a Porous Medium," in *Proceeding of the 2d ASME/JSME Thermal Engineering Joint Conference*, 401–406.

Woods, A. W., 1991, "Fluid Mixing During Melting," *Phys. Fluids*, A3, 1393–1404.

Worster, M. G., 1991, "Natural Convection in a Mushy Layer," *J. Fluid Mech.*, 224, 335–339.

Yortsos, Y. C., 1982, "Effect of Heat Losses on the Stability of Thermal Displacement Fronts in Porous Media," *AIChE J.*, 28, 480–486.

Nomenclature

An upper-case boldface letter indicates that the quantity is a second-order *tensor* and a lower-case boldface letter indicates that the quantity is a *vector* (or *spatial tensor*). Some symbols, which are introduced briefly through derivations and, otherwise are not referred to in the text, are locally defined in the appropriate locations and are not listed here. The mks units are used throughout.

a	phase distribution function
a_j	$j = 1, 2, \ldots$, constants
A	area, cross section (m^2)
A_o	volumetric (or specific) surface area (1/m)
A_{sf}	interfacial area between solid and fluid phases (m^2)
b	magnetic induction (Tesla, or V-s/m^2) or closure vector function (m)
Bo	Bond or Eötvös number $\rho_\ell g R^2/\sigma$ where ℓ stands for the wetting phase, also $\rho_\ell g K/\varepsilon\sigma$
c_o	speed of light in vacuum (m/s)
c_p	specific heat capacity at constant pressure (J/kg-K)
C	ratio of volumetric heat capacities $(\rho c_p)_s / (\rho c_p)_f$ or average interparticle clearance (m)
C_1	$= \varepsilon + (1-\varepsilon)C$, $C = (\rho c_p)_s/(\rho c_p)_f$
C_E	coefficient in the Ergun modification of the Darcy law
Ca	capillary number where ℓ stands for the wetting phase $\mu_\ell u_{D\ell}/\sigma$
d	pore-level linear length scale (m) or diameter (m)
d	displacement tensor (m) or electric displacement (V/m)
D	total thermal diffusion coefficient (m^2/s)
D^d	dispersion coefficient (m^2/s)
D_{Ki}	Knudsen mass diffusivity for component i (m^2/s)
D_m	binary mass diffusion coefficient (m^2/s)
D	total thermal diffusivity tensor (m^2/s) or rate of deformation tensor (1/s)
D^d	thermal dispersion tensor (m^2/s)
D$_m$	total binary mass diffusion tensor (m^2/s)
D$_{me}$	effective binary mass diffusion tensor (m^2/s)
D$_m^d$	mass dispersion tensor (m^2/s)
e	electric field intensity (V/m)
e	electric field intensity vector (V/m)
E	total radiation emissive power $\langle n^2 \rangle \sigma T^4$ (W/m^2)
E_s	modules of elasticity (Pa)

677

Nomenclature

E	strain tensor
f	frequency (Hz), force (N)
f_δ	van der Waals force (N/m^2)
f	force vector (m/s^2)
F	scattering matrix or free energy (J)
$F_{0-\lambda T}$	fraction of blackbody emission power in spectral region $0 - \lambda T$
g	gravitational constant (m/s^2)
g	gravitational acceleration vector (m/s^2)
h	gap size (m) or heat transfer coefficient (W/m^2-K) or height (m) or magnetic field intensity (A/m)
h_{sf}	interstitial convection heat transfer coefficient (W/m^2-K)
h_P	Planck constant 6.626×10^{-34} (J-s)
h	magnetic field intensity vector (A/m)
H	mean curvature of the meniscus $(1/2)(1/r_1 + 1/r_2)$ (m) where r_1 and r_2 are the two principal radii of curvature
i	volumetric enthalpy (J/m^3) or specific enthalpy (J/kg) or $(-1)^{1/2}$ or radiation intensity (W/m^2)
I	radiation intensity (W/m^2)
I	second-order identity tensor
j$_e$	current density (A/m^2)
J	Leverett function
J_i	Bessel function
Ja	Jakob number $c_p \Delta T / \Delta i_{\ell g}$
k	thermal conductivity (W/m-K), or wave number (1/m)
k_B	Boltzmann constant 1.381×10^{-23} (J/K)
k_e	effective thermal conductivity (W/m-K)
k_K	Kozeny constant
k_p	equilibrium partition ratio
K	permeability (m^2)
K_{rg}	nonwetting phase relative permeability
$K_{r\ell}$	wetting phase relative permeability
Kn	Knudsen number, λ(mean free path)/C(average interparticle clearance)
K	permeability tensor (m^2)
K$_e$	effective thermal conductivity tensor (W/m-K)
ℓ	linear length scale for representative elementary volume or unit-cell length (m)
ℓ	length of a period vector (m)
L, L_1, L_2	system dimension, linear length scale (m)
L$_t^*$	tortuosity tensor
m	complex refraction index $n - i\kappa$ or mobility ratio $(K_{r\ell}\mu_g)/(K_{rg}\mu_\ell)$
$\dot{m}$	mass flux (kg/m^2-s)
$\dot{m}^*_{ad}$	adsorption capacity
ṁ	mass flux vector (kg/m^2-s)
M	molecular weight (kg/kg·mol)

Ma	Marangoni number $(\partial\sigma/\partial T)(\Delta T)R/(\alpha\mu)$
n	index of refraction, also number of molecules per unit volume (molecules/m^3)
$\dot{n}_i$	volumetric rate of production of component i (kg/m^3-s)
n_r	number of components in the gas mixture
$\mathbf{n}$	normal unit vector
N_A	Avogadro number 6.0225×10^{23} (molecules/mol)
Nu_x	Nusselt number $qx/k\Delta T$
p	pressure (Pa)
p_c	capillary pressure (Pa)
P	probability density function
P_i	Legendre polynomial
Pe_x	Peclet number $Re_x Pr = ux/\alpha$
Pr	Prandtl number $\nu/\alpha = (\mu c_p)_f/k_f$
Pr_e	effective Prandtl number $\nu/\alpha_e = (\mu c_p)_f/k_e$
q	heat flux (W/m^2)
$\mathbf{q}$	heat flux vector (W/m^2)
Q	heat flow rate (W)
r	radial coordinate axis (m), separation distance (m)
$\mathbf{r}$	radial position vector (m)
R	radius (m)
R_g	universal gas constant 8.3144 (kJ/kg·mol-K) $= k_B N_A$
Ra	Rayleigh number $g\beta\Delta T R^3/(\alpha\nu)$
Re_x	Reynolds number ux/ν
s	saturation
s_1	surface saturation
$\dot{s}$	volumetric heat generation (W/m^3)
s_{ir}	immobile (or irreducible) wetting-phase saturation
$s_{\text{ir }g}$	immobile (or irreducible) nonwetting-phase saturation
s_g	nonwetting phase saturation $\varepsilon_g/\varepsilon$
$\mathbf{s}$	unit vector
S	reduced (or effective) saturation $(s - s_{\text{ir}})/(1 - s_{\text{ir}} - s_{\text{ir }g})$ or path length (m)
$\mathbf{S}$	shear component of stress tensor (Pa)
Sc	Schmidt number ν/D
Sh_x	Sherwood number $\dot{m}x/D_m\Delta\rho$
t	time (s)
$\mathbf{t}$	tangential unit vector
T	temperature (K)
$\mathbf{T}$	stress tensor (Pa)
T_r	transmissivity
u, v, w	components of velocity vector in x-, y-, and z-directions (m/s)
$\mathbf{u}$	velocity vector (m/s)
$\mathbf{u}_D$	Darcean (or superficial) velocity vector (m/s)
u_F	front velocity (m/s)

$\mathbf{u}_p$	pore (interstitial or fluid intrinsic) velocity vector (m/s)
V	volume (m³)
x, y, z	coordinate axes (m)
$\mathbf{x}$	position vector (m)
w	width (m)
W	work (J)
We	Weber number where ℓ stands for the wetting phase $\rho_\ell u_{D\ell}^2 d/\sigma$
x_F	front location (m)
Y_i	mass fraction of species i

Greek

α	thermal diffusivity (m²/s) or slip coefficient
α_e	effective thermal diffusivity $k_e/(\rho c_p)_f$ (m²/s)
α_R	radiation size parameter $2\pi R/\lambda$
α_T	temperature slip coefficient
β	volumetric expansion coefficient (1/K)
δ	boundary layer thickness (m) or liquid film thickness (m)
δ_j	$j = 1, 2, \ldots$, linear dimension of microstructure (m)
Δ	surface roughness (m)
ΔE_a	specific heat of activation or adsorption energy (kJ/kg), also given on molar basis
Δi_c	specific heat of combustion (kJ/kg), also given on molar basis
$\Delta i_{\ell g}$	specific heat of evaporation (J/kg)
Δi_o	specific formation energy (kJ/kg), also given on molar basis
ε	porosity
ε_ℓ	dielectric constant (relative permeability)
ε_o	permittivity of free space
ε_r	emissivity
η	scattering efficiency
η_λ	specular efficiency
θ	polar angle (rad)
θ_c	contact angle (rad) measured through the wetting fluid
θ_o	angle between incident and scattered beam (rad)
κ	compressibility (1/Pa) or index of extinction
λ	wavelength (m) or mean free path (m)
μ	dynamic viscosity (kg/m-s) or $\cos\theta$ (used in radiation or chemical potential) (J/mol)
μ_m	magnetic permeability (N/A²)
μ_p	Poisson ratio
ν	kinematic viscosity μ/ρ (m²/s)
ρ	density (kg/m³) or electrical resistivity (ohm/m) or reflectivity
σ	Stephan-Boltzmann constant 5.6696×10^{-8} (W/m²-K⁴) or surface tension (N/m)
σ_a	absorption coefficient (1/m)

σ_e	electric conductivity (mho/m)
σ_{ex}	extinction coefficient $\sigma_a + \sigma_s$ (1/m)
σ_s	scattering coefficient (1/m)
τ	optical thickness, $(\sigma_s + \sigma_a)\Delta S$ or shear stress (Pa)
τ	tangent unit vector
ϕ	azimuthal angle (rad)
φ	scalar
Φ	particle scattering phase function
ω	angular frequency (rad/s)
ω_a	scattering albedo $\sigma_s/(\sigma_s + \sigma_a)$
Ω	solid angle (sr)

Superscripts

$\hat{}$	Fourier Laplace or other transformation
—	average value
†	transpose
*	dimensionless quantity
′	deviation from volume-averaged value or directional quantity
d	dispersion component
ℓ	liquid
f	fluid
fs	fluid-solid
g	gas
s	solid
sf	solid-fluid

Subscripts

0	reference or surface
1	surface condition
a	absorption or ambient
ad	adsorbed
b	blackbody radiation
B	boundary
c	condensible
cr	critical
d	particle
D	Darcy
e	effective or emission
ex	extinction
f	fluid-phase
fs	solid-fluid interface
g	gas-phase
$g\ell$	gas (or nonwetting phase) liquid interface

$g\ell1, g\ell2$	gas-liquid drag
$g\Delta\sigma$	gas-phase surface tension gradient
gi	gas-phase inertia
h	hydraulic
H	curvature
i	interfacial or incident or component i or inside
ind	independent
j	component j
K	Knudsen or Kozeny
ℓ	liquid (or wetting phase), representative elementary volume
ℓf	liquid film
ℓg	liquid-gas
$\ell g1, \ell g2$	liquid-gas drag
ℓi	liquid-phase inertia
$\ell\Delta\sigma$	liquid-phase surface tension gradient
L	system
m	molecular or mass or mean or wetting
n	normal
o	reference or surface
p	pore
pa	porous-ambient
r	radiation or reflected or relative
s	solid or scattering or symmetric or saturation
s_b	bounding solid surface
$s\ell$	solid-liquid interface
sf	solid-fluid
sg	solid-gas interface
t	transmitted
T	temperature
v	vapor
x, y	x-, y-component
β	liquid or wetting phase or displacing phase
λ	wavelength dependent
$\perp$	axial (or longitudinal) component
$\|$	lateral (or transverse) component

Others

[ ]	matrix
$\langle\ \rangle$	volume average
$o(\ )$	order of magnitude

Glossary

Absorption coefficient: Inverse of the mean free path that a *photon* travels before undergoing *absorption*. The *spectral* absorption coefficient a_λ is found from $I_\lambda(x) = I_\lambda(0) \exp\left[-\int_0^x \sigma_{\lambda\,a}(x)\,dx\right]$, where the beam is traveling along x.

Adsorption: Enrichment of one or more components in an interfacial layer.

Adsorption isotherm: Variation of the extent of enrichment of one component (amount adsorbed) in the solid-gas *interfacial layer* with respect to the gas pressure and at a constant temperature. For porous media, the amount adsorbed can be expressed in terms of the gram of gas adsorbed per gram of solid.

Brinkman screening length: A distance $o\left(K^{1/2}\right)$ over which the velocity disturbances, caused by a source, decay; the same as the boundary-layer thickness.

Bulk properties: Quantities measured or assigned to the matrix/fluid system without consideration of the existence of boundaries (due to finiteness of the system). Some properties take on different values than their bulk values at or adjacent to these boundaries.

Capillary pressure: Local pressure difference between the *nonwetting* phase and *wetting* phase (or the pressure difference between the concave and convex sides of the meniscus).

Channeling: In packed beds made of nearly spherical particles, the packing near the boundaries is not uniform and the local porosity (if a meaningful representative elementary volume could be defined) is larger than the bulk porosity. When the packed bed is confined by a solid surface and a fluid flows through the bed, this increase in the local porosity (a decrease in the local flow resistance) causes an increase in the local velocity. This increase in the velocity adjacent to the solid boundary is called channeling.

Coordination number: The number of contact points between a sphere (or particle of any regular geometry) and adjacent spheres (or particles of the same geometry).

Darcean flow: A flow that obeys $\mathbf{u}_D = -(\mathbf{K}/\mu) \cdot \nabla p$.

Dispersion: In the context of heat transfer in porous media and in the presence of a net Darcean motion and a temperature gradient, dispersion is the *spreading* of heat accounted for *separately* from the Darcean convection and the effective (collective) molecular conduction. It is a result of the simultaneous existence of temperature and velocity gradients within the pores. Due to the volume averaging over the pore space, this contribution is not included in the Darcean convection, and because of its dependence on $\nabla \langle T \rangle$, it is included in the *total effective thermal diffusivity tensor*.

Drainage: Displacement of a wetting phase by a nonwetting phase. Also called *desaturation* or *dewetting*. A more *restrictive* definition requires that the only force present during draining must be the capillary force.

Dupuit-Forchheimer velocity: Same as *pore* or *interstitial velocity*, defined as u_D/ε where u_D is the filter (or *superficial or Darcy*) velocity and ε is porosity.

Effective porosity: The interconnected void volume divided by the total (solid plus total void) volume. The effective porosity is smaller than or equal to porosity.

Effective thermal conductivity: Local volume-averaged thermal conductivity used for the fluid-filled matrices along with the assumption of local thermal equilibrium between the solid and fluid phases. The effective thermal conductivity is *not* only a function of porosity and the thermal conductivity of each phase, but is very sensitive to the microstructure.

Extinction of radiation intensity: Sum of the *absorbed* and *scattered* radiation energy, as the incident beam travels through a particle or a collection of particles.

Extinction coefficient: Sum of the *scattering* and *absorption* coefficients $\sigma_s + \sigma_a = \sigma_{\text{ex}}$.

Formation factor: Ratio of electrical resistivity of fully *saturated matrix* (with an electrolyte fluid) to the electrical resistivity of the *fluid*.

Funicular state: Or *funicular flow regime*. The flow regime in two-phase flow through porous media, where the wetting phase is continuous. The name funicular is based on the concept of a continuous wetting phase flowing on the outside of the nonwetting phase and over the solid phase [this two-phase flow arrangement is not realized in practice; instead each phase flows through its individual network of interconnected channels (Dullien, 1979, p. 252)].

Hydrodynamic dispersion: In the presence of both a net and nonuniform fluid motion and a gradient in temperature that portion of diffusion or spreading of heat caused by the nonuniformity of velocity within each pore. Also called *Taylor-Aris dispersion*.

Hysteresis: Any difference in behavior associated with the past state of the phase distributions. Examples are the hysteresis loop in the capillary pressure-saturation, relative permeability-saturation, or adsorption isotherm-saturation curves. In these curves, depending on whether a given saturation is reached through *drainage* (reduction in the wetting phase saturation) or by *imbibition* (increase in the wetting phase saturation) a *different* value for the dependent variable is found.

Imbibition: Displacement of a nonwetting phase with a wetting phase. Also called *saturation, free,* or *spontaneous imbibition.* A more restrictive definition requires that the only force present during imbibition be the capillary force.

Immiscible displacement: Displacement of one phase (wetting or nonwetting) by another phase (nonwetting or wetting) *without* any mass transfer at the interface between the phases (diffusion or phase change). In some cases a displacement or *front* develops and right downstream of it the saturation of the phase being displaced is the largest, and behind the front, the saturation of the *displacing* phase is the largest.

Immobile or irreducible saturations: s_{ir}, the reduced volume of the *wetting* phase retained at the *highest* capillary pressure. For very smooth surfaces the wetting phase saturation does not reduce any further as the capillary pressure increases. However, for rough and etched surfaces the irreducible saturation can be zero. The *nonwetting* phase immobile saturation $s_{ir\,g}$ is found when the capillary pressure is nearly zero and yet some of the nonwetting phase is trapped.

Infiltration: Displacement of a wetting (or nonwetting) phase by a nonwetting (or wetting) phase.

Intrinsic phase average: For any quantity ψ, the intrinsic phase average over any phase ℓ is defined as

$$\langle \psi_\ell \rangle^\ell = \frac{1}{V_\ell} \int_{V_\ell} \psi_\ell \, dV.$$

If ψ_ℓ is a constant, then $\langle \psi_\ell \rangle^\ell = \psi_\ell$. The intrinsic average is useful in analysis of multiphase flow and in dealing with the energy equation.

Knudsen diffusion: When the Knudsen number Kn satisfies

$$Kn = \frac{\lambda \, (\text{mean free path})}{C \, (\text{average interparticle clearance})} > 10$$

then the gaseous mass transfer in porous media is by the molecular or Knudsen diffusion. In this *regime*, the intermolecular collisions do *not* occur as frequent as the molecule-wall (matrix surface) collisions, i.e., the motion of molecules is *independent* of all the other molecules present in the gas.

Laplace or Young-Laplace equation: Equation describing the capillary pressure in terms of the liquid-fluid interfacial curvature, $\langle p_c \rangle = \langle p \rangle^g - \langle p \rangle^\ell = 2\sigma H$.

Local thermal equilibrium: When the temperature difference between the solid and fluid phases is much smaller than the smallest temperature difference across the system at the level of representative elementary volume, i.e., $\Delta T_\ell \ll \Delta T_L$.

Macroscopic behavior: System-level (over the entire volume of the porous medium) variations in velocity, temperature, pressure, concentration, and porosity.

Matrix or solid matrix: A solid structure with *distributed* void space in its interior as well as its surface; the solid structure in the porous medium.

Mean penetration distance of radiation: $(\sigma_a + \sigma_s)^{-1}$ is the inverse of the sum of the absorption and scattering coefficients.

Mechanical dispersion: That portion of diffusion or spreading of heat that is due to the presence of the matrix (mechanical dispersion is present only for matrices with random structures) independent of molecular diffusion and in the presence of both a net fluid motion and a gradient in temperature. The tortuous path the fluid particle takes in disordered porous media, as it moves through the matrix, makes it continuously branch out into neighboring conduits, causing spreading of its heat content when a temperature gradient exists.

Microscopic behavior: Pore-level variations in velocity, temperature, pressure, and concentration. This is different than the micro- or molecular-level variations used in statistical mechanics.

Mobility ratio: The ratio of flow conductivity of the displacing phase to that of the displaced phase $m = (K_{r\ell}\mu_g)/(K_{rg}\mu_\ell)$. For miscible displacement, the mobility ratio is the viscosity ratio.

Molecular diffusion: Diffusion or spreading of heat content in the presence of a temperature gradient and absence of a net fluid motion. This molecular diffusion is caused by the *Brownian motion* of the fluid particles.

Optical thickness: $\tau = (\sigma_a + \sigma_s)d$ or the *number* of mean penetration distances a photon encounters as it passes through the particle of diameter d (or as it passes through a finite length d). The optical path length is $\tau(x) = \int_0^x (\sigma_a + \sigma_s)\,dx$.

Partial saturation: When both liquid and gaseous phases occupy the pores *simultaneously*, each occupying a portion of the representative elementary volume.

Pendular state: Or *pendular stage*. The phase distribution at very low saturations (wetting phase), where the wetting phase is distributed in the pores as discrete masses. Each mass is a ring of liquid wrapped around the contact point of adjacent elements of the solid matrix (Dullien, 1979, p. 29).

Phase average: For any quantity ψ, the phase average over any phase ℓ is defined as

$$\langle \psi_\ell \rangle = \frac{1}{V} \int_{V_\ell} \psi_\ell \, dV.$$

If ψ_ℓ is a constant, then $\langle \psi_\ell \rangle = \varepsilon_\ell \psi_\ell$, where $\varepsilon_\ell = V_\ell/V$. In dealing with the single-phase flow, the phase average suffices, otherwise the intrinsic phase average is used.

Plain media: The domain where no solid matrix is present, i.e., the ordinary fluid domain.

Porosity: Ratio of void volume to total (solid plus void) volume.

Reduced or effective saturation: The wetting phase saturation normalized using the immobile saturations,

$$\frac{s - s_{\text{ir}}}{1 - s_{\text{ir}} - s_{\text{ir }g}}.$$

Representative elementary volume: The smallest differential volume that results in statistically *meaningful* local average properties such as local porosity, saturation, and capillary pressure. When the representative elementary volume is appropriately chosen, the limited addition of extra pores around this local volume will not change the values of local properties.

Saturated matrix: A matrix fully filled with one fluid.

Saturation: The volume fraction of the void volume occupied by a fluid phase, $s = \varepsilon_\ell/\varepsilon$, $0 \leq s \leq 1$. Generally the wetting phase saturation $\varepsilon_\ell/\varepsilon$ is used.

Scattering: Interaction between a photon and one or more particles where the photon does *not* lose its entire energy.

Scattering albedo: The scattering coefficient divided by the sum of the scattering and absorption coefficients, $\sigma_s(\sigma_s + \sigma_a)^{-1}$. For purely scattering media, the albedo is one and for purely absorbing media it is zero.

Scattering by diffraction: The change in the direction of motion of a photon as it passes *near* the edges of a particle.

Scattering by reflection: The change in the direction of a photon as it collides with a particle and is reflected from the particle surface.

Scattering by refraction: The change in the direction of a photon as it penetrates *through* and then escapes from a particle.

Scattering coefficient: Inverse of the mean free path that a photon travels before undergoing scattering. The spectral scattering coefficient $\sigma_{\lambda s}$ is found from $I_\lambda(x) = I_o(x) \exp(-\int_0^x \sigma_{\lambda s}\, dx)$, where the beam travels along x.

Scattering phase function: Scattered intensity in a direction (θ, ϕ), divided by that intensity corresponding to isotropic scattering. This includes the *reflected*, *refracted*, and *diffracted* radiation scattered in any direction.

Specific surface area: Or *volumetric surface area*. The surface area of pores (interstitial surface area or surface area between the solid and fluid) per unit volume of the matrix. Direct and inferred methods of measurement are discussed in Scheidegger (1974). In some specific applications, the volume is taken as the volume of the *solid* phase.

Spectral or monochromatic: Indicates that the quantity is for a specific wavelength.

Superficial velocity: Same as Darcy or the filter velocity. It is the volumetric flow rate divided by the surface area (both solid and fluid), so it can be readily used to calculate flow rates.

Thermal transpiration: When a temperature gradient exists in a porous medium, the gas saturating it flows due to this temperature gradient. The coefficient L in $\rho \mathbf{u}_D = -(L/T)\nabla T$ depends on the gas and local temperature.

Tortuosity: Traditionally the length of the actual path line between two ports that the fluid particle travels divided by the length of a straight line between these ports. This path is taken by a *diffusion (Brownian) motion* and is independent of the net velocity. In the modern usage the tortuosity is found from $\varepsilon(1 + L_t^*) = k_e/k_f$ for $k_s = 0$. L_t^* is also called the tortuosity (Carbonell and Whitaker, 1983). The tortuosity tensor is designated by $\mathbf{L}_t^*$.

Total thermal diffusivity tensor: $\mathbf{D}$, the sum of the *effective thermal diffusivity tensor*, $\mathbf{K}_e/(\rho c_p)_f$, where

$$\mathbf{K}_e/k_f = \mathbf{K}_e/k_f\left(k_s/k_f, \varepsilon, \text{structure}\right),$$

and the *dispersion tensor*

$$\mathbf{D}^d = \mathbf{D}^d\left(\frac{k_s}{k_f}, \varepsilon, \text{structure}, Re, Pr, \text{and } \frac{(\rho c_p)_s}{(\rho c_p)_f}\right),$$

i.e., $\mathbf{D} = \mathbf{K}_e/(\rho c_p)_f + \varepsilon \mathbf{D}^d$.

Void ratio: Ratio of void fraction (porosity) to solid fraction, i.e., $\varepsilon(1-\varepsilon)^{-1}$.

Wetting phase: The phase that has a smaller contact angle (the contact angle is measured through a perspective phase).

Citation Index

Ackerberg, R. C., et al., 587
Acrivos, A., 128, 190
Acrivos, A., et al., 208
Adler, P. M., 21, 28, 495
Afgan, N. M., et al., 624
Ahuja, S., et al., 668
Altobelli, S. A., 53, 93
Antoniou, A. A., 518
Aris, R., 158, 200
Arpaci, V. S., 279
Auriault, J.-L., 45
Awschalom, D. D., et al., 539, 542
Azad, F. H., 317

Bachmat, Y., 61
Baldi, G., 555, 564
Bankoff, G., 596
Barber, D. J., 534
Basel, M. D., 631
Batchelor, G. K., 127
Batycky, R. P., et al., 158
Bau, H. H., 616, 621, 623
Bayazitoğlu, Y., 318
Bear, J., 54, 61, 101, 193, 459, 464, 470
Beasley, D. E., 102
Beavers, G. S., 72, 150
Beavers, G. S., et al., 72, 85, 102, 104
Beckermann, C., 658, 668, 669
Behl, S., 70, 98, 108
Behrendt, D. R., 655
Bejan, A., 11
Benenati, R. F., 103
Bennon, W. D., 658, 662
Bensabat, J., 459, 464, 470
Bensoussan, A., et al., 61
Beran, M., 136
Bergles, A. E., 624

Beveridge, G. S. G., 141
Bird, R. B., et al., 378, 642
Bischoff, K. B., 104
Bohren, G. F., 269, 289
Bongiorno, V., 510
Born, M., 279
Brady, J. F., 52, 183, 191, 200, 205, 385
Brenner, H., 33, 183, 205, 495
Bressler, R. G., 573
Bretherton, F. P., 441
Brewster, M. Q., 260, 295, 307, 313
Brinkman, H. C., 40, 52, 66, 92
Brosilow, C. B., 103
Bruno, C., et al., 383
Buchin, J.-M., 616
Buckius, R. O., 331, 383
Buckley, S. E., 497
Burelbach, J. P., et al., 459

Cahn, J. W., 510
Carbonell, R. G., 33, 120, 167, 227, 243, 383, 392, 398, 466, 488, 596
Carlson, B. G., 322, 325
Carman, P. C., 32
Carslaw, H. S., 224, 567
Cartigny, J. D., et al., 294
Catton, I., 196, 458, 573, 575, 578
Cess, R. D., 113, 313
Chai, J. C., et al., 327
Chan, C. K., 333
Chandrasekhar, B. C., 106
Chandrasekhar, S., 260, 279, 313, 322
Chang, H.-C., 128
Chang, S. L., 297
Chang, S.-H., 124
Chappuis, J., 374, 432, 434, 448, 452
Chatwin, P. C., 233

689

Chen, C. K., 128
Chen, J. C., 305, 315
Chen, J.-D., 495
Cheng, P., 11, 98, 247, 253, 318, 610
Chern, B.-C., et al., 346
Chikhliwala, E. D., 500
Chikhliwala, E. D., et al., 500
Chou, W. T.-H., 639
Chu, C. F., 104
Chu, C. M., 328
Chu, C. M., et al., 308
Chui, E. H., 313, 327
Chung, J. N., et al., 611
Churaev, N. V., 455
Churchill, S. W., 305, 315, 328
Chyu, M. C., 624
Clark, J. A., 102
Cole, M. W., 539, 541
Combarnous, M. A., 158, 213, 220
Cook, R., et al., 573
Corey, A. T., 488
Corson, D. R., 268, 275
Cox, R. G., 453
Cramer, K. R., 113
Crawford, M. E., 407, 587, 593
Crine, M., 561, 569
Crine, M., et al., 561
Crosbie, A. L., 277, 331
Cunningham, G. R., 128
Cunningham, R. E., 365, 371, 374

Darcy, H., 17
Dash, S. K., 669
Davidson, G. W., 277
Davis, L. B., 139
Davison, B., 313, 330
Davison, G. W., 331
de Gennes, P. G., 448, 452
de Gennes, P. G., et al., 448, 453
De Josselin De Jong, G., 196
De Souza, J. F. C., 219
Defay, R., 429, 432, 507, 511, 519, 530
Deissler, R. G., 317
Delshad, M., 488
Demond, A. H., et al., 477
Derjaguin, B. V., 455
Dhir, V. K., 442, 466, 481, 488, 616
Dixon, A. G., 147

Drolen, B. L., 294, 307, 334, 340, 341, 346
du Plessis, J. P., 48
Dullien, F. A. L., 20–22, 27, 29, 442, 445, 471, 474, 475, 481, 484, 497
Dullien, F. A. L., et al., 471
Durlofsky, L., 52
Dussan, E. B., 452, 453
Dybbs, A., 48, 67, 104
Dzyaloskinskii, I. E., et al., 456, 523

Eckert, E. R. G., et al., 616
Edwards, R. V., 48, 67, 104
Eidsath, A., et al., 38, 172
Elrick, D. E., 223
Ene, H. I., 61, 71, 89
Epstein, N., 385
Ergun, S., 46
Erle, M. A., et al., 435
Ernst, W. R., 383
Evans, R., et al., 533, 538

Fakheri, A., 383
Fand, R. M., et al., 617
Fatehi, M., 88, 388
Felske, J. D., 279
Firdaouss, M., 45
Fisher, D. J., 658
Fiveland, W. A., 322, 324, 356
Fogler, H. S., 387
Fortinn, A., 153
Frickle, J., 128, 134, 145
Fried, J. J., 158, 213, 220
Fukusako, S., et al., 616

Ganesan, S., 661
Gauvin, W. H., 51
Georgiadis, J. G., 196
Giedt, W. H., 134
Gill, N. M., 669
Gill, W. N., 158
Gilver, R. C., 53, 93
Glassman, I., 387, 585
Glatzmaier, G. C., 121
Glicksman, L., et al., 308, 310, 333
Goodier, J. N., 132
Goplen, S. P., 573
Gordon, D. T., 481

Grassmann, P., 237
Gray, W. A., et al., 55
Gray, W. G., 459
Greenkorn, R. A., 197
Gregg, S. J., 523, 525
Grosser, K., et al., 442, 466
Grosshandler, W. H., 309
Guermond, J.-L., 45
Gunn, D. J., 174, 180, 213, 217, 219, 222
Gunn, R. D., 376
Gupte, A. R., 33

Hadley, G. R., 128
Hall, M. J., 112
Han, N.-W., et al., 160, 220
Hanamura, K., 645, 651
Hanratty, T. J., 51
Hansen, J. E., 280
Happel, J., 33
Haring, R. E., 197
Harrison, B. K., 383
Hashimoto, K., et al., 555, 564
Hashin, Z., 136
Hasimoto, H., 183
Hassanizadeh, S. M., 459
Haughey, D. P., 141
Hendicks, T. J., 331
Hendricks, T. J., 309, 326
Hiatt, J. P., 112
Higdon, J. J. L., 41, 82
Higenyi, J., 318
Hilliard, J. E., 510
Hills, R. N., 666
Hills, R. N., et al., 658
Hocking, L. M., 453
Holm, F. W., 573
Homsy, G. M., 495, 496, 575
Horn, F. J. M., 200
Horn, R. G., 450
Hottel, H. C., et al., 294, 307, 312, 322, 346
Howell, J. R., 260, 268, 281, 283, 309, 313, 318, 337
Hrubesh, L. W., 128, 134
Hsieh, C. K., 289
Hsieh, J.-S., 511
Hsu, C. T., 247, 253
Hsu, C. T., et al., 145

Huang, A. B., 500
Huffman, D. R., 269, 289
Hupport, H. E., 658

Ilic, M., 639
Incropera, F. P., 658, 662
Ishii, M., 491
Ishimaru, A., 294, 307, 346
Israelachvili, J. N., 374, 432, 448, 450

Jackson, R., 365, 374
Jaeger, J. C., 224, 567
Jamaluddin, A. S., 322
Jeans, J. H., 318
Jeffrey, D. J., 206, 213
Jennings, J. D., 621
Jerauld, G. R., 484
Jerauld, G. R., et al., 500
Joanny, J. F., 452
Jodrey, W. S., 334
Jolls, K. R., 51
Jones, I. P., 87
Jones, S. W., et al., 618, 623
Joseph, D. D., 72, 150
Joseph, J. H., et al., 330

Kaguei, S., 365, 401, 403
Kamiuto, K., 348, 357
Kanury, A. M., 366
Kaviany, M, 410
Kaviany, M., 11, 38, 42, 59, 68, 79, 88, 106, 113, 125, 149, 233, 235, 241, 245, 254, 297, 321, 334, 342, 345, 346, 388, 404, 416, 438, 583, 590, 596, 610, 639, 645, 651, 656, 658
Kaviany. M, 175
Kays, W. M., 407, 587, 593
Kee, R. J., et al., 382
Kee, R. J., et al., 370, 372
Kerker, M., 268
Kerker, M., et al., 279
Kesten, A. S., 383
Khalid, M., 217
Kim, C.-J., 658
Kim, I. C., 29
King, C. J., 376
Kocamustafaogullari, G., 491
Koch, D. L., 183, 191, 200, 205, 385

Koch, D. L., et al., 183, 190, 385
Koh, J. C. Y., 153
Konev, S. K., et al., 624
Koplik, J., et al., 93
Kovalev, S. A., et al., 624
Kruger, E. H., 371
Krupiczka, R., 144
Krylov, V. S., 439
Ku, J. C., 279
Kudo, K., et al., 334
Kuga, Y., 294, 307, 346
Kumar, S., 294, 346
Kumar, S., et al., 322
Kunii, D., 144, 146, 242
Kuo, K. K. Y., 382
Kuo, S. M., 247
Kurz, W., 658
Kurzweg, U. H., 233
Kurzweg, U. H., et al., 237
Kyan, C. P., et al., 34

Ladd, T., 42
Larson, R. E., 41, 82
Lathrop, K. D., 322, 325
Lawson, D. W., 223
Lee, C. K., et al., 166
Lee, H., 302, 331
Lee, S. C., et al., 302
Lefebvre du Prey, E. J., 483
Leith, J. R., 145
Lenormand, R., 495
Levec, J., 392
Levec, J., et al., 481, 487
Leverett, M. C., 471, 473, 497
Levich, V. G., 439
Levy, Th., 71
Lienhard, J .H., 517
Lienhard, J. H., 139, 366, 371
Ling, S. C., 587
Liou, K.-N., 280
Loeffler, A. L., Jr., 34
Lorrain, P., 268, 275
Loudun, R., 534
Luikov, A. V., 639
Luikov, A. V., et al., 134
Lundgren, T. S., 52, 67, 70, 92
Lykoudis, P. S., 113

Macdonald, I. F., et al., 46

Majumdar, A., 611
Marshall, H., et al., 38, 44, 502
Marteney, P. J., 383
Martin, H., 103, 227
Martinez, M. J., 440
Matsuura, A., et al., 554, 564
Mazza, G. D., et al., 339
Mei, C., 166
Mei, C. C., 45
Melanson, M. M., 147
Melrose, J. C., 445, 477
Menegus, D. K., 631
Mengüc, M. P., 259, 305, 318
Meres, A. W., 468, 478
Mickley, H. S., et al., 51
Miller, C. A., 496, 631
Miller, M., 136
Mills, A. F., 375
Minkowycz, W., 11
Mirzamoghadam, A., 458, 573, 575, 578
Mittal, M., 639
Modest, M. F., 260, 317
Monterio, S. L. P., 309
Moosman, S., 575
Morel-Seytoux, H. J., 497, 499
Morgan, J. T., 481
Morrow, N. R., 446
Müller, U., 462, 488
Muskat, M., 468, 478
Myers, J. E., 555, 564, 566
Myshkis, A. D., et al., 434, 453

Nader, W. K., 40, 93, 95, 189, 384
Naik, A. S., 442, 616
Nayak, L., 139
Neale, G. H., 40, 93, 95, 189, 384
Nelson, H. F., et al., 308
Newel, R., 104
Ng, K. M., 104, 443, 481
Ng, K. M., et al., 447
Ngan, C. G., 453
Nguyen, V. V., et al., 386
Ni, J., 668, 669
Nield, D. A., 11
Nimick, F. B., 145
Nozad, I., 145
Nozad, I., et al., 124
Nunge, R. J., 158

O'Brien, R. K., 127
Ochoa-Tapia, J. A., et al., 127, 145
Ofuchi, K., 146, 242
Ogniewicz, Y., 128
Okusu, N. M., 497
Oliver, F. W. J., 235
Orell, A., 596
Osoba, J. S., et al., 479, 481
Ozisik, M. N., 260, 311, 313

Pai, S.-I., 113
Palik, E. D., 289
Papini, M., 308
Parmentier, E. M., 610, 614
Payatakes, A. C., 495
Pekala, R. W., 128, 134
Penndorf, R. B., 291
Pesaran, A. A., 375
Peterson, B. K., et al., 533
Pfefferle, L. D., 383
Pfefferle, W. C., 383
Plumb, O. A., 588, 593, 611, 656, 657
Plumb, O. A., et al., 639
Poirer, D. R., et al., 658
Poirier, D. R., 661
Polisevski, D., 61
Poots, G., 113
Pope, G. A., 488
Porter, H. F., et al., 7
Prakash, C., 668
Prasad, V., et al., 145
Prat, M., 148
Prescott, P. J., et al., 668
Prigogine, I., 429, 432, 507, 511, 519, 530
Pryce, C., 174, 180, 213

Quiblier, J. A., 28
Quintard, M., 127, 158, 174, 392, 416
Quintard, M., et al., 402

Raithby, G. D., 313, 327
Ramesh, P. S., 621
Ramirez, W. F., 121
Rappaz, M., 658, 663, 665, 666
Ratzel, A. C., 318
Reed, A. W., 616
Reid, R. C., et al., 371

Reinelt, D. A., 440
Reis, J. F. G., et al., 212
Renken, K. J., et al., 604
Rhee, K. T., 297
Richardson, S., 78
Rish, J. W., 322
Rivers, A. D., 453
Roberts, P. H., 666
Rogers, J. A., 438, 639
Romig, M. F., 112
Rose, J. W., 608
Rossow, V. J., 113
Roux, J. A., 322
Rubenstein, J., 65
Rumpf, H., 33
Ryan, D., et al., 126, 175, 383, 384

Saam, W. F., 539, 541
Saez, A. E., 434, 466, 488, 558, 596
Saez, A. E., et al., 483, 554, 558, 560, 569
Saffman, P. G., 75, 197, 208, 212, 440
Sahimi, M., et al., 194
Sahraoui, M., 38, 42, 79, 106, 125, 149, 175, 241, 245, 254, 410, 416
Saleh, S., et al., 101
Salter, S. J., 484
Sanchez-Palencia, E., 61, 71, 89
Sangani, A. S., 70, 98, 108, 128, 190
Scheidegger, A. E., 19, 21, 22, 28, 29, 102, 193, 442, 470, 484, 495
Schertz, W. W., 104
Schlichting, H., 235
Schlünder, E. U., 144, 227, 580
Schmidt, F. W., 405
Schulenberg, T., 462, 488
Schulgasser, K., 138
Schwartz, C. E., 104
Scriven, L. E., 510
Selamet, A., 279
Shekarriz, A., 611
Sherwood, T. K., et al., 445
Shonnard, D. R., 126
Shtrikman, S., 136
Siegel, R., 260, 268, 281, 283, 313, 337

Sing, K. S. W., 523, 525
Singh, B. P., 297, 334, 342, 345, 346
Singh, M., 655
Slattery, J. C., 55, 66, 121, 124, 167, 377, 439, 447, 459
Slejko, F. L., 524
Smith, J. M., 104, 144
Smith, P. J., 322
Somerton, W. H., et al., 554
Sondergeld, C. H., 616, 620, 621
Soo, G. O., 447
Sparrow, E. M., 34, 113, 313
Specchia, V., 555, 564
Springer, G. S., 134
Standish, N., 104
Stanek, V., 47
Stanish, M. A., et al., 639
Stark, C., 128, 134, 145
Stewart, L. D., 631
Stewart, L. D., et al., 631
Stokes, J. P., et al., 495
Strieder, W., 135, 340
Strieder, W. C., 134
Stubos, A. K., 616
Styrikovich, M. A., et al., 624
Su, K. C., 289
Sullivan, R. R., 37
Swathi, P. S., 318
Szekely, J., 47

Tao, Y.-X., 583, 596
Taylor, G. I., 78, 158, 163
Thévoz, Ph., 663, 665
Thévoz, Ph., 666
Thiyagaraja, R., 94
Tien, C., 502
Tien, C.-L., 11, 66, 68, 128, 139, 247, 294, 302, 307, 313, 316, 333, 334, 340, 341, 346, 366, 371, 610, 611
Timoshenko, S. P., 132
Tobis, J., 247
Todorovic, P., 196
Tong, T. W., 302, 316, 318
Tong, T. W., et al., 302
Torquato, S., 29, 65, 138
Torrance, K. E., 616, 621, 623
Tory, E. M., 334
Truelove, J. S., 322

Truong, J. G., 456, 458, 521, 573
Tseng, J. W. C., et al., 334
Tsung, V. X., et al., 617
Tuma, M., 237
Tung, V. X., 444, 481, 488
Turcotte, D. L., 616, 621
Turner, I. W., 639
Tutu, N. K., et al., 488

Udell, K. S., 440, 497, 517, 616, 618, 621, 622, 631

Vafai, K., 11, 66, 68, 94, 98, 105, 138
van Brakel, J., 571
van de Hulst, H. C., 278, 302, 337
van der Merwe, D. F., 51
van Genuchten, M. Th., 478
Vidali, G., 539
Vincenti, W. G., 371
Viskanta, R., 259, 305, 318, 404, 658, 668
Vodak, F., et al., 658
Voller, V. R., 658
Voller, V. R., et al., 668
Vortmeyer, D., 98, 106, 247, 313, 340

Wakao, N., 365, 401, 403
Wang, C. Y., 668
Wang, C.-C., 588, 593
Wang, K. Y., 302
Ward, J. C., 48
Wardlaw, N. C., 474, 476
Watson, E. J., 233
Wayner, P. C., 456, 458, 521, 573, 574
Wayner, P. C., et al., 573
Weekman, V. W., 555, 564, 566
Whitaker, S., 33, 55, 67, 120, 126, 127, 158, 167, 174, 383, 392, 416, 459, 462, 545, 639
White, S. M., 610
Wilkinson, D., 495
Williams, F. A., 382
Williams, R. J. J., 365, 371, 374
Willis, D. R., 134
Willmott, A. J., 405
Wiscombe, W. J., 329
Wolf, E., 279
Wolf, J. R., 134

Wolf, J. R., et al., 346
Wooding, R. A., 66, 497, 499
Woods, A. W., 658
Worster, M. G., 658
Wyatt, P. W., 573
Wyllie, M. R. J., 480

Xia, Y., 340

Yagi, S., 242
Yagi, S., et al., 232
Yamada, Y., et al., 307
Yang, Y. S., et al., 333
Yortsos, Y. C., 500, 631

Yovanovich, M. M., 128
Yu, L., 474, 476
Yuan, Z., et al., 158
Yunis, L. B., 404

Zanotti, F., 392, 398
Zarcone, C., 495
Zehnder, P., 144
Zeng, S. Q., et al., 134
Zheng, L., 134
Zhu, H., 247
Zimmerman, W., 145
Ziolkowski, D., 247

Subject Index

Abrasives, 449
absolute permeability, 478
absorbed layer, 576
absorption coefficient, 261
accelerated recalescence, 668
activation energy, 374, 382
adiabatic flame temperature, 418
adjacent to bounding surfaces, 252
adsorbed film thickness, 457
adsorbed surfactant, 439
adsorption, 507, 523
    and surface flux, 373
    by solid surface, 524
    capacity, 524
    isotherm, 374, 523, 524
    of gases, 450
    resistance, 375
adsorption/desorption, 379
advancing
    contact angle, 451
    drop experiment, 454
aerogels, 134, 502
aerosols, 501
analogy with electro- and magneto-hydrodynamics, 112
angle of reflection, 336
angle of refraction, 281, 337
angular distribution functions, 277
anisotropic, 47, 313
anisotropic media, 19
anisotropy, 44, 553, 557, 561
approximation, 317, 322
areal contact, 128
Arrhenius relation, 412
asymmetry factor, 277
asymptotic behavior, 177, 181
asymptotic limit, 225
attenuating factor, 338

attractive
    fluid-fluid potential, 533
average incident intensity, 332
axial dispersion for a periodic structure, 558

Back scattered, 307
Beavers-Joseph slip coefficient, 75
binary
    solution, 520
    system, 654
binary mixture, 376, 388
blackbody emissive power, 261, 343
blackbody intensity, 261
blob, 447
Bond number, 440, 453, 483, 616
boundary effect, 107
boundary layer, 212
boundary-layer analyses, 11
boundary-layer thickness, 94
bounding surfaces, 545, 603
bounds for effective conductivity, 136, 138
Brinkman
    equation, 210
    model, 95
    screening distance, 9
    screening length, 41
    superposition of bulk and boundary effects, 52
Brownian diffusion, 502
Brunauer-Emmett-Teller equation, 527, 542
bubble/dispersed-bubble regimes, 443
Buckley-Leverett
    equation, 498
Buckley-Leverett front, 497, 501
bulk permeability, 74

697

bulk transverse dispersion coefficient, 251
bulk-mixed gas temperature, 413
buoyancy-influenced dendritic crystal growth, 662
buoyant liquid and crystal motions, 668
buoyant motion of a bubble in a tube, 440

Capillarity, 497
capillary
    condensation, 507, 523, 529
    condensation/evaporation, 530
    equilibrium, 446
    force, 455
    meniscus, 455
    models, 29
    number, 440, 447, 453, 495
    phenomena, 438
    pressure, 7, 427, 429, 440, 471
    pressure hysteresis, 471
Carman-Kozeny
    equation, 33, 208
    theory, 32
centrifuge method, 445
ceramic foams, 309
ceramics, 128
change in freezing temperature, 517
change in triple-point temperature, 518
channeling, 21, 101, 107
characteristic equation, 500
charge-polarity-based forces, 531
chemical
    equilibrium, 512
    kinetic, 382
    potential, 519, 521
    reaction, 1, 380, 409
classification
    of adsorption isotherms, 525
    of particle size, 7
    of pore size, 526
Clausius-Clapeyron equation, 515, 576
closure, 59, 161
    constitutive equation, 124, 168
    equations, 551
coalescing, 528
cocurrent downflow, 442
cocurrent flow, 484
coefficients in momentum equations, 468
coherent addition, 294
collimated irradiation, 327
collision diameters, 534
collision-frequency factor, 382
columnar dendritic growth, 658
combustion, 297, 383
complete chemical potential, 509
complete wetting, 453
completely backward scattering, 277
completely forward scattering, 277
completely nonwetting, 455
complex amplitude scattering matrix, 271
compressibility work, 508
compressive force, 128
condensation, 380, 494, 603
    vertical impermeable bounding surfaces, 603
condensation front, 649
condensation front moving into dry porous media, 644
condensed phase, 603
condensible, 549
conditional ensemble average, 206
conserved scattering, 328
consolidated, 7
constant drying rate, 582
contact angle, 429, 435, 481
contact angle hysteresis, 451, 452
contact line, 448
continuous phase distribution, 442
convection-diffusion layer, 585
coordination number, 139
countercurrent flows, 444
creeping flow over cylinders, 34
critical mobility, 501
critical temperature, 510
crossed diffusion, 370
current density, 275
curvature, 427, 507
curvature effect, 438, 523
cyrogenic, 297

Darcean regime, 466
Darcy
    deviation from, 45

equation, 17
flow regime, 48
law, 17
degree of polarization, 277
delta-M approximation, 329
dendrictic temperature, 654
dendritic growth, 658, 663
dendritic mass fractions, 654
density of displaced and displacing fluids, 494
density ratio, 483, 494
dependence-included discrete ordinates method, 347, 350
dependent scattering, 297, 302
dependent scattering theory, 348
desaturation, 471, 473
desorption, 515, 523, 530
deviation, 161
  radial, 161
devolatization, 380
dielectric constant, 269
dielectric susceptibilities, 457
diffraction, 283
diffuse reflection, 283
diffusely scattering, 336
diffusion, 317
diffusion layer, 584
diffusion-limited aggregation, 495
diffusional resistance, 387
dimensionless solid conductivity, 344
direct simulation, 7, 411
directional, spectral specular reflectivity, 281
discontinuous
  nonwetting phase, 447
  phase distributions, 445
  wetting phase, 446
discrete-ordinates, 322
  approximated radiation model, 309
  equation, 323
disjoining pressure, 455
dispersed element, 663
dispersed-element model, 662
dispersion
  adjacent to surfaces, 237
  force per unit area, 457
  in a tube, 157
  in disordered structures, 192, 205
  in mass transfer, 376, 380
  in oscillating flow, 232
  in porous media, 164
  near a bounding channel flow, 247
dispersion coefficient
  closed-form expressions, 227
  correlations, 227
  correlations for two-phase flow, 568
  experiment, 217
dispersion near bounding surface, 569
dispersion tensor, 550
  properties of, 215
  variation near surfaces, 240
distinct phase change, 655
distribution functions, 428
disturbance of the internal field, 294
divergence
  of radiative heat flux, 263
  theorem, 56
dominance of conduction, 246
downstream region, 650
drag models, 34
drag theory, 492
drainage, 446, 471, 495
dusty-gas model for transition flows, 376
dyad product, 124
dynamic contact angle, 448
Dzyaloskinskii-Lifshitz-Pitaevskii theory, 456

Eddington approximation, 330
eddy viscosity, 250
effect
  of curvature, 511
  of surface tension gradient, 464
  of thin liquid film dispersion forces, 523
effect of
  solid conductivity, 343
effective
  binary mass diffusivity, 377
  conductivity, 657
  conductivity tensor, 550
  emissivity, 339
  mass diffusion tensor, 376, 380
  mass diffusivity, 383

Subject Index   699

porosity, 20
radiative properties, 259
reflectivity, 339
spectral absorption and scattering coefficients, 309
thermal conductivity, 129, 135, 144, 551
thermal conductivity tensor, 124, 545
transmissitivity, 339
viscosity, 52, 92
Einstein formula, 52
electric
conductivity, 269
dipole scattering, 279
displacement vector, 269
field intensity, 269
electric potential, 438
electrolytic polishing, 450
electrostatic forces, 374, 502
emitting particles, 338
empirical slip coefficient, 240
empirical treatments, 2
energy minimization, 435
ensemble averaging, 182, 195
entrance
effect, 111, 225
length, 111, 225
entropy, 508
Eötvös number, 438
equation of radiative transfer, 262
equiaxial dendritic growth, 658
equilibrium partition ratio, 654
equilibrium treatment of solidification, 658
equivalence ratio, 412
equivalent particle diameter, 487
equivalent thermal-circuit model, 414
Ergun coefficient, 48
Euler-Lagrange equation, 437
evaporation, 603
  horizontal impermeable bounding surfaces, 615
  thin porous layer coated surfaces, 624
  vertical impermeable bounding surfaces, 614
evaporation from heated liquid film, 571

excess
  potential, 456
  resistance, 480
  temperature, 418
extreme incident angles, 280

Fabricated, 7
far-field effects, 293, 297
fiberglass insulation, 308
fingers, 495
finite-volume approximation, 345
finite-volume method, 327
first and second laws of thermodynamics, 508
first-order phase transitions, 654
first-order transition, 538
flame location, 418
flames, 387
flooding, 442, 445
flooding limit, 430
flow
  development, 67
  maldistributions, 567
  recirculation, 175
  regimes, 48
  structure, 51
flow regimes
  in two-phase flow, 442
  transitions, 443
fluid behavior in small pores, 532
fluid-phase distribution function, 55
fluid-solid two-phase flow, 427, 501
foametals, 74
formation of a monolayer, 524
four length scales, 9
Fraunhofer diffraction pattern, 280
Fresnel coefficients, 337
fritted-glass filter, 377
front saturation, 499
front velocity, 409, 499
full Mie solution, 284
fully irrigated, 561
funicular flow regime, 442
funicular regime, 639

Ganglion, 447
gas-liquid transition, 537
Gauss quadrature, 323
Gaussian distribution, 195

general transport theorem, 56
geometric
    calculations, 291
    range, 297
    scattering, 279
    treatment, 283
geometric, layered model, 339
geometric-optics, 279
geometrical model, 251
    mixing length, 251
Gibbs-Duhem equation, 507, 510, 512
glory, 292
grand potential, 535, 542
grand potential functional, 533
gravity, 497
    drainage, 445
    force, 455
gray body, 308
grinding, 450

Hagen-Poiseuille profile, 232
Hagen-Rubens law, 289
Hamaker constant, 449
hard sphere fluid, 533
heat
    of adsorption, 528
    of desorption, 515
heat and momentum transfer, 163
    analogy between, 163
heat of extension of the surface, 510
Helmholtz free energy, 508
Helmholtz function, 508
hemispherical isotropy, 313
hemispherical specular reflectivity, 283
Henyey-Greenstein approximation, 331
heterogeneous
    reaction, 386
    reaction rate coefficient, 383
    reactions, 382
    system, 2
high Reynolds number flows, 45
holdup effect, 212
homogeneous
    nucleation limit, 517
    reaction, 386
    reactions, 382
homogenization
    method, 61
    theory, 166

Horn method of moments, 200
hydraulic
    continuity, 472
    diameter, 32
    radius model, 32
hydraulically isolated elements, 446
hydrodynamics of two-phase flow, 427
hydrosols, 502
hysteresis, 471, 481, 507, 530, 532, 538
hysteresis loop, 523

Ideal solution, 519
imbibition, 447, 471
immiscible
    displacement, 427, 492
    solid phase, 654
immobile, 501
    nonwetting phase saturation, 447
    saturation, 446
impermeable spheres, 384
in-scattering, 262
incident
    beam, 260, 262
    radiation, 264
    ray, 354
increase in heat of evaporation, 514
independent scattering, 295, 308, 350, 353
index of refraction, 267
inertial
    flow regime, 49
    force, 430
    impaction, 502
    regime, 465
integral
    concept, 254
intensity of the scattered radiation, 273
intensity parameter, 273
interaction potential, 534
interatomic potential, 448
interception, 502
interconnected pores, 20
interfacial
    boundary conditions, 71
    contact separation, 449
    convection heat transfer coefficient, 391

convective heat and mass transfer, 571
convective heat transfer coefficient, 401
convective transfer coefficient, 397
drag, 462, 465, 466
drag coefficients, 469
effective viscosity, 17
free energy, 510
heat transfer coefficient, 212
instabilities, 495
layer, 507, 520
location, 74
mass transfer, 460, 494
normal force balance, 462
Nusselt number, 416
tension between two liquids, 434
tension liquid-vapor, 432
tensions, 427
velocity, 73
interfacial-averaged Nusselt and Sherwood numbers, 669
interference, 293
intermolecular force, 455
intermolecular forces, 448
internal energy, 508
internal reflections, 280, 312
interparticle spacing, 366
intrinsic contact angle, 450
invading nonwetting phase, 495
irreducible
    nonwetting phase saturation, 473
    saturation, 446
    wetting phase saturation, 472
isobaric flow, 388
isosteric heat of adsorption, 529
isosteric molar heat of adsorption, 527
isotropic media, 19, 384
isotropic scattering, 307, 329

Jump in saturation, 500

Kelvin
    equation, 513, 522
    equation for multicomponent systems, 521
kinematic condition, 75

kinetic theory, 371
kinetic-diffusion controlled dendritic crystal growth, 662
Knudsen
    diffusion, 368, 370
    diffusion coefficient, 373
    diffusion for tube flows, 372
    flow, 366, 367, 377
    number, 366
    regime, 430
    transition flows, 376
Kozeny constant, 33, 173

Laguerre polynomials, 323
Lambert cosine law, 338
Langmuir equation, 527
Laplace equation, 511, 512
large Bond number asymptote, 609
large Knudsen number model, 580
layered models, 243
Legendre polynomials, 262, 302
length scales, 494
Lennard-Jones type potential, 534
Leverett
    J-function, 477
    reduced function, 476
    reduced relationship, 473
Lewis number, 412
limited solubility, 655
linear dimension, 7
linear isotherm for adsorption, 375
linear stability theory, 500
liquid
    compressibility, 512
    holdup, 434, 438
    region, 633, 649
    superheat, 516
liquid-film region, 606
liquid-gas interfacial drag, 427, 466, 490
liquid-gas interfacial tension, 507
liquidus, 654
local
    chemical nonequilibrium, 387
    effective thermal conductivity, 243
    porosity, 55
    thermal and chemical noinequilibrium, 668

thermal equilibrium, 7, 120, 548
thermal nonequilibrium, 418
thermodynamic equilibrium, 658
volume average, 55
volume-averaging method, 53
long-wavelength excitations, 539
longitudinal total diffusivity, 175
lowering of the vapor pressure, 523
lubricants, 449

Machine-finished, 449
macroscopic
    boundary layer, 69
    entrance length, 70
    viscous shear stress, 67
macroscopic inertial terms, 466
magnetic
    field intensity, 269
    induction vector, 269
    permeability, 275
marginal
    stability, 501
    stability criterion, 501
    states, 500, 501, 532, 541
mass
    Chapman-Enskog, 372
    diffusivity, 372
    dispersion tensor, 385
    dynamic viscosity, 372
    fraction, 412
    transfer from a single strip, 587
    transfer in gases, 365
matrix structure, 494
maximum mobility, 500
Maxwell equations, 275
mean
    curvature of the interface, 429
    free path of molecules, 366
    molecular speed, 371
    radius of curvature, 445, 511
mechanical equilibrium, 511
Melrose function, 445, 477
melt-front speed, 657
melt-front thickness, 657
melting, 380, 603, 655, 656
    and solidification, 654
    temperature, 657
metallic solids, 267
metallic surface, 449

metals, 128, 267
metastable
    films, 532
    gas, 536
    gas phase, 535
    state, 532
metric coefficient, 210
mica, 450
microscopic
    fluid dynamics, 48
    inertial coefficients, 469, 488
Mie
    scattering, 278
    solution, 284
    theory, 280, 283, 295
miscible liquids, 654
mixing length, 254
mixing-length theory, 247
mobility ratio, 495
model for lateral dispersion, 561
modifications to energy equation, 388
modified
    Arrhenius model, 382
    Ergun relation, 67
    Kelvin equation, 523
    mean free path, 372
molar
    energy of adsorption, 528
    heat of adsorption, 374
    integral enthalpy of adsorption, 528
    internal energy, 528
molar isosteric enthalpy of adsorption, 529
molecular
    fluctuations, 537
    interaction theories, 507
    mass flux, 371
    slip, 365, 368
molecular thermal conductivity, 372
momentum equations
    for two-phase flow, 466
monolayer capacity, 527
Monte Carlo, 259, 301
    method, 343, 348
    simulation, 333
moving
    contact angle, 427
    contact line, 453

evaporation or condensation front, 627
moving-interface regime, 644
multicomponent
    gas mixtures, 365
    gas phase, 551
    systems, 507, 519, 658
multilayers, 524
multiple scattering, 293, 299, 301, 350
multiple surface sources, 588
multiple-medium treatment, 658
mushy phase, 658
mushy region, 655

Naturally formed, 7
near-field effects, 297
net isosteric heat of adsorption, 529
net mass transport, 371
no-slip temperature boundary condition, 245
nonabsorbing sphere, 282
nonconsolidated, 7
nonequilibrium treatment of solidification, 662
nonhygroscopic, 11
nonorthogonal boundaries, 327
nonpolarized incident radiation, 282
nonpolarized irradiation, 337
nonuniformities near the boundaries, 21
normal intensity, 309
normalized number density distribution, 533
number density, 449
Nusselt number, 413

Off-diagonal
    elements, 209
off-principal
    axes flows, 181, 191
one-point correlation, 137
onset of film evaporation, 623
opaque
    particles, 335
    spheres, 301
    spherical particles, 343
optical
    properties, 264, 267, 283, 292

thickness, 327, 328, 330, 347, 352
thicknesses, 308
optical thickness, 309
ordinary chemical potential, 509
orthotropic media, 19
overall convection transfer coefficient, 398, 399
oxide fragments, 449

Parallel/series conduit, 30
partially
    nonwetting, 455
    saturated surfaces, 571
    wetting, 455
pendular element, 446
pendular state, 446
Penndorf extension, 291
percolation concepts, 497
Percus-Yevick approximation, 534
Percus-Yevick model, 346
perfect conductors, 276
perfect gas behavior, 512
perfectly wetting liquids, 427, 521
periodically restricting tube, 483
permeability, 9, 18, 28, 208
    tensor, 19
permeable solid, 1
perturbed field, 500
phase
    change, 1, 603
    distributions, 430
    distributions for two-phase flow, 442
    function, 262, 302, 327, 348
    mass fractions, 661
    permeabilities, 427
    permeability coefficients, 468
    shift, 280
    transition, 532
phase transition in small pores, 533
phase-change transition theory, 532
phenomenological chemical kinetic expressions, 382
physical adsorption, 524
physioadsorption, 533
plane-parallel geometry, 260
point scattering, 293
points of contact, 139

Poiseuille-Couette flow, 73
polarization, 337
polarization forces, 374
polymers, 128
polyurethane insulations, 309
pore
    linear length scale, 21
    structure, 24
porosity, 20
    local variation, 102
    variation near surfaces, 101
porous diaphragm tensiometer, 471
precipitation, 380
preexponential factor, 382
premixed gaseous reactions, 409
pressure slip, 89
principal axes, 19, 181
probability density function, 206, 350
production rate, 380
pulsing flow, 430, 442, 566
pulsing flow regime, 443

Quadrature, 322
    directions, 325
    points, 322
quantum hydrodynamic theory, 539
quantum-mechanical forces, 374

Radiant conductivity, 340, 343
radiation
    heat flux, 332
    intensity, 260
    properties of a single particle, 264
radiative
    equilibrium, 263, 325
    exchange factor, 340
    heat flux, 263, 339, 344
    heat transfer scaling, 327
    properties, 354
    properties dependent, independent, 292
    properties experimental, 305
    transfer equation, 355
radius of curvature, 7
rainbow, 292
random
    numbers, 335
    porous media, 182
    walk model, 193
randomly arranged spheres, 251
rapture instability, 459
rate of condensation, 527
rate of evaporation, 527
Rayleigh
    limit, 291
    scattering, 279
Rayleigh theory, 280, 283
Rayleigh-Penndorf, 292
reaction, 1
reaction front, 409
recalescence, 666
receding contact angle, 451
reduction of chemical potential, 513
reflection, 280
    angle, 270
refraction, 280
relative
    index of refraction, 260
    permeabilities, 466, 494
    permeability, 478
    turbulent intensity, 51
    velocity, 490
remobilization, 447
representative elementary volume, 7, 21
Reynolds number, 453
Reynolds number effects, 213
rhombohedral arrangement, 181, 230
rhombohedral lattice arrangement, 295

Saffman-Taylor stability, 501
saturation, 428, 471, 473, 501
saturation temperature, 549, 576
scales, 7
scaling, 327, 348
scaling factor, 348, 349
scattered beam, 262
scattered electric field intensity, 272
scatterers, 259, 340, 354
scattering, 348
    albedo, 328
    coefficient, 261
    efficiency, 277, 282
Schuster-Schwarzchild approximation, 313
secondary flows, 45

sedimentation, 502
semiheuristic
    momentum equation, 66
semiheuristic momentum equation, 465
semitransparent
    boundaries, 312
    particles, 300, 337, 350
    spheres, 350
Sherwood number, 388
short range intermolecular forces, 510
significance of macroscopic forces, 68
silica gel, 375
similarity, 327, 329
simultaneous heat and mass transfer, 599
single scatterers, 264
single scattering, 293
single-component
    capillary systems, 511
    system, 521
    systems, 654, 656
single-medium treatment, 415, 658
size parameter, 264, 279, 283
size parameters, 297
skin depth, 276
slip
    boundary condition, 72, 240
    coefficient, 17, 73, 453
    coefficient, average, 86
    model, 453
    self-diffusivity, 373
    two-dimensional structure, 79
slip velocity, 464
slip-boundary condition, 368
small Bond number approximation, 611
small size parameters, 291
Snell law, 281, 338
solid matrix, 2
solid-fluid interfacial forces, 507
solidification, 603, 655
solidus, 654
solution methods
    equation of radiative transfer, 313
sorption coefficient, 374
spatial deviation component, 122
specific interfacial area, 491

specific surface area, 32
spectral
    absorption efficiency, 278
    extinction cross section, 278
    extinction efficiency, 278
    scattering cross section, 278
    scattering efficiency, 277
spherical harmonics, 332
spray flow regime, 443
stability
    of Buckley-Leverett front, 500
    of fingers, 495
stability of liquid film in small pores, 539
staggered arrangement, 182
static
    contact angle, 427, 448, 450
    equilibrium, 455
    equilibrium at liquid-gas interface, 434
stepped isotherm, 528
stoichiometric coefficient, 366
stoichiometric reaction, 366
stoichiometric reaction equation, 412
Stokes
    drag force, 41
    number, 405
Stokes flow, 17
structural parameters, 133
subcooled liquid, 655
superheated liquid, 656
superheated solid, 655
superpolishing, 450
surface, 452
    adsorption, 452
    catalytic reaction, 379
    convective heat transfer coefficient, 217
    energy, 448
    flux, 374
    heterogeneity, 452
    mass flux, 375
    molecular forces, 374
    reaction, 379
    roughness, 452
    saturation, 571, 589
    structure, 88
    viscosity, 439

surface tension, 432, 433, 483, 494, 511
   gradient-induced shear, 427
   of solid, 448
   of solution, 519
   thermodynamic definition, 433
surface-bulk surfactant, 439
surface-tension gradient coefficients, 470
symmetry factor, 330

Tangential interfacial stress continuity, 75
tangential interfacial velocity, 72
temperature exponent, 382
temperature of the adsorbed layer, 576
temperature-slip model, 240
tension of adhesion, 450
theorems, 56
theory of independent scattering, 259
thermal
   dispersion, 555
   dispersion tensor, 545
   eddy diffusivity, 250
   equilibrium, 508, 511
thermo- and diffuso-capillarity, 438
thermodynamic stability, 523
thermodynamics, 382, 507
thickness of
   adsorbed layer, 458
   interfacial layer, 510, 511
thin
   adsorbed film, 455
   extension of meniscus, 455
   fluid layer, 496
   liquid extension, 455
   liquid film region, 613
   liquid film stability theory, 532
three-phase
   systems, 427, 545
three-phase systems, 11
throat dimension, 478
tilt angle, 181, 182
time periodic behavior, 275
time scales, 9
   convection, 9
   diffusion, 9
tortuosities, 465

tortuosity, 32
   tensor, 168
tortuosity factor, 385
total
   axial diffusion coefficient, longitudinal, 162
   conductivity tensor, 550
   convective velocity, 397, 400
   diffusivity tensor, 169
   effective thermal diffusivity, longitudinal, 174
   effective thermal diffusivity, transverse, 174
total energy of extension of the surface, 510
total extinction coefficient, 309
total mobility, 500
total thermal diffusivity tensors, 397
transformation, 161
   tensor, 59
   tensors, 461
   vector, 60, 124
transformations, 124
transformed kernel, 328
transient two-phase flows, 428
transition, 45, 466
   flow, 366
   layer, 455, 508
   region, 455
transitional layer, 507
transitional relative pressure, 529
transmitted radiation, 308
transparent boundaries, 311
transparent spheres, 300
transport, 1
transport through bounding surfaces, 571
transportation, 352
transportation effect, 301
transversability conditions, 437
trickle bed, 430
trickling flow, 442
trickling flow regime, 443
turbulence, 51
   closure, 163
   spectra, 51
   transition to, 51
two-equation models, 391
two-flux

approximations, 313
two-medium treatment, 9, 391, 418
two-phase flow, 6
two-phase region, 608, 621, 631, 648
type
    I isotherm, 526
    II isotherm, 527
    III isotherm, 528
    IV isotherm, 528
    V isotherm, 528
    VI isotherm, 528

Ultramicropores, 7, 526
unit cell, 410
unit cells, 6
unit-cell linear dimension, 410
unit-cell model, 663
unsteady and chaotic flow regime, 49
unsteady laminar flow regime, 49
upstream region, 647

van de Hulst diagram, 284
van der Waals
    dispersion forces, 456, 521, 531
    forces, 458
    interfacial-layer forces, 427
van Driest wall function, 249, 252
vapor channels, 442
vapor pressure reduction, 513, 520
vapor-film region, 621
variable velocity zone, 499
velocity
    deviation, 172, 252
    mass-averaged, 365
    nonuniformities, 104, 106
    of displacing fluid, 494
    slip, 73, 453
view factor, 339
viscosity of displacing fluid, 494
viscosity ratio, 494
viscous flow, 366
void distribution function, 55
volumetric
    evaporation rate, 549
    rate of production, 378
    size distribution function, 294

Wall effect, 242, 247
wall heat transfer coefficient, 243
wave numbers, 500
wavelengths, 500
weakly anisotropic, 327
Weber number, 447
well depths, 534
wettability, 429, 455
    of displaced fluid, 494
    of invading phase, 495
wetting fluid, 495
Womersley number, 405
work
    of adhesion, 448
    of surface formation, 507, 508
    of wetting, 450

Young equation, 435, 436, 438
Young-Laplace equation, 429, 435, 437, 438